D. K. Linedall

IEE Electromagnetic Waves Series 19

Series Editors: Professors P.J.B. Clarricoats
E.D.R. Shearman and J.R. Wait

Microwave Antenna Theory and Design

Previous volumes in this series

Volume 1	Geometrical theory of diffraction for electromagnetic waves Graeme L. James
Volume 2	Electromagnetic waves and curved structures Leonard Lewin, David C. Chang and Edward F. Kuester
Volume 3	Microwave homodyne systems Ray J. King
Volume 4	Radio direction-finding P.J.D. Gething
Volume 5	ELF communications antennas Michael L. Burrows
Volume 6	Waveguide tapers, transitions and couplers F. Sporleder and H.G. Unger
Volume 7	Reflector antenna analysis and design P.J. Wood
Volume 8	Effects of the troposphere on radio communications Martin P.M. Hall
Volume 9	Schuman resonances in the earth-ionosphere cavity P.V. Bliokh, A.P. Nikolaenko and Y.F. Filippov
Volume 10	Aperture antennas and diffraction theory E.V. Jull
Volume 11	Adaptive array principles J.E. Hudson
Volume 12	Microstrip antenna theory and design J.R. James, P.S. Hall and C. Wood
Volume 13	Energy in Electromagnetism H.G. Booker
Volume 14	Leaky feeders and subsurface radio communications P. Delogne
Volume 15	The Handbook of Antenna Design Volume 1 Editors: A.W. Rudge, K. Milne, A.D. Olver, P. Knight
Volume 16	The Handbook of Antenna Design Volume 2 Editors: A.W. Rudge, K. Milne, A.D. Olver, P. Knight
Volume 17	Surveillance Radar Performance Prediction P. Rohan
Volume 18	Corrugated horns for microwave antennas P.J.B. Clarricoats and A.D. Olver

Microwave Antenna Theory and Design

Edited by Samuel Silver

Peter Peregrinus Ltd
On behalf of The Institution of Electrical Engineers

Published by Peter Peregrinus Ltd., London, UK

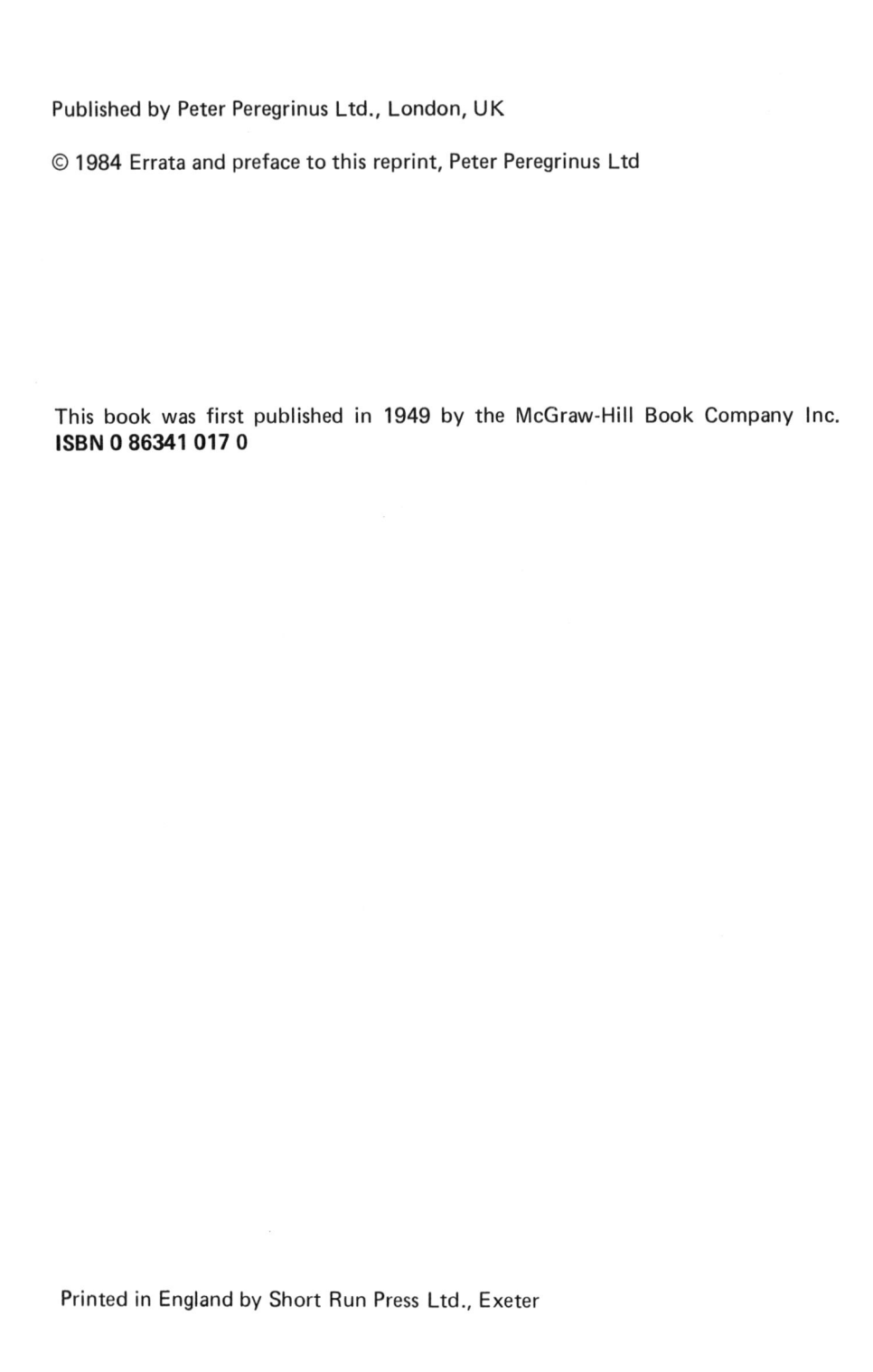

© 1984 Errata and preface to this reprint, Peter Peregrinus Ltd

This book was first published in 1949 by the McGraw-Hill Book Company Inc.
ISBN 0 86341 017 0

Printed in England by Short Run Press Ltd., Exeter

Microwave Antenna Theory and Design

Editorial Staff

SAMUEL SILVER
HUBERT M. JAMES

Contributing Authors

J.E. Eaton	R.M. Redheffer
L.J. Eyges	J.R. Risser
T.J. Keary	S. Silver
H. Krutter	O.A. Tyson
G.G. Macfarlane	L.C. Van Atta

Foreword

The tremendous research and development effort that went into the development of radar and related techniques during World War II resulted not only in hundreds of radar sets for military (and some for possible peacetime) use but also in a great body of information and new techniques in the electronics and high-frequency fields. Because this basic material may be of great value to science and engineering, it seemed most important to publish it as soon as security permitted.

The Radiation Laboratory of MIT, which operated under the supervision of the National Defense Research Committee, undertook the great task of preparing these volumes. The work described herein, however, is the collective result of work done at many laboratories, Army, Navy, university and industrial, both in this country and in England, Canada, and other Dominions.

The Radiation Laboratory, once its proposals were approved and finances provided by the Office of Scientific Research and Development, chose Louis N. Ridenour as Editor-in-Chief to lead and direct the entire project. An editorial staff was then selected of those best qualified for this type of task. Finally the authors for the various volumes or chapters or sections were chosen from among those experts who were intimately familiar with the various fields, and who were able and willing to write the summaries of them. This entire staff agreed to remain at work at MIT for six months or more after the work of the Radiation Laboratory was complete. These volumes stand as a monument to this group.

The volumes serve as a memorial to the unnamed hundreds and thousands of other scientists, engineers, and others who actually carried on the research, development, and engineering work the results of which are herein described. There were so many involved in this work and they worked so closely together even though often in widely separated laboratories that it is impossible to name or even to know those who contributed to a particular idea or development. Only certain ones who wrote reports or articles have even been mentioned. But to all those who contributed in any way to this great cooperative development enterprise, both in this country and in England, these volumes are dedicated.

L.A. DuBridge.

Preface

The need that arose during the war for utilizing the microwave region of the radio frequency spectrum for communications and radar stimulated the development of new types of antennas. The problems and design techniques, lying as they do in the domain of both applied electromagnetic theory and optics, are quite distinct from those of long-wave antennas. It is the aim of the present volume to make available to the antenna engineer a systematic treatment of the basic principles and the fundamental microwave antenna types and techniques. The elements of electromagnetic theory and physical optics that are needed as a basis for design techniques are developed quite fully. Critical attention is paid to the assumptions and approximations that are commonly made in the theoretical developments to emphasize the domain of applicability of the results. The subject of geometrical optics has been treated only to the extent necessary to formulate its basic principles and to show its relation as a short wavelength approximation to the more exact methods of field theory. The brevity of treatment should not be taken as an index of the relative importance of geometrical optics to that of electromagnetic theory and physical optics. It is in fact true that the former is generally the starting point in the design of the optical elements (reflectors and lenses) of an antenna. However, the use of ray theory for microwave systems presents no new problems over those encountered in optics—on which there are a number of excellent treatises—except that perhaps the law of the optical path appears more prominently in microwave applications.

In the original planning of the book it was the intention of the editors to integrate all of the major work done in this country and in Great Britain and Canada. This proved, however, to be too ambitious an undertaking. Many subjects have regrettably been omitted completely, and others have had to be treated in a purely cursory manner. It was unfortunately necessary to omit two chapters on rapid scanning antennas prepared by Dr. C.V. Robinson. The time required to revise the material to conform with the requirements of military security and yet to represent an adequate exposition of the subject would have unduly delayed the publication of the book. Certain sections of Dr. Robinson's material have been incorporated into Chaps. 6 and 12.

I take pleasure in expressing here my appreciation to Prof. Hubert M. James who, as Technical Editor, shared with me much of the editorial work and the attendant responsibilities. The scope of the book, the order of presentation of the material, and the sectional division within chapters were arrived at by us jointly in consultation with the authors. I am personally indebted to Professor James for his editorial work on my own chapters.

The responsibility for the final form of the book, the errors of omission and commission, is mine. A word of explanation to the authors of the various chapters is in order. After the close of the Office of Publications and the dispersal of the group, I have on occasions made use of my editorial prerogative to revise their presentations. I hope that the results meet with their approval. The policy of assignment of credit also needs explanation. The interpretation of both Professor James and myself of the policy on credit assignment formulated by the Editorial Board for the Technical Series has been to the effect that no piece of work discussed in the text would be associated with an individual or individuals. Radiation Laboratory reports are referred to in the sense that they represent source material for the chapter rather than individual acknowledgements. References to unpublished material of the Radiation Laboratory notebooks have been assiduously avoided, although such material has been drawn upon extensively by all of us. In defense of this policy it may be stated that the work at the Radiation Laboratory was truly a cooperative effort, and in only a few instances would it have been possible to assign individual credit unequivocally.

The completion of the book was made possible through the efforts of a number of people; on behalf of the editorial staff and the authors I wish to acknowledge their assistance and contributions. Mrs. Barbara Vogel and Mrs. Ellen Fine of the Radiation Laboratory served as technical assistants; the production of figures and photographs was expedited by Mrs. Frances Bourget and Mrs. Mary Sheats. It proved impossible to finish the work by the closing date of the Office of Publications; the Naval Research Laboratory accepted the work as one of the projects of the newly formed Antenna Research Section and contributed generously in personnel and facilities. Special thanks are due to A. S. Dunbar, I. Katz, and Dr. I. Maddaus for their editorial assistance; to Queenie Parigian and Louise Beltramini for preparation of the manuscript; and to Betty Hodgkins who prepared almost all the figures. The editors are indebted to Dr. G.G. Macfarlane of the Telecommunications Research Establishment, Great Britain, for his critical review of several of the theoretical chapters and his contribution on the theory of slot radiators in Chap. 9. John Powell of the Radiation Laboratory prepared material on lenses that was used in Chap. 11. The National Research Council of Canada and the British Central Radio Bureau have graciously granted us permission to take material from Canadian and British reports in accordance with current security regulations. The Bell Telephone Laboratory supplied the photographs of metal lens antennas.

The publishers have agreed that ten years after the date on which each volume of this series is issued, the copyright thereon shall be relinquished, and the work shall become part of the public domain.

Samuel Silver.

Naval Research Laboratory,
Washington, D.C.,
  April, 1947.

Preface to this reprint

This volume is an unabridged reprint of Microwave Antenna Theory and Design which was published by McGraw Hill as Volume 12 of the MIT Radiation Laboratory Series in 1949. The Editor of the Volume, the late Professor Samuel Silver contributed extensively to the text as chapter authorship shows, he subsequently became one of the best known figures in the world of Radio Science including in his life-time service as the President of the International Union of Radio Science (URSI).

Although many of the volumes in the MIT Series have helped several generations of microwave and radar engineers to acquire the foundations of their subject, probably no text in the series has stood the test of time better than Volume 12. It remains today essential reading for all those engaged in the microwave antenna field. Since 1949, only a few major treatises on microwave antennas have appeared, the most recent being the Handbook of Antenna Design (Volumes 15 and 16 in this series). However, the authors of these texts have generally assumed the reader's familiarity with material in Silver's volume but access to it has become increasingly difficult with the passage of time, as the McGraw Hill text and the 1965 Dover reprint have long been out-of-print. The decision to reprint the volume in the Electromagnetic Wave Series was taken after consultation with leading workers in the antenna field on both sides of the Atlantic. No attempt has been made to alter the text although a list of corrections compiled by Dr. John Bennett of the University of Sheffield, England is included. It is not claimed that the list is comprehensive but if readers would communicate with us should other corrections be found, the Errata Sheet could then be made more complete in the future, thus benefiting future generations of antenna research workers.

Peter Clarricoats

Queen Mary College,
University of London.
January 1984.

Errata

Page		Correction
73	Eqn (62)	$j(\omega t - k\mathbf{r}\cdot\mathbf{s})$
76	sentence after (73)	... of order zero.
78	last paragraph	cylindrical
117	Eqn (34)	$\dfrac{1}{F}\dfrac{\partial F}{\partial \theta}\theta_1$
141	Eqn (40)	$(u,v,w;\ x,y,z)$
141	Eqn (41)	faintly printed s_0 and r
146	Eqn (57b)	$\left(\dfrac{\epsilon}{\mu}\right)^{1/2}$
146	Eqn (57b)	$(\mathbf{s}_1 \times \mathbf{E}_r)$
165	Para before (115)	Eqs (114) *not* Eqs (110)
165	Eqn (115)	u_p not clearly printed
168	Eqn (127)	$\mathbf{H}_1 - \mathbf{E}_2 = \mathbf{G}$
174	Eqn (11b)	$\iint_A$
180	Eqn (30)	$\left(\dfrac{\pi b}{\lambda}\sin\theta\sin\phi\right)$
197	Fig. 6.9	label P missing
343	Eqn (22) for E_ϕ	$\cos\theta$ not $\cos\phi$ inside first [ ]
344	Eqn (23b)	$\dfrac{-\beta_{10}}{k}\Bigg]$
357) 358)	Eqns (45a) and (45b)	term in $-\left(\dfrac{\mu}{\epsilon}\right)^{1/2}$ should read $+\left(\dfrac{\mu}{\epsilon}\right)^{1/2}$
418	Eqn (7)	minus sign in exponent missing
420	Eqn (12a) exponent should read	$-jkp[1 + \cos\psi\cos\theta$ $-\sin\psi\sin\theta\cos(\xi-\phi)]$

Contents

FOREWORD BY L. A. DuBridge. vi

PREFACE. xi

Chap. 1. SURVEY OF MICROWAVE ANTENNA DESIGN PROBLEMS . 1

    1·1. The Wavelength Region 1
    1·2. Antenna Patterns 2
    1·3. Types of Microwave Beams 6
    1·4. Microwave Transmission Lines 7
    1·5. Radiating Elements 8
    1·6. A Survey of Microwave Antenna Types 9
    1·7. Impedance Specifications 13
    1·8. Program of the Present Volume 14

Chap. 2. CIRCUIT RELATIONS, RECIPROCITY THEOREMS 16

    2·1. Introduction 16
    2·2. The Four-terminal Network 17
    2·3. The Rayleigh Reciprocity Theorem 19
    2·4. Thévenin's Theorem and the Maximum-power Theorem . . . 20
    2·5. The Two-wire Transmission Line 21
    2·6. The Homogeneous Transmission Line 23
    2·7. The Lossless Line 26
    2·8. Transformation Charts 29
    2·9. The Four-terminal Network Equivalent of a Section of Transmission Line 36

Transmitting and Receiving Antennas 37

    2·10. The Antenna as a Terminating Impedance 37
    2·11. The Receiving Antenna System 40
    2·12. The Transmitter and Receiver as a Coupled System 45
    2·13. Reciprocity between the Transmitting and Receiving Patterns of an Antenna 48
    2·14. The Average Cross Section for a Matched System 50
    2·15. Dependence of the Cross Section on Antenna Mismatch . . . 51
    2·16. The Four-terminal Network Representation 53
    2·17. Development of the Network Equations 56
    2·18. The Reciprocity Relation between the Transfer Impedance Coefficients 59

CONTENTS

Chap. 3. RADIATION FROM CURRENT DISTRIBUTIONS. 61

    3·1. The Field Equations. 61
    3·2. The Constitutive Parameters; Linearity and Superposition . . . 65
    3·3. Boundary Conditions. 66
    3·4. The Field Equations for Harmonic Time Dependence 68
    3·5. Poynting's Theorem 69
    3·6. The Wave Equations. 71
    3·7. Simple Wave Solutions. 73
    3·8. General Solution of the Field Equations in Terms of the Sources, for a Time-periodic Field. 80
    3·9. Field Due to Sources in an Unbounded Region 84
    3·10. Field in a Region Bounded by Surfaces of Infinitely Conductive Media 86
    3·11. The Far-zone Fields 87
    3·12. Polarization. 91
    3·13. The Electric Dipole 92
    3·14. The Magnetic Dipole 95
    3·15. The Far-zone Fields of Line-current Distributions 96
    3·16. The "Half-wave Dipole". 98
    3·17. Superposition of Fields. 99
    3·18. The Double-dipole System 101
    3·19. Regular Space Arrays 104

Chap. 4. WAVE FRONTS AND RAYS 107

    4·1. The Huygens-Green Formula for the Electromagnetic Field. . . 107
    4·2. Geometrical Optics: Wavefronts and Rays 110
    4·3. Curvature of the Rays in an Inhomogeneous Medium 111
    4·4. Energy Flow in Geometrical Optics 112
    4·5. Geometrical Optics as a Zero-wavelength Limit 114
    4·6. The Huygens-Fresnel Principle and Geometrical Optics: The Far-zone Approximation 116
    4·7. The Principle of Stationary Phase 119
    4·8. Fermat's Principle. 122
    4·9. The Law of the Optical Path 125

Chap. 5. SCATTERING AND DIFFRACTION. 129

    5·1. General Considerations. 129
    5·2. Boundary Conditions 130
    5·3. Reflection by an Infinite Plane Surface; the Principle of Images . 132

    Approximate Methods for Reflectors of Arbitrary Shape. 137

    5·4. The Geometrical-optics Method. 138
    5·5. Calculation of the Scattered Field 139
    5·6. Superposition of the Source Field and the Scattered Field. . . 143
    5·7. The Current-distribution Method 144
    5·8. Calculation of the Scattered Field 146
    5·9. Application to Point-source and Line-source Feeds. 149
    5·10. Reaction of a Reflector on a Point-source Feed 155
    5·11. The Aperture-field Method 158
    5·12. The Fraunhofer Region. 160

CONTENTS

DIFFRACTION... 162
    5·13. General Considerations on the Approximate Methods 162
    5·14. Reduction to a Scalar Diffraction Problem 164
    5·15. Babinet's Principle for the Electromagnetic Field 167

CHAP. 6. APERTURE ILLUMINATION AND ANTENNA PATTERNS . 169
    6·1. Primary and Secondary Patterns 169
    6·2. The Diffraction Field 169
    6·3. Fourier Integral Representation of the Fraunhofer Region ... 174
    6·4. General Features of the Secondary Pattern 175
    6·5. The Rectangular Aperture 180
    6·6. Two-dimensional Problems 182
    6·7. Phase-error Effects....................................... 186
    6·8. The Circular Aperture 192
    6·9. The Field on the Axis in the Fresnel Region 196

CHAP. 7. MICROWAVE TRANSMISSION LINES 200
    7·1. Microwave and Long-wave Transmission Lines 200
    7·2. Propagation in Waveguides of Uniform Cross Section 201
    7·3. Orthogonality Relations and Power Flow 207
    7·4. Transmission Line Considerations in Waveguides 209
    7·5. Network Equivalents of Junctions and Obstacles 214
    7·6. TEM-mode Transmission Lines 216
    7·7. Coaxial Lines: TEM-mode 217
    7·8. Coaxial Lines: TM- and TE-modes 219
    7·9. Cascade Transformers: TEM-mode 221
    7·10. Parallel Stubs and Series Reactances...................... 223
    7·11. Rectangular Waveguides: TE- and TM-modes 226
    7·12. Impedance Transformers for Rectangular Guides 229
    7·13. Circular Waveguide: TE- and TM-modes.................. 233
    7·14. Windows for Use in Circular Guides 235
    7·15. Parallel-plate Waveguide................................. 235
    7·16. Design Notes 238

CHAP. 8. MICROWAVE DIPOLE ANTENNAS AND FEEDS.......... 239
    8·1. Characteristics of Antenna Feeds 239
    8·2. Coaxial Line Terminations: The Skirt Dipole 240
    8·3. Asymmetric Dipole Termination.......................... 242
    8·4. Symmetrically Energized Dipoles: Slot-fed Systems 245
    8·5. Shape and Size of the Dipole 248
    8·6. Waveguide-line-fed Dipoles............................... 250
    8·7. Directive Dipole Feeds................................... 250
    8·8. Dipole-disk Feeds 251
    8·9. Double-dipole Feeds 253
    8·10. Multi-dipole Systems.................................... 256

CHAP. 9. LINEAR ARRAY ANTENNAS AND FEEDS................ 257
    9·1. General Considerations................................... 257

xvi CONTENTS

PATTERN THEORY 258

    9·2. General Array Formula. 258
    9·3. The Associated Polynomial 261
    9·4. Uniform Arrays 264
    9·5. Broadside Beams 267
    9·6. End-fire Beams 274
    9·7. Beam Synthesis 279

RADIATING ELEMENTS 284

    9·8. Dipole Radiators 284
    9·9. Slots in Waveguide Walls 286
    9·10. Theory of Slot Radiators 287
    9·11. Slots in Rectangular Waveguide; TE_{10}-mode 291
    9·12. Experimental Data on Slot Radiators 295
    9·13. Probe-fed Slots 299
    9·14. Waveguide Radiators 301
    9·15. Axially Symmetrical Radiators 303
    9·16. Streamlined Radiators 310

ARRAYS 312

    9·17. Loaded-line Analysis 313
    9·18. End-fire Array 316

BROADSIDE ARRAYS 318

    9·19. Suppression of Extraneous Major Lobes 318
    9·20. Resonant Arrays 321
    9·21. Beacon Antenna Systems 327
    9·22. Nonresonant Arrays 328
    9·23. Broadband Systems with Normal Beams 331

CHAP. 10. WAVEGUIDE AND HORN FEEDS 334

    10·1. Radiation from Waveguide of Arbitrary Cross Section 334
    10·2. Radiation from Circular Waveguide 336
    10·3. Radiation from Rectangular Guide 341
    10·4. Waveguide Antenna Feeds 347
    10·5. The Double-slot Feed 348
    10·6. Electromagnetic Horns 349
    10·7. Modes in E-plane Sectoral Horns 350
    10·8. Modes in H-plane Sectoral Horns 355
    10·9. Vector Diffraction Theory Applied to Sectoral Horns 357
    10·10. Characteristics of Observed Radiation Patterns from Horns of
           Rectangular Cross Section 358
    10·11. Admittance of Waveguide and Horns 366
    10·12. Transformation of the E-plane Horn Admittance from the Throat
           to the Uniform Guide 369
    10·13. Admittance Characteristics of H-plane Sectoral Horns . . . 374
    10·14. Compound Horns 376
    10·15. The Box Horn 377
    10·16. Beam Shaping by Means of Obstacles in Horn and Waveguide
           Apertures 380
    10·17. Pressurizing and Matching 383

CONTENTS

CHAP. 11. DIELECTRIC AND METAL-PLATE LENSES 388

    11·1. Uses of Lenses in Microwave Antennas. 388

DIELECTRIC LENSES. 389

    11·2. Principles of Design 389
    11·3. Simple Lenses without Zoning. 390
    11·4. Zoned Dielectric Lenses 395
    11·5. Use of Materials with High Refractive Indexes 398
    11·6. Dielectric Losses and Tolerances on Lens Parameters. 399
    11·7. Reflections from Dielectric Surfaces 401

METAL-PLATE LENSES. 402

    11·8. Parallel-plate Lenses. 402
    11·9. Other Metal-lens Structures. 406
    11·10. Metal-plate Lens Tolerances 407
    11·11. Bandwidth of Metal-plate Lenses; Achromatic Doublets 408
    11·12. Reflections from Surfaces of Parallel-plate Lenses 410

CHAP. 12. PENCIL-BEAM AND SIMPLE FANNED-BEAM ANTENNAS 413

PENCIL-BEAM ANTENNAS. 413

    12·1. Pencil-beam Requirements and Techniques 413
    12·2. Geometrical Parameters 415
    12·3. The Surface-current and Aperture-field Distributions. 417
    12·4. The Radiation Field of the Reflector. 420
    12·5. The Antenna Gain. 423
    12·6. Primary Pattern Designs for Maximizing Gain 433
    12·7. Experimental Results on Secondary Patterns 433
    12·8. Impedance Characteristics 439
    12·9. The Vertex-plate Matching Technique 443
    12·10. Rotation of Polarization Technique 447
    12·11. Structural Design Problems. 448

SIMPLE FANNED-BEAM ANTENNAS. 450

    12·12. Applications of Fanned Beams and Methods of Production . . . 450
    12·13. Symmetrically Cut Paraboloids 451
    12·14. Feed Offset and Contour Cutting of Reflectors 453
    12·15. The Parabolic Cylinder and Line Source 457
    12·16. Parallel-plate Systems 459
    12·17. Pillbox Design Problems 460

CHAP. 13. SHAPED-BEAM ANTENNAS. 465

    13·1. Shaped-beam Applications and Requirements 465
    13·2. Effect of a Directional Target Response 468
    13·3. Survey of Beam-shaping Techniques. 471
    13·4. Design of Extended Feeds 487
    13·5. Cylindrical Reflector Antennas 494
    13·6. Reflector Design on the Basis of Ray Theory 497
    13·7. Radiation Pattern Analysis. 500
    13·8. Double Curvature Reflector Antennas 502
    13·9. Variable Beam Shape 508

CONTENTS

CHAP. 14. ANTENNA INSTALLATION PROBLEMS 510

GENERAL SURVEY OF INSTALLATION PROBLEMS. 510

14·1. Ground Antennas 510
14·2. Ship Antennas 511
14·3. Aircraft Antennas 512
14·4. Scanning Antennas on Aircraft 513
14·5. Beacon Antennas on Aircraft 521

RADOME DESIGN PROBLEMS AND PROCEDURES 522

14·6. Relation of the Radome to System Performance. 523
14·7. Radome Wall Design. 528
14·8. Normal Incidence Radomes 537
14·9. Streamlined Radomes 540

CHAP. 15. ANTENNA MEASUREMENTS—TECHNIQUES 543

15·1. Introduction 543

IMPEDANCE MEASUREMENTS 544

15·2. Transmission-line Relations. 544
15·3. Standing-wave Ratios 545
15·4. Measurement of Voltage Standing-wave Ratio 547
15·5. Determination of Electrical Length 550
15·6. Calibration of Detection System 552
15·7. Probe Reflections 556

PRIMARY FEED PATTERN MEASUREMENTS. 557

15·8. Primary Pattern Apparatus for Point-source Feeds. 557
15·9. Intensity Measurements 561
15·10. Phase Determinations 564
15·11. Line-source Primary Pattern 570
15·12. Magic T 572
15·13. Beacon Azimuth Patterns. 573

SECONDARY PATTERN MEASUREMENTS. 574

15·14. Siting Considerations. 574
15·15. Pattern Measurements. 578
15·16. Gain Measurements 580
15·17. Directive Gain 580
15·18. Gain Comparison 581
15·19. Primary Gain Standard Determination. 582
15·20. Reflection Method for Gain Determination 585
15·21. Secondary Gain Standards 586
15·22. Interaction between Antennas. 587

CHAP. 16. ANTENNA MEASUREMENTS—EQUIPMENT. 593

16·1. Survey of Equipment Requirements 593
16·2. Sources of R-F Power 594
16·3. Detectors. 601
16·4. Amplifiers 604
16·5. Recorders 609

INDEX. 615

CHAPTER 1

SURVEY OF MICROWAVE ANTENNA DESIGN PROBLEMS

By S. Silver

1·1. The Wavelength Region.—The designation of the boundaries of the microwave region of the electromagnetic spectrum is purely arbitrary. The long-wavelength limit has been set variously at 25 or 40 cm, even at 100 cm. From the point of view of antenna theory and design techniques, the 25-cm value is the most appropriate choice. The short-wavelength limit to which it is possible to extend the present techniques has not yet been reached; it is in the neighborhood of 1 mm. Accordingly we shall consider the microwave region to extend in wavelength from 0.1 to 25 cm, in frequency from 3×10^5 to 1200 Mc/sec.

This is the transition region between the ordinary radio region, in which the wavelength is very large compared with the dimensions of all the components of the system (except perhaps for the large and cumbersome antennas), and the optical region, in which the wavelengths are excessively small. Long-wave concepts and techniques continue to be useful in the microwave region, and at the same time certain devices used in the optical region such as lenses and mirrors are employed. From the point of view of the antenna designer the most important characteristic of this frequency region is that the wavelengths are of the order of magnitude of the dimensions of conventional and easily handled mechanical devices. This leads to radical modification of earlier antenna techniques and to the appearance of new and striking possibilities, especially in the construction and use of complex antenna structures.

It follows from elementary diffraction theory that if D is the maximum dimension of an antenna in a given plane and λ the wavelength of the radiation, then the minimum angle within which the radiation can be concentrated in that plane is

$$\theta \approx \frac{\lambda}{D}. \qquad (1)$$

With microwaves one can thus produce highly directive antennas such as have no parallel in long-wave practice; if a given directivity is desired, it can be obtained with a microwave antenna which is smaller than the equivalent long-wave antenna. The ease with which these small antennas can be installed and manipulated in a restricted space contributes greatly to the potential uses of microwaves. In addition, the convenient size of microwave antenna elements and of the complete antenna structure makes it feasible to construct and use antennas of elaborate structure for special purposes; in particular, it is possible to introduce mechanical motions of parts of the antenna with respect to other parts, with consequent rapid motion of the antenna beam.

The microwave region is a transition region also as regards theoretical methods. The techniques required range from lumped-constant circuit theory, on the low-frequency side, through transmission-line theory, field theory, and diffraction theory to geometrical optics, on the high-frequency side. There is frequent need for using several of these theories in parallel—combining field theory and transmission-line theory, supplementing geometrical optics by diffraction theory, and so on. Optical problems in the microwave antenna field are relatively complex, and some are of quite novel character: For instance, the optics of a curved two-dimensional domain finds practical application in the design of rapid-scanning antennas.

1·2. Antenna Patterns.—Before undertaking a survey of the more important types of microwave antenna, it will be necessary to state precisely the terms in which the performance of an antenna will be described.

The Antenna as a Radiating Device: The Gain Function.—The field set up by any radiating system can be divided into two components: the induction field and the radiation field. The induction field is important only in the immediate vicinity of the radiating system; the energy associated with it pulsates back and forth between the radiator and near-by space. At large distances the radiation field is dominant; it represents a continual flow of energy directly outward from the radiator, with a density that varies inversely with the square of the distance and, in general, depends on the direction from the source.

In evaluating the performance of an antenna as a radiating system one considers only the field at a large distance, where the induction field can be neglected. The antenna is then treated as an effective point source, radiating power that, per unit solid angle, is a function of direction only. The directive properties of an antenna are most conveniently expressed in terms of the "gain function" $G(\theta,\phi)$. Let θ and ϕ be respectively the colatitude and azimuth angles in a set of polar coordinates centered at the antenna. Let $P(\theta,\phi)$ be the power radiated per unit solid angle in direction θ, ϕ and P_t the total power radiated. The gain function is defined as the ratio of the power radiated in a given direction per unit solid angle to the average power radiated per unit solid angle:

$$G(\theta,\phi) = \frac{P(\theta,\phi)}{\dfrac{P_t}{4\pi}}. \qquad (2)$$

Thus $G(\theta,\phi)$ expresses the increase in power radiated in a given direction by the antenna over that from an isotropic radiator emitting the same total power; it is independent of the actual power level. The gain function is conveniently visualized as the surface

$$r = G(\theta,\phi) \tag{3}$$

distant from origin in each direction by an amount equal to the gain function for that direction. Typical gain-function surfaces for microwave antennas are illustrated in Fig. 1·1.

The maximum value of the gain function is called the "gain"; it will be denoted by G_M. The gain of an antenna is the greatest factor by which the power transmitted in a given direction can be increased by using that antenna instead of an isotropic radiator.

The "transmitting pattern" of an antenna is the surface

$$r = \frac{G(\theta,\phi)}{G_M}; \tag{4}$$

it is thus the gain-function surface normalized to unit maximum radius. A cross section of this surface in any plane that includes the origin is called the "polar diagram" of the antenna in this plane. The polar diagram is sometimes renormalized to unit maximum radius.

When the pattern of an antenna has a single principal lobe, this is usually referred to as the "antenna beam." This beam may have a wide variety of forms, as is shown in Fig. 1·1.

The Antenna as a Receiving Device: The Receiving Cross Section.—The performance of an antenna as a receiving device can be described in terms of a receiving cross section or receiving pattern.

A receiving antenna will pick up energy from an incident plane wave and will feed it into a transmission line which terminates in an absorbing load, the detector. The amount of energy absorbed in the load will depend on the orientation of the antenna, the polarization of the wave, and the impedance match in the receiving system. In specifying the performance of the antenna, we shall suppose that the polarization of the wave and the impedance characteristics of the detector are such that maximum power is absorbed. The absorbed power can then be expressed as the power incident on an effective absorbing area, called the "receiving cross section," or "absorption cross section" A_r of the antenna. If S is the power flux density in the incident wave, the absorbed power is

$$P_r = SA_r \tag{5}$$

The receiving cross section will depend on the direction in which the plane wave is incident on the antenna. We shall write it as $A_r = A_r(\theta,\phi)$, where θ and ϕ are the spherical angles, already defined, of the direction of incidence of the wave. This function, like the gain function, is represented conveniently as the surface

$$r = A_r(\theta,\phi). \tag{6}$$

The "receiving pattern" of an antenna is defined, analogously to the transmitting pattern, as the above surface normalized to unit maximum radius:

$$r = \frac{A_r(\theta,\phi)}{A_{rM}}. \tag{7}$$

It is a consequence of the reciprocity theorem to be discussed in Chap. 2 that the receiving and transmitting patterns of an antenna are identical:

$$\frac{G(\theta,\phi)}{G_M} = \frac{A_r(\theta,\phi)}{A_{rM}}. \tag{8}$$

It will also be shown that the ratio A_{rM}/G_M is a constant for all matched antennas:

$$\frac{A_{rM}}{G_M} = \frac{\lambda^2}{4\pi}. \tag{9}$$

Thus for any matched receiving system

$$A_r(\theta,\phi) = \frac{\lambda^2}{4\pi} G(\theta,\phi). \tag{10}$$

Coverage Pattern, One Way.—The characteristics of an antenna may also be described in terms of the performance of a radio or radar system of which it is a part. It is necessary to distinguish between the case of one-way transmission, in which a given antenna serves for transmission or for reception only, and the case of radar or two-way transmission, in which a single antenna performs both functions.

We consider first a transmitting antenna and a receiving antenna separated by a large distance R. Let G_t and G_r be the respective gain functions of the two antennas for the direction of transmission. If the total power transmitted is P, the power radiated in the direction of the receiver, per unit solid angle, will be $(1/4\pi)PG_t$. The receiving antenna will present a receiving cross section $(1/4\pi)G_r\lambda^2$ to the incident wave; it will, in effect, subtend a solid angle $G_r\lambda^2/4\pi R^2$ at the transmitter. The power absorbed at the receiver will thus be

$$P_r = P\frac{G_t G_r \lambda^2}{16\pi^2 R^2}. \tag{11}$$

The maximum operating range is determined by the signal-to-noise ratio of the detector system. If P_{rm} is the minimum detectable signal for the receiver, the maximum operating range is

$$R_{\max} = \left(\frac{P}{P_{rm}}\right)^{\frac{1}{2}} \frac{\lambda}{4\pi} (G_t G_r)^{\frac{1}{2}} \tag{12}$$

Thus, if it is possible to ignore the effect of the earth on the propagation of the wave and if G_r is constant, it will be possible to operate the receiving system satisfactorily everywhere within the surface

$$r = \left(\frac{P}{P_{rm}}\right)^{1/2} \frac{\lambda}{4\pi} G_r^{1/2} [G_t(\theta,\phi)]^{1/2}, \tag{13}$$

where the transmitter is taken to be at the origin. This surface will be called the "free-space coverage pattern for one-way transmission."

Coverage Pattern, Two Ways.—In most radar applications the same antenna is used for transmission and reception. One is here interested in detecting a target, which may be characterized by its "scattering cross section" σ. This is the actual cross section of a sphere that in the same position as the target would scatter back to the receiver the same amount of energy as is returned by the target. For this fictitious isotropic scatterer, the effective angle subtended at the transmitter is σ/R^2 and the total power intercepted is

$$P_i = \frac{1}{4\pi} P G_t \frac{\sigma}{R^2}. \tag{14}$$

Scattered isotropically, this power would appear back at the transmitter as a power flux, per unit area,

$$S = \frac{P_i}{4\pi R^2} = \frac{P G_t \sigma}{(4\pi)^2 R^4}. \tag{15}$$

Actually, the scattering of most targets is not uniform. The scattering cross section of the target will in any case be defined by Eq. (15), but it will usually be a function of the orientation of the target.

The power absorbed by the receiver from the scattered wave will be

$$P_r = A_r S = \frac{P \sigma \lambda^2 G_t^2}{(4\pi)^3 R^4} \tag{16}$$

since here $G_t = G_r$. If the effect of the earth on transmission of the waves can be neglected, it will be possible to detect the target only when it lies within the surface

$$r = \left[\frac{P}{P_{rm}} \frac{\sigma \lambda^2}{(4\pi)^3}\right]^{1/4} [G_t(\theta,\phi)]^{1/2} \tag{17}$$

about the transmitter as an origin. This surface will be called the "free-space coverage pattern for two-way transmission."

The extent of the coverage patterns is determined by characteristics of the system and target—output power, receiver sensitivity, target size —that are not under the control of the antenna designer. The form of the coverage patterns is determined by but is not the same as the form of the antenna transmitting and receiving patterns; in the coverage patterns, r is proportional to $[G_t(\theta,\phi)]^{1/2}$ rather than to $G_t(\theta,\phi)$. The desired form of the coverage pattern is largely determined by the use to be made of the system. From it, one can derive the required form of the transmitting or receiving pattern of the antenna; it is usually in terms of this type of pattern that antenna performance is measured and specified.

It is to be emphasized that the discussion of coverage patterns given

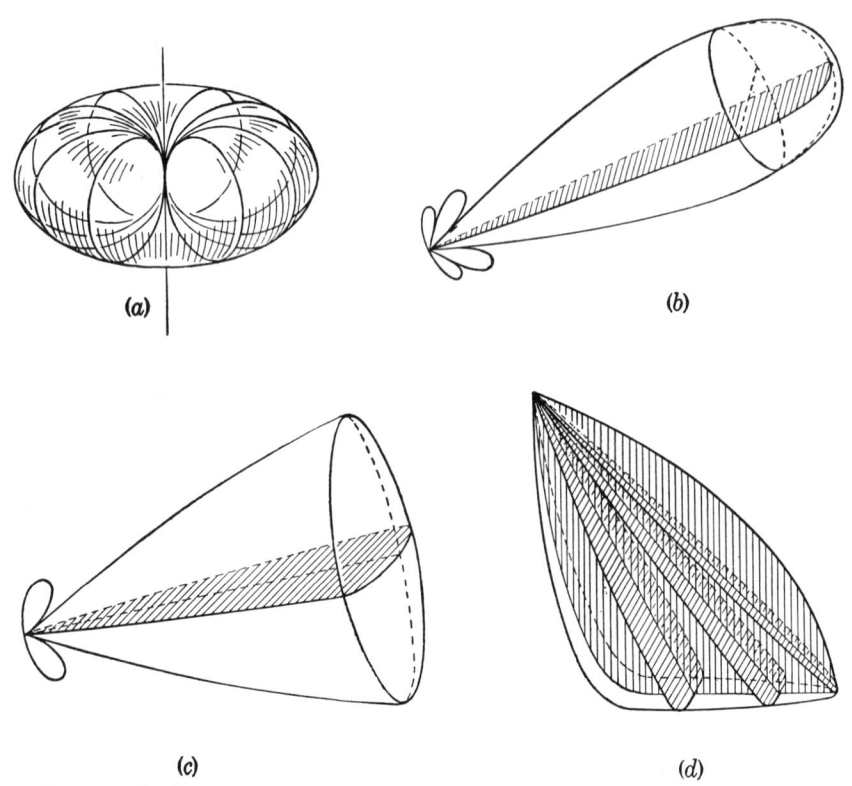

Fig. 1·1.—Typical gain-function surfaces for microwave antennas. (a) Toroidal (omnidirectional) pattern; (b) pencil-beam pattern; (c) flat-top flared beam; (d) asymmetrically flared beam.

here assumes free-space conditions. In many important applications, coverage is affected by interference and diffraction phenomena due to the earth, by meteorological conditions, and by other factors. A detailed account of these factors, which may be of considerable importance in determining the antenna transmitting pattern required for a given application, will be found in Vol. 13 of the Radiation Laboratory Series.

1·3. Types of Microwave Beams.—The most important types of microwave beams are illustrated in Fig. 1·1.

The least directive beam is the "toroidal beam,"[1] which is uniform in

---

[1] Such a beam is also referred to as "omnidirectional." (IRE Standards and Definitions, 1946.)

azimuth but directive in elevation. Such a beam is desirable as a marker for an airfield because it can be detected from all directions.

The most directive type of antenna gives a "pencil beam," in which the major portion of the energy is confined to a small cone of nearly circular cross section. With the high directivity of this beam goes a very high gain, often as great as 1000. In radar applications such a beam may be used like a searchlight beam in determining the angular position of a target.

Although the pencil beam is useful for precise determination of radar target positions, it is difficult to use in locating random targets. For the latter purpose it is better to use a "fanned beam," which extends through a greater angle in one plane than it does in a plane perpendicular to that plane. The greater part of the energy is then directed into a cone of roughly elliptical cross section, with the long axis, for example, vertical. By sweeping this beam in azimuth, one can scan the sky more rapidly than with a pencil beam, decreasing the time during which a target may go undetected. Such a fanned beam still permits precise location of targets in azimuth, at the expense of loss of information concerning target elevation.

Other applications of microwave beams require the use of beams with carefully shaped polar diagrams. These include one-sided flares, such as is illustrated in Fig. 1·1d, in which the polar diagram in the flare plane is roughly an obtuse triangle, whereas in transverse planes the beam remains narrow. In radar use, such a beam at the same time permits precise location of targets in azimuth and assures most effective distribution of radiation within the vertical plane of the beam. Toroidal beams with a one-sided flare in elevation have also been developed.

No theoretical factors limit any of the above beam types to the microwave region, but many practical limitations are imposed on long-wave antennas by the necessary relationship between the dimensions of the antenna elements and the wavelengths.

1·4. Microwave Transmission Lines.—The form of microwave antennas depends upon the nature of the available radiating elements, and this in turn depends upon the nature of the transmission lines that feed energy to these elements. We therefore preface a survey of the main types of microwave antennas with a brief description of microwave transmission lines; a detailed discussion of these lines will be found in Chap. 7.

Unshielded parallel-wire transmission lines are not suitable for microwave use; if they are not to radiate excessively, the spacing of the wires must be so small that the power-carrying capacity of the line is severely limited.

Use of the self-shielding coaxial line is possible in the microwave region but is generally restricted to wavelengths of approximately 10 cm or more. For proper action as a transmission line, a coaxial line should transmit electromagnetic waves in only a single mode; otherwise the generator looks into an indeterminate impedance and tends to be erratic in operation. On this account it is necessary to keep the average circumference of inner and outer conductors less than the free-space wavelength of the transmitted waves. At wavelengths shorter than 10 cm this limitation on the dimensions of coaxial lines begins to limit their power-carrying capacity to a degree that makes them unsatisfactory for most purposes.

The most useful transmission line in the microwave region is the hollow pipe. Such pipes will support the propagation of an electromagnetic wave only when they are sufficiently large compared with its free-space wavelength. As guides for long-wave radiation, intolerably large pipes are required, but in the microwave region it becomes possible to use pipes of convenient size. Like the coaxial guide, there is also an upper limit imposed on the cross-sectional dimension of the pipe if it is to transmit the wave in only a single mode. However, in the absence of an inner conductor, this size limitation does not affect the power capacity so seriously as it does in the coaxial line.

1·5. Radiating Elements.—The nature of the radiating elements terminating a transmission line is to a considerable extent determined by the nature of the line itself. Typical long-wave radiating elements are the "dipole" antennas, such as the center-driven half-wave dipole, and loop antennas, such as the rhombic antenna, illustrations of which are given in Fig. 1·2a and b. It is evident that the parallel-wire and

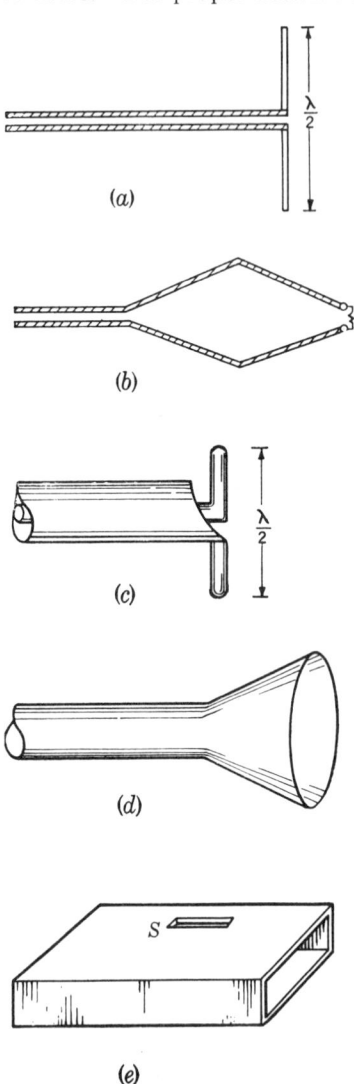

FIG. 1·2.—Transmission lines and radiating elements. (a) Center-driven half-wave dipole; (b) rhombic antenna terminating a two-wire line; (c) microwave dipole terminating a coaxial line; (d) conical horn fed by a circular waveguide; (e) slot radiator in the wall of a rectangular guide.

coaxial lines lend themselves to such terminations. Many long-wave antenna ideas have been carried over into the microwave region, particularly those connected with the half-wave dipole; the transition, however, is not merely a matter of wavelength scaling. In a microwave antenna the cross-sectional dimensions of the transmission line are comparable to the dimensions of the half-wave dipole, and consequently, the coupling between the radiator and the line becomes a more significant problem than in a corresponding long-wave system. The cross-sectional dimensions of the dipole element are also comparable to its length. A typical microwave dipole is shown in Fig. 1·2c; the analysis and understanding of such microwave dipoles is at best still in a qualitative stage.

The use of hollow waveguide lines leads to the employment of entirely different radiating systems. The simplest radiating termination for such a line is just the open end of the guide, through which the energy passes into space. The dimensions of the mouth aperture are then comparable to the wavelength; as a result of diffraction, the energy does not continue in a beam corresponding to the cross section of the pipe but spreads out considerably about the direction of propagation defined by the guide. The degree of spreading depends on the ratio of aperture dimensions to wavelength. On flaring or constricting the terminal region of the guide in order to control the directivity of the radiated energy, one arrives at electromagnetic horns based on the same fundamental principles as acoustic horns (Fig. 1·2d).

Another type of element that appears in microwave antennas is the radiating slot (Fig. 1·2e). There is a distribution of current over the inside wall of a waveguide associated with the wave that is propagated in the interior. If a slot is milled in the wall of the guide so as to cut across the lines of current flow, the interior of the guide is coupled to space and energy is radiated through the slot. (If the slot is milled along the line of current flow, the space coupling and radiation are negligible.) A slot will radiate most effectively if it is resonant at the frequency in question. The long dimension of a resonant slot is nearly a half wavelength, and the transverse dimension a small fraction of this; the perimeter of the slot is thus closely a wavelength.

1·6. A Survey of Microwave Antenna Types.—We are now in a position to mention briefly the principal types of antennas to be considered in this book.

Antennas for Toroidal Beams.—A toroidal beam may be produced by an isolated half-wave antenna. This is a useful antenna over a large frequency range, the limit being set by the mechanical problems of supporting the antenna and achieving the required isolation. The beam thus produced, however, is too broad in elevation for many purposes.

A simple system that maintains azimuthal symmetry but permits control of directivity in elevation is the biconical horn, illustrated in

Fig. 1·3. The primary driving element between the apexes of the cones is a stub fed from a coaxial line. The spread of the energy is determined by the flare angle and the ratio of mouth dimension to wavelength. Although this antenna is useful over a large frequency range, maximum directivity for given antenna weight and size is obtainable in the microwave region, where the largest ratio of aperture to wavelength can be realized.

Increased directivity in a toroidal beam can also be obtained with an array of radiating elements such as dipoles, slots, or biconical horns built up along the symmetry axis of the beam. The directivity of the array is determined by its length measured in wavelengths; high directivities are conveniently obtained by this method only in the microwave region. A typical microwave array of this type is shown in Fig. 1·4.

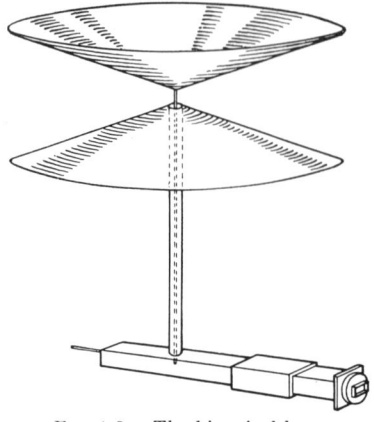

FIG. 1·3.—The biconical horn.

Pencil-beam Antennas.—Beams that have directivity both in elevation and azimuth may be produced by a pair of dipole elements or by a dipole with a reflecting plate. The major portion of the energy is contained in a cone with apex angle somewhat less than 180°.

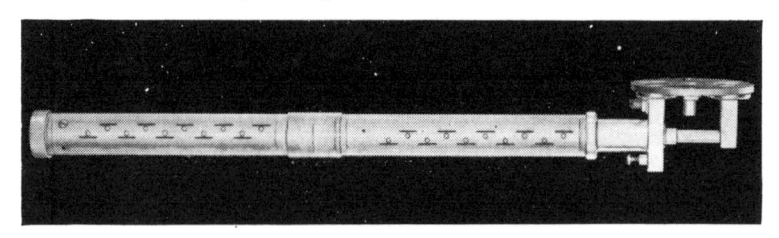

FIG. 1·4.—A microwave beacon array.

Similar beams are produced by horn antennas that permit control of the directivity through choice of the flare angle and the mouth dimensions. Horns are useful at lower frequencies as well as in the microwave region; indeed, the early work on horns was done for wavelengths ranging from 50 to 100 cm.

More directive beams—true pencil beams—can be produced by building up space arrays of the above systems. Two-dimensional arrays (mattress arrays) and multiunit horn systems are used at lower frequencies. Their directivity is severely limited, however, by the mechanical problems occasioned by the required ratio of dimensions to wavelengths. Such arrays have not been employed in the microwave region.

SEC. 1·6] A SURVEY OF MICROWAVE ANTENNA TYPES 11

At these wavelengths it becomes feasible, and indeed very convenient, to replace the two-dimensional array technique by the use of reflectors and lenses.

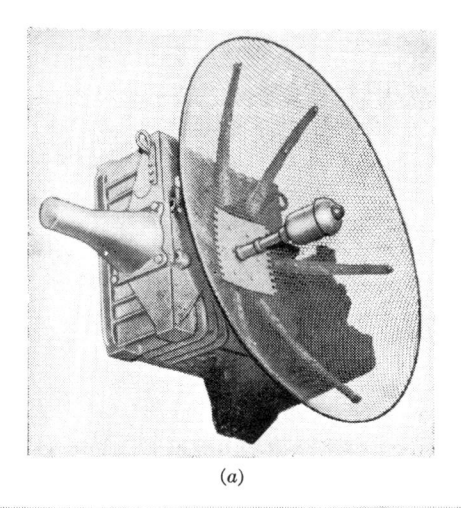

(a)

(b)

FIG. 1·5.—Pencil-beam antennas. (a) Paraboloidal mirror; (b) metal-plate lens. (*Metal-plate lens photograph courtesy of the Bell Telephone Laboratories.*)

Highly directive pencil beams are produced by placing a partially directive system such as the double-dipole unit, dipole-reflector unit, or horn at the focus of a paraboloidal reflector or a centrosymmetric lens. The use of these devices is based on the concepts of ray optics, according to which the reflector or lens takes the divergent rays from the point source at the focus and converts them into a beam of parallel rays. Despite the diffraction effects which limit the application of ray optics and are very important in the microwave region, it is practicable to make the apertures so large that extremely sharp beams can be produced. Conversely, it is possible to obtain good directivity with an antenna so small that aircraft installations are practical. Paraboloidal and parabolic reflectors are used at lower frequencies in some special cases, but in the required large sizes they tend to be less satisfactory than mattress arrays.

Plastic lenses are used in the microwave region in precisely the same way as glass lenses in the optical region. In addition, a new device, the metal lens, has been developed for microwaves. The wavelength of an electromagnetic wave in an air-filled waveguide is greater than that in free space; from the optical point of view the waveguide is a region of index of refraction less than unity. A stack of waveguides thus constitutes a refractive medium analogous to dielectric material, from which a metal lens can be fashioned. Figure 1·5 shows microwave pencil-beam antennas employing, respectively, a paraboloidal mirror and a metal lens as directive devices.

Antennas for Flared Beams.—Simple flared beams and one-sided flares are likewise produced by means of reflectors and lenses and by arrays of dipole-reflector units or radiating slots. Such arrays by themselves give beams that are highly directive in planes containing the array axis but are fairly broad in the transverse plane. In order to gain greater directivity in the transverse plane the array may be used as a line source along the focal line of a parabolic cylindrical reflector; this focuses radiation from a line source in the same way that a reflector in the form of a paraboloid of revolution focuses radiation from a point source. By suitable shaping of the cross section of the cylinder, one can produce beams with carefully controlled one-sided flares and other useful special characteristics. Typical microwave antennas of this type are shown in Fig. 1·6.

Except for a few types of linear array, all microwave antennas use primary sources of radiation together with reflectors and lenses. The radiating element, which extracts power directly from the transmission line, is spoken of as the "primary feed," the "antenna feed," or simply the "feed"; its radiation pattern as an isolated unit is known as the "primary pattern" of the antenna. In combination with the optical elements of the antenna, the feed produces the over-all pattern of the antenna, often referred to as the "secondary pattern" of the antenna.

One of our major problems will be to establish the relationships among the primary pattern of the antenna feed, the properties of the optical elements, and the secondary pattern.

(a)

(b)

Fig. 1·6.—Antennas for producing flared beams. (a) Simple flared-beam antenna; (b) one-sided flared-beam system.

1·7. Impedance Specifications.—The achievement of a satisfactory antenna pattern is by no means the only problem to be considered by the antenna designer. It is important that the antenna pick up maximum power from an incident wave and that it radiate the power delivered to it by a transmission line without reflecting an appreciable portion of it back into the transmitter. In other words, it is important that the antenna have satisfactory impedance characteristics.

The impedance problem in microwave antenna design takes on a somewhat special character because of the characteristics of other elements of the system, particularly the transmitting tubes. Conventional triode-tube oscillators are not generally useful in the microwave region. This is due to inherent limitations in the tube itself and to the fact that elements in the tank circuit no longer behave like lumped impedances. The self-resonant frequency of the ordinary tube is considerably below the microwave range, and it is therefore impossible to design a practical circuit that will oscillate at the required high frequency. A modified triode has been designed for use down to 10 cm. It has limited power capacity and is used where low power is acceptable. More generally, magnetrons and klystrons are used, the former for very high power levels. The operating characteristics of these tubes are very sensitive to the impedance into which they are required to operate, the frequency varying rapidly with changes in this impedance. More serious than this "frequency pulling" is the fact that the magnetron will cease to oscillate without too much provocation. Closer tolerances are, therefore, imposed on the impedance of a microwave antenna than those which would be dictated by power considerations. Many tubes can be tuned over a frequency band, but at any frequency setting they must operate into the proper impedance. Thus it is customary to specify that a microwave antenna be satisfactorily matched to the transmission line within close tolerances, not simply at an intended operating frequency, but over a band of frequencies.

In rapid-scanning antennas the impedance problem is even more complex. The arrangement of the mechanical parts varies during a scan; it is necessary to make sure that the impedance properties of the antenna remain satisfactory in all parts of the scan, as well as for a given range of wavelengths. This element of the problem has an important bearing on the choice of schemes for rapid-scanning antennas.

Throughout this volume the impedance characteristics of antennas will be considered in parallel with their radiation patterns.

1·8. Program of the Present Volume.—This book falls into four main divisions: basic theory, theory and design of feeds, theory and design of complete antenna systems, and antenna-measuring techniques and equipment.

The following chapter summarizes certain parts of conventional circuit theory that are pertinent to antenna problems. In particular, it is shown that the antenna designer need make no distinction between transmitting and receiving antennas. Chapter 3 states the basic principles of field theory and applies them to the discussion of current distributions as sources of radiation fields. Chapters 4 to 6 then discuss electromagnetic waves without regard to their sources. Chapter 4 gives a brief treatment of wavefronts and rays. Chapter 5 deals with the interaction between electromagnetic waves and obstacles; the general theory of reflectors is here developed as a boundary-condition problem, and a discussion is given of the relation between this theory and conventional diffraction theory, which also finds application to microwave antenna problems. Finally, Chap. 6 applies this theory in treating one of the fundamental problems of antenna design—the relation between the field distribution over the aperture of an antenna (such as a lens or reflector) and its secondary pattern.

Chapter 7, on microwave transmission lines, serves as introduction to the chapters on antenna feeds: dipole feeds, linear arrays, and horns. Of these types all but the first have found applications also as complete antennas; these applications will be indicated in these chapters.

A chapter on lenses precedes the treatment of more complex antenna systems which is organized according to the type of beam to be produced: pencil beams, simple fanned beams, and more complexly shaped beams. When an antenna is installed on ground or a ship or airplane—generally, enclosed in a housing—its performance is modified from that in free space by its enclosure and neighboring objects. The subject of antenna-installation problems is discussed briefly to acquaint the engineer with the phenomena that may be expected to occur and some of the currently known solutions of the problems.

The concluding chapters provide a statement of the basic techniques of antenna measurements and a description of certain types of measuring equipment that have given satisfactory service in the Radiation Laboratory.

CHAPTER 2

CIRCUIT RELATIONS, RECIPROCITY THEOREMS

By S. Silver

2·1. Introduction.—The circuit theory considerations and techniques characteristic of low-frequency radio work do not carry over in a simple manner to the microwave region. Thus, for example, in treating a circuit element as a lumped impedance, it is assumed that the current (and voltage) at any given instant has the same value at every point in the element. This assumption is valid if the dimensions of the circuit element are small compared with the wavelength, with the result that the phase differences between separated points in the element are negligible. If, however, the wavelength becomes comparable to the dimensions of the element, these phase differences become significant; at a given instant the current at one point in the element may be passing through its maximum value, while at another point it is zero. In such cases the circuit element must be regarded as a system of distributed impedances.

The extension of conventional circuit theory to microwave systems is further complicated by the use of circuit elements such as waveguides, in which voltages and currents are not uniquely defined. The analysis of these elements must be approached from the point of view that they serve to guide electromagnetic waves; attention is centered on electric and magnetic fields rather than on voltage and current. The final result of the field theory analysis is that under suitable conditions—which are generally encountered in practice—a waveguide can be set into equivalence with a two-wire transmission line in which the fundamental quantities are voltage and current. The latter are directly related to the waveguide's electric and magnetic fields, respectively.[1] By means of this equivalence the concepts of impedance, impedance matching, and loaded lines are carried over to waveguides.

A waveguide can itself be treated as a system of distributed impedances. Distributed impedances are treated in the same way as lumped impedances, by use of Kirchhoff's current and voltage laws for networks. A system of distributed impedance can, in fact, be replaced by a network of lumped-impedance elements. The latter differ from the conventional radio-circuit elements in that their impedance is a transcendental func-

---

[1] The subject is treated in Chap. 7. A full treatment of the extension of circuit theory to waveguides will be found in Vol. 8 of this series.

tion of frequency rather than an algebraic function. By means of these equivalent lumped-element networks, the network theorems that are applicable to low-frequency lumped-element networks are carried over to systems with distributed impedance. The first part of this chapter will review several network theorems and the two-wire transmission-line theory that are used in microwave circuit theory. The subjects will be treated briefly, the reader being referred to standard texts[1] for more complete discussions and proofs of the results quoted here.

The relation between a transmitting and a receiving antenna also can be expressed in terms of an equivalent network. In this way one can arrive at a reciprocity theorem which relates the transmission characteristics of an antenna to its receiving characteristics. Of particular importance to antenna design is the fact, proved by use ef the reciprocity theorem, that the transmitting pattern of an antenna is the same as its receiving pattern.[2] The reciprocity theorem will be discussed in the latter part of this chapter.

2·2. The Four-terminal Network.—Let us consider an arbitrary network, free from generators, made up of linear bilateral elements. A linear bilateral element is one for which the relation between voltage and current is linear:

$$V = IZ, \qquad (1)$$

where the value of the impedance Z is independent of the direction of the

FIG. 2·1.—Four-terminal network.

voltage drop across the element.[3] For convenience the network will be pictured as enclosed in a box and presenting to the outside only a pair of input and a pair of output terminals. This is illustrated schematically in Fig. 2·1. A boxed network of this type is referred to as a four-terminal or two-terminal-pair network.

The network as a unit involves four quantities: the current i_1, the voltage drop V_1 from A to B, the current i_2, and the voltage drop V_2 from C to D. In consequence of the linear property [Eq. (1)] of each component element of the network, the relations between the voltages V_1, V_2 and the currents i_1, i_2 are linear:

$$\left.\begin{array}{l} V_1 = Z_{11}i_1 - Z_{12}i_2, \\ V_2 = Z_{21}i_1 - Z_{22}i_2. \end{array}\right\} \qquad (2)$$

---

[1] W. L. Everitt, *Communication Engineering*, McGraw-Hill, New York, 1937; E. A. Guillemin, *Communication Networks*, Vols. I, II, Wiley, New York, 1931; T. E. Shea, *Transmission Networks and Wave Filters*, Van Nostrand, New York, 1929.

[2] See Chap. 1 for the definitions of these patterns.

[3] It is assumed that we are dealing with a single frequency, that both the voltage and current depend on time through the same factor $e^{j\omega t}$.

The impedance coefficient Z_{11} is the input impedance at AB when CD is open-circuited ($i_2 = 0$); similarly Z_{22} is the input impedance at CD when AB is open-circuited. The quantities Z_{12} and Z_{21} are known as the transfer impedance coefficients of the network. As a result of the bilateral property of the component elements of the network, the transfer impedance coefficients satisfy the reciprocity relation[1]

$$Z_{12} = Z_{21}. \tag{3}$$

As an alternative to the relations expressed by Eq. (2), the currents may be expressed as linear functions of the voltage:

$$\begin{aligned} i_1 &= Y_{11}V_1 - Y_{12}V_2, \\ i_2 &= Y_{21}V_1 - Y_{22}V_2. \end{aligned} \tag{4}$$

The admittance coefficient Y_{11} is the input admittance at AB when the terminals CD are short-circuited; Y_{22} is the admittance at CD when AB is short-circuited; and Y_{12}, Y_{21} are the transfer admittance coefficients. The latter coefficients satisfy a reciprocity relation

$$Y_{12} = Y_{21} \tag{5}$$

in the case of bilateral elements. The impedance and admittance coefficients of the network are related:

$$Y_{11} = \frac{Z_{22}}{\Delta}; \quad Y_{22} = \frac{Z_{11}}{\Delta}; \quad Y_{21} = Y_{12} = \frac{Z_{12}}{\Delta}, \tag{6}$$

where

$$\Delta = Z_{11}Z_{22} - Z_{12}Z_{21}. \tag{7}$$

By virtue of the reciprocity relations, [Eqs. (3) and (5)], the network has only three independent parameters. Consequently it can be replaced by a network of three lumped-impedance elements arranged in the form of either a T- or Π-section as shown in Fig. 2·2. The impedance elements of the T-section are designated by Z_1, Z_2, Z_3. In the case of the Π-section it is more convenient to use admittances; the elements are designated by $Y_A = 1/Z_A$, $Y_B = 1/Z_B$, $Y_C = 1/Z_C$. The relations between the elements of the reduced networks and the coefficients of Eqs. (2) and (4) are

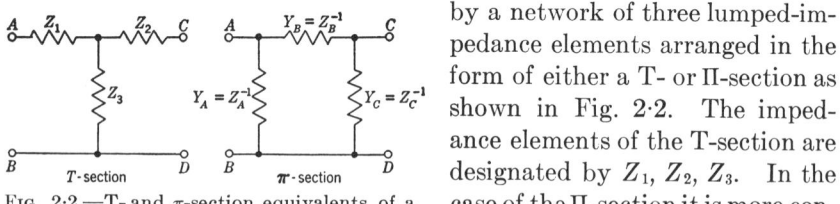

FIG. 2·2.—T- and π-section equivalents of a four-terminal network.

a. T-section:

$$\begin{aligned} Z_1 &= Z_{11} - Z_{12}, \\ Z_2 &= Z_{22} - Z_{12}, \\ Z_3 &= Z_{12}, \end{aligned} \tag{8}$$

---

[1] E. A. Guillemin, *op. cit.*, Vols. I, II, Wiley, New York, 1931, particularly Vol. I, Chap. IV.

b. Π-section:

$$\left.\begin{array}{l}Y_A = Y_{11} - Y_{12}, \\ Y_C = Y_{22} - Y_{12}, \\ Y_B = Y_{12}.\end{array}\right\} \quad (9)$$

The relations between the T- and Π-section elements for one and the same four-terminal network are given by

$$Z_A = \frac{\Delta}{Z_2}, \quad Z_B = \frac{\Delta}{Z_3}, \quad Z_C = \frac{\Delta}{Z_1}, \quad (10)$$

where the quantity Δ is that defined in Eq. (7).

The network can also be characterized by any three of the following measurable quantities: the input impedance at AB when CD is short-circuited, the input impedance at AB when CD is open-circuited, the input impedances at CD when AB is open-circuited or short-circuited. The relations between these quantities and the impedance coefficients or the T- and Π-section elements can easily be derived from Eqs. (2) and (8) or (9); they are given explicitly by Everitt.[1]

2·3. The Rayleigh Reciprocity Theorem.—The reciprocity relation between the transfer impedance coefficients given in Eq. (3) is fundamental to the various reciprocity theorems pertaining to networks. All of these theorems are variants of the general theorem derived by Rayleigh.[2] The particular form of the theorem as it applies to a four-terminal network will be discussed here.

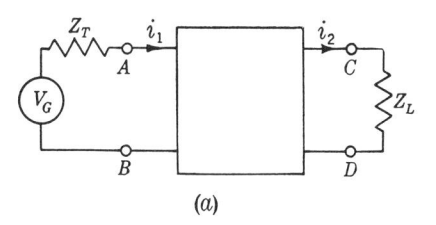

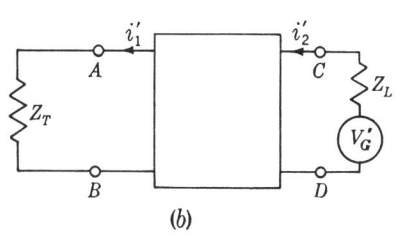

FIG. 2·3.—Reciprocity theorem for the four-terminal network.

In Fig. 2·3, i_1 and i_2 are the currents in the network terminals when a generator of emf V_G is applied to the terminals AB through an impedance Z_T to feed a load Z_L across the terminals CD; i_1' and i_2' are the corresponding currents at the terminals when a generator of emf V_G' is applied to the terminals CD through an impedance Z_L to feed a load Z_T across AB. The generator in each case is assumed to have zero internal impedance. The *reciprocity theorem* states that

$$V_G i_1' = V_G' i_2. \quad (11)$$

[1] W. L. Everitt, *op. cit.*, Chap. II.
[2] Rayleigh, *Theory of Sound*, Vol. I, Secs. 105–111, Macmillan, New York, reprinted by Dover Publications, New York, 1945.

Using Eqs. (2), we find for Case a of Fig. 2·3

$$i_2 = \frac{Z_{21}V_G}{(Z_{11} + Z_T)(Z_{22} + Z_L) - Z_{12}Z_{21}}.$$

For Case b, remembering that the role of input and output terminals must be interchanged in Eqs. (2), we have

$$i_1'' = \frac{Z_{12}V_G'}{(Z_{11} + Z_T)(Z_{22} + Z_L) - Z_{12}Z_{21}}.$$

Multiplying the first of these by V_G' and the second by V_G, one finds that the reciprocity theorem in Eq. (11) holds provided that $Z_{12} = Z_{21}$. Conversely, if a four-terminal network is linear in the sense of Eq. (2)

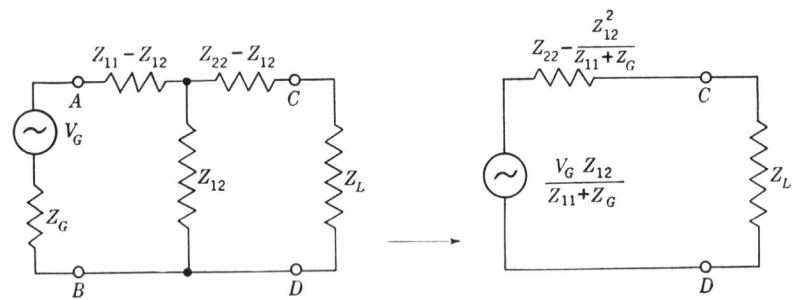

FIG. 2·4.—Thévenin's theorem and the maximum-power transfer condition.

and if the reciprocity theorem [Eq. (11)] holds for the network, then the transfer impedance coefficients satisfy the reciprocity relation of Eq. (3).

2·4. Thévenin's Theorem and the Maximum-power Theorem.—Consider a network made up of linear bilateral elements and containing a system of generators. Thévenin's theorem states that the current through any impedance Z_L across a pair of terminals C, D of the network is the same as the current in an impedance Z_L connected across a generator whose emf is the open-circuit voltage across CD (the voltage with Z_L removed) and whose internal impedance is the input impedance measured at CD looking into the passive network (the network with generators replaced by their respective internal impedances).[1] The theorem is illustrated diagrammatically in Fig. 2·4.

Thévenin's theorem is useful in discussing the conditions for maximum-power transfer from a generator through a network to a load impedance Z_L. As is well known, when a load impedance is connected directly to a generator of internal impedance Z_G, maximum-power transfer is effected with a load impedance that is the complex conjugate of the generator impedance:

$$Z_L = Z_G^*.$$

[1] W. L. Everitt, op. cit., p. 47.

Consider then the case in which the load Z_L is fed by the generator through a four-terminal network, the generator emf being V_G and ts internal impedance Z_G (Fig. 2·4). The four-terminal network may be replaced by its T-section equivalent as shown. By Thévenin's theorem the system is equivalent to a generator of emf $V_G Z_{12}/(Z_{11} + Z_G)$ and internal impedance $Z_{22} - Z_{12}^2/(Z_{11} + Z_G)$ feeding the load impedance Z_L directly. It follows then that maximum-power transfer will be achieved with a load that is the complex conjugate of the internal impedance of the effective generator:

$$Z_L = Z_{22}^* - \frac{(Z_{12}^*)^2}{Z_{11}^* + Z_G^*}. \quad (12)$$

2·5. The Two-wire Transmission Line.—One of the most important distributed-impedance systems from the point of view of antenna theory is the two-wire transmission line.[1] For the present the line will be considered in its conventional form, as a pair of linear conductors in a plane,

FIG. 2·5.—Two-wire line.

which support the propagation of a wave of wavelength small compared with the length of the lines The problem of interest is the distribution of voltage and current along the line for a wave of single frequency, in which the voltage and current vary with $e^{j\omega t}$.

The line is shown schematically in Fig. 2·5 as a pair of parallel wires. In general, however, the spacing between the wires may vary along the line; the only restriction imposed is that the line have an axis of symmetry. Position along the line is specified by the coordinate z along the symmetry axis. It is further assumed that the line is isolated from perturbing objects, so that at any position along the line the currents at every instant may be equal and opposite in the two component lines. The properties of the line are specified by its distributed parameters: (1) the series impedance per unit length,

$$\mathfrak{Z}(z) = R(z) + j\omega L(z), \quad (13a)$$

where $R(z)$ is the series resistance and $L(z)$ the series inductance per unit length, taking both component lines together, and (2) the shunt

---

[1] W. L. Everitt, *op. cit.* For a very complete treatment the reader is referred to R. W. King, H. R. Mimno, A. H. Wing, *Transmission Lines, Antennas, and Wave Guides*, McGraw-Hill, New York, 1945, Chap. 1.

admittance per unit length,

$$\mathfrak{N}(z) = G(z) + j\omega C(z), \tag{13b}$$

where $G(z)$ is the transverse conductance and $C(z)$ the capacitance per unit length between the component members of the line. These parameters may be functions of position because of variations in the conductors, in the spacing between the latter, or in the structure of the surrounding dielectric medium.

Taking either conductor for reference, let $i(z)$ be the current at the point z and $V(z)$ the voltage drop from the reference conductor to the other member at the same point. To obtain the space dependence of $i(z)$ and $V(z)$, consider a section of line of length dz about the point z. Applying Ohm's law, we have

$$V(z + dz) - V(z) = -i(z)\mathfrak{Z}(z)\, dz$$

and

$$i(z + dz) - i(z) = -V(z)\mathfrak{N}(z)\, dz$$

for, respectively, the series and shunt relations across the element of line. The terms on the left-hand side, by use of Taylor's theorem, become $(dV/dz)\, dz$ and $(di/dz)\, dz$ respectively. Thus the differential equations of the line are found to be

$$\frac{dV}{dz} = -\mathfrak{Z}(z)i(z), \tag{14a}$$

$$\frac{di}{dz} = -\mathfrak{N}(z)V(z). \tag{14b}$$

Second-order differential equations for voltage and current alone are obtained by eliminating voltage or current from one or the other of these equations:

$$\frac{d^2V}{dz^2} - \left[\frac{d}{dz}(\ln \mathfrak{Z})\right]\frac{dV}{dz} - (\mathfrak{Z}\mathfrak{N})V = 0, \tag{15a}$$

$$\frac{d^2i}{dz^2} - \left[\frac{d}{dz}(\ln \mathfrak{N})\right]\frac{di}{dz} - (\mathfrak{Z}\mathfrak{N})i = 0. \tag{15b}$$

From a generalized point of view, Eqs. (14) can be regarded as the definition of a "two-wire" transmission line. That is, given a physical system supporting a wave with time dependence $e^{j\omega t}$, the propagation of which is expressible in terms of a single coordinate z and two quantities (i, V) related by equations of the form of Eqs. (14), it is possible to set up a two-wire line representation for the system. The voltage and current of the equivalent line are directly proportional to the wave quantities entering the differential equations, and the series impedance and shunt admittance per unit length of the equivalent line are proportional to the coefficients of the wave quantities in the differential equations.

The generalized concept of a transmission line will be made use of in the discussion of waveguides in Chap. 7, where it will be seen that the electric and magnetic field vectors satisfy transmission-line equations.

2·6. The Homogeneous Transmission Line.—Equations (15) are the general equations for a line whose parameters $\mathfrak{Z}$ and $\mathfrak{R}$ are functions of position. We shall be concerned mainly with lines for which the parameters are independent of position, and the subsequent discussion will be confined to the so-called homogeneous line. For such a line the coefficients of dV/dz and di/dz in Eqs. (15) vanish; consequently, voltage and current satisfy the same differential equation. The voltage equation becomes

$$\frac{d^2V}{dz^2} - (\mathfrak{Z}\mathfrak{R})V = 0. \tag{16}$$

Defining the complex number γ by

$$\gamma = \alpha + j\beta = (\mathfrak{Z}\mathfrak{R})^{1/2} \tag{17}$$

with the square root taken to be such that both α and β are positive quantities, we find the solution of Eq. (16) to be

$$V(z) = A_1 e^{-\gamma z} + A_2 e^{\gamma z} \tag{18}$$

or

$$V(z) = A_1 e^{-\alpha z} e^{-j\beta z} + A_2 e^{\alpha z} e^{j\beta z}. \tag{18a}$$

The current $i(z)$ has the same form but is not independent of the voltage. The relation between them is established by Eq. (14a). On inserting Eq. (18) into this equation, it is found that

$$i(z) = \frac{1}{Z_0}(A_1 e^{-\gamma z} - A_2 e^{\gamma z}). \tag{19}$$

The constant Z_0 is known as the *characteristic impedance* of the line; it is given by

$$Z_0 = \left(\frac{\mathfrak{Z}}{\mathfrak{R}}\right)^{1/2}. \tag{20}$$

If Eq. (18a) is multiplied through by the time factor $e^{j\omega t}$, it will be seen that the right-hand side is the sum of two waves: The term $e^{-j\beta z}$ represents a wave traveling in the positive z-direction, whereas $e^{j\beta z}$ represents a wave traveling in the negative z-direction. The wavelength of propagation is related to the *phase constant* β by

$$\lambda = \frac{2\pi}{\beta}. \tag{21}$$

The amplitude of each component wave undergoes attenuation along the direction of propagation as represented by the factors $e^{-\alpha z}$ and $e^{\alpha z}$ respectively; α is known as the *voltage attenuation constant*. It is seen from

Eq. (17) that α may be different from zero, that is, the line may be lossy if one or both of the distributed parameters $\mathfrak{Z}$ and $\mathfrak{N}$ are complex, and that the line is nonlossy, $\alpha = 0$, if the distributed parameters are both pure imaginary quantities of the same sign. In the case of the two-wire line for which the distributed parameters are given by Eqs. (13) this means that the line is nonlossy if the series resistance and shunt conductance are zero, that is, if the distributed impedance along the line is purely reactive.

The amplitudes A_1 and A_2 of the component waves are determined by the excitation conditions at the input end of the line and the nature of the termination of the line. Consider a line of total length L, fed by a generator of emf V_G and internal impedance Z_G, and terminated in a load impedance Z_L as shown in Fig. 2·5. In this case the component waves are interpreted simply as a wave of amplitude A_1 incident on the load Z_L and a wave of amplitude A_2 reflected by it. Let the origin $z = 0$ be taken at the termination; the generator is thus located at $z = -L$. The impedance at any point z along the line looking toward the termination is the ratio $Z(z) = V(z)/i(z)$, which is, by Eqs. (18) and (19),

$$Z(z) = Z_0 \left(\frac{A_1 e^{-\gamma z} + A_2 e^{\gamma z}}{A_1 e^{-\gamma z} - A_2 e^{\gamma z}} \right). \tag{22}$$

At the terminal point, $z = 0$, this must be equal to the terminating impedance Z_L; we have then

$$\frac{A_1 + A_2}{A_1 - A_2} = \frac{Z_L}{Z_0}. \tag{23}$$

Thus the ratio of the amplitudes A_2/A_1 is determined solely by the termination. This shows also the significance of the characteristic impedance: If $Z_L = Z_0$, then $A_2 = 0$; there is no reflected wave. A line terminated in an impedance equal to its characteristic impedance thus behaves as though it extended to infinity.

A second relation between the amplitudes is obtained from the conditions at the input end of the line. The input impedance Z_{in} to the line is obtained from Eq. (22) by setting $z = -L$, and the current at the point is obtained from Eq. (19) by the same substitution. We have then

$$V_G = i_{(z=-L)}(Z_G + Z_{in}),$$

whence we obtain

$$\left(1 + \frac{Z_G}{Z_0}\right) A_1 e^{\gamma L} + \left(1 - \frac{Z_G}{Z_0}\right) A_2 e^{-\gamma L} = V_G. \tag{24}$$

From Eqs. (23) and (24) we finally get

$$A_1 = \frac{V_G Z_0 (Z_L + Z_0)}{(Z_G + Z_0)(Z_L + Z_0)e^{\gamma L} - (Z_G - Z_0)(Z_L - Z_0)e^{-\gamma L}} \tag{25a}$$

and
$$A_2 = \frac{V_G Z_0 (Z_L - Z_0)}{(Z_G + Z_0)(Z_L + Z_0)e^{\gamma L} - (Z_G - Z_0)(Z_L - Z_0)e^{-\gamma L}}. \quad (25b)$$

It should be noted that these expressions give the amplitudes of the incident and reflected waves at the termination, or more specifically at $z = 0$. The respective amplitudes $A_1'(z)$ and $A_2'(z)$ at an arbitrary point z are given in terms of the above by

$$A_1'(z) = A_1 e^{-\gamma z}; \qquad A_2'(z) = A_2 e^{\gamma z}. \quad (26)$$

The ratio of the amplitudes of the waves at any given point is known as the voltage reflection coefficient $\Gamma(z)$ at that point. We have

$$\Gamma(z) = \frac{A_2'(z)}{A_1'(z)} = \frac{A_2}{A_1} e^{2\gamma z} \quad (27)$$

or

$$\Gamma(z) = \Gamma(0) e^{2\gamma z}, \quad (27a)$$

where $\Gamma(0)$ is the reflection coefficient at the point $z = 0$. On making use of Eqs. (26) and (27) together with (22), we find that the relation between $\Gamma(z)$ and the impedance $Z(z)$ is

$$\Gamma(z) = \frac{Z(z) - Z_0}{Z(z) + Z_0}. \quad (28)$$

It is convenient for many purposes to introduce the *normalized impedance* $\zeta(z)$:

$$\zeta(z) = \frac{Z(z)}{Z_0}. \quad (29)$$

The relations between the reflection coefficient and the normalized impedance are then

$$\Gamma(z) = \frac{\zeta(z) - 1}{\zeta(z) + 1}; \qquad \zeta = \frac{1 + \Gamma}{1 - \Gamma}. \quad (30)$$

Equation (27a) expresses the transformation property of a transmission line. It is readily seen that Eq. (27a) can be generalized to the form

$$\Gamma(z \pm l) = \Gamma(z) e^{\pm 2\gamma l} = \Gamma(z) e^{\pm 2\alpha l} e^{\pm j 2\beta l}. \quad (31)$$

The phase of the reflection coefficient has a space periodicity of $\lambda/2$. The amplitude of the reflection coefficient is independent of position in a nonlossy line. In a lossy line it decreases as we move along the line toward the generator from the load, corresponding to the increase in the amplitude of the incident wave and the attenuation of the reflected wave. The transformation property of the line applies to the impedance likewise. From Eqs. (28) and (31) it follows that the impedance at a point $z - l$ is related to the impedance at the point z by

$$Z(z - l) = Z_0 \left[\frac{Z(z) + Z_0 \tanh (\gamma l)}{Z_0 + Z(z) \tanh (\gamma l)} \right]. \tag{32}$$

or, in terms of the normalized impedance,

$$\zeta(z - l) = \frac{\zeta(z) + \tanh (\gamma l)}{1 + \zeta \tanh (\gamma l)}. \tag{32a}$$

A section of line of length l thus serves as an impedance-transformation device, converting an impedance $Z(z)$ at the output end into an impedance $Z(z - l)$ at the input end. The impedance transformation is associated with the reflected wave; if the terminal impedance is equal to the characteristic impedance, the reflection coefficient vanishes and the input impedance at any point on the line (looking toward the termination) is equal to Z_0. If the reflection coefficient is zero, the termination is said to be matched to the line; otherwise, it is said to be mismatched.

The properties of the line can be discussed in terms of admittance as well as impedance. The corresponding relations are obtained by replacing Z by $1/Y$. The admittance transformation effected by a section of line is

$$Y(z - l) = Y_0 \left[\frac{Y(z) + Y_0 \tanh (\gamma l)}{Y_0 + Y(z) \tanh (\gamma l)} \right]. \tag{33}$$

where the characteristic admittance is defined to be

$$Y_0 = \frac{1}{Z_0}. \tag{34}$$

A normalized admittance $\eta(z)$ is defined in a similar manner as the normalized impedance

$$\eta(z) = \frac{Y(z)}{Y_0}. \tag{35}$$

and the relations between it and the reflection coefficient are

$$\Gamma = \frac{1 - \eta}{1 + \eta}; \qquad \eta = \frac{1 - \Gamma}{1 + \Gamma}. \tag{36}$$

2·7. The Lossless Line.—The further discussion of the transmission line will be particularized to the case of a lossless line. The microwave lines to be treated in Chap. 7 can be considered to be lossless over the length of line that enters into the problem of the design of an antenna. If the line is lossless, $\alpha = 0$ and the propagation constant γ is a pure imaginary,

$$\gamma = j\beta.$$

The voltage and current relations in this case are

$$V(z) = A_1 e^{-j\beta z} + A_2 e^{j\beta z}, \tag{37a}$$

$$i(z) = \frac{1}{Z_0} (A_1 e^{-j\beta z} - A_2 e^{j\beta z}), \tag{37b}$$

and the impedance and admittance transformation formulas become

$$\zeta(z - l) = \frac{\zeta(z) + j \tan \beta l}{1 + j\zeta \tan \beta l}, \quad (38)$$

$$\eta(z - l) = \frac{\eta(z) + j \tan \beta l}{1 + j\eta \tan \beta l}. \quad (39)$$

The transformations have a space periodicity of a half wavelength:

$$\zeta\left(z \pm \frac{\lambda}{2}\right) = \zeta(z);$$

$$\eta\left(z \pm \frac{\lambda}{2}\right) = \eta(z);$$

the impedance and admittance take on the same values at intervals of a half wavelength. The reflection coefficient is likewise periodic; if in Eq. (31) α is set equal to zero, we get

$$\Gamma(z \pm l) = \Gamma(z)e^{\pm 2j\beta l}. \quad (40)$$

Since Γ passes through a complete cycle of phase over a half-wavelength section of line, there are two points within every such interval at which Γ is a real number. It follows from Eq. (30) that at these points the impedance and admittance are real numbers. The magnitude of Γ does not vary along the line. Consequently, at every point the reflection coefficient is a measure of the power loss arising from the impedance mismatch at the termination. The power carried by the incident wave is proportional to $|A_1|^2$, and that carried by the reflected wave is proportional to $|A_2|^2$. The magnitude of Γ, is given by

$$|\Gamma| = \left|\frac{A_2}{A_1}\right|; \quad (41)$$

hence $|\Gamma|^2$ is the fraction of the incident power reflected by the termination, and $1 - |\Gamma|^2$ is the fraction of the incident power extracted by the termination.

In measurements on a transmission line the significant quantity is the square of the magnitude of the voltage averaged over a time cycle, given directly by $\frac{1}{2}|V(z)|^2$. In computing this from Eq. (37a) it must be remembered that the amplitudes A_1 and A_2 are in general complex. Writing

$$A_1 = |A_1|e^{j\phi_1}, \quad A_2 = |A_2|e^{j\phi_2}, \quad (42)$$

in Eq. (37a), we obtain

$$\tfrac{1}{2}|V(z)|^2 = \tfrac{1}{2}VV^* = \tfrac{1}{2}[|A_1|^2 + |A_2|^2 + 2|A_1A_2|\cos(2\beta z - \phi_1 + \phi_2)]. \quad (43)$$

Thus the time average $|V|^2$ takes the form of a standing-wave pattern along the line. The maxima and minima occur at those points for which

$$2\beta z - \phi_1 + \phi_2 = \pm 2n\pi$$

and

$$2\beta z - \phi_1 + \phi_2 = \pm(2n+1)\pi, \qquad n = 0, 1, 2, \cdots,$$

respectively, the distance between a maximum and adjacent minimum being a quarter wavelength. The maximum and minimum values are

$$|V|^2_{max} = (|A_1| + |A_2|)^2,$$
$$|V|^2_{min} = (|A_1| - |A_2|)^2.$$

The ratio of the maximum to minimum value is known as the *power standing-wave ratio*, designated here by r^2:

$$r^2 = \frac{|V|^2_{max}}{|V|^2_{min}} = \left(\frac{|A_1| + |A_2|}{|A_1| - |A_2|}\right)^2.$$

The square root of power standing-wave ratio r is known as the *voltage standing-wave ratio*. It follows from Eq. (41) that

$$r = \frac{1 + |\Gamma|}{1 - |\Gamma|}, \tag{44a}$$

with the inverse relation

$$|\Gamma| = \frac{r - 1}{r + 1}. \tag{44b}$$

The magnitude of Γ may be determined from the measured standing-wave ratio by means of Eq. (44b). The phase of Γ can be deduced from the positions of the maxima and minima. On inserting Eq. (42) into Eq. (27a) and replacing γ by $j\beta$, we have

$$\Gamma(z) = |\Gamma|e^{j(2\beta z - \phi_1 + \phi_2)}. \tag{45}$$

Thus Γ takes on real values at the points where the standing-wave pattern takes on maximum and minimum values. The phase of Γ may be taken to be zero at a maximum point, with

$$\Gamma = |\Gamma| = \frac{r - 1}{r + 1}; \tag{46a}$$

then at a minimum point the phase of Γ will correspondingly be π, and

$$\Gamma = |\Gamma|e^{j\pi} = \frac{1 - r}{1 + r}. \tag{46b}$$

The phase of Γ at any other point, taking a maximum or minimum position as a reference point, is then readily deduced by means of Eq. (40).

The impedance at any point can likewise be deduced from measured values of the standing-wave ratio and the positions of maxima or minima.

It was noted previously that the impedance and admittance is real at the points where Γ is a real number; hence the impedance is real at the maximum and minimum points of the standing-wave pattern. Making use of Eq. (30) together with Eqs. (46), we find in fact that the impedance takes on the following values at those points:

$$\zeta = \frac{1+|\Gamma|}{1-|\Gamma|} = r \quad \text{(max)}; \tag{47a}$$

$$\zeta = \frac{1-|\Gamma|}{1+|\Gamma|} = \frac{1}{r} \quad \text{(min)}. \tag{47b}$$

Given the value of ζ at any one such point, the value at any other point is obtained by means of the transformation formula [Eq. (38)]. Similar considerations apply to the admittance values. At the points of maxima and minima the admittance is a pure conductance with the values

$$\eta = \frac{1}{r} \quad \text{(max)}, \tag{48a}$$

$$\eta = r \quad \text{(min)}. \tag{48b}$$

2·8. Transformation Charts.—The impedance and reflection coefficient transformations along a line can be presented graphically in forms that are very useful in experimental work. There are many types of charts, of which two, the so-called circle diagrams, will be discussed here. They are especially suited to lossless lines.

The Reflection Coefficient (Smith) Charts.[1]—Consider first the reflection coefficient transformation along a lossless line as expressed by Eq. (45):

$$\Gamma(z) = |\Gamma|e^{j(2\beta z - \phi_1 + \phi_2)}. \tag{2·45}$$

Let us set up a complex plane, as shown in Fig. 2·6, with the real and imaginary axes associated

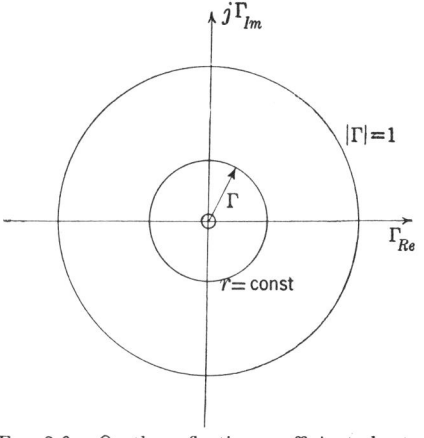

FIG. 2·6.—On the reflection coefficient chart.

with corresponding components of Γ, designated by Γ_{Re} and Γ_{Im}. Γ is then represented by a vector from the origin. The magnitude of Γ can never exceed unity because the amplitude of the reflected wave must be less than that of the incident wave; consequently we are confined to the portion of the complex plane circumscribed by the unit circle. It is evident that polar coordinates in the complex plane are more appropriate than the cartesian coordinates Γ_{Re}, Γ_{Im} for discussing

[1] P. H. Smith, *Electronics*, January, 1944.

the line transformation of Eq. (45). The family of circles centered about the origin correspond to curves $|\Gamma|$ = constant or, by virtue of Eq. (44a), to curves of constant voltage standing-wave ratio. The curves of constant phase of Γ are the family of radial lines from the origin. The line transformation given by Eq. (45) corresponds to a rotation of Γ about the origin without change in length: displacement along the line in the direction of increasing z, that is, away from the generator, produces an increase in the phase of Γ, thus rotating Γ in the positive sense (counterclockwise), whereas a displacement along the line toward the generator rotates Γ in the negative sense.

The polar coordinate curves are of such simple form that usually they are not drawn in explicitly on the chart. Instead, another pair of families of curves are introduced, the circles of constant resistive and reactive components of the impedance, R and X respectively. Writing

$$\zeta = \left(\frac{R}{Z_0}\right) + j\left(\frac{X}{Z_0}\right)$$

and $\Gamma = \Gamma_{\text{Re}} + j\Gamma_{\text{Im}}$, in Eq. (30),

$$\zeta = \frac{1 + \Gamma}{1 - \Gamma}, \tag{2.30}$$

and separating real and imaginary parts, one finds

$$\left.\begin{array}{l} \dfrac{R}{Z_0} = \dfrac{1 - (\Gamma_{\text{Re}}^2 + \Gamma_{\text{Im}}^2)}{(1 - \Gamma_{\text{Re}})^2 + \Gamma_{\text{Im}}^2}, \\[2mm] \dfrac{X}{Z_0} = \dfrac{2\Gamma_{\text{Im}}}{(1 - \Gamma_{\text{Re}})^2 + \Gamma_{\text{Im}}^2}. \end{array}\right\} \tag{49}$$

These can be written as

$$\left(\Gamma_{\text{Re}} - \frac{\frac{R}{Z_0}}{1 + \frac{R}{Z_0}}\right)^2 + \Gamma_{\text{Im}}^2 = \frac{1}{\left(1 + \frac{R}{Z_0}\right)^2} \tag{50a}$$

$$(\Gamma_{\text{Re}} - 1)^2 + \left(\Gamma_{\text{Im}} - \frac{Z_0}{X}\right)^2 = \left(\frac{Z_0}{X}\right)^2, \tag{50b}$$

respectively. It is seen from Eq. (50a) that the curve R/Z_0 = constant is a circle with its center on the positive real axis at $(R/Z_0)/(1 + R/Z_0)$ and radius $1/(1 + R/Z_0)$. Every such circle is tangent to the line $\Gamma_{\text{Re}} = 1$ at its point of intersection with the real axis. The circle corresponding to $R/Z_0 = 1$ passes through the origin and encloses all the circles for which $R/Z_0 > 1$.

Similarly Eq. (50b) shows that the curves X/Z_0 = constant are a family of circles. For a given value of X/Z_0, the center of the circle is

Sec. 2·8] TRANSFORMATION CHARTS 31

at the point $(1, Z_0/X)$ and its radius is $|Z_0/X|$. Every such circle is tangent to the real axis at the point $\Gamma_{Re} = 1$. The curves lying in the upper half of the plane correspond to positive (inductive) reactance, and those in the lower half plane to negative (capacitive) reactance. It can be shown that the circles $X/Z_0 =$ constant are orthogonal to the circles $R/Z_0 =$ constant.

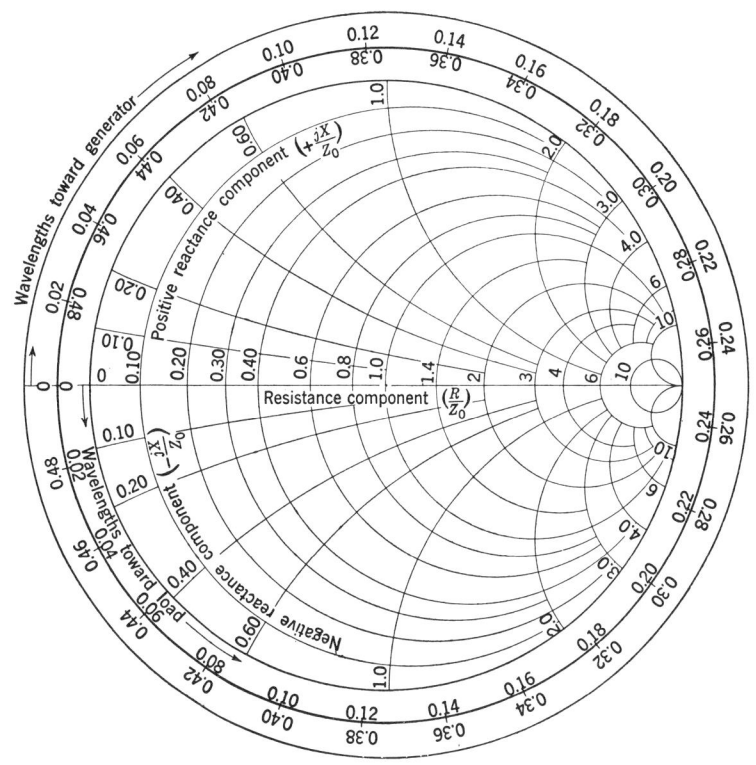

Fig. 2·7.—The Smith chart.

The Smith chart consists of the circles just described. A typical chart is shown in Fig. 2·7, the circles being labeled with the corresponding values of the parameters R/Z_0, X/Z_0. These curves serve as a system of coordinate lines. The terminal point of the vector Γ associated with the complex number $\zeta = (R/Z_0) + j(X/Z_0)$ is located at the intersection of the circles R/Z_0 and X/Z_0. The distance from the origin to the intersection of the circle R/Z_0 with the real axis is equal to the magnitude of the vector Γ that corresponds to a standing-wave ratio

$$r = \frac{R}{Z_0} \quad \text{if} \quad \frac{R}{Z_0} > 1,$$
$$r = \frac{Z_0}{R} \quad \text{if} \quad \frac{R}{Z_0} < 1.$$

This follows from the fact that Γ is real when ζ is real and from the relations of Eq. (47) between the value of ζ when it is a real number and the standing-wave ratio.

To illustrate these relationships let us suppose that the standing-wave ratio r has been measured on a given line, together with the position of a voltage minimum; the reflection coefficient and impedance are desired at a point a distance l from the minimum position away from the generator. It will be recalled [Eq. (46b)] that at a maximum position the phase of Γ is equal to π; Γ is then directed along the negative real axis. The impedance at this point is real, being $R/Z_0 = 1/r$. The vector Γ thus extends from the origin to the circle corresponding to $R/Z_0 = 1/r$. Counterclockwise rotation of this vector through an angle $2\beta l$ carries us to the desired point on the chart; the components of ζ at that point are read off from the pair of intersecting circles. It will be noted on Fig. 2·7 that the periphery of the chart carries a phase angle scale with the phase designated by the ratio of line length to wavelength.

The Smith chart can also be used to study the admittance transformation. First it should be noted that there are two conventions for the definition of admittance. The convention adopted in this book defines the normalized admittance $\eta = (G/Y_0) + j(B/Y_0)$ to be the reciprocal of the normalized impedance $\zeta = (R/Z_0) + j(X/Z_0)$; positive susceptance B thus corresponds to negative (capacitive) reactance. The other convention defines the admittance to be the conjugate of the reciprocal impedance, in order that positive susceptance (like positive reactance) should be inductive. The use of the latter convention changes the use of the chart in ways which the reader can easily develop.

Equation (36) gives the relation between the admittance and the voltage reflection coefficient:

$$\eta = \frac{1 - \Gamma}{1 + \Gamma}.$$

Let us define a new coefficient

$$\Upsilon = -\Gamma \tag{51}$$

and associate with it a complex plane with axes Υ_{Re} and Υ_{Im}. (Actually the same complex plane serves for both Γ and Υ, the two vectors making an angle of 180° with each other.) The vector Υ is, in fact, the *current reflection coefficient*, expressing the ratio of the amplitude of the reflected current wave to the amplitude of the incident current wave. The law of transformation of Υ along the line is precisely the same as that given for Γ by Eq. (49). On substituting Eq. (51) into the relation between η and Γ, we obtain

$$\eta = \frac{1 + \Upsilon}{1 - \Upsilon}. \tag{52}$$

This is the same as the relation between ζ and Γ; it follows that the curves

G/Y_0 = constant are a family of circles that coincide with the constant R/Z_0 family in the ζ-Γ transformation and that the curves B/Y_0 = constant coincide with the X/Z_0 circles. With respect to the latter it should be noted (in using the chart for admittance) that the curves lying in the upper half plane represent capacitive susceptance. The distinction that need be made between the use of the chart for impedance and admittance can be made clear by considering the problem of finding the admittance at a point distant l from a voltage minimum in the direction away from the generator, the standing-wave ratio again being r. At a voltage minimum Γ lies along the negative real axis; hence Υ extends along the positive real axis to the circle

$$\frac{G}{Y_0} = r.$$

The starting point thus lies on the positive real axis, instead of on the negative axis. Moving along the line away from the generator again rotates Υ in the positive sense (counterclockwise) through an angle $2\beta l$. The admittance at the new point is determined from the pair of intersecting coordinate curves, just as in the case of the impedance. It should be clear that the admittance and impedance points on the Smith chart for one and the same point on the line are diametrically opposite to one another.

The Smith chart is particularly suited to the study of an impedance mismatch that arises from the superposition of reflections. For example, there may be a series of discontinuities on a transmission line; the overall reflection coefficient at a given point is, to a good approximation, the vector sum of the reflection coefficients that would be produced at the point by the individual discontinuities acting separately in the absence of all the others. The vector addition of the component coefficients yields good results if the components are small. The subject will be considered further in later chapters in connection with specific problems.

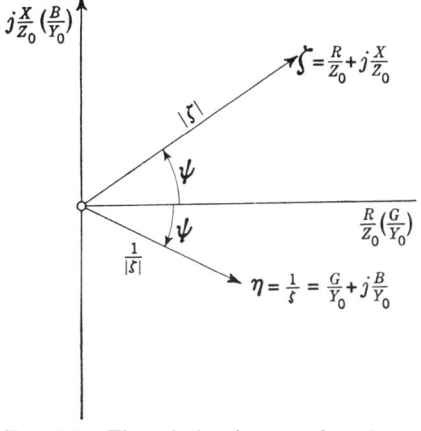

Fig. 2·8.—The relation between impedance and admittance.

The Bipolar Charts.—A complex plane can be set up of which the real and imaginary axes are associated with corresponding components of the normalized impedance ζ (or normalized admittance η) just as in the case of the reflection coefficients. Since the real parts R/Z_0 of the impedance and G/Y_0 of the admittance can never be negative, only the half plane containing the positive real axis comes under consideration. The impedance (admittance) is represented in this plane by a vector from the origin. With reference to the admittance we note again that it is taken here to be the reciprocal of the impedance. One and the same plane serves for both impedance and admittance; Fig. 2·8 shows the relation between the impedance and admittance points in the plane for a given point on a transmission line.

The impedance transformation

$$\zeta(z - l) = \frac{\zeta(z) + j \tan \beta l}{1 + j\zeta \tan \beta l} \qquad (2 \cdot 30)$$

does not take so simple a form in the ζ-plane as did the reflection coefficient transformation in the Γ-plane. Displacement along the line produces a change in both the magnitude and phase of the impedance. The geometrical transformation is simplified by introducing two families of circles: the curves $|\Gamma|$ = constant and the curves Γ-phase = constant. These curves are obtained from the Γ-ζ transformation $\Gamma = (\zeta - 1)/(\zeta + 1)$ of Eq. (30). Writing $\Gamma = |\Gamma|e^{j\Phi}$, we find that

$$|\Gamma|^2 = \frac{\left(\dfrac{R}{Z_0} - 1\right)^2 + \left(\dfrac{X}{Z_0}\right)^2}{\left(\dfrac{R}{Z_0} + 1\right)^2 + \left(\dfrac{X}{Z_0}\right)^2}$$

and

$$\tan \Phi = \frac{2\left(\dfrac{X}{Z_0}\right)}{\left(\dfrac{R}{Z_0}\right)^2 + \left(\dfrac{X}{Z_0}\right)^2 - 1}.$$

These can be rewritten as

$$\left(\frac{R}{Z_0} - \frac{1 + |\Gamma|^2}{1 - |\Gamma|^2}\right)^2 + \left(\frac{X}{Z_0}\right)^2 = \left(\frac{2|\Gamma|}{1 - |\Gamma|^2}\right)^2, \qquad (53)$$

$$\left(\frac{R}{Z_0}\right)^2 + \left(\frac{X}{Z_0} - \cot \Phi\right)^2 = \csc^2 \Phi, \qquad (54)$$

respectively. It will be seen that the curves $|\Gamma|$ = constant and Φ = constant are circles. The circle for a given $|\Gamma|$ has its center on the real axis at a distance $(1 + |\Gamma|^2)/(1 - |\Gamma|^2)$ from the origin; its radius is $2|\Gamma|/(1 - |\Gamma|^2)$. Curves of constant $|\Gamma|$ are also curves of constant standing-wave ratio. By Eq. (43b), we find that the center of the circle is at $(r^2 + 1)/2r$ and that its radius is $(r^2 - 1)/2r$. The circle intersects the real axis at the points $1/r$ and r, corresponding to the values that we obtained previously [Eq. (47)] for the impedance at these points on the line where it is real. These two points on the chart thus correspond to points on the line at which the voltage minima and voltage maxima,

SEC. 2·8]　　　　　TRANSFORMATION CHARTS　　　　　35

respectively, occur. The family of circles $|\Gamma|$ = constant is shown in Fig. 2·9, where they are labeled according to the *power standing-wave ratio* r^2.

A circle of constant phase, Φ = constant, has its center on the imaginary axis at the point $(0, \cot \Phi)$, and has a radius $|\csc \Phi|$. This second family of circles is orthogonal to the first, just as in the Γ-plane the curves of constant $|\Gamma|$ and constant phase are orthogonal. In the ζ-plane all

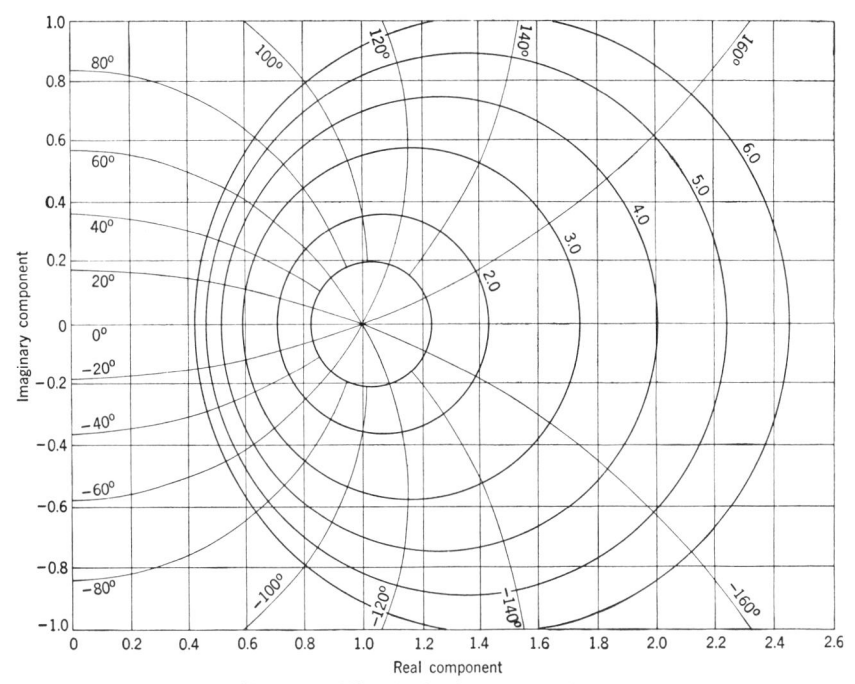

FIG. 2·9.—The bipolar impedance chart.

the constant-phase circles intersect in a point (1, 0), corresponding to the intersection of all the constant phase lines at the origin in the Γ-plane. The two families of curves in the ζ-plane, taken together with their image families in the left-hand portion of the plane, constitute a system of curvilinear coordinates known as the bipolar coordinates; hence the name of the chart.

The constant-phase curves are labeled in Fig. 2·9 so as to give directly the change in the phase of Γ corresponding to a displacement along the line from a voltage-minimum point. All voltage-minimum points must be on the segment of the real axis between zero and unity; this is therefore taken as the zero-phase line. The separation between a voltage minimum and the adjacent maximum on a line is $l = \lambda/4$, which corresponds to a phase shift $2\beta l = 180°$. All voltage-maximum points must lie on the real axis between 1 to ∞; hence this segment of the real axis is taken as the phase line $\Phi = 180°$.

By means of the bipolar curves the line transformation can be followed easily. It is clear that displacement along a given transmission line causes the impedance point in the ζ-plane to move around a circle of constant standing-wave ratio. Displacement in the counterclockwise sense corresponds to the same sense of rotation in the Γ-plane. A half wavelength of line produces a phase shift of $2\beta l = 360°$ and hence a complete revolution around the $r =$ constant circle. This periodic property of the impedance transformation was noted previously (Sec. 2·7). To illustrate the use of the chart, consider again a line in which a standing-wave ratio r has been measured and a voltage minimum point has been located. It is desired to find the impedance at a distance l from the minimum point away from the generator. The starting point is the intersection between the r^2-circle and the real axis on the segment (0, 1). We then move counterclockwise on the r^2-circle until it intersects the constant phase circle $\Phi = -2\beta l$; this is the desired impedance point.

The same families of bipolar curves serve for the admittance diagram likewise. In using the chart for admittance it must be noted that voltage minimum points are on the segment of the real axis (1, ∞) while voltage maximum points lie on the segment (0, 1). If the voltage minimum is retained as a zero-phase reference point, the real axis segment (1, ∞) must be taken as the zero-phase line and the segment (0, 1) as the 180° line. The sense of rotation about a circle $r =$ constant remains the same.

It should be kept in mind that the normalized impedance is discontinuous across a junction between lines of different characteristic impedances; the impedance itself is continuous. On moving across such a junction the point in both the reflection coefficient and the bipolar charts in general will move from one circle $r =$ constant to another. If we pass from a line of characteristic impedance Z_{0_1} to a line of characteristic impedance Z_{0_2}, the normalized impedance undergoes a change given by

$$\zeta_2 = \frac{Z_{0_1}}{Z_{0_2}} \zeta_1.$$

2·9. The Four-terminal Network Equivalent of a Section of Transmission Line.—For many purposes, in the analysis of systems involving transmission lines it is convenient to replace a section of line by its equivalent four-terminal network. The elements of the network will be derived here for the case of the lossy homogeneous line. Consider a section of line of length l, and take the origin $z = 0$ at the input end; let V_1, i_1 be the voltage and current at this end, and let V_2, i_2 be the voltage and current at the output end $z = l$. From the line equations [Eqs. (18) and (19)] we have then

$$z = 0: \quad V_1 = A_1 + A_2,$$
$$i_1 = \frac{1}{Z_0}(A_1 - A_2);$$
$$z = l: \quad V_2 = A_1 e^{-\gamma l} + A_2 e^{\gamma l},$$
$$i_2 = \frac{1}{Z_0}(A_1 e^{-\gamma l} - A_2 e^{\gamma l}).$$

Using the two current equations to solve for A_1 and A_2 in terms of i_1 and i_2 and substituting into the voltage equations, we obtain

$$\left. \begin{array}{l} V_1 = Z_{11} i_1 - Z_{12} i_2, \\ V_2 = Z_{21} i_1 - Z_{22} i_2, \end{array} \right\} \quad (55)$$

with

$$Z_{11} = Z_{22} = Z_0 \coth(\gamma l), \quad (56a)$$
$$Z_{12} = Z_{21} = Z_0 \operatorname{csch}(\gamma l). \quad (56b)$$

We thus find directly that the network is linear and that the transfer impedance coefficients satisfy the reciprocity relations. Since the line is homogeneous, the network is symmetrical with respect to its two ends; hence $Z_{11} = Z_{22}$. For a nonlossy line $\gamma = j\beta$; on substitution into the above, the network parameters are found to be

$$Z_{11} = Z_{22} = -jZ_0 \cot \beta l, \quad (57a)$$
$$Z_{12} = Z_{21} = -jZ_0 \csc \beta l. \quad (57b)$$

TRANSMITTING AND RECEIVING ANTENNAS

2·10. The Antenna as a Terminating Impedance.—The impedance relations between a transmitting or receiving antenna and its transmission line are of particular interest. In the following sections several general ideas that are associated with the analysis of these relations will be discussed. Let us consider first the case of a line feeding a transmitting antenna. It will be assumed for the present that the antenna is isolated—in particular, that it is removed from all other antennas—so that interactions with other systems need not be considered. The antenna functions like a dissipative load on the line in that it extracts power from it; part of this energy is radiated into space, and part is dissipated into heat in the antenna structure. In general, the antenna does not absorb all of the power incident on it from the line but gives rise to a reflected wave in the line; in effect the line is terminated by an impedance different from its characteristic impedance. However, the definition of the terminal impedance representing the antenna is not free from ambiguity and requires some consideration.

It is to be noted first that the definition of a terminal impedance implies the identification of a driving point, or set of input terminals, for the antenna. In some cases, such as the half-wave dipole or rhombic loop antennas fed from a two-wire line as illustrated in Fig. 1·2, the structural discontinuity between the line and the radiator suggests a driving point. This, however, is not enough; it is necessary that the current distribution in the line be that characteristic of a transmission line up to the assigned driving point. At long wavelengths this condition is realized with the antennas cited above: the interaction between the antenna and line can be represented by a lumped reactive impedance across the driving terminals in parallel with the impedance characteristic of the antenna itself. At short wavelengths, however, the interaction between the radiating system and the line causes a perturbation of the current distribution on the latter that may extend back over an appreciable distance; electrically there is no point of transition from transmission-line currents to antenna currents. This is a particularly cogent point in the case of microwave systems that make use of waveguide lines, in which the electromagnetic fields exist in the form of a number of modes.[1] A waveguide is equivalent to a two-wire line only when it is supporting propagation of a wave in a single mode. Microwave lines are, in fact, generally so designed that they can support free propagation of only one mode. Nevertheless, though a single mode is incident on the antenna, the antenna itself excites other modes, in addition to giving rise to a reflected wave in the incident mode. It is only at points so far from the antenna that the other modes have been attenuated to negligible amplitudes that a waveguide is equivalent to a two-wire line. Attention should also be called to the absence of a unique driving point in cases where the transition from the line to the radiator is effected by a continuous structural transition. An example of this is a waveguide flaring gradually into a horn without structural discontinuities in the walls. In these cases, again, the transition from transmission-line currents to antenna currents cannot be localized to a point.

The action of an arbitrary antenna as a terminal load on the line can be specified in terms of the reflection coefficient Γ measured in the transmission line, at a point so far from the antenna that its only effect is the production of the reflected transmission-line wave. At any point in the transmission-line region an impedance (or admittance) can be determined from the measured Γ, by means of Eq. (30); this can be taken as the load impedance terminating the line at that point. Furthermore, any such point may be regarded as the junction between the line and the input terminals to the antenna in so far as the practical analysis of the system is concerned.

This raises the question of the representation of an antenna by an equivalent network. There is no unique network associated with a

[1] See Chap. 7.

given value of Γ. The load to be associated with Γ at a point taken arbitrarily as the input to the antenna may be represented by an impedance in the form of a series combination of a resistance and reactance or equally well by an admittance made up of a resistance and reactance in parallel. In either case the resistance measures the power dissipated in the region beyond the input terminals to the antenna; this, if the line is lossless, is the power dissipated by the antenna in radiation and ohmic losses. If $\bar{P}$ is the total power (averaged over a cycle) dissipated by the antenna and $\bar{\imath}$ and $\bar{V}$ are respectively the effective current and voltage at the input terminals, the resistance of the impedance representation is given by
$$\bar{P} = \bar{\imath}^2 R \tag{58a}$$
and that of the admittance representation is given by
$$\bar{P} = \frac{\bar{V}^2}{R} \tag{58b}$$

It is tempting to carry over the concept of radiation resistance, used so extensively in the long-wavelength region. The total power dissipated by the antenna is the sum of the radiated power $\bar{P}_r$ and the power $\bar{P}_o$ dissipated in ohmic losses in the antenna structure. Correspondingly, the resistive component of the impedance representing the antenna would be taken as the sum of two elements: an ohmic resistance R_o and a radiation resistance R_r. Each element would be given in terms of the power component by a defining relation such as Eq. (58), for example, the radiation resistance by
$$\bar{P}_r = \bar{\imath}^2 R_r. \tag{59}$$
In the long-wavelength region this resolution is possible because one can define uniquely a driving point at which the antenna network can be dissociated from the line and because it is possible, on the basis of field theory, to set up an unambiguous network that is characteristic of the antenna itself. In the case of the dipole and loop antennas referred to earlier the network is a series combination of a resistance and reactance. However, in the general case, where the driving point is merely an arbitrary reference point on the line, the antenna network cannot be dissociated from the line, and either an impedance or an admittance representation can be used. In the admittance representation the resolution of R into an ohmic component R_o and a radiation component R_r (if it is to be made at all) must place the two components in parallel. In view of the transformation properties of the line, it is evident that these resistances will be functions of the position of toe reference point.

The practical significance of the reference point and of the "antenna impedance" Z_A determined from the measured value of Γ at that point may be illustrated by reference to the matching problem. Let l be the length of line from the reference point (regarded now as the terminal point of the line) to the generator. It follows from Secs. 2·4 and 2·9, the line will transfer maximum power from the generator to a terminal load of impedance

$$Z_L = jZ_0 \cot \beta l + \frac{Z_0^2 \csc^2 \beta l}{Z_G^* + jZ_0 \cot \beta l}, \qquad (60)$$

where Z_G is the internal impedance of the generator and Z_0 is the characteristic impedance of the line. If the "antenna impedance" Z_A is different from Z_L, it is possible to introduce a reactive network between the input terminals of Z_A and the line, which (at one frequency at least) transforms Z_A into Z_L; this network will effect maximum-power transfer to the antenna.

It is to be noted that in microwave systems another matching problem exists: The characteristics of the generator are such that the reflected wave in the line must be eliminated. This requires that the antenna impedance Z_A be transformed into Z_0—in general a different transformation from that required by the maximum-power-transfer condition. In these systems the generator must be independently matched to the line; the generator internal impedance Z_G is transformed into Z_0 with the result [cf. Eq. (60)] that the maximum-power condition then coincides with the condition for eliminating the reflected wave in the line.

2·11. The Receiving Antenna System.—The equivalent circuit representations used in discussing receiving antennas also need examination. Consider an arbitrary antenna—it may be a center-driven dipole, a horn, or a combination of such elements with reflectors and lenses—feeding into one end of a transmission line that at the other end is terminated in a passive load impedance. (That is, the receiving circuit is free from generators.) When an external electromagnetic field falls on the receiving-antenna system, the interaction between the antenna and the field gives rise to a wave in the line. The antenna may be regarded as a device that transforms energy carried by a free wave in space into energy carried by a guided wave on the transmission line. From the point of view of the terminal load, however, the antenna functions as a generator, and it is customary to replace it by a generator in discussing the efficiency of the receiving system as it depends on the antenna, line, and load impedances. It is our purpose to discuss the nature of the equivalent generator. In this connection the problem of modes in microwave systems again arises. The field excited in the line by the antenna always consists of a number of the modes that are possible in the given line. It will be assumed that the line is designed to support free propagation of a single mode and that the length of line between the antenna and load is more than sufficient to attenuate the other modes to negligible amplitudes; there will then be an appreciable region over which the guide is equivalent to a two-wire line.

Before discussing the equivalent generator representation, it will be well to consider briefly the physical processes of the interaction between the receiving system and the external field. For this purpose it will be assumed that an essentially plane wave from a very distant source is falling on the receiving antenna. In the neighborhood of the receiving antenna the incident wavefront may be regarded as a plane surface, over which the electric and magnetic field intensities are sensibly constant in magnitude; furthermore, the electric and magnetic field vectors lie in the plane, normal to the direction of propagation of the wavefront.[1] We shall assume for the moment that the load impedance terminating the line is equal to the characteristic impedance of the line. Under the action of the incident wave a distribution of currents and charges is excited in the antenna structure; the currents are communicated to the transmission line and give rise to a wave in this which proceeds toward the load. Since the load is matched to the line, this wave is completely absorbed by the load. The current and charge distribution existing on the antenna under this matched-load condition will be designated as the primary induced distribution.

Consider now an arbitrary load impedance. This will absorb only part of the wave excited by the primary induced distribution and will give rise to a reflected wave, which will proceed to the antenna and excite there a charge and current distribution, as if the system were a transmitting system. This new distribution of charges and currents will be termed a secondary induced distribution. The reaction of the antenna to the reflected wave depends on the impedance of the antenna relative to the line, as discussed in the preceding section. If the antenna impedance is equal to the line characteristic impedance, there will exist in the line only the two component waves already mentioned. On the other hand, if the antenna is mismatched, there will occur a process of multiple reflection between the antenna and the load. The resultant secondary induced distribution on the antenna is the sum of the component distributions arising from the multiple reflections between the antenna and load; its magnitude and phase relative to the primary distribution are determined by the antenna and load impedances and the length of line between them. It will be recognized that since the component waves are all of the same frequency, the net result inside the line is two waves, one—the resultant incident wave—traveling toward the load, and the second—the resultant reflected wave—traveling away from it. Their relative amplitudes are given by the reflection coefficient corresponding to the impedance mismatch between the load and the line.

Since the primary and secondary induced distributions on the antenna both vary with time (with a frequency equal to that of the incident wave), they radiate and set up an electromagnetic wave in space. This

[1] A general treatment of electromagnetic fields is given in Chap. 3.

wave is known as the scattered wave. The interaction between the receiving system and the incident wave is completely expressed in the relation between the scattered wave and the incident wave fields. There are two interaction effects: (1) energy is taken from the incident wave and dissipated in heat in the antenna, the line and the load, being thus completely lost to the field in space, and (2) energy is taken from the incident wave and reradiated into all directions about the antenna. The first effect is known as *absorption;* the second as *scattering.* If the dimensions of the antenna are large compared with the wavelength, the interaction between the scattered wave field and the incident wave is such as to give rise to a rather sharply defined shadow region behind the antenna, that is, on the side of the antenna away from the source of the incident wave. In this direction the scattered wave set up by the induced distribution on the antenna is out of phase with the incident wave; the destructive interference between the two fields results in the removal of energy from the incident wave. This energy includes both the *absorbed* and *scattered* energy.[1] If the dimensions of the antenna are of the order of magnitude of the wavelength or are small compared with it, there is no sharply defined shadow region. The fundamental process is the same, however, in that destructive interference between the scattered wave and the incident wave in various directions removes energy from the latter wave; this energy is in part absorbed and in part scattered by the antenna.

The interaction between the antenna and the incident wave may be visualized by thinking of the antenna as presenting a certain interception area or cross section to the incident wave and removing from it all the energy incident on the cross section. The total interception area is resolved into two parts: the *absorption cross section* and the *scattering cross section.* Reference was made to cross sections in Sec. 1·2. To repeat: Let S be the power intensity, that is, power flow per unit area of the incident wave, P_{abs} and P_{scat} the absorbed and scattered powers, and A_r and A_s the corresponding cross sections; then

$$P_{abs} = A_r S, \qquad (61a)$$
$$P_{scat} = A_s S. \qquad (61b)$$

The cross sections are functions of the aspect presented by the antenna to the incident wave. The reader is referred to Sec. 1·2 for the definition of the receiving pattern.

The definition and measurement of the absorbed power is unambiguous in principle. In microwave systems the power dissipation in the antenna and line is generally small compared with that in the load; hence the

---

[1] The significance of the shadow has been discussed in great detail for the case of a plane wave incident on a sphere by L. Brillouin, "On Light Scattering by Spheres," *Applied Math. Panel Reports,* NDRC, Columbia University, 87.1, December 1943, and 87.2, April 1944.

absorption cross section—or receiving cross section—can be evaluated with small error from the power absorbed by the load. The scattered power, however, is not directly measurable, and its theoretical evaluation is subject to ambiguities. Although electromagnetic fields are additive, their energies are not additive, the resultant energy being modified by the interaction between the fields. Consequently the energy flow computed for the scattered wave field, regarded as isolated from the incident wave field, does not necessarily represent the energy removed from the latter and reradiated in all other directions. This is particularly true when the antenna dimensions are comparable to the wavelength and the interaction between the scattered and incident waves, which results in removal of energy from the latter, cannot be localized to a well-defined shadow region.

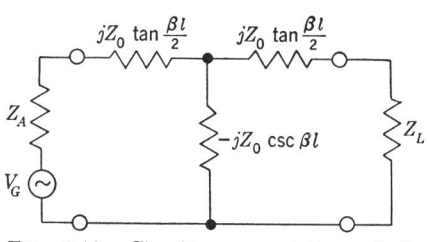

FIG. 2·10.—Circuit representation of the receiving antenna system.

The equivalent circuit representation of the receiving system is based on the fact that the antenna functions like a generator in so far as the load is concerned. In replacing the antenna by an equivalent generator it is generally assumed (1) that for a given aspect of the antenna toward the incident wave, the emf of the generator is proportional to the field intensity of the wave and (2) that the generator has an internal impedance equal to the input impedance which the antenna presents to the line when used as a transmitter. The complete circuit is shown in Fig. 2·10, where the line, assumed to be nonlossy, is replaced by its equivalent T-section; Z_A and Z_L are the antenna and the load impedance respectively. It is evident that this circuit representation involves the same difficulties as the representation of the antenna by a load impedance—the definition of Z_A and of the input terminals to the antenna. When a driving point can be localized in the transmission problem, the same point also serves for the output terminals of the generator feeding the line in the receiver problem. More generally, when the input terminals to the antenna can be defined only as an arbitrary reference point on the line, the generator voltage must be a function of the position of that point; it is not *a priori* evident that the power relations between the antenna and load calculated on the basis of the equivalent circuit are independent of the choice of antenna terminals. It will be shown in a later section that the results for the absorption cross section are independent of that choice.

It will be noted that in Fig. 2·10 power is dissipated both in the load impedance Z_L and in the internal impedance of the generator. The power dissipated in Z_L is interpreted as the power absorbed from the incident wave by the antenna and delivered to the load. The power dissipated in Z_A is frequently interpreted as the scattered power—the power absorbed by the antenna (dissipated in its ohmic resistance) plus the power reradiated. Neglecting the ohmic losses, the power dissipated in Z_A would thus measure the scattering cross section. It will, however, be seen in Sec. 2·12 that the power dissipated in the internal impedance of the equivalent generator has no direct relation to the energy reradiated by the antenna and in general cannot be used in discussing the scattering cross section. Two important cases in which the above interpretation is valid are that of the dipole antenna and the small (compared with wavelength) loop antenna. In these antennas, the current distributions induced by the incident wave under conditions of matched load terminations are the same as the currents excited on the antennas when they are driven by the line in transmission.

The equivalent circuit representation can thus in general be used only for the treatment of absorption. It is readily found that the power delivered to the load by the generator is given by

$$P_{abs} = \frac{1}{2} \frac{|V_G|^2}{D^2} \operatorname{Re} \left\{ jZ_0 \tan \frac{\beta l}{2} - \left[\frac{jZ_0 \csc \beta l \left(Z_L + jZ_0 \tan \frac{\beta l}{2} \right)}{Z_L + jZ_0 \tan \frac{\beta l}{2} - jZ_0 \csc \beta l} \right] \right\} \quad (62)$$

where

$$D^2 = \left| Z_A + jZ_0 \tan \frac{\beta l}{2} - \left[\frac{jZ_0 \csc \beta l \left(Z_L + jZ_0 \tan \frac{\beta l}{2} \right)}{Z_L + jZ_0 \tan \frac{\beta l}{2} - jZ_0 \csc \beta l} \right] \right|^2 \quad (62a)$$

The condition for maximum-power transfer from the generator to the load in the equivalent circuit gives the impedance relations required for maximum absorption cross section: the load impedance Z_L must be such that its impedance, transformed through the T-network of Fig. 2·10, is equal to the complex conjugate of Z_A. It was noted before that if a conjugate impedance relationship exists across any point in the line, it exists at all points on the line; consequently the load impedance determined by the conjugate condition is independent of the arbitrary point taken to be the input terminals of the antenna.

It follows from Eq. (62) that the absorption cross section is zero when the line is terminated in either a short circuit ($Z_L = 0$) or an open circuit ($Z_L = \infty$). In each case the reflection coefficient of the termination has the magnitude unity, and all power incident on the termination is reflected. It is of interest for these cases to compute the power dissipated in Z_A on the basis of the circuit representation. We find

$$P_{op} = \frac{|V_G|^2}{2|Z_A - jZ_0 \cot \beta l|^2} \operatorname{Re} Z_A, \quad (Z_L = \infty), \quad (63a)$$

$$P_{cl} = \frac{|V_G|^2}{2|Z_A + jZ_0 \tan \beta l|^2} \operatorname{Re} Z_A, \qquad (Z_L = 0). \tag{63b}$$

In both cases there are certain lengths of line

$$\left[\frac{n\lambda}{2} \text{ for } Z_L = \infty, \quad \left(\frac{n\lambda}{2}\right) - \left(\frac{\lambda}{4}\right) \text{ for } Z_L = 0, \, n \text{ being an integer} \right]$$

for which the power given by these equations is equal to zero. For cases in which the dissipation in Z_A may be interpreted as scattered power, this means that the scattering cross section vanishes for the stated terminations and associated line lengths. This can be understood readily from physical considerations. Since the reflection coefficient of the load is unity, the voltage impressed across the driving point of the antenna by the reflected wave in the line is equal in magnitude to that impressed by the external incident wave. The current distributions excited on the antenna by the two waves are the same except for phase; hence, by suitable adjustment of the line length, the primary and secondary induced distributions on the antenna can be put 180° out of phase, with the result that they give rise to no resultant scattered wave. The absorption and scattering cross sections are then both equal to zero. Similar phenomena can be observed with more general types of antennas. The phase between the primary and secondary induced antenna distributions is determined by the load impedance and the line length. If the load reflection coefficient is unity, the component distributions on the antenna will be comparable in magnitude, and by suitable adjustment of the line length their relative phase can be adjusted to give a minimum scattering cross section.

2·12. The Transmitter and Receiver as a Coupled System.—The preceding sections treat the transmitting and receiving antennas as isolated systems and neglect the significant feature of the interaction between them. Any discussion of a transmitting pattern implies the presence of a receiving antenna to explore the field; conversely, a discussion of a receiving antenna assumes the existence of a radiating system. The interaction between the transmitter and receiver is a result of scattering. Consider a transmitting antenna that, when completely isolated, is matched to its line. When a receiving antenna is introduced into the field of this transmitting antenna, it gives rise to a scattered wave. This, when intercepted by the transmitting antenna, in turn gives rise to a wave transmitted down the feed line of that antenna. The net effect is that the transmitting antenna no longer presents a matched impedance to its line. The transmitting antenna also in turn gives rise to a scattered wave that is partly absorbed by the receiving system and partly rescattered. The interaction between the two antennas is thus due to multiple scattering and absorption.

From the point of view of the transmission lines, the antennas and the external space form a network that couples the lines together. In Fig. 2·11, A and B represent the transmitter and receiver respectively, and O and O' are arbitrary but fixed reference points on the respective lines. It will be assumed that there is no activated generator other than the one feeding the transmitter A; the network between O and O'

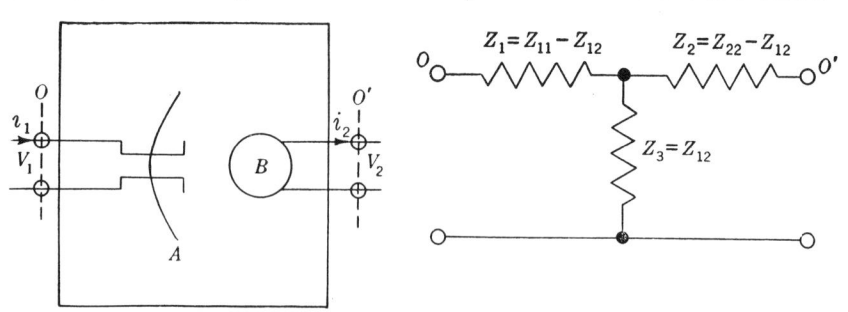

Fig. 2·11.—Four-terminal network representation of the coupled transmitter-receiver.

is passive. It will also be assumed that the network is a four-terminal network in the sense of Sec. 2·2. Thus the voltages and currents V_1, i_1 at O and V_2, i_2 at O', are linearly related:

$$V_1 = Z_{11}i_1 - Z_{12}i_2;$$
$$V_2 = Z_{21}i_1 - Z_{22}i_2;$$

and the transfer impedance coefficients obey the reciprocity condition $Z_{12} = Z_{21}$. The transfer impedance expresses the coupling between the antennas. The basis for these assumptions concerning the properties of the network is discussed in Secs. 2·16 to 2·18.

The network may be replaced by an equivalent T-section in the manner discussed in Sec. 2·2. This has been indicated in Fig. 2·11. The impedance coefficients are functions of the antennas, their relative configurations, the properties of the external medium and of the transmission lines, and the distance between the antennas. In the case of waveguide lines, the reference points O and O' defining the network terminals must be at such distances from the antennas that all modes other than that for which the line is designed have negligible amplitudes.

As the distance R_{AB} between the antennas increases, the importance of multiple scattering diminishes. The amplitude of the wave returning to a given antenna as a result of a single scattering process is attenuated by a distance factor $(R_{AB})^{-2}$; that due to stage multiple scattering process is attenuated by a factor $(R_{AB})^{-4}$. In the limit $R_{AB} = \infty$ the coupling between the two antennas vanishes—the terminals O and O' are isolated from each other. In this limit the impedance arm Z_{12} of the T-section becomes a short circuit:

$$\lim_{R_{AB} \to \infty} Z_{12} = 0. \tag{64}$$

Also, in this limiting case, Z_1 and Z_2 reduce to the input impedances Z_1^0 and Z_2^0 (referred to O and O' respectively) of antennas A and B in their isolated states. When R_{AB} is large but not infinite and A is transmitting, the scattered wave from B has a small amplitude when it reaches A; the input impedance of A is but slightly different from Z_1^0. If the impedance at O is sensibly independent of the position and orientation of antenna B, we have one of the requisite conditions under which B, acting as a receiver, may be considered to be measuring the transmission pattern of A. In this situation the antennas are weakly coupled; the transfer impedance is negligible in its effect on the transmitting antenna. As concerns the receiver, however, the transfer impedance is not negligible, for it represents the transfer of energy from the transmitting antenna to the receiving system. The same considerations apply when B is trans-

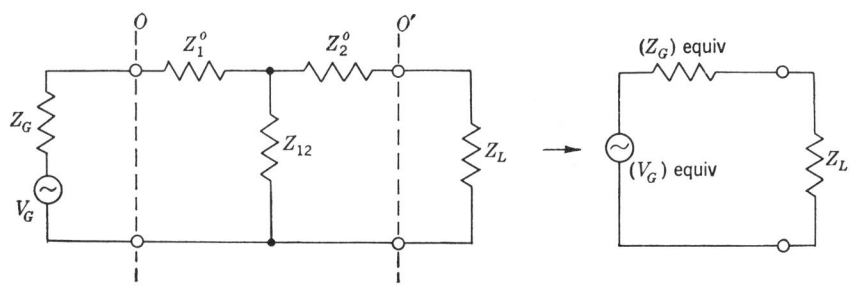

FIG. 2·12.—On the receiving system circuit.

mitting and A is the receiver. For the weakly coupled case we may then set Z_1 and Z_2 equal to the respective values at $R_{AB} = \infty$ and to a first approximation write

$$Z_{11} \approx Z_1^0 + Z_{12}, \tag{65a}$$
$$Z_{22} \approx Z_2^0 + Z_{12}. \tag{65b}$$

This coupled network representation provides the correct approach to the equivalent circuit of the receiving system discussed in Sec. 2·11. That case was actually one of a weakly coupled transmitter-receiver system. Without loss of generality we may consider a generator of emf V_G and internal impedance Z_G to be applied directly to the terminals at O and a load impedance Z_L to be applied directly at O' (Fig. 2·12). By Thévenin's theorem (Sec. 2·4) the system is equivalent to one in which the load is connected to a generator producing an emf

$$(V_G)_{\text{equiv}} = \frac{V_G Z_{12}}{Z_G + Z_1^0 + Z_{12}}, \tag{66}$$

and having an internal impedance

$$(Z_G)_{\text{equiv}} = Z_2^0 + \frac{Z_{12}(Z_1^0 + Z_G)}{Z_{12} + Z_1^0 + Z_G}. \tag{67}$$

In obtaining these results the weak-coupling approximations for Z_{11} and

Z_{22} given by Eqs. (65) have been used. The receiving antenna is thus represented by an equivalent generator; the emf of the generator is proportional to the amplitude of the incident wave (which is proportional to V_G). The effect of the orientation of the antenna with respect to the wave is contained in the functional dependence of the transfer coefficient Z_{12} on orientation. The internal impedance differs from Z_2^0 by the small quantity Z_{12}; neglecting the latter, we have the result (assumed previously) that the equivalent generator impedance is equal to the input impedance of the antenna when it is transmitting. The present analysis shows explicitly that the equivalent circuit applies only to absorption, for Thévenin's theorem is applicable only to the treatment of the power transferred to Z_L. In general the power dissipation computed for the equivalent generator impedance is not equal to the power dissipated in the network between V_G and the load; hence it cannot be interpreted as scattered power.

2·13. Reciprocity between the Transmitting and Receiving Patterns of an Antenna.—The four-terminal network analysis leads to the very important theorem that the transmitting and receiving patterns of an antenna are the same. In this connection the meaning of a pattern must be understood from the practical standpoint of the coupled system. One condition has already been stated: In the case of the transmission pattern, the distance from the transmitter to the receiver must be so large that the former is not affected (within the limits of measurements) by the wave scattered from the latter. In addition, however, one must consider the interactions between the receiving antenna and objects in its immediate neighborhood. Multiple reflection and scattering will take place between the receiver and such objects; the receiving antenna consists, in fact, of the antenna proper together with all neighboring objects with which its interactions are significant. If the receiving antenna is to measure the field at a point, its directive properties must be such that all such interactions are negligible. These interactions at the receiving antenna are similar to but are to be distinguished from the interactions between the transmitter and surrounding objects such as ground. The receiver measures the resultant of the field produced by the transmitter and any neighboring objects that interact with it; these together form, in fact, an extended radiating system.

In Fig. 2·13, A represents the antenna under consideration. In taking a transmitting pattern a receiver B is, in principle, moved over a large sphere about A, and the relative amounts of power absorbed by the load terminating the line B in successive positions give the transmitting pattern of A. Conversely, the receiving pattern of A is obtained as the relative amounts of power absorbed by a load terminating A when it is receiving from the antenna B at successive positions on the sphere. In accordance with the usual experimental conditions, no restrictions are made as to the generator impedance or load impedance; the only requirement is that they remain constant in the course of taking a given pattern. The load in the receiving system will again be taken to be applied directly to the reference point O or O'.

There is an equivalent four-terminal network between O and O' for every position of B. Consider the transmitting pattern. If Z_L is the load impedance at O', the network equations give (without approximations)

$$i_2 Z_L = V_2 = Z_{12} i_1 - Z_{22} i_2$$

or

$$i_2 = \frac{Z_{12}}{Z_{22} + Z_L} i_1. \quad (68)$$

The currents have the usual significance, indicated in Fig. 2·11. The power absorbed in the load is

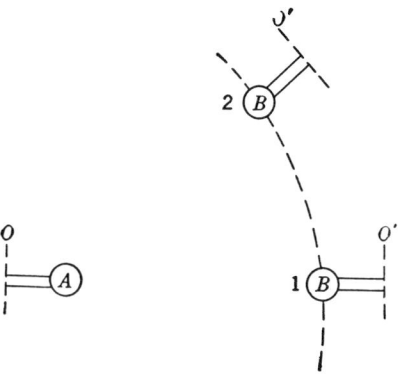

Fig. 2·13.—On the pattern reciprocity theorem.

$$P_{abs} = \frac{1}{2} |i_1|^2 \left| \frac{Z_{12}}{Z_{22} + Z_L} \right|^2 \operatorname{Re} Z_L. \quad (69)$$

Since the coupling is weak, the dependence of the input current i_1 on the position of antenna B is negligible. In the denominator of Eq. (69), the coefficient Z_{22} may be replaced by Z_2^0, for it follows from the weak-coupling approximation of Eq. (65) that this introduces an error of the magnitude $(\operatorname{Re} Z_{12})^3$. For two successive positions of B the ratio of the absorbed powers is given by

$$\frac{(P_{abs})_1}{(P_{abs})_2} = \frac{|Z_{12}|_1^2}{|Z_{12}|_2^2}. \quad (70)$$

The transmitting pattern is thus determined by the transfer impedance coefficient alone.

If now B is transmitting and the power absorbed by a fixed load terminating A at the point O is measured, the result should be the same as in Eq. (69) except that i_1 is replaced by the input current i_2 at O' and Z_{22} is replaced by Z_{11}. The variation in power with the position of B (assuming again weak-coupling conditions) is then likewise given by the transfer impedance alone—in fact, by Eq. (70). Hence, *subject to the condition that the transfer impedance coefficients obey the reciprocity relation, it is found that the transmitting and receiving patterns of an antenna are the same.* If then $G(\theta,\phi)$ is the gain function of the antenna as a transmitter in the direction θ, ϕ, the absorption cross section $A_r(\theta,\phi)$

presented by the antenna to a plane wave incident from the direction θ, ϕ is

$$A_r(\theta,\phi) = G(\theta,\phi)\bar{A}_r, \qquad (71)$$

where

$$\bar{A}_r = \frac{1}{4\pi} \int\int A_r(\theta,\phi) \sin\theta \, d\theta \, d\phi \qquad (72)$$

is the average cross section over all aspects. The practical result of the reciprocity theorem is that no distinction need be made between the transmitting and receiving functions of an antenna in the analysis of design problems.

2·14. The Average Cross Section for a Matched System.—In consequence of the reciprocal relation between the transfer impedance coefficients $Z_{12} = Z_{21}$, the four-terminal network representation of the transmitter-receiver system obeys the Rayleigh theorem of Eq. (11). This, taken together with the pattern reciprocity theorem established in the preceding section, leads to a further important result: *The average absorption cross section of receiving system in which the load is matched to the antenna impedance is a universal constant.* The demonstration given here applies strictly to the case in which the ohmic losses in the antenna and line are negligible.

Consider again a weakly coupled transmitter-receiver system made up of antennas A and B, with input terminals at assigned reference points O and O' as in Fig. 2·11. Let the input impedances of the respective antennas be

$$Z_1^0 = R_A + jX_A,$$
$$Z_2^0 = R_B + jX_B.$$

For a weakly coupled system these are but negligibly different from the input impedances at O and O' when the respective antennas are transmitting. Let us apply a generator of emf V_G and internal impedance Z_1^{0*}, equal to the conjugate of the impedance of antenna A, across the terminals at O. The receiving system is assumed to be so matched that the load impedance across O' is Z_2^{0*}. If i_2 is the current at O', the power absorbed by the receiver is

$$P_{AB} = \tfrac{1}{2}|i_2|^2 \operatorname{Re} Z_2^{0*} = \tfrac{1}{2}|i_2|^2 R_B. \qquad (73)$$

This power can be computed in another way. Let P_A be the total power radiated by the antenna A; the power radiated per unit solid angle in the direction of B is $(P_A/4\pi)G_{AB}$, G_{AB} being the gain function of A in the direction AB. The absorption cross section presented by B to the wave from A is by Eq. (71) equal to $G_{BA}\bar{A}_{rB}$, G_{BA} being the gain function of B in the direction of A. The solid angle subtended by the cross section at A is $G_{BA}A_{rB}/R_{AB}^2$, whence the power absorbed by B is

$$P_{AB} = \frac{P_A}{4\pi} \frac{G_{AB}G_{BA}\bar{A}_{rB}}{R_{AB}^2}. \tag{74}$$

However, P_A is equal to the power supplied to antenna A by the generator:

$$P_A = \frac{|V_G|^2}{8R_A}. \tag{75}$$

Collecting these results, we obtain

$$G_{AB}G_{BA}\bar{A}_{rB} = \frac{16\pi|i_2|^2 R_{AB}^2 R_A R_B}{|V_G|^2}. \tag{76}$$

If the situation is reversed so that B transmits and A receives, with a generator of emf V_G and internal impedance Z_2^{0*} applied across O' and a load impedance Z_1^{0*} across O, we obtain by the same calculation as before:

$$G_{AB}G_{BA}\bar{A}_{rA} = \frac{16\pi|i_1'|^2 R_{AB}^2 R_A R_B}{|V_G|^2}. \tag{77}$$

In this case i_1' is the current at the terminals at O. By the Rayleigh theorem we have

$$i_1' = i_2; \tag{78}$$

hence, on comparing Eqs. (76) and (77), we find

$$\bar{A}_{rA} = \bar{A}_{rB}; \tag{79}$$

The average cross sections of the two antennas are equal. Since the antennas are purely arbitrary, this means that the average cross section of a matched system is a universal constant.

The evaluation of the constant requires at least one detailed analysis of the interaction between an antenna and a plane wave on the basis of electromagnetic field theory. The reader is referred to Slater[1] for such a treatment of the electric dipole antenna. It is shown there that the value of the constant is

$$\bar{A}_r = \frac{\lambda^2}{4\pi}. \tag{80}$$

The cross section $A_r(\theta,\phi)$ presented by an antenna to a plane wave incident from the direction θ, ϕ is therefore

$$A_r(\theta,\phi) = G(\theta,\phi) \frac{\lambda^2}{4\pi}. \tag{81}$$

2·15. Dependence of the Cross Section on Antenna Mismatch.—The matched-impedance condition between the antenna and the load—that the load impedance be the conjugate of that of the antenna—is the same

[1] J. C. Slater, *Microwave Transmission*, McGraw-Hill, New York, 1942. Chap. VI.

as the condition for maximum-power transfer from a generator to a load. This condition can be realized by separately matching the antenna and load to the characteristic impedance of the transmission line if the characteristic impedance is real, as it is for a nonlossy line. The line-matched system is of particular interest in the study of microwave antennas and is generally taken as a reference system, since transmitting antennas are required to be matched to the line. Consequently, it is of interest to determine the effect of a mismatch between the antenna and the line on the absorption cross section.

The functional dependence of the cross section on line mismatch is of considerable importance in the measurement of the gain of microwave antennas. It may be desired, for example, to study the dependence of the gain on configurational parameters, such as the relative positions of a radiator and a reflector in a scanning antenna. It is impractical in such investigations to match the antenna to the line in each configuration; rather, a line-matched detector is used throughout, and the results are corrected for the antenna mismatch of the given configurations.

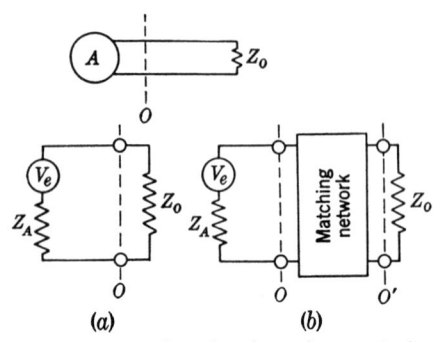

Fig. 2·14.—On the dependence of the absorption cross section on mismatch: (a) the mismatched system; (b) the line-matched system in which a network transforms Z_A into Z_0.

Consider the receiving system in Fig. 2·14, composed of an antenna A feeding a line terminated in a load equal to the characteristic impedance Z_0 of line. Let Γ be the reflection coefficient of the antenna (in transmission) at a given reference point O and $Z_A = R_A + jX_A$ the associated impedance. We may replace the antenna by an equivalent generator of internal impedance Z_A; the emf of the generator will be designated by V_e. Consider now two cases: (1) Fig. 2·14a, in which the antenna is mismatched and feeds directly to the line at O, and (2) Fig. 2·14b—the line-matched system—in which a lossless network has been introduced between the antenna terminals at O and the line to transform the antenna impedance into Z_0 at the output terminals O'. It is readily verified that such a network which transforms the impedance Z_A at O into Z_0 at O' transforms the impedance Z_0 at O' into the complex conjugate Z_A^* at O. Case b therefore meets the conditions of Sec. 2·14. The power absorbed in the load in the two cases is

Case a:

$$(P_{abs})_a = \frac{|V_e|^2}{2} \frac{Z_0}{|Z_0 + Z_A|^2}, \qquad (82)$$

Case b:

$$(P_{abs})_b = \frac{1}{2} \frac{|V_e|^2}{|Z_A + Z_A^*|^2} \operatorname{Re} Z_A$$
$$= \frac{|V_e|^2}{2} \cdot \frac{1}{4R_A}. \tag{83}$$

The ratio of the power absorbed in the two cases is the ratio of the respective absorption cross sections:

$$\frac{(A_r)_{mis}}{(A_r)_0} = \frac{4Z_0 R_A}{|Z_0 + Z_A|^2}. \tag{84}$$

Here $(A_r)_0$ designates the cross section of the matched system.

The antenna impedance can be evaluated in terms of the reflection coefficient Γ. Thus

$$Z_A = Z_0 \left(\frac{1 + \Gamma}{1 - \Gamma}\right).$$

and

$$R_A = \frac{1}{2}(Z_A + Z_A^*) = \frac{Z_0(1 - |\Gamma|^2)}{|1 - \Gamma|^2}.$$

Substituting into Eq. (83), we obtain the desired result:

$$(A_r)_{mis} = (A_r)_0(1 - |\Gamma|^2). \tag{85}$$

The decrease in cross section—or reception efficiency—is precisely the same as the reflection loss introduced by the mismatch on transmission. Also it will be noted that the mismatch depends only on $|\Gamma|^2$; hence the result is independent of the choice of the reference point O taken as the input terminals to the antenna.

2·16. The Four-terminal Network Representation.—This and the following sections summarize the considerations underlying the postulate (Sec. 2·12) that the transmitter-receiver system is equivalent to a four-terminal network between the respective transmission lines. Use will be made of results proved later in Chaps. 3 and 7. The treatment is formulated primarily for microwave systems in which the transmission lines are waveguides. The systems are assumed to be ideal, in the sense that ohmic losses in the lines and the antennas are negligible.

Consider a pair of antennas A and B, each of which is fed from a waveguide, as shown in Fig. 2·15. It is assumed that the guides are designed to support free propagation of a single mode only. The reference planes O and O' which serve as the input terminals to the antennas are perpendicular to the respective guide axes and are taken in the transmission-line region of the guides, where only the freely propagated mode has an amplitude significantly different from zero. We shall consider the closed surface S made up of the surface O inside the guide A, the interior surface of the guide, the surfaces of the conductors comprising the antenna, and finally the exterior surface of guide A; this encloses the A-system completely. A similar surface S' encloses the B-system. We shall be concerned with the electromagnetic field in the region V bounded by a sphere of infinite radius and by the surfaces S and S'.

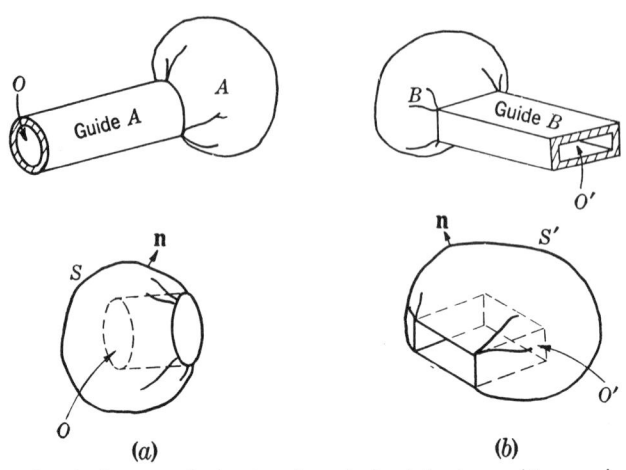

Fig. 2·15.—On the four-terminal network analysis of the transmitter-receiver system.

It will be assumed that there are no generators in the region V. As regards antennas A and B, either we may have the one transmitting and the other receiving or generators may be applied to both antennas simultaneously. However, the particular case involved is of no concern, since we are interested in the general nature of the relation between the tangential components $\mathbf{E}_1$, $\mathbf{H}_1$ of the field over the plane O in guide A and the tangential components $\mathbf{E}_2$, $\mathbf{H}_2$ over the plane O' in B.

The magnitudes of the tangential electric and magnetic fields are determined by voltage and current parameters V and i, respectively, which are analogous to the voltage and current in a balanced two-wire line. In order to set up a four-terminal network representation, we must show that the relation between the voltage and current parameters V_1, i_1 at the plane O and the parameters V_2, i_2 at O' is linear:

$$\left. \begin{array}{l} V_1 = Z_{11}i_1 - Z_{12}i_2; \\ V_2 = Z_{21}i_1 - Z_{22}i_2. \end{array} \right\} \tag{86}$$

To validate the various reciprocity theorems developed in Secs. 2·13 to 2·15 we must then show that the transfer impedance coefficients satisfy the reciprocity relation

$$Z_{12} = Z_{21}. \tag{87}$$

The remainder of this section will concern itself with the definition of the voltage and current parameters and an exposition of certain of

SEC. 2·16]   *THE FOUR-TERMINAL NETWORK REPRESENTATION*   55

their properties that are needed in developing the proof of the four-terminal network representation. The latter subject proper will be treated in the following section, and in Sec. 2·18 the reciprocity relation between the transfer impedance coefficients will be established.

The fields in a waveguide are functions of position both over the cross section of the guide and along its axis. It will be shown in Chap. 7 that the tangential components of the field over any cross section of a guide, for a given mode, have the form

$$\begin{aligned} \mathbf{E}_{\text{tang}} &= V\mathbf{g}(x,y), \\ \mathbf{H}_{\text{tang}} &= i\mathbf{h}(x,y), \end{aligned} \quad (88)$$

where the coordinates x, y refer to position on the cross-section plane. The functions $\mathbf{g}(x,y)$ and $\mathbf{h}(x,y)$ are characteristic of the given mode and satisfy the relation

$$\int_{\text{cross section}} \mathbf{i}_z \cdot [\mathbf{g}(x,y) \times \mathbf{h}(x,y)] \, dS = 1. \quad (88a)$$

The quantities V and i—the voltage and current parameters, respectively—are functions of position along the guide axis. If position along the latter is designated by z, the voltage and current parameters for a general field of a given mode take the form

$$V = V_+ e^{-i\beta z} + V_- e^{+i\beta z}, \quad (89a)$$
$$i = \Upsilon_0 (V_+ e^{-i\beta z} - V_- e^{+i\beta z}); \quad (89b)$$

that is, the general field is made up of two waves traveling in opposite directions along the guide axis, the subscript ± in Eqs. (89) referring to the direction of propagation of the component wave with respect to the positive z-direction. The quantity Υ_0 is a constant, characteristic of the given mode. Thus the voltage and current parameters obey the same equations as do the voltage and current in a two-wire line, of characteristic admittance Υ_0. As in the case of the two-wire line the amplitudes V_+ and V_- are determined by the boundary conditions at the input and terminal points in the guide.

If V_α and V_β are the voltage parameters of two fields of the same mode, for different boundary conditions on the line, and i_α and i_β are the respective current parameters, it follows from Eqs. (89) that the field with a voltage parameter

$$V_\gamma = m_\alpha V_\alpha + m_\beta V_\beta \quad (90a)$$

has a current parameter

$$i_\gamma = m_\alpha i_\alpha + m_\beta i_\beta. \quad (90b)$$

This leads at once, by virtue of Eqs. (88), to the corresponding property of the electric and magnetic fields: Let $\mathbf{E}_\alpha$, $\mathbf{H}_\alpha$ and $\mathbf{E}_\beta$, $\mathbf{H}_\beta$ be two linearly independent fields, of the same mode; then, if we construct the field

$$\mathbf{E}_\gamma = m_\alpha \mathbf{E}_\alpha + m_\beta \mathbf{E}_\beta, \tag{91a}$$

where m_α and m_β are both different from zero, the magnetic field $\mathbf{H}_\gamma$ associated with $\mathbf{E}_\gamma$ is correspondingly

$$\mathbf{H}_\gamma = m_\alpha \mathbf{H}_\alpha + m_\beta \mathbf{H}_\beta. \tag{91b}$$

This relation between the fields is of fundamental importance to the discussion in the following section.

2·17. Development of the Network Equations.—We may now pass to the details of the four-terminal network problem. The procedure is to consider the relation between the fields within the respective guides and the fields in the external space, thereby arriving at a relation between the fields in the two guides A and B. For this purpose the interior regions of the guides are thought of as connected with external space to form a composite region V bounded by the surfaces S and S', as was outlined in the previous section and illustrated schematically in Fig. 2·15.

Every set of values of electric and magnetic fields $\mathbf{E}_1$, $\mathbf{H}_1$ over O and $\mathbf{E}_2$, $\mathbf{H}_2$ over O' (and hence voltage and current parameters V_1, i_1, V_2, i_2) is associated with a field $\mathbf{E}$, $\mathbf{H}$ in the region V. Consider three such fields that are not simple multiples of one another:

$$(\mathbf{E}_{1\alpha}, \mathbf{H}_{1\alpha}; \mathbf{E}_{2\alpha}, \mathbf{H}_{2\alpha}; \mathbf{E}_\alpha, \mathbf{H}_\alpha),$$
$$(\mathbf{E}_{1\beta}, \mathbf{H}_{1\beta}; \mathbf{E}_{2\beta}, \mathbf{H}_{2\beta}; \mathbf{E}_\beta, \mathbf{H}_\beta),$$
$$(\mathbf{E}_{1\gamma}, \mathbf{H}_{1\gamma}; \mathbf{E}_{2\gamma}, \mathbf{H}_{2\gamma}; \mathbf{E}_\gamma, \mathbf{H}_\gamma).$$

It follows from Eq. (88) that over the planes O and O' the successive fields differ from each other only in their voltage parameters. (Only a single mode exists in each guide in the regions of the reference planes.) Any one of the three fields can be obtained as a linear combination of the other two, with coefficients m_α and m_β which satisfy the relations

$$\left. \begin{array}{l} V_{1\gamma} = m_\alpha V_{1\alpha} + m_\beta V_{1\beta}, \\ V_{2\gamma} = m_\alpha V_{2\alpha} + m_\beta V_{2\beta}. \end{array} \right\} \tag{92a}$$

By virtue of Eq. (88) the voltage parameters can be replaced by the electric fields $\mathbf{E}_{1\alpha} \ldots \mathbf{E}_{2\gamma}$. By Eqs. (91), the associated magnetic fields follow the same law of resolution:

$$\left. \begin{array}{l} \mathbf{H}_{1\gamma} = m_\alpha \mathbf{H}_{1\alpha} + m_\beta \mathbf{H}_{1\beta}, \\ \mathbf{H}_{2\gamma} = m_\alpha \mathbf{H}_{2\alpha} + m_\beta \mathbf{H}_{2\beta}. \end{array} \right\} \tag{92b}$$

This resolution can be effected regardless of the behavior of the fields throughout the region V. However, it is meaningful only if the field $\mathbf{E}_\gamma$, $\mathbf{H}_\gamma$ is the same linear combination of the fields $\mathbf{E}_\alpha$, $\mathbf{H}_\alpha$ and $\mathbf{E}_\beta$, $\mathbf{H}_\beta$ throughout V as it is over the reference planes, that is, if

$$\mathbf{E}_\gamma = m_\alpha \mathbf{E}_\alpha + m_\beta \mathbf{E}_\beta. \tag{92c}$$

Proof of Eq. (92c) follows from the uniqueness theorem of the electromagnetic field.[1] The application of the theorem, however, involves restrictions on the fields. The medium in the region V is characterized by three constitutive parameters: the conductivity σ, the electric inductive capacity ϵ, and the magnetic permeability μ.[2] These in general vary from point to point and are functions of frequencies. In special cases (such as ferromagnetic media) they are functions of the field intensities; such nonlinear regions are excluded in the formulation of the uniqueness theorem. Since the region V includes virtually all space, ferromagnetic media cannot be simply excluded; we must instead impose the restriction that the fields set up by the antennas be such that their amplitudes are negligible in the neighborhood of such media. Subject to this condition, the uniqueness theorem states that in a region V which is free from generators the field is determined completely by the values of $\mathbf{n} \times \mathbf{E}$ over the boundary surfaces S and S'. The reader is referred to Stratton for the proof. The same technique that is employed in the development of the uniqueness theorem leads to the following superposition principle: If $\mathbf{E}_a$ is the field in V corresponding to the boundary condition $\mathbf{n} \times \mathbf{E} = \mathbf{F}_a$ over S and S' and $\mathbf{E}_b$ the field with the boundary condition $\mathbf{n} \times \mathbf{E} = \mathbf{F}_b$, then the field $\mathbf{E}_c$ associated with the boundary conditions

$$\mathbf{n} \times \mathbf{E} = m_a \mathbf{F}_a + m_b \mathbf{F}_b,$$

m_a and m_b being constants, is

$$\mathbf{E}_c = m_a \mathbf{E}_a + m_b \mathbf{E}_b.$$

It will be noted that since the waveguides and antennas are all ideal conductors, all fields $\mathbf{E}$, $\mathbf{H}$, with which we are concerned in the region V, satisfy the same boundary conditions

$$\mathbf{n} \times \mathbf{E} = 0$$

over the surfaces S and S' exclusive of the cross sections O and O'. Over the regions O and O' the tangential component of $\mathbf{E}$ assumes prescribed values $\mathbf{E}_1$ and $\mathbf{E}_2$ respectively. Hence the resolution of $\mathbf{E}_{1\gamma}$ and $\mathbf{E}_{2\gamma}$ in Eq. (92) becomes, in fact, a resolution of the tangential components of the field $\mathbf{E}_\gamma$ over S and S' in terms of a pair of linearly independent fields:

$$\mathbf{n} \times \mathbf{E}_\gamma = m_\alpha (\mathbf{n} \times \mathbf{E}_\alpha) + m_\beta (\mathbf{n} \times \mathbf{E}_\beta).$$

From the superposition theorem we have then that everywhere in V

$$\mathbf{E}_\gamma = m_\alpha \mathbf{E}_\alpha + m_\beta \mathbf{E}_\beta,$$

which was the desired result stated in Eq. (92c). Thus given any pair

---

[1] See for example, J. A. Stratton, *Electromagnetic Theory*, McGraw-Hill, New York, 1941, Sec. 9·2.

[2] *Cf.* Chap. 3, Sec. 3·2.

of linearly independent fields over the reference planes O and O', all other fields may be expressed as a linear combination of the two, the law of combination holding for all points in the region V.

It is convenient to take as the basic set of linearly independent fields the two fields corresponding to short-circuit terminations over the plane O' and the plane O respectively. Consider first the short-circuit termination over O', and let the fields over O be designated by $\mathbf{E}_{1\alpha}$, $\mathbf{H}_{1\alpha}$, the fields over O' by $\mathbf{E}_{2\alpha}$, $\mathbf{H}_{2\alpha}$; let $V_{1\alpha} \ldots i_{2\alpha}$ be the corresponding voltage and current parameters. Since the short-circuit means that O' is the surface of the perfect conductor, we must have $\mathbf{E}_{2\alpha} = 0$, and hence $V_{2\alpha} = 0$. Of the three remaining quantities, one may be regarded as an independent variable, being adjustable, for example, by a generator applied over the surface O. Let $V_{1\alpha}$ be the independent variable. From Eqs. (89) it follows that for fixed conditions in V, that is, a prescribed termination in antenna B and hence a fixed terminal condition in guide A, the current parameter $i_{1\alpha}$ varies directly with the voltage parameter $V_{1\alpha}$:

$$V_{1\alpha} = a_{11} i_{1\alpha}, \tag{93}$$

where a_{11} is a constant independent of the field amplitude. Furthermore since O' is short-circuited, the field in V must satisfy the condition $\mathbf{n} \times \mathbf{E} = 0$ over *all* of S', for all values of $\mathbf{E}_{1\alpha}$. From the superposition principle it follows then that the field at all points in V is proportional to the magnitude of $\mathbf{E}_{1\alpha}$; in particular, then, the current $i_{2\alpha}$ is proportional to $V_{1\alpha}$:

$$i_{2\alpha} = b_\alpha V_{1\alpha} = b_\alpha a_{11} i_{1\alpha}, \tag{94}$$

with b_α also a constant independent of the field amplitude.

Similar relations are obtained for the case of a short-circuit termination over O. Letting $V_{1\beta}$, $i_{1\beta}$, $V_{2\beta}$, $i_{2\beta}$ be the voltages and currents over O and O' respectively, we have in this case

$$V_{1\beta} = 0, \tag{95a}$$
$$V_{2\beta} = a_{22} i_{2\beta}, \tag{95b}$$
$$i_{1\beta} = b_\beta V_{2\beta} = b_\beta a_{22} i_{2\beta}. \tag{95c}$$

The general field can be written as a linear combination of this basic set:

$$\left. \begin{array}{l} V_1 = m_\alpha V_{1\alpha} + m_\beta V_{1\beta} = m_\alpha V_{1\alpha} = a_{11}(m_\alpha i_{1\alpha}), \\ V_2 = m_\alpha V_{2\alpha} + m_\beta V_{2\beta} = m_\beta V_{2\beta} = a_{22}(m_\beta i_{2\beta}), \end{array} \right\} \tag{96}$$

and

$$\left. \begin{array}{l} i_1 = m_\alpha i_{1\alpha} + m_\beta i_{1\beta} = (m_\alpha i_{1\alpha}) + b_\beta a_{22}(m_\beta i_{2\beta}), \\ i_2 = m_\alpha i_{2\alpha} + m_\beta i_{2\beta} = b_\alpha a_{11}(m_\alpha i_{1\alpha}) + (m_\beta i_{2\beta}). \end{array} \right\} \tag{97}$$

Solution of Eqs. (97) for $m_\alpha i_{1\alpha}$ and $m_\beta i_{2\beta}$ and substitution into Eq. (96) give the linear relation between the voltages and currents in the two

SEC. 2·18]   *THE RECIPROCITY RELATION*   59

guides:
$$V_1 = Z_{11}i_1 - Z_{12}i_2, \\ V_2 = Z_{21}i_1 - Z_{22}i_2,$$ (98)

where

$$Z_{11} = \frac{a_{11}}{A}, \quad Z_{22} = -\frac{a_{22}}{A},$$
$$Z_{12} = \frac{a_{11}a_{22}b_\beta}{A}, \quad Z_{21} = -\frac{a_{11}a_{22}b_\alpha}{A},$$
$$A = 1 - b_\alpha b_\beta a_{11} a_{22}.$$ (98a)

It is necessary to observe sign conventions in using Eqs. (98) to relate the fields over O' to the fields over O. The convention will be adopted here to correspond to that used in Sec. 2·2: regarding O as the input terminals to the four-terminal network, the positive z-direction in guide A is toward the antenna, and i_1 is the positive current entering the network; at the input terminals O', the positive current leaves the network, the positive z-direction in the second guide being away from the antenna.

2·18. The Reciprocity Relation between the Transfer Impedance Coefficients.—Equations (98) establish a four-terminal network representation for the coupled transmitter-receiver system. The final problem to be considered is the justification of the assumption that the transfer impedance coefficients satisfy the reciprocity relation

$$Z_{12} = Z_{21}.$$

We shall make use of the Lorentz reciprocity theorem:[1] Let $\mathbf{E}_\alpha$, $\mathbf{H}_\alpha$ and $\mathbf{E}_\beta$, $\mathbf{H}_\beta$ be two linearly independent fields in the region V; then

$$\oint_{S+S'} (\mathbf{n} \times \mathbf{E}_\alpha) \cdot \mathbf{H}_\beta \, dS = \oint_{S+S'} (\mathbf{n} \times \mathbf{E}_\beta) \cdot \mathbf{H}_\alpha \, dS.$$ (99)

The conditions for the validity of the Lorentz theorem are the same as those stipulated for the uniqueness theorem and superposition principle in the preceding section.

Let us apply the theorem to the two basic fields employed in the preceding section. The relation (99) in this case reduces to

$$\int_O (\mathbf{n} \times \mathbf{E}_\alpha) \cdot \mathbf{H}_\beta \, dS = \int_{O'} (\mathbf{n} \times \mathbf{E}_\beta) \cdot \mathbf{H}_\alpha \, dS.$$ (100)

Making use of Eq. (88) and taking into account the sign conventions on the current parameters, we obtain

---

[1] See the article by A. Sommerfeld in Frank and V. Mises, *Die Differential- und Integralgleichungen der Mechanik und Physik*, Vol. II, p. 953, reprinted by Mary S. Rosenberg, New York, 1943.

$$-V_{1\alpha}i_{1\beta}\int_{O}\mathbf{i}_z\cdot[\mathbf{g}_1(x,y)\times\mathbf{h}_1(x,y)]\,dS$$
$$=V_{2\beta}i_{2\alpha}\int_{O'}\mathbf{i}_z\cdot[\mathbf{g}_2(x,y)\times\mathbf{h}_2(x,y)]\,dS. \quad (101)$$

By virtue of the property of the functions $\mathbf{g}, \mathbf{h}$ of Eq. (88a) it follows that

$$-V_{1\alpha}i_{1\beta}=V_{2\beta}i_{2\alpha}. \quad (101a)$$

If now the currents are expressed in terms of the voltages by means of Eqs. (94) and (95c), it is seen that the coefficients b_α and b_β of the previous section are related:

$$b_\alpha=-b_\beta.$$

It then follows from Eqs. (98a) that the transfer impedance coefficients obey the reciprocity relation

$$Z_{12}=Z_{21}.$$

CHAPTER 3

RADIATION FROM CURRENT DISTRIBUTIONS

By S. Silver

The fundamental approach to an understanding of microwave antennas is necessarily based on electromagnetic theory. This chapter therefore begins with a discussion of the field equations and the general properties of an electromagnetic field; the treatment is necessarily cursory, being intended as a summary of material that is familiar to the reader.[1] This theory is then applied to the simplest problem of antenna theory, the calculation of the radiation fields due to known current distributions. A discussion of certain idealized current distributions illustrates the principles of superposition and interference and furnishes a theoretical guide to the design of various antenna feeds.

3·1. The Field Equations.—The field equations relate the electric field vectors **E** and **D** and the magnetic field vectors **B** and **H** to each other and to the sources of the field, the electric charges and currents.

Sources of the Field.—The sources will be specified in terms of density functions.

The excess of positive over negative charge in a volume V is

$$Q = \int_V \rho \, dv, \tag{1}$$

where ρ is the charge density per unit volume.

The rate of transport of charge across a surface S, that is, the net current passing through S, is

$$I = \int_S \mathbf{J} \cdot \mathbf{n} \, dS, \tag{2}$$

where **J** is the current density and **n** is the unit normal to the surface S in the direction defined as positive. The current **J** has the direction of flow of positive charge, a negative charge moving in one direction being equivalent to a positive charge moving in the opposite direction.

In the rationalized meter-kilogram-second (mks) system of units,[2]

---

[1] The reader is referred to J. A. Stratton, *Electromagnetic Theory*, McGraw-Hill, New York, 1941, for a more detailed treatment of many of the subjects covered in this chapter.

[2] Stratton, *op. cit.*, pp. 16, 602.

which is used in this book, the charge density is measured in coulombs per cubic meter and the current density in amperes per square meter.

As a consequence of the conservation of charge, the charge density and current density are subject to an important relation. The total current passing out of a closed surface S must equal the rate of decrease of positive charge in the enclosed volume. That is,

$$\oint_S \mathbf{J} \cdot \mathbf{n}\, dS = -\frac{\partial}{\partial t} \int_V \rho\, dv, \tag{3}$$

where $\mathbf{n}$ is the unit vector normal to the surface and directed out from the region V. By the divergence theorem[1]

$$\oint_S \mathbf{J} \cdot \mathbf{n}\, dS = \int_V \mathbf{\nabla} \cdot \mathbf{J}\, dv. \tag{4}$$

Substitution of this into Eq. (3) gives

$$\int_V \left(\mathbf{\nabla} \cdot \mathbf{J} + \frac{\partial \rho}{\partial t} \right) dv = 0. \tag{5}$$

This must hold for any arbitrary volume, no matter how small; consequently the integrand itself must be zero:

$$\mathbf{\nabla} \cdot \mathbf{J} + \frac{\partial \rho}{\partial t} = 0. \tag{6}$$

This is the so-called "equation of continuity."

Finite charges and currents are sometimes limited to surfaces of discontinuity. In such cases the excess of positive over negative charge on a surface S is

$$Q = \int_S \eta\, dS, \tag{7}$$

where η is the charge density per unit area. Similarly if we let C be a curve on the surface of discontinuity and $\mathbf{n}_1$ a unit vector normal to C in the tangent plane, then the total current crossing C, that is, the rate of transport of charge across C, is

$$I = \int_C \mathbf{K} \cdot \mathbf{n}_1\, ds, \tag{8}$$

where $\mathbf{K}$ is the surface-current density. The surface-current density $\mathbf{K}$ and the charge distribution η on the boundary of an infinitely conducting medium must satisfy an equation of continuity analogous to the volume

---

[1] A treatment of the divergence theorem and Stokes's and Green's theorems, which are used subsequently, may be found in any text on vector analysis. See for example, H. B. Phillips, *Vector Analysis*, Wiley, New York, 1933.

Sec. 3·1]   *THE FIELD EQUATIONS*   63

distributions. This equation of continuity, in integral form, is

$$\oint_C \mathbf{K} \cdot \mathbf{n}_1 \, ds = -\int_S \frac{\partial \eta}{\partial t} \, dS, \tag{9}$$

where C is any closed curve enclosing an area S.

Another form of this relation is

$$\boldsymbol{\nabla}_s \cdot \mathbf{K} + \frac{\partial \eta}{\partial t} = 0, \tag{10}$$

where the "surface divergence" of $\mathbf{K}$, $\boldsymbol{\nabla}_s \cdot \mathbf{K}$, is defined by

$$\boldsymbol{\nabla}_s \cdot \mathbf{K} = \lim_{A \to 0} \frac{1}{A} \oint_C \mathbf{K} \cdot \mathbf{n}_1 \, ds, \tag{11}$$

A being the area circumscribed by the curve C.

Definitions of the Field Quantities.—The field vectors $\mathbf{E}$ and $\mathbf{B}$ measure the forces exerted on charges and currents respectively. The force on a stationary charge q at any point in the field is

$$\mathbf{F} = \mathbf{E}q. \tag{12}$$

The total force on a current distribution through a volume V of space is

$$\mathbf{F} = \int_V \mathbf{J} \times \mathbf{B} \, dv, \tag{13}$$

the integrand being the vector product of $\mathbf{J}$ and $\mathbf{B}$. The vector $\mathbf{E}$ is measured in volts per meter and $\mathbf{B}$ in webers per square meter.

The field vectors $\mathbf{D}$ and $\mathbf{H}$ are determined by the field sources and are independent of the medium. The net outward flux of $\mathbf{D}$ through a closed surface S is a direct measure of the enclosed charge Q:

$$\oint_S \mathbf{D} \cdot \mathbf{n} \, dS = Q, \tag{14}$$

where $\mathbf{n}$ is the unit vector normal outward from the enclosed region. The magnetic field $\mathbf{H}$ is related to the current. If I is the net current passing through a surface S bounded by a curve C, then

$$\oint_C \mathbf{H} \cdot d\mathbf{s} = I. \tag{15}$$

The integral on the left is the line integral of the tangential component of $\mathbf{H}$ along the curve C; the direction of integration is such that an observer traversing the curve in that direction will have on his left the positive normal $\mathbf{n}$ used in defining the current I.

The Field Equations.—The field equations expressing the relations between the field vectors and the sources may be set up either in differential or integral form.

The differential relationships, Maxwell's equations, are

$$\nabla \times \mathbf{E} + \frac{\partial \mathbf{B}}{\partial t} = 0, \tag{16a}$$

$$\nabla \times \mathbf{H} = \mathbf{J} + \frac{\partial \mathbf{D}}{\partial t}, \tag{16b}$$

$$\nabla \cdot \mathbf{B} = 0, \tag{16c}$$

$$\nabla \cdot \mathbf{D} = \rho. \tag{16d}$$

Equation (16c) may be derived from Eq. (16a) by taking the divergence of the latter. Similarly, Eq. (16d) may be derived by taking the divergence of Eq. (16b) and comparing the result with the equation of continuity

$$\nabla \cdot \mathbf{J} + \frac{\partial \rho}{\partial t} = 0. \tag{16e}$$

Equations (16a) to (16e) must be obeyed simultaneously by the field components and sources of any electromagnetic field.

The corresponding integral relations are the following. Let C be a closed curve spanned by an arbitrary surface S; then

$$\oint_C \mathbf{E} \cdot d\mathbf{s} = -\frac{\partial}{\partial t}\int_S \mathbf{B} \cdot \mathbf{n}\, dS, \tag{17a}$$

$$\oint_C \mathbf{H} \cdot d\mathbf{s} = \int_S \left(\mathbf{J} + \frac{\partial \mathbf{D}}{\partial t}\right) \cdot \mathbf{n}\, dS, \tag{17b}$$

the positive direction of integration around the curve C being that defined previously. The first of these relations is Faraday's law of electromagnetic induction, and the second is the generalization of Ampère's law in which the current density $\mathbf{J}$ due to charge is supplemented by the "displacement-current density" $\partial \mathbf{D}/\partial t$. These equations can be derived from Eqs. (16a) and (16b) by the use of Stokes's theorem. By application of the divergence theorem to Eqs. (16c) and (16d) one obtains two more integral relations:

$$\oint_S \mathbf{B} \cdot \mathbf{n}\, dS = 0, \tag{17c}$$

$$\oint_S \mathbf{D} \cdot \mathbf{n}\, dS = \int_V \rho\, dv = Q, \tag{17d}$$

where the integrals extend over the closed surface S of a volume V.

Equivalent Magnetic Charge and Current.—Equations (16c) and (17c) express the fact that there exist no free magnetic charges and corresponding magnetic currents. However, it is at times convenient to introduce equivalent distributions of such charges and currents. A simple example is provided by the infinitesimal current loop. This is equivalent to a magnetic dipole normal to the plane of the loop. If the current in the loop varies with time, the dipole strength varies likewise; the effect is that of a magnetic-current element.

In diffraction theory, equivalent magnetic-charge and magnetic-current distributions are introduced in a more general way. In the presence of a magnetic-charge distribution of density ρ_m and a magnetic-current distribution of density $\mathbf{J}_m$, Maxwell's equations assume the more symmetrical form

$$\nabla \times \mathbf{E} = -\mathbf{J}_m - \frac{\partial \mathbf{B}}{\partial t}, \tag{18a}$$

$$\nabla \times \mathbf{H} = \mathbf{J} + \frac{\partial \mathbf{D}}{\partial t}, \tag{18b}$$

$$\nabla \cdot \mathbf{B} = \rho_m, \tag{18c}$$

$$\nabla \cdot \mathbf{D} = \rho, \tag{18d}$$

with two equations of continuity

$$\nabla \cdot \mathbf{J} + \frac{\partial \rho}{\partial t} = 0, \tag{18e}$$

$$\nabla \cdot \mathbf{J}_m + \frac{\partial \rho_m}{\partial t} = 0. \tag{18f}$$

It is to be emphasized that the magnetic-source densities are mere formalisms. We introduce them here to avoid later repetition of certain mathematical developments. They will be different from zero only under very special circumstances.

3·2. The Constitutive Parameters; Linearity and Superposition.— There exist between the various field vectors further relations that depend on the medium.

In isotropic media the vectors $\mathbf{D}$ and $\mathbf{E}$ have the same direction at any given point, as do the vectors $\mathbf{B}$ and $\mathbf{H}$. The ratios of their magnitudes are constitutive parameters of the medium:

$$\epsilon = \frac{D}{E}, \tag{19a}$$

the electric inductive capacity, and

$$\mu = \frac{B}{H}, \tag{19b}$$

the magnetic inductive capacity. These quantities may be functions of the field intensities and the frequencies. They depend on the field intensities only for a small group of substances which we shall exclude from our discussion. The frequency dependence is a very general property. In vacuo these parameters are constants and have the values

$$\epsilon_0 = 8.85 \times 10^{-12} \text{ farad/meter},$$
$$\mu_0 = 4\pi \times 10^{-7} \text{ henry/meter}$$

The constitutive parameters are more commonly specified in terms of the specific inductive capacities

$$k_e = \frac{\epsilon}{\epsilon_0}, \tag{20a}$$

$$k_m = \frac{\mu}{\mu_0}. \tag{20b}$$

The quantity k_e is known as the dielectric constant; k_m as the magnetic permeability. These ratios are dimensionless and independent of the units. For practically all materials of interest in antenna work k_m is but negligibly different from unity and will be taken equal to unity unless otherwise indicated.

It is important to note that although **D** and **E** are in the same direction, they are not necessarily in phase. Such phase differences depend on the molecular structure of the medium and are connected with dissipation of electromagnetic energy in the medium. They are conveniently taken into account by expressing ϵ as a complex number,

$$\epsilon = \epsilon_r - j\epsilon_i. \tag{21}$$

The energy losses associated with the imaginary part of ϵ are to be distinguished from the conduction loss associated with conduction currents.

Two types of currents may contribute to the source function **J**: convection currents and conduction currents. In the present volume we shall be concerned only with conduction currents, for which the current density is proportional to the electric field vector **E**:

$$\mathbf{J} = \sigma \mathbf{E}. \tag{22}$$

The constant σ is the conductivity of the medium. Like the other constitutive parameters it may be frequency dependent. A conducting medium cannot support a free volume-charge density ρ; if the conductivity is at all appreciable, ρ may be taken to be zero at all times.

If the constitutive parameters are independent of the field strength, all relations between the field vectors—Maxwell's equations and the constitutive relations [Eqs. (19a), (19b), and (22)]—are linear. Under such circumstances the superposition principle applies. This states that if a set of field vectors $\mathbf{E}_1, \ldots, \mathbf{H}_1$ and source functions ρ_1 and $\mathbf{J}_1$ satisfies the field equations and a second set of field vectors $\mathbf{E}_2, \ldots, \mathbf{H}_2$ and source functions ρ_2 and $\mathbf{J}_2$ does so also, then the sum of these two solutions $\mathbf{E}_1 + \mathbf{E}_2, \ldots, \rho_1 + \rho_2, \mathbf{J}_1 + \mathbf{J}_2$ also satisfies the field and constitutive equations and describes a possible electromagnetic field.

3·3. Boundary Conditions.—In addition to the field equations, which give the relations between the elements of the field in a medium with continuously varying properties, we must know the relations that exist at a boundary where the properties of the medium change discontinuously. The derivation of these boundary conditions starts from the integral forms of the field equations; the procedure is standard and will be found in any text on electromagnetic theory; we shall simply state the results.

Let us consider the boundary surface between two media with constitutive parameters ϵ_1, μ_1, σ_1, and ϵ_2, μ_2, σ_2, respectively. Let the positive unit vector **n** normal to the boundary surface be directed from medium 1 into medium 2. If $\mathbf{E}_1$, $\mathbf{E}_2$, ..., $\mathbf{H}_1$, $\mathbf{H}_2$ are the field vectors at contiguous points on either side of the boundary, the boundary conditions are the following:

1. The tangential component of the electric field intensity is continuous across the boundary:

$$\mathbf{n} \times (\mathbf{E}_2 - \mathbf{E}_1) = 0. \tag{23}$$

It can be shown that a field penetrates into a conducting medium a distance inversely proportional to the square root of the conductivity.[1] Thus if $\sigma_1 = \infty$, $\mathbf{E}_1$ must be zero; this boundary condition then reduces to

$$\mathbf{n} \times \mathbf{E}_2 = 0 \qquad (\sigma_1 = \infty). \tag{24}$$

2. There is a discontinuity in the normal component of **D** at the boundary if there exists a surface layer of charge:

$$\mathbf{n} \cdot (\mathbf{D}_2 - \mathbf{D}_1) = \mathbf{n} \cdot (\epsilon_2 \mathbf{E}_2 - \epsilon_1 \mathbf{E}_1) = \eta, \tag{25}$$

the charge density per unit area being η. Such layers of charge occur, in general, only when one of the media has infinite conductivity.

3. The normal component of **B** varies continuously across a boundary:

$$\mathbf{n} \cdot (\mathbf{B}_2 - \mathbf{B}_1) = \mathbf{n} \cdot (\mu_2 \mathbf{H}_2 - \mu_1 \mathbf{H}_1) = 0. \tag{26}$$

4. A discontinuity in the tangential component of **H** occurs only when there is a surface-current sheet on the boundary

$$\mathbf{n} \times (\mathbf{H}_2 - \mathbf{H}_1) = \mathbf{K}, \tag{27a}$$

K being the surface-current density. Such current sheets exist only if one of the media, say the first, is infinitely conducting. In this case, however, the field cannot penetrate the medium; $\mathbf{H}_1$ must be zero. We have then

$$\mathbf{n} \times \mathbf{H}_2 = \mathbf{K} \qquad (\sigma_1 = \infty) \tag{27b}$$

and likewise

$$\mathbf{n} \cdot \mathbf{B}_2 = 0 \qquad (\sigma_1 = \infty). \tag{28}$$

[1] J. A. Stratton, *Electromagnetic Theory*, McGraw-Hill, New York, 1941, p. 504.

Under all other conditions **K** is zero, and the tangential component of **H** as well as the normal component of **B** is continuous.

These boundary conditions apply to fields that satisfy Maxwell's equations [Eqs. (11)] everywhere. We shall have occasion in diffraction problems to consider a boundary surface between two regions of the same medium. From solutions of Maxwell's equations in these two regions we shall form functions that are solutions of Maxwell's equations everywhere except on this surface, where they are discontinuous. These discontinuities can be formally associated with distributions of magnetic charges and currents on the boundary surface by equations that can be obtained from the Maxwell equations [Eqs. (13)] in which magnetic sources have been introduced:

$$\mathbf{n} \times (\mathbf{E}_2 - \mathbf{E}_1) = -\mathbf{K}_m, \tag{29}$$

and

$$\mathbf{n} \cdot (\mathbf{B}_2 - \mathbf{B}_1) = \eta_m, \tag{30}$$

respectively, where $\mathbf{K}_m$ is the density of the fictitious magnetic-current sheet over the boundary and η_m is the density of the fictitious surface layer of magnetic charge. As in the case of electric current and charge, the magnetic-source functions must satisfy a surface equation of continuity,

$$\boldsymbol{\nabla}_s \cdot \mathbf{K}_m + \frac{\partial \eta_m}{\partial t} = 0, \tag{31}$$

where as before $\boldsymbol{\nabla}_s\cdot$ is the surface-divergence operator.

3·4. The Field Equations for Harmonic Time Dependence.—It will be sufficient for most of our purposes to consider fields having a harmonic time dependence. In such cases we shall take all field and source distributions to depend on time through the same factor $e^{j\omega t}$. The real and imaginary parts of these complex solutions of the field equations will themselves be solutions of the field equations and will describe real fields. The assumption of harmonic time dependence will not greatly affect the generality of our results because an arbitrary field and source distribution can be resolved into harmonic components.

With the restriction of the time dependence to the time factor $e^{j\omega t}$, the field equations may be written as

$$\boldsymbol{\nabla} \times \mathbf{E} + j\omega\mu\mathbf{H} = -\mathbf{J}_m, \tag{32a}$$
$$\boldsymbol{\nabla} \times \mathbf{H} = (\sigma + j\omega\epsilon)\mathbf{E}, \tag{32b}$$
$$\boldsymbol{\nabla} \cdot (\mu\mathbf{H}) = \rho_m, \tag{32c}$$
$$\boldsymbol{\nabla} \cdot (\epsilon\mathbf{E}) = \rho, \tag{32d}$$
$$\boldsymbol{\nabla} \cdot \mathbf{J} + j\omega\rho = 0, \tag{32e}$$
$$\boldsymbol{\nabla} \cdot \mathbf{J}_m + j\omega\rho_m = 0. \tag{32f}$$

These equations apply equally to the field quantities and their space-dependent factors. Equations (19a), (19b), and (22) have been applied in this formulation. Equations (32c) and (32d) have been written in the general form, for inhomogeneous media in which ϵ and μ are functions of position. It should be noted that the equation of continuity determines the charge density directly from the current.

3·5. Poynting's Theorem.—Discussions of the energy relations in an electromagnetic field are usually based on Poynting's theorem. From the first two of Maxwell's equations, (16a) and (16b), we obtain

$$\mathbf{H} \cdot \nabla \times \mathbf{E} - \mathbf{E} \cdot \nabla \times \mathbf{H} = -\mathbf{H} \cdot \frac{\partial \mathbf{B}}{\partial t} - \mathbf{E} \cdot \frac{\partial \mathbf{D}}{\partial t} - \mathbf{E} \cdot \mathbf{J}. \qquad (33)$$

The quantity on the left is equal to $\nabla \cdot (\mathbf{E} \times \mathbf{H})$. On use of the constitutive relations [Equations. (19a) and (19b)], Eq. (33) becomes

$$\nabla \cdot (\mathbf{E} \times \mathbf{H}) + \mathbf{E} \cdot \mathbf{J} = -\frac{\partial}{\partial t}\left(\frac{\epsilon E^2}{2} + \frac{\mu H^2}{2}\right). \qquad (34)$$

This is Poynting's theorem. Formally, Poynting's theorem resembles the equations of continuity previously considered; it expresses the conservation of energy, rather than that of charge. The *Poynting vector*

$$\mathbf{S} = \mathbf{E} \times \mathbf{H} \qquad (35)$$

is interpreted as the intensity of flow of energy, that is, the rate of flow of energy per unit area normal to the direction of $\mathbf{S}$. The quantities $\epsilon E^2/2$ and $\mu H^2/2$ represent the densities of electric and magnetic energy, respectively. The term $\mathbf{E} \cdot \mathbf{J}$ measures the rate of dissipation or production of electromagnetic energy per unit volume. If $\mathbf{E} \cdot \mathbf{J}$ is positive, it is a dissipation term; if it is negative, it represents production of electromagnetic energy.

The analogy of Poynting's theorem to the equation of continuity is brought out more clearly in the corresponding integral form. Let us integrate Eq. (34) over a volume V enclosed by a surface S:

$$\int_V \nabla \cdot (\mathbf{E} \times \mathbf{H}) \, dv + \int_V \mathbf{E} \cdot \mathbf{J} \, dv = -\frac{\partial}{\partial t} \int_V \left(\frac{\epsilon E^2}{2} + \frac{\mu H^2}{2}\right) dv. \qquad (36)$$

Making use of the divergence theorem, we can transform the first integral into a surface integral over the boundary, obtaining

$$\oint_S \mathbf{S} \cdot \mathbf{n} \, dS + \int_V \mathbf{E} \cdot \mathbf{J} \, dv = -\frac{\partial}{\partial t} \int_V \left(\frac{\epsilon E^2}{2} + \frac{\mu H^2}{2}\right) dv. \qquad (37)$$

With the interpretations of the integrands given above, Eq. (37) states that the net rate of flow of energy out through the boundary surface plus the rate of dissipation of electromagnetic energy within the volume (or minus the rate of production) is equal to the rate of decrease of electromagnetic energy stored in the volume V.

Equation (34) is quite general in its applications. We have now to express Poynting's theorem in a form applicable to fields varying periodically with time. In this connection it must be noted that the complex exponential representation of periodic fields can be carried through all linear operations but that in nonlinear operations (such as formation of the products occurring in the Poynting theorem) the real expressions for the field quantities must be used. The complex field vectors may be expressed as

$$\mathbf{E} = (\mathbf{E}e^{j\omega t}) = (\mathbf{E}_r + j\mathbf{E}_i)e^{j\omega t}, \qquad (38a)$$
$$\mathbf{H} = (\mathbf{H}e^{j\omega t}) = (\mathbf{H}_r + j\mathbf{H}_i)e^{j\omega t}. \qquad (38b)$$

The corresponding real fields are

$$\text{Re } \mathbf{E} = (\mathbf{E}_r \cos \omega t - \mathbf{E}_i \sin \omega t), \qquad (39a)$$
$$\text{Re } \mathbf{H} = (\mathbf{H}_r \cos \omega t - \mathbf{H}_i \sin \omega t). \qquad (39b)$$

The Poynting vector is thus

$$\mathbf{S} = \text{Re } \mathbf{E} \times \text{Re } \mathbf{H} \qquad (40)$$
$$= [\mathbf{E}_r \times \mathbf{H}_r \cos^2 \omega t + \mathbf{E}_i \times \mathbf{H}_i \sin^2 \omega t - (\mathbf{E}_r \times \mathbf{H}_i + \mathbf{E}_i \times \mathbf{H}_r) \sin \omega t \cos \omega t]$$

In general we are not interested in the instantaneous flow but in the energy flow averaged over a cycle. That is, we wish to know

$$\bar{\mathbf{S}} = \frac{1}{\tau} \int_0^{\tau} \mathbf{S} \, dt, \qquad \left(\tau = \frac{2\pi}{\omega}\right), \qquad (41)$$

the overline denoting the time-average value. Now the time average of $\sin \omega t \cos \omega t$ vanishes, and the time average of both $\cos^2 \omega t$ and $\sin^2 \omega t$ is $\frac{1}{2}$. Hence

$$\bar{\mathbf{S}} = \tfrac{1}{2}(\mathbf{E}_r \times \mathbf{H}_r + \mathbf{E}_i \times \mathbf{H}_i). \qquad (42)$$

It will be observed that except for the factor $\frac{1}{2}$, the right-hand side of Eq. (42) is the real part of $\mathbf{E} \times \mathbf{H}^*$, where $\mathbf{H}^*$ represents the complex conjugate of $\mathbf{H}$. We have then

$$\bar{\mathbf{S}} = \tfrac{1}{2}\text{Re } (\mathbf{E} \times \mathbf{H}^*). \qquad (43)$$

We shall seldom be concerned with the instantaneous Poynting vector. Unless explicitly stated otherwise, all future reference to the Poynting vector will be to the time-average value given by Eq. (43); the overline will be omitted hereafter except where a distinction must be made.

It is of interest to formulate Poynting's theorem in terms of time-averaged quantities. Since the divergence is a linear operator, involving space derivatives only,

$$\overline{\boldsymbol{\nabla} \cdot \mathbf{S}} = \boldsymbol{\nabla} \cdot \bar{\mathbf{S}} = \boldsymbol{\nabla} \cdot \text{Re } \tfrac{1}{2}(\mathbf{E} \times \mathbf{H}^*) = \tfrac{1}{2} \text{Re } \boldsymbol{\nabla} \cdot (\mathbf{E} \times \mathbf{H}^*). \qquad (44)$$

In the absence of magnetic charges or currents one has, for a field with harmonic time dependence,

$$\frac{1}{2}\nabla \cdot (\mathbf{E} \times \mathbf{H}^*) = \frac{1}{2}(\mathbf{H}^* \cdot \nabla \times \mathbf{E} - \mathbf{E} \cdot \nabla \times \mathbf{H}^*)$$

$$= -\frac{1}{2}(\sigma - j\omega\epsilon^*)\mathbf{E} \cdot \mathbf{E}^* - \frac{j\omega\mu}{2}\mathbf{H} \cdot \mathbf{H}^*. \quad (45)$$

Taking the real part of Eq. (45), with due regard for the complex form of ϵ^* [Eq. (21)], we obtain the modified Poynting's theorem

$$\nabla \cdot \mathbf{S} = -\tfrac{1}{2}(\sigma + \omega\epsilon_i)\mathbf{E} \cdot \mathbf{E},^* \quad (46)$$

or, in integral form,

$$-\oint_S \mathbf{S} \cdot \mathbf{n}\, dS = \int_V \frac{1}{2}(\sigma + \omega\epsilon_i)\mathbf{E} \cdot \mathbf{E}^*\, dv. \quad (47)$$

Since the unit normal $\mathbf{n}$ is directed outward from the region enclosed by the surface S, the term on the left of Eq. (47) is the net average power flow across S *into* the region V. In view of the harmonic time dependence of the field, there can be no average increase in the energy stored; the terms on the right must be interpreted as electromagnetic energy dissipated within the region V. Thus, the imaginary component of the electric inductive capacity, like conductivity of the material, results in energy dissipation. A material with a complex dielectric constant is called a "lossy dielectric." By Eq. (47), if a medium is neither a conductor nor a lossy dielectric, the net power flow across a closed surface S into the region enclosed by it is zero.

3·6. The Wave Equations.—We turn now to a consideration of the wave equations satisfied by electromagnetic fields. We begin with Maxwell's equations in the form [Eqs. (18)] that includes magnetic sources but confine our discussion to linear homogeneous media; ϵ and μ are constants independent of position.

Taking the curl of Eq. (18a), eliminating the magnetic vector $\mathbf{B}$ by means of Eqs. (18b) and (19b), we obtain

$$\nabla \times \nabla \times \mathbf{E} + \mu\epsilon \frac{\partial^2 \mathbf{E}}{\partial t^2} = -\mu \frac{\partial \mathbf{J}}{\partial t} - \nabla \times \mathbf{J}_m. \quad (48)$$

Similarly, interchanging the roles of Eqs. (18a) and (18b), we get

$$\nabla \times \nabla \times \mathbf{H} + \mu\epsilon \frac{\partial^2 \mathbf{H}}{\partial t^2} = -\epsilon \frac{\partial \mathbf{J}_m}{\partial t} + \nabla \times \mathbf{J}. \quad (49)$$

We now make use of the vector identity

$$\nabla \times \nabla \times \mathbf{P} = \nabla(\nabla \cdot \mathbf{P}) - \nabla^2 \mathbf{P}. \quad (50)$$

On application of this to both the previous equations and replacement of $\nabla \cdot \mathbf{E}$ and $\nabla \cdot \mathbf{H}$ by ρ/ϵ and ρ_m/μ respectively, Eqs. (48) and (49) become

$$\nabla^2 \mathbf{E} - \mu\epsilon \frac{\partial^2 \mathbf{E}}{\partial t^2} = \mu \frac{\partial \mathbf{J}}{\partial t} + \nabla \times \mathbf{J}_m + \frac{1}{\epsilon} \nabla \rho, \qquad (51a)$$

$$\nabla^2 \mathbf{H} - \mu\epsilon \frac{\partial^2 \mathbf{H}}{\partial t^2} = \epsilon \frac{\partial \mathbf{J}_m}{\partial t} - \nabla \times \mathbf{J} + \frac{1}{\mu} \nabla \rho_m. \qquad (51b)$$

On the left sides of these equations are the familiar differential terms of the wave equation; the terms on the right represent the effects of distributions of sources. In a source-free medium these equations reduce to the homogeneous wave equations

$$\nabla^2 \mathbf{E} - \mu\epsilon \frac{\partial^2 \mathbf{E}}{\partial t^2} = 0, \qquad (52a)$$

$$\nabla^2 \mathbf{H} - \mu\epsilon \frac{\partial^2 \mathbf{H}}{\partial t^2} = 0, \qquad (52b)$$

with the speed of propagation of the wave given by

$$v = \frac{1}{\sqrt{\mu\epsilon}}, \qquad (53)$$

The speed of propagation in free space is a constant, independent of frequency:

$$c = \frac{1}{\sqrt{\mu_0 \epsilon_0}} = 3 \times 10^8 \text{ meters/sec.} \qquad (54)$$

The index of refraction of a medium is defined as

$$n = \frac{c}{v} = \sqrt{k_m k_e}. \qquad (55a)$$

For most media the magnetic permeability k_m is unity, and

$$n = \sqrt{k_e}. \qquad (55b)$$

The wave equations simplify for fields with time dependence $e^{j\omega t}$, in that the time can be totally eliminated from the equations. There result the so-called "vector Helmholtz" equations for the space dependence of the fields:

$$\nabla \times \nabla \times \mathbf{E} - k^2 \mathbf{E} = -j\omega\mu \mathbf{J} - \nabla \times \mathbf{J}_m, \qquad (56a)$$
$$\nabla \times \nabla \times \mathbf{H} - k^2 \mathbf{H} = -j\omega\epsilon \mathbf{J}_m + \nabla \times \mathbf{J}, \qquad (56b)$$

where

$$k^2 = \omega^2 \mu\epsilon. \qquad (57)$$

The constant k is known as the propagation constant. In nonlossy media it is real and is related to the wavelength by

$$k = \frac{2\pi}{\lambda}. \qquad (58)$$

If ϵ is complex, both the speed of propagation defined by Eq. (53) and the propagation constant are complex. The attenuation of a wave as it propagates in a lossy medium is directly connected with the imaginary part of the propagation constant.

Applying Eq. (50) to Eqs. (56) yields

$$\nabla^2 \mathbf{E} + k^2 \mathbf{E} = j\omega\mu\mathbf{J} + \nabla \times \mathbf{J}_m + \frac{1}{\epsilon}\nabla\rho, \qquad (59a)$$

$$\nabla^2 \mathbf{H} + k^2 \mathbf{H} = j\omega\epsilon\mathbf{J}_m - \nabla \times \mathbf{J} + \frac{1}{\mu}\nabla\rho_m. \qquad (59b)$$

In a source-free medium these reduce to the homogeneous equations

$$\nabla^2 \mathbf{E} + k^2 \mathbf{E} = 0, \qquad (60a)$$
$$\nabla^2 \mathbf{H} + k^2 \mathbf{H} = 0. \qquad (60b)$$

It should be emphasized that Eqs. (60) imply that each rectangular component of the field vectors E_x, E_y, . . . , H_z satisfies the *scalar Helmholtz equation*

$$\nabla^2 \psi + k^2 \psi = 0. \qquad (61)$$

Though all fields that satisfy Maxwell's equations necessarily satisfy the wave equations, the converse is not true. A set of field vectors $\mathbf{E}$ and $\mathbf{H}$ that satisfy the wave equations constitute an admissible electromagnetic field only if at the same time they satisfy Maxwell's equations. Furthermore, the fields must behave properly at the boundaries of the region concerned in accordance with the boundary conditions formulated in Sec. 3·3. If the region is infinite in extent, separate attention must be paid to the behavior at infinity.

3·7. Simple Wave Solutions.—General considerations relative to wave propagation will be developed in the next chapter. We shall consider here several simple waveforms, solutions of Eqs. (60), that recur frequently in general antenna theory. These are (1) the homogeneous plane wave, (2) the circularly symmetrical cylindrical waves, and (3) the isotropic spherical wave. In each case the medium is assumed to be homogeneous, nonconducting, and free from sources.

Plane Waves.—The plane wave is mathematically the simplest type of electromagnetic wave; its propagation is essentially one-dimensional. Let us attempt to find a field such that the directions and magnitudes of the field vectors are constant over any plane normal to the direction of a vector **s** (Fig. 3·1) but vary periodically along lines parallel to **s**. In the case of the electric field vector **E**, the conditions stated above will be satisfied if the field has the form

$$\mathbf{E}(x,y,z,t) = \mathbf{E}_0 e^{(j\omega t - k\mathbf{r}\cdot\mathbf{s})}, \qquad (62)$$

when **r** is the position vector from the origin to the field point (x,y,z)

and k is the propagation constant defined by Eq. (57); the amplitude $\mathbf{E}_0$ is independent of position and time. Since the planes normal to the unit vector $\mathbf{s}$ are defined by $\mathbf{r} \cdot \mathbf{s} = $ constant, this field must be uniform over every such plane. These planes are equiphase surfaces for the wave, and its propagation can be visualized as a continuous progression of one equiphase surface into the contiguous one. It is seen further that at any instant the field has the same magnitude over each of the family of parallel planes

$$\mathbf{r} \cdot \mathbf{s} = C \pm \frac{2\pi n}{k} = C \pm n\lambda,$$
$$n = 0, 1, 2, \cdots . \quad (63)$$

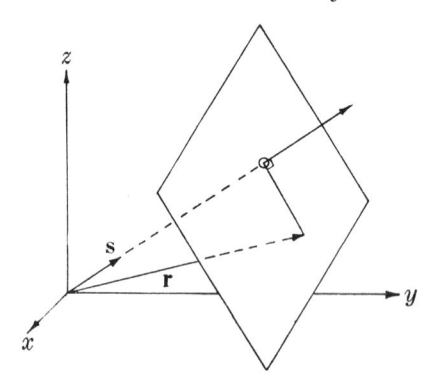

Fig. 3·1.—The plane wave.

It is readily verified that the electric field vector defined by Eq. (62) satisfies the wave equation [Eq. (52a)]. Obviously a similar expression for the magnetic field vector $\mathbf{H}(x,y,z,t)$ is a solution of Eq. (52b). However, if these field vectors are to describe an electromagnetic field, they must be so related as to satisfy Maxwell's equations. The required relation is Eq. (32a):

$$\mathbf{H} = \frac{j}{\omega\mu} \nabla \times \mathbf{E}. \quad (64)$$

On introduction of Eq. (62) this becomes

$$\mathbf{H} = \sqrt{\frac{\epsilon}{\mu}} (\mathbf{s} \times \mathbf{E}) = \sqrt{\frac{\epsilon}{\mu}} (\mathbf{s} \times \mathbf{E}_0) e^{j(\omega t - k\mathbf{s}\cdot\mathbf{r})}. \quad (65)$$

The space-time dependence of $\mathbf{H}$ is the same as that of $\mathbf{E}$, but the direction of $\mathbf{H}$ is normal to that of both $\mathbf{s}$ and $\mathbf{E}_0$. Equation (32d) requires $\nabla \cdot \mathbf{E}$ to be zero in a source-free medium. Thus

$$\nabla \cdot \mathbf{E} = -jk\, \mathbf{s} \cdot \mathbf{E} = 0; \quad (66)$$

that is, $\mathbf{E}$ is normal to $\mathbf{s}$. To satisfy Maxwell's equations, the electric and magnetic field vectors must thus lie in the plane normal to $\mathbf{s}$. It follows at once that the energy flow, that is, the Poynting vector,

$$\mathbf{S} = \frac{1}{2} \operatorname{Re} (\mathbf{E} \times \mathbf{H}^*) = \frac{1}{2} \left(\frac{\epsilon}{\mu}\right)^{1/2} |E_0|^2 \mathbf{s}, \quad (67)$$

is in the direction of propagation of the wave, normal to the equiphase surfaces.

It is of interest to determine whether or not there can exist a plane wave of the form of that in Eq. (62) if the magnitude of $\mathbf{E}_0$ is an arbitrary function of position over an equiphase plane. Without loss of generality we can take the direction of propagation along the z-axis and the direction of $\mathbf{E}_0$ along the x-axis. We are thus considering the field

$$E_x = E_0(x,y)e^{-jkz}, \qquad E_y = E_z = 0 \tag{68}$$

(omitting the time factor $e^{j\omega t}$). If this is to be a possible field, E_x must satisfy the scalar Helmholtz equation [Eq. (61)]. This will be true only if $E_0(x,y)$ is a solution of the two-dimensional Laplace equation:

$$\frac{\partial^2 E_0}{\partial x^2} + \frac{\partial^2 E_0}{\partial y^2} = 0. \tag{69}$$

Since there are no sources, $E_0(x,y)$ must be finite and continuous over the infinite x,y-domain. However, being a solution to Laplace's equation, E_0 can have no maxima or minima in this infinite region. Consequently, $E_0(x,y)$ must be a constant; arbitrary amplitude distributions and *infinite* plane equiphase surfaces are incompatible.

It should be noted that the infinite plane wave is impossible physically because the total energy transported across an equiphase surface is infinite. The practical importance of the plane wave lies in its use in the analysis of other waves. There are two parameters characterizing the plane wave: its angular frequency ω and the direction of propagation $\mathbf{s}$. By superposing time-periodic plane waves, all traveling in the same direction but with various values of ω and amplitudes $E_0(\omega)$, it is possible to build up a plane wave of more general time dependence—a pulse-modulated or otherwise modulated wave. By superposing plane waves with the same frequency ω but with various directions of propagation and amplitudes $E_0(\mathbf{s})$, it is possible to synthesize a time-periodic wave with a more general type of equiphase surface. Because each component wave satisfies Maxwell's equations, the resultant obtained by superposition likewise satisfies the field equations.

Cylindrical Waves.—Circularly symmetrical cylindrical waves are the elementary forms of two-dimensional propagation. The equiphase surfaces of these waves are coaxial circular cylinders; the wave is propagated along the radii of the phase surfaces.

Cylindrical coordinates, as defined in Fig. 3·2, are appropriate for the analysis. The z-axis is taken as the axis of symmetry, and r and θ are polar coordinates in a plane normal to the z-axis. At each point we define unit vectors $\mathbf{i}_r$, $\mathbf{i}_\theta$, $\mathbf{i}_z$ in the direction of increasing r, θ, and z, respectively; the field vectors may, on occasion, be resolved into components in these directions.

We shall now seek solutions of the field equations in which the field vectors are everywhere tangential to the cylindrical equiphase surfaces and have constant amplitude over each such surface (that is, the amplitudes are functions of r only). We shall seek solutions of two different types, distinguished with reference to the directions of the field vectors:

Case a:
$$H_z = 0, \quad E_z(r) \neq 0.$$
Case b:
$$E_z = 0, \quad H_z(r) \neq 0.$$

In each solution, of course, $H_r = E_r = 0$.

We begin by determining the form of the z-component of the field as a solution of the wave equation; later we shall determine the remaining field components by means of the field equations.

Since $E_z(r)$ and $H_z(r)$ are components of the respective field vectors in a rectangular coordinate system, they must satisfy the scalar Helmholtz equation [Eq. (61)]. In cylindrical coordinates this becomes

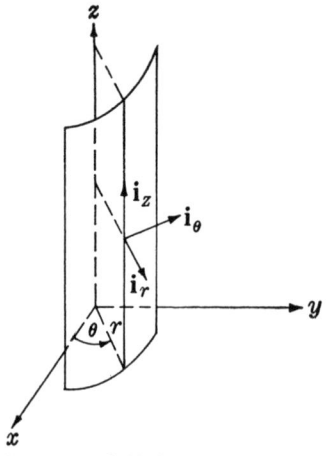

Fig. 3·2.—Cylindrical coordinates.

$$\frac{1}{r}\frac{\partial}{\partial r}\left(r\frac{\partial \psi}{\partial r}\right) + \frac{1}{r^2}\frac{\partial^2 \psi}{\partial \theta^2} + \frac{\partial^2 \psi}{\partial z^2} + k^2\psi = 0, \tag{70}$$

where ψ may represent either E_z or H_z. Since ψ is independent of θ and z, this reduces to

$$\frac{d^2\psi}{dr^2} + \frac{1}{r}\frac{d\psi}{dr} + k^2\psi = 0. \tag{71}$$

On introduction of
$$\xi = kr, \tag{72}$$
this becomes
$$\frac{d^2\psi}{d\xi^2} + \frac{1}{\xi}\frac{d\psi}{d\xi} + \psi = 0 \tag{73}$$

This is the differential equation satisfied by the Bessel functions or, more generally speaking, by the cylinder functions or order zero.[1] Of the many solutions of this equation which we might identify with the functions E_z or H_z, those of immediate interest here are the Hankel functions $H_0^{(1)}(\xi)$ and $H_0^{(2)}(\xi)$. The nature of these functions is most evident in their asymptotic behavior for large values of $\xi = kr$:

$$H_0^{(1)}(kr) \approx \sqrt{\frac{2}{\pi kr}}\, e^{j\left(kr - \frac{\pi}{4}\right)}, \tag{74a}$$

$$H_0^{(2)}(kr) \approx \sqrt{\frac{2}{\pi kr}}\, e^{-j\left(kr - \frac{\pi}{4}\right)}, \tag{74b}$$

$(kr \gg 1).$

[1] G. N. Watson, *Theory of Bessel Functions*, Macmillan, 1944.

The second of these functions, multiplied by the time factor $e^{j\omega t}$, represents a wave traveling in the positive r-direction; the phase quantity $\omega t - kr$ is the analogue of the quantity $\omega t - kx$ for a plane wave traveling in the positive x-direction. Thus $H_0^{(2)}(kr)$ represents a cylindrical wave diverging from a line source on the z-axis. Similarly, $H_0^{(1)}(kr)$ represents a wave traveling in the negative r-direction, that is, a wave from infinity converging to a line focus along the z-axis.

Restricting attention to the diverging wave function $H_0^{(2)}(kr)$, we consider first Case a. We assume

$$\mathbf{E} = H_0^{(2)}(kr)e^{j\omega t}\mathbf{i}_z \tag{75}$$

and use the field equations to determine the associated magnetic field. The curl of a vector $\mathbf{P}$, expressed in cylindrical coordinates, is

$$\nabla \times \mathbf{P} = \left(\frac{1}{r}\frac{\partial P_z}{\partial \theta} - \frac{\partial P_\theta}{\partial z}\right)\mathbf{i}_r + \left(\frac{\partial P_r}{\partial z} - \frac{\partial P_z}{\partial r}\right)\mathbf{i}_\theta + \frac{1}{r}\left[\frac{\partial}{\partial r}(rP_\theta) - \frac{\partial P_r}{\partial \theta}\right]\mathbf{i}_z. \tag{76}$$

Taking the curl of the vector $\mathbf{E}$ and making use of Eq. (32a), we obtain

$$\mathbf{H} = -\frac{j}{\omega\mu}\left[\frac{d}{dr}H_0^{(2)}(kr)\right]e^{j\omega t}\mathbf{i}_\theta. \tag{77}$$

It is left to the reader to verify that the field vectors $\mathbf{E}$ and $\mathbf{H}$ defined by Eqs. (75) and (77) satisfy the other field equations. Over the cylindrical surfaces of constant r, $\mathbf{E}$ and $\mathbf{H}$ are perpendicular to each other at every point and lie in the tangent plane to the surface; as in the case of a plane wave, $\mathbf{E}$ and $\mathbf{H}$ are normal to the direction of propagation, and the Poynting vector is normal to the equiphase surface. As the radius of the equiphase surface becomes large, it becomes sensibly plane in the neighborhood of any point. We must, therefore, expect that as $r \to \infty$, the relationship between $\mathbf{E}$ and $\mathbf{H}$ approaches that existing in a plane wave. The asymptotic form for $\mathbf{H}$ may be obtained by introducing Eq. (74b) into Eq. (77). Neglecting terms of higher order in $1/r$, we find

$$\frac{d}{dr}H_0^{(2)}(kr) \approx -jk\sqrt{\frac{2}{\pi kr}}e^{-j\left(kr-\frac{\pi}{4}\right)}, \tag{78}$$

whence

$$\mathbf{H} \approx -\frac{k}{\omega\mu}H_0^{(2)}(kr)e^{j\omega t}\mathbf{i}_\theta \qquad (kr \gg 1). \tag{79}$$

Thus, in the limit as $r \to \infty$,

$$\mathbf{H} = \sqrt{\frac{\epsilon}{\mu}}(\mathbf{i}_r \times \mathbf{E}) \tag{80}$$

as was to be expected.

The derivation of the field for Case b proceeds in a similar manner. We assume

$$\mathbf{H} = H_0^{(2)}(kr)e^{j\omega t}\mathbf{i}_z; \qquad (81)$$

the associated electric field follows by application of Eq. (32b), which, in the case at hand, becomes

$$\mathbf{E} = \frac{1}{j\omega\epsilon} \nabla \times \mathbf{H}. \qquad (82)$$

It follows that

$$\mathbf{E} = \frac{j}{\omega\epsilon}\left[\frac{d}{dr}H_0^{(2)}(kr)\right]e^{j\omega t}\mathbf{i}_\theta. \qquad (83)$$

The general remarks concerning Case a apply to the present case also. It is easily verified that here too the relationships approach those in a plane wave as the radius of the cylindrical phase surface becomes very large; that is,

$$\lim_{r \to \infty} \mathbf{H} = \sqrt{\frac{\epsilon}{\mu}}(\mathbf{i}_r \times \mathbf{E}). \qquad (84)$$

We have thus obtained two independent field distributions with cylindrical equiphase surfaces. We shall refer to these as cylindrical modes of free-space propagation. The first field, Case a, can arise from a linear distribution of electric current along the z-axis and will be spoken of as a field of the electric type; Case b can be associated with a linear distribution of magnetic current along the z-axis and is correspondingly referred to as a field of the magnetic type. These are the simplest cylindrical modes of free-space propagation. A treatment of the general theory of cylindrical waves will be found in Stratton.[1]

Isotropic Spherical Waves.—Next we shall consider the isotropic spherical wave with equiphase surfaces that are concentric spheres and field amplitudes that are constant in magnitude over each equiphase surface.

The spherical coordinates r, θ, and ϕ, illustrated in Fig. 3·3, are appropriate for this discussion. With the spherical coordinates are associated a set of orthogonal unit vectors $\mathbf{i}_r$, $\mathbf{i}_\theta$, $\mathbf{i}_\phi$ at each point in space, in the directions of increasing r, θ, and ϕ respectively.

Let the center of the family of equiphase spheres be at the origin of the coordinate system. An attempt to construct a field that is a function of r alone, as in the case of cylindical waves, will fail. For example, suppose that we try to construct a field in which the field vectors have only the components

$$\left.\begin{array}{l}\mathbf{E} = E(r)\mathbf{i}_\theta, \\ \mathbf{H} = H(r)\mathbf{i}_\phi.\end{array}\right\} \qquad (85)$$

[1] J. A. Stratton, *Electromagnetic Theory*, McGraw-Hill, New York, 1941, Chap VI.

Sec. 3·7]   SIMPLE WAVE SOLUTIONS   79

It will be seen that there is an essential ambiguity in the directions of these vectors at all points for which $\theta = 0$ or π. The ambiguity can be rendered trivial only by making the magnitudes of the fields vanish for $\theta = 0$ and π; the field can then be independent of θ only if it vanishes identically.

The isotropic spherical wave is, in general, a possible waveform only for scalar fields such as are encountered in acoustics. However, it is often useful for reference and comparison with electromagnetic waves.

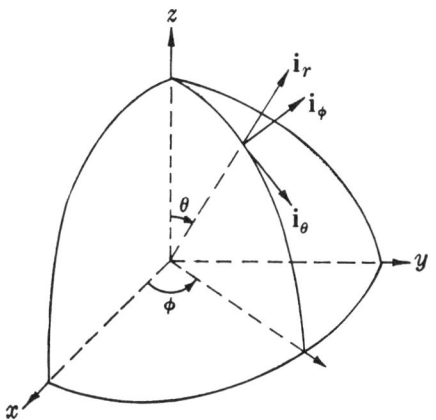

Fig. 3·3.—Spherical coordinates.

Accordingly we shall note briefly the spherically symmetrical solutions of the scalar Helmholtz equation [Eq. (61)]. In spherical coordinates, the Laplacian ∇^2 is

$$\nabla^2 = \frac{1}{r^2}\frac{\partial}{\partial r}\left(r^2 \frac{\partial}{\partial r}\right) + \frac{1}{r^2 \sin \theta}\frac{\partial}{\partial \theta}\left(\sin \theta \frac{\partial}{\partial \theta}\right) + \frac{1}{r^2 \sin^2 \theta}\frac{\partial^2}{\partial \phi^2}. \quad (86)$$

When ψ is a function of r only, the Helmholtz equation becomes

$$\frac{1}{r^2}\frac{d}{dr}\left(r^2 \frac{d\psi}{dr}\right) + k^2 \psi = 0. \quad (87)$$

It is readily verified that

$$\left.\begin{array}{l} \psi_+ = \dfrac{e^{jkr}}{r}, \\[6pt] \psi_- = \dfrac{e^{-jkr}}{r} \end{array}\right\} \quad (88)$$

are solutions of this equation. The solution ψ_-, multiplied by the time factor $e^{j\omega t}$, represents a wave diverging from a source at the origin, while $\psi_+ e^{j\omega t}$ represents a spherical wave converging to a point focus at the origin.

3·8. General Solution of the Field Equations in Terms of the Sources, for a Time-periodic Field.

—The plane and cylindrical waves discussed in the preceding section are solutions of the homogeneous field equations which apply in regions of space free from charge and current distributions. In deriving the form of these fields, no attention was paid to their ultimate sources, which lay outside the domain of validity of the solution. Our present task is the more exacting one of determining what fields will arise from a prescribed set of sources in a homogeneous medium.

For reference the complete set of field equations is repeated here. Magnetic charge and current distributions are included for later use.

$$\nabla \times \mathbf{E} + j\omega\mu\mathbf{H} = -\mathbf{J}_m, \quad (3\cdot32a)$$
$$\nabla \times \mathbf{H} - j\omega\epsilon\mathbf{E} = \mathbf{J}, \quad (3\cdot32b)$$
$$\nabla \cdot \mathbf{H} = \frac{\rho_m}{\mu}, \quad (3\cdot32c)$$
$$\nabla \cdot \mathbf{E} = \frac{\rho}{\epsilon}, \quad (3\cdot32d)$$
$$\nabla \cdot \mathbf{J} + j\omega\rho = 0, \quad (3\cdot32e)$$
$$\nabla \cdot \mathbf{J}_m + j\omega\rho_m = 0, \quad (3\cdot32f)$$

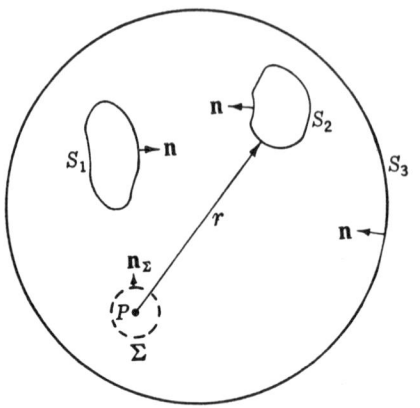

FIG. 3·4.—Notation for Green's theorem.

also the pair of vector Helmholtz equations,

$$\nabla \times \nabla \times \mathbf{E} - k^2\mathbf{E} = -j\omega\mu\mathbf{J} - \nabla \times \mathbf{J}_m, \quad (89)$$
$$\nabla \times \nabla \times \mathbf{H} - k^2\mathbf{H} = -j\omega\epsilon\mathbf{J}_m + \nabla \times \mathbf{J}. \quad (90)$$

The integration of these equations is based on a vector Green's theorem:[1] Consider the region V, illustrated in Fig. 3·4, bounded by the surfaces $S_1, \ldots, S_n$. Let $\mathbf{F}$ and $\mathbf{G}$ be two vector functions of position in this region, each continuous and having continuous first and second derivatives everywhere within V and on the boundary surfaces. Then, if $\mathbf{n}$ is the unit vector normal to a bounding surface, directed into the region V,

$$\int_V (\mathbf{F} \cdot \nabla \times \nabla \times \mathbf{G} - \mathbf{G} \cdot \nabla \times \nabla \times \mathbf{F}) \, dv$$
$$= -\int_{S_1+S_2+\cdots+S_n} (\mathbf{G} \times \nabla \times \mathbf{F} - \mathbf{F} \times \nabla \times \mathbf{G}) \cdot \mathbf{n} \, dS. \quad (91)$$

As indicated, the surface integral extends over all boundary surfaces.

Let us suppose that there exists in a volume V, such as that considered above, an electromagnetic field such that $\mathbf{E}$ and $\mathbf{H}$ meet the condi-

---

[1] The procedure adopted here is due to J. A. Stratton and L. J. Chu, *Phys. Rev.*, **56**, 99 (1939). A proof of the Green's theorem is given in this paper.

SEC. 3·8] GENERAL SOLUTION OF THE FIELD EQUATIONS 81

tions of continuity required of the vector function **F** of the Green's theorem. We shall now see, with the aid of this theorem, how one can express the field at an arbitrary point P in the volume V in terms of the field sources within this volume and the values of the field itself over the boundaries of the region.

We define the vector function of position

$$\mathbf{G} = \frac{e^{-jkr}}{r}\mathbf{a} = \psi\mathbf{a}, \tag{92}$$

where r is the distance from P to any other point in the region and **a** is an arbitrary but otherwise constant vector. This will satisfy the continuity conditions required of the function **G** in the Green's theorem everywhere, except at P, where it has a singularity. Accordingly, we surround P by a sphere Σ of radius r_0 and consider that portion V' of V which is bounded by the surfaces $S_1, \ldots, S_n$ and Σ; in this restricted region, **G** as defined by Eq. (92) and $\mathbf{F} = \mathbf{E}$ of the electromagnetic field satisfy the conditions required for application of the Green's theorem. We have then

$$\int_{V'} (\psi\mathbf{a} \cdot \nabla \times \nabla \times \mathbf{E} - \mathbf{E} \cdot \nabla \times \nabla \times \psi\mathbf{a})\, dv$$

$$= -\int_{S_1 + \cdots + S_n + \Sigma} (\mathbf{E} \times \nabla \times \psi\mathbf{a} - \psi\mathbf{a} \times \nabla \times \mathbf{E}) \cdot \mathbf{n}\, dS. \tag{93}$$

As the first step in the manipulation of this equation, we shall transform the volume integral involving the electric field into an equivalent integral involving only the field sources. Introduction of the vector $\psi\mathbf{a}$ into the vector identity [Eq. (50)] and use of the facts that ψ satisfies the scalar Helmholtz equation and **a** is a constant vector will suffice to show that

$$\nabla \times \nabla \times \psi\mathbf{a} = \nabla(\mathbf{a} \cdot \nabla\psi) + k^2\psi\mathbf{a}. \tag{94}$$

Taking this in conjunction with Eq. (89), we obtain

$$\psi\mathbf{a} \cdot \nabla \times \nabla \times \mathbf{E} - \mathbf{E} \cdot \nabla \times \nabla \times \psi\mathbf{a} = \mathbf{a} \cdot (-j\omega\mu\mathbf{J}\psi - \psi\nabla \times \mathbf{J}_m)$$
$$- \mathbf{E} \cdot \nabla(\mathbf{a} \cdot \nabla\psi). \tag{95}$$

A few additional transformations are necessary:

$$\mathbf{E} \cdot \nabla(\mathbf{a} \cdot \nabla\psi) = \nabla \cdot [\mathbf{E}(\mathbf{a} \cdot \nabla\psi)] - (\mathbf{a} \cdot \nabla\psi)\nabla \cdot \mathbf{E} \tag{96}$$
$$= \nabla \cdot [\mathbf{E}(\mathbf{a} \cdot \nabla\psi)] - \frac{\rho}{\epsilon}\mathbf{a} \cdot \nabla\psi,$$

and

$$\psi\nabla \times \mathbf{J}_m = \nabla \times \mathbf{J}_m\psi + \mathbf{J}_m \times \nabla\psi. \tag{97}$$

By use of these, Eq. (93) can be given the desired form:

$$\mathbf{a} \cdot \int_{V'} \left(j\omega\mu \mathbf{J}\psi + \mathbf{J}_m \times \nabla\psi - \frac{\rho}{\epsilon}\nabla\psi \right) dv + \mathbf{a} \cdot \int_{V'} \nabla \times \psi \mathbf{J}_m \, dv$$
$$+ \int_{V'} \nabla \cdot [\mathbf{E}(\mathbf{a} \cdot \nabla\psi)] \, dv$$
$$= \int_{S_1 + \cdots + \Sigma} [(\mathbf{E} \times \nabla \times \psi\mathbf{a}) \cdot \mathbf{n} - (\psi\mathbf{a} \times \nabla \times \mathbf{E}) \cdot \mathbf{n}] \, dS. \quad (98)$$

We can now bring each term in Eq. (98) into the form of a scalar product with the vector **a** and then completely eliminate this vector from the problem. The second and third volume integrals can be transformed into surface integrals:

$$\mathbf{a} \cdot \int_{V'} \nabla \times \psi\mathbf{J}_m \, dv = -\mathbf{a} \cdot \int_{S_1 + \cdots + \Sigma} \psi \mathbf{n} \times \mathbf{J}_m \, dS, \quad (99)$$

$$\int_{V'} \nabla \cdot [\mathbf{E}(\mathbf{a} \cdot \nabla\psi)] \, dv = -\int_{S_1 + \cdots + \Sigma} (\mathbf{n} \cdot \mathbf{E})(\mathbf{a} \cdot \nabla\psi) \, dS$$
$$= -\mathbf{a} \cdot \int_{S_1 + \cdots + \Sigma} (\mathbf{n} \cdot \mathbf{E}) \nabla\psi \, dS. \quad (100)$$

To the surface integrals on the right-hand side of Eq. (98) we apply the following transformations:

$$[\mathbf{E} \times (\nabla \times \psi\mathbf{a})] \cdot \mathbf{n} = [\mathbf{E} \times (\nabla\psi \times \mathbf{a})] \cdot \mathbf{n} = [(\mathbf{n} \times \mathbf{E}) \times \nabla\psi] \cdot \mathbf{a}, \quad (101)$$
$$\psi(\mathbf{a} \times \nabla \times \mathbf{E}) \cdot \mathbf{n} = -j\omega\mu\psi(\mathbf{a} \times \mathbf{H}) \cdot \mathbf{n} - \psi(\mathbf{a} \times \mathbf{J}_m) \cdot \mathbf{n}$$
$$= j\omega\mu\psi\mathbf{a} \cdot (\mathbf{n} \times \mathbf{H}) + \psi\mathbf{a} \cdot (\mathbf{n} \times \mathbf{J}_m). \quad (102)$$

Collecting these results, we obtain finally

$$\mathbf{a} \cdot \int_{V'} \left(j\omega\mu\psi\mathbf{J} + \mathbf{J}_m \times \nabla\psi - \frac{\rho}{\epsilon}\nabla\psi \right) dv$$
$$= \mathbf{a} \cdot \int_{S_1 + \cdots + \Sigma} [-j\omega\mu\psi(\mathbf{n} \times \mathbf{H}) + (\mathbf{n} \times \mathbf{E}) \times \nabla\psi + (\mathbf{n} \cdot \mathbf{E}) \nabla\psi] \, dS. \quad (103)$$

Since Eq. (103) must hold for *every* vector **a**, the integrals themselves must be equal. That is,

$$\int_{\Sigma} [-j\omega\mu\psi(\mathbf{n} \times \mathbf{H}) + (\mathbf{n} \times \mathbf{E}) \times \nabla\psi + (\mathbf{n} \cdot \mathbf{E}) \nabla\psi] \, dS$$
$$= \int_{V'} \left(j\omega\mu\psi\mathbf{J} + \mathbf{J}_m \times \nabla\psi - \frac{\rho}{\epsilon}\nabla\psi \right) dv - \int_{S_1 + \cdots + S_n} [-j\omega\mu\psi(\mathbf{n} \times \mathbf{H})$$
$$+ (\mathbf{n} \times \mathbf{E}) \times \nabla\psi + (\mathbf{n} \cdot \mathbf{E}) \nabla\psi] \, dS, \quad (104)$$

where for convenience we have split off the integral over the sphere Σ. In the limit as Σ shrinks down on P, this integral will depend only on the

Sec. 3·8] *GENERAL SOLUTION OF THE FIELD EQUATIONS* 83

field at P. Thus we have a relation between the field at P and a volume integral over the sources of the field, plus surface integrals involving the field itself.

Next let us consider the integral over Σ. On the surface of this sphere we have

$$(\nabla \psi) = \left[\frac{d}{dr}\left(\frac{e^{-jkr}}{r}\right)\right]_{r=r_0} \mathbf{n} = -\left(jk + \frac{1}{r_0}\right)\frac{e^{-jkr_0}}{r_0}\mathbf{n}. \quad (105)$$

The normal $\mathbf{n}$ is directed along the radius out from P. Let $d\Omega$ be the solid angle subtended at P by an element of surface dS on Σ; the surface integral can then be written

$$\int_\Sigma [\ ]\, dS = -jr_0 e^{-jkr_0} \int_\Sigma \{\omega\mu(\mathbf{n} \times \mathbf{H}) + k[(\mathbf{n} \times \mathbf{E}) \times \mathbf{n} + (\mathbf{n} \cdot \mathbf{E})\mathbf{n}]\}\, d\Omega$$

$$- e^{-jkr_0} \int_\Sigma [(\mathbf{n} \times \mathbf{E}) \times \mathbf{n} + (\mathbf{n} \cdot \mathbf{E})\mathbf{n}]\, d\Omega$$

$$= -j4\pi r_0 e^{-jkr_0}(\omega\mu\overline{\mathbf{n} \times \mathbf{H}} + k\overline{\mathbf{E}}) - 4\pi e^{-jkr_0}\overline{\mathbf{E}}, \quad (106)$$

where the overline denotes the mean value of the function over the surface of the sphere. If now we let the sphere shrink to zero, the term containing r_0 vanishes because by hypothesis the field vectors are finite in the neighborhood of P. At the same time $\overline{\mathbf{E}}$ approaches $\mathbf{E}_P$, the value of the field vector at P. Thus

$$\lim_{r_0 \to 0} \int_\Sigma [\ ]\, dS = -4\pi\mathbf{E}_P. \quad (107)$$

In this limit the region V' comes to include the whole of the region V, and Eq. (104) becomes

$$\mathbf{E}_P = -\frac{1}{4\pi}\int_V \left(j\omega\mu\psi\mathbf{J} + \mathbf{J}_m \times \nabla\psi - \frac{\rho}{\epsilon}\nabla\psi\right) dv$$

$$+ \frac{1}{4\pi}\int_{S_1+S_2+\ldots+S_n} [-j\omega\mu\psi(\mathbf{n} \times \mathbf{H}) + (\mathbf{n} \times \mathbf{E}) \times \nabla\psi + (\mathbf{n} \cdot \mathbf{E})\nabla\psi]\, dS. \quad (108)$$

The analysis follows the same course for the magnetic vector $\mathbf{H}$, with the corresponding result:

$$\mathbf{H}_P = -\frac{1}{4\pi}\int_V \left(j\omega\epsilon\psi\mathbf{J}_m - \mathbf{J} \times \nabla\psi - \frac{\rho_m}{\mu}\nabla\psi\right) dv$$

$$+ \frac{1}{4\pi}\int_{S_1+\ldots-S_n} [j\omega\epsilon(\mathbf{n} \times \mathbf{E})\psi + (\mathbf{n} \times \mathbf{H}) \times \nabla\psi + (\mathbf{n} \cdot \mathbf{H})\nabla\psi]\, dS. \quad (109)$$

The fields at the observation point P have thus been expressed as the sum of contributions from the sources distributed through the region V

and from fields existing on the bounding surfaces. These latter surface integrals represent contributions to the field from sources lying outside V; specifically, the surface integral over a surface S_i enclosing an exterior volume V_i represents the effect of sources within V_i.

Each of the three terms in the surface integral can be correlated with a corresponding term in the volume integral according to the way in which the function

$$\psi = \frac{e^{-jkr}}{r} \qquad (110)$$

is involved. In Eq. (108), for example, $(\mathbf{n} \times \mathbf{H})$, $(\mathbf{n} \times \mathbf{E})$, and $(\mathbf{n} \cdot \mathbf{E})$ enter the surface integral exactly as the electric-current density $\mathbf{J}$, magnetic-current density $\mathbf{J}_m$, and the charge density ρ, respectively, enter the volume integral; a similar correspondence will be observed in Eq. (109). Thus the effects of sources lying in an exterior region V_i, bounded by the surface S_i, are represented formally as arising from a surface distribution of charges and currents on the boundary S_i, with surface densities

$$\left.\begin{array}{l} \mathbf{K} = (\mathbf{n} \times \mathbf{H}), \\ \mathbf{K}_m = -(\mathbf{n} \times \mathbf{E}), \\ \eta = \epsilon(\mathbf{n} \cdot \mathbf{E}), \\ \eta_m = \mu(\mathbf{n} \cdot \mathbf{H}), \end{array}\right\} \qquad (111)$$

$\mathbf{E}$ and $\mathbf{H}$ being the fields existing over that surface.

3·9. Field Due to Sources in an Unbounded Region.—We have now to consider the case in which the region V is unbounded and the sources of the field are confined to a region of finite extent. There is then only one boundary surface S_n, which we shall at first take to be a sphere of large radius R about the point P, enclosing all sources of the field. Equations (108) and (109) then reduce to a single surface integral over this large sphere $S(R)$.

Let $\mathbf{R}_1$ be a unit vector directed *out* along the radius of this sphere; that is, let $\mathbf{R}_1 = -\mathbf{n}$. On introduction of this vector and the explicit form of ψ, the surface integral of Eq. (108) becomes

$$\frac{1}{4\pi} \int_{S(R)} [-j\omega\mu\psi(\mathbf{n} \times \mathbf{H}) + (\mathbf{n} \times \mathbf{E}) \times \nabla\psi + (\mathbf{n} \cdot \mathbf{E}) \nabla\psi]\, dS$$

$$= \frac{1}{4\pi} \int_{S(R)} \left\{ j\omega\mu(\mathbf{R}_1 \times \mathbf{H}) - \left(jk + \frac{1}{R}\right)[\mathbf{R}_1 \times (\mathbf{R}_1 \times \mathbf{E}) \right.$$

$$\left. - (\mathbf{R}_1 \cdot \mathbf{E})\mathbf{R}_1] \right\} \frac{e^{-jkR}}{R} dS = \frac{1}{4\pi} \int_{S(R)} \left\{ j\omega\mu \left[(\mathbf{R}_1 \times \mathbf{H}) + \left(\frac{\epsilon}{\mu}\right)^{\frac{1}{2}} \mathbf{E} \right] \right.$$

$$\left. + \frac{\mathbf{E}}{R} \right\} \frac{e^{-jkR}}{R} dS. \quad (112)$$

SEC. 3·9] *FIELD DUE TO SOURCES IN AN UNBOUNDED REGION* 85

If we now let the radius R become infinite, the surface of the sphere increases as R^2. The surface integral will vanish as $R \to \infty$ if the fields satisfy the conditions

$$\lim_{R \to \infty} R\mathbf{E} \text{ is finite,} \tag{113a}$$

$$\lim_{R \to \infty} R \left[(\mathbf{R}_1 \times \mathbf{H}) + \left(\frac{\epsilon}{\mu}\right)^{1/2} \mathbf{E} \right] = 0. \tag{113b}$$

In the case of Eq. (109), the surface integral will vanish if

$$\lim_{R \to \infty} R\mathbf{H} \text{ is finite,} \tag{113c}$$

$$\lim_{R \to \infty} R \left[\left(\frac{\epsilon}{\mu}\right)^{1/2} (\mathbf{R}_1 \times \mathbf{E}) - \mathbf{H} \right] = 0. \tag{113d}$$

Conditions (113a) and (113c) require that at large distances from the sources, the magnitudes of the field vectors decrease at least as rapidly as R^{-1}. Conditions (113b) and (113d), the so-called "radiation conditions," ensure that all radiation across the bounding sphere consist of waves diverging to infinity. This may be seen as follows: Taking the scalar product of Eq. (113b) with $\mathbf{R}_1$, we obtain

$$\lim_{R \to \infty} (R\mathbf{E}) \cdot \mathbf{R}_1 = 0. \tag{114}$$

The component of $\mathbf{E}$ in the direction $\mathbf{R}_1$ thus diminishes more rapidly than R^{-1}; we may say that $\mathbf{E}$ is perpendicular to $\mathbf{R}_1$, to terms of the order of R^{-1}. On the other hand, Eq. (113d) states that

$$\lim_{R \to \infty} R\mathbf{H} = \left(\frac{\epsilon}{\mu}\right)^{1/2} (\mathbf{R}_1 \times \mathbf{E}R). \tag{115}$$

It follows that to terms of the order of R^{-1}, $\mathbf{H}$ is perpendicular to both $\mathbf{E}$ and $\mathbf{R}_1$ and $\mathbf{E}$ and $\mathbf{H}$ are related in the same manner as in a plane wave progressing away from the center of the sphere $S(R)$.

If (as will be shown in Sec. 3·11 to be the case) the fields arising from sources confined to a finite region of space satisfy Eqs. (113) at infinity, then the surface integrals over the infinite spheres vanish and the field vectors in the unbounded region are given by

$$\mathbf{E}_P = -\frac{1}{4\pi} \int_V \left[j\omega\mu \mathbf{J} \frac{e^{-jkr}}{r} + \mathbf{J}_m \times \nabla\left(\frac{e^{-jkr}}{r}\right) - \frac{\rho}{\epsilon} \nabla\left(\frac{e^{-jkr}}{r}\right) \right] dv, \tag{116a}$$

$$\mathbf{H}_P = -\frac{1}{4\pi} \int_V \left[j\omega\epsilon \mathbf{J}_m \frac{e^{-jkr}}{r} - \mathbf{J} \times \nabla\left(\frac{e^{-jkr}}{r}\right) - \frac{\rho_m}{\mu} \nabla\left(\frac{e^{-jkr}}{r}\right) \right] dv. \tag{116b}$$

The fields are expressed here entirely in terms of the sources.

These fields can be expressed in terms of the current distributions alone by use of the equations of continuity [Eqs. (32e) and (32f)], which relate the charge densities to the current distributions. Thus Eq. (116a) becomes

$$E_P = \frac{j}{4\pi\omega\epsilon} \int_V \left[-k^2 \mathbf{J} \frac{e^{-jkr}}{r} + (\boldsymbol{\nabla} \cdot \mathbf{J})\boldsymbol{\nabla} \frac{e^{-jkr}}{r} + j\omega\epsilon \mathbf{J}_m \times \boldsymbol{\nabla}\left(\frac{e^{-jkr}}{r}\right) \right] dv. \quad (117)$$

Let $\mathbf{i}_\alpha$, $\alpha = 1, 2, 3$ be unit vectors in the x-, y-, and z-directions, respectively. Then

$$(\boldsymbol{\nabla} \cdot \mathbf{J})\boldsymbol{\nabla} \frac{e^{-jkr}}{r} = \sum_\alpha (\boldsymbol{\nabla} \cdot \mathbf{J}) \frac{\partial}{\partial x_\alpha}\left(\frac{e^{-jkr}}{r}\right) \mathbf{i}_\alpha \quad (118a)$$

$$= \sum_\alpha \mathbf{i}_\alpha \boldsymbol{\nabla} \cdot \left(\mathbf{J} \frac{\partial}{\partial x_\alpha} \frac{e^{-jkr}}{r}\right) - (\mathbf{J} \cdot \boldsymbol{\nabla})\boldsymbol{\nabla}\left(\frac{e^{-jkr}}{r}\right) \quad (118b)$$

Now

$$\int_{V(R)} \boldsymbol{\nabla} \cdot \left(\mathbf{J} \frac{\partial}{\partial x_\alpha} \frac{e^{-jkr}}{r}\right) dv = -\int_{S(R)} \mathbf{n} \cdot \mathbf{J} \frac{\partial}{\partial x_\alpha} \frac{e^{-jkr}}{r} dS = 0 \quad (119)$$

as soon as R is taken so large that $S(R)$ lies outside the region to which the current distribution is confined. It follows that the first terms on the right of Eq. (118b) contribute nothing to the integral in Eq. (117). Thus we obtain

$$\mathbf{E}_P = -\frac{j}{4\pi\omega\epsilon} \int_V [(\mathbf{J} \cdot \boldsymbol{\nabla})\boldsymbol{\nabla} + k^2\mathbf{J} - j\omega\epsilon \mathbf{J}_m \times \boldsymbol{\nabla}] \frac{e^{-jkr}}{r} dv. \quad (120)$$

Similarly, for the magnetic field we obtain

$$\mathbf{H}_P = -\frac{j}{4\pi\omega\mu} \int_V [(\mathbf{J}_m \cdot \boldsymbol{\nabla})\boldsymbol{\nabla} + k^2\mathbf{J}_m + j\omega\mu \mathbf{J} \times \boldsymbol{\nabla}] \frac{e^{-jkr}}{r} dv. \quad (121)$$

3·10. Field in a Region Bounded by Surfaces of Infinitely Conductive Media.—A second case of importance is that in which the region V is bounded by surfaces S_i which are the surfaces of bodies of infinite conductivity and by the surface S_∞ at infinity. We again assume that the fields at infinity satisfy the condition of Eq. (113). The integrals over S_∞ in Eqs. (108) and (109) then vanish, and we have to consider only the integrals over the surfaces of the conductors. At the surface of an infinitely conducting body the boundary conditions of Sec. 3·3 are

$$\left.\begin{array}{ll} \mathbf{n} \times \mathbf{E} = 0, & \mathbf{n} \cdot \mathbf{H} = 0, \\ \mathbf{n} \cdot \mathbf{E} = \dfrac{\eta}{\epsilon}, & \mathbf{n} \times \mathbf{H} = \mathbf{K}, \end{array}\right\} \quad (122)$$

η and $\mathbf{K}$ being the surface distributions of electric charge and current. Thus Eqs. (108) and (109) become

$$\mathbf{E}_P = -\frac{1}{4\pi}\int_V \left(j\omega\mu\mathbf{J} - \frac{\rho}{\epsilon}\boldsymbol{\nabla} + \mathbf{J}_m \times \boldsymbol{\nabla}\right)\frac{e^{-jkr}}{r}\,dv$$

$$-\frac{1}{4\pi}\int_{S_i}\left(j\omega\mu\mathbf{K} - \frac{\eta}{\epsilon}\boldsymbol{\nabla}\right)\frac{e^{-jkr}}{r}\,dS, \quad (123)$$

$$\mathbf{H}_P = -\frac{1}{4\pi}\int_V \left(j\omega\epsilon\mathbf{J}_m - \frac{\rho_m}{\mu}\boldsymbol{\nabla} - \mathbf{J} \times \boldsymbol{\nabla}\right)\frac{e^{-jkr}}{r}\,dv$$

$$+\frac{1}{4\pi}\int_{S_i}\mathbf{K}\times\boldsymbol{\nabla}\left(\frac{e^{-jkr}}{r}\right)\,dS. \quad (124)$$

It will be observed that the expressions for the fields due to surface currents and charges could have been obtained from the volume integrals as limiting forms, on considering that the volume distribution passes into a surface-layer distribution.

The results of this section will form the basis for the general theory of reflectors to be developed in Chap. 5.

3·11. The Far-zone Fields.—Let us now return to the case of the unbounded region and examine in more detail the relations between the field solutions

$$\mathbf{E}_P = -\frac{j}{4\pi\omega\epsilon}\int_V [(\mathbf{J}\cdot\boldsymbol{\nabla})\boldsymbol{\nabla} + k^2\mathbf{J} - j\omega\epsilon\mathbf{J}_m\times\boldsymbol{\nabla}]\frac{e^{-jkr}}{r}\,dv, \quad (3\cdot120)$$

$$\mathbf{H}_P = -\frac{j}{4\pi\omega\mu}\int_V [(\mathbf{J}_m\cdot\boldsymbol{\nabla})\boldsymbol{\nabla} + k^2\mathbf{J}_m + j\omega\mu\mathbf{J}\times\boldsymbol{\nabla}]\frac{e^{-jkr}}{r}\,dv \quad (3\cdot121)$$

and the radiation conditions developed in Sec. 3·9.

These solutions are based on the assumption that the sources are confined to a finite region of space. Let us choose an origin in the neighborhood of these sources, and let ϱ be the vector from the origin to the source element at the point x, y, z (Fig. 3·5). The vector from the origin to the field point P we shall write as $R\mathbf{R}_1$, $\mathbf{R}_1$ being a unit vector; similarly, $r\mathbf{r}_1$ will be the vector from the source element to the point P.

In the integrands of Eqs. (120) and (121), the operator $\boldsymbol{\nabla}$ acts on the coordinates of the source element, whereas the point P is treated as a fixed origin. For example,

$$\boldsymbol{\nabla}\left(\frac{e^{-jkr}}{r}\right) = \left(jk + \frac{1}{r}\right)\frac{e^{-jkr}}{r}\mathbf{r}_1, \quad (125)$$

and

$$(\mathbf{J}\cdot\boldsymbol{\nabla})\boldsymbol{\nabla}\left(\frac{e^{-jkr}}{r}\right) = \left[-k^2(\mathbf{J}\cdot\mathbf{r}_1)\mathbf{r}_1 + \frac{3}{r}\left(jk + \frac{1}{r}\right)(\mathbf{J}\cdot\mathbf{r}_1)\mathbf{r}_1 \right.$$
$$\left. -\frac{\mathbf{J}}{r}\left(jk + \frac{1}{r}\right)\right]\frac{e^{-jkr}}{r}. \quad (126)$$

Thus the integrands in these equations are power series in r^{-1}; for the first-degree terms in Eq. (120) we have

$$\frac{-j}{4\pi\omega\epsilon} \int_V [k^2\mathbf{J} - k^2(\mathbf{J}\cdot\mathbf{r}_1)\mathbf{r}_1 + k\omega\epsilon\mathbf{J}_m \times \mathbf{r}_1] \frac{e^{-jkr}}{r} dv.$$

In evaluating these integrals we must take into account the variation of r and of the unit vector $\mathbf{r}_1$ with the position of the source element. In general this offers serious difficulties, but simplifications can be effected if the field point is at a very great distance from the current distribution and the origin. First, the angle between the vectors $\mathbf{r}_1$ and $\mathbf{R}_1$, which decreases with R^{-1}, can be neglected; $\mathbf{r}_1$ can be replaced by $\mathbf{R}_1$ in the integrals. Next, the factor r^{-1} in the integrand can be replaced by the constant R^{-1}, from which it differs by the terms of the second order in R^{-1}. The variation of r cannot be neglected wholly in the phase factor. Here, making use of the fact that $\mathbf{r}_1$ and $\mathbf{R}_1$ are effectively parallel, we write

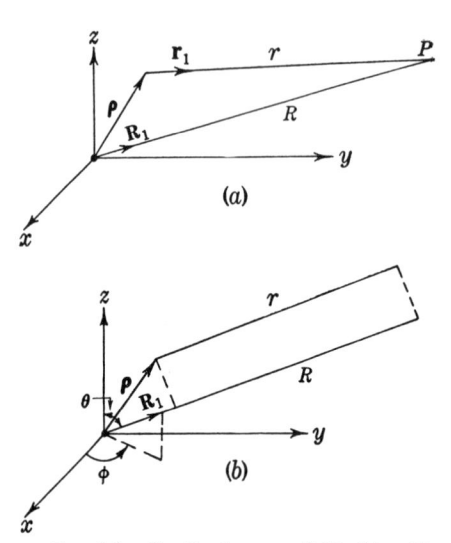

$$r = R - \boldsymbol{\varrho}\cdot\mathbf{R}_1. \quad (127)$$

Fig. 3·5.—On the far-zone field: (a) arbitrary field point P; (b) simplifying relationships for a point in the far zone.

With these approximations, Eqs. (120) and (121) take on forms valid for the far-zone fields:

$$\mathbf{E}_P = \frac{-j\omega\mu}{4\pi R} e^{-jkR} \int_V \left[\mathbf{J} - (\mathbf{J}\cdot\mathbf{R}_1)\mathbf{R}_1 + \left(\frac{\epsilon}{\mu}\right)^{1/2} \mathbf{J}_m \times \mathbf{R}_1\right] e^{+jk\boldsymbol{\varrho}\cdot\mathbf{R}_1} dv + O\left(\frac{1}{R^2}\right), \quad (128)$$

and

$$\mathbf{H}_P = -\frac{j\omega\epsilon}{4\pi R} e^{-jkR} \int_V \left[\mathbf{J}_m - (\mathbf{J}_m\cdot\mathbf{R}_1)\mathbf{R}_1 - \left(\frac{\mu}{\epsilon}\right)^{1/2} \mathbf{J} \times \mathbf{R}_1\right] e^{jk\boldsymbol{\varrho}\cdot\mathbf{R}_1} dv + O\left(\frac{1}{R^2}\right). \quad (129)$$

The calculation of the terms of order R^{-2} is tedious but straightforward and will be left to the reader.

The integrals in Eqs. (128) and (129) are independent of r. Thus it is evident that $R\mathbf{E}_P$ and $R\mathbf{H}_P$ remain finite as $R \to \infty$, as required by the boundary conditions [Eqs. (113a) and (113c)]. It is further evident that the field vectors are transverse to the unit vector $\mathbf{R}_1$; the $\mathbf{J}$ term in the integrand has a component in the direction of $\mathbf{R}_1$, but this is always canceled by subtraction of the second term $(\mathbf{J} \cdot \mathbf{R}_1)\mathbf{R}_1$. A simple calculation shows that the radiation conditions [Eqs. (113b) and (113d)] are satisfied; for example,

$$\mathbf{R}_1 \times \mathbf{H} + \left(\frac{\epsilon}{\mu}\right)^{1/2} \mathbf{E} = \frac{e^{-jkR}}{4\pi R} \int_V \{-j\omega\epsilon(\mathbf{R}_1 \times \mathbf{J}_m) + j\omega(\mu\epsilon)^{1/2}\mathbf{R}_1 \times (\mathbf{J} \times \mathbf{R}_1) - j\omega(\epsilon\mu)^{1/2}[\mathbf{J} - (\mathbf{J} \cdot \mathbf{R}_1)\mathbf{R}_1] - j\omega\epsilon(\mathbf{J}_m \times \mathbf{R}_1)\} e^{+jk\boldsymbol{\rho}\cdot\mathbf{R}_1} dv = 0. \quad (130)$$

Thus, $\mathbf{E}$ and $\mathbf{H}$ are related as in a plane wave, being mutually perpendicular and in a plane normal to $\mathbf{R}_1$.

We must now examine the integrals of Eqs. (128) and (129) in a little more detail. We introduce the system of spherical coordinates R, θ, ϕ, defined in Fig. 3·5, with polar axis along the z-axis. Let $\mathbf{i}_\theta$ and $\mathbf{i}_\phi$ be unit vectors having the directions of increasing θ and ϕ at the point P; $\mathbf{R}_1$ is, of course, the unit vector in the radial direction. In terms of Cartesian components

$$\boldsymbol{\rho} = x\mathbf{i}_x + y\mathbf{i}_y + z\mathbf{i}_z, \quad (131)$$
$$\mathbf{R}_1 = \sin\theta\cos\phi\,\mathbf{i}_x + \sin\theta\sin\phi\,\mathbf{i}_y + \cos\theta\,\mathbf{i}_z; \quad (132)$$

thus

$$\boldsymbol{\rho} \cdot \mathbf{R}_1 = (x\cos\phi + y\sin\phi)\sin\theta + z\cos\theta. \quad (133)$$

The components of the electric field vector along $\mathbf{i}_\theta$ and $\mathbf{i}_\phi$ are easily found to be

$$E_\theta = -\frac{j\omega\mu}{4\pi R} e^{-jkR} \int_V \left[\mathbf{J}\cdot\mathbf{i}_\theta + \left(\frac{\epsilon}{\mu}\right)^{1/2}\mathbf{J}_m\cdot\mathbf{i}_\phi\right] e^{jk\boldsymbol{\rho}\cdot\mathbf{R}_1} dv$$
$$= -\frac{j\omega\mu}{4\pi R} e^{-jkR} F_1(\theta,\phi) \quad (134a)$$

and

$$E_\phi = -\frac{j\omega\mu}{4\pi R} e^{-jkR} \int_V \left[\mathbf{J}\cdot\mathbf{i}_\phi - \left(\frac{\epsilon}{\mu}\right)^{1/2}\mathbf{J}_m\cdot\mathbf{i}_\theta\right] e^{jk\boldsymbol{\rho}\cdot\mathbf{R}_1} dv$$
$$= -\frac{j\omega\mu}{4\pi R} e^{-jkR} F_2(\theta,\phi). \quad (134b)$$

As indicated, the integrals are functions of only the angular coordinates θ and ϕ. The components of the electric field and the resultant far-field vector have the form to be expected for a source located at the origin. However, the far field is only a quasi-point-source field; the equiphase surfaces are not the family of spheres of constant R because the space factors F_1 and F_2 are in general complex. This is to be expected because the choice of origin was purely arbitrary.

The point-source character of the far field becomes more evident on considering the power flow in the far zone. The Poynting vector is

$$\mathbf{S} = \frac{1}{2} \operatorname{Re} (\mathbf{E} \times \mathbf{H}^*) = \frac{1}{2} \left(\frac{\epsilon}{\mu}\right)^{1/2} \operatorname{Re} [\mathbf{E} \times (\mathbf{R}_1 \times \mathbf{E}^*)]$$
$$= \frac{1}{2} \left(\frac{\epsilon}{\mu}\right)^{1/2} (|E_\theta|^2 + |E_\phi|^2) \mathbf{R}_1 \qquad (135)$$

or

$$\mathbf{S} = \frac{1}{8\lambda^2 R^2} \left(\frac{\mu}{\epsilon}\right)^{1/2} \Psi(\theta,\phi) \mathbf{R}_1, \qquad (136)$$

where

$$\Psi(\theta,\phi) = |F_1(\theta,\phi)|^2 + |F_2(\theta,\phi)|^2. \qquad (137)$$

The power flow is radially outward from the origin, with an intensity of flow that falls off with the square of R and depends also upon θ and ϕ; with respect to power flow the current distribution is, in effect, a directive point source at the origin.

In discussing the power flow it is convenient to use, instead of the Poynting vector, the power $P(\theta,\phi)$ radiated per unit solid angle in the direction θ, ϕ. This is given by

$$P(\theta,\phi) = R^2 |\mathbf{S}| = \frac{1}{8\lambda^2} \left(\frac{\mu}{\epsilon}\right)^{1/2} \Psi(\theta,\phi), \qquad (138)$$

which is independent of the radial distance R. The angular distribution of the power flow may be represented graphically by a three-dimensional plot in spherical coordinates, in which the angular coordinates θ and ϕ are those of the direction of observation and the radial coordinate is proportional to $P(\theta,\phi)$. It is customary to normalize the maximum of the power pattern to unity. The resulting figure is spoken of as the "polar diagram" or "radiation pattern" of the current distribution.

The power distribution is also specified in terms of a gain function $G(\theta,\phi)$ with respect to an isotropic radiator, as defined in Chap. 1; in terms of $P(\theta,\phi)$ we have

$$G(\theta,\phi) = \frac{P(\theta,\phi)}{\frac{1}{4\pi} \int_0^{2\pi} \int_0^{\pi} P(\theta,\phi) \sin\theta\, d\theta\, d\phi}$$
$$= \frac{4\pi \Psi(\theta,\phi)}{\int_0^{2\pi} \int_0^{\pi} \Psi(\theta,\phi) \sin\theta\, d\theta\, d\phi} \qquad (139)$$

The maximum value of the gain function is termed the "absolute gain." In design specifications this is generally quoted in decibels above the gain of an isotropic radiator (which is unity):

$$\text{Gain in db} = 10 \log_{10} [G(\theta,\phi)]_{\max}. \qquad (140)$$

3·12. Polarization.

In the preceding section we have considered the separate components E_θ and E_ϕ of the electric field vector in the far zone; we have now to note some properties of the resultant field vector.

The factors $F_1(\theta,\phi)$ and $F_2(\theta,\phi)$ in the expressions for E_θ and E_ϕ are in general complex quantities, which we may write thus:

$$F_1(\theta,\phi) = A_1(\theta,\phi)e^{-j\gamma_1(\theta,\phi)}, \quad (141a)$$
$$F_2(\theta,\phi) = A_2(\theta,\phi)e^{-j\gamma_2(\theta,\phi)}. \quad (141b)$$

Here the A's and γ's are real, and γ_1 and γ_2 are in general not equal. The vector $\mathbf{E}_P$ is thus the resultant of a pair of time-periodic vectors $E_\theta \mathbf{i}_\theta$ and $E_\phi \mathbf{i}_\phi$ at right angles to each other, with relative amplitude and

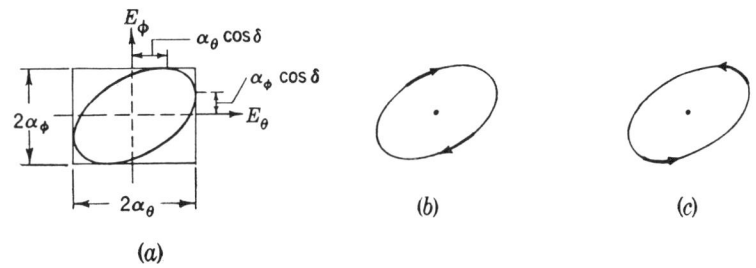

Fig. 3·6.—Elliptical polarization: (a) orientation of the ellipse; (b) right-handed polarization; (c) left-handed polarization, with direction of propagation toward the reader.

phase which vary with θ and ϕ. This resultant vector $\mathbf{E}_P$ simultaneously rotates in space and varies in magnitude in such a way that its terminal point describes an ellipse; the radiation field is elliptically polarized. To show this we note that the real parts of $E_\theta e^{+j\omega t}$ and $E_\phi e^{j\omega t}$, as given by Eq. (134), are the real E_θ- and E_ϕ-components of the electric field. These become, on use of Eqs. (141),

$$E_\theta = \frac{\omega\mu A_1(\theta,\phi)}{4\pi R}\sin(\omega t - kR - \gamma_1) = \alpha_\theta \sin(\omega t - kR - \gamma_1), \quad (142a)$$

$$E_\phi = \frac{\omega\mu A_2(\theta,\phi)}{4\pi R}\sin(\omega t - kR - \gamma_2) = \alpha_\phi \sin(\omega t - kR - \gamma_1 - \delta), \quad (142b)$$

where $\delta = \gamma_2 - \gamma_1$ is the phase of E_ϕ with respect to E_θ. Expanding the sine term in E_ϕ and eliminating the terms involving $\omega t - kR - \gamma_1$, we obtain a relation between E_ϕ and E_θ that holds at all times:

$$\frac{E_\theta^2}{\alpha_\theta^2} + \frac{E_\phi^2}{\alpha_\phi^2} - 2\cos\delta\,\frac{E_\theta}{\alpha_\theta}\frac{E_\phi}{\alpha_\phi} = \sin^2\delta. \quad (143)$$

This is the equation of an ellipse traced out by the terminus of the vector $\mathbf{E}_P$. The relation of the ellipse to the component vectors is shown in Fig. 3·6. The sense of polarization is defined for an observer watching the oncoming wave: The polarization is termed "right-handed" or "left-handed" according as the terminus of the vector $\mathbf{E}_P$ traces out the ellipse in the clockwise or counterclockwise sense, respectively.

If the phase difference δ is an odd multiple of $\pi/2$ and the amplitudes are equal, the ellipse becomes a circle; right-handed and left-handed circular polarization are defined in the same manner as for elliptical polarization. If the phase δ is an integral multiple of π, the ellipse degenerates into a straight line traced out by a linearly polarized resultant.

As θ and ϕ are varied, both $\delta(\theta,\phi)$ and E_θ/E_ϕ will vary; the polarization of the radiation from an extended source may change from linear to elliptical to circular and back again as one changes the direction of the observation.

3·13. The Electric Dipole.—In the preceding sections we have seen how a radiation field arises from a distribution of time-varying currents. We now turn to a discussion of some special idealized current distributions and their associated electromagnetic fields, leaving aside the question of their physical realizability.

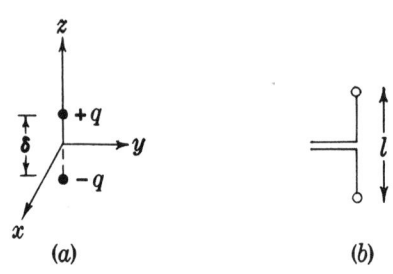

Fig. 3·7.—The electric dipole: (a) mathematical dipole; (b) antenna representation of a dipole, $l < < \lambda$.

The most elementary form of idealized radiator is the oscillating electric dipole (Fig. 3·7). A dipole consists mathematically of a pair of equal and opposite charges, each of magnitude q, separated by an infinitesimal distance δ. If the vector $\boldsymbol{\delta}$ is directed from $-q$ to $+q$, the dipole moment of the dipole is defined to be the vector

$$\mathbf{p} = q\boldsymbol{\delta}. \tag{144}$$

An antenna equivalent to a dipole is shown in Fig. 3·7. It consists of thin wires terminated in small spheres, the over-all dimensions of the structure being very small compared with a wavelength. The spheres form the capacitive element of the structure, and the charge at any instant can be considered to be localized on them. If the antenna is energized by a harmonic emf applied across the gap at the center, the charges on the spheres are given by

$$q = q_0 e^{j\omega t}; \tag{145}$$

the magnitude of the dipole moment of the antenna is

$$p = q_0 l e^{j\omega t} = p_0 e^{j\omega t}, \tag{146}$$

with amplitude

$$p_0 = q_0 l.$$

Since $l \ll \lambda$, the current at any instant may be taken to be the same at all points along the wings of the antenna. The current I is related to the charge q by $I = dq/dt = j\omega q$ and to the magnitude of the dipole moment by

$$p = \frac{Il}{j\omega}. \tag{147}$$

The electromagnetic field set up by a dipole is best described in spherical coordinates with the origin at the center and the polar axis along the axis of the dipole (Fig. 3·8). The derivation of the field will be found in any text on electromagnetic theory;[1] we shall simply state the results:

$$E_r = \frac{1}{2\pi\epsilon}\left(\frac{1}{r^3} + \frac{jk}{r^2}\right)\cos\theta\, p_0 e^{j(\omega t - kr)}, \tag{148a}$$

$$E_\theta = \frac{1}{4\pi\epsilon}\left(\frac{1}{r^3} + \frac{jk}{r^2} - \frac{k^2}{r}\right)\sin\theta\, p_0 e^{j(\omega t - kr)}, \tag{148b}$$

$$H_\phi = \frac{j\omega}{4\pi}\left(\frac{1}{r^2} + \frac{jk}{r}\right)\sin\theta\, p_0 e^{j(\omega t - kr)}. \tag{148c}$$

As a consequence of the axial symmetry of the radiator, the field is independent of ϕ. It can be resolved into three partial fields according to the dependence on r: (1) the "static field" varying inversely with r^3, (2) the "induction field" varying inversely with r^2, and (3) the "radiation field" varying inversely with r. The static field is, in fact, that which would be computed for a static dipole with fixed moment $p_0 e^{j(\omega t - kr)}$. The induction field is the quasi-stationary-state field commonly observed in the neighborhood of circuit elements at low frequencies; the magnetic component of the induction field is that which would be calculated on the basis of the Biot-Savart law for stationary currents. At small distances from the dipole the static and induction fields predominate. At a distance,

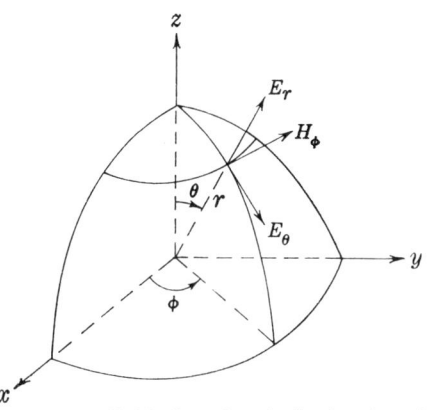

Fig. 3·8.—Field of an electric dipole oriented along the z-axis.

$$r > \frac{1}{k} = \frac{\lambda}{2\pi},$$

the radiation field becomes the leading term, and at sufficiently large distances the static and induction fields become negligible relative to

[1] For example, J. A. Stratton, *Electromagnetic Theory*, McGraw-Hill, New York, 1941, Chap. VIII.

the radiation field. However, it is only at distances much greater than $r = \lambda/2\pi$ that one can entirely neglect the static and induction fields.

The radiation field represents a flow of energy away from the dipole. There is no corresponding energy loss in the static and induction fields; the energy associated with these fields pulsates periodically back and forth between space and the antenna and its associated circuit just as do the energies in capacitances and inductances at low frequencies. The far-field Poynting vector computed by Eq. (43) arises entirely from the r^{-1} terms in the fields. It is

$$\mathbf{S} = \frac{\omega k^3}{32\pi^2 \epsilon} |p_0|^2 \frac{\sin^2 \theta}{r^2} \mathbf{i}_r, \quad (149)$$

where $\mathbf{i}_r$ is the unit vector in the outward radial direction.

The dipole is a true point source because the equiphase surfaces are spheres with centers at the origin; it is directive because the intensity of the field varies with the direction of observation. The power pattern of the dipole is independent of azimuth ϕ and is sufficiently represented by a cut in any one meridian plane, like that shown in Fig. 3·9. In design

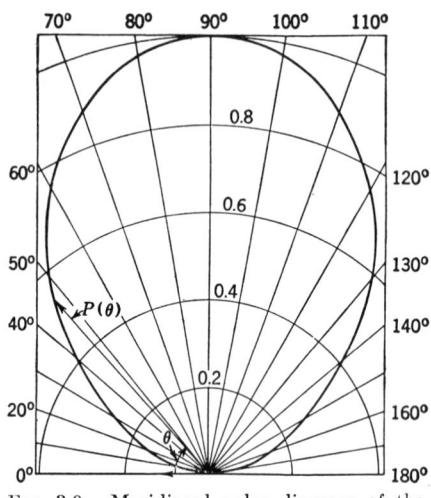

Fig. 3·9.—Meridional polar diagram of the power pattern of an electric dipole.

specifications it is customary to characterize such cuts in the three-dimensional polar diagram by two widths if they exist: (1) the "half-power width" Θ, which is the full angle in that cut between the two directions in which the power radiated is one-half the maximum value, and (2) the "tenth-power width" $\Theta\left(\frac{1}{10}\right)$, the angle between the directions in which the power radiated is one-tenth of the maximum. These widths for the meridional polar diagram of an electric dipole are

$$\Theta = 90°,$$
$$\Theta(\tfrac{1}{10}) = 146°.$$

Since the pattern is uniform in azimuth, the polar diagram in a cut taken normally to the dipole axis is a circle. The gain function of the dipole [Eq. (139)] is

$$G(\theta,\phi) = \tfrac{3}{2} \sin^2 \theta, \quad (150)$$

and the absolute gain is

$$G_m = (\tfrac{3}{2}) = 1.76 \text{ db.} \quad (151)$$

The impedance presented by the dipole to its feed line consists of a resistive component and a reactive component. We shall here consider only the resistive component, which corresponds to the power dissipated by the dipole. There are two elements in the power dissipation: (1) ohmic losses in the conductors of the dipole structure and (2) power radiated to space. In the idealized case, to which we restrict ourselves, the dipole consists of perfectly conducting elements. There is then only radiation loss to consider; the resistive component of the impedance is its radiation resistance. Let $\bar{P}$ be the average power radiated per unit time. The radiation resistance R_r is defined by

$$\bar{P} = \tfrac{1}{2}|I_0|^2 R_r, \tag{152}$$

where I_0 is the maximum value of the current. The radiated power $\bar{P}$ is computed by integrating the Poynting vector [Eq. (149)] over a complete sphere. By use of Eqs. (147) and (152) the radiation resistance is then found to be

$$R_r = \frac{k^2 l^2}{6\pi} \left(\frac{\mu}{\epsilon}\right)^{1/2}. \tag{153}$$

3·14. The Magnetic Dipole.—The magnetic counterpart of the electric dipole antenna is a current loop with radius small compared with the wavelength (Fig. 3·10). Such a current loop is equivalent to a magnetic dipole along the axis normal to the plane of the loop; this axis has been taken to be the z-axis in the figure. If I is the current in the loop and **A** is a vector normal to the loop, with magnitude equal to its area, the magnetic moment at any instant is

$$\mathbf{m} = I\mathbf{A}. \tag{154}$$

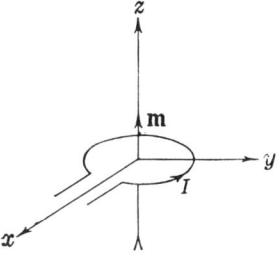

Fig. 3·10.—Magnetic dipole antenna: current loop and equivalent magnetic dipole.

If I_0 is the amplitude of the time-periodic current and m_0 the corresponding amplitude of the magnetic moment, the magnitude of the magnetic moment is given by

$$m = I_0 A e^{j\omega t} = m_0 e^{j\omega t}. \tag{155}$$

The direction of the dipole in relation to the direction of the current is shown in Fig. 3·10.

The field of the magnetic dipole, like that of the electric dipole, is most conveniently described in spherical coordinates. The field components are

$$E_\phi = \frac{k^2}{4\pi} \left(\frac{\mu}{\epsilon}\right)^{1/2} \left(\frac{1}{r} - \frac{j}{kr^2}\right) \sin\theta \; m_0 e^{j(\omega t - kr)}, \tag{156a}$$

$$H_r = \frac{1}{2\pi} \left(\frac{1}{r^3} + \frac{jk}{r^2} \right) \cos\theta \; m_0 e^{j(\omega t - kr)}, \tag{156b}$$

$$H_\theta = \frac{1}{4\pi} \left(\frac{1}{r^3} + \frac{jk}{r^2} - \frac{k^2}{r} \right) \sin\theta \; m_0 e^{j(\omega t - kr)}. \tag{156c}$$

As with the electric dipole, the field is independent of the azimuth angle ϕ. Comparison of Eqs. (156) with the electric dipole field [Eqs. (148)] will show that the roles of **E** and **H** are interchanged. With minor revisions required by this interchange, the discussion of the electric dipole as a directive point source can be carried over to the magnetic dipole. The power patterns are identical, and the absolute gain of the magnetic dipole, like that of the electric dipole, is 1.76 db. The radiation resistance of the loop is found to be

$$R_r = \frac{1}{6\pi} \left(\frac{\mu}{\epsilon} \right)^{1/2} k^4 A^2. \tag{157}$$

The reader should note that the far-zone fields of the electric and magnetic dipoles show the general properties mentioned in Sec. 3·11. In particular, he should note that in the far zone (and there only)

$$\mathbf{H} = \left(\frac{\epsilon}{\mu} \right)^{1/2} (\mathbf{i}_r \times \mathbf{E}); \tag{158}$$

E and **H** are mutually perpendicular and lie in a plane transverse to the direction of propagation.

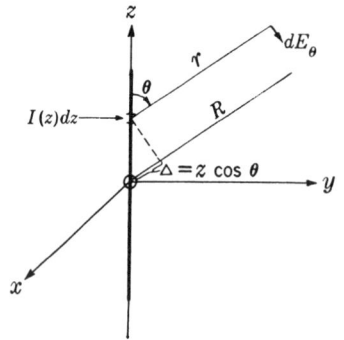

Fig. 3·11.—Far-zone field of a line-current distribution.

3·15. The Far-zone Fields of Line-current Distributions.—We shall next compute the far-zone fields due to a time-periodic current in a thin straight wire extending along the z-axis from $z = -l/2$ to $z = +l/2$, that is, along the polar axis of the r, θ, ϕ coordinate system. We shall allow the length of the wire to be comparable to a wavelength or even equal to a number of wavelengths. The phase differences between the currents at separated points on the wire will then be significant, and we shall need to consider the current to be a function of position along the wire:

$$\mathbf{I} = I(z)e^{j\omega t}\mathbf{i}_z, \tag{159}$$

Since the properties of the field in the far zone are those of a plane wave, it will be sufficient to calculate the electric field intensity. In Eqs. (134) we can first of all discard the magnetic-current density $\mathbf{J}_m$.

We note further that $\mathbf{J} \cdot \mathbf{i}_\phi$ is zero; consequently,

$$E_\phi = 0. \tag{160}$$

By Eq. (133), with $x = y = 0$, $\boldsymbol{\varrho} \cdot \mathbf{R}_1$ is simply $z \cos \theta$, and the volume integral for E_θ degenerates into a line integral:

$$E_\theta = -\frac{j\omega\mu}{4\pi R} e^{-jkR} \int_{-l/2}^{l/2} I(z)\mathbf{i}_z \cdot \mathbf{i}_\theta e^{jkz \cos \theta}\, dz, \tag{161a}$$

or

$$E_\theta = +\frac{j\omega\mu}{4\pi R} e^{-jkR} \int_{-l/2}^{l/2} I(z) \sin\theta\, e^{jkz \cos \theta}\, dz$$
$$= \frac{j\omega\mu}{4\pi R} e^{-jkR} F(\theta). \tag{161b}$$

(As usual, the time factor $e^{j\omega t}$ is understood implicitly.)

Again, because of the axial symmetry of the radiator, the field is independent of the azimuth angle ϕ. As with an electric dipole, the electric-field vector lies in the meridian plane; the magnetic-field vector is at right angles to this, parallel to $\mathbf{i}_\phi$. The function $F(\theta)$, known as the "form factor" of the field pattern, will in general be complex; the equiphase surfaces are not spheres of constant R.

The integral expression in Eq. (161b) admits of an interesting interpretation. On comparing the integrand with the far field of an electric dipole [Eq. (148)] it will be noted that the integral can be interpreted as a sum of the fields of a distribution of dipoles along the wire, the dipole moment $d\mathbf{p}$ associated with the element of conductor dz at the point z (Fig. 3·11) being given by

$$d\mathbf{p} = \frac{1}{j\omega} I(z)\, dz\, \mathbf{i}_z. \tag{162}$$

In superposing the component fields at the field point one must, of course, take account of the phase differences between the contributions from different dipole elements, due to the differences in path length to the field point. If Δ is the path difference between two elements, the phase difference is $2\pi\Delta/\lambda = k\Delta$. Taking the origin as a reference point for path length, the path difference corresponding to a point z on the wire is $\Delta(z) = z \cos \theta$; hence the phase factor $e^{jkz \cos \theta}$ in the integrand. It will be noted that Eq. (162) is essentially the relation between the current and dipole moment set down in Eq. (147).

The precise form of the current function $I(z)$ can be controlled by changing the point at which the driving voltage is applied to the wire and the way in which the wire is terminated. We shall now consider the case in which the wire is driven at the center, for example, by a parallel-wire line feeding across a small gap at the origin, and there is no load at the ends of the wire. In this case the current is necessarily zero at the ends of the wire; its distribution along the wire can always be expressed as a sum of standing waves, each of which vanishes at the ends. Such standing waves have the form

$$I_m(z) = I_0(m) \cos \frac{m\pi z}{l}, \qquad m = 1, 3, 5, \cdots,$$
$$I_m(z) = I_0(m) \sin \frac{m\pi z}{l}, \qquad m = 2, 4, 6, \cdots, \quad (163)$$

where $I_0(m)$ is the value of the current at a current antinode. In general the current will consist of a superposition of standing waves. It will, however, consist of a single standing wave if $l = m\lambda/2$; this is the case which we shall treat. Substituting the corresponding $I(z)$ into Eq. (161b), one finds with little difficulty

$$E_\theta = j \left(\frac{\mu}{\epsilon}\right)^{1/2} \frac{I_0(m) \sin \frac{m\pi}{2}}{2\pi R} e^{-jkR} \left[\frac{\cos\left(\frac{m\pi}{2} \cos \theta\right)}{\sin \theta}\right],$$
$$m = 1, 3, 5, \cdots, \quad (164a)$$

$$E_\theta = \left(\frac{\mu}{\epsilon}\right)^{1/2} \frac{I_0(m) \cos \frac{m\pi}{2}}{2\pi R} e^{-jkR} \left[\frac{\sin\left(\frac{m\pi}{2} \cos \theta\right)}{\sin \theta}\right].$$
$$m = 2, 4, \cdots. \quad (164b)$$

The term "form factor" is here applied to the terms in brackets. The surfaces of constant R are equiphase surfaces; the far-zone field of a standing-wave current is that of a true point source at the center of the current distribution. The field intensity in the equatorial plane $\theta = \pi/2$ is zero when m is an even integer because the current distribution is antisymmetrical with respect to the origin; the contributions to the field from current elements at $+z$ and $-z$ are 180° out of phase at points in the equatorial plane and there annul each other.

3·16. The "Half-wave Dipole."—The most important line-current distribution in microwave antenna theory is that with $l = \lambda/2$. This is usually called the "half-wave dipole"—a misnomer due, perhaps, to its diminutive structure at microwave frequencies and here retained because of its convenience. On setting $m = 1$ in Eq. (164a) we obtain the field pattern of this radiator:

$$E_\theta = j \left(\frac{\mu}{\epsilon}\right)^{1/2} \frac{I_0}{2\pi R} e^{-jkR} \left[\frac{\cos\left(\frac{\pi}{2} \cos \theta\right)}{\sin \theta}\right]. \quad (165)$$

The corresponding power pattern is

$$P(\theta,\phi) = \left(\frac{\mu}{\epsilon}\right)^{\frac{1}{2}} \frac{I_0^2}{8\pi^2} \left[\frac{\cos\left(\frac{\pi}{2}\cos\theta\right)}{\sin\theta}\right]^2. \tag{166}$$

The pattern differs only slightly from that of the electric dipole; it is uniform in azimuth and has its single maximum in the equatorial plane.

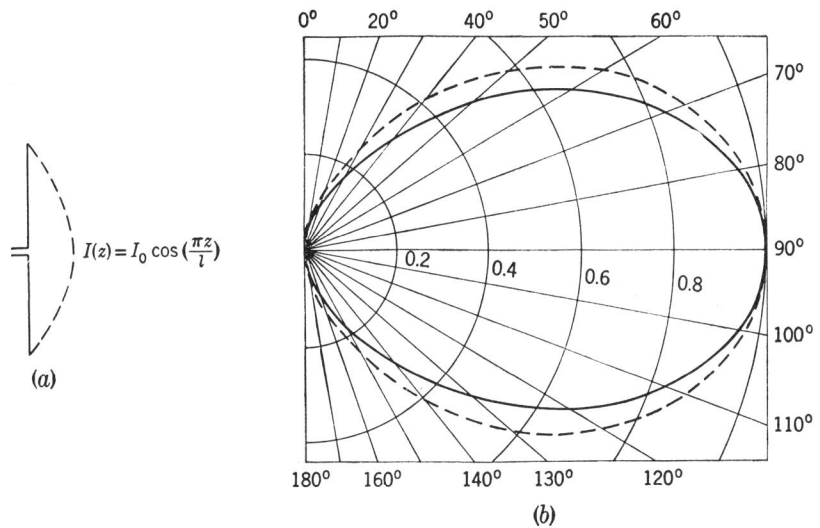

Fig. 3·12.—The half-wave dipole: (a) current distribution along the wire; (b) meridional polar diagram compared with that of the electric dipole: — $\frac{\lambda}{2}$ dipole, --- infinitesimal dipole.

Figure 3·12 shows the meridional polar diagram in comparison with that of the dipole. The gain of the half-wave dipole is

$$G_m = (1.65) = 2.17 \text{ db}. \tag{167}$$

The slight increase in directivity over that of the electric dipole arises from the fact that at points off the equatorial plane there is partial destructive interference between contributions from different portions of the wire, which lie at different distances from the point of observation; this leaves the radiation in the equatorial plane relatively stronger.

3·17. Superposition of Fields.—We shall often have occasion to deal with sources that consist of a number of separate current distributions. As long as the total system is confined to a finite region of space—the only practical case—this problem is in principle covered adequately by the general theory of Secs. 3·9 to 3·11. It will, however, be useful to reconsider it from the point of view of the superposition principle stated in Sec. 3·2. The total field is the sum of the component fields due to each component current system. We shall confine our attention to the far-zone fields, existing at field points far removed from every source in the total system.

The notation to be employed is illustrated in Fig. 3·13. We choose an origin O within the neighborhood of the sources; a primary system of rectangular coordinates x, y, and z; and an associated spherical system r, θ, and ϕ. The distance from the origin to the field point will, as before, be R; $\mathbf{R}_0$ is a unit vector in that direction. In connection with any of the component radiating systems, say the ith, we use a secondary coordinate

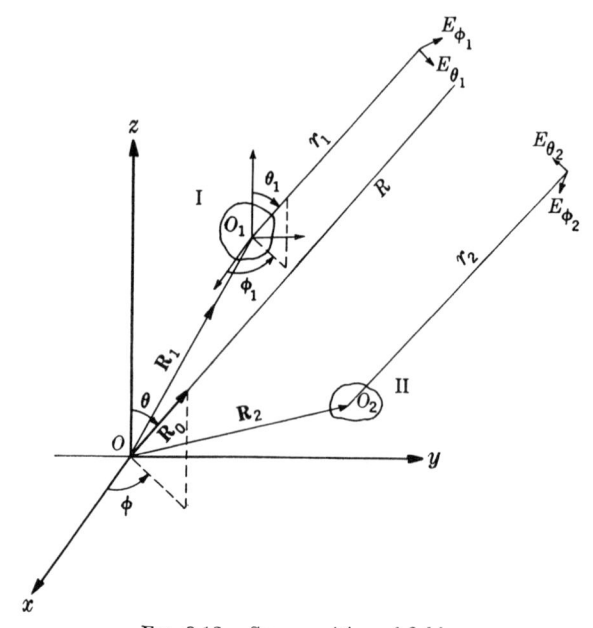

FIG. 3·13.—Superposition of fields.

system, with axes parallel to those of the primary system and origin O_i within that source distribution at the vector position $\mathbf{R}_i$ with respect to O. The polar coordinates of the field point P in this secondary coordinate system will be denoted by r_i, θ_i, ϕ_i.

As in the general discussion of far-zone fields, we may consider all the O_iP to be parallel to OP and all the θ_i, ϕ_i to be equal to θ and ϕ respectively. Furthermore, the field due to the ith radiating system can be expressed in terms of an equivalent quasi–point source at O_i. That is, the component fields are, by Eqs. (134),

$$E_{\theta_i} = -\frac{j\omega\mu}{4\pi r_i} e^{-jkr_i} \int_{V_i} \left[\mathbf{J}_i \cdot \mathbf{i}_\theta + \left(\frac{\epsilon}{\mu}\right)^{\frac{1}{2}} \mathbf{J}_{mi} \cdot \mathbf{i}_\phi \right] e^{jk\varrho \cdot \mathbf{R}_0} dv$$

$$= -\frac{j\omega\mu}{4\pi r_i} e^{-jkr_i} F_{1i}(\theta,\phi), \qquad (168a)$$

and

$$E_{\phi_i} = -\frac{j\omega\mu}{4\pi r_i} e^{-jkr_i} \int_{V_i} \left[\mathbf{J}_i \cdot \mathbf{i}_\phi - \left(\frac{\epsilon}{\mu}\right)^{\frac{1}{2}} \mathbf{J}_{mi} \cdot \mathbf{i}_\theta \right] e^{jk\boldsymbol{\rho}\cdot\mathbf{R}_0} dv$$
$$= -\frac{j\omega\mu}{4\pi r_i} e^{-jkr_i} F_{2i}(\theta,\phi). \tag{168b}$$

The total field is obtained by summing the component fields. We note, however, as in the discussion in Sec. 3·11, that we can replace r_i^{-1} by R^{-1}, with an error of the order of R^{-2}; in the phase factors we can similarly write

$$r_i = R - \mathbf{R}_0 \cdot \mathbf{R}_i. \tag{169}$$

The total field is, therefore, given by

$$E_\theta = \sum_i E_{\theta_i} = -\frac{j\omega\mu}{4\pi R} e^{-jkR} \mathfrak{F}_1(\theta,\phi), \tag{170a}$$

where

$$\mathfrak{F}_1(\theta,\phi) = \sum_i F_{1i}(\theta,\phi) e^{jk\mathbf{R}_0\cdot\mathbf{R}_i}, \tag{170b}$$

and

$$E_\phi = \sum_i E_{\phi_i} = -\frac{j\omega\mu}{4\pi R} e^{-jkR} \mathfrak{F}_2(\theta,\phi), \tag{171a}$$

where

$$\mathfrak{F}_2(\theta,\phi) = \sum_i F_{2i}(\theta,\phi) e^{jk\mathbf{R}_0\cdot\mathbf{R}_i}. \tag{171b}$$

The space factors $\mathfrak{F}_1$ and $\mathfrak{F}_2$ are complex, and the discussion of polarization in Sec. 3·12 applies without change.

The problem is thus reduced to the superposition of quasi-point-source fields arising from sources O_i and described by the space factors F_{1i} and F_{2i}. The composition of the over-all space factors $\mathfrak{F}_1$ and $\mathfrak{F}_2$ in terms of these and the phase differences arising from the relative positions of the sources is a procedure useful in many other fields—for example, the theory of X-ray diffraction.

3·18. The Double-dipole System.—The radiation patterns of compound systems are usually more directive than the patterns of the component systems; destructive interference between the fields of the component systems takes place in certain directions, constructive interference in others, with the consequence that the total power density changes more rapidly with angle and reaches more extreme values than does the power density for any component system.

An important compound system with wide application to microwave antennas is obtained by superposing two half-wave dipoles. We shall here restrict ourselves to the case in which the dipole axes are parallel and the currents are of equal strength, though of arbitrary relative phase. We consider, then, two half-wave dipoles with centers at $(0, -a/2, 0)$ and $(0, a/2, 0)$ and axes parallel to the z-axis, carrying currents of amplitudes I_0 and $I_0 e^{-j\psi}$ (Fig. 3·14). Since neither source gives rise to an

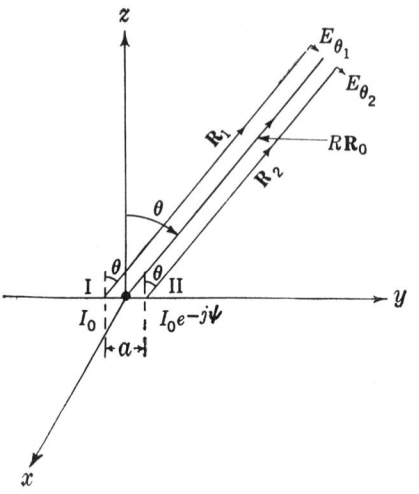

Fig. 3·14.—The double-dipole system.

E_ϕ-component in the far field, the total field can have no such component. The space factors of the dipoles are alike, except for the current phase term $e^{-j\psi}$. Combining Eqs. (165) and (170b), we find the resultant field:

$$E_\theta = j\left(\frac{\mu}{\epsilon}\right)^{\frac{1}{2}} \frac{I_0 e^{-jkR}}{2\pi R} \left[\frac{\cos\left(\frac{\pi}{2}\cos\theta\right)}{\sin\theta}\right] [e^{jk\mathbf{R}_0\cdot\mathbf{R}_1} + e^{j(k\mathbf{R}_0\cdot\mathbf{R}_2 - \psi)}]. \quad (172)$$

On making the substitutions

$$\mathbf{R}_0 \cdot \mathbf{R}_1 = -\frac{a}{2}\sin\theta\sin\phi,$$
$$\mathbf{R}_0 \cdot \mathbf{R}_2 = \frac{a}{2}\sin\theta\sin\phi, \quad (173)$$

we obtain finally

$$E_\theta = j\left(\frac{\mu}{\epsilon}\right)^{\frac{1}{2}} \frac{I_0 e^{-j[kR+(\psi/2)]}}{\pi R} \left[\frac{\cos\left(\frac{\pi}{2}\cos\theta\right)}{\sin\theta}\right] \cos\left(\frac{\pi a}{\lambda}\sin\theta\sin\phi - \frac{\psi}{2}\right). \quad (174)$$

This is a dipole field modified by the presence of the last factor. The spheres of constant R are the equiphase surfaces; at large R the field is that of a directive point source at the origin midway between the dipoles. The pattern is symmetrical in ϕ about $\phi = \pi/2$ and in θ about $\theta = \pi/2$;

SEC. 3·18]   THE DOUBLE-DIPOLE SYSTEM   103

that is, it is symmetrical with respect to the yz-plane, which contains the dipoles, and the xy-plane, to which they are perpendicular. These planes of symmetry are known, respectively, as the principal E-plane and the principal H-plane of the radiation pattern. Since the pattern is a function of both θ and ϕ, a three-dimensional polar diagram is required

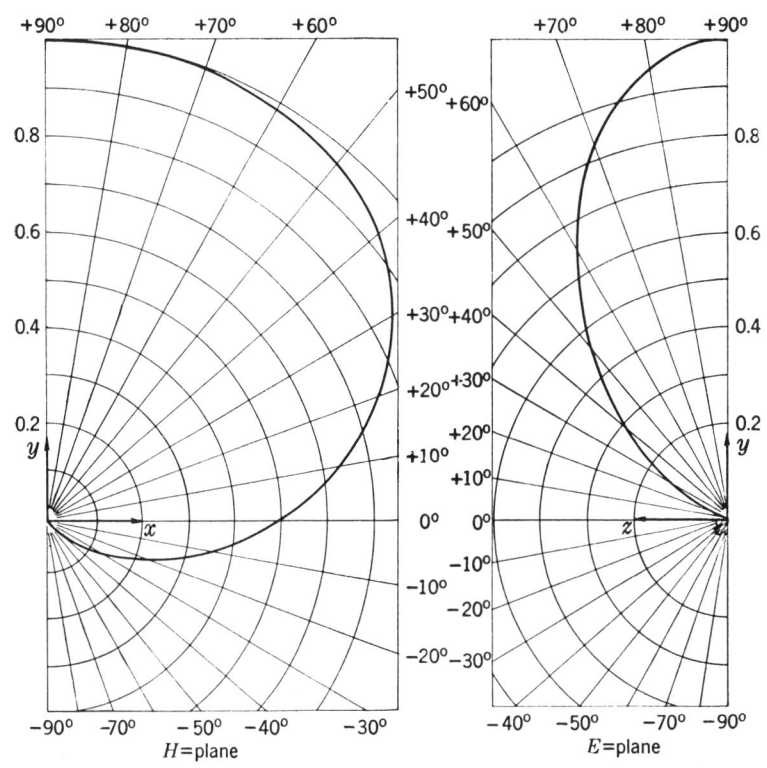

FIG. 3·15.—E- and H-plane polar diagrams in the power pattern of the double-dipole system.

for a complete presentation of its properties. However, in practice it is usually sufficient to consider the principal E- and H-plane cuts.

The details of the pattern depend on the precise values of a and ψ. We shall here consider one special case, in which $a = \lambda/4$, and $\psi = \pi/2$. The form factor is then (except for constant terms)

$$\mathcal{F}(\theta,\phi) = \frac{\cos\left(\dfrac{\pi}{2}\cos\theta\right)}{\sin\theta} \cos\left[\frac{\pi}{4}(1 - \sin\theta \sin\phi)\right]. \quad (175)$$

The principal E- and H-plane cuts of the power pattern (proportional to the square of the form factor) are shown in Fig. 3·15. Only a small fraction of the power is radiated in the hemisphere to the left of the

xz-plane; no power is radiated in the negative y-direction; maximum power in the positive y-direction. In the negative y-direction the radiation from the dipole at $y = +a/2$ must travel a distance greater by a quarter wavelength than the radiation from $y = -a/2$ with resulting phase retardation of 90°. Since the current in the first dipole is 90° behind that in the other dipole, the fields from the two dipoles are 180° out of phase and annul each other. In the positive y-direction, the phase retardation in the field from the dipole at $y = -a/2$, due to the additional path length traversed, is just compensated by the 90° phase lead of the radiating current; the fields from the two dipoles are in phase and reinforce each other. Since each dipole has maximum field intensity in the xy-plane, this has the consequence that the maximum in the total field intensity lies in the $+y$-direction.

As a measure of the directivity of the power pattern, we may take the half- and tenth-power widths of the polar diagram in each of the principal planes. These are designated by Θ_E and $\Theta_E(\frac{1}{10})$ for the E-plane half- and tenth-power widths respectively; corresponding notation applies in the H-plane. For the system under consideration

$$\left. \begin{array}{ll} \Theta_E = 76°, & \Theta_H = 180°, \\ \Theta_E(\tfrac{1}{10}) = 130°, & \Theta_H(\tfrac{1}{10}) = 252°. \end{array} \right\} \qquad (176)$$

3·19. Regular Space Arrays.—The double-dipole system is the simplest possible example of an important class of directive systems: regular space arrays of similar radiators. Let us consider a system of current distributions, identical in structure but perhaps differing from one another in over-all amplitude and phase. The radiating units need not be simple dipoles; they may be double-dipole systems or more complex current systems, but all must have the same orientation in space and be described by similar space factors $F_1(\theta,\phi)$ and $F_2(\theta,\phi)$, with respect to similarly situated origins O_i. Now let these radiating units and their origins O_i be arranged into a space array at the intersection points of a three-dimensional rectangular lattice (see Fig. 3·16). Let $\mathbf{a}_1$, $\mathbf{a}_2$, and $\mathbf{a}_3$ be the basis vectors of the lattice in the x-, y-, and z-directions, respectively, and let the extents of the lattice in these directions be $N_1 a_1$, $N_2 a_2$, and $N_3 a_3$. Choosing one of the corner elements of this lattice as a reference point, we can specify the position of an arbitrary lattice point O_i by the relative-position vector

$$\mathbf{R}_i = n_1 \mathbf{a}_1 + n_2 \mathbf{a}_2 + n_3 \mathbf{a}_3, \qquad (177)$$

where n_1, n_2, and n_3 are integers less than or equal to N_1, N_2, and N_3, respectively. We shall let the amplitude of the ith system be $A_{n_1 n_2 n_3}$ and shall admit the possibility of a progressive phase delay in each of the three basis directions of the lattice: the phase of the ith radiating system, relative to the reference system, shall be

$$\psi_{n_1 n_2 n_3} = n_1\psi_1 + n_2\psi_2 + n_3\psi_3. \tag{178}$$

We need consider only the space factors $\mathfrak{F}_1$ and $\mathfrak{F}_2$ defined in Eqs. (170b) and (171b). The space factors F_{1i} and F_{2i} are independent of i except for

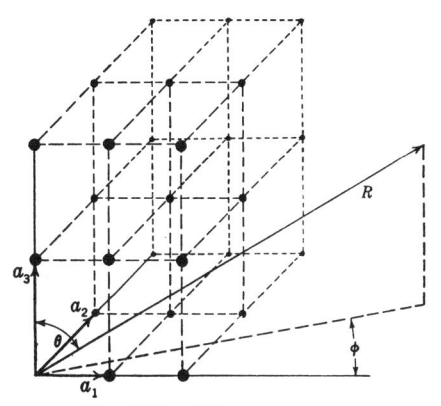

FIG. 3·16.—The space array.

the constant multipliers $A_{n_1 n_2 n_3} e^{-j\psi_{n_1 n_2 n_3}}$. Accordingly, the space factors for the system as a whole are given by

$$\left.\begin{array}{l} \mathfrak{F}_1(\theta,\phi) = F_1(\theta,\phi)\Lambda(\theta,\phi), \\ \mathfrak{F}_2(\theta,\phi) = F_2(\theta,\phi)\Lambda(\theta,\phi), \end{array}\right\} \tag{179}$$

where

$$\Lambda(\theta,\phi) = \sum_{n_1=0}^{N_1} \sum_{n_2=0}^{N_2} \sum_{n_3=0}^{N_3} A_{n_1 n_2 n_3} \exp\left[j \sum_{i=1}^{3} n_i(k\mathbf{R}_0 \cdot \mathbf{a}_i - \psi_i)\right]. \tag{180}$$

Here

$$\mathbf{R}_0 \cdot \mathbf{a}_1 = a_1 \sin\theta \cos\phi = a_1 u_1, \tag{181a}$$
$$\mathbf{R}_0 \cdot \mathbf{a}_2 = a_2 \sin\theta \sin\phi = a_2 u_2, \tag{181b}$$
$$\mathbf{R}_0 \cdot \mathbf{a}_3 = a_3 \cos\theta \quad\quad\, = a_3 u_3. \tag{181c}$$

The total space factor is thus a product of the space factor for a radiating unit by a lattice factor. The lattice factor, it will be noted, is itself the space factor of a lattice array of isotropic radiators with relative amplitudes $A_{n_1 n_2 n_3}$ and relative phases $\psi_{n_1 n_2 n_3}$.

If the radiating units all have the same amplitude, say equal to unity, the sums in Eq. (180) can be evaluated. The term on the right becomes a product of three factors:

$$\Lambda(\theta,\phi) = \Lambda_1\Lambda_2\Lambda_3, \tag{182}$$

where

$$\Lambda_i = \sum_{n_i=0}^{N_i} e^{jn_i(ka_iu_i-\psi_i)}. \tag{183}$$

This geometric series is easily summed. One finds

$$\Lambda_i = \frac{e^{j(N_i+1/2)(ka_iu_i-\psi_i)}}{e^{j\frac{1}{2}(ka_iu_i-\psi_i)}} \left[\frac{\sin \frac{N_i+1}{2}(ka_iu_i-\psi_i)}{\sin \frac{1}{2}(ka_iu_i-\psi_i)} \right]. \tag{184}$$

The power pattern of the space array is proportional to

$$|\mathfrak{F}_1|^2 + |\mathfrak{F}_2|^2$$

and is consequently given by

$$P(\theta,\phi) = P_0(\theta,\phi)|\Lambda(\theta,\phi)|^2 \tag{185}$$

except for the multiplicative constants. The second factor is the product of three factors,

$$|\Lambda_i|^2 = \left[\frac{\sin (N_i+1)\left(\frac{\pi a_i u_i}{\lambda} - \frac{\psi_i}{2}\right)}{\sin \left(\frac{\pi a_i u_i}{\lambda} - \frac{\psi_i}{2}\right)} \right]^2, \quad (i = 1, 2, 3), \tag{186}$$

each of the form $\sin Nx/\sin x$. Such a function has principal maxima at $x = h\pi$, h being an integer; if N is large, the maxima are very sharp, the function being only slightly different from zero between successive peaks. The composite lattice factor will then have its principal maxima only for those values of u_i for which the three factors simultaneously achieve their maximum values, that is, when

$$u_1 = \sin \theta \cos \phi = \left(h_1 + \frac{\psi_1}{2\pi}\right)\frac{\lambda}{a_1}, \tag{187a}$$

$$u_2 = \sin \theta \sin \phi = \left(h_2 + \frac{\psi_2}{2\pi}\right)\frac{\lambda}{a_2}, \tag{187b}$$

$$u_3 = \cos \theta \quad = \left(h_3 + \frac{\psi_3}{2\pi}\right)\frac{\lambda}{a_3}, \tag{187c}$$

h_1, h_2, h_3 being positive or negative integers. These conditions cannot be satisfied simultaneously by any choice of θ and ϕ for arbitrarily chosen h_1, h_2, h_3; the possibility of simultaneously satisfying the three conditions is determined by the values of the phases ψ_i and the lattice dimensions a_i/λ. Except when $P_0(\theta,\phi)$ has a zero in direction θ, ϕ determined by the above conditions, the lattice space factor of a very large lattice determines, essentially completely, the direction of the principal maxima in the total radiation pattern.

CHAPTER 4

WAVEFRONTS AND RAYS

By S. Silver

The preceding chapter dealt with radiation fields in their direct relation to the sources. It was found that the field represents a flow of energy outward from the region of the sources; also it was demonstrated separately that the energy flow in a time-varying field is a wave phenomenon. We now turn our attention to the study of wave propagation and the associated energy flow, without direct reference to the sources. Several simple waveforms have already been discussed: plane, cylindrical, and spherical waves. In each case the wave was described by a family of equiphase surfaces or wavefronts, and the propagation of the wave was visualized as a progression of each wavefront into a contiguous one; furthermore, the energy flow at every point was in a direction normal to the wavefront. The main subject of this chapter is the extension of these ideas to general waveforms.

4·1. The Huygens-Green Formula for the Electromagnetic Field.—We have now to consider the following problem: Given the values of the electric and magnetic field vectors over an equiphase surface, how can we determine the field vectors at a specified field point?

The solution to this problem is, in fact, contained in the general integral of the field equations obtained in Sec. 3·8. Let the fields be specified over an equiphase surface S (Fig. 4·1) which encloses all sources of the field, and let P be the field point at which the vectors $\mathbf{E}$ and $\mathbf{H}$ are to be determined. We now apply the general relations of Eqs. (3·108) and (3·109) to the region bounded by S and the sphere at infinity. Since the sources of the field lie outside this region, the volume integrals vanish and we have

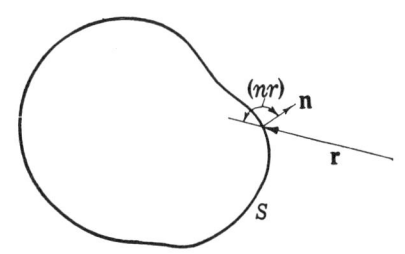

Fig. 4·1.—On the Huygens-Green relation.

$$E_P = \frac{1}{4\pi} \int_S [-j\omega\mu(\mathbf{n} \times \mathbf{H})\psi + (\mathbf{n} \times \mathbf{E}) \times \nabla\psi + (\mathbf{n} \cdot \mathbf{E})\nabla\psi]\, dS \quad (1a)$$

and

$$H_P = \frac{1}{4\pi} \int_S [j\omega\epsilon(\mathbf{n} \times \mathbf{E})\psi + (\mathbf{n} \times \mathbf{H}) \times \nabla\psi + (\mathbf{n} \cdot \mathbf{H}) \nabla\psi] \, dS, \quad (1b)$$

where $\psi = e^{-jkr}/r$ and $\mathbf{n}$ is the unit vector normal to S indicated in Fig. 4·1. These equations provide the solution of the stated problem.

Equations (1) may be regarded as an analytical formulation of the *Huygens-Fresnel* principle, which serves generally as a basis for the study of wave propagation. The Huygens-Fresnel principle states that each point on a given wavefront can be regarded as a secondary source which gives rise to a spherical wavelet; the wave at a field point is to be obtained by superposition of these elementary wavelets, with due regard to their phase differences when they reach the point in question. Equations (1) specify the nature of the wavelets arising at the various points on the equiphase surface;[1] as was pointed out in connection with Eq. (3·111), the sources of the wavelets can be regarded as surface layers of electric and magnetic currents and charges.

For the further purposes of this chapter it is desirable to write the surface integrals in somewhat different form. By means of a rather laborious calculation they can be transformed into

$$\mathbf{E}_P = -\frac{1}{4\pi} \int_S \left(\psi \frac{\partial \mathbf{E}}{\partial n} - \mathbf{E} \frac{\partial \psi}{\partial n} \right) dS \quad (2a)$$

and

$$\mathbf{H}_P = -\frac{1}{4\pi} \int_S \left(\psi \frac{\partial \mathbf{H}}{\partial n} - \mathbf{H} \frac{\partial \psi}{\partial n} \right) dS, \quad (2b)$$

respectively.[2] Relations of the same form must, of course, hold for the

---

[1] A comprehensive treatment of Huygens' principle and its application to scalar and vector waves has been given by Baker and Copson, *The Mathematical Theory of Huygens' Principle*, Oxford, New York, 1939. It should be noted that the integral expression for the fields, and hence the interpretation of the sources, is not unique; it is possible to add to Eqs. (1) any surface integral that is equal to zero. This is actually done in making the transformation from Eqs. (1) to Eqs. (2) in this chapter.

[2] This transformation can be effected only if the surface S is completely closed; otherwise additional terms appear. The results can be obtained by a simpler and for our purposes more useful procedure than by direct transformation of Eqs. (1). It was shown in Sec. 3·6 that in a source-free region each rectangular component of a field vector satisfies the scalar Helmholtz equation

$$\nabla^2 u + k^2 u = 0.$$

The integration of this equation can be performed by means of Green's theorem in a manner analogous to that by which we integrated the field equations. The scalar Green's theorem states that given two continuous scalar functions F and G having continuous first and second derivatives in a region V such as was illustrated in Fig. 3·4, then components of **E** and **H** in any rectangular coordinate system. We can therefore develop most of our considerations in terms of the scalar relation

$$u_P = -\frac{1}{4\pi} \int_S \left(\psi \frac{\partial u}{\partial n} - u \frac{\partial \psi}{\partial n} \right) dS, \qquad (3)$$

where u will stand for any one of the rectangular components of **E** or **H**.

Equation (3) can be regarded as the mathematical expression of Huygens' principle for a scalar wave; the resultant wave amplitude at P is again expressed as a sum of contributions for the elements of surface dS. The first part of the integral is a summation of terms of the form $(e^{-jkr}/r)(\partial u/\partial n)\,dS$—a summation of the amplitudes of isotropic spherical wavelets arising from sources of strength proportional to $(\partial u/\partial n)\,dS$ on the surface elements dS. The second part of the integral can be interpreted similarly. We note that

$$\frac{\partial \psi}{\partial n} = \frac{d}{dr}\left(\frac{e^{-jkr}}{r}\right)\cos(\mathbf{n},\mathbf{r}) = -\left(jk + \frac{1}{r}\right)\frac{e^{-jkr}}{r}\cos(\mathbf{n},\mathbf{r}), \qquad (4)$$

because the field point P is the origin in the integral formulation. The second part of the integral is thus a summation of anisotropic wavelets from sources of strength proportional to $u\,dS$ on the surface elements dS. The directivity of the sources is expressed by the factor $\cos(\mathbf{n},\mathbf{r})$; each wavelet includes a term for which the amplitude falls off with r^{-2}, like the induction field of a dipole source. Substituting this result into Eq. (3), we obtain

$$u_P = -\frac{1}{4\pi} \int_S \frac{e^{-jkr}}{r} \left[u\left(jk + \frac{1}{r}\right)\cos(\mathbf{n},\mathbf{r}) + \frac{\partial u}{\partial n} \right] dS. \qquad (5)$$

Despite the arbitrary feature of the integral formulations pointed out in the footnote on page 108, we shall consider the Huygens-Green relations [Eqs. (1) and (5)] as the analytical formulations of the Huygens-Fresnel principle for electromagnetic and scalar waves respectively. It

---

$$\int_V (F\nabla^2 G - G\nabla^2 F)\,dv = -\int_{S_1 \cdots S_n} \left(F\frac{\partial G}{\partial n} - G\frac{\partial F}{\partial n} \right) dS.$$

The co vention as to the direction of **n** is the same as shown in the figure. Let F be the spherical wave function $\psi = e^{-jkr}/r$ and G the function u satisfying the Helmholtz equation for the same value of k. The field point P is again surrounded by a sphere Σ, the radius of which later is allowed to approach zero. In the region bounded by $S_1 \ldots S_n$ and Σ the volume integral vanishes. The details of the limiting process that is then applied to Σ follow very closely those for the vector case; the result gives the value of u at the field point P, namely,

$$u_P = -\frac{1}{4\pi} \int_{S_1 \cdots S_n} \left(\psi \frac{\partial u}{\partial n} - u \frac{\partial \psi}{\partial n} \right) dS.$$

should be emphasized that according to the Huygens-Fresnel principle, there is no one-to-one correspondence between the field at the point P and the field at any point on the wave surface; the field at P is an integrated effect of contributions from every point on the wave surface.

4·2. Geometrical Optics: Wavefronts and Rays.—The Huygens-Fresnel principle, as expressed by the Green's theorem integrals, gives a rigorous solution of the wave equation. It is frequently convenient, however, to approach the subject of wave propagation from the less rigorous point of view of geometrical optics, in which attention is focused on the successive positions of equiphase surfaces, or wavefronts, and an associated system of rays.

Let the wavefront at time t_0 be the surface $L(x,y,z) = L_0$ of Fig. 4·2 and the new wavefront after passage of a very short time δt be the surface $L(x,y,z) = L_0 + \delta L$. Geometrical optics is then concerned not only with the form of these surfaces but also with a point-to-point transformation from one wavefront into the succeeding one. This is, of course, in fundamental contrast to the point of view of the Huygens-Fresnel principle.[1] The point-to-point correlation of the wavefronts is established by the "rays," a family of curves having at each point the direction of the energy flow in the field. In the case of electromagnetic waves, a ray can be traced out by proceeding at each point in the direction of the Poynting vector at that point. The rays are nearly normal to the wavefront—exactly normal in the wave systems to be discussed in this volume—and pass through corresponding points in successive wavefronts.

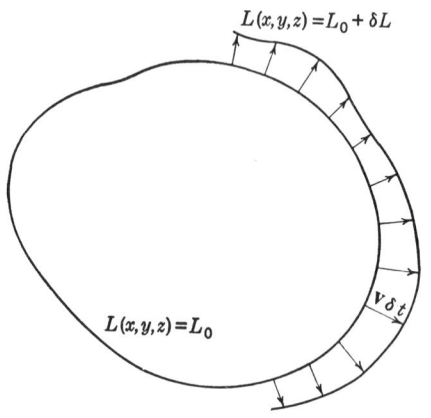

FIG. 4·2.—On the propagation of a wavefront in geometrical optics.

In an arbitrary medium the wave field is characterized by a ray velocity and a wave velocity at every point. The ray velocity is the velocity of energy propagation; it is represented at each point by a vector that is tangent to the ray passing through that point. The wave velocity, on the other hand, is always normal to the wavefront; it is the rate of displacement of the wavefront in the direction normal to that surface. Thus if $\mathbf{v}(x,y,z)$ is the wave velocity at a point (x,y,z) of the first wave-

---

[1] Treatments of geometrical optics as a self-contained theory are given by J. L. Synge, *Geometrical Optics*, Cambridge, London, 1937, and by Ph. Frank and V. Mises, *Differential-gleichungen der Physik*, Vol. II, Chap. 1, reprinted by Mary S. Rosenberg, New York, 1943.

front in Fig. 4·2, the vector **v** δt will extend from that point to the corresponding point on the second wavefront. The case illustrated is that of an inhomogeneous medium in which **v**(x,y,z) is a function of position.

In an anisotropic medium the ray velocity and wave velocity differ, in general, both in magnitude and direction; in isotropic media the ray velocity and wave velocity are identical. We shall here restrict our attention to isotropic but possibly inhomogeneous media; more general discussions will be found in the references of the footnote on page 110. As a result of the identity of the ray velocity and wave velocity, the rays in an isotropic medium make up a family of curves orthogonal to the family of wavefronts; the energy flow at any point is normal to the wavefront passing through that point.

The form of the wavefronts and rays can be determined as soon as the function $L(x,y,z)$ is given. This function is not uniquely determined by the foregoing remarks. We shall, in addition, require that it be chosen so that the wavefront $L(x,y,z) = L_0$ shall be one of constant phase $(\omega/c)L_0$ relative to the phase at some chosen point. The function $L(x,y,z)$ thus defined is of basic importance in the analytic theory of geometrical optics. It satisfies a differential equation which we shall now derive.

The phase increment between the two successive surfaces of Fig. 4·2 is $(\omega/c)\,\delta L$. Moreover, since the wave proceeds from one surface to the next in time δt while the phase at any fixed position changes at the rate ω, this phase difference must be $\omega \delta t$. Finally if δs_n is the distance between the surfaces at (x,y,z) and v is the wave velocity at that point, we have

$$\frac{\omega}{c}\,\delta L = \omega\,\delta t = \omega\,\frac{\delta s_n}{v}. \tag{6}$$

However, we must also have

$$\delta L = |\boldsymbol{\nabla} L|\,\delta s_n. \tag{7}$$

It follows that

$$|\boldsymbol{\nabla} L| = \frac{c}{v} = n, \tag{8}$$

where n is the index of refraction—in general a function of position in the medium. The function L must therefore satisfy the differential equation

$$|\boldsymbol{\nabla} L|^2 = \left(\frac{\partial L}{\partial x}\right)^2 + \left(\frac{\partial L}{\partial y}\right)^2 + \left(\frac{\partial L}{\partial z}\right)^2 = n^2. \tag{9}$$

4·3. Curvature of the Rays in an Inhomogeneous Medium.—In a homogeneous medium the rays are straight lines; in an inhomogeneous medium they have a curvature that we shall now compute. Let **s** be a unit vector in the direction of the ray at a chosen point. This is normal to the wavefront and must have the direction of $\boldsymbol{\nabla} L$; so by Eq. (8) we

have then
$$\mathbf{s} = \frac{\nabla L}{n}. \tag{10}$$

Let $\mathbf{N}$ be a unit vector in the direction of the radius of curvature of the ray at the same point and ρ the radius of curvature; the vector curvature of the ray is then $\mathbf{N}/\rho$. This curvature, however, is also given by $d\mathbf{s}/ds$, where s is distance measured along the ray. By the vector identity

$$\frac{d\mathbf{s}}{ds} = (\mathbf{s} \cdot \nabla)\mathbf{s} = -\mathbf{s} \times (\nabla \times \mathbf{s}) \tag{11}$$

we have then

$$\frac{\mathbf{N}}{\rho} = -\mathbf{s} \times (\nabla \times \mathbf{s}). \tag{12}$$

On taking the scalar product with $\mathbf{N}$, Eq. (12) becomes

$$\frac{1}{\rho} = -\mathbf{N} \cdot (\mathbf{s} \times \nabla \times \mathbf{s}) = -(\mathbf{N} \times \mathbf{s}) \cdot (\nabla \times \mathbf{s}). \tag{13}$$

Using Eq. (10) to compute $\nabla \times \mathbf{s}$, and replacing $\nabla\left(\frac{1}{n}\right)$ by $-(1/n)\nabla(\ln n)$ we obtain finally

$$\frac{1}{\rho} = \mathbf{N} \cdot \nabla(\ln n). \tag{14}$$

Since the radius of curvature is an essentially positive quantity, it follows from Eq. (14) that the rate of change of the refractive index in the direction of the radius of curvature is positive; that is, the ray bends toward the region of higher index of refraction. In a homogeneous medium where n is independent of position, the right-hand side of Eq. (14) is zero, the radius of curvature is infinite, and the rays are straight lines. From Eq. (12) it follows also that in a homogeneous medium the vector field of the rays satisfies the condition

$$\nabla \times \mathbf{s} = 0. \tag{15}$$

This is a sufficient condition for the existence of a family of surfaces orthogonal to the field of vectors $\mathbf{s}$.

4.4. Energy Flow in Geometrical Optics.—Consideration of the rays leads to a simple hydrodynamic picture of the energy flow. It was pointed out previously that the rays are lines of flow of energy. Let us consider the two wave surfaces L_1 and L_2 of Fig. 4·3 and a tube of rays that cuts out elements dA_1 and dA_2 on the respective surfaces. No power will flow across the sides of the tube; the flow across any section normal to the tubes will be constant. If S is the rate of flow per unit area, the condition of constant power flow through the tube is

$$S_1\,dA_1 = S_2\,dA_2. \tag{16}$$

SEC. 4·4]   ENERGY FLOW IN GEOMETRICAL OPTICS   113

In the case of electromagnetic waves the quantity S is the magnitude of the Poynting vector; we shall assume that as in the case of plane and cylindrical waves (Sec. 3·7)

$$S = \frac{1}{2}\left(\frac{\epsilon}{\mu}\right)^{1/2}|E|^2. \qquad (17)$$

If the permeability μ is independent of position, the relation between the electric amplitudes at dA_1 and dA_2 is

$$\epsilon_1^{1/2}|E_1|^2\, dA_1 = \epsilon_2^{1/2}|E_2|^2\, dA_2. \qquad (18)$$

In terms of the refractive index $n = (\epsilon/\epsilon_0)^{1/2}$ we have

$$n_1|E_1|^2\, dA_1 = n_2|E_2|^2\, dA_2. \qquad (19)$$

Unlike the Huygens-Fresnel principle, geometrical optics sets up a one-to-one correspondence between the amplitude at one field point and the amplitude at another.

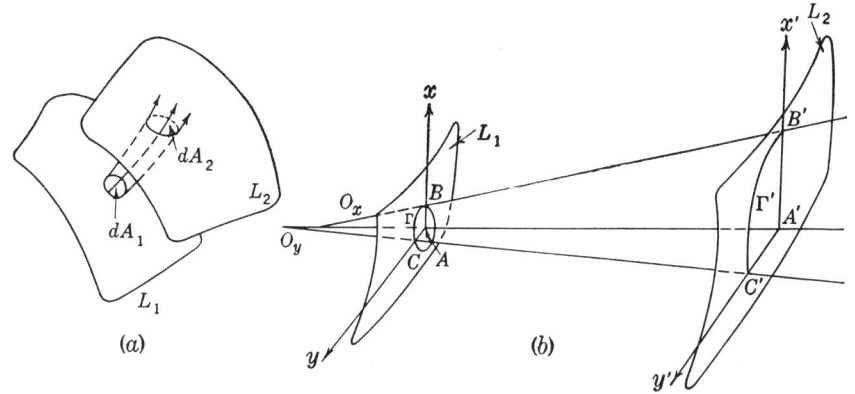

FIG. 4·3.—Energy relations in geometric optics: (a) tube of rays in an inhomogeneous medium; (b) relations between wavefronts in a homogeneous medium.

It will be of interest to apply Eq. (19) to the case of a homogeneous medium in which the rays consist of straight lines. The segments of rays between the wavefronts L_1 and L_2, as shown in Fig. 4·3b, will have equal lengths p. Let the ray through the point A on surface L_1 be the z-axis, and let the xz- and yz-planes coincide with the principal planes of L_1 at A. A ray through an adjacent point B lying in L_1 and the xz-plane will intersect the ray through A at the point O_x, at a distance R_1 which is one of the principal radii of curvature of L_1 at point A; a ray through an adjacent point C in the yz-plane will similarly intersect the ray through A at the point O_y, at a distance R_2 which is the second principal radius of curvature of L_1 at A. The radii of curvature will be considered to be positive if the centers of curvature lie on the negative z-axis, as shown.

The point A' on the surface L_2 lies on the ray through A. It can be shown that the principal planes of L_2 at A' are coincident with those of L_1; through A' we can pass coordinate axes x', y' which correspond to the axes of x, y, respectively. It is obvious that the principal radii of curvature of the surface L_2 at the point A' are $R_1 + p$ and $R_2 + p$.

Let us now consider an element of area dA_1 which includes A and is bounded by the curve Γ. The rays through the curve intersect L_2 in the curve Γ', which bounds an element of area dA_2 around the point A'. These areas are given by

$$\left. \begin{aligned} dA_1 &= \oint_\Gamma (x\,dy - y\,dx), \\ dA_2 &= \oint_{\Gamma'} (x'\,dy' - y'\,dx'). \end{aligned} \right\} \quad (20)$$

It is evident from the figure that the coordinates of corresponding points (x,y) and (x',y'), near A and A' respectively, are thus related:

$$x' = \left| \frac{R_1 + p}{R_1} \right| x \quad (21a)$$

$$y' = \left| \frac{R_2 + p}{R_2} \right| y. \quad (21b)$$

Substitution of these relations into Eq. (20) gives the relation between the cross sections of the tube of rays at L_1 and L_2:

$$dA_2 = \left| \frac{(p + R_1)(p + R_2)}{R_1 R_2} \right| dA_1. \quad (22)$$

Inserting this result into Eq. (19) and recalling that in the present case $n_1 = n_2$, we obtain the relation

$$|E_2| = |E_1| \left(\left| \frac{R_1 R_2}{(R_1 + p)(R_2 + p)} \right| \right)^{1/2}. \quad (23)$$

When R_1 and R_2 are both finite and the surface L_2 is so far from L_1 that $p \gg R_1$ and R_2, this reduces to

$$|E_2| \approx \frac{|E_1|}{p} |R_1 R_2|^{1/2}. \quad (24)$$

This last relation will be of use to us in the discussion of scattering of radiation by curved surfaces.

4·5. Geometrical Optics as a Zero-wavelength Limit.—We shall now investigate the relation between geometrical optics and the field equations,[1] taking up in the succeeding section its connection with the

---

[1] The subject is treated from the point of view of the scalar wave equation by P. Debye, *Polar Molecules*, Chap. 8, reprint by Dover Publications, New York, 1945; also in the article by A. Sommerfeld in Ph. Frank and V. Mises, *Differentialgleichungen der Physik*, Chap. 20, reprint by Mary Rosenberg, New York, 1943.

Huygens-Fresnel principle; the analysis will be confined to homogeneous media.

A careful review of the ideas of the preceding sections will make it evident that geometrical optics is based on the idea that a wavefront behaves locally like a plane wave. The corresponding solution to the scalar wave equation is

$$u = A(x,y,z)e^{j[\omega t - k_0 L(x,y,z)]}, \tag{25}$$

where $k_0 = 2\pi/\lambda_0$, $A(x,y,z)$ is the amplitude of the wave (usually a function of position) and $L(x,y,z)$ is the characteristic function defining surfaces of equal phase. We are here concerned with a linearly polarized electromagnetic field and must consider the vector counterparts of this solution. Let us then investigate the possibility of satisfying the field equations by electric- and magetic-field vectors having the form

$$\mathbf{E} = \boldsymbol{\alpha}(x,y,z)e^{-jk_0 L(x,y,z)}, \tag{26a}$$
$$\mathbf{H} = \boldsymbol{\beta}(x,y,z)e^{-jk_0 L(x,y,z)}. \tag{26b}$$

The amplitude vectors $\boldsymbol{\alpha}$ and $\boldsymbol{\beta}$ may be complex, but their phases in that case must be independent of position.

On substituting these expressions into the homogeneous forms of the field equations [Eqs. (3·32)], it will be found that the amplitude and phase functions must satisfy the relations

$$\boldsymbol{\beta} = \frac{k_0}{\omega\mu}(\boldsymbol{\nabla}L \times \boldsymbol{\alpha}) - \frac{1}{j\omega\mu}(\boldsymbol{\nabla} \times \boldsymbol{\alpha}), \tag{27a}$$

$$\boldsymbol{\alpha} = -\frac{k_0}{\omega\epsilon}(\boldsymbol{\nabla}L \times \boldsymbol{\beta}) + \frac{1}{j\omega\epsilon}(\boldsymbol{\nabla} \times \boldsymbol{\beta}). \tag{27b}$$

On eliminating $\boldsymbol{\beta}$ from Eq. (27b) by means of Eq. (27a) and replacing k_0 by $\omega(\mu_0\epsilon_0)^{1/2}$ it will be found that $\boldsymbol{\alpha}$ must satisfy the equation

$$\boldsymbol{\alpha} = -\frac{1}{n^2}[\boldsymbol{\nabla}L(\boldsymbol{\alpha}\cdot\boldsymbol{\nabla}L) - \boldsymbol{\alpha}|\boldsymbol{\nabla}L|^2] + \frac{1}{jn^2k_0}[\boldsymbol{\nabla}L \times (\boldsymbol{\nabla}\times\boldsymbol{\alpha})$$
$$+ \boldsymbol{\nabla}\times(\boldsymbol{\nabla}L\times\boldsymbol{\alpha})] + \frac{1}{n^2k_0^2}[\boldsymbol{\nabla}\times(\boldsymbol{\nabla}\times\boldsymbol{\alpha})]; \tag{28}$$

n is again the index of refraction. Similarly, on eliminating $\boldsymbol{\alpha}$ from Eq. (27a) we find that $\boldsymbol{\beta}$ must also satisfy Eq. (28).

If $\boldsymbol{\nabla}L$ and the derivatives of $\boldsymbol{\alpha}$ and $\boldsymbol{\beta}$ are finite, the last two terms on the right are of the orders $1/k_0$ and $1/k_0^2$, respectively, as compared with the first. As λ goes to zero, k_0 approaches infinity and the last two terms approach zero. For Eq. (28) and the analogous equation in $\boldsymbol{\beta}$ to be satisfied under these conditions we must have

$$\boldsymbol{\alpha}\cdot\boldsymbol{\nabla}L = 0, \tag{29a}$$
$$\boldsymbol{\beta}\cdot\boldsymbol{\nabla}L = 0, \tag{29b}$$
$$|\boldsymbol{\nabla}L|^2 = n^2. \tag{29c}$$

The last of these conditions is the differential equation for the characteristic function that was developed in Sec. 4·2. The first two conditions state that α and β must be transverse to ∇L; it follows that α and β lie in a plane transverse to the direction of propagation. Furthermore, Eq. (27a) can be written as

$$\beta = \left(\frac{\epsilon}{\mu}\right)^{1/2} \left(\frac{\nabla L}{n} \times \alpha\right) - \frac{1}{jk_0 n} \left(\frac{\epsilon}{\mu}\right)^{1/2} (\nabla \times \alpha), \qquad (30)$$

the second term being of order $1/k_0$ compared with the first. In the limit $\lambda \to 0$ the second term vanishes. Since $\nabla L/n$ is a unit vector in the direction of propagation, we see that in the limit $\lambda \to 0$, β must be perpendicular to α as well as to the direction of propagation. It follows that the Poynting vector is normal to the wavefront and that its magnitude is

$$|S| \approx \left(\frac{\epsilon}{\mu}\right)^{1/2} |\alpha|^2, \qquad (31)$$

to terms of the order of $1/k_0$.

We have thus seen that the field vectors of geometrical optics [Eqs. (26)] possess the properties which were shown in Sec. 3·11 to be possessed by the far-zone fields. In this region, at least, we may expect geometrical optics to serve as a reasonable approximation to the exact theory.

It should be emphasized that the terms of order $1/k_0$ and $1/k_0^2$ in Eqs. (28) and (30) may be considered negligible for short wavelengths only if the derivatives entering into these terms are finite. In the neighborhood of a geometrical focal point the function L varies rapidly and its derivatives assume large values; at the boundary of a geometrical shadow the amplitude varies rapidly. In these regions the geometrical-optics approximation fails, and phenomena are observed that are not covered by the simple theory of wavefronts and rays.

4·6. The Huygens-Fresnel Principle and Geometrical Optics: The Far-zone Approximation.—It will be instructive to investigate the relation between the Huygens-Fresnel principle and geometrical optics to see under what conditions the point-to-point amplitude relation [Eq. (23)] that was obtained in Sec. 4·4 on the basis of the geometrical-optics concept of the flow of energy in tubes of rays can be derived from the Huygens-Fresnel principle in the limit of zero wavelength. The discussion will be restricted again to homogeneous media.

For our present purposes it is sufficient to consider any one scalar component of a field vector; we therefore take as our starting point the scalar integral formula [Eq. (5)]:

$$u_P = -\frac{1}{4\pi} \int_S \frac{e^{-jkr}}{r} \left[u\left(jk + \frac{1}{r}\right) \cos(\mathbf{n},\mathbf{r}) + \frac{\partial u}{\partial n}\right] dS, \qquad (4\cdot5)$$

where the surface S encloses all the sources of the field. In view of the results obtained in the previous section we confine our attention to the field far from the sources; the present section is directed toward the development of an approximation to Eq. (5) suitable for equiphase surfaces in this region.

In the far zone the field is a quasi-point-source field (Sec. 3·11); that is, the amplitude function takes the form

$$u = \frac{e^{-jk\rho}}{\rho} F(\theta,\phi) + O\left(\frac{1}{\rho^2}\right), \qquad (32)$$

where ρ, θ, ϕ are the spherical coordinates of a point in the far zone with respect to an arbitrary origin in the neighborhood of the sources. If $\mathbf{n}$ is the unit normal to S directed out from the region containing the sources (Fig. 4·1) and ϱ_1, θ_1, ϕ_1 are unit vectors in the directions of increasing ρ, θ and ϕ, respectively, at a point on S, the normal derivative of u on this surface is

$$\frac{\partial u}{\partial n} = \mathbf{n} \cdot \nabla u = \mathbf{n} \cdot \left(\frac{\partial u}{\partial \rho} \varrho_1 + \frac{1}{\rho}\frac{\partial u}{\partial \theta} \theta_1 + \frac{1}{\rho \sin \theta}\frac{\partial u}{\partial \phi} \phi_1\right). \qquad (33)$$

By use of Eq. (32) we obtain

$$\nabla u = -jku\varrho_1 + \frac{u}{\rho}\left(-\varrho_1 + \frac{1}{F}\frac{\partial F}{\partial \phi}\theta_1 + \frac{1}{F \sin \theta}\frac{\partial F}{\partial \phi}\phi_1\right) + O\left(\frac{1}{\rho^2}\right). \qquad (34)$$

In the far zone $\rho \gg \lambda$; consequently

$$\nabla u \approx -jku\varrho_1, \qquad (35)$$

and

$$\frac{\partial u}{\partial n} \approx -jku \cos (\mathbf{n},\varrho_1), \qquad (36)$$

providing also that the variation of the amplitude in the θ and ϕ directions is small compared to that in the radial direction. The integral relation thus becomes

$$u_P \approx \frac{1}{4\pi} \int_S \frac{e^{-jkr}}{r} \left\{jku[\cos (\mathbf{n},\varrho_1) - \cos (\mathbf{n},\mathbf{r})] - \frac{u}{r} \cos (\mathbf{n},\mathbf{r})\right\} dS. \qquad (37)$$

Finally, if we consider only field points P such that $r \gg \lambda$ for all points on S, the last term in the integrand is negligible with respect to the first. We then have, as an approximation valid in the far zone,

$$u_P \approx \frac{j}{2\lambda} \int_S u[\cos (\mathbf{n},\varrho_1) - \cos (\mathbf{n},\mathbf{r})] \frac{e^{-jkr}}{r} dS, \qquad r \gg \lambda. \qquad (38)$$

In the limit $\lambda \to 0$ this equation can be applied with virtually no restriction as to the location of the field point P.

Equation (38) applies to any surface in the far-zone region that encloses the sources of the field. Let us now consider the surface S to be an equiphase surface and assume on the basis of the preceding section that the field can be expressed in the form of Eq. (25),

$$u = A(x,y,z)e^{-jk_0 L(x,y,z)}. \tag{4.25}$$

It was seen that in the limit $\lambda \to 0$ this leads to a solution of the field equations such that the Poynting vector is normal to the equiphase surface. On the other hand, in the investigation of the far-zone fields in Sec. 3·11 it was found that neglecting terms of order $1/\rho^3$ the Poynting vector is in the direction of ϱ_1 independent of the choice of the origin in the neighborhood of the sources. Consequently, if S is an equiphase surface, we have as an approximation valid for short wavelengths

$$\cos(\mathbf{n},\varrho_1) \approx 1. \tag{39}$$

The integral relation [Eq. (38)] in this case reduces to

$$u_P \approx \frac{j}{2\lambda} \int_S u[1 - \cos(\mathbf{n},\mathbf{r})] \frac{e^{-jkr}}{r} dS. \tag{40}$$

It will be recognized that Eq. (39) is tantamount to assuming that the equiphase surfaces do not differ widely from spheres about the source distribution. Also in view of the condition associated with Eq. (35) that $\rho \gg \lambda$, the assumption is implied that the radii of curvature of the equiphase surfaces are large compared with the wavelength.

A consideration of the normal derivative $\partial u/\partial n$ in terms of field expression of Eq. (25) shows an additional assumption, concerning the amplitude $A(x,y,z)$, which underlies the use of the far-zone integral [Eq. (40)]. Taking Eq. (25), we have for the normal derivative of u on an equiphase surface

$$\frac{\partial u}{\partial n} = u\left(-jk + \frac{1}{A}\frac{\partial A}{\partial n}\right). \tag{41}$$

In obtaining this result use is made of Eq. (29c). Substituting Eq. (41) into Eq. (5) shows that we pass from the latter to Eq. (40) under the condition that

$$\frac{1}{A}\frac{\partial A}{\partial n} \ll \frac{2\pi}{\lambda}. \tag{42}$$

This is satisfied, of course, in the limit $\lambda \to 0$ provided that $(1/A)(\partial A/\partial n)$ is finite everywhere. In the practical case, where the wavelength is small but not equal to zero, the contribution of $(1/A)(\partial A/\partial n)$ can be neglected to a good approximation if the fractional change in amplitude over a distance equal to the wavelength is small compared with unity.

4·7. The Principle of Stationary Phase.—Equation (40) still expresses the field at point P as a superposition of spherical wavelets arising from every point on the equiphase surface. The transition to the geometrical optics result of Sec. 4·4 is carried through on the basis of the principle of stationary phase which we shall discuss in this section.

Let the surface S of Fig. 4·4 be the equiphase surface and P be the field point. There are at least two points on S at which the normal to the surface lies along the line passing through P. Let N be the nearest of such points to the latter, and let NP be the z-axis of our coordinate system; the x- and y-axes are taken in the principal planes of curvature

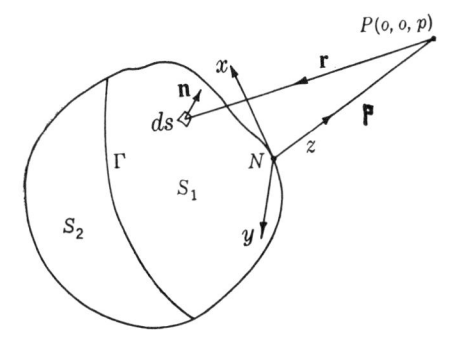

Fig. 4·4.—On the principle of stationary phase.

of the surface at the point N. The surface is divided into segments S_1 and S_2 by the curve Γ along which the tangent planes to the surface are parallel to the z-axis. We shall assume that in each segment there is only one point at which the line of the normal passes through P. This condition implies that P is not a focal point of the rays associated with the surface S.

Denoting the distance NP by p, the integral of Eq. (40) can be rewritten as

$$u_P = \frac{j}{2\lambda} e^{-jkp} \iint_{S_1+S_2} \frac{u[1 - \cos(\mathbf{n},\mathbf{r})]}{r \cos(\mathbf{n},z)} e^{-jk(r-p)} \, dx \, dy. \tag{43}$$

The integral is a sum of vector elements and can be treated graphically by the customary procedures of vector addition. The magnitude of the vector element contributed by an arbitrary element of surface dS is $(u/r)[1 - \cos(\mathbf{n},\mathbf{r})] \, dS$, and the angle between it and the vector from the element of area at the point N is $(2\pi/\lambda)(r - p)$. Consider now the contribution from an arbitrary portion of the surface as a function of the wavelength. If the wavelength is large, the angle between vectors from adjacent surface elements is small; the vector diagram in this case takes the form of a gradual curve, as is illustrated in Fig. 4·5a, and we may in general expect a resultant vector u_R significantly different from zero. On the other hand, if the wavelength is small, the angle between adjacent elements is large and the vector diagram takes the form of a tightly wound curve as is shown in Fig. 4·5b. In the latter case the resultant vector u_R may in general be expected to be virtually zero, the more so as $\lambda \to 0$. Thus, as a result of the rapid variation in the phase of the integrand of Eq. (43), we have destructive interference and virtually complete cancellation between the spherical wavelets from an arbitrary portion of the phase surface.

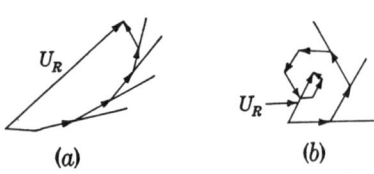

FIG. 4·5.—The vector representation of the Huygens-Fresnel integral: (a) the long-wavelength case; (b) the short-wavelength case.

The situation is different, however, for those portions of the surface in the neighborhood of the point N on the segment S_1 and the corresponding point on S_2. It is observed that the phase function

$$\phi(x,y) = r - p = [x^2 + y^2 + (p - z)^2]^{1/2} - p \tag{44}$$

is stationary in the neighborhood of these points; at these points

$$\frac{\partial \phi}{\partial x} = \frac{\partial \phi}{\partial y} = 0. \tag{45}$$

Consequently, in the neighborhood of these points the phase varies slowly, despite the short wavelength, and the vector diagrams representing the contributions of the areas around these points take the form of Fig. 4·5a rather than that of Fig. 4·5b. The stationary phase areas yield contributions to the integral [Eq. (43)] compared with which the contributions of other portions of the surface are negligible. We are thus led to the principle of stationary phase: For short wavelengths, the integral of Eq. (43) representing the effect of the whole surface S is negligibly different from the sum of the contributions of the areas about those points on S at which the phase has a stationary value.

It will be observed further that at the stationary point on the segment S_2, $\cos(\mathbf{n},\mathbf{r}) = 1$ and that the $\cos(\mathbf{n},\mathbf{r})$ will not be very different from unity over the area in the neighborhood of the stationary point in view of our earlier assumptions (Sec. 4·6) as to the nature of the surface. The contribution from this area is again zero, since $1 - \cos(\mathbf{n},\mathbf{r})$ vanishes, and we are left then solely with the contribution of the area around the point N. The amplitude of the integrand of Eq. (43) may be considered constant—equal to its value at the point N— over this area, and Eq. (43) then reduces to

$$u_P \approx \frac{j}{\lambda} u_N \frac{e^{-jkp}}{p} \int\int_{\delta_N} e^{-jk\phi} dx\, dy, \tag{46}$$

where δ_N is a small area around the point N.

The equation of the surface in the neighborhood of N is

$$z = -\left(\frac{x^2}{2R_1} + \frac{y^2}{2R_2}\right) + \cdots, \qquad (47)$$

R_1 and R_2 being the principal radii of curvature. Inserting this into Eq. (44), we find that to second-order terms the phase function over the area δ_N is

$$\phi = \frac{1}{2}\left(\frac{R_1 + p}{R_1 p}\right)x^2 + \frac{1}{2}\left(\frac{R_2 + p}{R_2 p}\right)y^2; \qquad (48)$$

this is to be inserted into Eq. (46). We may now, however, reverse the application of the stationary phase principle and argue that the integral of Eq. (46) may be extended over the infinite (x,y)-domain with negligible error. We thus obtain

$$u_P \approx \frac{j}{\lambda} u_N \frac{e^{-jkp}}{p} \int \int_{-\infty}^{\infty} e^{-j\frac{k}{2}[\alpha x^2 + \beta y^2]} dx\, dy, \qquad (49)$$

with

$$\alpha = \left(\frac{R_1 + p}{R_1 p}\right); \qquad \beta = \left(\frac{R_2 + p}{R_2 p}\right). \qquad (50)$$

The integral of Eq. (49) can be transformed to Fresnel integrals,

$$\int_{-\infty}^{\infty} e^{-j\frac{\pi}{2}\xi^2} d\xi = \sqrt{2}\, e^{-j\frac{\pi}{4}},$$

with the final result[1]

[1] The argument may be applied in general to integrals

$$u = \int\int_R F(u,v) e^{-jk\phi(u,v)}\, du\, dv$$

over a region R in which $F(u,v)$ has bounded variation in each variable. If (u_0, v_0) is a stationary point of the function ϕ in the region R, and if the coefficients α and β of the canonical form of $d^2\phi$ at that point,

$$d^2\phi = \tfrac{1}{2}(\alpha\xi^2 + \beta\eta^2),$$

are both different from zero, the asymptotic value of the integral for large k is

$$u \approx F(u_0,v_0) e^{-jk\phi(u_0,v_0)} \int_{-\infty}^{\infty}\int_{-\infty}^{\infty} e^{-j\frac{k}{2}(\alpha\xi^2 + \beta\eta^2)}\, d\xi\, d\eta;$$

or,

$$u \approx \frac{2\pi}{k|\alpha\beta|^{1/2}} F(u_0,v_0) e^{-jk\phi(u_0,v_0)} e^{-j\frac{\pi}{4}\left(\frac{\alpha}{|\alpha|} + \frac{\beta}{|\beta|}\right)}.$$

If ϕ has more than one stationary point in the region, the total value of the integral is obtained by summing the latter expression over the stationary points.

The principle was formulated by Lord Kelvin, *Math. Phys. Papers* **IV**, 303–306 (1910), for one-dimensional integrals; the latter has been discussed recently in a

$$u_P \approx u_N \left| \frac{R_1 R_2}{(R_1 + p)(R_2 + p)} \right|^{1/2} e^{-jkp}. \tag{51}$$

The amplitude relation is seen to be identical with that obtained on the energy flow basis. The factor e^{-jkp} simply represents the phase change corresponding to the displacement of the wavefront S along the optical rays to the wavefront containing the point P.

4·8. Fermat's Principle.—We shall now return to the discussion of the methods of geometrical optics and shall consider several principles that underlie the design of reflectors and lenses. The first of these is Fermat's principle, which is often taken as the basic postulate in the development of the general theory.[1]

Before stating Fermat's principle we must introduce the idea of "optical path length." The optical path length ΔL along a curve Γ between points P_1 and P_2 is defined by a line integral along this curve:

$$\Delta L = \int_\Gamma n |ds|, \tag{52}$$

where n is the refractive index at the line element ds.

This concept is intimately connected with the ideas discussed in Sec. 4·2. Between two adjacent phase surfaces $L(x,y,z) = L_0$ and $L(x,y,z) = L_0 + \delta L$, there is an increment in the value of the characteristic function L which is, by Eq. (6),

$$\delta L = \frac{c}{v} \delta s_n = n\, \delta s_n. \tag{53}$$

The distance δs_n between the two surfaces is a function of position, but the quantity $\delta L = n\, \delta s_n$ is a constant; this, it will be noted, is the optical path length along any ray between the two surfaces. It follows immediately that the optical path length, as given by Eq. (51), is the same for every ray between any two wavefronts $L(x,y,z) = L_0$ and $L(x,y,z) = L_1$; it is, in fact,

$$\Delta L = |L_1 - L_0|. \tag{54}$$

Thus the characteristic function $L(x,y,z)$ can be interpreted as the optical path length along a ray from the wavefront $L(x,y,z) = 0$ to the wavefront in which the point (x,y,z) lies.

---

rigorous manner by A. Wintner, *J. Math. Phys.*, **24**, 127 (1945). As yet, rigorous extension to the two-dimensional case has not been made. The convergence of the integrals that is required for the process outlined here to be valid is assured in the case that α and β are both positive and k has a small negative imaginary component k_i; the final result is then to be interpreted as the limit (after integration) as $k_i \to 0$.

[1] *Cf.* J. L. Synge, *Geometrical Optics*, Cambridge, London, 1937.

SEC. 4·8] FERMAT'S PRINCIPLE 123

The idea of optical path length is not restricted in its application to rays. One can determine the optical path length between two points P_1 and P_2 along any curve Γ whatever; its value will in general vary with the choice of Γ. Fermat's principle provides a method for using these values in selecting possible ray paths from P_1 to P_2 from all other paths. It may be stated thus:

Fermat's principle: The optical ray or rays from a source at a point P_1 to a point of observation P_2 is the curve along which the optical path length is stationary with respect to infinitesimal variations in path.

Usually the optical path along a ray has a maximum or minimum value with respect to neighboring paths. The inclusion of the plural possibility "rays" in the above formulation of Fermat's principle is

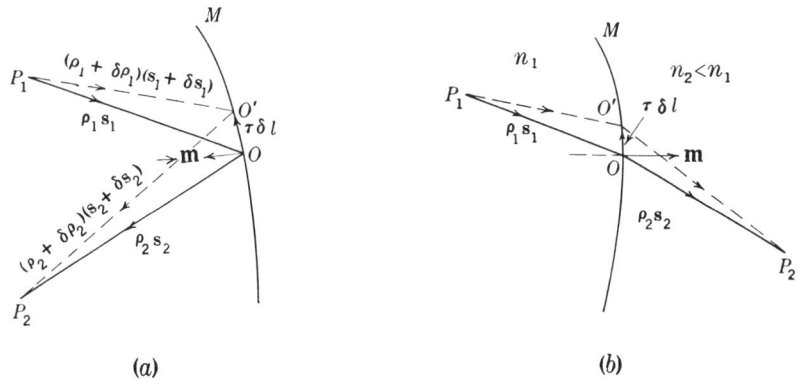

FIG. 4·6.—Notation for the derivation of Snell's laws: (a) reflection; (b) refraction.

required to cover situations in which the point P_2 may be reached by rays from P_1 by a direct path or by reflection from surfaces at which there are discontinuities in the index of refraction.

It follows directly from Fermat's principle that in a homogeneous medium (n = constant) the rays are straight lines. The optical path length is in this case proportional to the geometrical path length, and a straight line gives a minimum value for both.

Fermat's principle can also be used in deriving Snell's laws of reflection and refraction at the interface between two homogeneous media. Let us consider first the laws of reflection. Let the point O of Fig. 4·6a be the point on the reflecting surface M for which the optical path length from P_1 to O to P_2 has a stationary value. The optical path must consist of straight line segments from P_1 to O and O to P_2, since these paths are in a homogeneous medium. The optical path length is then certainly stationary with respect to neighboring curved paths from P_1 to P_2 by way of M which leave the point O unchanged; but by our postulate it is stationary also with respect to straight-line paths with near-by reflection points O'. Let us then consider a neighboring point O', displaced with respect to O by $\boldsymbol{\tau}\,\delta l$, $\boldsymbol{\tau}$ being a unit vector in the tangent plane. We shall now compute the variation in optical path length as O' is changed. Let $\mathbf{s}_1$ and $\mathbf{s}_2$ (Fig. 4·6) be unit vectors in the direction P_1O and OP_2, respectively, and $\mathbf{m}$ the unit vector normal to the surface at O. Then we may write the vector P_1O as $\rho_1\mathbf{s}_1$ and the vector OP_2 as $\rho_2\mathbf{s}_2$. Similarly let $\mathbf{s}_1 + \delta\mathbf{s}_1$ and $\mathbf{s}_2 + \delta\mathbf{s}_2$ be unit vectors along the lines P_1O' and $O'P_2$, respectively, and $\rho_1 + \delta\rho_1$ and $\rho_2 + \delta\rho_2$ the lengths of these lines. The variation in optical path by way of O' with respect to the path by way of the point O is

$$\delta L = n(\delta\rho_1 + \delta\rho_2); \tag{55}$$

by our postulate this must vanish to terms of the first order in δl. From Fig. 4·6 it is clear that

$$(\rho_1 + \delta\rho_1)(\mathbf{s}_1 + \delta\mathbf{s}_1) = \rho_1\mathbf{s}_1 + \boldsymbol{\tau}\,\delta l, \tag{56a}$$
$$(\rho_2 + \delta\rho_2)(\mathbf{s}_2 + \delta\mathbf{s}_2) = \rho_2\mathbf{s}_2 - \boldsymbol{\tau}\,\delta l. \tag{56b}$$

To terms of the first order we have then

$$\delta\rho_1\mathbf{s}_1 + \rho_1\,\delta\mathbf{s}_1 = \boldsymbol{\tau}\,\delta l, \tag{57a}$$
$$\delta\rho_2\mathbf{s}_2 + \rho_2\,\delta\mathbf{s}_2 = -\boldsymbol{\tau}\,\delta l, \tag{57b}$$

whence

$$\delta\rho_1 = \mathbf{s}_1 \cdot \boldsymbol{\tau}\,\delta l, \tag{58a}$$
$$\delta\rho_2 = -\mathbf{s}_2 \cdot \boldsymbol{\tau}\,\delta l, \tag{58b}$$

since $\mathbf{s}_1 \cdot \delta\mathbf{s}_1 = 0$. By Fermat's principle, then,

$$\delta L = n(\mathbf{s}_1 - \mathbf{s}_2) \cdot \boldsymbol{\tau}\,\delta l = 0 \tag{59}$$

for all variations δl; hence

$$(\mathbf{s}_1 - \mathbf{s}_2) \cdot \boldsymbol{\tau} = 0 \tag{60}$$

for every unit vector $\boldsymbol{\tau}$ in the tangent plane. This gives immediately the two laws of reflection:

1. The incident ray, the reflected ray, and the normal to the reflecting surface all lie in the same plane. (The plane defined by $\mathbf{s}_1$ and $\mathbf{s}_2$ is normal to the tangent plane.)
2. The incident and reflected rays make equal angles with the normal. [$\cos(\mathbf{s}_1,\boldsymbol{\tau}) = \cos(\mathbf{s}_2,\boldsymbol{\tau})$; that is, the angles $(\mathbf{s}_1,\boldsymbol{\tau})$ and $(\mathbf{s}_2,\boldsymbol{\tau})$ are equal.]

The law of refraction is derived in a similar manner in Fig. 4·6b. The variations in actual length of the paths P_1O' and $O'P_2$ are given again by Eqs. (58); the optical path variation is, however,

$$\delta L = n_1\,\delta\rho_1 + n_2\,\delta\rho_2 = (n_1\mathbf{s}_1 - n_2\mathbf{s}_2) \cdot \boldsymbol{\tau}\,\delta l. \tag{61}$$

By Fermat's principle this again vanishes for every δl, whence

$$(n_1\mathbf{s}_1 - n_2\mathbf{s}_2) \cdot \boldsymbol{\tau} = 0 \tag{62}$$

for every vector $\boldsymbol{\tau}$ in the tangent plane. This implies the two laws of refraction:

1. The incident ray, reflected ray, and the normal lie in a plane.
2. $$n_1 \cos(\mathbf{s}_1,\boldsymbol{\tau}) = n_2 \cos(\mathbf{s}_2,\boldsymbol{\tau}), \tag{63a}$$

or in terms of the angles between the rays and the normal

$$n_1 \sin(\mathbf{m},\mathbf{s}_1) = n_2 \sin(\mathbf{m},\mathbf{s}_2). \tag{63b}$$

Snell's laws of reflection and refraction are again an expression of the fundamental assumption of geometrical optics that the wavefront behaves locally like a plane wave; in addition, they assume that the boundary surface can be treated locally like its tangent plane. In field theory Snell's laws derive rigorously from application of the boundary conditions of Sec. 3·3 only for the case of an infinite plane wave incident upon an infinite plane boundary.[1] They follow in a good approximation from these boundary relations if the radii of curvature of the two wavefronts (incident and reflected or refracted) and of the boundary are large compared with the wavelength.

4·9. The Law of the Optical Path.—Fermat's principle provides an independent formulation of optical rays from the method of the characteristic function $L(x,y,z)$ and equiphase surfaces developed in Sec. 4·2. It was shown in the latter section that the rays are orthogonal to the equiphase surfaces, and it was observed further in the preceding section that the optical path along the rays between a pair of equiphase surfaces is a constant. The treatment of Sec. 4·2 applies, however, only to media in which the index of refraction is a continuous function of position. We shall now show that the system of rays arising by refraction or reflection (in accordance with Snell's laws) at a boundary of discontinuity in the refractive index have associated with them a family of equiphase surfaces, so that the law of constant optical path holds for any pair of wavefronts, one a member of the incident system and one of the refracted (reflected) system.

It was seen that in a homogeneous medium the rays are straight lines. A family of straight lines for which there exists a family of orthogonal surfaces is said to constitute a normal congruence. Thus, the rays defined in Sec. 4·2—the normals to the surfaces $L(x,y,z) = \text{constant}$—constitute a normal congruence. Let us now consider the problem of refraction or reflection, it being given that the incident system of rays form a normal congruence associated with a family of equiphase surfaces

[1] See M. Born, *Optik*, p. 15, reprint by Edwards Bros., Ann Arbor, Mich., 1943.

$L(x,y,z)$ = constant. The first question is whether the refracted (or reflected) system of rays is likewise a normal congruence. This is answered in the affirmative by the theorem of Malus which we state without proof:[1]

Theorem of Malus: A normal congruence after any number of reflections and refractions is again a normal congruence. The system of refracted rays thus has associated therewith a family of orthogonal surfaces.

We shall now investigate the optical path along the rays between a member of the incident wavefronts and a member of the surfaces orthogonal to the refracted system of rays. Let L_1 of Fig. 4·7 be a wavefront

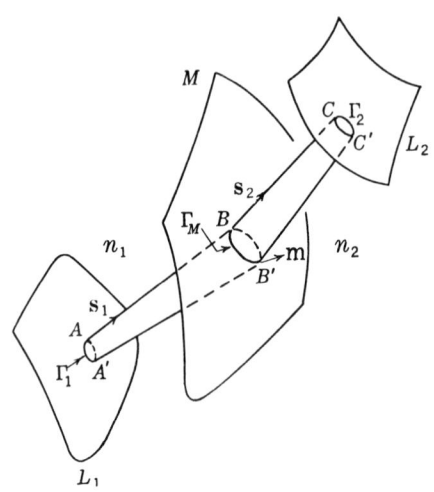

FIG. 4·7.—On the law of the optical path.

in the incident system and L_2 one of the orthogonal surfaces of the refracted system of rays and consider an incident tube of rays passing through the closed curve Γ_1 on L_1; let Γ_m be the curve of intersection of the tube with the refracting surface M and Γ_2 the curve of intersection of the refracted tube of rays with the surface L_2.

We shall evaluate the optical path from L_1 to L_2 along any pair of rays, say the paths ABC and $A'B'C'$ shown in Fig. 4·7. Let us consider first the integrals

$$\oint_{(I)} n_1\mathbf{s}_1 \cdot d\mathbf{l} = \int_A^B n_1\mathbf{s}_1 \cdot d\mathbf{l} + \int_{B\,(\Gamma_m)}^{B'} n_1\mathbf{s}_1 \cdot d\mathbf{l} + \int_{B'}^{A'} n_1\mathbf{s}_1 \cdot d\mathbf{l} + \int_{A'\,(\Gamma_1)}^{A} n_1\mathbf{s}_1 \cdot d\mathbf{l}, \quad (64)$$

---

[1] See for example R. K. Luneberg, *Mathematical Theory of Optics*, Lectures in Applied Mathematics, Brown University, 1944.

and
$$\oint_{(\text{II})} n_2 \mathbf{s}_2 \cdot d\mathbf{l} = \int_B^C n_2 \mathbf{s}_2 \cdot d\mathbf{l} + \int_{C \atop (\Gamma_2)}^{C'} n_2 \mathbf{s}_2 \cdot d\mathbf{l} + \int_{C'}^{B'} n_2 \mathbf{s}_2 \cdot d\mathbf{l} + \int_{B' \atop (\Gamma_m)}^{B} n_2 \mathbf{s}_2 \cdot d\mathbf{l}, \quad (65)$$

where $\mathbf{s}_1$ and $\mathbf{s}_2$ are unit vectors along the incident and refracted rays, respectively. Since $\mathbf{s}_1$ and $\mathbf{s}_2$ are both normal congruences, they each satisfy the equation [see Eq. (15)]

$$\nabla \times \mathbf{s} = 0;$$

therefore, the line integrals around the closed paths $\oint_{(\text{I})}$ and $\oint_{(\text{II})}$ are zero. Furthermore the integrals over Γ_1 and Γ_2 are zero, since $\mathbf{s}_1$ and $\mathbf{s}_2$ are normal to these respective curves. Adding the above integrals (I) and (II) and transposing suitable terms, we then obtain

$$\int_A^B n_1 \mathbf{s}_1 \cdot d\mathbf{l} + \int_B^C n_2 \mathbf{s}_2 \cdot d\mathbf{l} = \int_{A'}^{B'} n_1 \mathbf{s}_1 \cdot d\mathbf{l} + \int_{B'}^{C'} n_2 \mathbf{s}_2 \cdot d\mathbf{l} + \int_{B' \atop (\Gamma_m)}^{B} (n_1 \mathbf{s}_1 - n_2 \mathbf{s}_2) \cdot d\mathbf{l}. \quad (66)$$

The last integral of Eq. (66) vanishes as a result of Snell's law of refraction. The left-hand side is the optical path ABC, while the first two integrals on the right-hand side constitute the optical path $A'B'C'$. We have, therefore,

$$\int_{ABC} n \, ds = \int_{A'B'C'} n \, ds. \quad (67)$$

The optical path and hence the phase increment are constant along all rays from the equiphase surface L_1 in the incident system to the surface L_2 in the refracted region. The family of surfaces orthogonal to the refracted rays thus constitutes the refracted system of equiphase surfaces.

The law of the optical path often provides a simpler approach to the determination of reflecting or refracting surfaces than do Snell's laws. As an example, let us design a reflector that transforms a spherical wave into a plane wave. It is evident that the surface is a surface of revolution and that it is sufficient to consider a plane section containing the axis of revolution. In Fig. 4·8 let F be a point source, the center of curvature of the spherical wave; M the reflecting surface; and L_0 any one of the family of plane wavefronts into which the spherical waves are to be transformed. The optical path from F to the wavefront L_0 is

$$FP + AP = \text{const.} = f + d. \quad (68)$$

The constant may be evaluated by considering the path along the axis; if the distance $OF = f$, the optical path is equal to $f + d$. When FP and AP are evaluated in terms of ρ and ψ, Eq. (68) becomes the equation

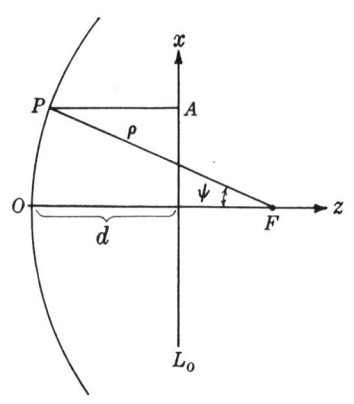

Fig. 4·8.—Application of the law of the optical path.

of the surface in polar form

$$\rho = \frac{2f}{1 + \cos \psi}. \tag{69}$$

This is the equation of a parabola of focal length f.

In contrast to the above calculation, application of Snell's laws would lead to the setting up of the differential equation of the surface; it would then be necessary to integrate this equation. Further examples of the application of the law of the optical path will be discussed in later chapters in the design of mirrors and lenses.

CHAPTER 5

SCATTERING AND DIFFRACTION

By S. Silver

The introduction of an obstacle into the path of a wave gives rise to phenomena that are not covered by the geometrical theory of wavefronts and rays developed in the preceding chapter. These phenomena —scattering and diffraction—are of fundamental importance in microwave antennas, for they underlie the formation of antenna patterns by reflectors and lenses. In the present chapter the theory of scattering and diffraction is developed with reference to general techniques; the specific problems associated with antenna patterns will be taken up in Chap. 6.

5·1. General Considerations.—The discussion of the scattering problem will be restricted to the case of an obstacle of infinite conductivity. The problem with which we are concerned is the following: Given a primary system of sources that produces an electromagnetic field E_0, H_0; an infinitely conducting body is introduced into the field, and it is required to find the new field E, H.

In practice the primary sources are distributions of currents and charges over a system of conductors activated by generators. We shall refer to the latter system of conductors and generators as the source system, in distinction to the currents and charges over the obstacles.

The solution to our problem is based on the superposition principle of Sec. 3·2. On introducing the body into the field of the sources a distribution of current and charge is induced over its surface. We then have two component fields: one arising from the induced distribution over the body and the second arising from the currents and charges in the source system. The total field E, H results from the superposition of the component fields. It should be noted, however, that the field of the body reacts on the source system with a resulting perturbation of its current distribution, so that the component field of the latter differs from the original field E_0, H_0.

The interaction between the body and the source system—and the total field E, H—can be analyzed as a superposition of multiple scattering processes. First we consider the interaction of the body with the original field E_0, H_0, assuming no change in the source currents. The body sets up a scattered wave E'_s, H'_s, arising from an induced distribution over its surface. The scattered wave falling on the source-system conductors induces a current distribution in the latter that gives rise to a secondary scattered wave $\mathbf{E}_0'$, $\mathbf{H}_0'$. The interaction of the secondary wave with the body is again a scattering process leading to an induced distribution over the body and a scattered wave $\mathbf{E}_s''$, $\mathbf{H}_s''$, and so on. The total induced distribution over the body is the sum of the distributions associated with the component scattered waves $\mathbf{E}_s'$, $\mathbf{E}_s''$, ... , and the resultant distribution in the source system is the sum of the distributions associated with $\mathbf{E}_0$, $\mathbf{E}_0'$, ... , respectively.

If the distance R between the source system and the body is large compared with the dimensions of either, the scattering processes of order higher than the first can generally be neglected; for example, in general the ratio E_0'/E_0 evaluated at the body is of order $1/R^2$ and the ratio E_0''/E_0 is of the order $1/R^4$. Also, in special cases, where, although the distance R is not large, the geometry of the body is such that the amplitude of the scattered wave $\mathbf{E}_s$, $\mathbf{H}_s$ at the source system is small, multiple scattering may be neglected in the analysis of the total field $\mathbf{E}$, $\mathbf{H}$.

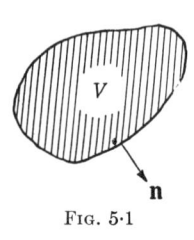

Fig. 5·1

These conditions are usually met in microwave antennas, and the multiple scattering will be neglected in the study of the antenna pattern.

5.2. Boundary Conditions.—With attention restricted to a single scattering process, our problem is that of finding the scattered field $\mathbf{E}_1$, $\mathbf{H}_1$ set up by an infinitely conducting body when it is introduced into an initial field $\mathbf{E}_0$, $\mathbf{H}_0$; the total field is then

$$\mathbf{E} = \mathbf{E}_0 + \mathbf{E}_1, \qquad (1a)$$
$$\mathbf{H} = \mathbf{H}_0 + \mathbf{H}_1. \qquad (1b)$$

It is assumed that the initial field is prescribed for all space.

Let V in Fig. 5·1 be the region occupied by the body; $\mathbf{n}$ is a unit vector normal to the boundary surface S of V, directed outward into the surrounding space. Since the conductivity of the body is infinite, the total field $\mathbf{E}$, $\mathbf{H}$ is zero everywhere inside the region V; according to the boundary conditions of Sec. 3·3 there is a distribution of charge and current over the surface S:

$$\eta = \epsilon(\mathbf{n} \cdot \mathbf{E}), \qquad (2a)$$
$$\mathbf{K} = \mathbf{n} \times \mathbf{H}, \qquad (2b)$$

respectively. $\mathbf{E}$ and $\mathbf{H}$ are the total fields just outside V, and ϵ and μ are the constitutive parameters of the surrounding medium at the boundary surface. These charge and current distributions are the sources of the scattered wave $\mathbf{E}_1$, $\mathbf{H}_1$.

From Eqs. (1) it is seen immediately that at all points in the interior of the body the scattered wave is out of phase with the original field:

$$\mathbf{E}_1 = -\mathbf{E}_0, \qquad \mathbf{H}_1 = -\mathbf{H}_0, \qquad (3)$$

since the total field is zero. Accordingly we need concern ourselves only with the region exterior to V. Here the scattered field must be determined as a solution of Maxwell's equations that satisfies appropriate boundary conditions at infinity and over the surface S. The boundary conditions to be imposed at infinity are the radiation conditions [Eqs. (3·113)], since the field arises from a current distribution confined to a finite region of space. Over the surface S, the scattered field must be such that the total field satisfies the boundary conditions [Eqs. (3·24) and (3·28)]:

$$\mathbf{n} \times \mathbf{E} = 0, \qquad (4a)$$
$$\mathbf{n} \cdot \mathbf{H} = 0. \qquad (4b)$$

From Eqs. (1) we have that the corresponding boundary conditions on $\mathbf{E}_1$ and $\mathbf{H}_1$ are

$$\mathbf{n} \times \mathbf{E}_1 = -\mathbf{n} \times \mathbf{E}_0, \qquad (5a)$$
$$\mathbf{n} \cdot \mathbf{H}_1 = -\mathbf{n} \cdot \mathbf{H}_0. \qquad (5b)$$

Since the field $\mathbf{E}_0$, $\mathbf{H}_0$ is known, Eqs. (5) prescribe the tangential component of $\mathbf{E}_1$ and the normal component of $\mathbf{H}_1$ as known functions over S.

The boundary conditions [Eqs. (4a) and (4b) or (5)] are not independent. If the field satisfies Maxwell's equations and one of the boundary conditions, it necessarily satisfies the other. Let us assume, for example, that condition (4a) is satisfied by the total field. Applying the integral relation between the field vectors [Eq. (3·17a)] to any area on S bounded by an arbitrary curve Γ, we have

$$\oint_\Gamma \mathbf{E} \cdot d\mathbf{s} = -\frac{d}{dt} \int_S \mathbf{B} \cdot \mathbf{n} \, dS = 0, \qquad (6)$$

since $\mathbf{E} \cdot d\mathbf{s} = 0$ by virtue of the boundary condition (4a). The result holds for an arbitrary area, no matter how small; consequently $\mathbf{n} \cdot \mathbf{B} = 0$ over the surface. Therefore only one of the boundary conditions need be considered in selecting the appropriate solutions of Maxwell's equations.

The problem can be approached from another point of view. We shall restrict ourselves at this point to an $e^{j\omega t}$ time dependence and to homogeneous media. It is evident that if the surface distributions [Eqs. (2)] are known, the scattered field is obtained directly by the methods of Secs. 3·9 and 3·10. It can be verified readily that the surface distributions [Eqs. (2)], satisfy the equation of continuity [Eq. (3·9)], over the surface ($\mathbf{E}$, $\mathbf{H}$ being required to satisfy Maxwell's equations); as a result the field vectors $\mathbf{E}_1$, $\mathbf{H}_1$ can be expressed in terms of the current distribution alone, as was done in Sec. 3·9. In fact, the appropriate expressions are obtained from Eqs. (3·120) and (3·121) by passing from volume to surface integrals. The scattered wave is then

$$\mathbf{E}_1 = -\frac{j}{4\pi\omega\epsilon} \int_S [(\mathbf{K} \cdot \boldsymbol{\nabla})\boldsymbol{\nabla} + k^2\mathbf{K}] \frac{e^{-jkr}}{r} dS, \tag{7}$$

$$\mathbf{H}_1 = \frac{1}{4\pi} \int_S (\mathbf{K} \times \boldsymbol{\nabla}) \frac{e^{-jkr}}{r} dS, \tag{8}$$

where r is the distance from the field point to the element of surface dS.

The fields given by Eqs. (7) and (8) necessarily satisfy Maxwell's equations and the radiation conditions at infinity. To determine the current density $\mathbf{K}$ on the boundary surface S we must use condition (5a) or (5b). Letting $\mathbf{n}'$ denote the unit vector normal to S at the point of observation, we have

$$\mathbf{n}' \times \mathbf{E}_0 = \frac{j}{4\pi\omega\epsilon} \int_S \mathbf{n}' \times [(\mathbf{K} \cdot \boldsymbol{\nabla})\boldsymbol{\nabla} + k^2\mathbf{K}] \frac{e^{-jkr}}{r} dS. \tag{9}$$

The left-hand side is a known function, and Eq. (9) is an integral equation for the determination of the unknown current distribution $\mathbf{K}$. The scattering problem is thus transformed to the problem of solving the integral equation rather than Maxwell's equations.

It will be observed that the current distribution which satisfies the integral equation leads through Eq. (7) to an electric field that satisfies the requisite boundary conditions over S and at infinity. It was pointed out earlier that the electric and magnetic fields [Eqs. (7) and (8)] satisfy Maxwell's equations. Solution of the integral equation (9) thus yields the unique solution of the problem.[1]

5·3. Reflection by an Infinite Plane Surface: the Principle of Images. The simplest obstacle problem is that of an infinite plane conductor. Here the solution can be obtained on the basis of geometrical considerations. Two cases will be discussed: (1) the initial field is a plane wave, and (2) the initial field arises from a dipole source.

Reflection of a Plane Wave.—Although the reflection of a plane wave by a plane surface has been treated frequently elsewhere, it will be of interest to treat the problem here in terms of the general ideas set forth in the preceding section.

Let us consider a plane wave, of the type discussed in Sec. 3·7, traveling in the direction defined by the unit vector $\mathbf{s}_0$. The initial field is then [Eq. (3·62)]

$$\mathbf{E}_i = \mathbf{E}_0 e^{j(\omega t - k \mathbf{s}_0 \cdot \mathbf{r})}. \tag{10}$$

An infinite plane conducting sheet is now introduced into the field. For convenience the conductor will be taken to lie in the xy-plane (Fig. 5·2). The unit vector $\mathbf{n}$, normal to the sheet, is taken to be in the positive

---

[1] For a discussion of the uniqueness theorem see J Stratton, *Electromagnetic Theory*, McGraw-Hill, New York, 1941, Chap. 9, Sec. 2.

SEC. 5·3] *REFLECTION BY AN INFINITE PLANE SURFACE* 133

z-direction, and the angle of incidence, which is the acute angle between the lines of direction of $\mathbf{s}_0$ and $\mathbf{n}$, is designated as θ.

The field set up by the current and charge distribution over the surface of the conductor must be such as to produce zero resultant field in the negative z-region. The scattered field in this region is therefore a

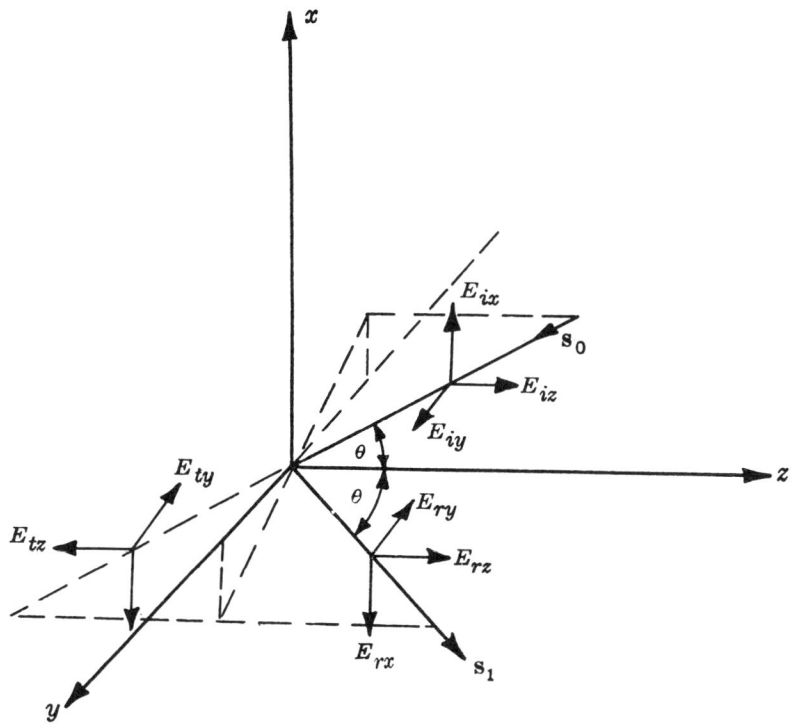

FIG. 5·2.—Reflection of a plane wave.

plane wave traveling in the same direction as $\mathbf{E}_i$ but 180° out of phase with it; denoting the former by $\mathbf{E}_t$, we have then

$$\mathbf{E}_t = -\mathbf{E}_0 e^{j(\omega t - k\mathbf{s}_0 \cdot \mathbf{r})}. \tag{11}$$

It is evident, however, that the infinite plane current sheet sets up in the positive z-region a field that is the mirror image of that in the negative z-region. Hence the scattered field in the region of interest is a plane wave

$$\mathbf{E}_r = \mathbf{E}_1 e^{j(\omega t - k\mathbf{s}_1 \cdot \mathbf{r})}, \tag{12}$$

traveling in the direction $\mathbf{s}_1$ which is the mirror image of $\mathbf{s}_0$, with an amplitude $\mathbf{E}_1$ bearing the following relations to the amplitude of $\mathbf{E}_t$ and thereby to the incident wave amplitude $\mathbf{E}_0$: (1) Their magnitudes are equal,

$$|E_1| = |E_0|; \tag{13}$$

(2) their respective components parallel to the xy-plane are equal in magnitude and direction

$$\mathbf{n} \times \mathbf{E}_1 = \mathbf{n} \times (-\mathbf{E}_0) = -\mathbf{n} \times \mathbf{E}_0; \tag{14}$$

(3) their components normal to the xy-plane are equal in magnitude but opposite in direction,

$$\mathbf{n} \cdot \mathbf{E}_1 = -\mathbf{n} \cdot (-\mathbf{E}_0) = \mathbf{n} \cdot \mathbf{E}_0. \tag{15}$$

It is seen that as a result of Eq. (14) the boundary conditions [Eqs. (5)] are satisfied.

It follows from the image relation between $\mathbf{s}_0$ and $\mathbf{s}_1$ that the vectors $\mathbf{s}_0$, $\mathbf{n}$, and $\mathbf{s}_1$ all lie in the same plane and that

$$\mathbf{s}_0 \cdot \mathbf{n} = -\mathbf{s}_1 \cdot \mathbf{n}. \tag{16}$$

The relations between these vectors can also be expressed as

$$\mathbf{s}_1 = \mathbf{s}_0 - 2(\mathbf{n} \cdot \mathbf{s}_0)\mathbf{n}, \tag{17a}$$
$$\mathbf{s}_0 = \mathbf{s}_1 - 2(\mathbf{n} \cdot \mathbf{s}_1)\mathbf{n}. \tag{17b}$$

From the point of view of geometrical optics the unit vectors $\mathbf{s}_0$ and $\mathbf{s}_1$ define the directions of the rays in the waves $\mathbf{E}_i$ and $\mathbf{E}_r$, respectively. It will be recognized that the relations among $\mathbf{s}_0$, $\mathbf{n}$, and $\mathbf{s}_1$ are just the laws of reflection derived in Chap. 4. It is thus seen that in this case the scattering reduces to geometrical reflection of the initial wave.

The magnetic-field vectors are obtained from the respective electric-field vectors by the plane wave relation of Eq. (3·65). Letting $\mathbf{H}_i$ and $\mathbf{H}_r$ be the magnetic vectors of the incident and reflected waves, respectively, we have

$$\mathbf{H}_i = \left(\frac{\epsilon}{\mu}\right)^{\frac{1}{2}} (\mathbf{s}_0 \times \mathbf{E}_i), \tag{18a}$$

$$\mathbf{H}_r = \left(\frac{\epsilon}{\mu}\right)^{\frac{1}{2}} (\mathbf{s}_1 \times \mathbf{E}_r). \tag{18b}$$

The total magnetic field is $\mathbf{H} = \mathbf{H}_i + \mathbf{H}_r$, whence by Eq. (2b) the surface current density on the reflector is

$$\mathbf{K} = \mathbf{n} \times (\mathbf{H}_i + \mathbf{H}_r). \tag{19}$$

Either by symmetry considerations or by direct calculation, it can be shown that

$$\mathbf{n} \times \mathbf{H}_i = \mathbf{n} \times \mathbf{H}_r; \tag{20}$$

consequently, Eq. (19) becomes

$$\mathbf{K} = 2(\mathbf{n} \times \mathbf{H}_i) = 2\left(\frac{\epsilon}{\mu}\right)^{\frac{1}{2}} [\mathbf{n} \times (\mathbf{s}_0 \times \mathbf{E}_i)], \tag{21}$$

or, alternatively,

$$\mathbf{K} = 2(\mathbf{n} \times \mathbf{H}_r) = 2\left(\frac{\epsilon}{\mu}\right)^{1/2} [\mathbf{n} \times (\mathbf{s}_1 \times \mathbf{E}_r)]. \tag{22}$$

In the case of a linearly polarized wave it is convenient for some purposes to express the field amplitudes in another way. Let $\mathcal{E}$ be the amplitude of $\mathbf{E}_i$ in magnitude and phase at any given point on the surface. The vector amplitude is

$$\mathbf{E}_0 = \mathcal{E} \mathbf{e}_0, \tag{23}$$

where $\mathbf{e}_0$ is a unit vector that is constant over the reflecting surface. Similarly the vector amplitude of $\mathbf{E}_r$ at the same given point on the surface is

$$\mathbf{E}_r = \mathcal{E} \mathbf{e}_1, \tag{24}$$

with $\mathbf{e}_1$ likewise a unit vector. The unit vectors $\mathbf{e}_0$ and $\mathbf{e}_1$ are related by Eqs. (14) and (15):

$$\mathbf{n} \times (\mathbf{e}_0 + \mathbf{e}_1) = 0, \tag{25a}$$
$$\mathbf{n} \cdot \mathbf{e}_0 = \mathbf{n} \cdot \mathbf{e}_1. \tag{25b}$$

In terms of these the current density expressions [Eqs. (21) and (22)] become

$$\mathbf{K} = 2\left(\frac{\epsilon}{\mu}\right)^{1/2} [\mathbf{s}_0(\mathbf{n} \cdot \mathbf{e}_0) - \mathbf{e}_0(\mathbf{n} \cdot \mathbf{s}_0)] \mathcal{E} \tag{26}$$

and

$$\mathbf{K} = 2\left(\frac{\epsilon}{\mu}\right)^{1/2} [\mathbf{s}_1(\mathbf{n} \cdot \mathbf{e}_1) - \mathbf{e}_1(\mathbf{n} \cdot \mathbf{s}_1)] \mathcal{E}, \tag{27}$$

respectively.

Dipole Sources.—Let us now consider the case where the initial field is due to an infinitesimal electric dipole. The infinite plane reflector will again be taken to be the xy-plane, and the dipole is located on the z-axis at a distance a from the reflector as shown in Fig. 5·3. The orientation of the dipole axis with respect to the reflector is arbitrary.

The current on the dipole is, of course, changed by the presence of the reflector. In this case, however, the reaction of the reflector merely produces a new dipole moment M in the source. This is due to the fact that the current induced in the source by the reflector is necessarily that of an infinitesimal dipole of, say, moment M_1. The latter is along the same line as the original dipole moment M_0, and the superposition of these two is, therefore, again a simple dipole. The resultant moment of the source will be designated by M; the field of the dipole is given in Sec. 3·13.

As in the case of the plane wave, the current distribution over the surface of the conductor must be such that the total field is zero in the hemisphere of space of the negative z-axis. In so far as this region is concerned, the reflector is, therefore, equivalent to a dipole $-M$ coincident with the source. By symmetry, however, the reflector produces a field in the region of the positive z-axis that is the mirror image of its field in the negative z-region; with respect to the positive z-region the reflector is equivalent to a dipole located at a distance a on the negative

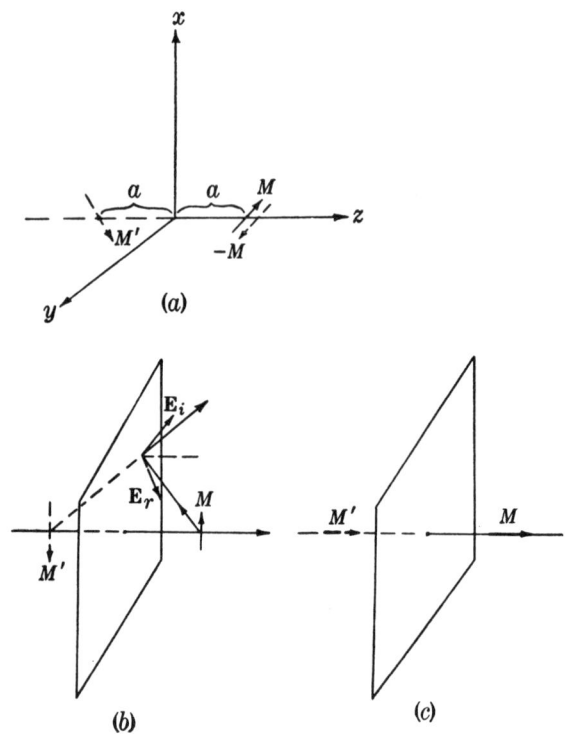

Fig. 5·3.—Dipole images: (a) arbitrary orientation; (b) dipole parallel to the reflector; (c) dipole normal to the reflector.

z-axis. The sense of the dipole with respect to the source is easily determined from the requirement that the fields of the image and the source must combine to give a zero resultant tangential electric field over the reflector. This leads at once to the result that the image dipole is obtained by reflection of $-M$ in the plane. The total field in the positive z-region is that of a double-dipole system made up of the source and the image dipole; the field is obtained by the methods discussed in Secs. 3·18 and 3·19.

The arbitrarily oriented dipole can always be resolved into a component parallel to the plane (Fig. 5·3b) and a component normal to the plane (Fig. 5·3c). The images for these two cases with respect to the source M are an antiphase dipole and a synphase dipole, respectively.

By considering the fields for these two cases, the reader can verify that the image sources correspond to geometrical reflection of the spherical wave from the source by the conducting plane; at each point on the latter the reflection takes place as though the incident wave were an infinite plane wave.

The image sources for magnetic dipoles are easily arrived at either by direct consideration of magnetic dipole fields (Sec. 3·14) or by considering the image of a small rectangular current loop, which is equivalent to a magnetic dipole normal to its plane. The image of a current loop can be obtained by regarding it as an array of electric dipoles. It is then found that the image of a magnetic dipole is obtained by direct reflection of the source in the plane: images for dipoles parallel and normal to the plane are synphase and antiphase, respectively.

The method of images can be applied to any source distribution. If only the radiation field is desired, the source distribution can be considered as a system of electric dipoles, the dipole moment distribution being given in terms of the current density **J** by

$$d\mathbf{M} = \frac{1}{j\omega} \mathbf{J} \, dv, \tag{28}$$

dv being an element of volume in the source distribution [*cf.* Eq. (3·162)]. Every dipole moment is resolved into a parallel and a normal component with respect to the reflector, and the total field is the sum of the component fields of the dipole elements and their images. With arbitrary current distributions, however, it must be kept in mind that the reflector plays an important part in determining the distribution. Only in special cases, such as a half-wave dipole radiator of negligible thickness, does the reaction of the reflector produce a change in the magnitude and phase of the amplitude of the distribution as a whole without affecting the relative magnitude and phase throughout the entire distribution. The half-wave dipole can be treated on the same basis as the infinitesimal dipole, substituting for the field of the latter the field of the half-wave radiator given in Sec. 3·16.

APPROXIMATE METHODS FOR REFLECTORS OF ARBITRARY SHAPE

Exact solutions of the scattering problem have been obtained for only a limited number of cases involving simple primary fields and reflectors of simple geometry, such as spheres and cylinders. These problems are treated in standard works on electromagnetic theory, to which the reader is referred for the results.[1] In treating reflectors of arbitrary shape it is necessary to resort to approximation techniques. Several such methods,

---

[1] See, for example, J. A. Stratton, *Electromagnetic Theory*, McGraw-Hill, New York, 1941, Chap. 9.

which yield very good results at high frequencies, are discussed in the following sections.

5.4. The Geometrical-optics Method.—The first method to be considered belongs more properly to the field of geometrical optics than to that of electromagnetic theory. It is applicable to the case of a point source, which has a broad radiation pattern in the absence of a reflector, together with a defocusing reflector. A reflector of this type renders every divergent pencil of rays incident on it more divergent on reflection, as is illustrated below in Fig. 5.4. The scattering pattern of the reflector is, therefore, very broad, energy being scattered in almost every direction in space. In such a system the salient features of the total field, such

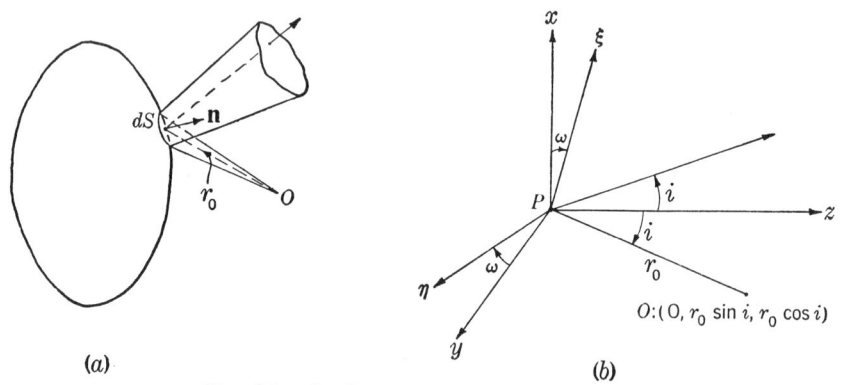

Fig. 5.4.—On the geometrical-optics method.

as the directions of zero and maximum amplitude, arise from the interaction between the scattered field and the primary source field. The finer details of the structure of the scattered field are of secondary interest, and therefore an analysis of the scattering on the basis of geometrical optics suffices.

Illustrative of the type of problem to which the method can be applied successfully is the analysis of the effects of the fuselage or wing structure of an airplane on the radiation pattern of a microwave beacon antenna mounted on it. The primary interest is in the lobe structure introduced into the beacon pattern by interaction with the scattered field from the aircraft structure, whereas the fine structure of the scattered field arising from deviations from geometrical optics is of negligible significance.

Let the primary source be located at the point O in Fig. 5.4a. The assumption that the source is a point radiator is justified in the practical case of a more general source system if the reflector is in the far-zone field of the former. It was shown in Sec. 3.11 that in so far as the far-zone field is concerned any current distribution reduces to a directive point source, and in Chap. 4 it was found that the far-zone field can be described adequately in terms of wavefronts and rays. We shall assume further that the wavefronts from the source differ negligibly from spheres about the point O.

The geometrical-optics analysis of the scattering assumes that at each point on the reflector the incident ray from the source is reflected by the tangent plane according to the laws of reflection developed in Sec. 4·7. The intensity of the scattered radiation in a given direction is obtained by applying the principle of conservation of energy to the total power contained in an incident cone of rays and the total power contained in the associated reflected pencil of rays. The use of the laws of reflection assumes that the reflector can be regarded locally as a plane surface and the incident wavefront can be regarded locally as a plane wave. It is, therefore, necessary to require that the radii of curvature of the reflector and of the incident wavefront be large compared with the wavelength. The latter condition, however, has already been assured by the fact that the reflector is in the far-zone field of the sources.

5·5. Calculation of the Scattered Field.—The procedure followed here[1] to determine the scattered power in a given direction is to consider the local transformation from the incident to the reflected wavefront at every point on the surface of the reflector. This determines the principal radii of curvature R_1 and R_2 of the reflected wavefront, together with the value of the field amplitude $\mathcal{E}_r$ at the point of reflection. The magnitude of the field amplitude $\mathcal{E}_p$ at a distance p along the reflected ray from a given point on the reflector is then obtained by means of Eq. (4·23):

$$|\mathcal{E}_p| = |\mathcal{E}_r| \left| \frac{R_1 R_2}{(R_1 + p)(R_2 + p)} \right|^{1/2}. \qquad (4·23)$$

We shall first investigate the amplitude transformation from the incident to the reflected wavefront. Let us consider an infinitesimal cone of rays from O incident on the reflector as shown in Fig. 5·4a; the cone intersects the reflector in an element of surface dS. The cone will be taken to have a circular cross section; the ray along the axis of symmetry is referred to as the central or principal ray. The vector **n** is a unit vector normal to dS at the point of incidence of the central ray; let i be the angle of incidence between the central ray and the normal. If $\mathcal{E}_i$ and $\mathcal{E}_r$ are the magnitudes at the surface of the reflector of the field amplitudes in the incident and reflected tubes of rays, respectively, and dS_1 and dS_2 are the cross-sectional areas of the respective tubes at the same point, the relation

$$\mathcal{E}_i^2 \, dS_1 = \mathcal{E}_r^2 \, dS_2 \qquad (29)$$

expresses the conservation of power in passing from the incident to the

---

[1] Alternative techniques have been developed by R. C. Spencer, "Reflections from Smooth Curved Surfaces," RL Report No. 661, January 1945; C. B. Barker and H. J. Riblet, "Reflections from Curved Surfaces," RL Report No. 976, February 1946.

reflected tubes of rays. From the law of reflection we have that the angle between the reflected principal ray and the normal is likewise i, so that
$$dS_1 = dS_2 = dS \cos i, \tag{30}$$
whence
$$|\mathcal{E}_i| = |\mathcal{E}_r|. \tag{31}$$

The transformation of the polarization on reflection is obtained directly from the results of the plane wave problem of Sec. 5·3. Let $\mathbf{E}_i$ be the incident electric-field vector at the surface and $\mathbf{E}_r$ the reflected-field vector; we have then from Eqs. (14) and (15)

$$\mathbf{n} \times (\mathbf{E}_r + \mathbf{E}_i) = 0, \tag{32a}$$
$$\mathbf{n} \cdot \mathbf{E}_r = \mathbf{n} \cdot \mathbf{E}_i, \tag{32b}$$
or,
$$\mathbf{E}_r = (\mathbf{n} \cdot \mathbf{E}_i)\mathbf{n} - (\mathbf{n} \times \mathbf{E}_i) \times \mathbf{n}. \tag{32c}$$

The determination of the radii of curvature of the reflected wavefront is a somewhat more difficult task. It will be necessary to make slight changes in notation: The point of incidence of the central ray on the reflector, at which the transformation of the wavefront is desired, will be designated by P, and the unit vector normal to the surface at that point by $\mathbf{n}_P$; the unit vector normal to the surface at any other point is $\mathbf{n}$ The point P is taken as the origin of the coordinate system (Fig. 5·4b) with the z-axis along $\mathbf{n}_P$ and the xy-plane tangent to the surface; the yz-plane is the plane of incidence (containing the central ray and $\mathbf{n}_P$). The axes ξ, η are the lines of intersection of the principal planes of curvature of the surface with xy-plane; the principal radii of curvature of the reflector at P will be designated by R_ξ and R_η, respectively. The plane of incidence makes an angle ω with one of the principal planes, say the plane containing the η-axis. Let r_0 be the distance OP; the distance from O to an arbitrary point x, y, z on the reflector is

$$r = [x^2 + y^2 + z^2 + r_0^2 - 2r_0(y \sin i + z \cos i)]^{1/2}. \tag{33}$$

Consider now the member of the family of reflected wavefronts that intersects the reflected central ray at a distance p from the reflector. Let u, v, w be the coordinates of a point on the wavefront, and let x, y, z be the coordinates of the point on the reflector for which the reflected ray passes through a given point (u,v,w) on the wavefront. By the law of the optical path (Sec. 4·8) the equation of the reflected wavefront is then

$$r + [(u - x)^2 + (v - y)^2 + (w - z)^2]^{1/2} = r_0 + p. \tag{34}$$

Now let $\mathbf{s}_0$ be a unit vector in the direction of an arbitrary incident ray and $\mathbf{s}$ a unit vector along the associated reflected ray. From the

law of reflection [Eq. (5·17a)] we have
$$\mathbf{s}_1 = \mathbf{s}_0 - 2(\mathbf{s}_0 \cdot \mathbf{n})\mathbf{n}. \tag{35}$$
If u,v,w are the coordinates of any point on the reflected ray, the unit vector $\mathbf{s}_1$ is given in component form by
$$s_{1x} = \frac{u - x}{[(u - x)^2 + (v - y)^2 + (w - z)^2]^{1/2}} \tag{36}$$
with corresponding expressions for s_{1y}, s_{1z}. Similarly, the components of $\mathbf{s}_0$ are
$$s_{0x} = \frac{x}{r}; \qquad s_{0y} = \frac{y - r_0 \sin i}{r}; \qquad s_{0z} = \frac{z - r_0 \cos i}{r}. \tag{37}$$
Let
$$z = F(x,y) \tag{38}$$
denote the equation of the reflector surface. The components of the normal $\mathbf{n}$ at an arbitrary point on the surface are then
$$n_x = -\frac{1}{\Delta}\frac{\partial F}{\partial x}; \qquad n_y = -\frac{1}{\Delta}\frac{\partial F}{\partial y}; \qquad n_z = \frac{1}{\Delta};$$
$$\Delta = \left[1 + \left(\frac{\partial F}{\partial x}\right)^2 + \left(\frac{\partial F}{\partial y}\right)^2\right]^{1/2} \tag{39}$$
Substitution of Eqs. (36), (37), and (39) into Eq. (35) gives
$$\left. \begin{array}{l} u = x + G_1(x,y)U(u,v,w;\, x,y,z), \\ v = y + G_2(x,y)U(u,v,w;\, c,y,z), \\ w = z + G_3(x,y)U(u,v,w;\, x,y,z), \\ U = [(u - x)^2 + (v - y)^2 + (w - z)^2]^{1/2}, \end{array} \right\} \tag{40}$$
where
$$\left. \begin{array}{l} G_1(x,y) = \dfrac{x}{r} + 2(\mathbf{s}_0 \cdot \mathbf{n})\dfrac{1}{\Delta}\dfrac{\partial F}{\partial x}, \\[4pt] G_2(x,y) = \dfrac{y - r_0 \sin i}{r} + 2(\mathbf{s}_0 \cdot \mathbf{n})\dfrac{1}{\Delta}\dfrac{\partial F}{\partial y}, \\[4pt] G_3(x,y) = \dfrac{z - r_0 \cos i}{r} - 2(\mathbf{s}_0 \cdot \mathbf{n})\dfrac{1}{\Delta}, \end{array} \right\} \tag{41}$$
and
$$\mathbf{s}_0 \cdot \mathbf{n} = r\frac{1}{\Delta}\left[-x\frac{\partial F}{\partial x} - (y - r_0 \sin i)\frac{\partial F}{\partial y} + (z - r_0 \cos i)\right]. \tag{42}$$

Equations (40) give the coordinates of arbitrary points on the system of reflected rays. If, in particular, we consider the family of points lying on the reflected wavefront that is defined by Eq. (34), the coordinates u, v, w of the system of Eqs. (40) must satisfy Eq. (34); in particular $U(u,v,w;\, x,y,z)$ must satisfy the latter equation. Substitution for U

into Eq. (40) then leads to

$$\left.\begin{array}{l}u = x + G_1(x,y)(r_0 - r + p), \\ v = y + G_2(x,y)(r_0 - r + p), \\ w = z + G_3(x,y)(r_0 - r + p).\end{array}\right\} \quad (43)$$

The coordinate z is eliminated from these equations by means of the equation $z = F(x,y)$ for the surface of the reflector. Equations (43) then become a set of parametric equations (x,y being the parameters) for the reflected wavefront that intersects the central reflected ray at a distance p from the reflector. On setting $p = 0$, we obtain

$$\left.\begin{array}{l}u = x + G_1(x,y)(r_0 - r), \\ v = y + G_2(x,y)(r_0 - r), \\ w = z + G_3(x,y)(r_0 - r),\end{array}\right\} \quad (44)$$

the parametric equations for a surface that intersects the reflector at the point P and there represents the reflected wavefront arising from the segment of the incident wavefront defined by a small cone of rays about the central ray.

The procedure for finding the principal radii of curvature of a surface from its parametric equations is straightforward and can be found in any book on differential geometry;[1] the details of the calculation will not be reproduced here. We are interested in the radii of curvature of the wave surface at the point $u = v = w = 0$; i.e., $x = y = 0$. In eliminating z from Eqs. (44) it is, therefore, necessary to use only the equation for the reflector surface in the neighborhood of the point P. Referred to the principal axes ξ, η, the equation of the surface of the reflector is

$$z = -\left(\frac{\xi^2}{2R_\xi} + \frac{\eta^2}{2R_\eta}\right) \cdots \quad (45)$$

By a simple transformation, the equation of the surface with respect to the x, y-axes is then found to be

$$z = \frac{1}{2}\left[\left(\frac{\cos^2 \omega}{R_\xi} + \frac{\sin^2 \omega}{R_\eta}\right)x^2 + \left(\frac{\sin^2 \omega}{R_\xi} + \frac{\cos^2 \omega}{R_\eta}\right)y^2 \right.$$
$$\left. - 2\sin\omega\cos\omega\left(\frac{1}{R_\xi} - \frac{1}{R_\eta}\right)xy\right] + \cdots ; \quad (46)$$

radii of curvature R_ξ, R_η are considered to be positive if the surface is convex with respect to the positive z-axis.

We are chiefly concerned with the over-all pattern produced by the reflector and the source system; hence we are interested in the scattered field at large distances from the reflector. Provided that neither one

---

[1] For example, L. P. Eisenhart, *A Treatise on the Differential Geometry of Curves and Surfaces*, Ginn, Boston, 1909.

SEC. 5·6]   SUPERPOSITION OF THE SOURCE FIELD   143

of the radii of curvature R_1, R_2 of the reflected wavefront is infinite, it is physically possible to consider distances p so large that $p \gg R_1$, $p \gg R_2$. In that case the calculation of the scattered field intensity is somewhat simplified; instead of Eq. (4·23) we can use Eq. (4.24),

$$|\mathcal{E}_P| \approx |\mathcal{E}_r| \frac{|R_1 R_2|^{1/2}}{p}.$$

The product of the radii of curvature of the reflected wave surface at the point P is

$$R_1 R_2 = \frac{\cos i}{\dfrac{4\cos i}{R_\xi R_\eta} + \dfrac{\cos i}{r_0^2} + \dfrac{2}{r_0}\left[\dfrac{\sin^2 \omega}{R_\xi} + \dfrac{\cos^2 \omega}{R_\eta} + \left(\dfrac{\cos^2 \omega}{R_\xi} + \dfrac{\sin^2 \omega}{R_\eta}\right)\cos^2 i\right]} \quad (47)$$

The result can be put into a more symmetrical form by introducing the angles θ_1 and θ_2 between the incident ray and the principal axes of the reflector ξ and η, respectively. The scattered field at a distance p from the reflector in the direction defined by the reflected ray is then given by

$$|\mathcal{E}_P| = \frac{r_0}{p}|\mathcal{E}_r|\left[\frac{R_\xi R_\eta \cos i}{(4r_0^2 + R_\xi R_\eta)\cos i + 2r_0(R_\xi \sin^2 \theta_1 + R_\eta \sin^2 \theta_2)}\right]^{1/2}. \quad (48)$$

The bracketed term is known as the divergence factor of the surface,

$$D = \frac{R_\xi R_\eta \cos i}{(4r_0^2 + R_\xi R_\eta)\cos i + 2r_0(R_\xi \sin^2 \theta_1 + R_\eta \sin^2 \theta_2)}; \quad (49)$$

it is the ratio of scattered power per unit solid angle in the direction of the reflected ray to the incident power per unit solid angle. By use of Eqs. (31) and (32c), together with Eq. (48), the scattered field can be obtained in magnitude, phase, and direction:

$$\mathbf{E}_P = \frac{r_0}{p}\{(\mathbf{n} \cdot \mathbf{E}_i)\mathbf{n} - (\mathbf{n} \times \mathbf{E}_i) \times \mathbf{n}\}D^{1/2}e^{-jkp}. \quad (50)$$

5·6. Superposition of the Source Field and the Scattered Field.—The method of superposing the scattered field on the original field of the sources is fundamentally the same as that used in Chap. 3 in treating the far-zone fields of current distributions. It will be assumed that the source field is linearly polarized. The fundamental elements are illustrated in Fig. 5·5. Let P_t be the total power radiated by the source, and let $G(\mathbf{s})$ be the gain function in the direction defined by the unit vector $\mathbf{s}$. The field of the source alone over a sphere of radius R is

$$\mathbf{E}_0 = \left[2\left(\frac{\mu}{\epsilon}\right)^{1/2}\frac{P_t}{4\pi}G(\mathbf{s})\right]^{1/2}\mathbf{e}_0\frac{e^{-jkR}}{R}. \quad (51)$$

The unit vector $\mathbf{e}_0$ describes the polarization of the field over a sphere with radius such that $kR = 2n\pi$, $n = 0, 1, 2, \ldots$. The total field in a given direction $\mathbf{s}$ is the sum of the scattered field produced in that direction by the reflector and the source field. In the far-zone treatment the ray from the source to the given field point is taken to be parallel to the ray from the reflector.

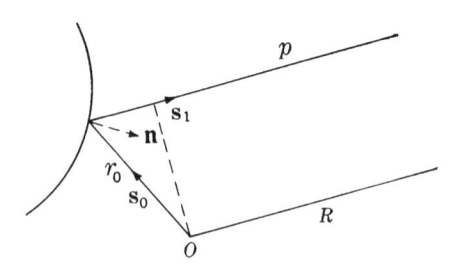

FIG. 5·5.—Superposition of the scattered field on the source field.

The field intensity incident on the reflector is

$$\mathbf{E}_i = \left[2\left(\frac{\mu}{\epsilon}\right)^{1/2}\frac{P_t}{4\pi}G(\mathbf{s}_0)\right]^{1/2} \mathbf{e}_0(\mathbf{s}_0)\frac{e^{-jkr_0}}{r_0}. \tag{52}$$

The scattered field is then, by Eq. (50),

$$\mathbf{E}_P = \frac{1}{R}\left[2\left(\frac{\mu}{\epsilon}\right)^{1/2}\frac{DP_t}{4\pi}G(\mathbf{s}_0)\right]^{1/2} [(\mathbf{n}\cdot\mathbf{e}_0)\mathbf{n} - (\mathbf{n}\times\mathbf{e}_0)\times\mathbf{n}]e^{-jk(r_0+p)}; \tag{53}$$

the distance p has been set equal to R in the expression for the amplitude. In so far as the phase is concerned, it is seen from the figure that

$$R\mathbf{s}_1 = r_0\mathbf{s}_0 + p\mathbf{s}_1, \tag{54}$$

whence

$$r_0 + p = R + r_0(1 + \cos 2i). \tag{55}$$

The total field in the direction $\mathbf{s}_1$ is, therefore,

$$\mathbf{E}(\mathbf{s}_1) = \left[2\left(\frac{\mu}{\epsilon}\right)^{1/2}\frac{P_t}{4\pi}\right]^{1/2}\frac{e^{-jkR}}{R}\{[G(\mathbf{s}_1)]^{1/2}\mathbf{e}_0(\mathbf{s}_1) + [DG(\mathbf{s}_0)]^{1/2}\mathbf{e}_r e^{-jkr_0(1+\cos 2i)}\}, \tag{56}$$

where

$$\mathbf{e}_r = [\mathbf{n}\cdot\mathbf{e}_0(\mathbf{s}_0)]\mathbf{n} - [\mathbf{n}\times\mathbf{e}_0(\mathbf{s}_0)]\times\mathbf{n}. \tag{56a}$$

5·7. The Current-distribution Method.—The geometrical-optics method discussed in the preceding sections can furnish no information on the structure of the scattered field that results from deviations from geometrical propagation of the reflected wavefront. By geometrical optics this wave is discontinuous (geometrical shadow behind the reflector), and it was pointed out in Sec. 4·5 that in the presence of a discontinuity geometrical optics does not give accurate results. The deviations decrease in significance as the wavelength goes to zero; the geometrical-optics method is to be regarded as a zero-wavelength approximation to the scattered field.

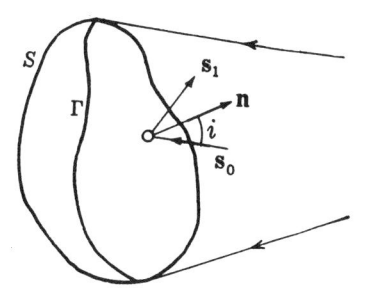

The current-distribution method which will be formulated in this section leads to a better approximation for the scattered field and also makes possible the analysis of secondary effects such as the reaction of the reflector on the sources. The cardinal feature of the method is that it attempts to approximate the current distribution over the surface of the reflector; the scattered field is obtained from the current distribution by Eqs. (7) and (8) and is thus an electromagnetic field that satisfies Maxwell's equations. We shall be interested primarily in the far-zone field of the current distribution in obtaining the composite pattern of the reflector and the sources.

The current distribution over the reflector is obtained on the basis of geometrical optics, which can be expected to yield good results only if the reflector is far enough from the sources for the field of these to be described adequately in terms of wavefronts and rays. On the basis of ray optics there is a sharply defined shadow region behind the reflector in which the total field is equal to zero.

FIG. 5·6.—On the current-distribution method.

In Fig. 5·6, S is the reflecting surface and Γ is the boundary curve between the geometrically illuminated area and the shadow area. According to the boundary condition [Eq. (2b)], since the *total field* is zero, the current distribution over the shadow area is zero. It is a matter of experience that the shadow region is more sharply defined the smaller the wavelength and the larger the ratio of the reflector dimensions to the wavelength. The first assumption of our approximation technique, then, is that there is no current over the shadow area of the reflector. The current distribution over the illuminated region of S is obtained on the assumption that at every point the incident field is reflected as though an infinite plane wave were incident on the infinite tangent plane. Let $\mathbf{E}_i$, $\mathbf{H}_i$ again be the initial field; let $\mathbf{s}_0$ be a unit vector in the direction of the Poynting vector, that is, along the incident ray. If $\mathbf{n}$ is the unit vector normal to the surface at the point of incidence and $\mathbf{s}_1$ a unit vector in the direction of the reflected ray, the surface current density, according to Eqs. (21) and (22), is

$$\mathbf{K} = 2(\mathbf{n} \times \mathbf{H}_i) = 2\left(\frac{\epsilon}{\mu}\right)^{\frac{1}{2}} [\mathbf{n} \times (\mathbf{s}_0 \times \mathbf{E}_i)], \qquad (57a)$$

or, in terms of the reflected field $\mathbf{E}_r$, $\mathbf{H}_r$ at the surface,

$$\mathbf{K} = 2(\mathbf{n} \times \mathbf{H}_r) = 2\left(\frac{\epsilon}{\mu}\right)^{1/2}[\mathbf{n} \times (\mathbf{s}_1 \times \mathbf{E})]. \tag{57b}$$

The surface charge density is obtained from the total field $\mathbf{E} \cdot \mathbf{E}_r$ by means of Eq. (2a); making use of the plane wave relations [Eq. (15)], we then find that the charge density is

$$\eta = 2\epsilon(\mathbf{n} \cdot \mathbf{E}_i) = 2\epsilon(\mathbf{n} \cdot \mathbf{E}_r). \tag{58}$$

From the discussion of Sec. 5·4 it is seen that the procedure for obtaining the current and charge distributions is based on the assumption that the radii of curvature of the incident wavefront are large compared with the wavelength as are also the radii of curvature of the reflector. On the other hand, in the present case there are no conditions imposed on the focusing or defocusing characteristics of the reflector. It is clear that Eqs. (57) and (58) represent high-frequency approximations to the actual currents and charges and may be expected to approach the latter in the limit of zero wavelength. The current method differs from the previous wavefront procedure in that a frequency dependence of the scattered field is introduced into the subsequent calculation of the field arising from the current and charge distributions. Also, the field at a given point in space is the resultant of contributions from all points on the illuminated area S_0 rather than from the point of geometrical reflection alone.

5·8. Calculation of the Scattered Field.—The expressions for the electric and magnetic fields in terms of the currents and charges were derived in Secs. 3·9 and 3·10. It will be recalled that the fields thus obtained satisfy Maxwell's equations only if the source-density functions satisfy the equation of continuity [Eqs. (3·6) and (3·9)]. The reader can readily verify that if the initial field $\mathbf{E}_i$, $\mathbf{H}_i$ satisfies Maxwell's equations, the current and charge distributions given by Eqs. (57a) and (58) do, in fact, satisfy the surface equation of continuity given in integral form by Eq. (3·9). The situation is different, however, at the boundary line Γ between the illuminated and shadow regions. The current distribution is discontinuous across the boundary, being zero over the shadow area; compatibility with the equation of continuity can be achieved only by introducing a line distribution of charge along the curve Γ.[1]

In Fig. 5·7 τ is a unit vector along the boundary curve Γ; $\mathbf{n}_1$ is a unit vector in the tangent plane normal to τ. The linear charge density along Γ will be denoted by σ. Considering a small area of sides ds and δl (the latter normal to Γ) and expressing the condition that the net current

---

[1] The discussion that follows parallels that given by Stratton and Chu in their treatment of diffraction; see J. A. Stratton, *Electromagnetic Theory*, McGraw-Hill, New York, 1941, Sec. 8·15.

flow from the area is equal to the rate of decrease of the charge enclosed, we obtain

$$-\mathbf{n}_1 \cdot \mathbf{K}\, ds = -\frac{\partial \sigma}{\partial t}\, ds; \tag{59}$$

the contributions from the sides δl vanish as $\delta l \to 0$ while ds remains fixed. We have then

$$\mathbf{n}_1 \cdot \mathbf{K} = \frac{\partial \sigma}{\partial t}. \tag{59a}$$

Substituting from Eqs. (57a) and (57b) for the current distribution, we find

$$\frac{\partial \sigma}{\partial t} = 2\mathbf{n}_1 \cdot (\mathbf{n} \times \mathbf{H}_i) = -2\boldsymbol{\tau} \cdot \mathbf{H}_i \tag{60a}$$

and

$$\frac{\partial \sigma}{\partial t} = -2\boldsymbol{\tau} \cdot \mathbf{H}_r. \tag{60b}$$

For time periodic fields these give the charge distribution directly; for we have $\partial \sigma/\partial t = j\omega\sigma$, whence

$$\sigma = -\frac{2}{j\omega}\boldsymbol{\tau} \cdot \mathbf{H}_i = -\frac{2}{j\omega}\boldsymbol{\tau} \cdot \mathbf{H}_r. \tag{61}$$

The scattered field is thus the sum of the contributions of three source distributions: (1) the surface currents over the illuminated area, (2) the surface charges over the same area, and (3) the line distribution of charge along the boundary curve Γ. We shall now restrict our analysis to time-periodic fields. Applying the results of Secs. 3·9 and 3·10, we find the scattered field to be

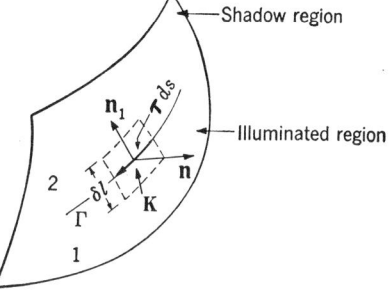

Fig. 5·7.—Calculation of the electric charge on the shadow boundary curve Γ.

$$\mathbf{E}_s = -\frac{1}{2\pi}\int_{S_0} [j\omega\mu(\mathbf{n} \times \mathbf{H}_i)\psi - (\mathbf{n} \cdot \mathbf{E}_i)\nabla\psi]\, dS$$

$$-\frac{1}{2\pi j\omega\epsilon}\oint_\Gamma \nabla\psi(\boldsymbol{\tau} \cdot \mathbf{H}_i)\, ds, \tag{62a}$$

$$\mathbf{H}_s = \frac{1}{2\pi}\int_{S_0} (\mathbf{n} \times \mathbf{H}_i) \times \nabla\psi\, dS, \tag{62b}$$

where $\psi = e^{\frac{jkr}{r}}$, with r the distance from the field point to the element of area dS on the reflector; S_0 designates the geometrically illuminated area; the sense of the line integral around Γ is such that the outward normal to S_0 is on the left. The fields can also be expressed in the same way in terms of the reflected fields $\mathbf{E}_r$, $\mathbf{H}_r$ at the surface S_0.

It was shown in Sec. 3·9 that if the current and charge distributions satisfy the equation of continuity, the fields can be expressed in terms of integrals involving the currents alone. In view of the introduction of the boundary line distribution it will be well to carry through the details of the transformation for this special case. It will be recalled (*cf.* Sec. 3·8) that the gradient operations in the integrands of Eqs. (62a) and (62b) are referred to the field point as an origin. Taking a rectangular system of coordinates with the origin at the field point, let the coordinates of a point on S_0 be x_α ($x_1 = x$, $x_2 = y$, $x_3 = z$), and let $\mathbf{i}_\alpha$ be unit vectors along the x_α-axes. The line integral of Eq. (62a) is then

$$\oint_\Gamma \nabla\psi(\boldsymbol{\tau} \cdot \mathbf{H}_i)\, ds = \sum_{\alpha=1}^{3} \mathbf{i}_\alpha \oint_\Gamma \boldsymbol{\tau} \cdot \left(\mathbf{H}_i \frac{\partial \psi}{\partial x_\alpha}\right) ds. \tag{63}$$

By Stokes' theorem each integral on the right-hand side transforms into a surface integral:

$$\oint_\Gamma \boldsymbol{\tau} \cdot \left(\mathbf{H}_i \frac{\partial \psi}{\partial x_\alpha}\right) ds = \int_{S_0} \left(\nabla \times \mathbf{H}_i \frac{\partial \psi}{\partial x_\alpha}\right) \cdot \mathbf{n}\, dS. \tag{64}$$

But

$$\nabla \times \left(\mathbf{H}_i \frac{\partial \psi}{\partial x_\alpha}\right) = \nabla \frac{\partial \psi}{\partial x_\alpha} \times \mathbf{H}_i + \frac{\partial \psi}{\partial x_\alpha} \nabla \times \mathbf{H}_i, \tag{65}$$

and

$$\left(\nabla \frac{\partial \psi}{\partial x} \times \mathbf{H}_i\right) \cdot \mathbf{n} = -(\mathbf{n} \times \mathbf{H}_i) \cdot \nabla \frac{\partial \psi}{\partial x_\alpha}, \tag{66}$$

$$\left(\frac{\partial \psi}{\partial x_\alpha} \nabla \times \mathbf{H}_i\right) \cdot \mathbf{n} = j\omega\epsilon \frac{\partial \psi}{\partial x_\alpha} (\mathbf{n} \cdot \mathbf{E}_i). \tag{67}$$

In the last of these use has been made of the field equation [Eq. (3·23b)]. Collecting these together, we have

$$\frac{1}{2\pi j\omega\epsilon} \oint_\Gamma \nabla\psi(\boldsymbol{\tau} \cdot \mathbf{H}_i)\, ds = \frac{1}{2\pi} \sum_\alpha \mathbf{i}_\alpha \int_{S_0} \left[\mathbf{n} \cdot \mathbf{E}_i \frac{\partial \psi}{\partial x_\alpha} \right.$$

$$\left. - \frac{1}{j\omega\epsilon}(\mathbf{n} \times \mathbf{H}_i) \cdot \nabla \frac{\partial \psi}{\partial x_\alpha}\right] dS,$$

or

$$\frac{1}{2\pi j\omega\epsilon} \oint_\Gamma \nabla\psi(\boldsymbol{\tau} \cdot \mathbf{H}_i)\, ds = \frac{1}{2\pi} \int_{S_0} \left[(\mathbf{n} \cdot \mathbf{E}_i) \nabla\psi \right.$$

$$\left. - \frac{1}{j\omega\epsilon}(\mathbf{n} \times \mathbf{H}_i) \cdot \nabla\nabla\psi\right] dS. \tag{68}$$

Substituting into Eq. (62a), we then obtain

$$\mathbf{E}_s = \frac{1}{2\pi j\omega\epsilon} \int_{S_0} \left[(\mathbf{n} \times \mathbf{H}_i) \cdot \nabla(\nabla\psi) + k^2(\mathbf{n} \times \mathbf{H}_i)\psi\right] dS. \tag{69}$$

SEC. 5·9] APPLICATION TO POINT-SOURCE FEEDS 149

It will be recognized that this is obtainable directly from Eq. (5·7) by inserting the value for the surface current density given by Eq. (57a). It was shown in Sec. 3·11 that the field integral, taking the form of Eq. (69), leads to a far-zone field in which the field vectors are transverse to the direction of propagation. The effect of the boundary line distribution is therefore to cancel the longitudinal field component introduced by the surface charge and current distributions. Subsequent calculations can be made on the basis of Eq. (69); the contributions of the charge distributions need not be evaluated explicitly.

Let ϱ be the vector from a given origin (see Fig. 5·8) to the element of surface dS; let $\mathbf{R}_1$ be a unit vector from the origin to the field point, the distance between them being R. The scattered field intensity in the far-zone is then, according to Eq. (3·128),

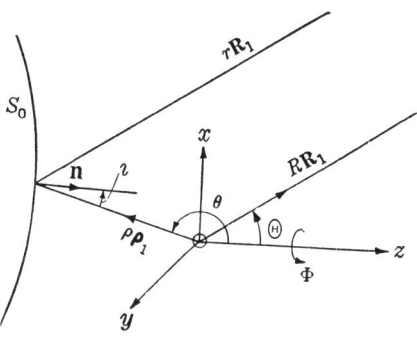

FIG. 5·8.—Calculation of the total field.

$$\mathbf{E}_S = -\frac{j\omega\mu}{2\pi R} e^{-jkR} \int_{S_0} \{\mathbf{n} \times \mathbf{H}_i - [(\mathbf{n} \times \mathbf{H}_i) \cdot \mathbf{R}_1]\mathbf{R}_1\} e^{jk\varrho \cdot \mathbf{R}_1} \, dS. \quad (70)$$

The magnetic field need not be calculated separately but is given by the far-zone relation

$$\mathbf{H}_S = \left(\frac{\epsilon}{\mu}\right)^{\frac{1}{2}} (\mathbf{R}_1 \times \mathbf{E}_S). \quad (71)$$

5·9. Application to Point-source and Line-source Feeds.—Two cases of major interest are those in which the initial field $\mathbf{E}_i$, $\mathbf{H}_i$ arises from a point-source system and a line-source system, respectively. Where the reflector enters into the problem by intention as a component of the antenna, the source system will be referred to as the *feed;* this term is used extensively in later chapters.

The Point-source Feed.—It was noted previously that at sufficiently large distances from any radiating system, the latter is equivalent to a directive point source. Microwave point-source feeds are specially designed so that the required distances are within practical ranges for use with a reflector.

Let the point O in Fig. 5·8 be the point-source equivalent of the feed; it will be assumed again that within the cone of illumination falling on the reflector the incident wavefronts differ negligibly from spheres about the point O. The reference system of coordinates will be taken with the origin at the source system. Spherical coordinates will be designated generally by ρ, θ, ϕ; the coordinates of a field point in the far-zone region of the system as a whole—reflector and feed—will be R, Θ, Φ. If $G_f(\theta,\phi)$ is the gain function of the feed and P is the total radiated power, the primary radiation field—of the feed alone—is

$$\mathbf{E}_i(\rho,\theta,\phi) = \frac{1}{\rho}\left[2\left(\frac{\mu}{\epsilon}\right)^{\frac{1}{2}}\frac{P}{4\pi}G_f(\theta,\phi)\right]^{\frac{1}{2}}\mathbf{e}_i(\theta,\phi)e^{-jk\rho}, \quad (72a)$$

$$\mathbf{H}_i = \left(\frac{\epsilon}{\mu}\right)^{\frac{1}{2}}(\boldsymbol{\varrho}_1 \times \mathbf{E}_i), \quad (72b)$$

where $\boldsymbol{\varrho}_1$ is a unit vector along ρ and $\mathbf{e}_i(\theta,\phi)$ is a unit vector defining the polarization of the electric field intensity. The current density $\mathbf{K}$ at a point ρ, θ, ϕ on the reflector is then

$$\mathbf{K} = 2(\mathbf{n} \times \mathbf{H}_i) = \frac{2}{\rho}\left[\left(\frac{\epsilon}{\mu}\right)^{\frac{1}{2}}\frac{P}{2\pi}G_f(\theta,\phi)\right]^{\frac{1}{2}}[\mathbf{n} \times (\boldsymbol{\varrho}_1 \times \mathbf{e}_i)]e^{-jk\rho}. \quad (73)$$

Substitution of the expression for $\mathbf{n} \times \mathbf{H}_i$ from Eq. (73) into Eq. (70) gives the scattered field in the far-zone. Equation (70) shows explicitly that there is no field component in the $\mathbf{R}_1$-direction. Let $\mathbf{i}_\Theta$ and $\mathbf{i}_\Phi$ be unit vectors in the direction of increasing Θ and Φ, respectively. The transverse components of the scattered field are then

$$\mathbf{E}_{s_\Theta} = -\frac{j\omega\mu}{2\pi}\frac{e^{-jkR}}{R}\left[\left(\frac{\epsilon}{\mu}\right)^{\frac{1}{2}}\frac{P}{2\pi}\right]^{\frac{1}{2}}\mathbf{i}_\Theta \cdot \mathbf{I}, \quad (74a)$$

$$\mathbf{E}_{s_\Phi} = -\frac{j\omega\mu}{2\pi}\frac{e^{-jkR}}{R}\left[\left(\frac{\epsilon}{\mu}\right)^{\frac{1}{2}}\frac{P}{2\pi}\right]^{\frac{1}{2}}\mathbf{i}_\Phi \cdot \mathbf{I}, \quad (74b)$$

$$\mathbf{I} = \int_{S_0}\frac{[G_f(\theta,\phi)]^{\frac{1}{2}}}{\rho}[\mathbf{n} \times (\boldsymbol{\varrho}_1 \times \mathbf{e}_i)]e^{-jk(\rho-\boldsymbol{\varrho}\cdot\mathbf{R}_1)}\,dS; \quad (74c)$$

the vector $\boldsymbol{\varrho} = \rho\boldsymbol{\varrho}_1$ is the radius vector from O to the element of surface dS. The total field at the point R, Θ, Φ is

$$\left.\begin{aligned}\mathbf{E}_\Theta &= \mathbf{E}_{i_\Theta} + \mathbf{E}_{s_\Theta} = \frac{e^{-jkR}}{R}\left[\frac{P}{2\pi}\left(\frac{\mu}{\epsilon}\right)^{\frac{1}{2}}\right]^{\frac{1}{2}}F_1(\Theta,\Phi),\\ F_1(\Theta,\Phi) &= \left\{[G_f(\Theta,\Phi)]^{\frac{1}{2}}\mathbf{i}_\Theta \cdot \mathbf{e}_i(\Theta,\Phi) - \frac{j}{\lambda}\mathbf{i}_\Theta \cdot \mathbf{I}\right\};\end{aligned}\right\} \quad (75)$$

$$\left.\begin{aligned}\mathbf{E}_\Phi &= \mathbf{E}_{i_\Phi} + \mathbf{E}_{s_\Phi} = \frac{e^{-jkR}}{R}\left[\frac{P}{2\pi}\left(\frac{\mu}{\epsilon}\right)^{\frac{1}{2}}\right]^{\frac{1}{2}}F_2(\Theta,\Phi),\\ F_2(\Theta,\Phi) &= \left\{[G_f(\Theta,\Phi)]^{\frac{1}{2}}\mathbf{i}_\Phi \cdot \mathbf{e}_i - \frac{j}{\lambda}\mathbf{i}_\Phi \cdot \mathbf{I}\right\}.\end{aligned}\right\} \quad (76)$$

The magnetic field is obtained by means of Eq. (71). The Poynting vector of the total field is $\mathbf{S} = \frac{1}{2}\operatorname{Re}(\mathbf{E} \times \mathbf{H}^*)$, and the power per unit solid angle $P(\Theta,\Phi)$, radiated by the system as a whole in the direction

(Θ,Φ), is $R^2|S|$; hence, the gain function of the composite system is

$$G(\Theta,\Phi) = \frac{4\pi P(\Theta,\Phi)}{P} = \frac{2\pi \left(\frac{\epsilon}{\mu}\right)^{\frac{1}{2}}\{|E_\Theta|^2 + |E_\Phi|^2\}R^2}{P} \quad (77a)$$

or

$$G(\Theta,\Phi) = |F_1|^2 + |F_2|^2. \quad (77b)$$

The Line-source Feed.—Line-source feeds are generally used with a cylindrical reflector, the generating element of which is parallel to the line source. The following analysis will be confined to such systems.

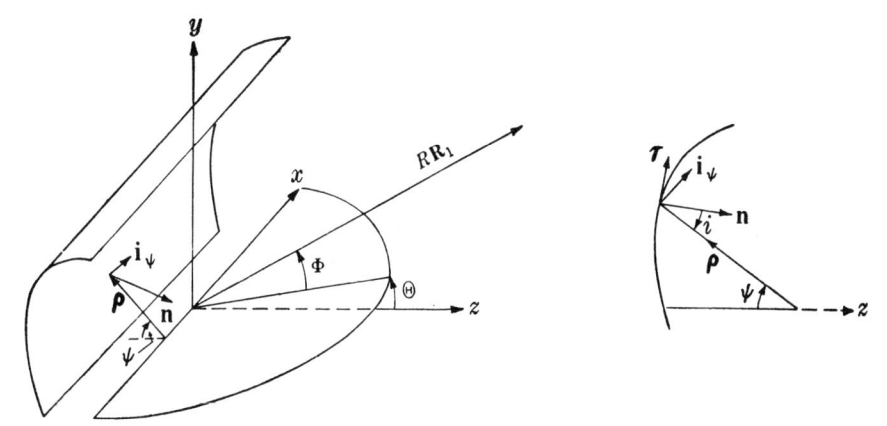

Fig. 5·9.—The cylindrical reflector with a line-source feed.

The line source may be a system of point-source radiators distributed along a line, such as the linear-array antennas discussed in Chap. 9, or it may take the form of a long, narrow, rectangular aperture through which energy is being radiated into space. It will be assumed that the length l of the source is large compared with the wavelength.

The reflector and source system are illustrated in Fig. 5·9, with the source along the x-axis. We shall assume that the maximum distance from the latter to the reflector does not exceed l^2/λ and that the minimum distance is large compared with the wavelength. Within such distances from the source its field is essentially in the form of a cylindrical wave (*cf.* Sec. 3·7). The wave incident on the reflector is, therefore, best discussed in terms of cylindrical coordinates. The x-axis in Fig. 5·9 serves as the axis for the cylindrical coordinate system, the polar coordinates of which, in the planes normal to the x-axis, are denoted by ρ and ψ.

The cylindrical-wave zone can be divided into two general regions: (1) a near-zone region in the immediate vicinity of the source and (2) a quasi-radiation zone at distances large compared with the wavelength but less than l^2/λ. In the latter region the predominant components of the field lie in the tangent plane of the cylindrical wavefront and are mutually perpendicular as in the case of the isotropic cylindrical wave discussed in Sec. 3·7. With a general line source, the field intensity is not uniform over the wavefront but varies both along the x-direction and about the cylinder axis. The radiation-zone field can be written

$$\mathbf{E}(\rho,\psi,x) = \mathbf{A}(\psi,x)\frac{e^{-jk\rho}}{\rho^{1/2}} \tag{78a}$$

and

$$\mathbf{H} = \left(\frac{\epsilon}{\mu}\right)^{1/2}(\boldsymbol{\varrho}_1 \times \mathbf{E}), \tag{78b}$$

where $\boldsymbol{\varrho}_1$ is a unit vector in the direction of increasing ρ. The radiation zone of the cylindrical wave field is to be distinguished from the general far-zone field with respect to which the line source behaves like a point source.

We shall confine our attention to fields in which the polarization is uniform over the wavefront. Two fundamental cases are to be considered: (1) longitudinal polarization in which the E-vector is parallel to the x-axis, so that

$$\mathbf{A}(\psi,x) = A(\psi,x)\mathbf{i}_x, \tag{79a}$$

and (2) transverse polarization in which the electric vector lies in the planes transverse to the x-axis, that is,

$$\mathbf{A}(\psi,x) = A(\psi,x)\mathbf{i}_\psi; \tag{79b}$$

the vectors $\mathbf{i}_x$ and $\mathbf{i}_\psi$ are the basis vectors of the cylindrical coordinate system. In most cases of interest the amplitude function $A(\psi,x)$ is separable in its dependence on the two coordinates. Referring to the power flow rather than the amplitude, we shall introduce a two-dimensional gain function $G(\psi)$. Let P be the total power radiated by the source; let $(P/l)F(x)\,dx$ be the total power in the cylindrical wave field between the planes $x = $ constant and $x + dx = $ constant. The power radiated per radian between these planes in the direction ψ is then

$$P(x,\psi)\,dx\,d\psi = \frac{1}{2\pi}\frac{P}{l}F(x)\,dx\,G(\psi)\,d\psi. \tag{80}$$

The gain function $G(\psi)$ must obviously satisfy the condition

$$\int_{-\pi}^{\pi} G(\psi)\,d\psi = 2\pi. \tag{81}$$

The function $F(x)$ expresses the distribution of intensity along the y-direction; it must satisfy the condition

$$\int_{-l/2}^{l/2} F(x)\,dx = l, \tag{82}$$

it being assumed that the origin of the coordinate system is at the center of the line source. The amplitude $A(x,\psi)$ may be evaluated in terms of the power-distribution functions as in the case of the point-source feed. The magnitude of the Poynting vector is

$$|S| = \frac{1}{2}|\text{Re }(\mathbf{E} \times \mathbf{H}^*)| = \frac{1}{2}\left(\frac{\epsilon}{\mu}\right)^{1/2}\frac{|A|^2}{\rho}.$$

We have then

$$\frac{1}{2}\left(\frac{\epsilon}{\mu}\right)^{1/2}|A|^2\,d\psi\,dx = \frac{1}{2\pi}\frac{P}{l}F(x)\,dx\,G(\psi)\,d\psi \qquad (83a)$$

or

$$|A(x,\psi)| = \left[\left(\frac{\mu}{\epsilon}\right)^{1/2}\frac{P}{l\pi}F(x)\,G(\psi)\right]^{1/2}. \qquad (83b)$$

The current densities, or rather $\mathbf{n} \times \mathbf{H}_i$, for the two types of polarization are the following:

1. Longitudinal polarization:

$$\mathbf{n} \times \mathbf{H}_i = \left[\left(\frac{\epsilon}{\mu}\right)^{1/2}\frac{P}{l\pi}F(x)\frac{G(\psi)}{\rho}\right]^{1/2}\mathbf{n} \times (\boldsymbol{\varrho}_1 \times \mathbf{i}_x)e^{-jk\rho}, \qquad (84a)$$

or

$$\mathbf{n} \times \mathbf{H}_i = \mathbf{i}_x\left[\left(\frac{\epsilon}{\mu}\right)^{1/2}\frac{P}{l\pi}F(x)\frac{G(\psi)}{\rho}\right]^{1/2}\cos i\,e^{-jk\rho}. \qquad (84b)$$

2. Transverse polarization:

$$\mathbf{n} \times \mathbf{H}_i = \left[\left(\frac{\epsilon}{\mu}\right)^{1/2}\frac{P}{l\pi}F(x)\frac{G(\psi)}{\rho}\right]^{1/2}\mathbf{n} \times (\boldsymbol{\varrho}_1 \times \mathbf{i}_\psi)e^{-jk\rho}, \qquad (85a)$$

or

$$\mathbf{n} \times \mathbf{H}_i = \left[\left(\frac{\epsilon}{\mu}\right)^{1/2}\frac{P}{l\pi}F(x)\frac{G(\psi)}{\rho}\right]^{1/2}\boldsymbol{\tau}e^{-jk\rho}. \qquad (85b)$$

The angle i is the angle of incidence, and $\boldsymbol{\tau}$ is a unit vector tangent to the cylinder in the cross-section plane. The positive directions of the angles and vectors are shown in Fig. 5·9.

The far-zone field is expressed in terms of spherical coordinates. Because of the geometry of the system it is convenient to use a set of spherical coordinates somewhat different from that used in the treatment of the point-source field, the x-axis being taken as the polar axis; the definition of the coordinates is given in Fig. 5·9. Let $\mathbf{i}_\Theta$ and $\mathbf{i}_\Phi$ again be unit vectors in the increasing Θ- and Φ-directions. For the case of the *longitudinally polarized* source, the scattered field is

$$E_{s_\Phi} = -\frac{j\omega\mu}{2\pi R}e^{-jkR}\left[\left(\frac{\epsilon}{\mu}\right)^{1/2}\frac{P}{l\pi}\right]^{1/2}\cos\Phi$$
$$\times \int_{S_0}\left[F(x)\frac{G(\psi)}{\rho}\right]^{1/2}\cos i\,e^{-jk[\rho-(\boldsymbol{\varrho}+x\mathbf{i}_x)\cdot\mathbf{R}_1]}\,ds\,dx]; \quad (86)$$

ds is the element of arc length along the cross section, or

$$ds = \left[1 + \left(\frac{1}{\rho}\frac{d\rho}{d\psi}\right)^2\right]^{1/2} \rho\, d\psi. \qquad (87)$$

The scattered field has no Θ-component, the electric vector lying entirely in the meridional plane passing through the x-axis. The scattered field for the *transversely polarized* source is

$$\mathbf{E}_{s_\Theta} = \frac{j\omega\mu}{2\pi R} e^{-jkR} \left[\left(\frac{\epsilon}{\mu}\right)^{1/2} \frac{P}{l\pi}\right]^{1/2} \int_{S_0} \left[F(x)\frac{G(\psi)}{\rho}\right]^{1/2} (\mathbf{i}_\Theta \cdot \boldsymbol{\tau}) \\ \times e^{-jk\{\rho - (\boldsymbol{\varrho}+x\mathbf{i}_x)\cdot\mathbf{R}_1\}}\, ds\, dx. \qquad (88)$$

In this case there is also a Φ-component proportional to sin Φ in magnitude. However, if $l \gg \lambda$, the beam is confined to the neighborhood of $\Phi = 0$ and the cross-polarization component is small.

On expressing the vectors $\boldsymbol{\varrho}$ and $\mathbf{R}_1$ in rectangular components, one finds the phase factor of the integrands of Eqs. (86) and (88) to be

$$\rho - (\boldsymbol{\varrho} + x\mathbf{i}_x)\cdot\mathbf{R}_1 = \rho[1 + \cos\Phi\cos(\psi + \Theta)] - x\sin\Phi. \qquad (89)$$

It is apparent that the integrands are factorable into a function of x and a function of ψ. Considering the central plane $\Phi = 0$, we see that the Θ-dependence of the field arises entirely from the integral over ψ; the field distribution is determined by the angular characteristic $G(\psi)$ of the source and the cross-section contour of the reflector. As regards the planes Θ = constant, both aspects of the reflector contribute to some degree. However, it will be shown in Chap. 6 that if the length l is large compared with the wavelength, the major portion of the field is confined within a small angular region Φ about the central plane. Over this region the variation of cos Φ in Eq. (89) is of second order compared with sin Φ; on setting cos $\Phi = 1$, separability is obtained, the field distribution in the planes Θ = constant being determined entirely by the linear characteristic of the source $F(x)$. The transverse distribution of the field is thus virtually the same for all transverse planes.

The primary field of the source alone must be added, of course, to the scattered field to obtain the total field. Here the far-zone field of the source (for which it is effectively a point source) must be used instead of the cylindrical wave field of the radiation zone. It will be assumed that this is known and expressed in a form similar to Eq. (72a), in terms, of course, of the spherical coordinates shown in Fig. 5·9. It is also assumed that the equivalent point source is located at the origin of the coordinate system, since the phase terms entering into the reflector field have been referred to that origin. The procedure for superposing the fields is exactly the same as that delineated in the previous case and need not be discussed further here.

Attention should be called to one point in the procedures discussed in this section that has been the cause of some concern in the past. It will be found in general that the radiation field of a current distribution—such as is given by Eqs. (72) and (78)—does not satisfy Maxwell's equations exactly. Consequently, except in special cases, the current and charge distributions on the surface of the reflector, as found by the methods already outlined, do not satisfy the equation of continuity exactly. However, the terms that are neglected, which would result in satisfying the required conditions exactly, are smaller in order of magnitude than the radiation field components and are in general in time quadrature with the latter; they therefore introduce a nonessential contribution to the scattered field and the scattered power pattern.

5·10. Reaction of a Reflector on a Point-source Feed.—One of the fundamental problems in the design of an antenna employing a reflector is the effect of the latter on the impedance characteristics of the antenna. The problem can be treated on the basis of re-radiation from the current distribution on the reflector;[1] the analysis will be carried out here for the case of a point-source feed.

The radiating system that constitutes the feed must be considered in its relation to a transmission line. The basic idea of the following analysis is that the interaction between the feed and the field of the current distribution on the reflector gives rise to a "reflected" wave in the transmission line and thus an impedance mismatch from the point of view of the line. Our object is to calculate the reflection coefficient—ratio of the reflected to incident wave amplitudes—in the transmission line due to the reflector. The field of the reflector, which is given in general by Eq. (69), is regarded as a superposition of spherical wavelets arising from every element of surface dS. The total reflected wave in the transmission line is then considered to be the sum of component waves arising from the interaction between the feed and the separate wavelets. The current element $\mathbf{K}\,dS$ is regarded as a dipole source, and only the radiation terms are retained for the individual wavelets. The interaction between one of these and the feed is evaluated on the assumption that the distance from the reflector to the feed is so large that the wavelet can be regarded as a plane wave over the effective area of the feed. This assumption is consistent with our previous condition that the reflector be in the far-zone of the feed system. Multiple scattering between the feed and the reflector is neglected; this is likewise consistent with the previous assumptions.

It will be assumed that in the absence of the reflector the feed is matched to the transmission line; there is then only an incident wave within the line. Let V_i be the voltage at some reference cross section

[1] S. Silver, "Analysis and Correction of the Impedance Mismatch Due to a Reflector," RL Report No. 810, September 1945.

of the line. The total power transported across the reference cross section is

$$P = \alpha |V_i|^2, \qquad (90)$$

where α is a constant characteristic of the line and the field distribution over the cross section of the line. If the dielectric and ohmic losses in the line and the radiating system are negligible, P is the total power radiated by the feed. By Eq. (73), the current density at a point ρ, θ, ϕ on the reflector, with its phase referred to that of the voltage at the reference cross section in the line, is

$$\mathbf{K} = 2(\mathbf{n} \times \mathbf{H}_i) = \frac{2V_i}{\rho}\left[\frac{\alpha}{2\pi}\left(\frac{\epsilon}{\mu}\right)^{\!\frac{1}{2}} G_f(\theta,\phi)\right]^{\!\frac{1}{2}} \mathbf{n} \times (\boldsymbol{\varrho}_1 \times \mathbf{e}_i)e^{-jk\rho}. \qquad (91)$$

Expanding the vector term, we have

$$\mathbf{n} \times (\boldsymbol{\varrho}_1 \times \mathbf{e}_i) = (\mathbf{n}\cdot\mathbf{e}_i)\boldsymbol{\varrho}_1 + \mathbf{e}_i \cos i, \qquad (92)$$

where i is the angle of incidence. The current is here resolved into one component along the incident ray and one component parallel to the polarization of the primary field.

The field of the wavelet arising from the current element $\mathbf{K}\,dS$ is the integrand of Eq. (69):

$$\frac{1}{2\pi j\omega\epsilon}\left[(\mathbf{n}\times\mathbf{H}_i)\cdot\boldsymbol{\nabla}(\boldsymbol{\nabla}\psi) + k^2(\mathbf{n}\times\mathbf{H}_i)\psi\right]\,dS.$$

Applying the results of Eqs. (3·125) and (3·126), one sees that the radiation field components arise only from the component of the current that is transverse to the direction of propagation of the wavelet. Consequently, to the order of approximation that all the other terms are neglected, the component of the current in the direction $\boldsymbol{\varrho}_1$ contributes nothing to the reaction on the feed. As regards the component in the direction $\mathbf{e}_i$ it is observed that this coincides with the polarization of the feed and therefore no polarization obliquity factor enters into the interaction with the feed. The field intensity of the spherical wavelet with which we are concerned is then

$$dE_r = \frac{-jk^2}{2\pi\omega\epsilon}(\mathbf{n}\times\mathbf{H}_i)\cdot\mathbf{e}_i\,dS$$

or

$$dE_r = \frac{-j\omega\mu}{2\pi\rho^2}V_i\left[\frac{\alpha}{2\pi}\left(\frac{\epsilon}{\mu}\right)^{\!\frac{1}{2}}G_f(\theta,\phi)\right]^{\!\frac{1}{2}}\cos i\ e^{-j2k\rho}\,dS. \qquad (93)$$

The magnitude of the Poynting vector of the spherical wavelet is

$$|S_r| = \frac{1}{2}\left(\frac{\epsilon}{\mu}\right)^{\!\frac{1}{2}}|dE_r|^2.$$

If the wavelet may be considered plane over the receiving cross section of the feed [see Eq. (2·80)], the power that would be extracted from the wavelet acting alone is

$$dP_r = |S_r| G_f(\theta,\phi) \frac{\lambda^2}{4\pi}. \quad (94)$$

The voltage dV_r of the reflected wave set up thereby in the line, at the given reference point, is

$$dV_r = \left(\frac{dP_r}{\alpha}\right)^{1/2} e^{-jk(2\rho+\delta)} = \left[\frac{\lambda^2}{8\pi\alpha}\left(\frac{\epsilon}{\mu}\right)^{1/2} G_f(\theta,\phi)\right]^{1/2} dE_r e^{-jk(2\rho+\delta)}. \quad (95)$$

The phase term δ is a constant determined by the feed and the choice of the reference point; we need not be concerned with its precise value. Substituting Eq. (93) into (95), we obtain the reflection coefficient contributed by the element of surface dS of the reflector:

$$d\Gamma = \frac{dV_r}{V_i} = \frac{1}{4\pi\rho^2} G_f(\theta,\phi) \cos i \, e^{-j(2k\rho+\delta')} \, dS. \quad (96)$$

The phase term δ' absorbs the $-j$ of Eq. (93). The reflection coefficient due to the entire reflector is, therefore,

$$\Gamma = e^{-jk\delta'} \int_{S_0} \frac{G_f(\theta,\phi)}{4\pi\rho^2} \cos i \, e^{-j2k\rho} \, dS. \quad (97)$$

Use will be made of this result in Chap. 12 to devise a method for eliminating the mismatch.

The principle of stationary phase (Sec. 4·7) may be applied to the integral of Eq. (97) to obtain an estimate of the mismatch for the case of short wavelengths. It will be recognized that the phase is stationary at those points on the reflector at which the rays from the feed strike at normal incidence. The essential contribution to Γ arises from the area in the immediate neighborhood of the stationary point. The calculation is hardly different from that used for Eqs. (4·46) to (4·51) and will not be given here. If G_n is the gain of the feed in the direction of normal incidence, ρ_n the feed-to-reflector distance, R_ξ and R_η the principal radii of curvature of the reflector at the stationary point, we have

$$\Gamma_n \approx \frac{G_n \lambda}{8\pi\rho_n} \left[\frac{R_\xi R_\eta}{(R_\xi + \rho_n)(R_\eta + \rho_n)}\right]^{1/2} e^{-j(2k\rho_n+\delta')}. \quad (98)$$

If there is more than one point of normal incidence, the total effect is obtained by summing the separate values of Γ_n.

The same result [Eq. (98)] can be obtained directly on the basis of geometrical optics.[1] The reflected field intensity at the feed is deter-

[1] S. Silver, "Contribution of the Dish to the Impedance of an Antenna," RL Report No. 442, September 1943.

mined by considering the dispersion of a small incident cone of rays by the reflector, making use of the techniques of Sec. 5·5.

5·11. The Aperture-field Method.—It was shown in Sec. 3·8 that the field at a point in space lying outside a surface that encloses all the sources of the field can be expressed in terms of integrals of the field vectors over the surface. Thus, if the scattered field $\mathbf{E}_s$, $\mathbf{H}_s$ is known over any surface Σ that surrounds the reflector completely, the scattered field at an external point P in space is given by Eqs. (3·108) and (3·109):

$$\mathbf{E}_s(P) = \frac{1}{4\pi} \int_\Sigma [-j\omega\mu(\mathbf{n} \times \mathbf{H}_s)\psi + (\mathbf{n} \times \mathbf{E}_s) \times \nabla\psi + (\mathbf{n} \cdot \mathbf{E}_s)\nabla\psi]\,dS; \quad (99a)$$

$$\mathbf{H}_s(P) = \frac{1}{4\pi} \int_\Sigma [j\omega\epsilon(\mathbf{n} \times \mathbf{E}_s)\psi + (\mathbf{n} \times \mathbf{H}_s) \times \nabla\psi + (\mathbf{n} \cdot \mathbf{H}_s)\nabla\psi]\,dS. \quad (99b)$$

In applications to antenna problems, the field over Σ may not be known. The aperture-field method formulates a high-frequency approximation to the field. The surface Σ is taken in the immediate vicinity of the reflector, and it is assumed that energy·passes to Σ from the reflector by propagation along the reflected rays. The field over Σ is then calculated by the methods of Sec. 5·5 in conjunction with Eq. (4·23); the same conditions must therefore be imposed on the radii of curvature of the incident wavefront and the reflecting surface.

The present method has no special advantages over the current-distribution method for the treatment of an arbitrary reflector. However, there is one class of reflectors for which it has decided advantages, both in ease of application and in establishing relations with other phenomena. The reflectors to which the method is particularly suited—and to which the subsequent discussion is restricted—have the property that the entire family of rays reflected from the illuminated area S_0 lie in one hemisphere of space, as shown in Fig. 5·10; also, in the neighborhood of the reflector it is possible in general to draw a finite curve Γ_A circumscribing the entire family of reflected rays. The shadow boundary Γ on the reflector then defines an aperture and serves as an exit pupil for the reflected rays, which can be regarded as arising from a distribution of image sources behind the reflector.

On the basis of the ray diagram it is to be expected that the scattered field will be concentrated largely in the hemisphere of space containing the reflected rays. Our discussion will pertain to points in this region, and the surface Σ will, therefore, be taken to be made up of an infinite plane containing a curve such as Γ_A, plus the hemispherical cap of infinite radius. The aperture of the system may be defined as the area A on the infinite plane circumscribed by the curve Γ_A obtained by projection of the shadow boundary Γ along the reflected rays. It may be noted at this point that since the scattered field must satisfy the radiation conditions [Eqs. (3·113) and (3·114)] at infinity, the hemispherical cap will make no contribution to the field integrals in Eqs. (99a) and (99b).

It is evident that the determination of the scattered field over the plane Σ by reference to the reflected rays leads to a discontinuous distribution, with a nonzero field over the area circumscribed by Γ_A and zero field over the area of Σ outside Γ_A. This introduces into the problem a feature that is equivalent to the discontinuity in the current distribution over the reflector at the shadow boundary in the previous method.

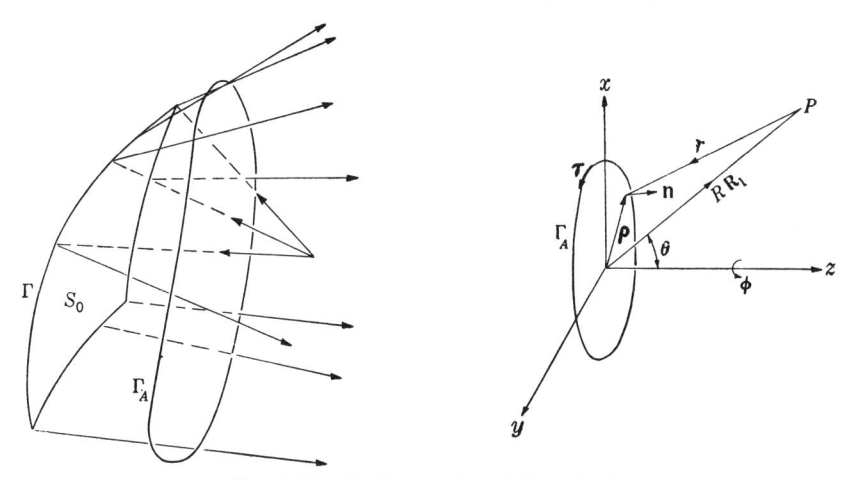

FIG. 5·10.—On the aperture-field method.

It was pointed out at the close of Sec. 3·8 that the terms entering into the integrands of Eqs. (99a) and (99b) can be set into correspondence with surface distributions of electric currents and charges and magnetic currents and charges. The electric and magnetic fields over the surface cannot be assigned arbitrarily; they must be assigned in such a way that the equivalent current and charge distributions satisfy the surface equation of continuity [Eq. (3·9)] if the integrals are to give field components that satisfy Maxwell's equations. In order to make the distributions over Σ compatible with the equation of continuity it is necessary to introduce line distributions of electric and magnetic charges along the boundary curve Γ_A.

The computation of the boundary charge distributions proceeds along exactly the same lines as in the case of the current-distribution methods.[1] With $\mathbf{E}_r$, $\mathbf{H}_r$ denoting components of the scattered field over Σ, the density σ_e of the boundary line distribution of electric charge and the

[1] See also J. A. Stratton, *Electromagnetic Theory*, McGraw-Hill, New York, 1941, Sec. 8·15.

density σ_m of the magnetic charge are

$$\sigma_e = -\frac{1}{j\omega}(\boldsymbol{\tau} \cdot \mathbf{H}_r); \qquad \sigma_m = \frac{1}{j\omega}(\boldsymbol{\tau} \cdot \mathbf{E}_r). \tag{100}$$

The unit vector $\boldsymbol{\tau}$ and the positive normal $\mathbf{n}$ to the surface Σ are defined in Fig. 5·10. The amended expressions for the fields are then

$$\mathbf{E}_s(P) = -\frac{1}{4\pi j\omega\epsilon}\oint_{\Gamma_A}\boldsymbol{\nabla}\psi(\boldsymbol{\tau}\cdot\mathbf{H}_r)\,ds$$
$$+\frac{1}{4\pi}\int_A[-j\omega\mu(\mathbf{n}\times\mathbf{H}_r)\psi+(\mathbf{n}\times\mathbf{E}_r)\times\boldsymbol{\nabla}\psi+(\mathbf{n}\cdot\mathbf{E}_r)\boldsymbol{\nabla}\psi]\,dS \tag{101a}$$

and

$$\mathbf{H}_s(P) = \frac{1}{4\pi j\omega\mu}\oint_{\Gamma_A}\boldsymbol{\nabla}\psi(\boldsymbol{\tau}\cdot\mathbf{E}_r)\,ds$$
$$+\frac{1}{4\pi}\int_A[j\omega\epsilon(\mathbf{n}\times\mathbf{E}_r)\psi+(\mathbf{n}\times\mathbf{H}_r)\times\boldsymbol{\nabla}\psi+(\mathbf{n}\cdot\mathbf{H}_r)\boldsymbol{\nabla}\psi]\,dS, \tag{101b}$$

where A is the area enclosed by Γ_A.

The integrals over the boundary Γ_A can be transformed into surface integrals by the same process used in transforming Eq. (5·62a). It is then found that the field expressions are

$$\mathbf{E}_s(P) = \frac{1}{4\pi j\omega\epsilon}\int_A[k^2(\mathbf{n}\times\mathbf{H}_r)\psi+(\mathbf{n}\times\mathbf{H}_r)\cdot\boldsymbol{\nabla}(\boldsymbol{\nabla}\psi)$$
$$+j\omega\epsilon(\mathbf{n}\times\mathbf{E}_r)\times\boldsymbol{\nabla}\psi]\,dS, \tag{102a}$$
$$\mathbf{H}_s(P) = -\frac{1}{4\pi j\omega\mu}\int_A[k^2(\mathbf{n}\times\mathbf{E}_r)\psi+(\mathbf{n}\times\mathbf{E}_r)\cdot\boldsymbol{\nabla}(\boldsymbol{\nabla}\psi)$$
$$-j\omega\mu(\mathbf{n}\times\mathbf{H}_r)\times\boldsymbol{\nabla}\psi]\,dS. \tag{102b}$$

The boundary line charges have the same effect here as in the case of the current distribution: They cancel the longitudinal field component of the far-zone field that arises from the surface current and charge distributions.

The discussion has been developed with reference to a plane area. This is not necessary for the application of Eqs. (99a) and (99b); the surface Σ may be any curved surface of infinite extent that divides the space into two regions—one of the reflector and one of the scattered field. The aperture area in that case will be a curved surface bounded by a curve Γ_A that is the projection of the shadow boundary along the reflected rays. There is no change in the final result; the integrals (102a) and (102b) apply to the curved aperture surface A.

5·12. The Fraunhofer Region.—We shall now carry through the reduction of the integrals for the far-zone field. The latter will be referred to henceforth as the Fraunhofer region because of the relation of the

SEC. 5·12]   THE FRAUNHOFER REGION   161

problem to optical diffraction problems. The latter will be discussed in a later section.

Let ϱ again be the vector from the origin of the coordinate system to the element dS of the aperture area and $\mathbf{R}_1$ a unit vector from the origin to the field point in the direction θ, ϕ. Applying the results of Sec. 3·11 and inserting therein the expressions for the electric and magnetic currents given in terms of the fields by Eqs. (3·111), we find that Eq. (102a) reduces to

$$\mathbf{E}_s(P) = \frac{-jk}{4\pi R} e^{-jkR} \mathbf{R}_1 \times \int_A \left[\mathbf{n} \times \mathbf{E}_r - \left(\frac{\mu}{\epsilon}\right)^{\frac{1}{2}} \mathbf{R}_1 \times (\mathbf{n} \times \mathbf{H}_r) \right] e^{jk\varrho \cdot \mathbf{R}_1} \, dS. \quad (103)$$

Let $\mathbf{s}$ be a unit vector along a ray through the aperture. In the geometrical-optics approximation, the electric and magnetic fields over the aperture are related by

$$\mathbf{H} = \alpha(\mathbf{s} \times \mathbf{E}). \quad (104)$$

[In free space $\alpha = (\epsilon/\mu)^{\frac{1}{2}}$. However, as we shall see later, the reflector is only one special case of an aperture problem; the theory can be applied to problems such as the radiation from horns in which α will have some other value.] Substituting for $\mathbf{H}_r$ in Eq. (103), we obtain

$$\mathbf{E}_s(P) = \frac{-jk}{4\pi R} e^{-jkR} \mathbf{R}_1 \times \int_A \left\{ \mathbf{n} \times \mathbf{E}_r - \alpha\left(\frac{\mu}{\epsilon}\right)^{\frac{1}{2}} [\mathbf{R}_1 \cdot (\mathbf{s} \times \mathbf{E}_r)\mathbf{n} - (\mathbf{s} \times \mathbf{E}_r)(\mathbf{n} \cdot \mathbf{R}_1)] \right\} e^{jk\varrho \cdot \mathbf{R}_1} \, dS. \quad (105)$$

For some purposes it may prove convenient to take as the aperture area A the wavefront of the system of rays. In that case the unit vectors $\mathbf{s}$ and $\mathbf{n}$ are identical, since the rays are normal to the wavefront. More generally, however, it is convenient to use a plane aperture; the vector $\mathbf{n}$ is then constant over the surface and in the direction of the polar axis of the spherical coordinate system. The field $\mathbf{E}_r$ over the aperture is generally specified in terms of the polarization, magnitude, and phase distribution $\Psi(x,y)$. If the wavefronts associated with the rays through the aperture are the surfaces $L(x,y,z) = $ constant (cf. Sec. 4·2), the phase distribution is

$$\Psi(x,y) = k_0 L(x,y,0), \quad (106)$$

where $k_0 = 2\pi/\lambda_0$ is the free-space propagation constant. From Eq. (4·10) it follows that the components of the vector $\mathbf{s}$ over the aperture plane are

$$s_x = \frac{1}{k}\frac{\partial \Psi}{\partial x}; \qquad s_y = \frac{1}{k}\frac{\partial \Psi}{\partial y}; \qquad s_z = (1 - s_x^2 - s_y^2)^{1/2}. \tag{107}$$

The total power passing through the aperture is the integral of the normal component of the Poynting vector:

$$P = \frac{1}{2}\int_A \text{Re } (\mathbf{E}_r \times \mathbf{H}_r^*) \cdot \mathbf{n}\, dS = \frac{\alpha}{2}\int_A |E_r|^2 s_z\, dS. \tag{108}$$

The gain function for the aperture is therefore

$$G = \frac{8\pi}{\alpha}\frac{R^2|E_s(P)|^2}{\displaystyle\int_A |E_r|^2 s_z\, dS}. \tag{109}$$

It is overlooked in many treatments of aperture problems that if there is a phase distribution over the aperture other than a constant phase, the field vectors $\mathbf{E}_r$, $\mathbf{H}_r$ do not lie in the aperture plane and the Poynting vector is not normal to the plane. In cases where the phase distribution $\Psi(x,y)$ represents small deviations from constant phase, these factors can be neglected without too serious an error. Subject to this approximation, Eq. (105) simplifies to

$$\mathbf{E}_s(P) = \frac{-jke^{-jkR}}{4\pi R}\mathbf{R}_1 \times \left[\left(\mathbf{n} + \alpha\left(\frac{\mu}{\epsilon}\right)^{1/2}\mathbf{R}_1\right)\times \mathbf{N}\right], \tag{110}$$

where the vector $\mathbf{N}$ is

$$\mathbf{N} = \int_A \mathbf{E}_r e^{jk(x\sin\theta\cos\phi + y\sin\theta\sin\phi)}\, dS. \tag{110a}$$

The expression for $\boldsymbol{\varrho}\cdot \mathbf{R}_1$ for the plane aperture has been inserted. In using these relations it must be kept in mind that the field vectors are assumed to lie in the aperture plane. The θ- and ϕ-components of Eq. (110) are

$$E_{s_\theta} = \frac{jke^{-jkR}}{4\pi R}\left[1 + \alpha\left(\frac{\mu}{\epsilon}\right)^{1/2}\cos\theta\right](N_x\cos\phi + N_y\sin\phi), \tag{111a}$$

$$E_{s_\phi} = \frac{-jke^{-jkR}}{4\pi R}\left[\cos\theta + \alpha\left(\frac{\mu}{\epsilon}\right)^{1/2}\right](N_x\sin\phi - N_y\cos\phi). \tag{111b}$$

DIFFRACTION

5·13. General Considerations on the Approximate Methods.—Both the current-distribution and aperture-field methods led to a calculation of the scattered field as arising from a distribution of sources over an open surface, the boundary of which is defined by the system of reflected rays. In contrast to the geometrical-optics method, the field at any point was found as the superposition of contributions from all elements of the source distribution. In general, therefore, the last two methods will lead to nonzero field intensities in the region of space not covered by the system of rays; also, in the region of the rays, the fields will differ from those obtained on the basis of geometrical scattering. These deviations from geometrical propagation of the scattered field are known as diffraction phenomena.

The diffraction effects are due fundamentally to the fact that the sources are distributed over an open surface; that is, the reflected wavefront is not a closed surface. The same effects will arise no matter by what process a field distribution is generated over a finite open area in space. Thus a lens illuminated by a point-source or line-source feed likewise defines an exit pupil for the system of rays incident in it and produces a segment of a wavefront in its aperture plane. Lenses and reflectors that have aperture areas of the same size and shape and produce the same field distributions over the apertures have field patterns that differ in no essential detail. The same phenomena are observed when a wave passes through an aperture in an infinite opaque screen or through the mouth of a horn into free space.

Experiment shows that whenever the dimensions of the aperture are large compared with the wavelength, the diffraction effects are small and the major portion of the field pattern is concentrated in the region covered by the rays from the aperture. On the basis of this fact a common high-frequency approximation technique is used for all problems of the type mentioned above. The mathematical details have already been developed in Secs. 5·11 to 5·12, and we need only summarize here the general ideas in the application of the results to the various types of problems. In each case the aperture area is associated with a surface Σ of infinite extent which divides all space into two separate regions. The problem is then equivalent to that of an aperture in an infinite screen on the surface Σ. It is assumed that the field over Σ is zero everywhere except over the aperture area; in effect, it is assumed that diffraction effects at wide angles with respect to the aperture-ray system are negligible. In the case of the reflector and lens it is assumed that the aperture field is produced by geometrical reflection or refraction of the rays from the primary feed. In the case of a horn the aperture field is taken to be that which would exist over the aperture area in a horn of infinite extent—possibly after correction is made for reflection from the opening. In the infinite screen problem, the aperture field is taken to be that which exists over the area in the unperturbed wave in the absence of the screen.

As was pointed out in Sec. 5·11, the calculation of the diffraction field is based on the integrals of the field equations obtained in Sec. 3·8 by means of Green's theorem. However, the application of Green's theorem was predicated on certain assumptions concerning the continuity of the distribution over the surface—assumptions that are not fulfilled by the distribution over Σ in the approximation technique. The method should rather be considered to be based on the Huygens-Fresnel principle which postulates that each point on a wave surface is a source of elementary fields (cf. Sec. 4·1), and the results of the Green's theorem integration are to be regarded as furnishing the appropriate identification for the sources as stated in Eqs. (3·111). The requirement that the source distribution must satisfy the equation of continuity then leads to the addition of a line distribution of electric and magnetic charge along the boundary of the aperture surface. Thus Eqs. (101a) and (101b) and hence Eqs. (102a) and (102b) derived from them apply to all diffraction problems in the high-frequency approximation method; the fields $\mathbf{E}_r$, $\mathbf{H}_r$ are to be interpreted quite generally as the fields over the aperture surface.

5·14. Reduction to a Scalar Diffraction Problem.—In many antennas the field over the aperture is almost completely linearly polarized, only a small fraction of the energy being in the cross-polarization component of the field. If the latter is neglected, the calculation of the diffraction field is simplified; by a further approximation, consistent with the high-frequency approximations made already, the problem can be reduced to a scalar diffraction problem.

The analysis will be restricted to a plane aperture; the aperture will be taken in the xy-plane as in Fig. 5·10, and the electric field will be taken to be polarized in the x-direction.

It was pointed out in Sec. 4.1 that the integrals of Eqs. (99a) and (99b) can be transformed into

$$\mathbf{E}_s(P) = -\frac{1}{4\pi} \int_\Sigma \left[\psi \frac{\partial \mathbf{E}}{\partial n} - \mathbf{E} \frac{\partial \psi}{\partial n} \right] dS, \qquad (112a)$$

$$\mathbf{H}_s(P) = -\frac{1}{4\pi} \int_\Sigma \left[\psi \frac{\partial \mathbf{H}}{\partial n} - \mathbf{H} \frac{\partial \psi}{\partial n} \right] dS, \qquad (112b)$$

provided that Σ is a closed surface over the whole of which $\mathbf{E}$ and $\mathbf{H}$ are continuous. Equations (112a) and (112b) are each a set of three equations for the three cartesian components of the field vectors; the normal derivative $\partial/\partial n$ is applied component by component. If Σ is an open surface, as in the case of the diffraction problem, a similar transformation can be effected; additional terms appear that vanish in the former case of integration over a closed surface. The reader can verify that for the aperture the integrals transform as follows:

$$\frac{1}{4\pi} \int_A [-j\omega\mu(\mathbf{n} \times \mathbf{H})\psi + (\mathbf{n} \times \mathbf{E}) \times \nabla\psi + (\mathbf{n} \cdot \mathbf{E}) \nabla\psi] \, dS$$

$$= -\frac{1}{4\pi} \int_A \left[\psi \frac{\partial \mathbf{E}}{\partial n} - \mathbf{E} \frac{\partial \psi}{\partial n} \right] dS + \frac{1}{4\pi} \oint_{\Gamma_A} \psi \mathbf{E} \times \boldsymbol{\tau} \, ds; \quad (113)$$

SEC. 5·14] REDUCTION TO A SCALAR DIFFRACTION PROBLEM 165

a similar expression holds for Eq. (99b). The line integral around the boundary is different from that of the line distribution of charge but arises from the transformation of the surface integrals into one another. The diffraction field is the sum of the contribution of the surface integral and the line distribution of charges on the boundary. The complete expression is then

$$\mathbf{E}(P) = -\frac{1}{4\pi} \int_A \left(\psi \frac{\partial \mathbf{E}}{\partial n} - \mathbf{E} \frac{\partial \psi}{\partial n} \right) dS + \frac{1}{4\pi} \oint_{\Gamma_A} \psi (\mathbf{E} \times \boldsymbol{\tau}) \, ds \\ - \frac{1}{4\pi j\omega\epsilon} \oint_{\Gamma_A} \nabla\psi (\boldsymbol{\tau} \cdot \mathbf{H}) \, ds. \quad (114)$$

If the field over the aperture is linearly polarized with, say,

$$E_y = E_z = 0,$$

the surface integral contributes only to $E_x(P)$. As is seen in Fig. 5·11, this leads to a component of the field in the direction of propagation.

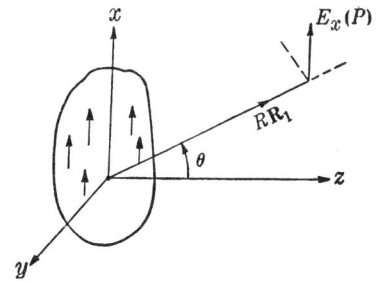

FIG. 5·11.—Reduction to a scalar diffraction problem.

For a given angle θ, the order of magnitude of this longitudinal component is at most $E_x(P) \sin \theta$. The vector $\mathbf{E} \times \boldsymbol{\tau}$ is normal to the aperture. It therefore gives rise only to a component $E_z(P)$; the contribution of the latter to the transverse field is again proportional to $\sin \theta$. Now the high-frequency approximation method is based on the assumption that the diffraction field is contained almost entirely in the region of small values of θ; therefore in the significant region of the field the longitudinal and transverse components arising from the surface integral and $\mathbf{E} \times \boldsymbol{\tau}$ respectively are negligible, and the surface integral may be taken alone to calculate the transverse field. As regards the longitudinal component introduced by $\mathbf{E} \times \boldsymbol{\tau}$, it will be recalled that the third integral of Eq. (110) was such as just to cancel the longitudinal component of the field introduced by the first two terms. Therefore, the last two integrals of Eqs. (110) virtually cancel each other for small angles θ.

The diffraction field is thus given by the scalar integral formula

$$u_p = -\frac{1}{4\pi} \int_A \left(\psi \frac{\partial u}{\partial n} - u \frac{\partial \psi}{\partial n} \right) dS, \quad (115)$$

where u stands for the particular component of the field involved. It will be recognized that this is the Kirchhoff diffraction formula used in physical optics.[1]

In the geometrical-optics approximation the field in the region of the aperture has the form (Sec. 4·5)

$$u = A(x,y,z)e^{-jk_0 L(x,y,z)}, \qquad (116)$$

where $A(x,y,z)$ is the amplitude and $L(x,y,z) = $ constant are the equiphase surfaces. Then

$$\frac{\partial u}{\partial n} = \mathbf{n} \cdot \nabla u = -jk_0 u \mathbf{n} \cdot \nabla L + u \frac{1}{A} \frac{\partial A}{\partial n}. \qquad (117)$$

If the wavelength is short, k_0 is large and the second term may be neglected in comparison with the first:

$$\frac{\partial u}{\partial n} \approx -jk_0 u \mathbf{n} \cdot \nabla L. \qquad (118)$$

The field over the aperture is usually given in terms of the amplitude $A(x,y,z)$ and the phase distribution $\Psi(x,y) = k_0 L(x,y,0)$. If $\mathbf{s}$ is the unit vector in the direction of ray at a given point on the aperture, we have by Eq. (4·10)[2]

$$k_0 \nabla L = k\mathbf{s}; \qquad (119)$$

then

$$\frac{\partial u}{\partial n} \approx -jk u \mathbf{n} \cdot \mathbf{s}. \qquad (120)$$

The components of $\mathbf{s}$ in terms of the phase distribution $\Psi(x,y)$ have been given previously in Eq. (5·107).

With regard to $\partial \psi / \partial n$ it is observed that

$$\frac{\partial \psi}{\partial n} = \mathbf{n} \cdot \nabla \psi = \frac{d}{dr}\left(\frac{e^{-jkr}}{r}\right) \mathbf{n} \cdot \mathbf{r}_1$$

or

$$\frac{\partial \psi}{\partial n} = \left(jk + \frac{1}{r}\right) \mathbf{n} \cdot \mathbf{r}_1 \frac{e^{-jkr}}{r}, \qquad (121)$$

where $\mathbf{r}_1$ is a unit vector from the point on the aperture to the field point. Collecting the terms in Eq. (115), we obtain

$$u_p = \frac{1}{4\pi} \int_A u \frac{e^{-jkr}}{r} \left[\left(jk + \frac{1}{r}\right) \mathbf{n} \cdot \mathbf{r}_1 + jk \mathbf{n} \cdot \mathbf{s}\right] dS. \qquad (122)$$

[1] See, for example, M. Born, *Optik*, reprint by Edwards Bros., Ann Arbor, Mich., 1943.

[2] This covers the general case in which the wavelength in the region of the aperture differs from that in free space.

For the far-zone field the customary approximations are made with regard to e^{-jkr}/r. In addition, $\mathbf{n} \cdot \mathbf{r}_1 = \mathbf{n} \cdot \mathbf{R}_1 = \cos\theta$ is constant over the aperture, and $1/r$ is negligible compared with jk. The far-zone field is, therefore,

$$u_p \approx \frac{jk}{4\pi R} e^{-jkR} \int_A u e^{jk\mathbf{\varrho}\cdot\mathbf{R}_1} (\mathbf{n}\cdot\mathbf{s} + \cos\theta)\, dS, \qquad (123)$$

where ϱ is the vector from the origin to the surface element dS. Attention should be directed to the $\mathbf{n}\cdot\mathbf{s}$ term. Only if the phase distribution over the aperture represents a small deviation from constant phase can $\mathbf{n}\cdot\mathbf{s}$ be set equal to unity with little error. In that case, we have

$$u_p \approx \frac{jk}{4\pi R} e^{-jkR} (1 + \cos\theta) \int_A u e^{jk\mathbf{\varrho}\cdot\mathbf{R}_1}\, dS. \qquad (124)$$

5·15. Babinet's Principle for the Electromagnetic Field.—Considerable progress has been made during the past few years in obtaining rigorous solutions of diffraction problems. Discussion of these would carry us beyond the scope of this work. Attention must be called, however, to the relation of Babinet's principle to the electromagnetic field that results from the solution of the problem of the diffraction of a wave by a plane metal screen of infinite conductivity.[1]

It will be well to recall the form of Babinet's principle as it applies to a scalar wave field.[2] Suppose that we have a plane opaque screen in the xy-plane; A_m is the area covered by the screen and A_0 is the aperture area. The complementary screen is defined to be that covering the area A_0 and having aperture area A_m. In both cases let there be an initial field u_i arising from sources in the negative z-region of space, and let u_1 and u_2 be the diffraction field produced in the positive z-region by the respective screens. The optical Babinet's principle states that the sum of the two complementary fields at any point is equal to the initial wave amplitude at the point in the absence of any screen:

$$u_i = u_1 + u_2. \qquad (125)$$

This relates the problem of diffraction around a metal sheet of finite area to the diffraction of a wave through an aperture of the same size and shape in an infinite plane sheet.

The principle for the electromagnetic field is fundamentally different in that the initial fields are complementary as well as the screens. Let $\mathbf{E}_i = \mathbf{F}$, $\mathbf{H}_i = \mathbf{G}$ be the initial field arising from sources in the negative z-region in the case of one of the screens, and let $\mathbf{E}_1$, $\mathbf{H}_1$ be the diffraction

---

[1] H. G. Booker, "Babinet's Principle and the Theory of Resonant Slots," TRE (Great Britain) Report No. 29, December 1941; E. T. Copson, *Proc. Roy. Soc., A*, **186**, 100 (1946).

[2] M. Born, *Op. cit.*

field in the positive z-region. Let $\mathbf{E}_i = -\mathbf{G}$, $\mathbf{H}_i = \mathbf{F}$ be the initial field in the case of the complementary screen and $\mathbf{E}_2$, $\mathbf{H}_2$ its diffraction field. Then

$$\mathbf{E}_1 + \mathbf{H}_2 = \mathbf{F}, \qquad (126)$$
$$\mathbf{H}_1 - \mathbf{H}_2 = \mathbf{G}. \qquad (127)$$

The incident field for the complementary screen is rotated 90° with respect to the first field, and the complementary relation exists between

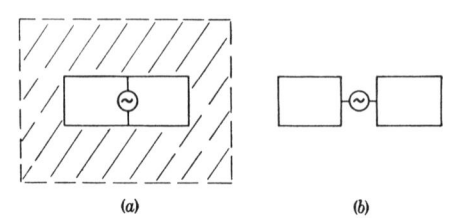

Fig. 5·12.—Relation between a slot and a dipole radiator.

the electric and magnetic field vector of the respective diffraction fields.

This principle leads to a useful relation between the radiation field of a slot and that of a dipole. Let S be a slot in an infinite plane conducting sheet, excited by a generator across its center as shown in Fig. 5·12a. The complementary dipole is a similar thin metal strip (Fig. 5·12b) energized by a generator across an infinitesimal gap at its center. The field vectors over the slot are perpendicular to the corresponding field vectors in the case of the dipole. It then follows from the Babinet's principle that the radiation field of the slot is the same as that for the dipole, but with the electric and magnetic field vectors interchanged; full details of the proof will be found in the paper by Booker.

CHAPTER 6

APERTURE ILLUMINATION AND ANTENNA PATTERNS

By S. Silver

6·1. Primary and Secondary Patterns.—The discussion of aperture systems will be continued in the present chapter with the object of developing in more detail the relations between the aperture field and the diffraction field. The results will furnish a basis for the design of the reflectors and lenses used in directive microwave antennas. The design considerations for such systems fall into two major groups: (1) transformation of the specifications that the radiation pattern of the antenna as a whole is required to meet into requirements on the aperture-field distribution, and (2) the design of the primary feed and reflector or lens to produce the required aperture field. The radiation pattern of the composite antenna will be referred to as the *secondary pattern* in distinction to the primary pattern of the feed system.

It must be kept in mind that strictly speaking the secondary pattern is a superposition of the diffraction field from the aperture and the field of the primary feed (*cf.* Sec. 5·1). Microwave feeds, however, are designed to have such directivity that the major portion of their energy is directed into illuminating the optical device. The overlapping of the field of the primary feed and the diffraction field gives rise, therefore, only to second-order effects; these will be treated in later chapters in the discussion of specific antennas. The secondary pattern can thus be resolved into two parts: (1) the diffraction field of the aperture and (2) the portion of the primary feed field that is not intercepted by the optical system. Specifications imposed on the secondary pattern (intensity level relative to peak intensity) in the second region are therefore requirements imposed on the primary feed pattern in addition to the requirements pertaining to the production of a desired aperture field.

6·2. The Diffraction Field.—The discussion will be restricted to a plane aperture and will be based on the scalar field approximation developed in Sec. 5·14. It is therefore being assumed that the field over the aperture is uniformly polarized in one direction, which, to fix our ideas, is, say, the x-direction, the aperture being taken in the xy-plane (Fig. 6·1).

Let the coordinates of a point in the aperture be ξ, η and those of a field point P be x, y, z. It will prove convenient to change the notation somewhat from that used in the preceding chapter. The field over the aperture will be designated by $F(\xi,\eta)$; $A(\xi,\eta)$ will be the amplitude distribution, and $\Psi(\xi,\eta)$ the phase distribution, so that

$$F(\xi,\eta) = A(\xi,\eta)e^{-j\Psi(\xi,\eta)}. \tag{1}$$

The method of determining the system of optical rays through the aperture associated with the phase distribution $\Psi(\xi,\eta)$ was discussed in Sec.

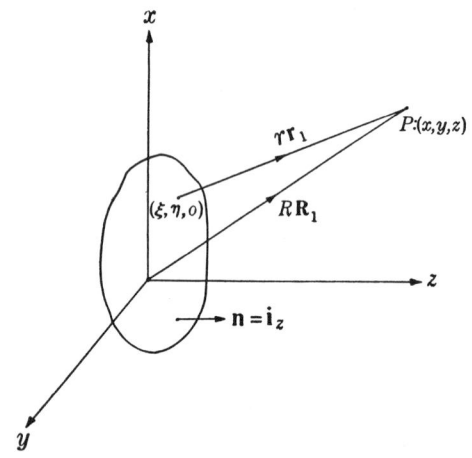

Fig. 6·1.—On the diffraction field.

5·12; the explicit relations between the components of a unit vector **s** along a ray and the phase function are given in Eq. (5·107). If $\mathbf{r}_1$ is a unit vector in the direction from the aperture point (ξ,η) to the field point P, then according to Eq. (5·122) the diffraction field U_p is given by

$$U_p = \frac{1}{4\pi} \int_A F(\xi,\eta) \frac{e^{-jkr}}{r} \left[\left(jk + \frac{1}{r} \right) \mathbf{i}_z \cdot \mathbf{r}_1 + jk \mathbf{i}_z \cdot \mathbf{s} \right] d\xi\, d\eta. \tag{2}$$

The diffraction field may be divided into three general zones which are determined mathematically by the nature of the approximations that may be made in the integral [Eq. (2)]. The three zones are also differentiated by the structure of the field, but it should be noted that the boundaries of these regions are not sharply defined.

First there is the *near-zone region* of points in the immediate neighborhood of the aperture for which no simplifying approximations can be made in Eq. (2). Although the dimensions of the aperture are large compared with the wavelength—an assumption that underlies the use of Eq. (2)—there is in general, for a given field point in this region, an appreciable area of the aperture for the points on which the $1/r$ term in the brackets of the integral is not negligible compared with $k = 2\pi/\lambda$. The region extends several wavelengths outward from the aperture, and it will be readily appreciated that this is not exactly infinitesimal for the wavelengths of the orders of magnitude of the microwave region. Also, for the near-zone region, the variation of $\mathbf{i}_z \cdot \mathbf{r}_1$ over the aperture for a given field point must be taken into account. The integrations are in general difficult to carry out, and a detailed study of the integral for this region is beyond the scope and purpose of our present discussion. In such cases where the field has been worked out in detail it has been found that the near-zone field is determined essentially by geometrical propagation along the aperture-ray system with fluctuations in intensity over the phase surface due to interference effects; the mean value of the intensity, however, differs little from that of the geometrically propagated field. The shadow region boundary is quite sharply defined.

Attention should be called to the fact that the scalar diffraction integral [Eq. (2)] can at best yield only qualitative results for the near field zone. In this region the contributions of the line integrals along the aperture boundary in Eq. (5·114) will make significant contributions to the field and must be taken into account if the results are to have a quantitative value.

From the near zone we pass into the region of the diffraction field which we shall call the optical-Fresnel field by virtue of its correspondence to the Fresnel region of optical diffraction problems. Several simplifying approximations are introduced; the orders of magnitudes of the errors involved must necessarily be evaluated for each case separately To start with, the term $1/r$ in the brackets of Eq. (2) is considered to be negligible with respect to k; at a distance of several wavelengths from the aperture this approximation is reasonable. Second, the variation of $(\mathbf{i}_z \cdot \mathbf{r}_1)$ over the aperture is neglected, and the term is replaced by the constant $\mathbf{i}_z \cdot \mathbf{R}_1 = \cos \theta$, where $\mathbf{R}_1$ is a unit vector directed from the origin to the field point. A third approximation in the same category is to neglect the variation of the $1/r$ term outside the brackets; it is set equal to the reciprocal distance $1/R$ from the origin to the field point.

The variation of r over the aperture must be treated more carefully in the phase term e^{-jkr}. We have in general

$$r = [(x - \xi)^2 + (y - \eta)^2 + z^2]^{1/2}. \tag{3}$$

If the field is concentrated in the region around the z-axis, a distance z from the aperture will be reached at which for the points in the significant region of the field $z \gg |x - \xi|, |y - \eta|$. Equation (3) can then be expanded as follows:

$$r \approx z + \frac{(x - \xi)^2}{2z} + \frac{(y - \eta)^2}{2z} + \cdots = z + r_a + \cdots. \tag{4}$$

Terms higher than the second order are neglected in the Fresnel field approximation. An alternative form of expansion is obtained by expressing the coordinates of the field point in spherical coordinates:

172 APERTURE ILLUMINATION AND ANTENNA PATTERNS [SEC. 6·2

$$x = R \sin \theta \cos \phi = R\alpha,$$
$$y = R \sin \theta \sin \phi = R\beta, \qquad (5)$$
$$z = R \cos \theta.$$

Introducing these into Eq. (3), we obtain

$$r \approx R - (\alpha\xi + \beta\eta) + \frac{\xi^2 + \eta^2 - (\alpha\xi + \beta\eta)^2}{2R} = R + r_b, \qquad (6)$$

neglecting terms of order higher than the second. It is seen that this assumes $\alpha\xi/R \ll 1$, $\beta\eta/R \ll 1$. The expansion in the form of Eq. (4) is suited for discussing the field over planes $z = $ constant, whereas Eq. (6) is best for discussing the field over a sphere of radius R about the origin. Since both expansions actually assume that the field is concentrated in the neighborhood of the z-axis, there is no significant difference between the results obtained with one or the other. The diffraction integral for the optical-Fresnel region thus becomes

$$U_p = \frac{j}{2\lambda} \frac{e^{-jkz}}{R} \int_A F(\xi,\eta) e^{-jkr_a} (\cos\theta + \mathbf{i}_z \cdot \mathbf{s})\, d\xi\, d\eta \qquad (7a)$$

or

$$U_p = \frac{j}{2\lambda} \frac{e^{-jkR}}{R} \int_A F(\xi,\eta) e^{-jkr_b} (\cos\theta + \mathbf{i}_z \cdot \mathbf{s})\, d\xi\, d\eta. \qquad (7b)$$

Eqs. (7a) and (7b) differ from the expressions for the Fresnel field generally found in the literature in the presence of the term $\mathbf{i}_z \cdot \mathbf{s}$ which arises from a nonuniform phase distribution over the aperture. It is to be noted that a phase distribution which represents wide deviations from constant phase has associated with it a highly dispersed system of rays. Under such conditions the assumption that the energy in the diffraction field is concentrated around the z-axis is not valid and the approximations entering into Eqs. (4) and (6) may not be justified. If, however, the phase distribution represents only small deviations from uniform phase, the deviation of the rays from a system of parallel rays that are normal to the aperture is small; the term $\mathbf{i}_z \cdot \mathbf{s}$ may then be treated as constant and equal to unity over the aperture.

The Fresnel region is characterized by the onset of diffusion of the field and the wavefront outside the boundaries defined by the extension of the rays through the aperture. The latter, however, still define the propagation of the major portion of the field; further details of the Fresnel region will be developed in later sections.

With increasing distance from the aperture we finally pass into the Fraunhofer or far-zone region of the field. This is the region with which the secondary pattern is concerned. The far-zone approximations have been discussed a number of times before. In the present connection it will be noted that the Fraunhofer region is differentiated from the Fresnel region by further approximations that are made in the phase term e^{-jkr}: the Fraunhofer field approximation neglects all terms in Eq. (6) above the first order in the aperture coordinates thereby considering in Fig. 6·1 that the unit vector $\mathbf{r}_1$ is parallel to $\mathbf{R}_1$. We have then

$$U_p = \frac{j}{2\lambda} \frac{e^{-jkR}}{R} \int_A F(\xi,\eta)(\cos\theta + \mathbf{i}_z \cdot \mathbf{s}) e^{jk\sin\theta(\xi\cos\phi + \eta\sin\phi)} \, d\xi \, d\eta. \quad (8)$$

Like all other far-zone fields encountered previously, the Fraunhofer field is a quasi-point source field. The field distribution is the same over all spheres about the origin; in a given direction θ, ϕ the amplitude varies monotonically as $1/R$ and the intensity as $1/R^2$.

Again if the phase error over the aperture—deviations from constant phase—are small, the $\mathbf{i}_z \cdot \mathbf{s}$ term may be replaced by unity. Equation (8) then becomes

$$U_p \approx \frac{j}{2\lambda R}(1 + \cos\theta)e^{-jkR}$$

$$\int_A F(\xi,\eta)e^{jk\sin\theta(\xi\cos\phi + \eta\sin\phi)} \, d\xi \, d\eta. \quad (8a)$$

It will be found that with nearly uniform phase over the aperture, almost all of the energy in the field is contained in a small angular region about the z-axis (corresponding to the geometrical property that the aperture rays are all parallel to the z-axis). The variation of $\cos\theta$ over the important region of the secondary pattern may then be neglected, and we have as our final approximation

$$U_p \approx \frac{j}{\lambda R} e^{-jkR} \int_A F(\xi,\eta)e^{jk\sin\theta(\xi\cos\phi + \eta\sin\phi)} \, d\xi \, d\eta. \quad (9)$$

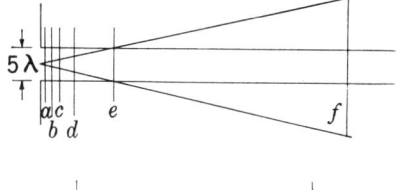

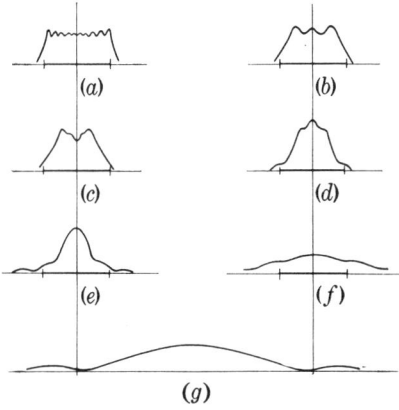

FIG. 6·2.—Transition from Fresnel to Fraunhofer diffraction for a slit; (a) ... (f) depict the field distribution across planes in the Fresnel region at increasing distances from the slit, showing progressive diffusion of the field into the shadow region; (g) is the Fraunhofer pattern. (*Reproduced from J. C. Slater and N. H. Frank, Introduction to Theoretical Physics, McGraw-Hill, New York, by courtesy of the authors.*)

Equation (9) is frequently used indiscriminately for both small and large phase errors over the aperture. This will be done in the present chapter and it should be remembered that for the latter cases the results have only qualitative value.

174   APERTURE ILLUMINATION AND ANTENNA PATTERNS   [Sec. 6·3

It was pointed out earlier that the boundaries of the three regions of the field cannot be sharply defined. It is clear that the passage from the Fresnel approximations to the Fraunhofer approximations is a gradual one and is determined to a large extent by the criteria of the acceptable error in the approximations that are made. In later sections we shall attempt to define an inner boundary for the Fraunhofer region on practical considerations for special types of apertures. The gradual transition of the physical characteristics of the field from one region to the next is illustrated very nicely in Fig. 6·2 taken from Slater and Frank.[1] The figures pertain to a slit over which the field is uniform in amplitude and phase. The near-zone pattern (Fig. 6·2b) is seen to consist essentially of the column of radiation propagated geometrically from the aperture. With increasing distance from the aperture the field diffuses into the shadow region, the system of parallel aperture rays finally passing over into a cone of rays in the Fraunhofer region.

6.3. Fourier Integral Representation of the Fraunhofer Region.—The final approximate expression that was obtained for the Fraunhofer region [Eq. (9)] has an interesting interpretation. Let us define

$$k_x = k \sin \theta \cos \phi, \qquad (10a)$$
$$k_y = k \sin \theta \sin \phi; \qquad (10b)$$

Eq. (9) then becomes

$$U_p = \frac{j}{\lambda R} e^{-jkR} g(k_x, k_y) \qquad (11a)$$

with

$$g(k_x, k_y) = \int_A F(\xi, \eta) e^{j(k_x \xi + k_y \eta)} \, d\xi \, d\eta. \qquad (11b)$$

Consider the plane $z = 0$. The aperture field can be regarded as the function $u(x,y)$ over the entire plane:

$$\begin{array}{ll} u(x,y) = F(x,y) & \text{inside } A, \\ u(x,y) = 0 & \text{outside } A. \end{array} \right\} \qquad (12)$$

The function $u(x,y)$ is stepwise continuous over the entire plane and can, therefore, be expressed as a Fourier expansion in the form of the Fourier integral:

$$u(x,y) = \frac{1}{(2\pi)^2} \int_{-\infty}^{\infty} \int_{-\infty}^{\infty} \int_{-\infty}^{\infty} \int_{-\infty}^{\infty} u(\xi,\eta) e^{jk_x(\xi-x)} e^{jk_y(\eta-y)} \, d\xi \, d\eta \, dk_x \, dk_y, \qquad (13)$$

or

$$u(x,y) = \frac{1}{2\pi} \int_{-\infty}^{\infty} \int_{-\infty}^{\infty} g(k_x, k_y) e^{-j(k_x x + k_y y)} \, dk_x \, dk_y, \qquad (13a)$$

[1] *Introduction to Theoretical Physics*, McGraw-Hill, New York, 1933, Chap. 27.

with

$$g(k_x, k_y) = \frac{1}{2\pi} \int_{-\infty}^{\infty} \int_{-\infty}^{\infty} u(\xi,\eta) e^{j(k_x\xi + k_y\eta)} \, d\xi \, d\eta,$$

$$= \frac{1}{2\pi} \int \int_A F(\xi,\eta) e^{j(k_x\xi + k_y\eta)} \, d\xi \, d\eta. \tag{13b}$$

It will be observed that except for the factor of $1/2\pi$ Eq. (13b) is identical with Eq. (11b).

Let us now examine Eq. (13a). If we define a vector **k**,

$$\mathbf{k}: \quad k_x; k_y; k_z = \pm \left(\frac{4\pi^2}{\lambda^2} - k_x^2 - k_y^2\right)^{\frac{1}{2}}, \tag{14}$$

the function $e^{-j\mathbf{k}\cdot\mathbf{r}}$ satisfies the wave equation and represents a plane wave of unit amplitude traveling in the direction of the vector **k**. Over the plane $z = 0$, the wave produces a distribution

$$e^{-j\mathbf{k}\cdot\mathbf{r}}\bigg]_{z=0} = e^{-j(k_x x + k_y y)}. \tag{15}$$

The integrand of Eq. (13a) is thus the distribution over the plane $z = 0$ produced by a plane wave in the direction **k** with an amplitude $g(k_x, k_y)$, and the arbitrary distribution $u(x,y)$ is given by Eq. (13a) as a superposition of plane waves traveling in all directions. Referring to Eq. (11a) it is then seen that the amplitude of the field in the Fraunhofer region in the direction defined by k_x and k_y [Eqs. (10)] is the amplitude of the plane wave component in that direction which enters into the synthesis of the arbitrary distribution over the aperture.

Equations (13a) and (13b) are referred to as the pair of mates of a Fourier transform. If the function $g(k_x, k_y)$ is given, that is, the Fraunhofer field is prescribed completely both as regards to amplitude and phase, Eq. (13a) serves to determine the field distribution over the plane $z = 0$ that is required to produce the prescribed secondary field pattern. In practice the use of the transform is limited by the fact that the secondary pattern is prescribed only in power; the phase of $g(k_x, k_y)$ can be assigned at will, and therefore the aperture distribution is not determined uniquely. Two different choices of the phase of $g(k_x, k_y)$ lead to two different aperture fields, one of which it may be physically possible to produce, whereas the other may not be realizable physically at all.

6·4. General Features of the Secondary Pattern.—The results of later sections will be anticipated here with a general summary of the relation between the secondary pattern and the aperture field. Considering Eq. (8) or (9) again from the point of view of the superposition of contributions from each element of surface on the aperture, the field at a given point is visualized as the resultant of a system of vector elements. The magnitude of the vector from the element of surface at a point ξ, η is $|F(\xi,\eta)|\,d\xi\,d\eta$; the angle that it makes with the element from the origin (which is taken as a reference) is determined by the intrinsic phase difference between them over the aperture and the phase difference arising from different path lengths to the field point. The absolute maximum value that the resultant of the system of vectors can have is equal to the sum of their magnitudes, obtained when the contributions are all in phase. If the phase over the aperture is constant, the absolute maximum is attained in the direction normal to the aperture, for in that direction the path length is the same from all aperture points to the field point. Since the path-length phase factor

$$k \sin \theta(\xi \cos \phi + \eta \sin \phi)$$

is a linear function of the coordinates on the aperture, the absolute maximum cannot be obtained in any direction in the case of arbitrary phase distributions over the aperture unless the distribution is a linear function of the aperture coordinates. In general, however, there will always exist directions in space for which the path phase factor makes the optimum compensation for the aperture phase differences between the elements as compared with neighboring directions. The secondary pattern thus has series of maxima and minima. If the phase distribution does not deviate too widely from constant phase over the aperture, there will in general be one maximum that is considerably greater in value than the others. The portion of the secondary pattern possessing this maximum and contained within the angular region bounded by the directions of the adjacent minima is known as the main lobe or sometimes as the main beam. The subsidiary maxima are referred to as side lobes. The line through the origin and the peak of the main beam is referred to as the beam axis.

From the practical point of view the pattern is specified by certain beam characteristics: the direction of the peak intensity; the gain, half-power, and tenth-power widths of the main lobe; and the magnitudes and positions of the side lobes. To define the beam widths consider any plane containing the axis of the beam; the half-power width Θ in that plane is the angular distance between the two directions about the axis in which the power radiated per unit solid angle is one-half the peak value; the tenth-power width $\Theta(\frac{1}{10})$ is defined correspondingly. If the aperture is symmetrical in shape and the field distribution over the aperture has certain symmetry elements in common with the aperture, the main lobe will reflect the symmetry of the field distribution. The symmetry elements are generally planes of symmetry; these are referred to as the principal planes of the pattern.

SEC. 6·4] *GENERAL FEATURES OF THE SECONDARY PATTERN* 177

Gain.—Let us consider first the relation between the gain and the aperture field. The power radiated per unit solid angle in a given direction is [*cf.* Eq. (5·77a)]

$$P(\theta,\phi) = \frac{1}{2}\left(\frac{\epsilon}{\mu}\right)^{1/2} R^2 |U_p|^2. \tag{16}$$

The total power P_a radiated by the aperture is equal to the power flow across the aperture, which is the integral of the normal component of the Poynting vector. The total power is then

$$P_a = \frac{1}{2}\left(\frac{\epsilon}{\mu}\right)^{1/2} \int_A |F(\xi,\eta)|^2 (\mathbf{i}_z \cdot \mathbf{s})\, d\xi\, d\eta, \tag{17}$$

and the gain function in the direction (θ,ϕ) is

$$G(\theta,\phi) = \frac{\pi}{\lambda^2} \frac{\left| \int_A F(\xi,\eta)(\cos\theta + s_z) e^{jk\sin\theta(\xi\cos\phi + \eta\sin\phi)}\, d\xi\, d\eta \right|^2}{\int_A |F(\xi,\eta)|^2 s_z\, d\xi\, d\eta}. \tag{18}$$

The more exact form [Eq. (8)] has been used in Eq. (18) for the field intensity U_p in the Fraunhofer region.

It was seen that if the phase is constant over the aperture, the secondary pattern attains the absolute maximum in the direction of the z-axis, $\theta = 0$. The aperture rays are parallel to the z-axis so that $s_z = 1$; the maximum value of the gain function, or simply the gain, is, therefore,

$$G_M = \frac{4\pi}{\lambda^2} \frac{\left| \int_A F(\xi,\eta)\, d\xi\, d\eta \right|^2}{\int_A |F(\xi,\eta)|^2\, d\xi\, d\eta}. \tag{19}$$

A case of especial interest is that of uniform illumination over the aperture. $F(\xi,\eta)$ is a constant; from Eq. (19) the gain G_0 for that case is found to be

$$G_0 = \frac{4\pi A}{\lambda^2}. \tag{20}$$

Consider now any other intensity distribution. Making use of the Schwartz inequality,

$$\left| \int fg\, d\xi\, d\eta \right|^2 \leq \int f^2\, d\xi\, d\eta \int g^2\, d\xi\, d\eta, \tag{21}$$

where f and g are any two functions; by taking $f = F(\xi,\eta)$ and $g = 1$, we find

$$\left| \int_A F(\xi,\eta)\, d\xi\, d\eta \right|^2 \leq A \int |F(\xi,\eta)|^2\, d\xi\, d\eta. \tag{22}$$

Hence,
$$G_M \le \frac{4\pi A}{\lambda^2}. \tag{23}$$

Thus, the uniform field distribution over the aperture gives the highest gain of all constant-phase distributions over the aperture. The ratio $g = G_M/G_0$, known as the gain factor, may be regarded as the efficiency of the aperture in concentrating the available energy into the peak intensity of the beam.

The proof given above that uniform illumination gives maximum gain is valid strictly for constant-phase distributions only, since Eq. (19) applies only to such distributions. A proof for the more general case must be based on Eq. (18); so far, to the author's knowledge, no such proof has been established. If the phase distribution represents a small deviation from constant phase, however, and is such that the peak intensity lies in the direction $\theta = 0$, it is certain that the gain is less than that of the uniform field. The value of the peak intensity is more sensitive to the interference effects between the vector elements from the aperture than is the value of the total power to the slight deviations of the aperture rays from the normal to the latter. The effect of such phase errors is, therefore, a reduction in the aperture efficiency.

The aperture efficiency can be given more pictorial significance by considering the performance of the antenna system on reception. Let us suppose for the moment that the primary feed is designed to illuminate the reflector or the lens but to have no radiation in other directions. In that case the secondary pattern arises entirely from the aperture, and the gain of the antenna is equal to the aperture gain G_M. If now a plane wave is incident on a matched antenna along the beam axis, by Eq. (2·80), the absorption cross section presented by the antenna to the plane wave is

$$A_{r_M} = \frac{\lambda^2}{4\pi} G_M. \tag{24}$$

From Eq. (20), it is then seen that if the aperture is uniformly illuminated, the absorption cross section is equal to the physical cross section presented by the antenna to the incident wave. In the case of any other type of constant-phase illumination we have

$$A_{r_M} = \frac{G_M}{G_0} A. \tag{25}$$

The effective area is reduced by the gain factor. The aperture efficiency may thus be regarded as measuring the effective aperture area presented by the antenna to the incident wave.

In the practical case the primary feed radiates in directions other than that required to illuminate the optical device. The energy not intercepted by the latter is referred to as the spill-over energy. Except for secondary effects the total power in the secondary pattern can be considered as consisting additively of the power in the Fraunhofer region of the aperture diffraction field and the spill-over energy of the feed. The over-all gain of the antenna must be referred to the total energy radiated in all directions. If P_m is the peak intensity in the secondary pattern and P is the total power radiated by the feed, the over-all gain is

$$G = \frac{4\pi P_m}{P}. \tag{26}$$

If the aperture intercepts a fraction α of the total power from the feed, the power radiated by the aperture is $P_a = \alpha P$, whence

$$G = \alpha \frac{4\pi P_m}{P_a} = \alpha G_M. \tag{27}$$

Taking again the uniformly illuminated aperture with an idealized feed as the reference, the over-all efficiency of the system is

$$\mathcal{G} = \frac{G}{G_0} = \alpha \frac{G_M}{G_0}. \tag{28}$$

The efficiency of the antenna is thus seen to be a product of two factors: (1) the fraction of the total power intercepted by the optical device and (2) the efficiency of the aperture in concentrating the available energy into the peak of the main lobe.

Beamwidths and Side Lobes.—The beamwidth and side-lobe characteristics are, of course, intimately related to the dependence of gain on the aperture field distributions. The following remarks are based on the results of the investigation of a number of special cases. Taking the constant-phase distribution first, we have seen that maximum gain is realized with uniform illumination. If the illumination over the aperture is modified so that the intensity is peaked in the central area of the aperture and tapered down in magnitude toward the aperture boundary, the diminution in gain is accompanied by an increase in beamwidth and a decrease in side-lobe intensity relative to the peak intensity of the main lobe. The prominence of the side lobes can be traced to the discontinuity at the edge of the aperture, considering the field distribution with regard to the plane $z = 0$ as a whole.

The effect of phase errors over the aperture, with the types of aperture fields that are commonly encountered, is in general to reduce the gain and broaden the main lobe. Side-lobe levels may be either raised or depressed depending on both the type of phase distortion and the intensity distribution over the aperture. Quite generally the sharpness of the minima is reduced and their levels are raised. Severe phase errors over the aperture may result in splitting of the main lobe and enhancement of the side lobes to such an extent that it is no longer possible to identify a major lobe.

6.5. The Rectangular Aperture.—A number of special problems associated with rectangular and circular apertures will be investigated to illustrate the general ideas formulated in the preceding section. The rectangular aperture will be treated first. Let the dimensions of the aperture be designated by a and b, the orientation of the aperture in the xy-plane being shown in Fig. 6·3. The secondary pattern depends only on the relative distribution over the aperture, and in the following discussion it will be assumed that the distribution $F(\xi,\eta)$ is normalized to have a maximum value of unity. For the present purposes the completely simplified expression for the Fraunhofer region [Eq. (9)] will be used; in so far as the pattern is concerned we need consider only the factor

$$g(\theta,\phi) = \int_{-a/2}^{a/2} \int_{-b/2}^{b/2} F(\xi,\eta) e^{jk \sin \theta (\xi \cos \phi + \eta \sin \phi)} \, d\xi \, d\eta. \tag{29}$$

Uniform Amplitude and Phase.—$F(\xi,\eta) = 1$ for a uniformly illuminated aperture; the integral of Eq. (29) is easily evaluated, and one finds that the secondary amplitude pattern is

$$g(\theta,\phi) = A \left[\frac{\sin\left(\frac{\pi a}{\lambda} \sin \theta \cos \phi\right)}{\frac{\pi a}{\lambda} \sin \theta \cos \phi} \right] \left[\frac{\sin\left(\frac{\pi b}{\lambda} \sin \theta \sin \phi\right)}{\frac{\pi b}{\lambda} \sin \theta \sin \phi} \right] \tag{30}$$

The patterns in the principal planes (the xz- and yz-planes) are of particular interest. For the xz-plane $\phi = 0$, Eq. (30) simplifies to

$$g(\theta,\phi) = A \frac{\sin\left(\frac{\pi a}{\lambda} \sin \theta\right)}{\frac{\pi a}{\lambda} \sin \theta}. \tag{31}$$

For the yz-plane $\phi = \pi/2$; the pattern in this plane is likewise given by Eq. (31) with a replaced by b. Both patterns are of the same form, $\sin u/u$, but are scaled in the angle θ according to the aperture dimensions in the respective planes. The secondary power pattern, normalized to a peak value of unity, is plotted in Fig. 6·3 on a logarithmic scale as a function of the reduced variable $u = \binom{a}{b} (\pi/\lambda) \sin \theta$. The minima in this case are equal to zero and occur at the points $u_n = n\pi$, $n = \pm 1$, $\pm 2, \cdots$. The full widths of the main lobe measured from null point

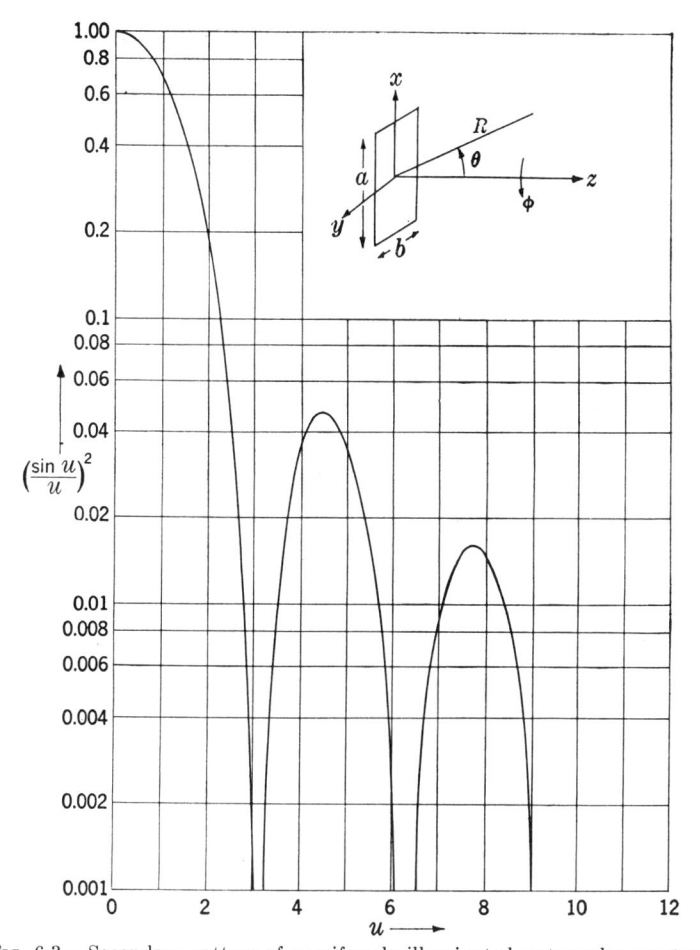

FIG. 6·3.—Secondary pattern of a uniformly illuminated rectangular aperture.

to null point on either side of the axis are

$$xz\text{-plane: } 2\sin^{-1}\left(\frac{\lambda}{a}\right) \approx \frac{2\lambda}{a},$$

$$yz\text{-plane: } 2\sin^{-1}\left(\frac{\lambda}{b}\right) \approx \frac{2\lambda}{b}.$$

The half-power point on the main lobe is very closely at $u = 1.39$; hence the half widths in the principal planes are

$$xz\text{-plane: } \Theta = 2\sin^{-1}\left(\frac{1.39\lambda}{\pi a}\right) \approx 0.88\frac{\lambda}{a}, \qquad (32a)$$

$$yz\text{-plane: } \Theta = 2\sin^{-1}\left(\frac{1.39\lambda}{\pi b}\right) \approx 0.88\frac{\lambda}{b}. \qquad (32b)$$

These illustrate two fundamental points: (1) In a plane of symmetry the width of the beam is determined by the aperture dimension in that plane, and (2) the diffraction pattern is confined to a smaller angular region the larger the dimensions of the aperture as measured in wavelengths.

The side lobes (peaks) are located at the points u_m that satisfy the relation $u = \tan u$. The first of these comes at $u_1 = 4.51$; the second at $u_2 = 7.73$. The side-lobe intensities relative to the peak intensities are readily found to be $1/(1 + u_m^2)$, which, from the values of u_1 and u_2, is seen to be very nearly equal to $1/u_m^2$. Referring to the amplitude expression [Eq. (31)], it is seen that $g(\theta)$ is positive over the entire main lobe, changing sign in passing through the first zero, returning to a positive value on passing through the second zero, and so on. The odd-numbered side lobes are, therefore, out of phase with the main lobe, and the even-numbered ones are in phase. Such phase reversals are characteristic of all power patterns in which the minima are equal to zero.

Separable Aperture-field Distributions.—A common type of aperture-field distribution is that which arises with a cylindrical reflector or lens and a line source (*cf.* Sec. 5·9) where the distribution over the aperture is separable into a product of two functions:

$$F(\xi,\eta) = F_1(\xi)F_2(\eta). \tag{33}$$

Substituting into Eq. (29), we find that the integral is likewise separable:

$$g(\theta,\phi) = \int_{-a/2}^{a/2} F_1(\xi)e^{jk\xi \sin\theta \cos\phi}\, d\xi \int_{-b/2}^{b/2} F_2(\eta)e^{jk\eta \sin\theta \sin\phi}\, d\eta. \tag{34}$$

If we consider again the principal plane patterns, we see that the pattern in a given plane is determined entirely by the field distribution along the corresponding aspect of the aperture. The principal plane patterns are

$$xz\text{-plane: } g(\theta) = \left[\int_{-b/2}^{b/2} F_2(\eta)\, d\eta\right] \int_{-a/2}^{a/2} F_1(\xi)e^{jk\xi \sin\theta}\, d\xi; \tag{35a}$$

$$yz\text{-plane: } g(\theta) = \left[\int_{-a/2}^{a/2} F_1(\xi)\, d\xi\right] \int_{-b/2}^{b/2} F_2(\eta)e^{jk\eta \sin\theta}\, d\eta. \tag{35b}$$

The effects of tapered illumination and phase errors on the principal plane patterns can thus be studied as two-dimensional problems, providing, of course, that the aperture field is separable in the form of Eq. (33) both in amplitude and in phase.

6·6. Two-dimensional Problems.—The remaining analysis of the secondary pattern of a rectangular aperture will be restricted to a separable distribution in which the field is uniform along, say, the y-direction; that is, $F_2(\eta) = 1$. The pattern in the plane $x = 0$ is just the $\sin u/u$ pattern,

$$g(\theta) = \frac{\sin\left(\dfrac{\pi b}{\lambda}\sin\theta\right)}{\dfrac{\pi b}{\lambda}\sin\theta},$$

and we are left with only the pattern in the plane $y = 0$,

$$g(\theta) = \int_{-a/2}^{a/2} F_1(\xi) e^{jk\xi\sin\theta}\, d\xi, \tag{36}$$

to consider. Multiplicative constants are being ignored in Eq. (36).

It is convenient to introduce new variables

$$x = \frac{2\xi}{a}, \qquad u = \frac{\pi a}{\lambda}\sin\theta; \tag{37}$$

the function $F_1(\xi)$ goes over into a function $f(x)$, and $g(\theta)$ becomes a function of u which to avoid difficulties of notation will be designated as $g(u)$. Equation (36) then becomes

$$g(u) = \frac{a}{2}\int_{-1}^{1} f(x) e^{jux}\, dx. \tag{38}$$

It is seen at once that if the same relative distribution, for example, $f(x) = [1 - (4\xi^2/a^2)] = (1 - x^2)$, is produced over two apertures of different size, the two apertures will produce the same secondary patterns when regarded as functions of u. The side-lobe intensities relative to the peak intensity will be the same in the two cases. However, since $\sin\theta = \lambda u/\pi a$, the angular distributions will differ; the diffraction field of the larger aperture will be contained in a smaller angular region than that of the smaller aperture, and in particular the main lobe will have a smaller beamwidth. The larger aperture will yield higher gain, corresponding to the fact that the pattern is confined to a smaller angular region in space. This can be seen directly from the expression for the gain [Eq. (19)]. For the present we shall consider only constant-phase distributions. Equation (19) reduces to

$$G_M = \frac{4\pi b}{\lambda^2} \frac{\left|\int_{-a/2}^{a/2} F_1(\xi)\, d\xi\right|^2}{\int_{-a/2}^{a/2} |F_1(\xi)|^2\, d\xi} \tag{39a}$$

for the separable type of distribution. On introducing the variable x, this becomes

$$G_M = \frac{2\pi ab}{\lambda^2} \frac{\left|\int_{-1}^{1} f(x)\, dx\right|^2}{\int_{-1}^{1} |f(x)|^2\, dx}, \tag{39b}$$

showing explicitly that the gain is proportional to the area of the aperture.

The distribution over the aperture can be characterized completely by its moments μ_m,

$$\mu_m = \int_{-1}^{1} x^m f(x)\, dx \qquad (m = 0, 1, 2, \cdots). \tag{40}$$

These are very useful in relating the properties of the secondary pattern to those of the aperture distribution.[1] Expanding the exponential in Eq. (38) we obtain

$$g(u) = \frac{a}{2} \sum_{m=0}^{\infty} \frac{(j)^m u^m}{m!} \int_{-1}^{1} x^m f(x)\, dx \tag{41}$$

or

$$g(u) = \frac{\mu_0 a}{2} \sum_{m=0}^{\infty} \frac{(j)^m \mu_m}{m!\, \mu_0} u^m. \tag{42}$$

The power pattern $p(u) = |g(u)|^2$ is then

$$p(u) = \left(\frac{\mu_0 a}{2}\right)^2 \left[1 - \left(\frac{\mu_2}{\mu_0} - \frac{\mu_1^2}{\mu_0^2}\right) u^2 + \left(\frac{\mu_4}{12\mu_0} - \frac{\mu_1 \mu_3}{3\mu_0^2} + \frac{\mu_2^2}{4\mu_0^2}\right) u^4 - \cdots \right]. \tag{43}$$

It is seen from Eq. (42) that an asymmetrical distribution over the aperture results in a $g(u)$ that is complex so that the equiphase surfaces in the Fraunhofer region are not spheres centered at the origin. If, however, the aperture distribution is symmetrical, that is, $f(x)$ is an even function, its odd moments vanish and $g(u)$ is real:

$$g(u) = \frac{\mu_0 a}{2} \left(1 - \frac{\mu_2}{2\mu_0} u^2 + \frac{\mu_4}{4!\mu_0} u^4 - \cdots \right). \tag{44}$$

For the latter case, convenient expressions for the beamwidth can be obtained by simple approximations. In the neighborhood of the beam axis, we shall approximate the pattern by neglecting all terms in Eq. (44) beyond the second:

$$g(u) \approx \frac{\mu_0 a}{2} \left(1 - \frac{\mu_2}{2\mu_0} u^2 \right). \tag{45}$$

The half-power point in the power pattern corresponds to point $\bar{u}$ at which the amplitude has fallen to $1/\sqrt{2}$ of its peak value. Hence

$$\bar{u} = \frac{\pi a}{\lambda} \sin \frac{\Theta}{2} \approx \left[(2 - \sqrt{2}) \frac{\mu_0}{\mu_2} \right]^{\frac{1}{2}},$$

---

[1] R. C. Spencer, "Fourier Integral Methods of Pattern Analysis," RL Report No. 762-1, Jan. 21, 1946.

and the half-power width is
$$\Theta \approx 2 \sin^{-1}\left\{\left[(2-\sqrt{2})\frac{\mu_0}{\mu_2}\right]^{1/2} \frac{\lambda}{\pi a}\right\}$$
or
$$\Theta \approx 0.48 \left(\frac{\mu_0}{\mu_2}\right)^{1/2} \frac{\lambda}{a}. \tag{46}$$

The effect of tapering the illumination down toward the edges of the aperture can be seen directly from this expression. Since the moment μ_2 is the average of the distribution function weighted by the factor x^2, peaking the function in the neighborhood of $x = 0$ decreases the second moment more rapidly than μ_0 which is the average of the function itself. The effect of such tapering is to increase the ratio μ_0/μ_2 and hence to increase the beamwidth.

A more accurate expression for the beamwidth has been obtained,[1] which can be used to construct the main lobe down to its tenth-power width. The results are applicable only to the cases of symmetrical aperture distributions. For the latter the expansion of the power pattern [Eq. (43)] reduces to

$$p(u) = 1 - \left(\frac{\mu_2}{\mu_0}\right) u^2 + \frac{1}{4}\left[\left(\frac{\mu_2}{\mu_0}\right)^2 + \frac{\mu_4}{3\mu_0}\right] u^4 - \cdots . \tag{47}$$

The factor $(\mu_0 a/2)^2$ has been dropped to normalize the pattern to a peak value of unity. The power drop in the pattern relative to the peak expressed in napiers is $N = -\ln p(u)$. Considering Eq. (47) to be of the form $1 - z$, the expansion for $\ln(1 - z)$ is used to obtain

$$N \approx \frac{\mu_2}{\mu_0} u^2 + \frac{1}{4}\left(\frac{\mu_2^2}{\mu_0^2} - \frac{\mu_4}{3\mu_0}\right) u^4. \tag{48}$$

Solving for u, we then get
$$u = AN^{1/2}(1 - BN), \tag{49}$$
where
$$A = \left(\frac{\mu_0}{\mu_2}\right)^{1/2}; \qquad B = \frac{1}{8}\left(1 - \frac{\mu_0 \mu_4}{3\mu_2^2}\right). \tag{49a}$$

Since $\ln p(u) = 2.303 \log p(u)$, the corresponding expression in terms of the decibel drop D is
$$u = A'D^{1/2}(1 - B'D), \tag{50}$$
where
$$A' = 1.518A; \qquad B' = 2.303B. \tag{50a}$$

[1] R. C. Spencer, op. cit.

The full angular width of the main lobe at a given decibel level is, therefore,

$$\Theta_D = 2 \sin^{-1}\left[\frac{A'D^{1/2}}{\pi}(1 - B'D)\frac{\lambda}{a}\right]. \tag{51}$$

The aperture-field–secondary-pattern relationships are further illustrated by the results in Table 6·1, in which the major secondary-pattern characteristics are given for several typical aperture distributions. The integration of Eq. (38) is easy to perform for each of these distributions, and the details need not be given here. The effect of reducing the discontinuity at the edge of the aperture is shown by the series of parabolic distributions $1 - (1 - \Delta)x^2$. It is seen that the gain decreases rapidly as Δ gets in the neighborhood of zero; and the beamwidth increases. The series $\cos^n(\pi x/2)$ shows the effect of a higher-order taper of illumination. All members of the series ($n = 1, 2, \ldots$) reduce to zero at the edge of the aperture, but in addition the nth member has $n - 1$ derivatives equal to zero at the edge of the aperture. The gain decreases and the beamwidth increases with increasing n; the side lobes appear at increasingly larger angles and with reduced intensity relative to the main lobe.

6.7. Phase-error Effects.—A phase-error distribution may arise over the aperture of an optical system from various causes such as a displacement of the primary feed from the focus or distortion of the reflector or lens, or it may be caused by phase error in the field of the primary feed; that is, the wavefront is not spherical or cylindrical as is presupposed in the design of the optical system.

It will again be assumed in the following that the aperture field is separable, the field being uniform in the y-direction and the phase error existing in the x-direction only. If $\Psi(2\xi/a) = \Psi(x)$ denotes the phase-error distribution, the expression for the secondary pattern [Eq. (38)] becomes

$$g(u) = \frac{a}{2}\int_{-1}^{1} f(x)e^{j[ux - \Psi(x)]}\,dx, \tag{52}$$

where $f(x)$ now denotes the amplitude of the aperture field. As in the preceding section, $f(x)$ is assumed to be normalized to unity.

The discussion will be limited to a consideration of special forms of $\Psi(x)$, specifically to the following:[1]

Linear error: $\Psi(x) = \beta x$.
Quadratic error: $\Psi(x) = \beta x^2$.
Cubic error: $\Psi(x) = \beta x^3$.

[1] The results presented here are taken largely from R. C. Spencer and P. M. Austin, "Tables and Methods of Calculation for Line Sources," RL Report No. 762-2, Mar. 30, 1946.

TABLE 6·1.—SECONDARY PATTERN CHARACTERISTICS PRODUCED BY VARIOUS TYPES OF APERTURE DISTRIBUTIONS.

	Gain factor $\mathcal{G}_n$	Full width at half power Θ, radians	Angular position θ of first zero	Intensity of first side lobe; db below peak intensity

Rectangular aperture, $-1 \le x \le +1$, height 1: $f(x) = 1 \quad |x| > 1$; $= 0 \quad |x| > 1$; $g(u) = \dfrac{\sin u}{u}$

| | 1 | $0.88 \dfrac{\lambda}{a}$ | $\dfrac{\lambda}{a}$ | 13.2 |

Parabolic taper: $f(x) = 1 - (1 - \Delta)x^2, \ |x| < 1$

$$g(u) = a\left[\frac{\sin u}{u} + (1-\Delta)\frac{d^2}{du^2}\left(\frac{\sin u}{u}\right)\right];$$

$$\mathcal{G}(\Delta) = \frac{(2+\Delta)^2}{9[1 - \frac{2}{3}(1-\Delta) + \frac{1}{5}(1-\Delta)^2]}$$

$\Delta = 1.0$	1	$0.88 \dfrac{\lambda}{a}$	$\dfrac{\lambda}{a}$	13.2
0.8	0.994	$0.92 \dfrac{\lambda}{a}$	$1.06 \dfrac{\lambda}{a}$	15.8
0.5	0.970	$0.97 \dfrac{\lambda}{a}$	$1.14 \dfrac{\lambda}{a}$	17.1
0.0	0.833	$1.15 \dfrac{\lambda}{a}$	$1.43 \dfrac{\lambda}{a}$	20.6

Cosine taper: $f(x) = \cos^n \dfrac{\pi x}{2} \quad |x| < 1^\circ$

$$g(u) = \frac{2a}{\pi} \frac{n!\cos u}{\prod_{k=0}^{\frac{n-1}{2}} \left[(2k+1)^2 - \dfrac{4u^2}{\pi^2}\right]} \quad n, \text{ odd};$$

$$g(u) = a \frac{n!}{\prod_{k=1}^{\frac{n}{2}}\left[(2k)^2 - \dfrac{4u^2}{\pi^2}\right]} \cdot \frac{\sin u}{u} \quad n, \text{ even}$$

$$\mathcal{G}_n = \frac{4}{\pi^2}\left[\frac{2 \cdot 4 \cdot 6 \cdots (n-1)}{1 \cdot 3 \cdot 5 \cdots n}\right]^2 \left(\frac{2 \cdot 4 \cdot 6 \cdots 2n}{1 \cdot 3 \cdot 5 \cdots 2n-1}\right) \quad n, \text{ odd}$$

$$\mathcal{G}_n = \left[\frac{1 \cdot 3 \cdot 5 \cdots (n-1)}{2 \cdot 4 \cdot 6 \cdots n}\right] \left[\frac{(n+2)(n+4)\cdots 2n}{(n+1)(n+3)\cdots 2n-1}\right] \quad n, \text{ even}$$

$n = 0$	1	$0.88 \dfrac{\lambda}{a}$	$\dfrac{\lambda}{a}$	13.2
1	0.810	$1.2 \dfrac{\lambda}{a}$	$1.5 \dfrac{\lambda}{a}$	23
2	0.667	$1.45 \dfrac{\lambda}{a}$	$2 \dfrac{\lambda}{a}$	32
3	0.575	$1.66 \dfrac{\lambda}{a}$	$2.5 \dfrac{\lambda}{a}$	40
4	0.515	$1.93 \dfrac{\lambda}{a}$	$3 \dfrac{\lambda}{a}$	48

Triangular taper: $f(x) = 1 - |x|, \ |x| < 1$

$$g(u) = 4a \left(\frac{\sin \dfrac{u}{2}}{\dfrac{u}{2}}\right)^2$$

| | 0.75 | $1.28 \dfrac{\lambda}{a}$ | $2 \dfrac{\lambda}{a}$ | 26.4 |

188 APERTURE ILLUMINATION AND ANTENNA PATTERNS [SEC. 6·7

Linear Error.—Inserting the appropriate expression for $\Psi(x)$ into Eq. (52) we obtain

$$g(u) = \frac{a}{2} \int_{-1}^{1} f(x) e^{j(u-\beta)x} \, dx. \tag{53}$$

It is directly evident that this is of the same form as Eq. (38) with u replaced by $(u - \beta)$. The pattern is, therefore, the same as that of the constant-phase distribution but displaced by an amount β. The peak intensity comes at $u = \beta$, that is, in the direction

$$\theta_0 = \sin^{-1} \frac{\beta\lambda}{a\pi}. \tag{54}$$

The pattern is thus the secondary pattern of a constant-phase aperture field rotated through the angle θ_0. The physical basis for this is very simple. On expressing the phase distribution in terms of the original aperture variable $\xi = ax/2$ and making use of Eq. (5·107), it will be found that the aperture rays form a system of parallel rays traveling in the direction θ_0 given by Eq. (54). The aperture field can be considered to have arisen from a plane wave incident on the aperture in the direction θ_0.

If the aperture is projected onto a plane normal to the aperture rays, a new aperture is obtained over which the field distribution has constant phase. The projected aperture dimension a' is

$$a' = a \cos \theta_0,$$

and therefore the gain G'_M in the case of linear-phase error will be related to the gain G_M of the constant-phase distribution by

$$G'_M = G_M \cos \theta_0. \tag{55}$$

This can be verified by a direct calculation on the basis of Eq. (18).

Quadratic Error.—The secondary amplitude pattern for this case is

$$g(u) = \frac{a}{2} \int_{-1}^{1} f(x) e^{j(ux - \beta x^2)} \, dx. \tag{56}$$

The evaluation of such integrals is generally laborious. For small phase errors, however, a convenient approximate method can be used. Expanding the exponential factor $e^{-j\beta x^2}$, we obtain

$$g(u) = \frac{a}{2} \sum_{m=0} \frac{(-j)^m \beta^m}{m!} \int_{-1}^{1} x^{2m} f(x) e^{jux} \, dx. \tag{57}$$

The integrals of Eq. (57) can be expressed as derivatives of the pattern $g_0(u)$ obtained in the absence of phase error ($\beta = 0$), for

$$\frac{d^k}{du^k}\int_{-1}^{1} f(x)e^{jux}\,dx = (j)^k \int_{-1}^{1} x^k f(x)e^{jux}\,dx. \qquad (58)$$

Hence

$$g(u) = \frac{a}{2}\sum_{m=0} \frac{(j)^m \beta^m}{m!} \frac{d^{2m}}{du^{2m}}[g_0(u)]. \qquad (59)$$

Retaining only the first two terms, we have

$$g(u) \approx \frac{a}{2}[g_0(u) + j\beta g_0''(u)]. \qquad (60)$$

If the amplitude distribution $F(x)$ is symmetrical, $g_0(u)$ is real; the power pattern $p(u)$ is then

$$p(u) \approx \frac{a^2}{4}\{[g_0(u)]^2 + \beta^2[g_0''(u)]^2\}. \qquad (61)$$

The effect of quadratic phase errors is illustrated in Fig. 6·4 for two types of illumination: uniform amplitude, $f(x) = 1$, and tapered illumination, $f(x) = \cos^2(\pi x/2)$. In both cases $\beta = \pi/2$, representing a path length deviation of $\lambda/4$ from constant phase at the edges of the aperture. The value of β is rather large for the use of Eq. (61) to be valid, but the qualitative features are not seriously affected by the errors involved. It is seen that the peak intensity still appears in the direction $\theta = 0$. Since the phase error is symmetrical with respect to the center of the aperture, the secondary pattern will always be symmetrical about the $\theta = 0$ axis. However, it will be found that when β gets sufficiently large the main lobe becomes bifurcated, with maxima appearing on either side of the $\theta = 0$ axis. The general effect of the phase error is to raise both the side-lobe level and the level of the minima. In the case of the tapered illumination these effects are so large that the first side lobe is almost completely absorbed into an extremely broadened main lobe. The effect of phase error on gain is exemplified by Fig. 6·5 which shows the gain relative to the constant-phase distribution for a uniformly illuminated aperture. The phase error is expressed in terms of path length deviation from constant phase. The loss in gain that can be tolerated in practice depends, of course, on the operational requirements on the antenna and the associated system.

Cubic Phase Errors.—The cubic phase errors can be treated by the same approximation technique as was employed in Eq. (60). The corresponding expression for the amplitude pattern is

$$g(u) = \frac{a}{2}[g_0(u) + \beta g_0'''(u)]. \qquad (62)$$

190 APERTURE ILLUMINATION AND ANTENNA PATTERNS [SEC. 6·7

In the case of a symmetrical distribution $f(x)$ over the aperture $g_0(u)$ is real and the power pattern is

$$p(u) = \frac{a^2}{4} [g_0(u) + \beta g_0'''(u)]^2. \tag{63}$$

If $f(x)$ is an even function, $g_0(u)$ is likewise even and hence $g_0'''(u)$ is odd. In the neighborhood of $u = 0$, $g_0'''(u)$ is positive for $u > 0$ and negative for $u < 0$. It is then directly evident from the form of $p(u)$ that the peak will occur at some value $u > 0$. The effect of the phase error is to

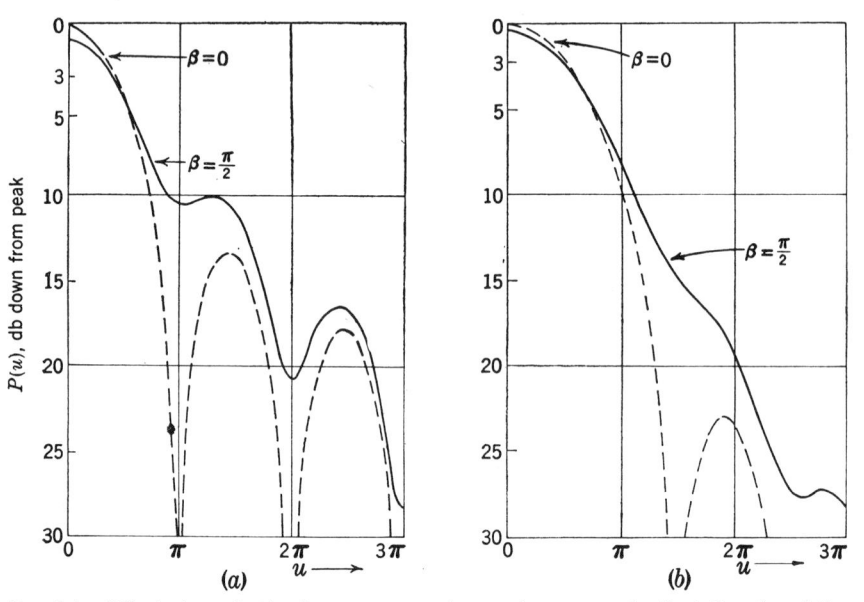

FIG. 6·4.—Effect of quadratic phase error; maximum phase error of $\pi/2$ at the edge of the aperture: (a) constant amplitude; (b) tapered illumination, $f(x) = \cos^2(\pi x/2)$.

tilt the beam as in the case of a linear error. In addition, however, the main lobe becomes asymmetrical and the side lobes increase on the side of the main lobe nearer $\theta = 0$ and decrease on the other side of the main lobe. The shift in the main lobe is also accompanied by a loss in gain.

Aperture Blocking.—The problem of an obstacle in the aperture is of interest in connection with the use of reflectors, for the primary feed is located in the path of the reflected rays, thus blocking out a portion of the aperture. The obstacle may be considered as a particular type of phase error. Assuming that over the exposed area the presence of the obstacle does not alter the distribution $f(x)$ which would exist in its absence, the obstacle can be regarded as producing a field 180° out of phase with $f(x)$ over the area that it covers.

Let us consider the particular case[1] illustrated in Fig. 6·6a of an obstacle located in the center of the aperture. The width w of the obstacle will be taken to be small compared with the total aperture width a. To express the pattern in terms of the variables of Eq. (37), we define the normalized width $2\delta = w/a$. Equation (38) for the pattern then becomes

$$g(u) = \frac{a}{2}\left[\int_{-1}^{-\delta} f(x)e^{jux}\,dx + \int_{+\delta}^{1} f(x)e^{jux}\,dx\right], \quad (64)$$

which can be rewritten as

$$g(u) = \frac{a}{2}\left[\int_{-1}^{1} f(x)e^{jux}\,dx - \int_{-\delta}^{+\delta} f(x)e^{jux}\,dx\right]. \quad (65)$$

It is seen explicitly that the obstacle can be regarded as an out-of-phase field superimposed on the original distribution.

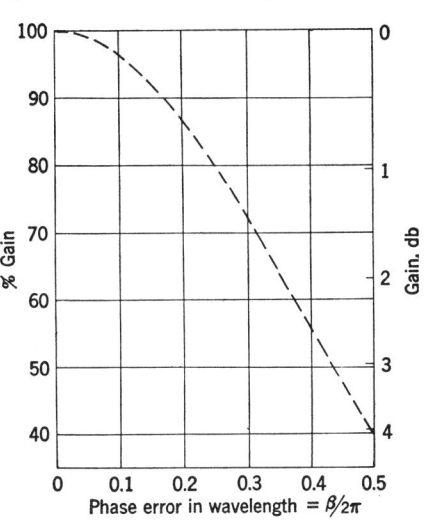

FIG. 6·5.—Loss of gain as a function of quadratic phase error over a uniformly illuminated aperture.

Over the region of the obstacle $f(x)$ may be considered constant and equal to unity; hence

$$g(u) = \frac{a}{2}\left[\int_{-1}^{1} f(x)e^{jux}\,dx - 2\delta\,\frac{\sin(u\delta)}{(u\delta)}\right]. \quad (66)$$

Since the width of the obstacle is small compared with the aperture width, the pattern produced by the former will be very broad compared with that of the aperture. For qualitative results the obstacle pattern may be regarded as constant over the region of the main lobe and near in side lobes of the aperture pattern. The effect of the obstacle is then simply that illustrated in Fig. 6·6b of subtracting a constant 2δ from the original amplitude pattern. If the peak amplitude of the original pattern is

$$a_0 = \int_{-1}^{1} f(x)\,dx$$

and the amplitude of the first side lobe is pa_0, the intensity of the first side lobe relative to the peak in the modified pattern is

$$p' = \frac{a_0 - 2\delta}{pa_0 + 2\delta} = \frac{1 - \dfrac{2\delta}{a_0}}{p + \dfrac{2\delta}{a_0}}. \quad (67)$$

[1] R. C. Spencer, "Fourier Integral Methods of Pattern Analysis," RL Report No. 761-1, Jan. 21, 1946.

The effect of the obstacle is to increase the magnitude of the first side lobe.

6.8. The Circular Aperture.—The fundamental considerations and results developed for the rectangular aperture pertaining to the relation between the aperture field and the secondary pattern apply in general to the circular aperture, but the quantitative details differ because of the difference in aperture geometry. In treating circular aperture problems

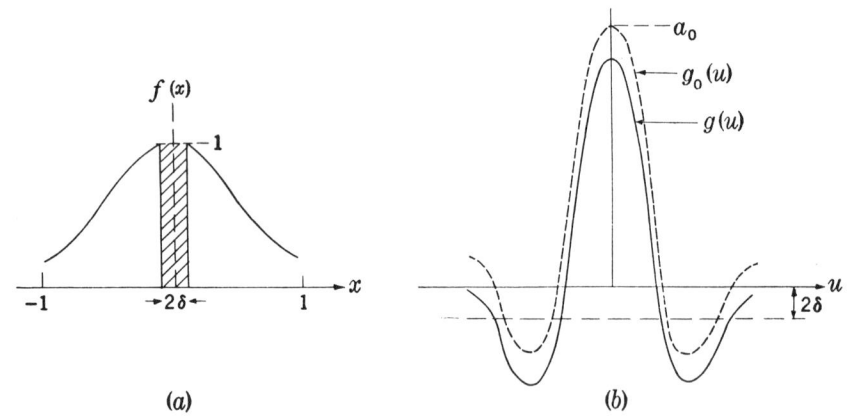

FIG. 6·6.—The effect of aperture blocking: (a) modified aperture distribution; (b) secondary pattern.

it is usually convenient to use polar coordinates ρ, ϕ' (Fig. 6·7), which are related to ξ, η by

$$\xi = \rho \cos \phi', \qquad \eta = \rho \sin \phi'. \tag{68}$$

Denoting the aperture field distribution by $F(\rho,\phi')$, the expression for the secondary pattern [Eq. (9)] becomes

$$g(\theta,\phi) = \int_0^{2\pi} \int_0^a F(\rho,\phi') e^{jk\rho \sin\theta \cos(\phi-\phi')} \rho \, d\rho \, d\phi', \tag{69}$$

where a is the radius of the aperture. Introducing the variables

$$r = \frac{\rho}{a}; \qquad u = \frac{2\pi a}{\lambda} \sin\theta = \frac{\pi D}{\lambda} \sin\theta, \tag{70}$$

the function $F(\rho,\phi')$ goes over into a function $f(r,\phi')$, and $g(\theta,\phi)$ goes over into a new function which we shall denote simply as $g(u,\phi)$. It will be assumed, as before, that $f(r,\phi')$ is normalized to unity. The pattern is then

$$g(u,\phi) = a^2 \int_0^{2\pi} \int_0^1 f(r,\phi') e^{jur \cos(\phi-\phi')} r \, dr \, d\phi'. \tag{71}$$

SEC. 6·8]   THE CIRCULAR APERTURE   193

It is observed that as in the case of the rectangular aperture, all apertures having the same relative distributions produce the same secondary patterns regarded as functions of u. The angular distribution in ϕ is the same for all, reflecting the symmetry of the distribution over the aperture; as seen from Eq. (70), the distribution in θ again scales by the

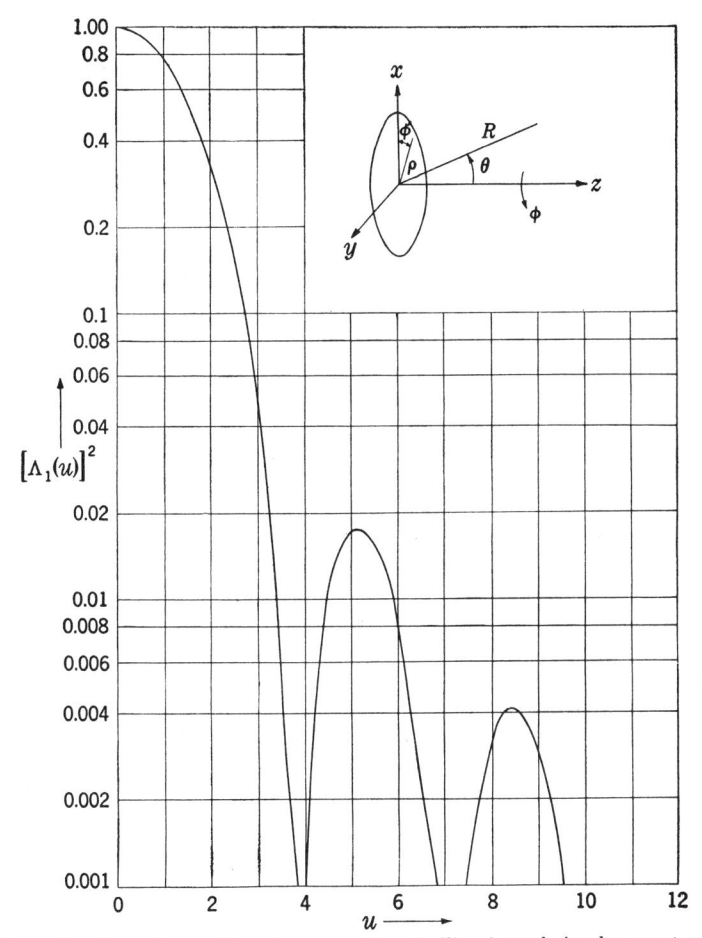

FIG. 6·7.—Secondary pattern from a uniformly illuminated circular aperture.

factor λ/D; the larger the diameter the smaller is the angular spread of the pattern about the ($\theta = 0$)-axis.

Uniform Phase and Amplitude.—Setting $f(r,\phi') = 1$ in Eq. (71) and carrying out the integration over ϕ', we obtain

$$g(u) = 2\pi a^2 \int_0^1 r J_0(ur) \, dr, \tag{72}$$

where $J_0(ur)$ is the Bessel function of order zero.[1] The integration over r leads to

$$g(u) = 2\pi a^2 \frac{J_1(u)}{u}. \tag{73}$$

The power pattern $p(u)$ normalized to unity is shown in Fig. 6·7, plotted on a logarithmic scale, as a function of u. The half width of the main lobe is

$$\Theta = 2 \sin^{-1}\left(0.51 \frac{\lambda}{D}\right) \approx 1.02 \frac{\lambda}{D}, \tag{74}$$

and the first side lobe is 17.5 db down from the peak. These are to be compared with the half width of $0.88\lambda/a$ and the 14-db side lobe of the secondary pattern of a rectangular aperture.

Tapered Illumination.—The effect of tapering the illumination down toward the edge is the same as with a rectangular aperture: reduction in gain, increase in beamwidth, and reduction in side lobes. The effects can be illustrated by considering the series of aperture field distributions $(1 - r^2)^p$, $p = 1, 2, \cdots$.[2] The secondary patterns are given by

$$g_p(u) = 2\pi a^2 \int_0^1 (1 - r^2)^p J_0(ur) r \, dr \tag{75}$$

or

$$g_p(u) = \pi a^2 \frac{2^p p! J_{p+1}(u)}{u^{p+1}} = \frac{\pi a^2}{p+1} \Lambda_{p+1}(u). \tag{75a}$$

The functions Λ_p are available in tabular form.[3] The major characteristics of the patterns are summarized in Table 6·2 and will not be discussed further.

The circular symmetry of the secondary pattern is associated of course with the corresponding symmetry of the field distribution. It is of interest to consider a distribution of the form

$$f(r,\phi') = 1 - r^2 \cos^2 \phi' \tag{76}$$

which is tapered in the plane $\phi' = 0$ and uniform in the plane $\phi' = \pi/2$. Substituting Eq. (76) into Eq. (71), we obtain

$$g(u,\phi) = 2\pi a^2 \left[\frac{J_1(u)}{u} - \int_0^{2\pi} \int_0^1 r^2 \cos^2 \phi' e^{jur \cos(\phi-\phi')} r \, dr \, d\phi' \right]. \tag{77}$$

---

[1] G. N. Watson, *Theory of Bessel Functions*, 2d ed., Macmillan, New York, 1945.

[2] R. C. Spencer, "Paraboloid Diffraction Patterns from the Standpoint of Physical Optics," RL Report T-7, Oct. 21, 1942.

[3] E. Jahnke and F. Emde, *Tables of Functions*, reprint by Dover Publications, New York, 1943.

TABLE 6·2.—SECONDARY PATTERN CHARACTERISTICS PRODUCED BY A DISTRIBUTION $(1 - r^2)^p$ OVER A CIRCULAR APERTURE

p	$\mathcal{G}$, gain factor	Θ, half-power width	ϑ, position of first zero	First side lobe, db below peak intensity
0	1.00	$1.02 \frac{\lambda}{D}$	$\sin^{-1} \frac{1.22\lambda}{D}$	17.6
1	0.75	$1.27 \frac{\lambda}{D}$	$\sin^{-1} \frac{1.63\lambda}{D}$	24.6
2	0.56	$1.47 \frac{\lambda}{D}$	$\sin^{-1} \frac{2.03\lambda}{D}$	30.6
3	0.44	$1.65 \frac{\lambda}{D}$	$\sin^{-1} \frac{2.42\lambda}{D}$	....
4	0.36	$1.81 \frac{\lambda}{D}$	$\sin^{-1} \frac{2.79\lambda}{D}$	....

This can be evaluated by means of the expansion[1]

$$e^{jur\cos(\phi-\phi')} = \sum_{n=-\infty}^{\infty} (j)^n J_n(ur) e^{jn(\phi-\phi')}. \quad (78)$$

Of particular interest are the patterns in the planes of symmetry, $\phi = 0$ and $\phi = \pi/2$. We find that these are

$\phi = 0$,

$$g(u) = 2\pi a^2 \left\{ \frac{J_1(u)}{u} - \frac{1}{2} \int_0^1 r^2 [J_0(ur) - J_2(ur)] r \, dr \right\}$$
$$= \frac{3\pi a^2}{4} \Lambda_2(u). \quad (79)$$

$\phi = \frac{\pi}{2}$,

$$g(u) = 2\pi a^2 \left\{ \frac{J_1(u)}{u} - \frac{1}{2} \int_0^1 r^2 [J_0(ur) + J_2(ur)] r \, dr \right\}$$
$$= \pi a^2 \left[\Lambda_1(u) - \frac{\Lambda_2(u)}{4} \right]. \quad (80)$$

The two patterns are shown in Fig. 6·8. It is seen that the beamwidth is greater in the plane $\phi = 0$ than in the plane $\phi = \pi/2$, corresponding to the fact that in the first principal plane the illumination over the aperture is tapered whereas in the second the aperture illumination is uniform.

[1] Watson, op. cit., Sec. 2·22.

6·9. The Field on the Axis in the Fresnel Region.—An important consideration in making measurements of secondary patterns is the minimum distance from the aperture at which the field may be regarded as being in the Fraunhofer region. To aid in arriving at a criterion it

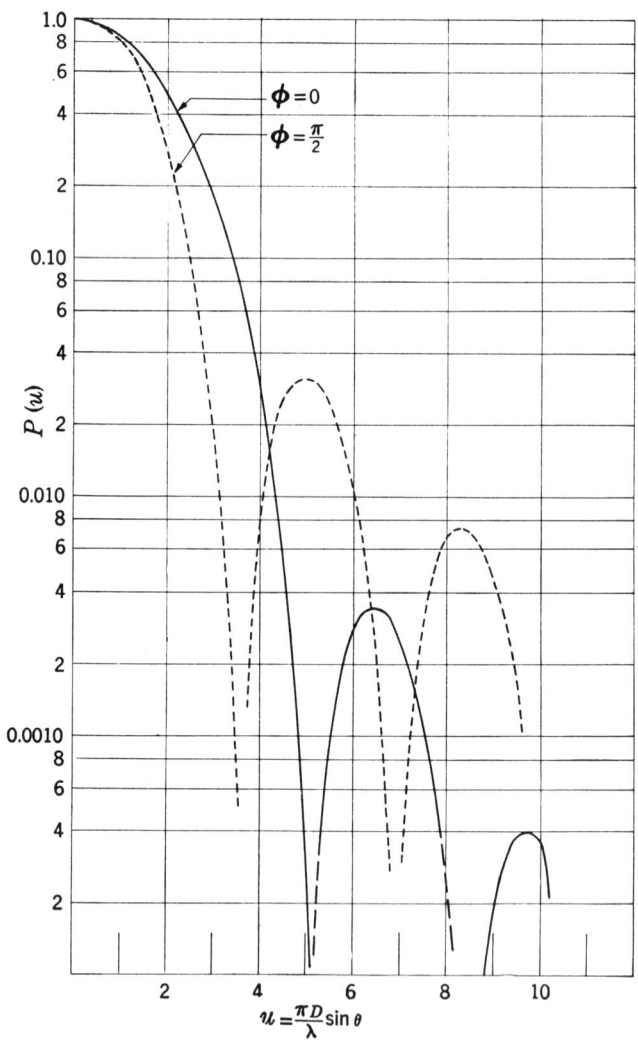

Fig. 6·8.—Principal plane patterns for the aperture distribution $f(r,\phi') = 1 - r^2 \cos^2 \phi'$.

will be well to discuss briefly the field on the axis in the Fresnel region and the transition to the Fraunhofer region. The aperture field will be taken to be uniform in amplitude and phase.

The method of Fresnel zones used extensively in optics affords a simple physical basis for understanding the effects that are observed in the Fresnel region. In Fig. 6·9 let the point P at distance R on the axis be the field point under consideration. Taking the point P as a center, we shall describe a family of spheres of radii $R + \lambda/2, R + 2(\lambda/2)$ .... Their intersections with the aperture divide the latter into annular regions; these are known as the Fresnel zones. The zones will be numbered as shown in the figure. Taking any two adjacent zones n and $n + 1$, it is seen from the method of construction that for every contribution to the field at P arising from an element of surface in the first zone

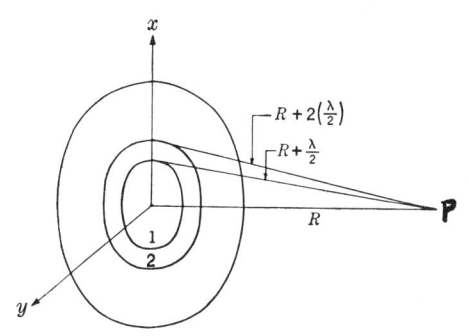

FIG. 6·9.—Division of the aperture into Fresnel zones.

there is a contribution from an element in the second zone 180° out of phase therewith, and the integrated contributions of the two zones are, therefore, very nearly 180° out of phase with one another. Denoting the magnitude of the contribution of the nth zone by S_n, the effect of the entire aperture is

$$\mathbf{S} = S_1 - S_2 + S_3 - S_4 + - \cdots . \tag{81}$$

The contributions S_n decrease slowly with increasing n; the resultant effect of pairs of adjacent zones is therefore virtually equal to zero. A careful analysis[1] shows that if the aperture contains a full number of zones N, the resultant is very closely equal to

$$\mathbf{S} = \tfrac{1}{2}(S_1 \pm S_N) \tag{82}$$

depending upon whether N is odd or even. As R increases, N decreases and $\mathbf{S}$ fluctuates between the values

$$\mathbf{S} = \tfrac{1}{2}(S_1 - S_N) \approx 0 \text{ for } N \text{ even} \tag{83a}$$

and

$$\mathbf{S} = \tfrac{1}{2}(S_1 + S_N) \approx S_1 \text{ for } N \text{ odd}. \tag{83b}$$

The amplitude of the field along the axis, therefore, passes through maxima and minima, the maxima coming at the points that subtend an odd number of Fresnel zones, the minima at the points subtending an

[1] See, for example, M. Born, *Optik*, p. 145, reprint by Edwards Bros., Ann Arbor, Mich., 1943.

even number. There will be no further fluctuations beyond the point on the axis for which the entire aperture consists of a single Fresnel zone; this distance is

$$R_f = \frac{D^2 - \lambda^2}{4\lambda} \approx \frac{D^2}{4\lambda}, \qquad (84)$$

D being the diameter of the aperture. However, the distance R_f cannot be taken as the beginning point of the Fraunhofer region, for at that point the contribution from the edge of the aperture is still 180° out of phase with that from the center; whereas the calculation of the Fraunhofer

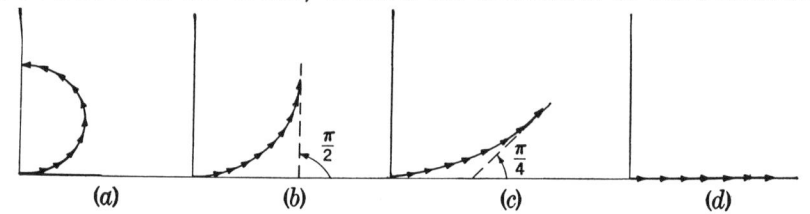

FIG. 6·10.—The transition region between the Fresnel and the Fraunhofer regions: (a) $R = R_f \approx D^2/4\lambda$; (b) $R \approx D^2/2\lambda$; (c) $R = D^2/\lambda$; (d) $R = \infty$.

region, on axis, assumes that the path differences between points on the aperture to the field point are negligible. Considering the aperture to be subdivided into small annular zones and resolving the resultant effect of the aperture into the superposition of the vector elements of these zones, one finds that the vector diagrams take the forms shown in Fig. 6·10 for distances greater than R_f.[1] The slope angle of the vector diagram at the terminal point is equal to the phase difference between the edge of the aperture and the center, corresponding to the difference in path length to the field point. At a distance greater than $D^2/2\lambda$ (Fig. 6·10b) there is no longer any cancellation between horizontal components of the vector elements; at a distance D^2/λ the resultant is a good approximation to the value for $R = \infty$.

To make a more quantitative evaluation we must consider the actual values of the field intensity and the gain. For this purpose we will start from the Fresnel approximation [Eq. (7a)], which in the present case takes the form

$$U_p = \frac{j}{\lambda} \frac{e^{-jkR}}{R} \int_0^{2\pi} \int_0^a F(\rho, \phi') e^{-jk\rho^2/2R} \rho \, d\rho \, d\phi'. \qquad (85)$$

It will be recognized that this is equivalent to the expression for the on-axis field intensity in the Fraunhofer region of an aperture having a quadratic phase error $\rho^2/2R$. Equation (78) is easily integrated for uniform illumination giving

$$U_p = 2j \sin\left(\frac{ka^2}{4R}\right) e^{-jkR}. \qquad (86)$$

[1] The vector diagrams depict the variation of only the form factors of the field—that is, the integrals of Eqs. (7)—with increasing distance R.

SEC. 6·9] *THE FIELD ON THE AXIS IN THE FRESNEL REGION* 199

This corresponds to radiated power per unit solid angle

$$P_m = \frac{1}{2}\left(\frac{\epsilon}{\mu}\right)^{1/2} 4R^2 \left[\sin\left(\frac{ka^2}{4R}\right)\right]^2$$

or

$$P_m = \frac{1}{2}\left(\frac{\epsilon}{\mu}\right)^{1/2} \frac{A}{\lambda} \left(\frac{\sin x}{x}\right)^2; \qquad x = \frac{ka^2}{4R}. \tag{87}$$

The total power radiated by the aperture is simply $\frac{1}{2}(\epsilon/\mu)^{1/2}A$, whence the gain is

$$G = \frac{4\pi A}{\lambda^2}\left(\frac{\sin x}{x}\right)^2. \tag{88}$$

The factor $(\sin x/x)^2$ expresses the ratio of the gain measured at a distance R to the gain G_0 of the true Fraunhofer field at infinity. The accepted values of the minimum distance at which pattern measurements may properly be made vary between $R = D^2/\lambda$ and $R = 2D^2/\lambda$. The values of the gain ratio of Eq. (89) for the two cases are

$$R = \frac{D^2}{\lambda}, \qquad \frac{G}{G_0} = 0.94; \tag{89}$$

$$R = \frac{2D^2}{\lambda}, \qquad \frac{G}{G_0} = 0.99; \tag{90}$$

There is a little difference between them; for the cases most commonly encountered the $2D^2/\lambda$ criterion is to be favored. Other considerations which are discussed in Chap. 15 also point in that direction.

CHAPTER 7

MICROWAVE TRANSMISSION LINES

BY S. SILVER

We have dwelt at considerable length on general theoretical considerations underlying the design and operation of microwave antennas as a whole. We now enter upon a program of studying the components of an antenna, starting with an investigation of microwave transmission lines.

Usually about a foot, or perhaps two, of the line immediately preceding the radiating system is at the disposal of the engineer for the insertion of matching devices to compensate for the impedance mismatch of his antenna; this section will be referred to as the feed line. The following discussion of feed lines will be confined to elementary transmission-line theory and problems; for more extensive treatments, in particular for the analysis of matching devices, the reader is referred to the sources indicated below.[1]

7·1. Microwave and Long-wave Transmission Lines.—A brief comparison of long-wave and microwave lines was made in Sec. 1·4. It was pointed out that the use of unshielded parallel wire lines becomes impractical at microwave frequencies largely because the power-carrying capacity is so sharply limited by the small interline spacing required if the line is not to radiate. The relation between the interline spacing and radiation follows from the ideas developed in Chap. 3. We may consider the alternating current in a wire as a line distribution of oscillating dipoles; corresponding points on a pair of wires carrying equal and opposite currents are occupied by similar dipoles in opposite phase. If the spacing between the dipoles is small compared with a wavelength, their radiation fields will be out of phase at all points in space and annul each other. On the other hand, if the spacing is comparable to the wavelength, the double-dipole system can radiate, there being directions in space for which path-length differences will compensate for the intrinsic phase difference of the members of the pair. In addition, it should be noted that large interline spacings can be used at long wavelengths, since the radiation-field intensity of a dipole varies inversely as the square of the wavelength [cf. Eqs. (3·148)].

[1] J. C. Slater, *Microwave Transmission*, McGraw-Hill, New York, 1941, Chaps. 3, 4; R. L. Lamont, *Wave Guides*, Methuen, London, 1942; Montgomery, Purcell, and Dickie, *The Principles of Microwave Circuits*, Vol. 8; and N. Marcuvitz, *The Waveguide Handbook*, Vol. 10, of this series.

Another hindrance in the use of unshielded lines is their susceptibility to interference, which would cause serious installation difficulties. If the line is not to radiate, its elements must be symmetrically disposed with respect to near-by conductors, in order that equal and opposite currents be maintained at all paired points on the line. Perturbation of the line balance would also give rise to impedance difficulties.

It accordingly becomes clear that shielded lines are required for transmission at microwave frequencies. Two types are in use: (1) two-conductor lines consisting of one conductor surrounded by a second, separated by dielectric, and (2) hollow metal tubes. These lines are to be considered as waveguides for electromagnetic waves in the enclosed dielectric rather than as transmission lines carrying current and voltage waves. In fact, in the hollow waveguide it is not possible to establish definitions of line current and line voltage that are comparable to the quantities defined in a parallel wire line; under special conditions current and voltage can be defined for the two-conductor line. However, while a new approach must be taken in the fundamental analysis, it is found that under suitable conditions transmission-line analogues of voltage and current can be defined and use can be made of the line theory summarized in Chap. 2.

7·2. Propagation in Waveguides of Uniform Cross Section.—We shall confine our discussion to lines of arbitrary but uniform cross section, that is, waveguides with cylindrical walls. The guide walls will be taken to have infinite conductivity; the dielectric in the interior will be assumed to be homogeneous, with dielectric constant ϵ, permeability μ, and zero conductivity; the dielectric will also be assumed to be free of charge. We shall consider electromagnetic fields in these waveguides which have a harmonic time dependence; they will satisfy the homogeneous field equations obtained from Eqs. (3·32) by setting the source functions and the conductivity equal to zero.

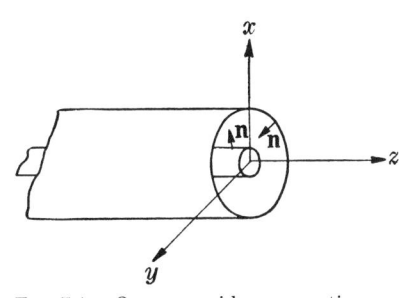

Fig. 7·1.—On waveguide propagation.

We shall take the z-axis of the coordinate system to be parallel to the generator of the cylindrical walls of the waveguide. Since the guide is homogeneous in structure along the z-direction, a wave of a single frequency will depend on z only through a phase factor and possibly a damping factor corresponding to progressive attenuation of the wave. That is, the z-dependence of all field components is of the form $e^{\mp \gamma z}$, where γ is possibly complex:

$$\gamma = \alpha + j\beta. \tag{1}$$

With the convention that α and β are both to be positive quantities, the upper sign in the exponential corresponds to propagation in the positive z-direction, the lower sign to propagation in the negative z-direction. Writing out the field equations in component form and taking into account the postulated form of the z-dependence, we obtain for a wave traveling in the positive z-direction

$$j\omega\mu H_x = -\frac{\partial E_z}{\partial y} - \gamma E_y, \qquad (2a)$$

$$j\omega\mu H_y = \frac{\partial E_z}{\partial x} + \gamma E_x, \qquad (2b)$$

$$j\omega\mu H_z = -\frac{\partial E_y}{\partial x} + \frac{\partial E_x}{\partial y}, \qquad (2c)$$

and

$$j\omega\epsilon E_x = \frac{\partial H_z}{\partial y} + \gamma H_y, \qquad (3a)$$

$$j\omega\epsilon E_y = -\frac{\partial H_z}{\partial x} - \gamma H_x, \qquad (3b)$$

$$j\omega\epsilon E_z = \frac{\partial H_y}{\partial x} - \frac{\partial H_x}{\partial y}. \qquad (3c)$$

For most purposes an alternative set of equations is more convenient to use. Taking Eqs. (2a), (2b), (3a), and (3b) we find, on substitution and rearrangement,

$$\kappa^2 E_x = -j\omega\mu \frac{\partial H_z}{\partial y} - \gamma \frac{\partial E_z}{\partial x}, \qquad (4a)$$

$$\kappa^2 E_y = j\omega\mu \frac{\partial H_z}{\partial x} - \gamma \frac{\partial E_z}{\partial y}, \qquad (4b)$$

and

$$\kappa^2 H_x = j\omega\epsilon \frac{\partial E_z}{\partial y} - \gamma \frac{\partial H_z}{\partial x}, \qquad (5a)$$

$$\kappa^2 H_y = -j\omega\epsilon \frac{\partial E_z}{\partial x} - \gamma \frac{\partial H_z}{\partial y}. \qquad (5b)$$

On substitution of these into Eqs. (2c) and (3c), we obtain

$$\frac{\partial^2 E_z}{\partial x^2} + \frac{\partial^2 E_z}{\partial y^2} + \kappa^2 E_z = 0, \qquad (6a)$$

$$\frac{\partial^2 H_z}{\partial x^2} + \frac{\partial^2 H_z}{\partial y^2} + \kappa^2 H_z = 0, \qquad (6b)$$

with

$$\kappa^2 = \omega^2 \mu\epsilon + \gamma^2 = k^2 + \gamma^2. \qquad (7)$$

The structure of this second set of equations shows that there are two independent field components E_z, H_z from which the others can be

derived by Eqs. (4) and (5). We can consequently classify waves in uniform guides into three fundamental types:
1. *TEM*-waves (transverse electromagnetic waves) with $E_z = H_z = 0$.
2. *TE*-waves (transverse electric waves) with $E_z = 0$, $H_z \neq 0$.
3. *TM*-waves (transverse magnetic waves) with $H_z = 0$, $E_z \neq 0$.

TEM-waves.—These are also known as the principal waves. The electric and magnetic field vectors lie in a plane transverse to the direction of propagation as they do in a plane wave. From Eqs. (4) and (5) it is evident that the principal wave will vanish identically unless $\kappa^2 = 0$. With this condition on κ, we find from Eq. (7)

$$\gamma = (\alpha + j\beta) = jk = j\omega(\mu\epsilon)^{1/2}. \tag{8}$$

If such a wave can exist, it will propagate without attenuation and with a phase constant $\beta = \omega(\mu\epsilon)^{1/2} = 2\pi/\lambda$ which is the same as that for a plane wave in an unbounded medium.

To obtain information about the field vectors we must return to Eqs. (2) and (3). It is directly evident from Eqs. (2a) and (2b) that

$$\mathbf{H} = \left(\frac{\epsilon}{\mu}\right)^{1/2} (\mathbf{i}_z \times \mathbf{E}). \tag{9}$$

Thus **E** and **H** are related as in a plane wave; they are mutually perpendicular and transverse to the direction of propagation. Equation (2c) becomes

$$\frac{\partial E_y}{\partial x} - \frac{\partial E_x}{\partial y} = 0,$$

whereas substitution of Eqs. (2a) and (2b) into Eq. (3c) yields

$$\frac{\partial E_x}{\partial x} + \frac{\partial E_y}{\partial y} = 0.$$

The first of these states that in the dependence on x and y the field is derivable from a potential function $U(x,y)$; that is, we can write

$$\mathbf{E} = e^{-\gamma z}\, \boldsymbol{\nabla} U(x,y). \tag{10}$$

It then follows from the second of the above equations that $U(x,y)$ must be a solution of the two-dimensional Laplace equation:

$$\frac{\partial^2 U}{\partial x^2} + \frac{\partial^2 U}{\partial y^2} = 0. \tag{11}$$

The electric vector of Eq. (10) is everywhere normal to the equipotential surfaces $U =$ constant. Since the conductivity of the guide walls is infinite, **E** must be normal to the walls by the boundary conditions formulated in Sec. 3·3; consequently, the walls of the guide must correspond to equipotential lines of U. This, however, raises an important distinction between single-conductor and two-conductor lines. In the case of a single-conductor line we are concerned with a solution of Laplace's equation in a simply connected region—a solution that assumes a constant value over the boundary. The only such solution is that for which U is a constant over the entire region; hence the gradient and the field vectors are zero. *There is no TEM-wave possible in a hollow waveguide.* In the two-conductor line we seek solutions of Laplace's equation in a multiply connected region. The required solution assumes one constant value over one boundary and another constant value over the second boundary, as in the electrostatic problem of two conductors at different potentials. Such solutions exist; consequently *a TEM-wave is possible in any two-conductor line. Furthermore since γ is a pure imaginary for all frequencies, the lossless two-conductor line supports free propagation of this wave type at all frequencies.*

TE-waves.—These are known also as H-waves. The electric field is wholly transverse to the direction of propagation, while the magnetic field has a longitudinal component H_z in the direction of propagation. It is clear from Eqs. (4) and (5) that all the other components can be derived from H_z. If we write

$$H_z = \psi(x,y)e^{-\gamma z}, \tag{12}$$

ψ must satisfy Eq. (6b):

$$\nabla^2 \psi + \kappa^2 \psi = 0. \tag{7.6b}$$

We must find solutions of Eq. (6b) that lead to field components satisfying appropriate boundary conditions at the guide walls. By Eqs. (3·24) and (3·28), these conditions are

$$(\mathbf{n} \times \mathbf{E}) = (n_x E_y - n_y E_x)\mathbf{i}_z = 0, \tag{I}$$
$$(\mathbf{n} \cdot \mathbf{H}) = n_x H_x + n_y H_y = 0, \tag{II}$$

where $\mathbf{n}$ is a unit normal to the boundary, directed into the interior of the guide. From Eqs. (5a) and (5b) it follows that (II) is equivalent to requiring

$$\frac{\partial H_z}{\partial n} = \frac{\partial \psi}{\partial n} = 0 \tag{13}$$

over the boundary. On inserting the values of E_x, E_y from Eqs. (4a) and (4b) into (I), one finds that condition (I) likewise reduces to Eq. (13). Thus, the boundary condition (13) is the only one that need be imposed on the solution.

Solutions to Eq. (6b) which satisfy Eq. (13) are possible only for definite values of κ. These are known as the characteristic values; we shall designate them by κ_{mn}. To each characteristic value there corresponds a set of wave types which are spoken of as modes of propagation; in most cases of interest there is only one mode for each value of κ. Any one mode is completely specified by giving the field configuration over a cross section of the line. The propagation constant γ_{mn} for a given mode is

$$\gamma_{mn} = (\kappa_{mn}^2 - k^2)^{\frac{1}{2}}. \tag{14}$$

It is immediately evident that if $\kappa_{mn}^2 < k^2$, then γ_{mn} is a pure imaginary and the wave is propagated without attenuation. Conversely, if $\kappa_{mn}^2 > k^2$, then γ_{mn} is real and the wave is attenuated. A wave of frequency $\nu = 2\pi/\omega$ will, therefore, be freely propagated only in those modes for which $\omega(\mu\epsilon)^{\frac{1}{2}} = 2\pi/\lambda > \kappa_{mn}$. The phase constant for a given mode in which free propagation takes place is

$$\beta_{mn} = \frac{2\pi}{\lambda_{g_{mn}}} = (k^2 - \kappa_{mn}^2)^{\frac{1}{2}}. \tag{15}$$

If we define the cutoff wavelength $\lambda_{mn}^{(c)}$ by the equation

$$\kappa_{mn} = \frac{2\pi}{\lambda_{mn}^{(c)}}, \tag{16}$$

then the wavelength in the guide is

$$\lambda_{g_{mn}} = \frac{\lambda}{\left[1 - \left(\frac{\lambda}{\lambda_{mn}^{(c)}}\right)^2\right]^{\frac{1}{2}}}. \tag{17}$$

When the wavelength in unbounded dielectric exceeds the cutoff wavelength, the wave cannot propagate in that particular mode. A hollow waveguide thus behaves like a high-pass filter, for there is a definite upper limit to the cutoff wavelength, corresponding to the smallest characteristic value κ_{mn}. In terms of the free-space wavelength λ_0 and the specific inductive capacity $k_e = \epsilon/\epsilon_0$, the guide wavelength is given by

$$\lambda_{g_{mn}} = \frac{\lambda_0/k_e^{\frac{1}{2}}}{\left[1 - \frac{1}{k_e}\left(\frac{\lambda_0}{\lambda_{mn}^{(c)}}\right)^2\right]^{\frac{1}{2}}}; \tag{17a}$$

the permeability μ of the medium is assumed to be negligibly different from that of free space, μ_0.

The wave type, or mode, corresponding to a characteristic value κ_{mn} is designated as TE_{mn}. It follows from Eqs. (5) and (12) that the transverse magnetic field is given by

$$\mathbf{H}_t = -\frac{\gamma}{\kappa^2} e^{-\gamma z} \nabla\psi.$$

The complete magnetic field is, therefore,

$$\mathbf{H} = e^{-\gamma z}\left(-\frac{\gamma}{\kappa^2}\nabla\psi + \psi \mathbf{i}_z\right). \tag{18}$$

The electric field is given by $\mathbf{E} = (1/j\omega\epsilon)\nabla \times \mathbf{H}$ or

$$\mathbf{E}_t = -\frac{\omega\mu}{j\gamma}(\mathbf{H} \times \mathbf{i}_z). \tag{19}$$

TM-waves.—The magnetic field is wholly transverse to the direction of propagation, whereas the electric field has a component E_z in the direction of propagation. These waves are known also as *E*-waves. If we write

$$E_z = \phi(x,y)e^{-\gamma z}, \tag{20}$$

$\phi(x,y)$ must satisfy Eq. (6a):

$$\nabla^2\phi + \kappa^2\phi = 0, \tag{7.6a}$$

which is the same equation as that for $\psi(x,y)$ in the case of *TE*-waves. The essential difference between the problems arises from the boundary conditions. The boundary condition (I) is a statement that at the walls the tangential electric field must be zero. We thus require

$$n_x E_y - n_y E_x = 0, \tag{21a}$$
$$E_z = 0; \quad \text{i.e.,} \quad \phi(x,y) = 0, \tag{21b}$$

over the walls. Substituting from Eqs. (4a) and (4b), we find that condition (21a) is equivalent to

$$(\mathbf{n} \times \nabla\phi) = 0. \tag{21c}$$

If condition (21b) is satisfied, the boundary corresponds to a curve of constant ϕ; hence $\nabla\phi$ is normal to the boundary, and condition (21c) is automatically satisfied. Further, from Eqs. (5a) and (5b) it follows that for *TM*-waves Eq. (21a) is equivalent to the boundary condition (II) on the magnetic field, stated previously. Again, therefore, we have a single boundary condition, namely, Eq. (21b), to impose on the solutions of Eq. (6a).

As in the case of *TE*-waves, it is found that solutions $\phi(x,y)$ of Eq. (6a) which satisfy the boundary condition exist only for certain characteristic values κ_{mn}; these are, of course, different from the *TE*-values. To each characteristic value there corresponds at least one wave type or *TM*-mode. The general remarks concerning the propagation constant γ_{mn} and the conditions for free propagation are equally applicable to the *TM*-mode; the guide wavelength is given again by Eqs. (17) and (17a). It follows from Eqs. (4a) and (4b) that the complete electric field, for a single mode, is given by

$$\mathbf{E} = e^{-\gamma z}\left(-\frac{\gamma}{\kappa^2}\nabla\phi + \phi\mathbf{i}_z\right); \quad (22)$$

the magnetic field obtained therefrom by $\mathbf{H} = -(1/j\omega\mu)(\nabla \times \mathbf{E})$ is

$$\mathbf{H}_t = \frac{\omega\epsilon}{j\gamma}(\mathbf{E} \times \mathbf{i}_z). \quad (23)$$

7·3. Orthogonality Relations and Power Flow.—Examination of Eqs. (18), (19), (22), and (23) shows that in a freely propagated mode, for which $\gamma = j\beta$ is a pure imaginary, the transverse electric and magnetic fields are in phase with each other and are in time quadrature with the longitudinal field component associated with the given mode. The functions ψ and ϕ are arbitrary to within a multiplicative constant; by a simple readjustment of constants which does not affect the relative magnitudes and phases of the field components, the latter can be put in the following form:

TE-waves:

$$H_z = jH_{az}e^{-j\beta_a z}; \quad H_{az} = \frac{\kappa_a^2}{\omega\mu}\psi_a, \quad (24a)$$

$$\mathbf{E}_t = \mathbf{E}_{at}e^{-j\beta_a z}; \quad \mathbf{E}_{at} = \nabla\psi_a \times \mathbf{i}_z, \quad (24b)$$

$$\mathbf{H}_t = \mathbf{H}_{at}e^{-j\beta_a z}; \quad \mathbf{H}_{at} = \frac{\beta_a}{\omega\mu}\nabla\phi_a. \quad (24c)$$

TM-waves:

$$E_z = jE_{az}e^{-j\beta_a z}; \quad E_{az} = \frac{\kappa_a^2}{\beta_a}\phi_a, \quad (25a)$$

$$\mathbf{E}_t = \mathbf{E}_{at}e^{-j\beta_a z}; \quad \mathbf{E}_{at} = \nabla\phi_a, \quad (25b)$$

$$\mathbf{H}_t = \mathbf{H}_{at}e^{-j\beta_a z}; \quad \mathbf{H}_{at} = \frac{\omega\epsilon}{\beta_a}\mathbf{i}_z \times \nabla\phi_a. \quad (25c)$$

where the functions H_{az}, E_{az}, $\mathbf{E}_{at}$, and $\mathbf{H}_{at}$ are all real. The subscript a represents the pair of mode indices m, n. Equations (24) and (25) are, of course, still to be multiplied by arbitrary constants determining the amplitudes of the waves. From these expressions it is seen that the Poynting vector $\mathbf{S} = \frac{1}{2}\text{Re}\,(\mathbf{E} \times \mathbf{H}^*)$ arises entirely from the transverse field components; the power flow is, therefore, entirely along the axis of the waveguide, no power flowing into the walls of the guide.

The same expressions [Eqs. (24) and (25)] with β_a replaced by $j\gamma_a$, γ_a being real, serve also for the modes that are beyond cutoff for the given operating wavelength. It is seen that in these modes the transverse electric and magnetic fields are in time quadrature; consequently, there is no energy flow along the axis of the guide. In fact, the Poynting vector $\mathbf{S} = \frac{1}{2}\text{Re}\,(\mathbf{E} \times \mathbf{H}^*)$ vanishes completely; the energy associated with these modes is stored in the waveguide in the neighborhood of the point of their excitation.

The modes possess important orthogonality properties.[1] The total power transported through any cross section of a guide that is supporting free propagation of several modes is the sum of the powers transported by the separate modes; there is no energy coupling between modes. For example, let us consider the power transport P_{ab} of the mixed Poynting vector $\mathbf{S}_{ab} = \frac{1}{2}(\mathbf{E}_{at} \times \mathbf{H}_{bt})$ of a pair of modes TE_a and TE_b; we have

$$P_{ab} = \frac{1}{2} \int_{\substack{\text{cross} \\ \text{section}}} (\mathbf{E}_{at} \times \mathbf{H}_{bt}) \cdot \mathbf{i}_z \, dS$$

$$= \frac{\beta_b}{2\omega\mu} \int [(\nabla \psi_a \times \mathbf{i}_z) \times \nabla \psi_b] \cdot \mathbf{i}_z \, dS = \frac{\beta_b}{2\omega\mu} \int \nabla \psi_a \cdot \nabla \psi_b \, dS. \quad (26)$$

The last integral transforms as follows:

$$\int \nabla \psi_a \cdot \nabla \psi_b \, dS = \int \nabla \cdot (\psi_a \nabla \psi_b) \, dS - \int \psi_a \nabla^2 \psi_b \, dS. \quad (27)$$

By means of Green's theorem, the first integral on the right-hand side transforms into a line integral over the boundary:

$$\int \nabla \cdot (\psi_a \nabla \psi_b) \, dS = - \oint \psi_a \frac{\partial \psi_b}{\partial n} \, ds, \quad (28)$$

the positive normal to the boundary being taken as shown in Fig. 7·1. Since the function ψ_b satisfies the boundary condition [Eq. (13)] for the TE-modes, the integral (28) is equal to zero. Making use of Eq. (6b) we have then

$$\int \nabla \psi_a \cdot \nabla \psi_b \, dS = \kappa_b^2 \int \psi_a \psi_b \, dS. \quad (29)$$

Interchanging the role of ψ_a and ψ_b in Eq. (28), we arrive in a similar manner to

$$\int \nabla \psi_a \cdot \nabla \psi_b \, dS = \kappa_a^2 \int \psi_a \psi_b \, dS. \quad (29a)$$

It is evident that if $a \neq b$, Eqs. (29) and (29a) can both be satisfied only if

$$\int \nabla \psi_a \cdot \nabla \psi_b \, dS = \int \psi_a \psi_b \, dS = 0. \quad (30)$$

We have thus found that

$$P_{ab} = 0, \quad a \neq b$$
$$= \frac{\beta_a}{2\omega\mu} \int |\nabla \psi_a|^2 \, dS, \quad a = b. \quad (31)$$

---

[1] H. A. Bethe, "Formal Theory of Waveguides of Arbitrary Cross Section," RL Report No. 43-26, March 1943.

It is readily seen that the proof applies without change to the case where one or both of the modes are beyond cutoff. Similar techniques lead to the result that there is no energy coupling between pairs of *TM*-modes or between a *TE*- and *TM*-mode. The power relation is only one of a number of orthogonality properties. The others are given without proof: if $a \neq b$,

$$\int E_{az}E_{bz}\,dS = \int H_{az}H_{bz}\,dS = 0, \tag{32a}$$

$$\int \mathbf{E}_{at} \cdot \mathbf{E}_{bt}\,dS = \int \mathbf{H}_{at} \cdot \mathbf{H}_{bt}\,dS = 0. \tag{32b}$$

7·4. Transmission-line Considerations in Waveguides.—We have concerned ourselves in the foregoing with a wave propagated in the positive z-direction; this is the physical situation which would exist in a waveguide extending to $z = +\infty$ with a generator at some remote point along the negative z-axis. It was found for every wave type that in a single mode there is a simple linear relationship between the transverse components of the electric and magnetic fields:

$$TEM\text{-mode: } \mathbf{H} = \frac{1}{Z^{(0)}}(\mathbf{i}_z \times \mathbf{E}), \quad Z^{(0)} = \left(\frac{\mu}{\epsilon}\right)^{\frac{1}{2}}, \tag{33a}$$

$$TE\text{-mode: } \mathbf{E} = Z_{mn}^{(0)}(\mathbf{H} \times \mathbf{i}_z), \quad Z_{mn}^{(0)} = \frac{j\omega\mu}{\gamma_{mn}}, \tag{33b}$$

$$TM\text{-mode: } \mathbf{H} = \frac{1}{Z_{mn}^{(0)}}(\mathbf{i}_z \times \mathbf{E}), \quad Z_{mn}^{(0)} = \frac{\gamma_{mn}}{j\omega\epsilon}. \tag{33c}$$

These are analogous to the current-voltage relationships in a single wave on an infinite two-wire transmission line. The quantity $Z_{mn}^{(0)}$ is known as the *transverse wave impedance*.

The general field for a single mode in a waveguide that does not extend to infinity consists of two waves, one propagating in the positive z-direction, the other in the negative z-direction. The field expressions for the latter are fundamentally the same as those given by the sets of Eqs. (24) and (25), but with $e^{j\beta_a z}$ replacing $e^{-j\beta_a z}$, and with the magnetic field components reversed in sign to give the proper direction to the Poynting vector of the wave. Consider, for example, the TE_a-mode. Let A_a and B_a be the amplitudes of the electric field in the waves propagating in the positive and negative z-directions, respectively; from Eqs. (24b) and (24c) we have then that the transverse fields are

$$\mathbf{E}_t = (A_a e^{-j\beta_a z} + B_a e^{j\beta_a z})(\nabla\psi_a \times \mathbf{i}_z), \tag{34a}$$

$$\mathbf{H}_t = \frac{1}{Z_a^{(0)}}(A_a e^{-j\beta_a z} - B_a e^{j\beta_a z})\nabla\psi_a. \tag{34b}$$

On considering the scalar factors that express the dependence of the fields on position along the waveguide axis, it is seen that the mode can be set into equivalence with a two-wire transmission line of characteristic impedance $Z_a^{(0)}$; the electric and magnetic fields are the analogues of the voltage and current, respectively. It must be noted that the characteristic impedance of the equivalent line differs from mode to mode; and consequently, a waveguide supporting free propagation of a number of modes cannot be set into correspondence with any one two-wire line.

The definition of the equivalent two-wire line for a given mode is arbitrary to a considerable extent. Given a function $\psi_a(x,y)$, we may define a pair of vector functions

$$\mathbf{g}_a(x,y) = c_1 \nabla \psi_a \times \mathbf{i}_z, \tag{35a}$$
$$\mathbf{h}_a(x,y) = c_2 \nabla \psi_a, \tag{35b}$$

where the constants c_1 and c_2 are required to be such that

$$\int_{\text{cross section}} (\mathbf{g}_a \times \mathbf{h}_a) \cdot \mathbf{i}_z \, dS = c_1 c_2 \int |\nabla \psi_a|^2 \, dS = 1. \tag{35c}$$

The constants c_1 and c_2 are arbitrary. In terms of the new vector functions, Eqs. (34a) and (34b) can then be written

$$\mathbf{E}_t = V_a \mathbf{g}_a(x,y) = [V_a^{(+)} e^{-j\beta_a z} + V_a^{(-)} e^{j\beta_a z}] \mathbf{g}_a(x,y), \tag{36a}$$
$$\mathbf{H}_t = I_a \mathbf{h}_a(x,y) = \frac{1}{Z_a^{(0)}} \frac{c_1}{c_2} [V_a^{(+)} e^{-j\beta_a z} - V_a^{(-)} e^{j\beta_a z}] \mathbf{h}_a(x,y). \tag{36b}$$

The quantities V_a and I_a will be named the voltage and current parameters of the mode, respectively. The voltage parameter is the sum of two "voltage" waves traveling in opposite directions, of amplitudes $V_a^{(+)}$ and $V_a^{(-)}$, respectively.

Equations (36a) and (36b) serve to emphasize the arbitrary feature of the two-wire line equivalent of a waveguide mode. The ratio c_1/c_2 can be chosen at will; given any ratio, the characteristic impedance of the equivalent line is

$$Z_0 = Z_a^{(0)} \frac{c_2}{c_1}; \tag{37}$$

the voltage and current parameters represent directly the voltage and current on the equivalent line. The voltage and current parameters possess one property that is unique, independent of the arbitrary choice of the constants c_1 and c_2, provided Eq. (35c) is satisfied. The net power passing through the cross section of the guide in the positive z-direction is

$$P = \tfrac{1}{2} \int [\text{Re } (\mathbf{E}_t \times \mathbf{H}_t^*)] \cdot \mathbf{i}_z \, dS = \tfrac{1}{2} \text{ Re } V_a I_a^*. \tag{38}$$

Thus, any choice of definition of the voltage and current parameters leads to a two-wire line representation in which the power flow computed on the basis of the equivalent voltage and current is equal to the power transport along the waveguide of the given mode.

One possible choice of the definition of the equivalent transmission line is to take $c_1 = c_2 = 1$. The function $\psi_a(x,y)$ is itself arbitrary to within a multiplicative constant; it can, therefore, be chosen so that it satisfies the normalization condition

$$\int_{\text{cross section}} |\nabla \psi_a|^2 \, dS = 1. \tag{39}$$

The characteristic impedance of the line in this case is equal to the transverse wave impedance. This definition has one shortcoming: It is possible to change the dimensions of the waveguide, other than by a scale factor, without changing the characteristic impedance of the equivalent line. Consider, for example, a pair of two-conductor lines, having different cross-sectional dimensions and configurations, joined together to form an infinite line. The wave impedance of the TEM-mode is independent of the cross-sectional dimensions, and on that basis alone the hybrid line is equivalent to an infinite homogeneous two-wire line. The treatment of the junction effect can be simplified considerably by a different choice of the definition of the characteristic impedance of the line, obtained by multiplying the wave impedance by a factor c_2/c_1 that is a function of the cross-section geometry. In Sec. 7·6 it will be shown that there is a natural physical definition of the voltage and current parameters for a TEM-mode which leads to a characteristic impedance having the desired properties. Similar considerations apply to the other modes; it is possible to choose the ratio c_2/c_1 in Eq. (37) to be a function of the cross-sectional dimensions of the waveguide in such a way as to simplify the analysis of problems involving junctions between waveguides of different cross section.[1]

The transmission-line analogy develops more fully if we consider the waveguide to undergo sudden changes in structure. Such changes may be produced by obstacles inserted at some point in the guide, a sharp transition in the properties of the dielectric medium, or a sudden transition to a waveguide of different cross section—to mention but a few. We shall consider in detail the simplest of these cases—a sharp transition in the dielectric in a guide of uniform cross section. For convenience the boundary between the two media will be taken to be in the plane $z = 0$, as shown in Fig. 7·2. Let the constants of the medium to the left of $z = 0$ be ϵ_1, μ_1 and those of the medium to the right of $z = 0$ be

---

[1] For further details see J. C. Slater, *Microwave Transmission*, McGraw-Hill, New York, 1942, Chap. 4.

ϵ_2, μ_2. As a typical case consider a TE-wave of a single mode to be incident on the boundary from the left. To the right of $z = 0$ we will have a transmitted wave and to the left a reflected wave, in addition to the incident wave. No other waves will occur, since the contours of the cross section are uniform and there is no necessary distortion of the field configuration at the boundary; these three waves will suffice to satisfy the boundary conditions on the fields at the discontinuity.

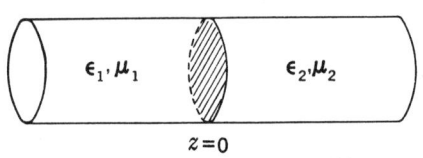

FIG. 7·2.—Discontinuity in waveguide structure.

The field vectors of each of the three waves are derived from a scalar function $\psi(x,y)$ according to Eqs. (24a) to (24c). Furthermore, the scalar functions for the three waves all satisfy the same differential equation [Eq. (6b)], and the same boundary conditions at the walls of the waveguide; hence, all three fields derive from the same scalar function. The ratio c_2/c_1 in Eq. (37) is of no immediate consequence in this case, because the cross section is uniform and may be chosen equal to unity; the function $\psi(x,y)$ may also be required to satisfy the normalization condition [Eq. (39)]. The only significant differences between the waves are the amplitudes and the transverse wave impedance. The field in region 1 is then[1]

$$\mathbf{E}_{t_1} = [V_1^{(+)} e^{-j\beta_1 z} + V_1^{(-)} e^{j\beta_1 z}]\mathbf{g}(x,y), \tag{40a}$$

$$\mathbf{H}_{t_1} = \frac{1}{Z_1^{(0)}} [V_1^{(+)} e^{j\beta_1 z} - V_1^{(-)} e^{j\beta_1 z}]\mathbf{h}(x,y); \tag{40b}$$

and in region 2,

$$\mathbf{E}_{t_2} = [V_2^{(+)} e^{-j\beta_2 z}]\mathbf{g}(x,y), \tag{41a}$$

$$\mathbf{H}_{t_2} = \frac{1}{Z_2^{(0)}} [V_2^{(+)} e^{-j\beta_2 z}]\mathbf{h}(x,y). \tag{41b}$$

According to the boundary conditions (Sec. 3·3) the transverse electric and magnetic fields must be continuous across the plane $z = 0$; we have then

$$V_1^{(+)} + V_1^{(-)} = V_2^{(+)}, \tag{42a}$$

$$\frac{1}{Z_1^{(0)}} [V_1^{(+)} - V_1^{(-)}] = \frac{1}{Z_2^{(0)}} V_2^{(+)}. \tag{42b}$$

As in the case of a two-wire line, these equations express the continuity of voltage and current at the junction of two lines of different characteristic impedance. We can also define an *electric-field reflection coefficient* $\Gamma(z)$,

[1] The mode subscript a will be dropped to simplify the notation.

$$\Gamma(z) = \frac{V_1^{(-)}}{V_1^{(+)}} e^{2i\beta_1 z} = \Gamma(0)e^{2i\beta_1 z}, \tag{43}$$

which corresponds to the voltage reflection coefficient of Eq. (2·27). From Eqs. (42a) and (42b) the value of the reflection coefficient $\Gamma(0)$ at $z = 0$ is found to be

$$\Gamma(0) = \frac{Z_2^{(0)} - Z_1^{(0)}}{Z_2^{(0)} + Z_1^{(0)}}. \tag{44}$$

It is evident that this is equivalent to the reflection coefficient of a line of characteristic impedance $Z_1^{(0)}$ terminated in an impedance $Z_2^{(0)}$. With respect to the terminal impedance, it will be noted that the line to the right, extending to infinity, is equivalent to a line terminated in its own characteristic impedance and hence presents an input impedance $Z_2^{(0)}$ at the plane $z = 0$.

The reflection coefficient $\Gamma(z)$ is to be regarded as the fundamental transmission-line quantity for a waveguide. Evidently it is free from the arbitrary factors entering into the definition of the voltage and current parameters and the characteristic impedance of the equivalent transmission line. It is apparent from Eq. (43) that it transforms along the line just like a voltage reflection coefficient. Also, on computing the Poynting vectors of the incident and reflected waves, it will be seen that the electric-field reflection coefficient bears the same relation to the incident and reflected power as the voltage reflection coefficient (Sec. 2·7). At any point along the line we can regard the section to the right as presenting an input impedance, normalized to the characteristic impedance of the mode,

$$\zeta(z) = \frac{1 + \Gamma(z)}{1 - \Gamma(z)}. \tag{45}$$

The normalized impedance is also independent of the choice of the definition of the equivalent transmission line. Making use of the transformation property of $\Gamma(z)$ expressed by Eq. (43) it is found that the normalized impedance transforms along the waveguide according to

$$\zeta(z \pm l) = \frac{\zeta(z) \mp j \tan \beta l}{1 \mp j\zeta(z) \tan \beta l}, \tag{46}$$

just as it does on a two-wire line. The normalized admittance can also be defined in the same manner as was done in Sec. 2·6,

$$\eta(z) = \frac{1}{\zeta(z)},$$

and it is evident that it also transforms along the waveguide according to Eq. (46). Thus, the entire discussion in Chap. 2 on impedance mismatch, standing-wave ratios, and line transformations can be carried over to the fields on any one mode in a waveguide.

7·5. Network Equivalents of Junctions and Obstacles.—The development of the problem considered above proceeds in a similar manner for TEM- and TM-modes and leads to equivalent two-wire line analogies for any single mode. The discontinuity that we have considered in that problem is equivalent to a junction between a pair of two-wire lines of different characteristic impedances, such that the capacitative and inductive effects due to the junction are negligible. At such a junction both the current and voltage are continuous, corresponding to the continuity in the transverse magnetic and electric fields, respectively, in the waveguide problem. As a second step in developing the transmission-line analysis we shall consider junction effects and the problem of obstacles inserted into a waveguide. The general theory of these problems is treated extensively in other volumes of this series.[1] We shall restrict ourselves here to several qualitative remarks.

As a specific problem let us consider a junction between two waveguides of the same cross-sectional shape but different dimensions, joined in the plane $z = 0$ (see Fig. 7·3). The dimensions of both guides are assumed to be such that they can support free propagation at the given frequency in one mode only; we shall refer to the latter as the dominant-mode wave. We shall assume the dominant-mode wave, set up by a generator at a remote point on the negative z-axis, to be incident on the junction. Since there is a change in cross section at the junction, we should certainly expect to find a reflected wave of the dominant mode on the left and a transmitted wave of that mode in the waveguide on the right. The fields must join in the plane $z = 0$ so as to satisfy the appropriate boundary conditions. Over the opening in the junction the transverse fields must be continuous; over the metal surface of the junction the transverse electric field and the normal component of the magnetic field must vanish. The latter conditions cannot be satisfied by the three dominant-mode waves alone; higher modes must be excited in both waveguides at the junction.

The generation of the higher modes arises from the necessary distortion of the electric and magnetic fields due to the edge of the junction and its metal surface. The electric-field lines must be normal to the latter—a condition that cannot be met by the dominant mode alone in the waveguide to the right. However, according to our assumptions as

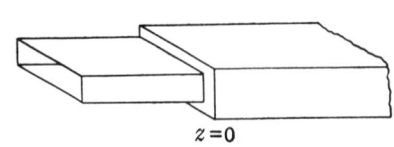

FIG. 7·3.—Junction effects in waveguides.

---

[1] *Principles of Microwave Circuits*, Vol. 8, and *The Waveguide Handbook*, Vol. 10.

to the waveguide dimensions, the higher modes cannot propagate; except within a short distance of the junction (of the order of a wavelength) the fields consist essentially of dominant-mode waves. The higher modes represent electric and magnetic energy stored at the junction. It is possible to represent these energies as energies stored in a reactive network equivalent for the junction.[1] In the general case the network takes the form of a T- or Π-section (see Sec. 2·2). The effect of the junction on the dominant-mode wave thus arises from two factors: (1) a discontinuity in characteristic impedance and (2) a reactive four-terminal network inserted between the lines. The precise values of the elements of the latter network again depend on the definition of the characteristic impedance of the equivalent line for the dominant mode. A number of junction networks are given in the *Waveguide Handbook*, Vol. 10 of this series; in each case, the definition of the characteristic impedance (or its reciprocal, the characteristic admittance) is given. The elements of the network can, of course, be expressed as normalized values with respect to the characteristic impedance of either guide.

In the waveguide to the left, at a short distance from the junction, we have only the incident and reflected dominant-mode waves. Here we can apply transmission-line concepts to the dominant mode and define the corresponding electric-field reflection coefficient. This reflection coefficient can be related to an effective impedance terminating the line at the junction. This impedance, in turn, may be expressed as due to a junction network across the output terminals of which there has been connected the characteristic impedance of the guide to the right. These procedures lead to consistent definitions of the junction impedance. The junction network necessary to represent the stored energies, when inserted between the transmission-line representations of the two waveguides, gives rise to a reflection coefficient in the region on the left corresponding to the electric-field reflection coefficient obtained on the basis of field-theory analysis.

The theory of obstacles develops along similar lines. It is found, as in the case of junctions, that an obstacle has the same effects on impedance and energy stored as a four-terminal network inserted between a pair of transmission lines whose characteristic impedances are the wave impedances of the dominant mode. It must be emphasized that each mode which can propagate has its own transmission-line analogue and that simple transmission-line theory applies to a waveguide only when it can support but one mode. Transmission theory alone can give no information as to the network equivalents of junctions and obstacles; these must be obtained by field-theory analysis. The equivalent network also depends on the particular dominant mode being considered. Once the equivalent network has been established, it can be expressed as a T-sec-

---

[1] See *Principles of Microwave Circuits*, Vol. 8 of this series.

tion, and the impedance transformation properties of such networks can be used in the conventional manner.

7·6. TEM-mode Transmission Lines.—We have pointed out earlier that in general it is not possible to set up unique definitions of voltage and current in waveguides, and we have therefore set up transmission-line analogues in terms of wave impedances and field-reflection coefficients. In the case of TEM-modes, however, it is possible to set up transmission-line quantities that are directly related to the two-wire line quantities discussed in Chap. 2.

It was found in Sec. 7·2 that the electric field over any cross section is derivable from a potential. Hence, over a cross section, the line integral of the electric field from the inner conductor C_1 to the outer conductor C_2 is independent of the path and indeed is equal to the difference between the values of the potential over the conductors. This defines the voltage:

$$Ve^{-\gamma z} = \int_{C_1}^{C_2} \mathbf{E} \cdot d\mathbf{r} = e^{-\gamma z} \int_{C_1}^{C_2} \mathbf{\nabla} U \cdot d\mathbf{r} = (U_2 - U_1)e^{-\gamma z}. \quad (47)$$

There exists also a relation between surface integrals over any closed region in a cross section:

$$\oint (\mathbf{\nabla} \times \mathbf{H}) \cdot \mathbf{i}_z \, dS = j\omega\epsilon \oint \mathbf{E} \cdot \mathbf{i}_z \, dS = 0. \quad (47a)$$

It follows that the line integral of $\mathbf{H}$ over a closed curve surrounding C_1 is independent of the choice of the curve. In particular, let us take a path along the boundary of C_1. $\mathbf{H}$ is tangential to C_1 and by the boundary condition (Sec. 3·3) is equal in magnitude to the surface current density $\mathbf{K}$. Hence the line integral of $\mathbf{H}$ gives the total current carried by C_1:

$$\oint_{C_1} \mathbf{H} \cdot d\mathbf{s} = Ie^{-\gamma z}. \quad (48)$$

The line integral of $\mathbf{H}$ along the boundary of C_2 gives the total current carried by the latter; by virtue of the equality of the line integrals the two currents are equal. On carrying through the details of the vector calculation, it will be found that the current on C_2 is opposite in direction to that on C_1. There is thus a direct two-wire line analogue with voltage V and current I. Corresponding to these we define a characteristic impedance,

$$Z_0 = \frac{V}{I} = \left(\frac{\mu}{\epsilon}\right)^{1/2} \frac{U_2 - U_1}{\oint_{C_1} |\nabla U| \, ds}. \quad (49)$$

This is, of course, different from the wave impedance for the mode.

The relationships between the Z_0 defined in Eq. (49) and the two-wire-line impedance become more evident on calculating the equivalent series inductance and shunt capacitance per unit length of the two-conductor system. The magnetic energy for unit volume is $\frac{1}{2}\mu|H|^2$, and therefore the magnetic energy per unit length of line is

$$W_m = \frac{\epsilon}{2} \iint_{\text{cross section}} |\nabla U|^2 \, dS. \tag{50}$$

If L is the equivalent inductance per unit length, then

$$W_m = \tfrac{1}{2} L I^2;$$

hence

$$L = \frac{\epsilon \iint |\nabla U|^2 \, dS}{I^2}. \tag{50a}$$

Similarly the electric energy per unit volume is $\frac{1}{2}\epsilon|E|^2$, and the electric energy per unit length is

$$W_e = \frac{\epsilon}{2} \iint_{\text{cross section}} |\nabla U|^2 \, dS. \tag{51}$$

The equivalent capacity C per unit length is then

$$W_e = \tfrac{1}{2} C V^2$$

or

$$C = \frac{\epsilon \iint |\nabla U|^2 \, dS}{V^2}. \tag{51a}$$

According to Eq. (2·20) the characteristic impedance of a lossless two-wire line is

$$Z_0 = \sqrt{\frac{L}{C}}.$$

Combining this with Eqs. (50a) and (51a), we obtain the quantity defined in Eq. (49).

For most practical purposes a two-conductor guide supporting the TEM-mode as its dominant wave can be treated from the voltage-current point of view. Applications of this fact will be made in Secs. 7·9 and 7·10 in discussing impedance transformations and matching devices for coaxial lines.

7·7. Coaxial Lines: TEM-mode.—The only type of two-conductor guide of major importance is the coaxial line formed by a pair of concentric circular cylinders. Let a be the radius of the inner conductor, b the radius of the outer conductor. Cylindrical coordinates r, θ, z are suited for the discussion of this system, r and θ being polar coordinates in a cross section of the line. We shall first consider the TEM-mode. The solution to the potential problem is well known from electrostatics:

$$U(x,y) = -\frac{V}{\left(\ln\frac{b}{a}\right)} \ln\left(\frac{r}{a}\right), \tag{52}$$

where V is the voltage across the line. The electric-field intensity is,

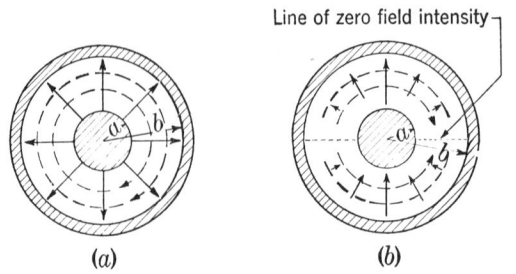

FIG. 7·4.—Coaxial-line modes: (a) TEM-mode (no cutoff wavelength); (b) TE_{11}-mode $[\lambda_{11}{}^{(c)} = (a + b)\pi]$. ———— electric field; – – – – magnetic field.

therefore,

$$\mathbf{E} = \frac{1}{r}\left(\frac{V}{\ln\frac{b}{a}}\right) \mathbf{i}_r e^{\mp j\beta z}, \tag{53}$$

and the magnetic-field intensity

$$\mathbf{H} = \pm \frac{1}{r}\left(\frac{\epsilon}{\mu}\right)^{\frac{1}{2}} \frac{V}{\ln\frac{b}{a}} \mathbf{i}_\theta e^{\mp j\beta z}, \tag{53a}$$

where $\mathbf{i}_r$ and $\mathbf{i}_\theta$ are unit vectors in the directions of increasing r and θ. Here γ has been replaced by $j\beta$, and the double sign indicates a wave traveling in either the positive or negative z-direction. The field configuration and the current distributions on the conductors are completely symmetrical about the z-axis; the former is shown in Fig. 7·4.

The current is given simply by $2\pi r H(r)$:

$$I = 2\pi \left(\frac{\epsilon}{\mu}\right)^{\frac{1}{2}} \frac{V}{\ln\left(\frac{b}{a}\right)}.$$

It follows then directly that the characteristic impedance, in the sense of the previous section, is

$$Z_0 = \frac{1}{2\pi}\left(\frac{\mu}{\epsilon}\right)^{\frac{1}{2}} \ln\left(\frac{b}{a}\right). \tag{54}$$

For most dielectrics of interest μ differs negligibly from the free-space value μ_0. On introduction of the specific inductive capacity $k_e = \epsilon/\epsilon_0$, the characteristic impedance becomes

$$Z_0 = \frac{60}{\sqrt{k_e}} \ln\left(\frac{b}{a}\right). \tag{54a}$$

The series inductance and shunt capacitance per unit length of the line, computed from Eqs. (50a) and (51a), are found to be

$$L = \frac{\mu}{2\pi} \ln\left(\frac{b}{a}\right), \tag{55a}$$

$$C = \frac{2\pi\epsilon}{\ln\left(\frac{b}{a}\right)}. \tag{55b}$$

7·8. Coaxial Lines: TM- and TE-modes.—In the study of the TE- and TM-modes we are concerned with the solutions of equations of the form

$$\frac{\partial^2 F}{\partial x^2} + \frac{\partial^2 F}{\partial y^2} + \kappa^2 F = 0, \tag{7·6a}$$

where F will stand for either of the functions $\psi(x,y)$ or $\phi(x,y)$ of Eqs. (12) and (20) respectively. On introduction of the polar coordinates r, θ, the differential equation becomes

$$\frac{\partial^2 F}{\partial r^2} + \frac{1}{r}\frac{\partial F}{\partial r} + \frac{1}{r^2}\frac{\partial^2 F}{\partial \theta^2} + \kappa^2 F = 0. \tag{56}$$

The equation is separable in the variables r and θ; in particular we shall write

$$F = R(r) \begin{cases} \cos m\theta \\ \sin m\theta \end{cases};$$

then $R(r)$ satisfies the equation

$$\frac{d^2 R}{dr^2} + \frac{1}{r}\frac{dR}{dr} + \kappa^2\left(1 - \frac{m^2}{\kappa^2 r^2}\right) R = 0. \tag{57}$$

This is the differential equation for the cylinder functions of order m, in the variable κr. The pair of linearly independent solutions suited to the finite region with which we are concerned here consists of the Bessel function $J_m(\kappa r)$ and the Neumann function $N_m(\kappa r)$. The latter becomes infinite at $r = 0$; however, since the origin is excluded by the inner conductor, the Neumann function is admissible as a solution. The general solutions of Eq. (56) are therefore

$$\left.\begin{array}{r}\psi(x,y) \\ \phi(x,y)\end{array}\right\} = [A J_m(\kappa r) + B N_m(\kappa r)](C \cos m\theta + D \sin m\theta).$$

The field must be single-valued in θ; as a consequence m can have only integral values. For any given value of m it is possible to eliminate one of the trigonometric functions by proper orientation of the x,y-axes. Without loss of generality we can set $D = 0$, taking as the solutions

$$\left.\begin{array}{l}\psi(x,y) \\ \phi(x,y)\end{array}\right\} = [AJ_m(\kappa r) + BN_m(\kappa r)]\cos m\theta. \tag{58}$$

a. *TM-modes*.—We must consider the TM- and TE-modes separately. In the case of the TM-modes we are concerned with the function $\phi(x,y)$ and the boundary condition of Eq. (21b); we require $\phi = 0$ for all values of θ at $r = a$ and $r = b$. This gives two homogeneous equations,

$$\left.\begin{array}{l}AJ_m(\kappa a) + BN_m(\kappa a) = 0, \\ AJ_m(\kappa b) + BN_m(\kappa b) = 0,\end{array}\right\} \tag{59}$$

for the determination of the ratio B/A. Solutions other than $A = B = 0$ exist only if the determinant of the coefficients vanishes:

$$\begin{vmatrix} J_m(\kappa a) & N_m(\kappa a) \\ J_m(\kappa b) & N_m(\kappa b) \end{vmatrix} = J_m(\kappa a)N_m(\kappa b) - J_m(\kappa b)N_m(\kappa a) = 0. \tag{60}$$

This in turn is satisfied only for a discrete set of values of κ; the latter are the characteristic values which, arranged in order of increasing magnitude, we shall designate by κ_{mn}. If we write $u = \kappa a$, $\alpha = b/a$, the equation appears in the standard form

$$J_m(u)N_m(\alpha u) - N_m(u)J_m(\alpha u) = 0. \tag{60a}$$

Roots of this equation are given in Jahnke and Emde.[1] For a given value of α the smallest value of u_{mn} occurs for $m = 0$; this gives the longest cutoff wavelength for these modes. Examination of the roots shows that for $1 \leq \alpha \leq 7$,

$$\frac{\pi a}{b-a} > u_{01} = a\kappa_{01} > \frac{3a}{b-a}.$$

Therefore the cutoff wavelength $\lambda_{01}^{(c)}$ for the mode is given approximately by

$$\lambda_{01}^{(c)} \approx 2(b-a). \tag{61}$$

We recall that propagation in a given mode can take place only if the wavelength in unbounded dielectric is shorter than $\lambda^{(c)}$. In all practical cases the spacing between the conductors is much smaller than the wavelength, and there is no need to be concerned about the simultaneous excitation of TM- and TEM-modes.

b. *TE-waves*.—Here we are concerned with the function $\psi(x,y)$ and the boundary condition of Eq. (13); for the case at hand the latter becomes $\partial\psi/\partial r = 0$ for $r = a$ and $r = b$. This leads to the conditions

---

[1] E. Jahnke and F. Emde, *Tables of Functions*, Fig. 204, Dover Publications Reprint, New York, 1943.

Sec. 7.9] CASCADE TRANSFORMERS: TEM-MODE

$$\left.\begin{array}{l}AJ'_m(\kappa a) + BN'_m(\kappa a) = 0, \\ AJ'_m(\kappa b) + BN'_m(\kappa b) = 0,\end{array}\right\} \quad (62)$$

on the constants A and B. Nontrivial solutions for the latter exist again only for the characteristic values κ_{mn} that satisfy

$$J'_m(\kappa a)N'_m(\kappa b) - J'_m(\kappa b)N'_m(\kappa a) = 0. \quad (63)$$

For $m = 0$ we have the relation $J'_0(z) = -J_1(z)$, and similarly for the Neumann function; the characteristic values of the TE_{0n}-modes are therefore given by the roots of

$$J_1(u)N_1(\alpha u) - J_1(\alpha u)N_1(u) = 0, \quad (63a)$$

where u, α have the same meanings as previously.

From what has been said about the roots of Eq. (60a), it is evident that the cutoff wavelength of the TE_{0n}-modes is shorter than that of the TM_{01}-mode and that the former are of no consequence as propagating modes in a practical case. The roots of Eq. (63) for $m > 0$ have been discussed by Truell.[1] For our immediate purposes we need concern ourselves only with the lowest mode of the series, the TE_{11}-mode. The field configuration for this mode is illustrated in Fig. 7·4. For this case it is found that the characteristic value is given very closely by

$$\kappa_{11} = \frac{2}{a+b}; \quad (64)$$

thus the cutoff wavelength is

$$\lambda_{11}^{(c)} = \pi(a+b). \quad (64a)$$

This is the mean circumference of the inner and outer conductors. *To prevent propagation of the TE_{11}-mode the mean circumference must be smaller than the operating wavelength.* This imposes limitations on the dimensions of the line and in particular on the spacing between the conductors; the latter in turn limits the power-carrying capacity of the line.

7·9. Cascade Transformers: TEM-mode.—The termination of the line in a radiating system in general gives rise to a reflected TEM-wave and to excitation of TM- and TE-modes. We shall assume that the line dimensions are such that the latter modes cannot propagate and confine our attention to the region of the line where only the incident and reflected TEM-waves exist. The reflected wave represents an impedance mismatch, and it is necessary to consider a correction for it. Perhaps the most useful device is a cascade transformer, a section of coaxial line of characteristic impedance different from that of the main line. Two such transformers are illustrated in Fig. 7·5: (a) the sleeve

[1] R. Truell, *Jour. Applied Phys.*, **14**, 350 (1943).

type with characteristic impedance smaller than the line impedance and (b) the undercut type with characteristic impedance larger than that of the line. As has been pointed out before, the junctions give rise to other modes; however, if the change in radius is small, the junction effect is small. Data on the latter will be given below.

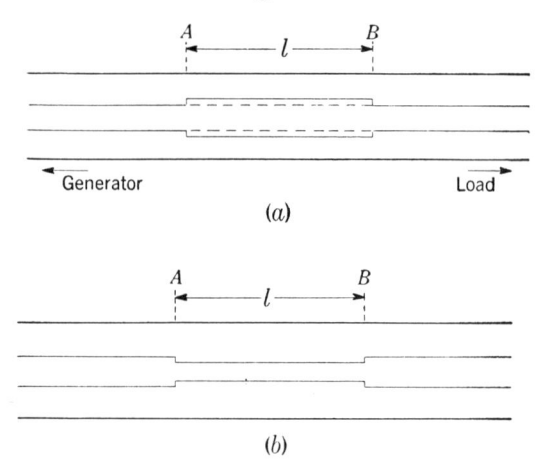

FIG. 7·5.—Cascade impedance transformers: (a) sleeve section; (b) undercut section.

The dimensions desired in a transformer can be determined as follows: Except for junction effects, the voltage and current and the input impedance looking toward the right have the same values at adjacent points on either side of the junction. Let Z_0' be the characteristic impedance of the transformer, Z_0 that of the line, and Z_1 the input impedance at B. Then, from Eq. (2·32), Sec. 2·6, the input impedance at A is

$$Z(A) = Z_0' \left(\frac{Z + jZ_0' \tan \beta l}{Z_0' + jZ \tan \beta l} \right). \tag{65a}$$

Impedance matching requires $Z(A) = Z_0$; that is,

$$Z_0 = Z_0' \left(\frac{Z + jZ_0' \tan \beta l}{Z_0' + jZ \tan \beta l} \right). \tag{65b}$$

Separation of real and imaginary parts gives two equations from which, for a given value of λ, one can obtain Z_0' and the length of the transformer that matches Z into Z_0; the dimensions of the transformers are obtained from Z_0' by means of Eq. (54a).

There are points along the line at which Z is real. These points are $\lambda/4$ apart, and the impedance is alternately rZ_0 and Z_0/r, where r is the voltage standing-wave ratio. If either of these is taken as the junction point B, it is found from Eq. (65b) that $l = \lambda/4$. The characteristic impedance of the quarter-wave section is found to be related to Z_0 as follows:

$$Z'_0 < Z_0 \quad \text{if} \quad Z = \frac{Z_0}{r},$$
$$Z'_0 > Z_0 \quad \text{if} \quad Z = rZ_0.$$

The first of these corresponds to a sleeve section; the second to an undercut section. In so far as matching is concerned, either can be used. The sleeve section has the advantage of simplicity of insertion, since it is necessary only to slip a piece of tubing over the inner conductor and to solder the seam to ensure good contact; it also has the advantage of strengthening the line mechanically. An undercut section requires machining and weakens the line. On the other hand, the sleeve section reduces the clearance between the conductors and consequently the power capacity. In both cases the edges of the junction increase the breakdown tendency; this difficulty can be minimized by rounding the edges of the junction without impairing the matching relations.

It must be emphasized that a single transformer matches properly at only one wavelength. In general the load impedance is a function of frequency. Matching over a frequency band, such that the standing-wave ratio remains less than a prescribed value, can often be achieved by a series of transformer sections of different lengths and characteristic impedances. It is difficult to carry the analysis through analytically for an arbitrary load $Z(\lambda)$. A method of rather limited applicability employing a tandem of quarter-wave sections has been developed by Fubini, Sutro, and Lewis.[1]

While the matching condition of Eq. (65b) always leads to a solution of the mathematical problem, it is not necessarily true that the transformer will be satisfactory. If a large change in radius is required at the junction, the junction effect becomes significant, and we must add to the equivalent transmission-line reactive networks at A and B corresponding to the junction effects. It is found[2] that the network consists of a capacity across the transmission line at the junction points. The junction effect can be studied experimentally by means of a half-wave section. From Eq. (65a) it is seen from transmisson-line considerations alone that if $l = \lambda/2$, then $Z(A) = Z$ regardless of the value of Z'_0; this means that the standing-wave ratio should be the same on either side of the transformer. Figure 7·6 shows experimental results obtained with a half-wavelength sleeve section on a 50-ohm coaxial line with inner diameter 0.375 in. It is seen that the deviation from simple transmission-line behavior increases rapidly with increasing diameter of the sleeve section.

7·10. Parallel Stubs and Series Reactances.—Another useful device in coaxial-line design is the parallel stub consisting of a section of coaxial

---

[1] "Frequency Characteristics of Wide-band Matching Sections," Radio Research Laboratory (Harvard University) Report No. 23, April 1943.

[2] *Waveguide Handbook*, Vol. 10 of this series.

line at right angles to the main line. The arrangement is shown schematically in Fig. 7·7a. The stub is terminated by a metal cap to prevent radiation. Electrically the stub is a shorted section of transmission line. If Z_0' is the characteristic impedance of the stub and l its length, then its input impedance, obtained from Eq. (65a) by setting $Z = 0$, is

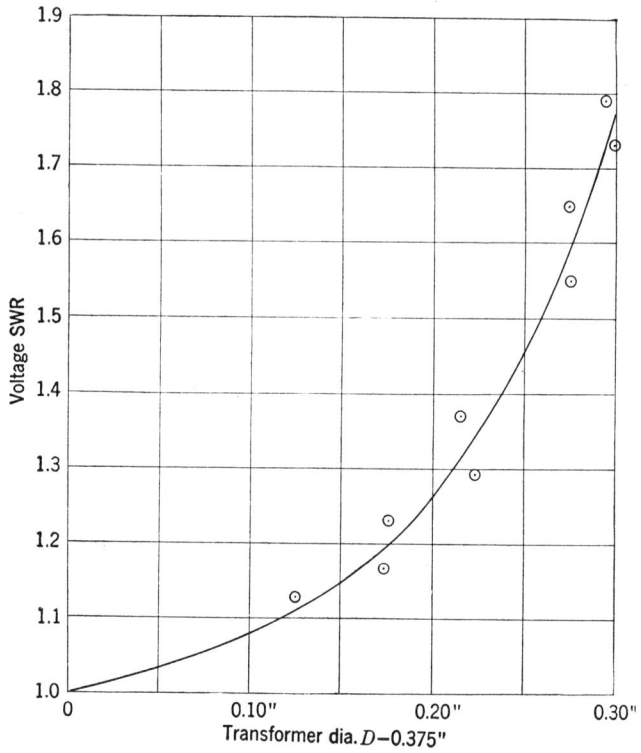

Fig. 7·6.—Junction effects with cascade transformers; mismatch of a $\lambda/2$ transformer as a function of diameter in a coaxial line of dimensions OD = 0.811, ID = 0.375 in.

$Z_s = jZ_0' \tan \beta l$. It is thus a reactive element. Consideration of the current division at A shows that, neglecting junction networks, the stub is to be regarded as a reactance shunted across the main line. If $l = \lambda/4$, then $Z = \infty$, and the stub introduces no change in impedance at A; such a quarter-wave stub is useful as a mechanical support for the inner conductor. We shall not consider here the refinements required to eliminate the frequency sensitivity.

The stub can also serve as a matching device. In this connection it is more convenient to speak in terms of admittances. Let $Y_0 = 1/Z_0$ be the characteristic admittance of the main line, Y the admittance seen to the right of A, and $Y_s = -jY_0' \cot \beta l$ the admittance of the stub. It is possible to locate the point A so that the admittance Y is $Y = Y_0 + jB$.

Sec. 7·10] PARALLEL STUBS AND SERIES REACTANCES

Insertion of the stub gives an admittance at the left of A equal to $Y + Y_s = Y_0 + j(B - Y'_0 \cot \beta l)$. For matching we require simply that
$$Y'_0 \cot \beta l = B. \tag{66}$$

The structure illustrated in Fig. 7·7b is less widely used but is worth consideration. The region between AB and the outer conductor C_2 acts

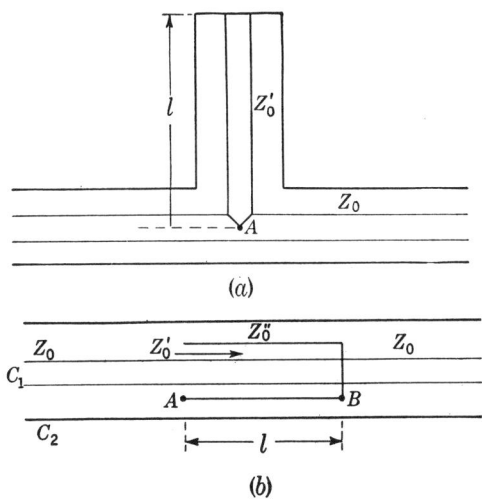

Fig. 7·7.—(a) Parallel stub reactance; (b) series reactance element.

as a cascade transformer. If Z is the impedance at B, the impedance just to the right of A in the transformer space is
$$Z_{AC_2} = Z''_0 \left(\frac{Z + jZ''_0 \tan \beta l}{Z''_0 + jZ \tan \beta l} \right),$$
where Z''_0 is the characteristic impedance of the transformer space. The region AB within the inner conductor acts as a shorted section of line which presents an impedance at A equal to $Z_{C_1A} = jZ'_0 \tan \beta l$, where Z'_0 is the characteristic impedance of the inner region. At A we have the voltage relation $V_{C_2C_1} = V_{C_1A} + V_{AC_2}$; the impedance just to the left of A is given by
$$Z(A) = Z_{C_1A} + Z_{AC_2}$$
$$= jZ'_0 \tan \beta l + Z''_0 \left(\frac{Z + jZ''_0 \tan \beta l}{Z''_0 + jZ \tan \beta l} \right). \tag{67}$$

The structure thus introduces a series impedance at A. It is of interest to note that the length of the inner region can be made shorter than the length of the outer region. If the latter is made equal to an integral number of half wavelengths, the effect of the transformer region is eliminated and at A we have simply an impedance Z_{C_1A} in series with Z.

It is found in practice that stubs and series reactance transformers with dimensions calculated on the basis of the transmission-line formulas given above do not quite meet the simple theoretical expectations. This is due to the junction effects neglected in the transmission-line arguments. The errors, however, are generally small and can be eliminated by small adjustments of the lengths of the structures. In the case of stubs, the shorting cap can be replaced by a sliding plunger in the experimental model to allow easy adjustment of the length. The use of series reactance transformers limits the power capacity of the line; the standing waves in the inner region produce intense fields at the open end and increase the tendency toward electrical breakdown and sparking. An alternative form, which mounts the transformer on the outer conductor where the electric field is weaker, is more satisfactory with respect to breakdown characteristic but is less desirable from assembly considerations.

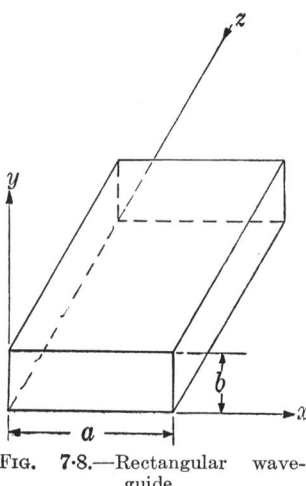

FIG. 7·8.—Rectangular waveguide.

7·11. Rectangular Waveguides: TE- and TM-modes.—The hollow guide of rectangular cross section is the most widely used line in microwave antennas. We shall take the x, y-axes to be oriented as shown in Fig. 7·8; a is the broad dimension of the guide; b the narrow dimension. The Helmholtz equation

$$\frac{\partial^2 F}{\partial x^2} + \frac{\partial^2 F}{\partial y^2} + \kappa^2 F = 0, \qquad F = \begin{cases} \psi(x,y) \\ \phi(x,y) \end{cases} \tag{7·6a}$$

is in this case separable in the form

$$F = X(x)Y(y).$$

Substitution into Eq. (6a) leads to the two equations

$$\frac{d^2 X}{dx^2} + \kappa_x^2 X = 0; \qquad \frac{d^2 Y}{dy^2} + \kappa_y^2 Y = 0, \tag{68}$$

with

$$\kappa_x^2 + \kappa_y^2 = \kappa^2. \tag{68a}$$

The solutions have the same form for both members of Eq. (68); for example,

$$X(x) = A \cos(\kappa_x x) + B \sin(\kappa_x x).$$

The general solution of Eq. (6a) is

SEC. 7·11] RECTANGULAR WAVEGUIDES: TE- AND TM-MODES 227

$$\left.\begin{array}{l}\psi(x,y)\\ \phi(x,y)\end{array}\right\} = [A \cos (\kappa_x x) + B \sin (\kappa_x x)][C \cos (\kappa_y y) + D \sin (\kappa_y y)]. \quad (69)$$

a. *TE-waves.*—The solution $\psi(x,y)$ must satisfy the boundary condition $\partial\psi/\partial n = 0$ over the walls. For the walls $x = 0$ and $x = a$, $\partial\psi/\partial n = \partial\psi/\partial x$; we thus require that for all values of y

$$\left.\frac{\partial\psi}{\partial x}\right]_{x=0} = \kappa_x BY(y) = 0, \quad (70a)$$

$$\left.\frac{\partial\psi}{\partial x}\right]_{x=a} = -\kappa_x[A \sin (\kappa_x a) - B \cos (\kappa_x a)]Y(y) = 0. \quad (70b)$$

This requires that $B = 0$ and that κ_x have the characteristic values

$$\kappa_x = \frac{m\pi}{a}, \quad m = 0, 1, 2, \cdots. \quad (71)$$

Over the walls $y = 0$ and $y = b$, $\partial\psi/\partial n = \partial\psi/\partial y$. This boundary condition requires that $D = 0$ and that κ_y have the characteristic values

$$\kappa_y = \frac{n\pi}{b}, \quad n = 0, 1, 2, \cdots. \quad (72)$$

The characteristic values κ_{mn} for the TE_{mn}-wave are therefore

$$\kappa_{mn}^2 = \left(\frac{m\pi}{a}\right)^2 + \left(\frac{n\pi}{b}\right)^2. \quad (73)$$

By use of Eqs. (18) and (19), the complete set of field components for the wave in the positive z-direction is found to be

$$\left.\begin{array}{l} H_z = \cos\dfrac{m\pi x}{a} \cos\dfrac{n\pi y}{b} e^{-\gamma_{mn}z}, \quad E_z = 0,\\[6pt] H_x = -\dfrac{\gamma_{mn}}{j\omega\mu} E_y = \dfrac{m\pi\gamma_{mn}}{\kappa_{mn}^2 a} \sin\left(\dfrac{m\pi x}{a}\right) \cos\left(\dfrac{n\pi y}{b}\right) e^{-\gamma_{mn}z},\\[6pt] H_y = \dfrac{\gamma_{mn}}{j\omega\mu} E_x = \dfrac{n\pi\gamma_{mn}}{\kappa_{mn}^2 b} \cos\left(\dfrac{m\pi x}{a}\right) \sin\left(\dfrac{n\pi y}{b}\right) e^{-\gamma_{mn}z}. \end{array}\right\} \quad (74)$$

The significance of the integers m, n, is directly apparent: They represent the number of sinusoids in the intensity of the field components E_y and E_x, respectively, over the cross section of the guide.

The cutoff wavelength, the guide wavelength, and the transverse wave impedance for a TE_{mn}-mode are respectively

$$\lambda_{mn}^{(c)} = \frac{1}{\left[\left(\dfrac{m}{2a}\right)^2 + \left(\dfrac{n}{2b}\right)^2\right]^{1/2}}, \quad (75)$$

$$\lambda_{g_{mn}} = \frac{\lambda}{\left\{1 - \left[\left(\frac{m\lambda}{2a}\right)^2 + \left(\frac{n\lambda}{2b}\right)^2\right]\right\}^{1/2}}, \quad (76)$$

$$Z^{(0)}_{mn} = \left(\frac{\mu}{\epsilon}\right)^{1/2} \frac{1}{\left\{1 - \left[\left(\frac{m\lambda}{2a}\right)^2 + \left(\frac{n\lambda}{2b}\right)^2\right]\right\}^{1/2}}. \quad (77)$$

The TE_{10}-mode ($m = 1$, $n = 0$) has the longest cutoff wavelength. It is by far the most important mode for antenna work. The electric field has but one component, E_y, which is uniform in the y-direction and varies sinusoidally along the x-direction with symmetry about the central section of the guide. The field configurations for this and several other TE-modes are shown in Fig. 7·9. It will be seen from Eq. (75) that to

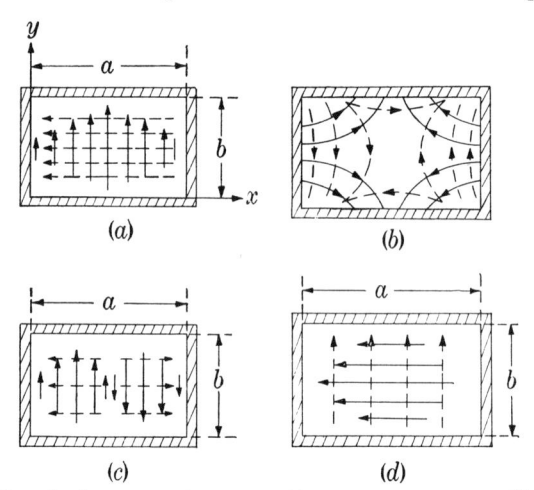

Fig. 7·9.—TE-modes in rectangular waveguides: (a) TE_{10}-mode $[\lambda_{10}{}^{(c)} = 2a]$; (b) TE_{11}-mode $[\lambda_{11}{}^{(c)} = 2ab/\sqrt{a^2 + b^2}]$; (c) TE_{20}-mode $[\lambda_{20}{}^{(c)} = a]$; (d) TE_{01}-mode $(\lambda_{01}{}^{(c)} = 2b)$. ——— electric field; - - - - - magnetic field.

ensure propagation of the TE_{10}-mode alone the dimensions of the guide must be such that

$$a < \lambda < 2a; \quad 2b < \lambda.$$

b. *TM-waves.*—The solution $\phi(x,y)$ must satisfy the boundary condition $\phi = 0$ over the walls. It is evident from Eq. (69) that we must set $A = C = 0$ to satisfy the condition over the surface $x = 0$ and $y = 0$. Over the walls $x = a$, $y = b$ the conditions are satisfied only for the characteristic values

$$\kappa_x = \frac{m\pi}{a}, \quad m = 0, 1, 2, \cdots. \quad (7\cdot71)$$

$$\kappa_y = \frac{n\pi}{b}, \quad n = 0, 1, 2, \cdots. \quad (7\cdot72)$$

Thus the characteristic value κ_{mn} for the TM_{mn}-mode line, like that of a TE_{mn}-wave, is given by

$$\kappa_{mn}^2 = \left(\frac{m\pi}{a}\right)^2 + \left(\frac{n\pi}{b}\right)^2. \tag{7.73}$$

The cutoff and guide wavelengths are given by Eqs. (75) and (76); the characteristic wave impedance, however, differs from that of the TE-wave. It is

$$Z_{mn}^{(0)} = \left(\frac{\mu}{\epsilon}\right)^{\frac{1}{2}} \left\{1 - \left[\left(\frac{m\lambda}{2a}\right)^2 + \left(\frac{n\lambda}{2b}\right)^2\right]\right\}^{\frac{1}{2}}. \tag{78}$$

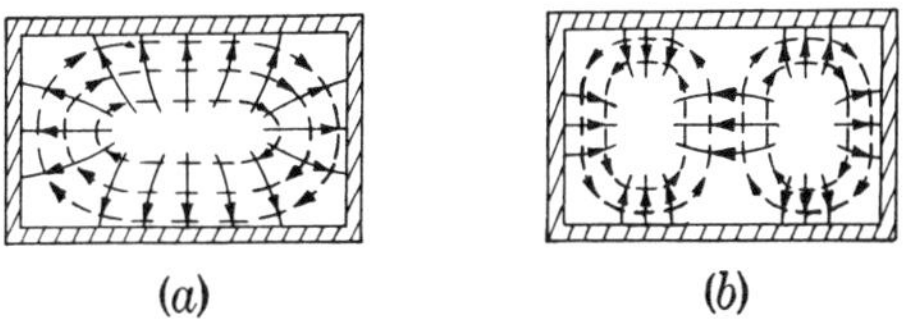

Fig. 7·10.—TM-modes in rectangular waveguides: (a) TM_{11}-mode $[\lambda_{11}{}^{(c)} = 2ab/\sqrt{a^2 + b^2}]$; (b) TM_{21}-mode $[\lambda_{21}{}^{(c)} = 2ab/\sqrt{a^2 + b^2}]$. ——— electric field; — — — magnetic field.

The complete set of field components obtained by means of Eqs. (22) and (23) is

$$\left.\begin{array}{l} E_z = \sin\left(\dfrac{m\pi x}{a}\right) \sin\left(\dfrac{n\pi y}{b}\right) e^{-\gamma_{mn}z}; \quad H_z = 0, \\[6pt] E_x = \dfrac{\gamma_{mn}}{j\omega\epsilon} H_y = -\dfrac{m\pi \gamma_{mn}}{\kappa_{mn}^2 a} \cos\left(\dfrac{m\pi x}{a}\right) \sin\left(\dfrac{n\pi y}{b}\right) e^{-\gamma_{mn}z}, \\[6pt] E_y = \dfrac{-\gamma_{mn}}{j\omega\epsilon} H_x = -\dfrac{n\pi \gamma_{mn}}{\kappa_{mn}^2 b} \sin\left(\dfrac{m\pi x}{a}\right) \cos\left(\dfrac{n\pi y}{b}\right) e^{-\gamma_{mn}z}. \end{array}\right\} \tag{79}$$

There is no mode for which either m or n is zero; the lowest is the TM_{11}-mode. It follows accordingly that a guide designed to cut off the TE-modes other than the TE_{10} will likewise not support free propagation of any of the TM-modes. The field configurations for several of the latter are shown in Fig. 7·10.

7·12. Impedance Transformers for Rectangular Guides.—Equivalent networks have been established for a number of types of obstacles in waveguides; these can serve to match out the reflected dominant mode wave set up by the line termination. We shall present here the pertinent data on elements designed for the TE_{10}-mode in rectangular guide and shall indicate their applicability. The simplest, from the point of view of the equivalent networks, are the *windows*: metal diaphragms inserted in the cross section of the guide. Typical forms are illustrated in Fig. 7·11. In the idealized case of infinite conductivity these elements behave like capacities or inductances shunted across the two-wire transmission-line representation of the TE_{10}-mode. Accordingly, in using these elements it is convenient to treat the line in terms of admittance rather than impedance. Let $Y_{10}^{(0)} = 1/Z_{10}^{(0)}$ be the characteristic wave admittance of the TE_{10}-mode line.[1] With any arbitrary termination, there exist points along the line, at quarter-wavelength intervals, at which the input admittance looking toward the load is alternately (1) $Y = Y_0 - jB$ and (2) $Y = Y_0 + jB$. At the points 1 the load susceptance is inductive and a parallel capacity is required for matching. Points (2), where the load susceptance is capacitative, require a parallel inductance. For the

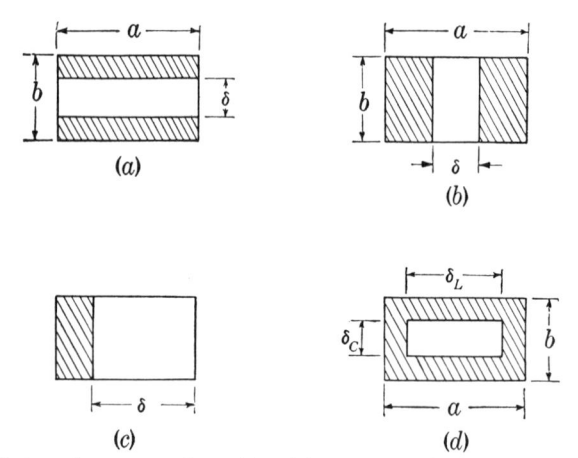

FIG. 7·11.—Windows for rectangular guides; (a) symmetrical capacitive; (b) symmetrical inductive; (c) asymmetrical inductive; (d) resonant.

former case the capacitative window (Fig. 7·11a), is suited, while for points 2 the inductive windows (Fig. 7·11b and c) are appropriate. Formulas and graphs for the susceptance of these and other windows, referred to the characteristic wave admittance of the TE_{10}-mode, are available in the literature.[2]

In practice the inductive windows are to be preferred, because the capacitative window, in presenting an edge across the electric-field lines, is more susceptible to electrical breakdown. Asymmetrical windows have experimental and design advantages in that only one side of the guide need be milled for an insertion.[3] This reduces the amount of machining required in making test runs on impedance and eliminates the

---

[1] We shall drop the mode notation hereafter and write simply Z_0 and Y_0 for the characteristic impedance and admittance respectively.

[2] *Microwave Transmission Design Data*, Sperry Gyroscope Company, 1944; "Waveguide Handbook," RL Group Report No. 43, Feb. 7, 1944; "Waveguide Handbook Supplement," RL Group Report No. 41, Jan. 23, 1945; *Waveguide Handbook*, Vol. 10 of this series.

[3] W. Sichak, "One-sided Inductive Irises and Quarter-wave Capacitative Transformers in Waveguide," RL Report No. 426, Nov. 17, 1943.

SEC. 7·12] IMPEDANCE TRANSFORMERS 231

problem of alignment of two halves of a symmetrical window. On the other hand, symmetrical windows lend themselves to use as pressurization devices; the two metal borders can serve as supports for a thin dielectric sheet. Such a sheet introduces an additional capacity in parallel with the window, the magnitude of which depends on the thickness and dielectric constant. No systematic design information seems to be available on this point at present, and the design of the pressurized window must be developed experimentally.

The circuit equivalents of the windows immediately suggest the possibility of combining a capacitative and inductive window to make the net susceptance zero, that is, to produce a resonant device that introduces no reflection in the guide. Such a resonant window is illustrated in Fig.

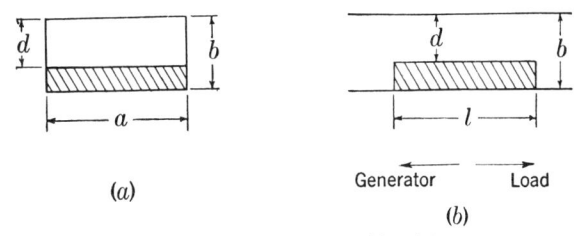

FIG. 7·12.—Step transformer in rectangular waveguide: (a) transverse cross section; (b) longitudinal cross section.

7·11d. To a first approximation, the dimensions δ_c, δ_L can be so chosen that the capacitative and inductive susceptances are equal in magnitude. The resonant window transmits all the incident power and, therefore, cannot be used as a matching device. It is useful as a pressurizing element to seal the waveguide; either the window frame serves as a support for a thin dielectric sheet, or the open area of the window is filled with a dielectric block. The dimensions of the window must be adjusted to compensate for the dielectric; this again must be determined empirically. It is obvious that true resonance behavior can be achieved at only one wavelength with a given window.

The use of windows at wavelengths shorter than 3 cm is rather limited. Several difficulties arise due to the decrease in the dimensions of the waveguide with decreasing wavelength. The most striking of these are (1) the increased liability to electrical breakdown in the neighborhood of a window, (2) errors in determining the position of the element, and (3) the machining and insertion of small parts. For wavelengths shorter than 3 cm the step transformer, illustrated in Fig. 7·12, is recommended. This is analogous to the cascade section discussed for coaxial lines. The characteristics of the step transformer can be expressed in terms of the input admittance presented at the generator side when the guide is terminated beyond the section in a matched load:[1]

[1] Sichak, *op. cit.*, p. 3.

$$1 - \frac{Y}{Y_0} = \frac{\left(\frac{b}{d}\right)^2 + \frac{4b^2 F}{\lambda_g d} \cot \frac{2\pi l}{\lambda_g} - \left(\frac{2b}{\lambda_g}\right)^2 F^2 + jA}{(1 + jA)},$$

$$A = \frac{2bF}{\lambda_g} - \frac{b}{d} \cot \frac{2\pi l}{\lambda_g}.$$

(80)

The parameters l, b, d are defined in the figure; λ_g is the wavelength in the waveguide; and F is a function of d/b alone, a few values of which are given in Table 7·1. When $l = \lambda_g/4$, Eq. (81) reduces to

TABLE 7·1.—F-FUNCTION FOR STEP TRANSFORMER

$1 - \frac{d}{b}$, %	F-function
0	0
10	0.020
20	0.063
30	0.130
40	0.235
50	0.395
60	0.598
70	0.820

$$\frac{Y}{Y_0} = \frac{\frac{b^2}{d^2} + j \frac{2bF}{\lambda_g} \left(1 - \frac{b^2}{d^2} + \frac{4b^2 F^2}{\lambda_g^2}\right)}{1 + \frac{4b^2 F^2}{\lambda_g^2}}.$$

(80a)

To design a transformer from either Eq. (80) or (80a) it is necessary to construct a graph, based on Table 7·1, from which the required values of F may be obtained. Over the range of useful values of b/d the section can be regarded as a quarter-wave transformer with a phase correction due to the capacitative effects at the junction. The phase correction makes itself felt in that the load end of the transformer is not placed at the point of a "voltage" minimum (the point of maximum load admittance) but is displaced slightly from that point toward the generator. To a first approximation the matching condition is that the conductance in Eq. (80a) be equal to the maximum normalized load admittance,

$$g_t = \frac{\frac{b^2}{d^2}}{1 + \frac{4b^2 F^2}{\lambda_g^2}}.$$

(81)

The latter directly equals the voltage standing-wave ratio due to the load. Accordingly it is suggested that the designer prepare for himself a set of charts of r or g_t against d/b over the range of λ_g with which he will be chiefly concerned. For a given case the transformer with dimensions determined in the indicated manner can be prepared to slide in the

SEC. 7·13]   *CIRCULAR WAVEGUIDE: TM- AND TE-MODES*   233

guide, the bottom of the transformer being tinned before insertion. The transformer is moved along the guide until the best matching position is located and then soldered into place by heating the outside of the guide.

7·13. Circular Waveguide: TM- and TE-Modes.—Let us consider next a hollow guide of circular cross section of radius a. As in the case of the coaxial line we are here concerned with solutions of the scalar Helmholtz equation in a circular region. The general solutions are the same as for the coaxial line:

$$\begin{matrix} \psi(x,y) \\ \phi(x,y) \end{matrix} = [AJ_m(\kappa r) + BN_m(\kappa r)] \cos m\theta. \qquad (7\cdot 58)$$

Here again r, θ are polar coordinates over the cross section, and m is an integer. In the present case, since there is no inner conductor, there are no sources in the interior and the fields must be finite at all points. The Neumann function, however, becomes infinite at $r = 0$; accordingly it must be removed from the solution: B must be equal to zero. The fundamental solutions are, therefore,

$$\begin{matrix} \psi(x,y) \\ \phi(x,y) \end{matrix} = AJ_m(\kappa r) \cos m\theta. \qquad (82)$$

a. TM-modes.—By the boundary condition of Eq. (21b) we require $\phi = 0$ at $r = a$ for all values of θ. This leads to characteristic values $\kappa_{mn}^\bullet$ which satisfy the relation[1]

$$J_m(\kappa_{mn}a) = 0.$$

The complete set of field components for the TM_{mn}-mode, obtained from Eqs. (22) and (23), are

$$\left. \begin{matrix} E_z = \kappa_{mn}^2 \cos m\theta J_m(\kappa_{mn}r)e^{-\gamma_{mn}z}; \quad H_z = 0, \\ E_r = \dfrac{\gamma_{mn}}{j\omega\epsilon} H_\theta = -\gamma_{mn}\kappa_{mn} \cos m\theta J'_m(\kappa_{mn}r)e^{-\gamma_{mn}z}, \\ E_\theta = -\dfrac{\gamma_{mn}}{j\omega\epsilon} H_r = m\gamma_{mn} \sin m\theta \dfrac{J_m(\kappa_{mn}r)}{r} e^{-\gamma_{mn}z}. \end{matrix} \right\} \qquad (83)$$

The field configurations for several of these modes, together with the cutoff wavelengths, are shown in Fig. 7·13.

b. TE-modes.—The function $\psi(x,y)$ is subject to the boundary condition of Eq. (13): $\partial\psi/\partial r]_{r=a} = 0$ for all θ. The characteristic values κ_{mn} satisfy the relation[2]

$$J'_m(\kappa_{mn}a) = 0. \qquad (84)$$

[1] For lower roots $z_{mn} = \kappa_{mn}a$ of this equation see E. Jahnke and F. Emde, *Tables of Functions*, Dover Publications Reprint, New York, 1943, p. 168.
[2] For the lower roots $z_{mn} = \kappa_{mn}a$ of this equation see *ibid*.

The field components for the general TE_{mn}-mode are found to be

$$\begin{aligned} H_z &= \kappa_{mn}^2 \cos m\theta J_m(\kappa_{mn}r)e^{-\gamma_{mn}z}; \qquad E_z = 0, \\ H_r &= -\frac{\gamma_{mn}}{j\omega\mu} E_\theta = -\kappa_{mn}\gamma_{mn} \cos m\theta J'_m(\kappa_{mn}r)e^{-\gamma_{mn}z}, \\ H_\theta &= \frac{\gamma_{mn}}{j\omega\mu} E_r = m\gamma_{mn} \sin m\theta \frac{J_m(\kappa_{mn}r)}{r} e^{-\gamma_{mn}z}. \end{aligned} \qquad (85)$$

These modes are illustrated in Fig. 7·14. On examination of the roots of the Bessel functions and their derivatives it will be seen that the lowest mode, that is, the mode with the longest cutoff wavelength, is the TE_{11}-

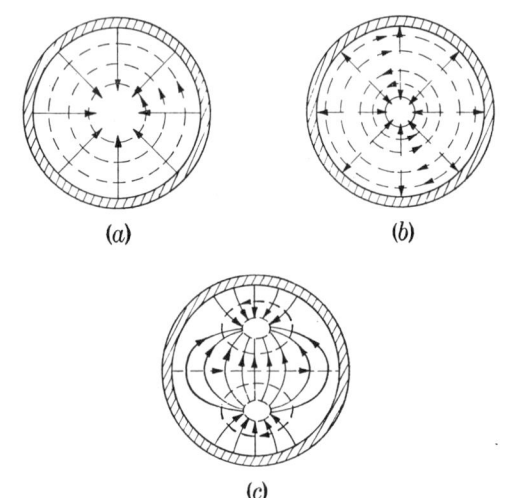

Fig. 7·13.—TM-mode in circular waveguide: (a) TM_{01}-mode $[\lambda_{01}{}^{(c)} = 1.31d]$; (b) TM_{02}-mode $[\lambda_{02}{}^{(c)} = 1.07d]$; (c) TM_{11}-mode $[\lambda_{11}{}^{(c)} = 0.82d]$. ——— electric field; − − − − − magnetic field.

mode. This is the mode generally utilized in antenna systems. It is the circular guide analogue of the TE_{10}-mode in rectangular waveguide.

The use of circular waveguide is limited by several factors, of which perhaps the most significant is instability in orientation of the field configurations. Since the guide has rotational symmetry, the field configuration can be rotated about the z-axis without violating boundary conditions; there is no preferred direction $\theta = 0$. Small irregularities in the wall of the guide or matching windows can cause such rotation of the fields giving rise to subsequent difficulties in designing the radiating system. In rectangular guide, on the other hand, the orientation of the field configuration is uniquely determined by the orientation of the cross section. Another difficulty in round guide is mode control over an appreciable frequency band. The radius is the only parameter available to determine the cutoff wavelength; in rectangular guide, both the dimensions a and b enter into the characteristic values of the higher modes.

Other comparative factors will be pointed out in the discussion of waveguide and horn feeds.

7·14. Windows for Use in Circular Guides.—As with rectangular guides, metal diaphragms can be inserted into circular guides to serve as matching devices for the TE_{11}-mode. The circuit equivalents of these windows are again reactive elements shunted across the two-wire line representation of the dominant mode. Capacitative windows cut across the E-lines, while inductive windows cut across the transverse magnetic field in the cross section. The admittance characteristics of such windows

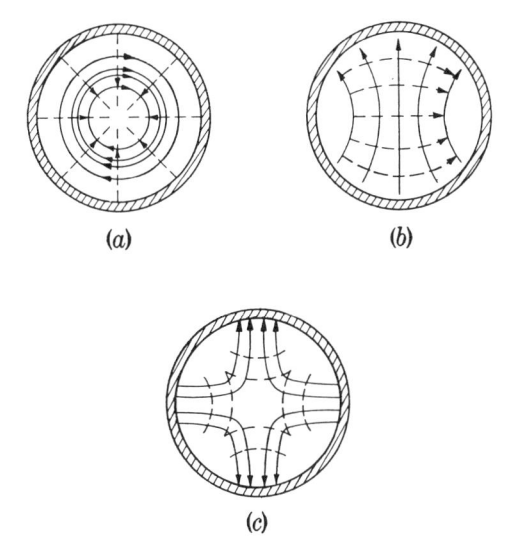

Fig. 7·14.—TE-modes in circular waveguides: (a) TE_{01}-mode [$\lambda_{01}^{(c)} = 0.82d$]; (b) TE_{11}-mode [$\lambda_{11}^{(c)} = 1.71d$]; (c) TE_{21}-mode [$\lambda_{21}^{(c)} = 1.03d$]. ——— electric field; - - - - - magnetic field.

may be found in the literature on the subject.[1] There is also available a resonant window which can be used as a frame to support a thin dielectric sheet to seal the waveguide.

7·15. Parallel-plate Waveguide.—Another type of waveguide that is used in microwave antennas is that formed by a pair of parallel plates. The modes can be derived, as in the previous sections, by a direct solution of the field equations, in the present case for a region bounded by a pair of parallel perfectly conducting surfaces of infinite extent. It will be instructive, however, to treat the parallel-plate system as a limiting case of the coaxial line and the rectangular waveguide.

The parallel-plate waveguide can be derived from the coaxial line

---

[1] *Microwave Transmission Design Data*, Sperry Gyroscope Company, 1944; "Waveguide Handbook," RL Group Report No. 43, Feb. 7, 1945; "Waveguide Handbook Supplement," RL Group Report No. 41, Jan. 23, 1945; *Waveguide Handbook*, Vol. 10 of this series.

by allowing the radii a and b of the inner and outer conductors to become infinite in such a way that the spacing $b - a$ between the conductors remains constant:

$$b - a = s. \tag{86}$$

It will be recalled that the *TEM*-mode is independent of the radii of the conductors and is supported by the line for all frequencies with a wavelength equal to that in free space. We thus arrive directly at the result that the parallel-plate guide supports free propagation of a *TEM*-mode at all frequencies. The electric vector is perpendicular to the plates,

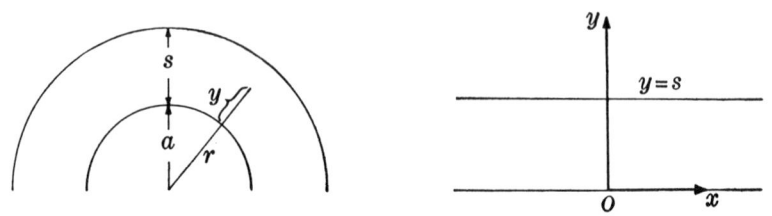

Fig. 7·15.—The parallel-plate waveguide as a limiting case of a coaxial line.

and the magnetic vector is parallel to the plates; neither field vector has a component in the direction of propagation. Taking Eq. (53) for the electric vector of the *TEM*-mode and writing $r = a + y$, $b = a + s$ (*cf.* Fig. 7·15), we find that the magnitude of the electric field is

$$E = \frac{1}{a+y} \frac{V}{\ln\left(1 + \frac{s}{a}\right)} = \frac{1}{(a+y)} \frac{V}{\frac{s}{a} - \frac{s^2}{2a^2} + \cdots}. \tag{87}$$

Letting a become infinite we obtain

$$\lim_{a \to \infty} |E| = \frac{V}{s}. \tag{88}$$

The magnitude of the electric-field vector is independent of position between the plates; the same result is obtained for the magnetic field. It will be recognized that Eq. (88) is the same expression as for the static electric field between a pair of plates at a difference of potential V.

Considering next the *TE*- and *TM*-modes of the coaxial line we note that as the radii become infinite, the periodicity condition disappears; that is, we need concern ourselves only with the modes of order $m = 0$ [Eq. (58)].

TM-modes.—The longitudinal component of the electric field [the function $\phi(x,y)$ in Eq. (58)] is

$$E_z = AJ_0(\kappa r) + BN_0(\kappa r). \tag{89}$$

Making use of the asymptotic forms of the Bessel functions[1] for large (κr), we get

[1] G. N. Watson, *Bessel Functions*, 2d ed., Macmillan, New York, 1945, Chap. 7.

$$E_z \approx \left(\frac{2}{\pi \kappa a}\right)^{1/2} \left[A \cos\left(\kappa y + \kappa a - \frac{\pi}{4}\right) + B \sin\left(\kappa y + \kappa a - \frac{\pi}{4}\right)\right].$$

We have here introduced again $r = a + y$. In the limit $a = \infty$, the solution takes the form

$$E_z = A' \cos(\kappa y + \tau). \tag{90}$$

Applying the boundary conditions $E_z = 0$ at $y = 0$ and $y = s$, we find that $\tau = (\pi/2) \pm 2m\pi$ and that the characteristic values of the modes are

$$\kappa_n = \frac{n\pi}{s}; \qquad n = 1, 2, \cdots. \tag{91}$$

Equation (90) can thus be rewritten as

$$E_z = A' \sin\left(\frac{n\pi y}{s}\right). \tag{92}$$

The cutoff wavelength for the TM_n-mode is [Eq. (16)]

$$\lambda_n^{(c)} = \frac{2s}{n}, \tag{93}$$

and the guide wavelength for the freely propagated mode [Eq. (17)] is

$$\lambda_{gn} = \frac{\lambda}{\left[1 - \left(\frac{n\lambda}{2s}\right)^2\right]^{1/2}}. \tag{94}$$

The transverse components of the field are obtained from E_z by means of the set of Eqs. (4) and (5):

$$E_y = -\frac{\gamma_n}{\kappa_n} A' \cos\left(\frac{n\pi y}{s}\right); \qquad E_x = 0, \tag{95a}$$

$$H_x = \frac{j\omega\epsilon}{\kappa_n} A' \cos\left(\frac{n\pi y}{s}\right); \qquad H_y = 0; \tag{95b}$$

the constant γ_n is defined by Eq. (14).

TE-modes.—The derivation of the *TE*-modes proceeds in a similar manner. Equation (90) in this case represents the longitudinal component of the magnetic field; that is,

$$H_z = A' \cos(\kappa y + \tau). \tag{96}$$

The boundary conditions $\partial H_z/\partial y = 0$ at $y = 0$ and $y = s$ lead to the result that $\tau = \pm 2m\pi$ and

$$\kappa_n = \frac{n\pi}{s}, \qquad n = 1, 2, \cdots. \tag{97}$$

The cutoff wavelength for the TE_n-mode is given by Eq. (93), and the guide wavelength by Eq. (94). The complete set of field components is

$$H_z = A' \cos\left(\frac{n\pi y}{s}\right), \tag{97a}$$

$$E_x = \frac{j\omega\mu}{\kappa_n} A' \sin\left(\frac{n\pi y}{s}\right), \qquad E_y = 0, \tag{97b}$$

$$H_y = \frac{\gamma_n}{\kappa_n} A' \sin\left(\frac{n\pi y}{s}\right); \qquad H_x = 0. \tag{97c}$$

It will be recognized that the field distributions and guide wavelengths correspond to TM- and TE-modes of the rectangular guide. The TM_n-modes of the parallel-plate system are the analogues of the $TM_{1,n}$-modes, and the TE_n-modes are the analogues of the $TE_{0,n}$-modes. As the broadside dimension a of the rectangular guide becomes infinite, the modes of the latter pass into parallel plate modes.

7·16. Design Notes.—Several remarks on design practice may prove of interest to the reader. These are particularly concerned with coaxial lines and circular waveguides. Unless an antenna is being developed as a single experimental model or for production in very limited numbers, some attention should be given to the production problem or the availability of parts. With respect to the coaxial lines and circular guides, dimensions should be chosen as near as possible to those of commercially standardized tubing. The primary considerations in the choice of dimensions are, of course, the characteristic impedance of the line and the control of higher modes; these, however, allow some latitude in design.

Special care should be taken in the inspection of tubing. Erratic results in standing-wave measurements on lines have frequently been traced to irregularities in the cross section of the line. Ridges and waves are found in the tube wall if the die through which the tubing was extruded is worn or if the driving unit is faulty. Such ridges and waves can be detected only by cutting the tube in half. It is recommended that a sample length of tubing from each new lot be cut down the middle for inspection before using the material. It is often useful to force a steel ball of proper diameter through the tubing under pressure, thus sizing and polishing the inside surface.

CHAPTER 8

MICROWAVE DIPOLE ANTENNAS AND FEEDS

By S. Silver

The early trends in microwave antenna design grew out of the practice of using dipole systems at longer wavelengths. Nevertheless, little systematic information has been obtained about microwave dipole systems. This is partly due to the greater difficulty in applying theory to practically useful microwave dipoles and partly to the urgent military needs which prevented systematic research during the early development in this field. More recently, attention has been concentrated on waveguide and horn radiators, which are more amenable to quantitative analysis. Consequently, the design of microwave dipole antennas is still in the empirical stage; quantitative data are available only with reference to particular systems.

8·1. Characteristics of Antenna Feeds.—The dipole systems that we shall consider in this chapter are, with a few exceptions, designed to serve as primary feeds to illuminate reflectors; it will be assumed throughout, unless the contrary is noted, that this is the end in view. The general design requirements and specifications imposed on primary feeds are the following:

Radiation Pattern.—It is evident that a primary feed radiation pattern must be directive, with the major fraction of the energy radiated toward the reflector. We have studied in Chap. 6 the relation between the radiation pattern of the antenna as a whole and the intensity and phase distribution over the aperture. The relation between the latter and the primary pattern will be developed in later chapters on the design problems of special types of antennas. It may be noted here, however, that the design of a reflector—or a lens—is generally based on the assumption that the feed is a point source. Deviations of the feed from a point-source radiator result in phase errors over the aperture of the antenna.

Particular attention must be paid to the phase. It was shown in Chap. 3 that many idealized radiating systems are effectively point sources in the sense that the equiphase surfaces constitute a family of concentric spheres. This situation is realized only approximately in the case of an actual feed. The pattern of the latter is usually specified in terms of the principal E- and H-plane patterns (Sec. 3·18). In each of these planes it should be possible to find an equivalent *center of feed*, with respect to which the equiphase lines are circular (to within a prescribed limit of error) over the region to be covered by the reflector. To minimize the problems of reflector design it is essential that the centers of feed for the principal planes be coincident. In general, it is desirable that on a sphere about the center of feed the phase shall be constant to within $\pm \pi/8$, corresponding to path differences of $\pm \lambda/16$; for some purposes, path differences of $\pm \lambda/8$ can be tolerated. The cone within which the feed is a point source in the sense of these criteria will be referred to as the *point-source cone*.

Impedance.—Impedance match is required over as broad a frequency band as possible. An antenna is generally considered to be usable throughout the frequency band in which the voltage standing-wave ratio is less than 1.4. Since interactions with the reflector tend to increase the total mismatch, it is desirable to keep the feed mismatch below the figure given above.

Power-carrying Capacity.—This is limited by electrical breakdown which may occur within the feed line and around the feed components under the peak voltage of a transmitted signal. The effect of matching devices on breakdown characteristics was noted in Sec. 7·9. The breakdown problem is particularly significant in antennas intended for aircraft, because the breakdown potential decreases with increasing altitude, due to the decrease in atmospheric pressure and the increase of free ion content. Feeds for high-altitude airborne systems must therefore be so designed that air can be held in the r-f line under pressure. The average requirement is 10 to 15-lb gauge pressure relative to sea level atmospheric pressure.

Weather Protection.—Antennas must be protected from the weather to prevent corrosion and consequent power dissipation in the antenna structure. Weatherization is an important consideration in shipborne antennas, which are exposed to sea-water sprays.

Mechanical Strength, Light Weight.—Antennas installed in aircraft and ships are subject to high stresses due to rapid changes in the motion of the airplane or oscillations of masts of the ship in a high wind. In aircraft systems, mechanical strength must be attained with economy of weight.

Reasonable Tolerances.—Tolerances should not be so close that production methods cannot be used effectively.

8·2. Coaxial Line Terminations: The Skirt Dipole.—The theoretical prototype of the dipole radiators is the half-wave dipole fed at the center from a balanced two-wire transmission line. The significant features of this system are the following:

1. The two wings of the dipole carry equal currents.

2. The current distribution is determined by the dipole structure, interaction between the dipole and the transmission line being negligible.
3. The dipole termination does not upset the balanced condition of the line.

While it is true that a coaxial line propagating the TEM-mode is equivalent to a balanced two-wire line, it is virtually impossible to make a

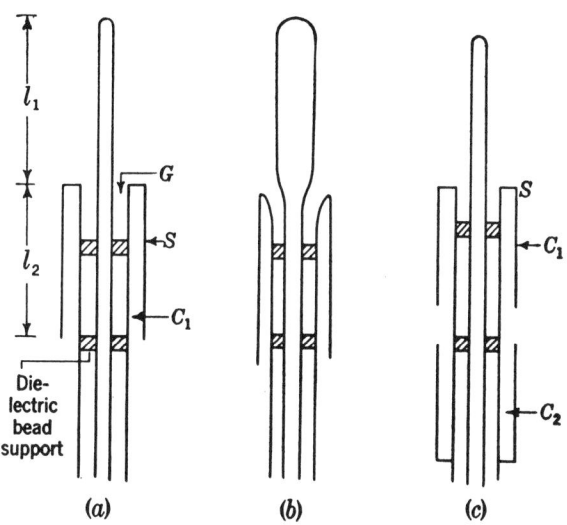

FIG. 8·1.—Skirt dipole: (a) simple form; (b) tapered gap to improve the impedance characteristics; (c) decoupling choke C_2 to prevent current leakage along the outer wall.

microwave dipole termination that behaves like the theoretical prototype.

The skirt dipole illustrated in Fig. 8·1 is an example of a coaxial-line termination that is used extensively at longer wavelengths and to a lesser extent in the microwave region. The two wings of the dipole consist of the unshielded section of the inner conductor and the folded-back section of the outer conductor (S in Fig. 8·1); we shall refer to the latter as the skirt. If the lengths l_1, l_2, of the respective elements are each about $\lambda/4$, the system approximates a center-driven half-wave dipole. This termination maintains the radial symmetry of the line; the current distribution over the wings is radially symmetric, and the radiation pattern has the axial symmetry of the idealized system.

It is to be expected, however, that the meridional pattern will differ from that of the line radiator. One reason for this is the fact that the current distribution is spread over a finite area instead of being confined to a line. The currents at different points on a circumference of the skirt are consequently at different distances from a field point and give contributions to the field that have correspondingly different phases. These phase differences are more significant with microwaves than with long waves, since they depend on the ratio of the skirt diameter to the wavelength. The pattern is also affected by the finite dimension of the gap at the driving point; this causes the current distribution along the length of the dipole to deviate from the sinusoidal distribution of the line dipole with an infinitesimal gap.

A further major factor is the coupling between the field of the dipole and the outside wall of the line, which produces a current distribution down the line beyond the skirt. This current distribution also radiates; the total pattern arises from superposition of this field and the dipole field. The pattern rapidly becomes less satisfactory as the current on the line increases; so the line current must therefore be kept as small as possible. It can be controlled in part by changing the cavity C_1 formed by the skirt and the outside wall of the line. This region constitutes a shorted section of line and as such presents at the open end of the skirt a reactive impedance in series with the dipole and the outer wall of the line; by making the depth $\lambda/4$, the reactive impedance can be made infinite. In practice it is found that best results are obtained with a skirt of length somewhat less than $\lambda/4$. Proper operation is obtained only at the design frequency, since the impedance of the choke C_1 varies rapidly with frequency. Improved over-all impedance characteristics have been obtained by shortening the skirt and compensating for the reduced physical length of C_1 by filling it with dielectric to bring the electrical length up to $\lambda/4$. It has also been found that more efficient decoupling between the dipole and the outer line can be effected by means of a second choke C_2 mounted as shown in Fig. 8·1c. The electrical depth of C_2 should again be a quarter wavelength, so that the choke presents an infinite impedance at the open end. Experimentally it is found that the decoupling is most complete when the separation of C_1 and C_2 is 0.15λ.

The structure of the gap G plays a significant part in determining the over-all impedance characteristics of the antenna. Because an abrupt discontinuity in structure gives rise to a reflected wave in the line, it is natural to replace the region G in Fig. 8·1a by the tapered structure shown in Fig. 8·1b. The increased diameter of the dipole stub also contributes to maintaining uniform impedance over a larger frequency band (*cf.* Sec. 8.5). Further methods of controlling the impedance characteristics, such as decreasing the length l_1 and loading the stub with a sphere (capacitative loading), will occur to the reader; we shall not dwell upon them here.

8·3. Asymmetrical Dipole Termination.—The asymmetric dipole terminations shown in Fig. 8·2 are designed to give a radiation pattern with peak intensity along the axis of the feed line. The dipole in Fig. 8·2a is center-fed from a two-wire line. The asymmetry of the termina-

SEC. 8·3] ASYMMETRICAL DIPOLE TERMINATION 243

tion unbalances the current distributions on the inner and outer conductors of the line, with the result that the two wings of the dipole are not equally excited. Also, strong coupling exists between the dipole system and the outer wall of the line, giving rise to radiating currents on the latter, just as in the case of the skirt dipole. The choke C remedies the situation to some extent; with a depth l_c of about $\lambda/4$ the choke presents at its open end an infinite impedance, in series between the outer wall of

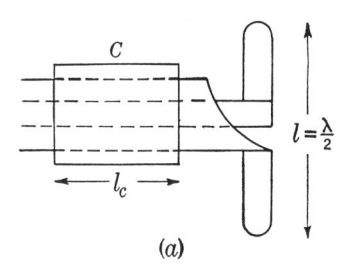

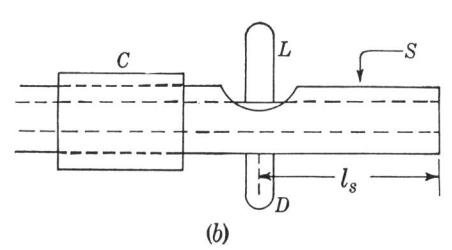

FIG. 8·2.—Asymmetric dipole terminations: (a) open-end termination; (b) stub-support termination.

the line to the left of the choke and the region of the line to the right. This serves to confine most of the outer-wall currents to the region between the dipole and the choke.

The open-ended termination has poor structural properties. In order to maintain alignment of the dipole wings it is necessary to fill the terminal region of the line with a dielectric plug. The latter gives rise to further problems of impedance mismatch and to poor contact between the dielectric and the conductors, which may lead to electrical breakdown; the seals generally deteriorate under exposure to moisture and thermal and mechanical stresses. In addition, radiation from the open-ended coaxial line distorts the dipole pattern. These defects are absent in the stub-support termination shown in Fig. 8·2b. The coaxial line is continued for a distance $l_s \approx \lambda/4$ beyond the dipole system and terminated there in a metal plate. The latter region, known as the terminating stub, is again a shorted section of line, presenting a reactive impedance at the input end. Consideration of the current division at the driving point of the dipole shows that the stub is equivalent to an impedance shunted across the gap between the dipole wings. With $l_s \approx \lambda/4$, this impedance is practically infinite; electrically the system is equivalent to an open-ended termination.

The disparity in the currents on the two wings of a stub termination is even greater than that in the open-ended termination. The dead wing (or stub) D is excited only by leakage currents which make their

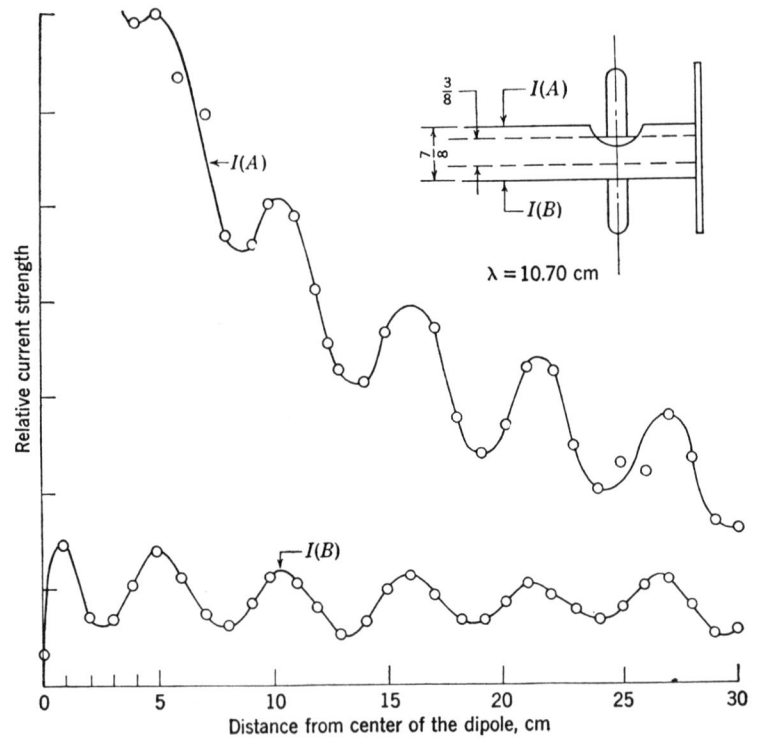

Fig. 8·3.—Leakage currents along the line; stub-supported dipole-disk feed without choke.

way through the opening in the outer wall and by coupling with the field of the live stub L. As in the case of the open-ended termination, coupling exists between the dipole system and the center wall of the line. A measure of the relative excitation of the dipole stubs is afforded by the intensity of the outer-wall line currents along lines in a plane containing the dipole axis. Figure 8·3 shows results of line-current studies made on a dipole system carrying a reflecting plate on the terminal stub. The standing-wave structure in the current is due to some obstruction on the outside surface at the input end of the coaxial line.

Control of the outer wall currents is achieved by means of the choke

C (Fig. 8·2b), as in the systems discussed previously. To present an infinite impedance at its open end the choke should, nominally, have a depth of $\lambda/4$. However, because of junction effects at the open end and coupling with the dipole system, the optimum value is somewhat less than $\lambda/4$. Figure 8·4 shows the line-current strength at a fixed point on the outer wall as a function of choke depth for the system studied in Fig. 8·3; the optimum depth is 0.23λ. Although this value is strictly significant only for the system illustrated, it has been found to give good results in other dipole systems employing chokes; it is a suitable value for the depth of the choke C_1 of the skirt dipole considered earlier.

It has been noted that the effect of the choke is to confine the outerwall current to the region between the choke and the end of the line. This current distribution

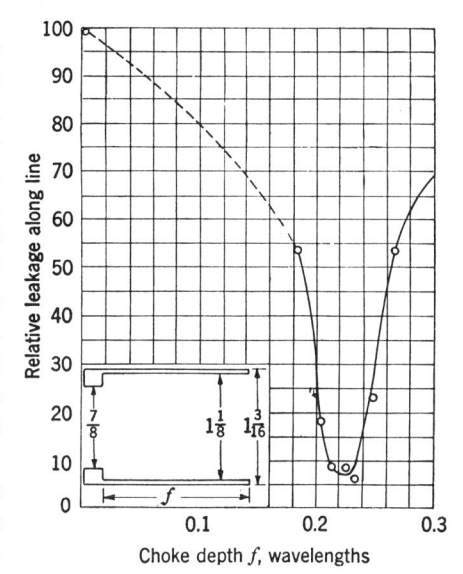

Fig. 8·4.—Leakage current as a function of choke depth.

serves as a linear radiator along the axis of the feed line. From the general considerations of Sec. 3·15 it will be evident that this radiates no energy in the direction of the line axis; it will, in general, give rise to a pattern with peak intensity on a cone having its axis coincident with the line axis. The phase of the line current with respect to the dipole current is determined by the position of the choke with respect to the dipole system. In combination with a paraboloidal mirror, in which the feed line lies on the axis of the mirror, the interaction between the dipole and line-current system produces a phenomenon known as *squint*, in which the over-all antenna beam is pointed, not along the axis of symmetry of the system, but in a direction making a small angle with that axis. Use is made of this phenomenon for scanning.

In closing the discussion of the asymmetric terminations, it should be noted that the input impedances of both the choke and the terminating stub vary rapidly with frequency. As a result, these structures are strong contributing factors in the frequency sensitivity of the impedance of these antennas. In addition, the cut-away region of the line introduces distributed capacities and inductances. These factors restrict the usability of the antenna to a narrow frequency band.

8·4. Symmetrically Energized Dipoles: Slot-fed Systems.—The shortcoming of unequal excitation of the dipole stubs, which characterizes the terminations discussed above, is eliminated in the slot-fed systems shown in Fig. 8·5. Both wings of the dipole are mounted on the outer conductor, in which a pair of slots S is milled in a plane normal to the dipole axis. The inner conductor is short-circuited to the outer conductor on one side by the post P, which usually is in the line of the dipole axis but may be inserted at any point along the line in the slotted region. Both open-ended and stub-terminated systems are used, ana-

(a)

(b)

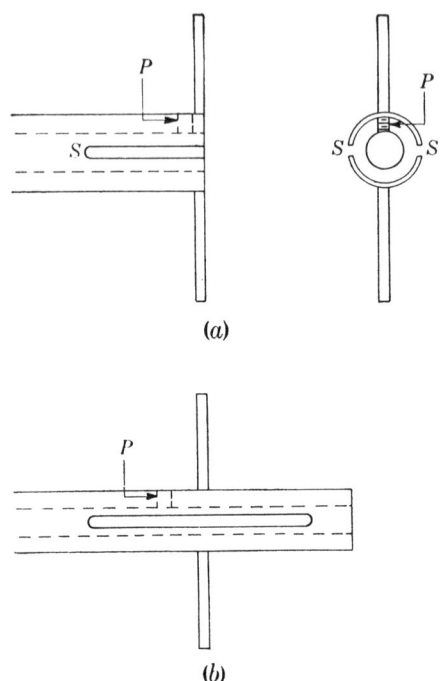

Fig. 8·5.—Slot-fed dipole terminations on coaxial line: (a) open-ended termination; (b) stub termination.

logous to the systems discussed in the preceding section. The open-ended type is used as a radiating element in linear arrays (cf. Sec. 9·8).

The operation of the dipole can be interpreted from various points of view. Perhaps the simplest picture is that the radiating system is energized by a voltage impressed across the slot. The origin of the voltage becomes evident on consideration of mode relationships in the slotted region. In the absence of the short-circuiting post P we would have the TEM-mode and possibly higher modes generated in the open-ended termination or, in the case of a wide slot, generated by the slot itself. All these modes, however, would be symmetric with respect to the plane containing the axes of the slots and give rise to no impressed field across the slot; under these conditions the dipole is not excited. With the insertion of the post, modes are generated that are symmetric with respect to the plane determined by the axis of the post and the axis of the inner conductor. These modes, when superposed on the preceding set, must give rise to a field such that the tangential electric field is zero over the surface of the post. In the case of a narrow slot we can ignore (for the qualitative picture) the modes generated by the slot itself; the prime effect of these modes is to relax the cutoff conditions and allow propagation within the slotted region of some of the modes generated by the post. The most significant of the latter is the TE_{11}-mode; Fig. 8·6 shows how superposition of the TE_{11}-mode on the TEM-mode leads

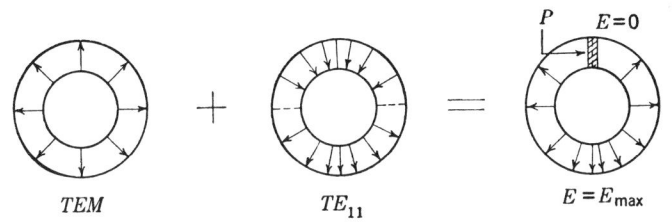

FIG. 8·6.—Superposition of the TEM- and TE_{11}-modes in the slotted region of the slot-fed dipole.

to a field configuration that satisfies the requisite boundary condition on the electric field in the case of a thin post. The resultant configuration gives a field that is zero along the post and increases with angle to a maximum value directly opposite to the post. It is readily seen that this impresses a voltage across the slot, with resulting excitation of the dipole structure.

The slotted dipole can also be analyzed from the transmission-line point of view.[1] The slotted region is conceived as a three-wire transmission line; this is the appropriate representation of a waveguide supporting simultaneous propagation of two modes, just as the two-wire line represents single-mode propagation. It will carry us too far afield to discuss the general theory of three-wire lines.[2] The equivalent circuit representations for the open-ended and stub-terminated systems are shown in Fig. 8·7a and b respectively, for the case in which the post lies along the dipole axis. Here Z_P is the impedance across the pair of lines connected by the post; Z_G the impedance at the gap opposite to the post; Z_A is the input impedance of a dipole having the same wing structure as in the given system, but center-fed from a balanced two-wire line; $l_{s'}$ is the length of the slot; and $l_{s''}$ the length of the terminating stub. At the end of the slot the outer lines are short-circuited, the three-wire line passing into the two-wire line.

[1] H. Riblet, "Slotted Dipole Impedance Theory," RL Report No. 772, Nov. 21, 1945.
[2] See S. O. Rice, "Steady State Solutions of Transmission Line Equations," *Bell System Tech. Jour.*, **20**, 131 (1941).

In the case of the open-ended termination with a narrow slot it is possible to reduce the system to a two-wire line with appropriate loading, as shown in Fig. 8·7c. The impedance Z_P has been taken to be zero; Z_s is the characteristic impedance of the three-wire line under the condition that no current is flowing in the central line. It will be seen that in this particular case the slot contributes only a susceptance, like a short-circuited section of two-wire line. This circuit representation indicates that the length of the slot can be so chosen as to match out the other reactive impedance elements involved in the termination.

The slot not only equalizes the excitation of the wings but also serves as a choke element to decouple the dipole system from the outer wall of the line. The resulting system is completely free from the squint phenomenon associated with the asymmetric termination. In the case of open-ended terminations it is possible to design units with high power capacity; these have found application in linear-array antennas. The stub-terminated units, on the other hand, are more limited in their power capacity than the corresponding asymmetrical terminations and have been used in place of the latter only where it is imperative to have a squint-free system and relatively lower power levels are acceptable.

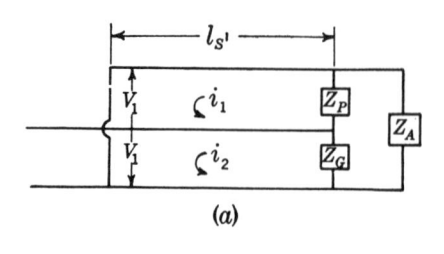

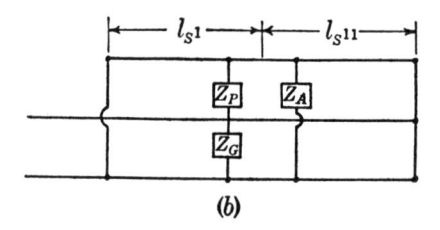

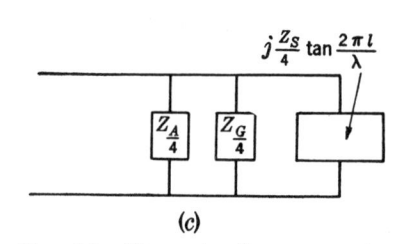

Fig. 8·7.—Three-wire line representation of the slot-fed dipole: (a) open-ended termination; (b) stub-terminated line; (c) reduced equivalent loading for Case a.

8·5. Shape and Size of the Dipole.—The impedance problem has been a troublesome one with dipole feeds, largely because of the frequency-sensitive elements—such as the choke, terminating stub, and slot—needed in making various types of terminations. A certain measure of adjustment is available in the size and shape of the dipole. The dependence of the impedance of a center-fed dipole on its size and shape has been the subject of considerable theoretical work.[1] All of the work

[1] S. A. Schelkunoff, *Electromagnetic Waves*, Van Nostrand, New York, 1943, Chap. 11; L. J. Chu and J. A. Stratton, *Jour. Applied Phys.*, **12**, 241 (1941); R. W. P. King and D. D. King, *Jour. Applied Phys.*, **16**, 445 (1945).

SHAPE AND SIZE OF THE DIPOLE

applies to an idealized system in which the dipole is driven from a balanced system across an infinitesimal gap; it is assumed that the coupling between the dipole and line plays no part in determining the current distribution of the wings. As we have noted above, this condition is never realized in microwave systems where the dimensions of the feedline cross section are comparable to those of the dipole structure.

The theoretical results, however, are helpful in a qualitative way. The various theories differ in quantitative details concerning the values of the impedance, but all show the same general qualitative features.

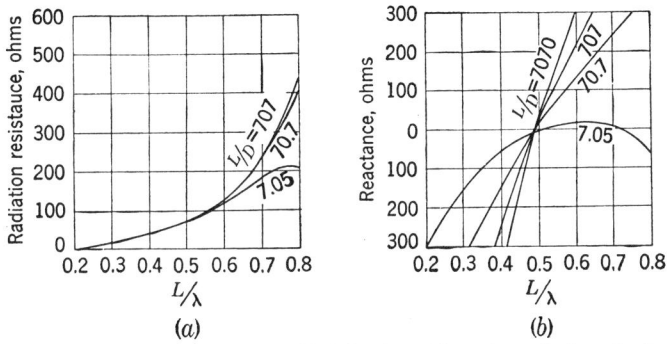

FIG. 8·8.—Input impedance of spheroidal dipoles with major axis L and minor axis D: (a) real component or radiation resistance; (b) imaginary component or reactance. (*From L. J. Chu and J. A. Stratton, J. Appl. Physics, by courtesy of the authors and the American Institute of Physics.*)

The curves shown in Fig. 8·8 are taken from the work of Chu and Stratton. They apply to spheroidal dipoles, the major axis of which is designated by L and the minor axis by D.

The curves show the dependence of the real and imaginary components of the impedance on wavelength for various values of the ratio L/D. It is observed that in the neighborhood of the resonant point, which corresponds closely to a length equal to $\lambda/2$, the resistive component is virtually independent of the value of L/D and is equal to about 70 ohms. The dependence of the resistance on wavelength does not become marked until the length is considerably larger than the resonant value. The reactive component, however, is seen to be a decided function of the frequency. The larger the ratio L/D, that is, the thinner the dipole, the more rapidly does the reactance vary and the sharper is the resonant point. Thus, a thin dipole is more frequency-sensitive than a fat dipole. The dipole dimensions can be chosen such that its reactive component balances the reactance which is associated with the termination; this in general will lead to better over-all impedance characteristics for the antenna than the choice of a dipole that alone has a flat reactance characteristic. The impedance characteristics of the dipole can also be controlled by such processes as top loading with a sphere or other structure in the same manner as is done at longer wavelengths. Here again the procedure is entirely empirical, and we shall not dwell upon it any further.

8·6. Waveguide-line-fed Dipoles.—It is much simpler to feed a dipole from a waveguide line than from a coaxial line. The technique of termination is shown in Fig. 8·9. The dipole is mounted on a web that fits into the mouth of the guide, parallel to the broad face of the guide and transverse to the electric vector in the dominant TE_{10}-mode. The E-vector is thus parallel to the dipole, which is driven by the radiation incident on it from the mouth of the guide. It is obvious that if the web is inserted symmetrically, the two wings of the dipole are excited equally. The taper shown in the diagram serves as an impedance-matching device; it also improves the radiation pattern in that it decouples the outer wall of the line from the dipole. The impedance of the system is also determined by the depth of insertion of the web and the position of the dipole with respect to the mouth.

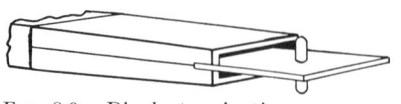

FIG. 8·9.—Dipole termination on waveguide.

8·7. Directive Dipole Feeds.—The design of directive feeds is based on the principle of interference between dipoles properly spaced and phased (Sec. 3·17) and on the principle of images (Sec. 5·3). Early designs utilized the skirt dipole with a reflecting plate and the open-ended asymmetric termination followed by a second dipole or a reflecting plate. These designs have very poor structural characteristics; they will not be discussed here. Stub-terminated coaxial systems and waveguide systems lend themselves admirably to the construction of directive feeds, the stub or web providing mechanical support for the system of dipoles involved or for the reflecting plate. These directive systems are designed to radiate maximum power back along the feed line; the reflector that is to be illuminated by the feed is then also mounted on the feed line. This rear-feed type of installation (examples of which are to be seen in Sec. 12·11) minimizes the length of line and the series of bends and joints required (factors of considerable importance for generator stability) and forms a compact and rugged system.

The directive system employing a reflecting plate, which may be termed a dipole-plate or dipole-disk feed, is based on the principle of images. In accordance with the general theory, to produce peak intensity along the feed line the reflecting plate is mounted a distance $\lambda/4$ behind the dipole. The principle of images assumes, of course, a reflecting plate of infinite extent. In the case of the feed system the plate must be kept as small as possible. Otherwise the feed will present too extended an obstruction in the path of the energy reflected from the large mirror;

the effects of such aperture blocking on the over-all antenna pattern are discussed in Sec. 6·7. It is thus necessary to sacrifice a certain measure of directivity, with the result that the primary feed has a back lobe, that is, radiation behind the reflector plate; this, too, has a significant effect on the over-all antenna pattern (*cf.* Sec. 12·5).

The coaxial-line-fed multidipole systems are usually designed so that only one dipole is excited directly from the line. The other members (dummy or parasitic dipoles), arranged in a linear array, are fed by coupling with the directly excited element. Microwave feeds have usually included a single dummy element to complete a double-dipole system such as that discussed in Sec. 3·18. In that section the case of $\lambda/4$ spacing and relative phase $\psi = \pi/2$ was considered in detail. However, by reference to Eq. (3·174) of Sec. 3·18, it may be seen that any pair of values of spacing a and phase which satisfy the relation

$$\frac{\pi a}{\lambda} - \frac{\psi}{2} = m\pi, \qquad m = 0, 1, 2, \cdots$$

will give peak intensity along the direction normal to both dipole axes, that is, along the feed line in the practical case. These other systems, however, unlike the $(\lambda/4, \pi/2)$ system, in general also give rise to a back lobe in the direction 180° away from the peak. In practice, the phase of the dummy relative to the driven element is controlled by the relative dimensions of the dipoles as well as by their spacing; from Sec. 8·5 we see that it is possible to make one dipole capacitative or inductive relative to the other, by proper choice of dimensions.

Directive feeds will be further discussed with reference to particular systems. In the following sections design data are presented on a number of feeds that have been developed in the Radiation Laboratory and used extensively. It is not to be assumed that the results given here represent the ultimate that can be achieved with these systems.

8·8. Dipole-disk Feeds.—Two dipole-disk systems have been developed, employing respectively the stub-terminated asymmetric dipole and the stub-terminated slot-fed dipole.

a. Asymmetric Dipole Termination.—Three such feeds have been designed[1] to illuminate paraboloidal mirrors, of focal length 10.6 in. and 30-in. aperture, at wavelengths of 9.1, 10.0, and 10.7 cm respectively. Details of the feed assembly are given in Fig. 8·10. The line has a characteristic impedance of 46 ohms; its dimensions are outer conductor, OD = 0.875 in. with wall thickness of 0.032 to 0.035 in.; inner conductor, OD = 0.375 in. Reasonable directivity was obtained with a reflector plate with diameter about 0.8λ. The principal E- and H-plane feed patterns are shown in Fig. 8·11. The peak intensity of the pattern is

---

[1] S. Breen and R. Hiatt, R L Report No. 54-23, June 21, 1943.

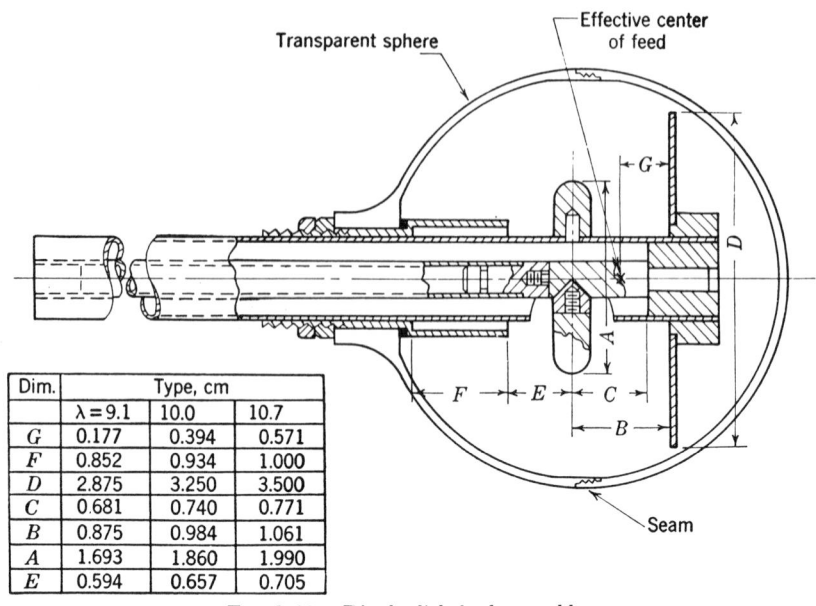

Dim.	Type, cm		
	λ = 9.1	10.0	10.7
G	0.177	0.394	0.571
F	0.852	0.934	1.000
D	2.875	3.250	3.500
C	0.681	0.740	0.771
B	0.875	0.984	1.061
A	1.693	1.860	1.990
E	0.594	0.657	0.705

FIG. 8·10.—Dipole-disk feed assembly.

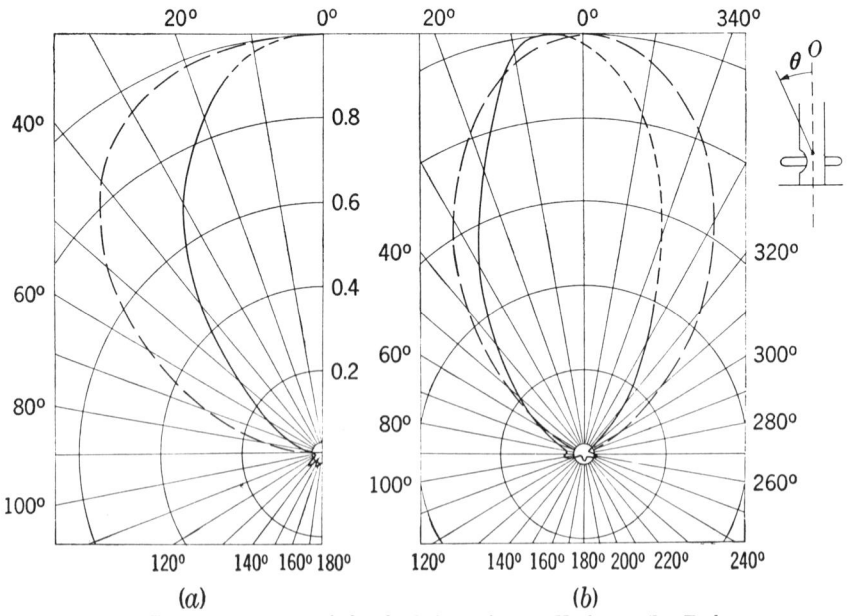

FIG. 8·11.—Primary pattern of dipole-disk feed: (a) H-plane; (b) E-plane. ——— dipole-disk feed of Fig. 8·10; – – – – theoretical pattern of a dipole at a distance $\lambda/4$ from an infinite plane.

directed along the feed line, and the data in this region are, therefore, somewhat uncertain. The dotted portion of the curves have been obtained by extrapolation. It is observed that the E-plane pattern is not symmetrical. This is due to the fact that one wing of the dipole is excited more strongly than the other in the asymmetric dipole termination, as was pointed out in Sec. 8·3. The peak appears on that side of the axis which corresponds to the dipole wing carrying the major portion of the current. The H-plane pattern, on the other hand, was found to be accurately symmetrical corresponding to the symmetry of the dipole structure in the plane. For comparison, there are plotted the theoretical patterns for the ideal system of a dipole placed $\lambda/4$ in front of an infinite reflecting plane. It is seen that the feed pattern is considerably more directive; the gain of the feed is found to be equal to 7.

The E- and H-plane centers of feed are coincident, lying between the dipole and the disk, somewhat nearer to the latter. The point-source cone is more than adequate to cover a mirror with dimensions given above. The unpressurized feed has a peak power capacity of 350 ± 35 kw. With suitable matching transformers it has been possible to realize an impedance characteristic for the composite system (feed and paraboloid) such that the standing-wave ratio r did not exceed 1.23 over a band of ±3 per cent about the matching frequency.

b. *Slot-fed Termination.*—A unit designed to operate at a wavelength of 9.1 cm with a paraboloidal mirror of 3.6-in. focal length and 12-in. aperture is illustrated in Fig. 8·12.[1] The line has a characteristic impedance of 45 ohms, with an inner conductor of $\tfrac{5}{16}$ in. diameter. The smaller line was used here to reduce weight, the power requirements on the feed having been smaller than in the preceding case. It will be noted that the disk diameter here is about 0.5λ. The system has a single center of feed for both principal planes and is completely free from squint. The composite antenna made up of the feed and the mirror indicated above has an impedance band of ±1.25 per cent about the design frequency over which $r \leq 1.23$.

8·9. Double-dipole Feeds. a. *Coaxial-line-fed System.*—Such a feed[2] is illustrated in Fig. 8·13; it is a lightweight unit employing a $\tfrac{5}{16}$-in. line like that discussed in Sec. 8·8b. The spacing between the dipoles is very nearly $\lambda/8$; correspondingly, the parasite element is longer than the driven element in order to produce the proper phase relationships. This system, like those discussed above, has a unique center of feed. An antenna consisting of this feed and a paraboloidal reflector of 3.6-in. focal length and 12-in. aperture has a standing-wave ratio $r \leq 1.23$ in a band of ±1 per cent about the design frequency.

[1] W. B. Nowak, RL Report No. 54-26, July 5, 1943.
[2] *Ibid.*

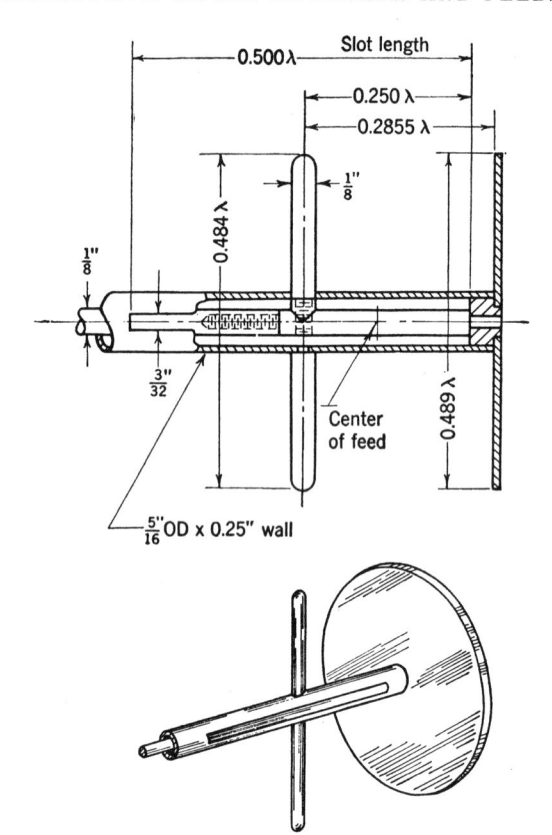

Fig. 8·12.—Dipole-disk feed.

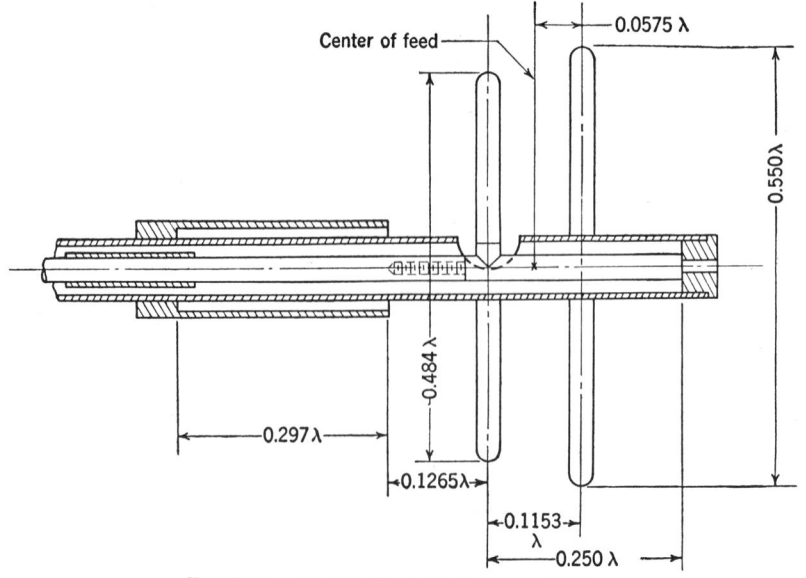

Fig. 8·13.—Double-dipole feed on coaxial line.

SEC. 8·9] DOUBLE-DIPOLE FEEDS

b. *Waveguide Systems.*—A double-dipole feed built up on a waveguide termination for use at a wavelength of 3.2 cm is shown in Fig. 8·14.[1] The two dipoles are mounted on the web so that their axes lie on the plane of symmetry of the guide. The spacing between the dipoles is

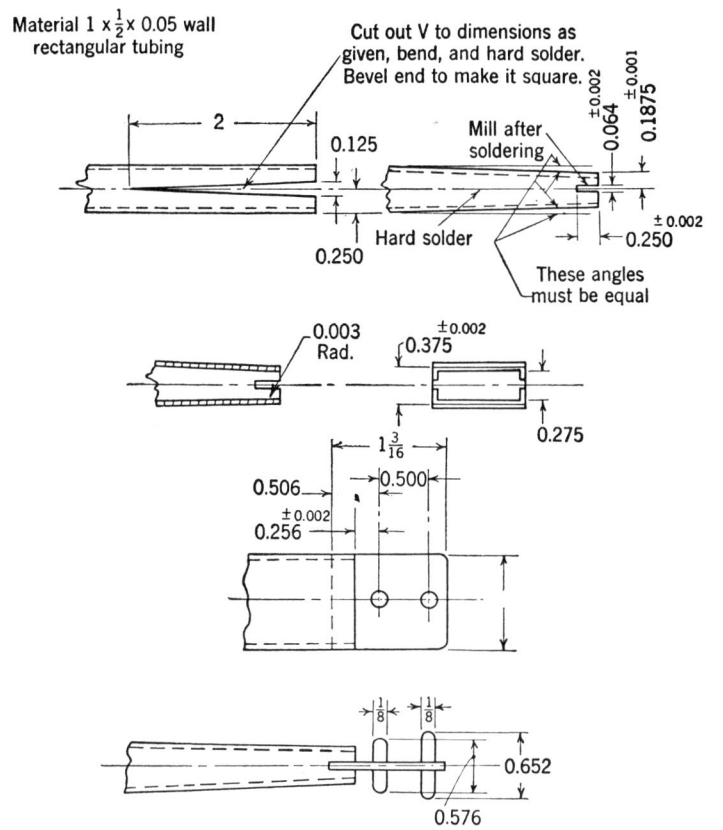

FIG. 8·14.—Waveguide double-dipole feed; $\lambda = 3.2$ cm.

about $\lambda/2.5$; again the coupling (and hence the relative phase) of the elements is adjusted by the suitable choice of their relative dimensions. The radiation pattern has an appreciable back lobe which is in some measure due to the guide itself; this is reduced by tapering the terminal region as shown in the figure. The E- and H- plane centers of feed are not concident; however, their separation is negligible for most purposes, and the equivalent center of feed can be taken to be located just behind the first dipole.

We have previously pointed out the dependence of the impedance on the taper, depth of insertion of the web, and the dipole factors. To obtain reproducible results, special care must be taken to remove excess

[1] W. Sichak, "Double Dipole Rectangular Wave Guide Antennas," RL Report No. 54-25, June 26, 1943.

solder at the base of the dipole and at the seams between web and waveguide. Antennas made up of this feed and paraboloidal mirrors of 18-in. aperture and focal length either 4.5 or 5.67 in. have a bandwidth of ± 1.5 per cent over which $r \leq 1.23$ if the antenna is matched by an inductive window at $\lambda = 3.2$ cm. The unpressurized antenna has a peak power capacity of 375 kw, corresponding to 50 kw at 50,000-ft altitude.

8·10. Multidipole Systems.—The web termination on a waveguide provides a convenient base on which to build multidipole systems in the form of two-dimensional arrays. Two such arrays have been designed

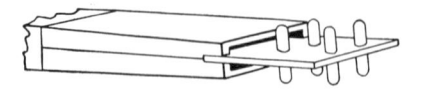

Fig. 8·15.—Four-dipole feed.

for the 3-cm band, one a triangular array of three dipoles, the other a rectangular array of four dipoles. Only the latter has been used in final antenna design. The four-dipole array shown schematically in Fig. 8·15 can be regarded as a pair of the double dipole units discussed in Sec. 8·9b, separated by a distance of approximately $\lambda/2$. Each double-dipole unit can be replaced by its equivalent point source, reducing the system to two directive sources in phase, spaced $\lambda/2$ apart. It is evident that no appreciable change is to be expected in the E-plane pattern. The H-plane pattern, however, must be multiplied by the directivity factor of two isotropic sources in phase and with $\lambda/2$ separation. This factor is readily found to be $[\cos(\pi/2 \cos \phi)]^2$ where ϕ is the angle with respect to the axis in the H-plane. Hence if $P_2(\phi)$ is the H-plane pattern of the double-dipole system, the pattern $P_4(\phi)$ of the four-dipole system is given closely by

$$P_4(\phi) = P_2(\phi) \left[\cos\left(\frac{\pi}{2} \cos \phi\right) \right]^2.$$

CHAPTER 9

LINEAR-ARRAY ANTENNAS AND FEEDS

BY J. E. EATON, L. J. EYGES, AND G. G. MACFARLANE

9·1. General Considerations.—The technique of producing directive beams by means of arrays of radiators that are suitably spaced and driven with appropriate relative amplitudes and phases has been used widely at the longer wavelengths. These arrays have generally been in the form of two-dimensional lattices with the possible addition of a reflecting surface to confine the radiation to a single hemisphere in space. In the microwave region, attention has been confined almost exclusively to the one-dimensional, that is, linear, arrays. The wavelength advantage becomes evident at once, for with economy in physical size it is still possible to have an array that is long measured in wavelengths and hence highly directive.

The arrays that have been designed to date can be grouped into two general classes: (1) end-fire arrays producing a beam directed along the axis of the array and (2) broadside arrays producing beams the peak intensity of which is in a direction normal to or nearly normal to the axis. End-fire arrays have proved to be particularly useful where it is necessary to mount an antenna close to an object; for example, such arrays have been mounted along a gun barrel in airplanes to furnish gunfire range information and to serve as gunfire directors. Axially symmetrical broadside arrays which produce beams symmetrical about the axis have been designed for use as beacons; installed both in ground or ship and on aircraft they provide a communication system between ground (or ship) and aircraft. The patterns of these arrays are axially symmetrical like the dipole patterns but have increased directivity in the meridional plane to give increased range. Other types of broadside arrays have been developed whose beams have a fair measure of directivity also in the plane perpendicular to the array axis. In a few cases, arrays of this type have been used as the terminal antenna system; more frequently these arrays have been used as line sources for illuminating cylindrical reflectors, in which case the reflector is placed sufficiently close to the array so as to be in its cylindrical wave zone.

While there is no fundamental difference in principle between long-wave and microwave arrays, the microwave arrays present problems of their own which are due to the wavelength region involved. In long-wave arrays it is possible to isolate to a large degree the feeding of one element of the array from another. Microwave arrays must be built on coaxial line or waveguides with the result that the feeding of the element becomes a mutual interaction problem. This type of feeding also requires special designs in the radiating elements of which there is quite a variety at microwave frequencies. The physical size of the radiating elements is generally small, and tolerance problems are associated with microwave arrays that are generally uncommon at longer wavelengths.

The problems and techniques of linear-array design have been divided in this chapter into three general parts. The first concerns itself with general pattern theory, that is, the relation between the far-zone pattern of an array and the amplitude and phase distribution among the elements and their spacing; in this section no attention is paid to the problem of realizing a given amplitude and phase distribution. The second part is a survey of the radiating elements that have been developed for microwave arrays. The final division treats the problems associated with combination of the elements into linear arrays and the techniques available to produce the desired amplitude and phase distributions.

PATTERN THEORY

9·2. General Array Formula.—A linear array is a specialization of the general space array discussed in Sec. 3·19. The space factor of the system can be obtained immediately from Eqs. (3·179) and (3·180) by imposing on those equations the simplifications gained in working with a one-dimensional rather than a three-dimensional complex. It may be instructive, however, to derive the space factor directly from the superposition of fields; we shall be concerned only with the far-zone field of the array.

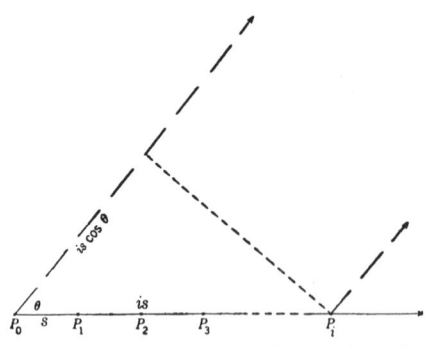

Fig. 9·1.—Difference in distance from the ith element and from the pole to a distant point in the direction θ, ϕ.

Suppose that there are n elements in the array under consideration, and let the reference line of the array be taken as the polar axis. The ordering of the elements $P_0, P_1, \ldots, P_{n-1}$ is shown in Fig. 9·1 with the element P_0 taken at the origin; the distance between two adjacent elements is s. Let us consider the field at a point (R, θ, ϕ) in the far zone. According to Eqs. (3·168a) and (3·168b) the field due to the ith element at a distance r_i from the element is

SEC. 9·2]   GENERAL ARRAY FORMULA   259

$$E_{\theta_i} = -\frac{j\omega\mu}{4\pi r_i} e^{-jkr_i} F_{1i}(\theta,\phi),$$

$$E_{\phi_i} = -\frac{j\omega\mu}{4\pi r_i} e^{-jkr_i} F_{2i}(\theta,\phi).$$

If the customary far-zone field approximations are made, r_i can be set equal to R in the denominator, while in the phase term we have

$$r_i = R - is\cos\theta \qquad (1)$$

as shown in Fig. 9·1. The component fields are then

$$E_{\theta_i} = -\frac{j\omega\mu}{4\pi R} e^{-jkR} F_{1i}(\theta,\phi)e^{j\psi_i}, \qquad (2a)$$

$$E_{\phi_i} = -\frac{j\omega\mu}{4\pi R} e^{-jkR} F_{2i}(\theta,\phi)e^{j\psi_i}, \qquad (2b)$$

where ψ_i is the phase difference between the ith element and the origin due to the difference in path length to the field point:

$$\psi_i = \frac{2\pi i s \cos\theta}{\lambda}. \qquad (3)$$

The elements of the array are identical in structure and carry similar current distributions. They differ only in the amplitude and phase. We can, therefore, write

$$F_{1i}(\theta,\phi) = a_i F_1(\theta,\phi), \qquad (4a)$$
$$F_{2i}(\theta,\phi) = a_i F_2(\theta,\phi). \qquad (4b)$$

The complex coefficients a_i express the amplitude and phase of the ith element with respect, say, to the zeroth element; they will be called the "feeding coefficients."

By the superposition principle, the field of the array is

$$E_\theta = -\frac{j\omega\mu}{4\pi R} e^{-jkR} F_1(\theta,\phi) \sum_{i=0}^{n-1} a_i e^{j(2\pi i s \cos\theta)/\lambda}, \qquad (5a)$$

$$E_\phi = -\frac{j\omega\mu}{4\pi R} e^{-jkR} F_2(\theta,\phi) \sum_{i=0}^{n-1} a_i e^{j(2\pi i s \cos\theta)/\lambda}. \qquad (5b)$$

The last two factors in each instance represent the corresponding space factor of the array. The power pattern is proportional to the sum of the squares of the absolute value of the two space factors; that is,

$$P(\theta,\phi) = \left| F_1(\theta,\phi) \sum_{i=0}^{n-1} a_i e^{j(2\pi i s \cos\theta)/\lambda} \right|^2 + \left| F_2(\theta,\phi) \sum_{i=0}^{n-1} a_i e^{j(2\pi i s \cos\theta)/\lambda} \right|^2.$$

Since the absolute value of a product is the product of the absolute values,

$$P(\theta,\phi) = [|F_1(\theta,\phi)|^2 + |F_2(\theta,\phi)|^2] \left| \sum_{i=0}^{n-1} a_i e^{j(2\pi i s \cos \theta)/\lambda} \right|^2,$$

or

$$P(\theta,\phi) = P_0(\theta,\phi) \left| \sum_{i=0}^{n-1} a_i e^{j(2\pi i s \cos \theta)/\lambda} \right|^2. \quad (6)$$

The first factor is the power pattern of an individual element of the array. The second factor depends on the number of elements in the array, their amplitudes and phases, and their spacing. It is formally independent of the type of element used, although in practice the value of the feeding coefficients a_i is intimately connected with the characteristics of the elements of the array. We shall call this function the "array factor" and denote it by $\Psi(\theta)$.

$$\Psi(\theta) = \left| \sum_{i=0}^{n-1} a_i e^{j(2\pi i s \cos \theta)/\lambda} \right|^2. \quad (7)$$

This factor is the power pattern of a similar array of isotropic radiators, for which $P_0(\theta,\phi) = 1$. Moreover, it is independent of ϕ as was to be expected.

If the feeding coefficient a_i is written as

$$a_i = |a_i| e^{j\chi_i},$$

the array factor is seen to be the square of the magnitude of the resultant of n vectors; the magnitude of the ith element vector is $|a_i|$ and the angle between it and the zeroth-element vector is $\chi_i + \psi_i$. The angles between the vectors vary with the angular position θ of the field point, with corresponding variation in the resultant vector. In general as θ covers the entire range from $\theta = 0$ to $\theta = \pi$, the magnitude of the resultant passes through maximum and minimum values. The absolute maximum value that could be attained by the resultant is the sum of the vectors when they are colinear and in the same direction. With arbitrary χ_i, however, there may be no angle θ for which this condition is realized and the maxima are less than the absolute maximum. Similarly, there may be no value of θ for which the minimum value of the resultant takes on the absolute minimum value of zero. However, with special relations between the χ_i it is possible to have directions θ for which the path-length phases ψ_i compensate for the intrinsic phase differences χ_i between the elements to bring all the component field vectors in phase; in this case, the absolute maximum resultant is attained.

A particularly simple and useful case is that in which the coefficients a_i are all real. This implies that the angle between any two vectors associated with adjacent elements is $(2\pi s \cos \theta)/\lambda$.

If the coefficients a_i are equal, it is readily apparent that the resultant vector is 0 whenever the vectors constituting the sum permute among themselves under a rotation of less than 2π. For then the resultant vector both rotates and remains unchanged and hence is 0. This occurs whenever $(2\pi s \cos \theta)/\lambda$ is any integral multiple of $2\pi/n$ less than n. When $(2\pi s \cos \theta)/\lambda = 2\pi$, the vectors obviously reinforce one another and an absolute maximum results.

Whenever $\Psi(\theta) = 0$, then $E_\theta = E_\phi = P(\theta,\phi) = 0$ [Eqs. (5a), (5b), and (6)] for all values of ϕ. The surface in spherical coordinates for which θ is constant is a right circular cone. The cones on which $\Psi(\theta) = 0$ are commonly called cones of silence.

9·3. The Associated Polynomial.—The vector representation of the array factor provides a method of rapidly analyzing the simple arrays frequently encountered in practice. Vector language is not, however, well suited to a more general study of arrays. An alternate method has been developed[1] that associates a polynomial with any linear array. The array factor may be completely analyzed in terms of properties of this polynomial.

Let z be the complex number $z = x + jy$. The polynomial associated with the linear array of elements having feeding coefficients a_i is

$$f(z) = a_0 + a_1 z + \cdots + a_{n-1} z^{n-1}.$$

The value of the polynomial for the complex number

$$z = \zeta = e^{j(2\pi s \cos \theta)/\lambda}$$

is the sum entering into Eqs. (6) and (7); the array factor is thus the norm[2] of the associated polynomial for $z = \zeta$,

$$\Psi(\theta) = |f(\zeta)|^2. \tag{8}$$

The complex number ζ is a vector from the origin in the complex plane, of magnitude unity, making an angle $\psi = (2\pi s \cos \theta)/\lambda$ with the real axis. As θ varies $z = \zeta$ describes a circle of unit radius about the origin. In the future we shall not distinguish between z and ζ; it is to be understood that z lies on the unit circle whenever $\Psi(\theta)$ is to be computed from the associated polynomial. When $\theta = 0$, $\psi = 2\pi s/\lambda$. As θ moves toward π, z moves along the unit circle clockwise toward the point where its angle $\psi = -2\pi s/\lambda$. In that interval z may traverse but a portion of

---

[1] S. A. Schelkunoff, "A Mathematical Theory of Linear Arrays," *Bell System Tech. Jour.*, **22**, 80 (1943).

[2] The norm of a complex number as used here is the square of its absolute value. It may have a more general meaning.

the unit circle or may complete several circuits of it depending on the value of s. Its path will be referred to as the range of z. In Fig. 9·2 the range of z is shown for three values of s. Since the angular distance traversed by z is $4\pi s/\lambda$, the range of z is exactly one circuit of the unit circle when $s = \lambda/2$, is less than one circuit when $s < \lambda/2$, and is more than one circuit when $s > \lambda/2$.

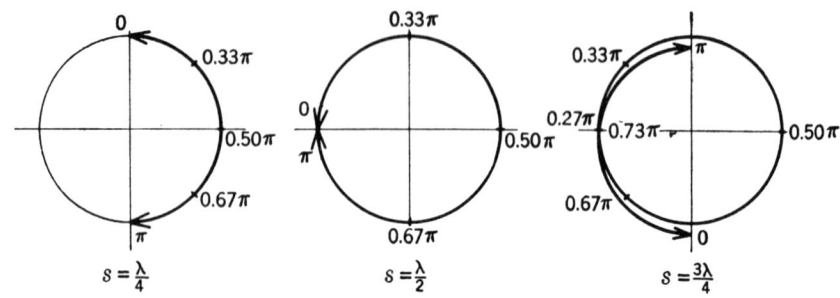

Fig. 9·2.—The portion of the unit circle in the complex plane that is the range of z. The real axis is horizontal. The figures on the perimeter show the corresponding values of θ for certain values of z.

Any polynomial can be expressed as a product of linear factors. In particular the associated polynomial may be written in the form

$$f(z) = a_{n-1}(z - z_1)(z - z_2) \cdots (z - z_{n-1}). \tag{9}$$

Since the feeding coefficients give only the relative phases and amplitudes of the elements of the array, a_{n-1} can be taken to be any convenient nonzero number. The complex numbers z_i (known as the "zeros" of the polynomial) are unaffected. Their values depend only on the set of ratios a_i/a_{n-1}. The factorization of $f(z)$ in Eq. (9) lends itself to a simple geometric interpretation of the array factor. Since the norm of a product is the product of the norms, Eq. (9) may be written

$$\Psi(\theta) = |z - z_1|^2 |z - z_2|^2 \cdots |z - z_{n-1}|^2$$

for z on the unit circle. The zeros of $f(z)$ are well-defined points in the complex plane but do not necessarily lie on the unit circle. For any value of z, $|z - z_i|^2$ is the square of the distance between the point z and the point z_i. The array factor is then the square of the product of the distances of $n - 1$ fixed points to a variable point moving on the unit circle. It is immediately obvious that $\Psi(\theta) = 0$ if and only if some z_i lies on the unit circle within the prescribed range of z. Shown in Fig. 9·3 is the range of z when $s = \lambda/4$. The zeros of $f(z)$ are shown for the case $n = 9$ and $a_0 = a_1 = \cdots = a_{n-1} = 1$. The array factor then vanishes for four values of θ and attains a maximum value at three points, each lying between an adjacent pair of nulls. The predominating influence on the value of $\Psi(\theta)$ is the distance from the corresponding value of z to the nearest zero of $f(z)$. In this connection it should be noted that the zeros of $f(z)$ lying outside the range of z have for the most part but small effect on the relative value of $\Psi(\theta)$. For the case illustrated, the point $z = 1$ (corresponding to $\theta = \pi/2$) is farthest from a zero of $f(z)$, and one would expect, as is the case, that $\Psi(\theta)$ has an absolute maximum at $\theta = \pi/2$. This is directly evident from the vector viewpoint of the previous section. The elements themselves are all in phase, and in the direction $\theta = \pi/2$ the distance is the same from all the elements to the field point. The contributions from the elements in that direction are, therefore, in phase, and the vectors are all in the same direction.

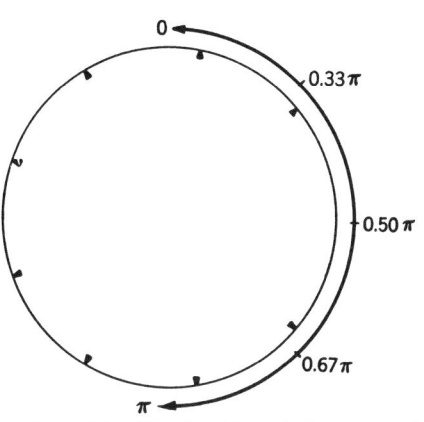

FIG. 9·3.—The location of the zeros of $f(z) = 1 + z + z^2 + \cdots + z^8$. The range of z for $s = \lambda/4$ is also shown together with a few of the corresponding values of θ.

It is frequently advantageous to separate from the relative phase χ_i of each element as expressed in the coefficients a_i a constant-phase delay ψ_0 between each pair of adjacent elements. This is equivalent to writing

$$a_i = \bar{a}_i e^{-j(i\psi_0)},$$

where now the angle of $\bar{a}_i$ is the deviation from $-i\psi_0$ of the difference in phase between the ith element and the element with index 0. Let

$$\bar{z} = e^{-j\psi_0} z; \tag{10}$$

then $f(z)$ transforms into the polynomial

$$\bar{f}(\bar{z}) = \sum_{i=0}^{n-1} \bar{a}_i \bar{z}. \tag{11}$$

When z lies on the unit circle, $\bar{z}$ given by Eq. (10) does likewise, and we have $f(z) = \bar{f}(\bar{z})$. Since Eq. (10) is equivalent to a rotation of the complex plane through an angle ψ_0 in the clockwise direction, the array factor may be computed from the zeros of $\bar{f}(\bar{z})$ in the same manner as before save that the range of $\bar{z}$ is the original range of z rotated clockwise through the angle ψ_0 (Fig. 9·4). Symbolically

$$\Psi(\theta) = |\bar{f}(\bar{z})|^2,$$

where

$$\bar{z} = e^{j[(2\pi s \cos \theta)/\lambda - \psi_0]}.$$

The association of a polynomial with any linear array of prescribed spacing provides a simple and elegant method for compounding array factors. Suppose $\Psi_1(\theta)$ and $\Psi_2(\theta)$ are the respective array factors of two arrays with the same spacing. If $f_1(z)$ and $f_2(z)$ are the associated polynomials of the arrays, then the array whose associated polynomial is

$$f(z) = f_1(z)f_2(z)$$

will have as its array factor

$$\Psi(\theta) = \Psi_1(\theta)\Psi_2(\theta), \quad (12)$$

for, as has already been observed, the norm of a product is the product of the norms. Explicit values of the feeding coefficients of an array whose array factor is given by Eq. (12) can thus be obtained by simply multiplying together the polynomials $f_1(z)$ and $f_2(z)$.

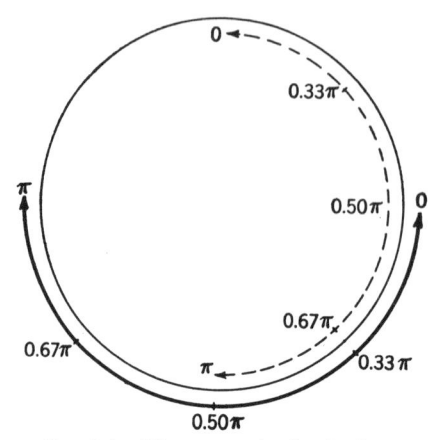

Fig. 9·4.—The range of $\bar{z}$ due to the constant phase delay $\psi_0 = \pi/2$ for $s = \lambda/4$. The dotted curve indicates the range of z associated with that of $\bar{z}$. The figures on the perimeter indicate the corresponding values of θ for certain values of z and $\bar{z}$.

9·4. Uniform Arrays.—A linear array that is made up of elements having equal amplitudes and a constant-phase difference between adjacent ones is of considerable importance. Such an array is called a "uniform array." Its feeding coefficients are

$$a_i = e^{-ji\psi_0}.$$

Although the associated polynomial of a uniform array will in general have complex coefficients, the related polynomial $\bar{f}(\bar{z})$ may, as was shown in the previous section, be used with equal effectiveness. Then

$$\bar{f}(\bar{z}) = 1 + \bar{z} + \bar{z}^2 + \cdots + \bar{z}^{n-1},$$

where

$$\bar{z} = e^{-j\psi_0}z. \quad (9 \cdot 10)$$

But

$$\bar{f}(\bar{z}) = \frac{\bar{z}^n - 1}{\bar{z} - 1} = \bar{z}^{(n-1)/2}\frac{\bar{z}^{n/2} - \bar{z}^{-n/2}}{\bar{z}^{1/2} - \bar{z}^{-1/2}}.$$

The array factor is then

$$\Psi(\theta) = |\bar{z}^{n-1}| \cdot \left|\frac{\bar{z}^{n/2} - \bar{z}^{-n/2}}{\bar{z}^{1/2} - \bar{z}^{-1/2}}\right|^2; \quad \bar{z} = e^{j\psi}$$

with

$$\psi = \frac{2\pi s}{\lambda}\cos\theta - \psi_0. \quad (13)$$

SEC. 9·4] UNIFORM ARRAYS 265

However, $|e^{j\psi}| = 1$ and $|e^{jr\psi} - e^{-jr\psi}|^2 = 4\sin^2 r\psi$. Thus the array factor of a uniform array is given by

$$\Psi(\theta) = \frac{\sin^2 \frac{n}{2}\psi}{n^2 \sin^2 \frac{1}{2}\psi} \tag{14}$$

together with Eq. (13). The number n^2 has been inserted in the denominator as a normalizing factor.

When $\psi = 2k\pi$ and $k = 0, \pm 1, \pm 2, \cdots$, $\Psi(\theta)$ is indeterminate. It can be readily shown, however, that it approaches the value unity at those points. The corresponding values of θ are given by

$$\cos\theta = \frac{\lambda}{2\pi s}(2k\pi + \psi_0) \qquad (k = 0, \pm 1, \pm 2, \cdots). \tag{15}$$

For every real value of θ satisfying Eq. (15) $\Psi(\theta)$ has an absolute maximum. In these directions the differences in phase between the vector contributions of successive elements that are due to differences in path to the field point just compensate for the intrinsic phase difference between the elements. The contributions are then all in phase, and we have, therefore, an absolute maximum equal to the sum of the lengths of the n vectors. For values of $s < \lambda/2$, however, there will be values of ψ_0 for which Eq. (15) has no real solution.

Since $\Psi(\theta)$ is never negative, its absolute minima will occur when $\Psi(\theta) = 0$, that is, for any value of θ satisfying

$$\cos\theta = \frac{\lambda}{2\pi s}\left(\frac{2k\pi}{n} + \psi_0\right) \qquad (k = 0, \pm 1, \pm 2 \cdots) \tag{16}$$

other than those satisfying Eq. (15); for at those points the numerator in Eq. (14) vanishes while the denominator does not. The points $\theta = 0$ and $\theta = \pi$ may also be minimum points. Certainly $\Psi(\theta)$ is an extremum at each of these values because it has the period 2π and is symmetrical with respect to the line $\theta = 0$.

No other minima of $\Psi(\theta)$ exist.[1] The maxima, other than those given by Eq. (15), will occur close to the point where the numerator in Eq. (14) reaches its maximum value of unity; for the numerator is changing much

[1] Differentiate $\Psi(\theta)$ with respect to θ.

$$\Psi'(\theta) = \frac{\sin\frac{n}{2}\psi}{n^2\sin^3\frac{1}{2}\psi}\left[\frac{n-1}{2}\sin\left(\frac{n+1}{2}\psi\right) - \frac{n+1}{2}\sin\left(\frac{n-1}{2}\psi\right)\right]\left(-\frac{2\pi s}{\lambda}\sin\theta\right).$$

The points at which $\sin n\psi/2$, and $\sin\theta$ vanish have already been examined. The only other critical values can arise from the factor

$$\frac{n-1}{2}\sin\left(\frac{n+1}{2}\psi\right) - \frac{n+1}{2}\sin\left(\frac{n-1}{2}\psi\right).$$

more rapidly than the denominator. An excellent approximation then for the remaining maximum points of $\Psi(\theta)$ is

$$\cos \theta = \frac{\lambda}{2\pi s}\left[\frac{(2k+1)}{n}\pi + \psi_0\right] \quad (k = \pm 1, \pm 2, \cdots)$$

in which values of k divisible by n are excluded.

Figure 9·5 shows Ψ as a function of ψ [Eq. (14)] for $n = 12$. For this functional dependence Ψ has the period 2π and is symmetrical with respect

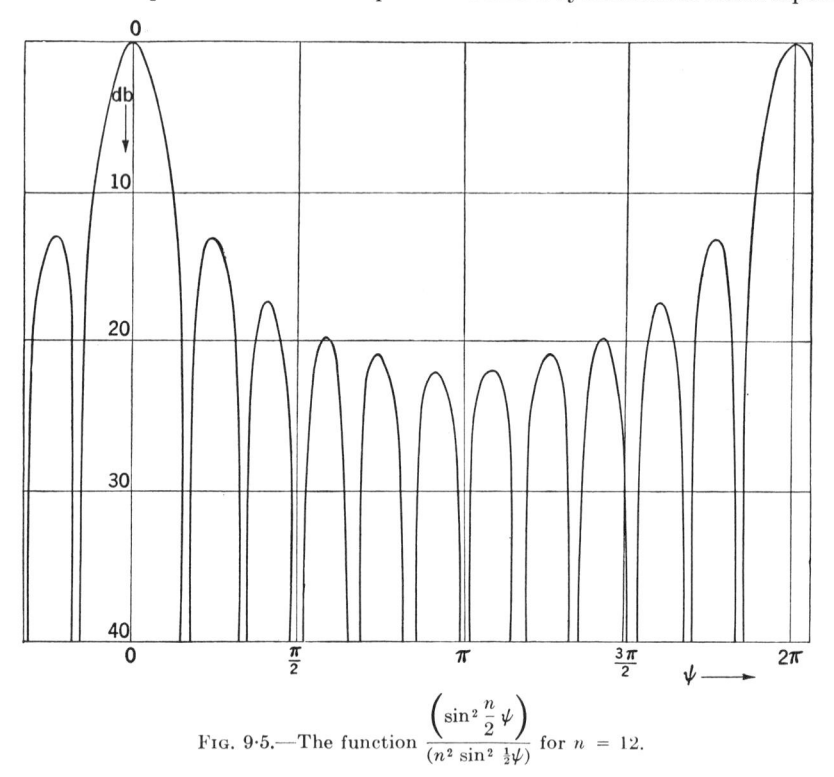

Fig. 9·5.—The function $\dfrac{\left(\sin^2 \dfrac{n}{2}\psi\right)}{(n^2 \sin^2 \frac{1}{2}\psi)}$ for $n = 12$.

to the line $\psi = 0$. An idea of the shape of the array factor for various values of s/λ and ψ_0 may be obtained from the graph. Because

$$\psi = \frac{2\pi s \cos \theta}{\lambda} - \psi_0, \tag{9.13}$$

This function, however, is monotone in any region between adjacent solutions of Eq. (16) because its derivative is

$$-\frac{n^2 - 1}{2} \sin \frac{n}{2}\psi \sin \frac{1}{2}\psi.$$

Thus it can vanish no more than once in the region whose end points are successive roots of Eq. (16).

the portion of Ψ that represents the array factor lies in the region for which

$$-\frac{2\pi s}{\lambda} - \psi_0 \leq \psi \leq \frac{2\pi s}{\lambda} - \psi_0.$$

Thus the values of ψ that determine the array factor extend over an interval of length $4\pi s/\lambda$; this may be less than the period of Ψ or several of its periods. In the language of Sec. 9·3 the range of z (whose angle is ψ) may be less than one circuit of the unit circle or several circuits.

9·5. Broadside Beams.—A linear array whose form factor has its absolute maxima only in directions normal to the axis of the array is known as a "broadside array." The array factor of such an array should then have a single absolute maximum in the direction $\theta = \pi/2$. The power pattern of the radiating element employed will, of course, determine whether, among other possibilities, the array will have a single direction of maximum intensity in the plane $\theta = \pi/2$ or its intensity will be maximum in every direction in that plane. Both of these types of arrays have widespread application; the latter is sometimes called an omnidirectional antenna.[1] In the microwave region the principal use of these antennas is as beacons, and in the following sections all such antennas will be referred to as beacons.

We have seen that the array factor for a uniform array has absolute maxima for the values θ satisfying the relation [Eq. (15)]

$$\cos\theta = \frac{k\lambda}{s} + \frac{\lambda\psi_0}{2\pi s} \qquad (k = 0, \pm 1, \pm 2, \cdots).$$

This will have the solution $\theta = \pi/2$ if $\psi_0 = 0$, and it will be the only such solution if $s < \lambda$. Arrays in which the elements all have the same amplitude and phase ($\psi_0 = 0$) are commonly referred to as *uniformly illuminated* arrays. For the moment let attention be restricted to the case $s = \lambda/2$. It will be shown later that this restriction is desirable. The array factor is then, from Eq. (14),

$$\Psi(\theta) = \frac{\sin^2\left(\frac{n\pi}{2}\cos\theta\right)}{n^2 \sin^2\left(\frac{\pi}{2}\cos\theta\right)}. \tag{17}$$

Equation (17) is plotted in Fig. 9·6 on a decibel scale for $n = 6$ and $n = 12$.

It will be observed that the side lobes (secondary maxima) on either side of the main beam decrease. Moreover, on the decibel scale used in Fig. 9·6, a straight line joining any two peaks on the same side of the main beam lies entirely above any intervening peak. That this is always true

[1] It should be remembered, however, should this usage be encountered, that "omnidirectional" means all directions in a plane.

may be verified by noting that the peaks of the side lobes lie approximately on the curve

$$\Psi_0 = \frac{1}{n^2 \sin^2\left(\frac{\pi}{2} \cos \theta\right)}.$$

The second derivative of $\ln \Psi_0$ with respect to θ is

$$\pi \left[\frac{\pi}{2} \csc^2\left(\frac{\pi}{2} \cos \theta\right) \sin^2 \theta + \cot\left(\frac{\pi}{2} \cos \theta\right) \cos \theta \right].$$

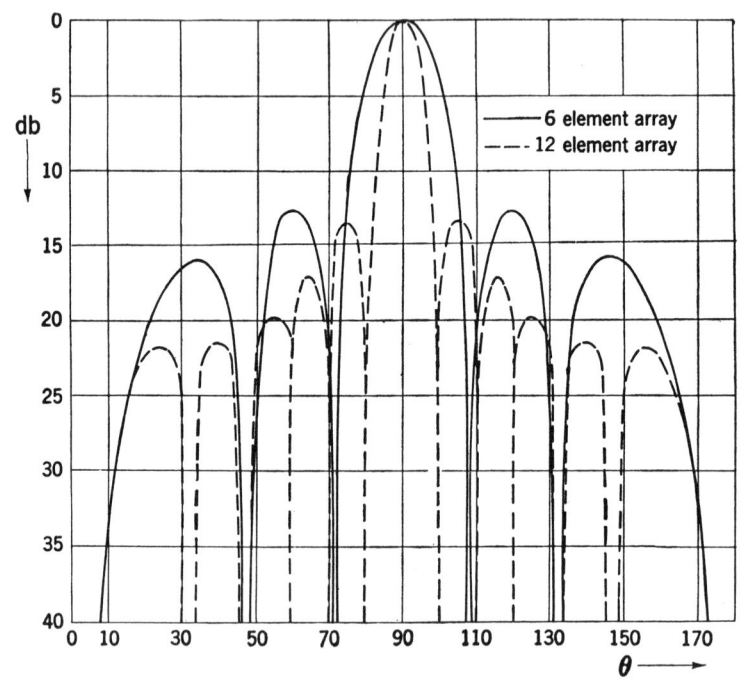

Fig. 9·6.—Graphs of $\dfrac{\left[\sin^2\left(\frac{n\pi}{2}\cos\theta\right)\right]}{\left[n^2 \sin^2\left(\frac{\pi}{2}\cos\theta\right)\right]}$ for $n = 6$ (full curve) and $n = 12$ (dotted curve).

This is positive in the interval on either side of the main beam; hence the peaks of the lobes on each side of the main beam lie on curves that are concave upward.

Direct computation shows that the height of the first side lobe, that is, the one nearest the main beam, varies from 0.056 for $n = 6$ to 0.047 for $n = 12$ and 0.045 for all sufficiently large n. The height of the last side lobe is $1/n^2$ for odd n and approximately that for even n.

If the sine appearing in the denominator of Eq. (17) is replaced by its argument, an approximation for the half-power width of the array factor may be obtained namely,

$$\Theta = \frac{101.8°}{n}.$$

An indication of the magnitude of the error is contained in Table 9·1 in which both the actual half-power widths and those computed by the approximate formula are given.

TABLE 9·1.—MAGNITUDE OF ERROR RESULTING FROM THE USE OF THE APPROXIMATION FOR THE HALF-POWER WIDTH

n	2	3	4	5	6	12	50
$\frac{101.8}{n}$	50.9	33.9	25.4	20.4	16.97	8.48	2.036
Θ	60.0	36.3	26.3	20.8	17.19	8.50	2.039

The relatively large height of the first side lobe is characteristic of a uniform array and at times may be annoying. Broadside arrays may readily be formulated, at least in theory, that have side lobes as small as desired. For example, consider an array whose associated polynomial is

$$f(z) = (1 + z + z^2 + \cdots + z^{m-1})^2$$

or

$$f(z) = 1 + 2z + \cdots + mz^{m-1} + \cdots + 2z^{2m-3} + z^{2m-2}.$$

The elements are all in phase, but their amplitudes decrease uniformly from the central element. This is a special case of what is commonly called a *gabled illumination*. Its array factor is the square of the array factor of a uniform array; hence its first side lobe will have a height of but ¼ per cent of the height of the main beam instead of the 5 per cent height of the uniform array. All of the other side lobes will be reduced in a similar fashion, but the main beam will be somewhat broader than the main beam of a uniform array with the same number of elements. The half-power widths of the gabled and uniform arrays are approximately $146°/n$ and $102°/n$, respectively, where n is the number of elements. Successively higher powers of the polynomial may be computed; the reduction in side lobes is accompanied by a rapid growth in beam width.

A general discussion of the problem of constructing high-gain broadside arrays with side lobes below a prescribed value will be given in Sec. 9·7. Attention may be called here to an array that eliminates side lobes completely. The feeding coefficients are equal to the binomial coefficients

$$C_{r,k} = \frac{r!}{k!(r-k)!}. \tag{18}$$

The array is derived from the two-element half-wavelength-spaced uniform array. The latter has an array factor

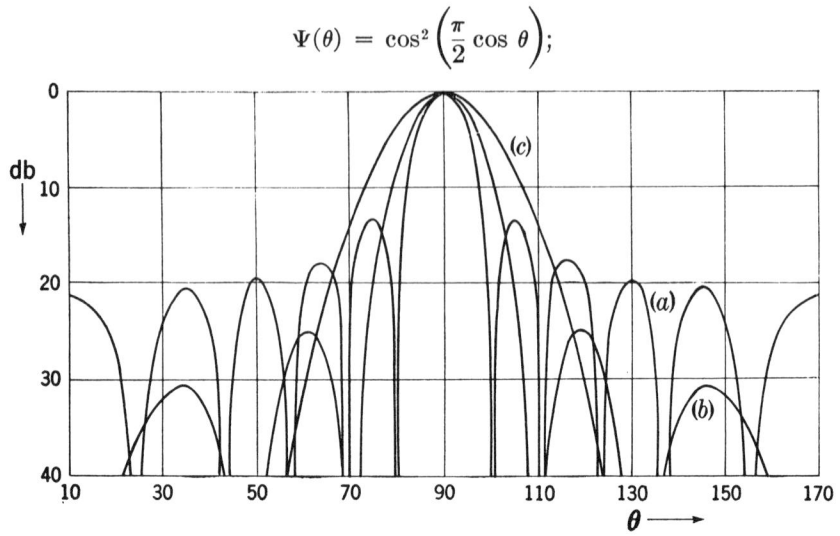

$$\Psi(\theta) = \cos^2\left(\frac{\pi}{2}\cos\theta\right);$$

Fig. 9·7.—Array factors for three 11-element $\lambda/2$-spaced arrays: (a) the uniform array $\left[\dfrac{\sin\left(\dfrac{11}{2}\pi\cos\theta\right)}{11\sin\left(\dfrac{\pi}{2}\cos\theta\right)}\right]^2$; (b) the gabled array $\left[\dfrac{\sin(3\pi\cos\theta)}{6\sin\left(\dfrac{\pi}{2}\cos\theta\right)}\right]^4$; (c) the "binomial" array $\left[\cos\left(\dfrac{\pi}{2}\cos\theta\right)\right]^{10}$.

this has no side lobes and has nulls at $\theta = 0$ and $\theta = \pi$. Its associated polynomial is

$$f(z) = 1 + z.$$

From Eq. (12) it follows then that the array whose polynomial is

$$f(z) = (1+z)^r = C_{r,0} + C_{r,1}z + C_{r,2}z^2 + \cdots + C_{r,r}z^r$$

has an array factor

$$\Psi = \cos^{2r}\left(\frac{\pi}{2}\cos\theta\right).$$

An inspection of the three space factors given in Fig. 9·7 shows that, of the three, the uniform broadside array concentrates the greatest percentage of the radiated energy in the direction normal to the array. It can be readily shown that of all arrays in which the elements have the same phase and the spacing is $\dfrac{\lambda}{2}$ the uniformly illuminated array has the maximum gain. Let

$$f(z) = \frac{\sum_{r=0}^{n-1} a_r z^r}{\sum_{r=0}^{n-1} a_r}$$

be the polynomial associated with an arbitrary array of n elements, normalized so that $f(1) = 1$.

Since the elements are spaced a half wavelength apart, $z = e^{j\pi \cos \theta}$. The gain G in the direction $\theta = \pi/2$ (which corresponds to $z = 1$) is given by

$$G = \frac{4\pi}{\int_0^{2\pi} \int_0^{\pi} |f(z)|^2 \sin \theta \, d\theta \, d\phi}, \qquad z = e^{j\pi \cos \theta}$$

or

$$G = \frac{2}{I}, \tag{19}$$

where

$$I = \int_0^{\pi} |f(z)|^2 \sin \theta \, d\theta. \tag{20}$$

To maximize G, I must be minimized. In Sec. 9·7 it is shown that there is no loss in generality by assuming that $f(z)$ has real coefficients. If we let $\psi = \cos \theta$, then Eq. (20) becomes, after expansion,

$$I = \left(\frac{1}{\sum_{r=0}^{n-1} a_r} \right)^2 \int_{-1}^{1} \left(\sum_{r=0}^{n-1} a_r^2 + \sum_{r=1}^{n-1} C_r \cos r\pi\psi \right) d\psi,$$

where the numbers C_r are combinations of the polynomial coefficients a_i. Hence

$$I = \frac{2 \sum_{r=0}^{n-1} a_r^2}{\left(\sum_{r=0}^{n-1} a_r \right)^2}, \tag{21}$$

and

$$\frac{\partial I}{\partial a_k} = 4 \frac{a_k \left(\sum_{r=0}^{n-1} a_r \right)^2 - \sum_{r=0}^{n-1} a_r \sum_{r=0}^{n-1} a_r^2}{\left(\sum_{r=0}^{n-1} a_r \right)^4} \qquad (k = 0, 1, \cdots, n-1).$$

We then have for a minimum the system of equations

$$a_k \sum_{r=0}^{n-1} a_r - \sum_{r=0}^{n-1} a_r^2 = 0 \qquad (k = 0, 1, \cdots, n - 1). \tag{22}$$

The difference between any two of Eqs. (22) is

$$(a_k - a_i) \sum_{r=0}^{n-1} a_r = 0.$$

Thus $a_k = a_i$ for all i and k. That these conditions actually yield a minimum may be shown by examining the second derivatives. Equations (19) and (21) show that the gain of such a uniform array is n, the number of elements.

The situation becomes more complicated if the restriction to half-wavelength spacing is removed. We shall attempt an answer only for broadside arrays having a total length (the distance between the first and last elements) of $\lambda/2$. In that event, the half-wavelength-spaced uniform array has the largest gain of all uniform arrays of the prescribed length. *However, the uniform array does not yield maximum gain of all arrays of a given length.*

The array factor of an n-element uniform broadside array $\lambda/2$ long is, from Eq. (14),

$$\Psi = \frac{\sin^2\left(\dfrac{n}{n-1}\dfrac{\pi}{2}\cos\theta\right)}{n^2 \sin^2\left(\dfrac{1}{n-1}\dfrac{\pi}{2}\cos\theta\right)}.$$

When $n = 2$, we have

$$\sqrt{\Psi} = \cos\left(\frac{\pi}{2}\cos\theta\right).$$

The two-element array will have maximum gain if

$$\cos\left(\frac{\pi}{2}\cos\theta\right) \leqq \frac{\sin\left(\dfrac{n}{n-1}\dfrac{\pi}{2}\cos\theta\right)}{n \sin\left(\dfrac{1}{n-1}\dfrac{\pi}{2}\cos\theta\right)}$$

in the interval $0 \leqq \theta < \pi/2$. This inequality, however, is equivalent to

$$\frac{\sin\left(\dfrac{n}{n-1}\dfrac{\pi}{2}\cos\theta\right)}{\dfrac{n}{n-1}\dfrac{\pi}{2}\cos\theta} \leqq \frac{\sin\left(\dfrac{n-2}{n-1}\dfrac{\pi}{2}\cos\theta\right)}{\dfrac{n-2}{n-1}\dfrac{\pi}{2}\cos\theta}.$$

Because $(\sin x)/x$ is a decreasing function in the first two quadrants, the inequality is established and the two-element array has the largest gain of all *uniform* arrays whose total length is $\lambda/2$.

To show that the uniform broadside array has not, however, the largest gain of all broadside arrays of the same length, we shall consider as a specific example the maximum gain of three-element quarter-wavelength-spaced arrays. Let

$$f(z) = \frac{a + bz + cz^2}{a + b + c}$$

be the polynomial associated with any such array. As before, a, b, and c are assumed to be real. Then from Eq. (20)

$$I = \frac{4}{\pi} \int_0^{\pi/2} \frac{(a^2 + b^2 + c^2) + 2(ab + bc) \cos \psi + 2ac \cos 2\psi}{(a + b + c)^2} d\psi$$

or

$$I = \frac{4}{\pi} \frac{\frac{\pi}{2}(a^2 + b^2 + c^2) + 2(ab + bc)}{(a + b + c)^2}.$$

The equations for minimizing I are

$$(a + b + c)^2(\pi a + 2b) - 2(a + b + c)N = 0,$$
$$(a + b + c)^2(\pi b + 2a + 2c) - 2(a + b + c)N = 0,$$
$$(a + b + c)^2(\pi c + 2b) - 2(a + b + c)N = 0,$$

where

$$N = \frac{\pi}{2}(a^2 + b^2 + c^2) + 2(ab + bc).$$

Thus

$$a = c,$$
$$b = \frac{\pi - 4}{\pi - 2} a = -0.7519a.$$

Hence $G = 2.4$ as contrasted to $G = 2.0$ for the two-element half-wavelength-spaced uniform array. In Fig. 9·8 are drawn the array factors of the uniform array and the array whose gain was just computed. Also shown is the array factor for the uniform continuous array, that is, one in which n has been allowed to increase without limit, subject only to the restriction $(n - 1)s = \lambda/2$.

In practice it is frequently desirable to avoid half-wavelength spacing because of the resonance that may occur at that spacing. As far as gain is concerned this is quite feasible; for the gain of a uniform array sufficiently long is nearly independent of the number of elements, provided only that the spacing does not greatly exceed a half wavelength. However, it is only at resonance that the requirement that the radiating elements be in phase can be readily met.[1] Nonetheless, if the spacing does not differ by much from a half wavelength, the progressive phase delay thereby introduced [ψ_0 in Eq. (13)] causes but a small deflection of the main beam from the normal to the array; the exact amount is given by

$$\delta = \sin^{-1} \frac{\psi_0 \lambda}{2\pi s}. \qquad (23)$$

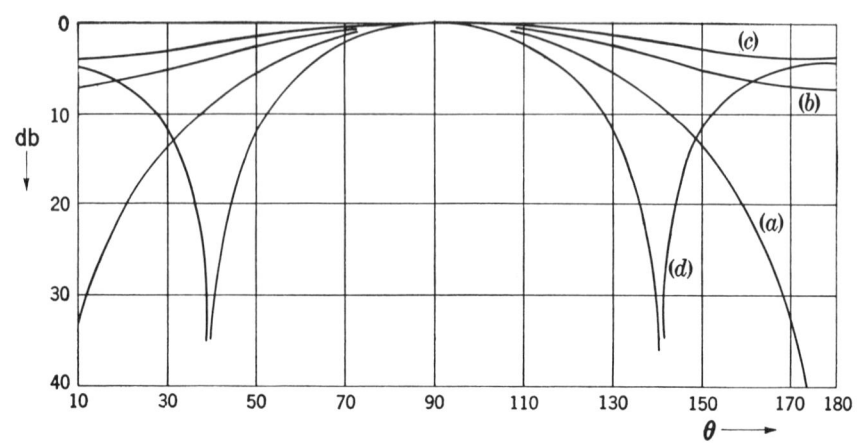

FIG. 9·8.—Array factors of four broadside arrays whose lengths are $\lambda/2$: (a) the two-element $\lambda/2$-spaced uniform array $\cos^2 \left(\frac{\pi}{2} \cos \theta \right)$; (b) the four-element $\lambda/6$-spaced uniform array $\dfrac{\sin^2 \left(\frac{2\pi}{3} \cos \theta \right)}{16 \sin^2 \left(\frac{\pi}{6} \cos \theta \right)}$; (c) the continuous uniform array $\dfrac{\sin^2 \frac{\pi}{2} \cos \theta}{\left(\frac{\pi}{2} \cos \theta \right)^2}$; and (d) the three-element $\lambda/4$-spaced array with maximum gain.

9·6. End-fire Beams.—The feeding coefficients of a linear array may be chosen so that the array factor has an absolute maximum along the axis of the array. If an element of the array produces a pattern having an absolute maximum in the same direction and if the product of the array factor and the element pattern has no other absolute maximum, the array is called an "end-fire" array.

If the elements of the array are not directive, the radiation pattern of the array is determined entirely by the array factor $\Psi(\theta)$. The pattern is the surface in spherical coordinates given by

$$r = \Psi(\theta)$$

and is therefore a surface of revolution symmetric with respect to the axis of the array. It is only by considering the three-dimensional picture that the great difference between end-fire and broadside arrays becomes apparent. The major lobe of an end-fire array is a pencil beam; thus a

[1] See Sec. 9·17.

one-dimensional configuration of sources produces radiation directive in two planes and does so without relying on any directivity of the individual sources. A broadside array on the other hand is directive in only one plane; it is omnidirectional in the plane perpendicular to the axis of the array.

Pencil beams whose half-power widths are in the region from about 15° to 35° can be produced quite readily by end-fire arrays that have lengths ranging from 3 to 18 wavelengths. The length of the array, however, varies inversely with the square of the beamwidth; narrow beams would require very long arrays. By properly choosing the feeding coefficients, end-fire arrays can be designed whose gains are almost double those of broadside arrays with the same length. This increase in gain has the greatest practical significance for arrays about 5 wavelengths long.

In order to eliminate from the array factor large lobes in any direction except $\theta = 0$, it is necessary to restrict the spacing of the elements. The necessary relations can be obtained from a study of the associated polynomial and an examination of the range of z on the unit circle in the complex plane. Since the angle ψ of z is given by

$$\psi = \frac{2\pi s}{\lambda} \cos \theta,$$

ψ assumes all values from $2\pi s/\lambda$ to $-2\pi s/\lambda$ as θ ranges from 0 to π.

Because the array factor has a period 2π as a function of ψ, the spacing s/λ must be such that over the range of θ the total variation of ψ is less than 2π; that is, the range of z in the unit circle is less than one revolution. In

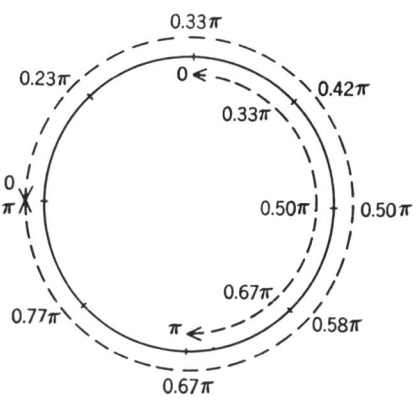

Fig. 9·9.—The range of ψ, the angle of z, as θ varies from 0 to π. The outer ring of figures shows corresponding values of θ at various points on the unit circle in the z-plane for $s = \lambda/2$. The inner set of figures shows values of θ for $s = \lambda/4$.

that case a principal maximum, which in the case of an end-fire occurs at $\psi = 2\pi s/\lambda$, will not be repeated. In Fig. 9·9 the mapping of θ on ψ is shown schematically. On the exterior of the unit circle in the z-plane are shown values of θ, corresponding to the indicated points on the circle for the spacing $s = \lambda/2$. The values shown on the interior are for the spacing $s = \lambda/4$. The array factor of a half-wavelength-spaced end-fire array will duplicate its value for $\theta = 0$ again at $\theta = \pi$. To suppress such an undesirable back lobe it is necessary to separate the end points of the range of z. The quarter-wavelength-spaced array will be examined as a typical array that satisfies this condition. Figure 9·9 shows that the range of ψ for the quarter-wavelength spacing of elements is from $\pi/2$ to $-\pi/2$. From the discussion in the previous sections it is seen that if such an array has equal feeding coefficients, it will have its principal maximum at $\theta = \pi/2$, that is, at $\psi = 0$.

We have seen, however (Sec. 9·3, Fig. 9·4), that if a given array is altered by having superimposed on it a constant-phase difference from element to element, the effect on the array factor is to rotate the range of z through an angle equal to that phase difference. If, in particular, a quarter-wavelength-spaced uniform array is adjusted to introduce a phase difference of $-\dfrac{\pi}{2}$ between each pair of adjacent elements, the principal maximum that occurs when $\psi = 0$ corresponds to $\theta = 0$, and an end-fire antenna results (Fig. 9·10). The array factor of such an array is obtained immediately from Eq. (14) and is

$$\Psi(\theta) = \frac{\sin^2 \dfrac{n\pi}{4}(\cos\theta - 1)}{n^2 \sin^2 \dfrac{\pi}{4}(\cos\theta - 1)}.$$

The factor n^2 has been inserted in the denominator so that $\Psi(0) = 1$.

The gain is more easily computed from the polynomial associated with the array

$$f(\bar{z}) = \frac{1}{n}\sum_{k=0}^{n-1} \bar{z}^k; \qquad \bar{z} = e^{j\pi(\cos\theta - 1)/2}.$$

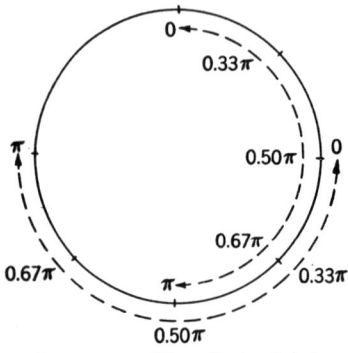

Fig. 9·10.—The effect of introducing a constant-phase difference $-\pi/2$ on the elements of a quarter-wavelength-spaced uniform array. The inner semicircle shows the original range of z and the corresponding values of θ. The outer semicircle is the range of z due to the phase difference; the corresponding values of θ are indicated.

Equation (20) becomes, after substituting and making the change of variable $\psi = \pi(\cos\theta - 1)/2$,

$$I = \frac{2}{\pi n^2}\int_{-\pi}^{0}\left[n + 2\sum_{k=1}^{n-1}(n - k)\cos k\psi\right]d\psi.$$

Thus from Eq. (19) $G = n$, the same gain as the longer half-wavelength-spaced uniform broadside array.

A uniform array with constant-phase difference between adjacent elements is one readily realized in practice. It is then well to inquire if the choice $-\pi/2$ for the phase difference is optimum for an end-fire array. If this difference is slightly less than $-\pi/2$, then the range of z is displaced slightly more than $\pi/2$. The direction $\theta = 0$ no longer represents the principal maximum of

$$\frac{\sin^2 \frac{n\psi}{2}}{\sin^2 \frac{\psi}{2}} \qquad (24)$$

but is displaced slightly from it (Fig. 9·11). However, the principal maximum of Expression (24) is not contained in the range of z. The end point of that range corresponds to $\theta = 0$ which then is a maximum of Expression (24) considered as a function of θ. The net effect is two-

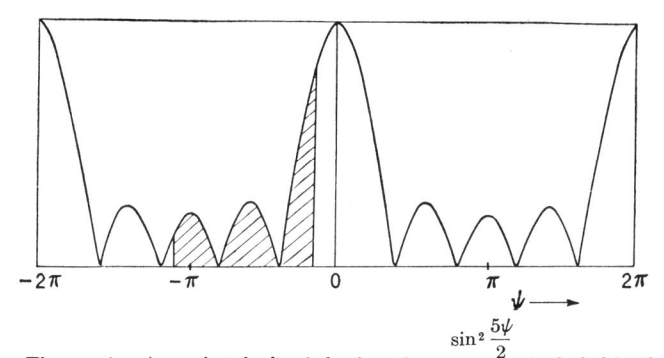

Fig. 9·11.—The portion (cross-hatched) of the function $\dfrac{\sin^2 \frac{5\psi}{2}}{25 \sin^2 \frac{\psi}{2}}$ included in the range of z due to a displacement slightly more than $\pi/2$.

fold: Because the value of Expression (24) at $\theta = 0$ has been reduced, the relative heights of the side lobes are increased and the gain tends to be reduced. On the other hand, the width of the main beam has been diminished, which has an opposite influence on the gain.

An estimate of the displacement yielding maximum gain may be made by approximate methods valid for large n. Suppose the phase difference between any two elements is $-(\pi/2) - \psi_0$; that is,

$$\psi = \frac{\pi}{2}(\cos \theta - 1) - \psi_0.$$

Then the array factor is

$$\Psi(\theta) = \frac{\sin^2 \frac{\psi_0}{2}}{\sin^2 \frac{n\psi_0}{2}} \frac{\sin^2 \left[\frac{n\pi}{4}(\cos \theta - 1) - \frac{n\psi_0}{2}\right]}{\sin^2 \left[\frac{\pi}{4}(\cos \theta - 1) - \frac{\psi_0}{2}\right]}.$$

The first factor has been inserted so that $\Psi(0) = 1$. An approximation of the gain may be obtained. If $y = -(n/2)\psi$, Eq. (20) becomes

$$I = \frac{4}{n\pi} \frac{\sin^2 \frac{\psi_0}{2}}{\sin^2 \frac{n\psi_0}{2}} \int_{n\psi_0/2}^{n(\pi+\psi_0)/2} \frac{\sin^2 y}{\sin^2 \frac{y}{n}} dy. \quad (25)$$

If n is large, the distant side lobes have little effect on the gain. We may replace $\sin^2 y/n$ by its argument, thus reducing the height of the distant side lobes. The range of integration may then be extended to ∞ and $\sin^2 \psi_0/2$ may be replaced by its argument with but a negligible effect on the value of I. Then Eq. (25) becomes

$$I = \frac{n}{\pi} \frac{\psi_0^2}{\sin^2 \frac{n\psi_0}{2}} \int_{n\psi_0/2}^{\infty} \frac{\sin^2 y}{y^2} dy.$$

By graphical methods[1] it has been shown that I is a minimum when $\psi_0 = 2.94/n$. Then the direction $\theta = 0$ corresponds to the point where the function given in Expression (24) is 46 per cent of its maximum value. The main beam is about half as broad as that of a uniform array with a phase shift of $-\pi/2$. On the other hand the heights of the side lobes have been more than doubled. The gain of such an array is

$$G = 1.82n.$$

Its half-power width is approximated by

$$\Theta = \frac{125°}{\sqrt{n}}.$$

For a more general spacing s between adjacent elements, maximum gain occurs when

$$\psi_0 = \frac{2.94\lambda}{4ns}.$$

Here ψ_0 still refers to the additional displacement of the range of z beyond $\pi/2$. The phase difference between adjacent elements is $-2\pi s/\lambda - \psi_0$. The gain for the general case is

$$G = 1.82 \frac{4ns}{\lambda}.$$

However, s is not completely arbitrary. We still must conform to our assumption that the distant side lobes have small effect on the gain. It has been suggested[2] that the approximations are valid for $s < \lambda/3$.

---

[1] W. W. Hansen and J. R. Woodyard, "A New Principle in Directional Antenna Design," *Proc. IRE*, **26**, 333 (1938).

[2] Hansen and Woodyard, *op. cit.*

An entirely different technique for increasing the gain of an end-fire array has been given.[1] Again we start with a quarter-wavelength-spaced uniform array with a phase delay $-\pi/2$ between adjacent elements. The polynomial associated with such an array is

$$f(z) = \sum_{k=0}^{n-1} (-jz)^k = (-j)^{n-1} \prod_{k=1}^{n-1} (z - j\omega^k) \qquad (26)$$

where ω is the nth root of unity with the smallest positive angle. The array factor is then

$$\Psi(\theta) = \prod_{k=1}^{n-1} |z - j\omega^k|^2.$$

The numbers $j\omega^k$ lie on the unit circle, and, it will be recalled, the array factor is formed by computing the square of the product of the distances from these numbers to the variable point z. It is apparent (Fig. 9·12) that the zeros of $f(z)$ lying outside the range of z add little to the directivity of the array. Thus, an array whose polynomial is

$$f(z) = \prod_{k=\left[\frac{n+1}{2}\right]}^{n-1} (z - j\omega^k),$$

retaining as it does only those zeros which lie on the range of z, will have the same nulls as formerly and at the same location. Its gain will have been reduced but little, while its number of elements has been almost halved.

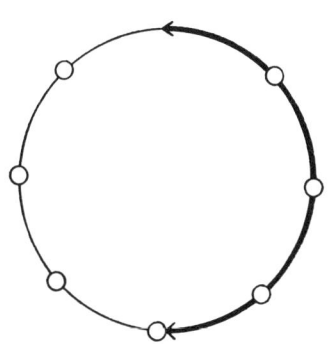

Fig. 9·12.—The location of the zeros of Eq. (26) with relation to the range of z for $n = 8$.

9·7. Beam Synthesis.—The preceding sections have dealt principally with the problem of analyzing the properties of the array factors of given linear arrays. The inverse problem, that of finding an array which will yield an array factor having prescribed characteristics, is far more difficult. The present section will treat some aspects of the synthesis problem.

The nature of the synthesis problem depends on the manner in which the desired pattern is specified. The latter may be prescribed as a complete function of θ over the physical range $0 \leq \theta \leq \pi$. In general a solution is sought that gives an acceptable approximation to the desired pattern. There can be no unique solution to such a problem, since the pattern is prescribed only as regards intensity distribution of the radia-

[1] S. A. Schelkunoff, "A Mathematical Theory of Linear Arrays," *Bell System Tech. Jour.*, **22**, 80 (1943).

tion field; the phase distribution is arbitrary, and each choice of such a distribution will lead to a different array. Only a partial solution to the problem will be given. It will be based on the general characterization of the array factor of an n-element array and the formulation of the properties that a pattern must possess in order to be the array factor of a linear array. All n-element arrays will be found that have a given array factor. The problem of finding the best approximation to a given pattern by a realizable array factor is beyond the scope of the present discussion.

The desired pattern may be specified with regard to general properties rather than as a complete function of θ. Examples of such synthesis problems are the design of a broadside array having minimum beamwidth for a given side-lobe level and the design of one having a minimum side-lobe level for a given beamwidth. These problems have exact solutions when the spacing of the elements of the array is $\lambda/2$ or greater; they will be discussed later in this section.

We shall consider first the characterization of the array factor of an n-element array. The array factor can be obtained from Eq. (8) by replacing ζ by $e^{j(2\pi s \cos \theta)/\lambda}$ and expanding. If the real numbers A_k and B_k are defined by

$$A_k + jB_k = \sum_{r=0}^{n-1-k} a_r a_{r+k}^*, \qquad (27)$$

Eq. (8) becomes

$$\Psi(\theta) = A_0 + 2 \sum_{k=1}^{n-1} \left[A_k \cos \left(k \frac{2\pi s}{\lambda} \cos \theta \right) + B_k \sin \left(k \frac{2\pi s}{\lambda} \cos \theta \right) \right]. \qquad (28)$$

Thus the array factor of any n-element linear array with spacing s is a trigonometric sum of order $n-1$ in the angle $\psi = (2\pi s \cos \theta)/\lambda$. The trigonometric sum is nonnegative for all real values of ψ. Conversely, every nonnegative trigonometric sum can be realized as the array factor of a linear array. It follows then that the necessary and sufficient condition that there exists a linear array having the prescribed pattern as its array factor is that the prescribed pattern can be expressed as a nonnegative trigonometric sum of a finite number of terms. Expressing the prescribed pattern as such a sum determines the coefficients A_k and B_k, and in principle the feeding coefficients a_r of the array can be determined from Eq. (27).

To find an n-element array that will approximate the prescribed pattern, the latter may be approximated[1] by the terms of order less than

---
[1] The method of approximation selected will depend on how the prescribed pattern is specified and what deviation from it is acceptable. For a general discussion of this problem see C. de la Vallee Poussin, *Leçons sur l' Approximation des Fonctions d' une Variable reéle*, Paris, 1919.

n in its Fourier series expansion in the angle $\psi = (2\pi s \cos \theta)/\lambda$. When $s > \lambda/2$, the periodicity of the Fourier series may present difficulty. If these terms form a trigonometric sum that is nonnegative for all ψ, the coefficients A_k and B_k may be used to determine the feeding coefficients of the n-element array.

A direct solution of Eq. (27) is, however, difficult. Instead, define an auxiliary polynomial $F(z)$ by

$$F(z) = \sum_{k=1}^{n-1} (A_k - jB_k)z^{n-1+k} + A_0 z^{n-1} + \sum_{k=1}^{n-1} (A_k + jB_k)z^{n-1-k}. \quad (29)$$

Then Eq. (28) becomes

$$\Psi(\theta) = |F(e^{j(2\pi s \cos \theta)/\lambda})|. \quad (30)$$

If z_k is a zero of $F(z)$, so also is its conjugate reciprocal $1/z_k^*$. The assumption that the trigonometric sum is nonnegative thus implies that the zeros lying on the unit circle, which are their own conjugate reciprocals, occur with even multiplicities. Hence the zeros of $F(z)$ may be grouped in pairs; and aside from a constant multiplier,

$$F(z) = \prod_{k=1}^{n-1} \left[(z - z_k) \left(z - \frac{1}{z_k^*} \right) \right]. \quad (31)$$

One zero in each pair may be selected as a zero of a new polynomial

$$f(z) = \prod_{k=1}^{n-1} (z - z_k). \quad (32)$$

For values of z on the unit circle

$$|f(z)|^2 = |F(z)| = \Psi(\theta),$$

where again a constant multiplier has been dropped. Equation (30) then implies that $\Psi(\theta)$ is the array factor of the array whose associated polynomial is given by Eq. (32). The separation of the zeros of $F(z)$ into two sets can in general be done in many ways. Each such partition will usually lead to two different arrays; all arrays will have the same array factor. When all the zeros lie on the unit circle (as, for example, in the uniform array), only one method of division is possible and the two sets obtained are the same. It should not be assumed that finding an array having a given array factor is an easy computational problem, even when n is as small as 5. It is necessary to find the zeros of a polynomial of degree $2n - 2$ and then perform the multiplications indicated in Eq. (32) to find the feeding coefficients. Simpler but less general methods have been devised.[1]

It is now possible to verify the assumption made in Sec. 9·5 that as far as the gain of broadside arrays is concerned, attention may be restricted to arrays whose elements are either in phase or out of phase by 180°, that is, to arrays whose feeding coefficients are real. The array factor of an arbitrary array is given by Eq. (28). If the sine terms are dropped, the resulting $\Psi(\theta)$ is still the array factor of some array. Moreover the gain in the direction $\theta = \pi/2$ is unchanged, since both the field intensity in the direction $\theta = \pi/2$ and the integral in Eq. (20) are unchanged by eliminating the sine terms. The corresponding polynomial $F(z)$ in Eq. (29) will have real coefficients, and its nonreal zeros occur in conjugate pairs. Hence in forming the associated polynomial $f(z)$, the pairing of conjugate zeros may be maintained and $f(z)$ will have real coefficients.

Let us consider next the problem of minimizing the side-lobe level of broadside arrays with a fixed beamwidth or maximizing the beamwidth for a given side-lobe level. The problem has received an exact solution[2] when the spacing between elements is at least $\lambda/2$ and sufficiently less than λ to eliminate any large end-fire lobe. For the present purposes a convenient definition of beamwidth is the angular difference between the position of the two nulls enclosing the main beam. Only those arrays will be considered whose main-beam nulls are symmetrically located with respect to the direction $\theta = \pi/2$.

The array factor having either the minimum beamwidth or the lowest side-lobe level may be expressed in terms of the Tchebyscheff polynomial $T_{2n}(x) = \cos(2n \cos^{-1} x)$. This polynomial falls between -1 and $+1$ in the interval $-1 \leq x \leq 1$, assumes the value $+1$ at the end points of the interval, increases steadily outside the interval, and is symmetric with respect to the line $x = 0$. The actual array factor is given by

$$\Psi(\theta) = \frac{1}{2}[1 + T_{2n}(ax)], \qquad x = \cos\left(\frac{\pi s \cos\theta}{\lambda}\right). \tag{33}$$

Figure 9·13 is a graph of $\frac{1}{2}[1 + T_{2n}(ax)]$ for $n = 4$ and $a = 1$. In Eq. (33) the direction $\theta = \pi/2$ corresponds to $x = 1$. If

$$\tfrac{1}{2}[1 + T_{2n}(a)] = r, \tag{34}$$

the relative height of each side lobe is $1/r$. The array factor with this

---

[1] S. A. Schelkunoff, "A Mathematical Theory of Linear Arrays," *Bell System Tech. Jour.*, **22**, 80 (1943); and Irving Wolff, "Determination of the Radiating System Which Will Produce a Specified Directional Characteristic," *Proc. IRE*, **25**, 630 (1937).

[2] C. L. Dolph, "A Current Distribution for Broadside Arrays Which Optimizes the Relationship between Beam Width and Side Lobe Level," *Proc. IRE*, **34**, 335 (1946). The results of Dolph have been generalized by Henry J. Riblet, *Proc. IRE*, **35**, 489 (1947).

side-lobe level and having the smallest beamwidth is given by Eq. (33) with a a solution of Eq. (34). The null nearest $x = 1$ occurs for

$$x_1 = \frac{1}{a} \cos \frac{\pi}{2n}. \qquad (35)$$

This, together with $x = \cos[(\pi s \cos \theta)/\lambda]$, gives the beamwidth. If the side-lobe level is to be minimized for a prescribed beam width, Eq. (35) is used to determine a and Eq. (34) to find the height of the side lobe.

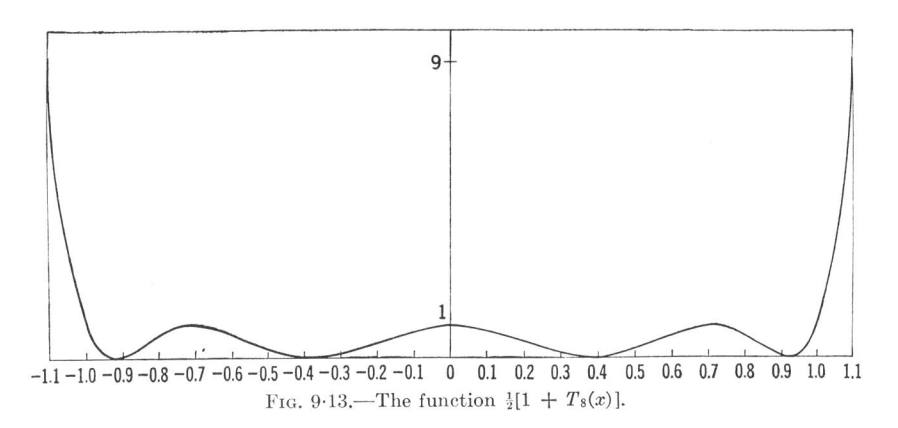

Fig. 9·13.—The function $\frac{1}{2}[1 + T_8(x)]$.

The substitution $x = \frac{1}{2}(z^{1/2} + z^{-1/2})$ transforms $\frac{1}{2}[1 + T_{2n}(ax)]$ into $z^{-n}F(z)$, where $F(z)$ is a polynomial of the form of Eq. (29) with $B_k = 0$ for all k. The symmetry of $T_{2n}(ax)$ ensures that the fractional powers of z in $F(z)$ are missing. The same substitution transforms

$$x = \cos\left[\frac{(\pi s \cos \theta)}{\lambda}\right]$$

into $z = e^{j(2\pi s \cos \theta)/\lambda}$. Hence this substitution transforms Eq. (33) into Eq. (30), and thus Eq. (33) represents an array factor of some linear array. The feeding coefficients are obtained most easily from the zeros of $\frac{1}{2}[1 + T_{2n}(ax)]$, for these transform into the zeros of $f(z)$, the associated polynomial of the array.

The optimum properties of the Tchebyscheff array are readily established. An argument similar to one used earlier in this section is sufficient to show that attention may be restricted to arrays whose associated polynomials have real coefficients. The array factor can be represented in the form of Eq. (30). The polynomial $F(z)$ defined in Eq. (29) has only real coefficients. Hence the substitution $x = \frac{1}{2}(z^{1/2} + z^{-1/2})$ transforms $z^{-n}F(z)$ into a polynomial $G(x)$ symmetric with respect to the line $x = 0$. Equation (30) is transformed into

$$\Psi(\theta) = G(x), \qquad x = \cos\left(\frac{\pi s \cos \theta}{\lambda}\right).$$

Suppose $G(x)$ is normalized so that $G(1) = r$ with r as in Eq. (34). It is impossible for $G(x)$ to have a zero for $x \geq x_1$ [Eq. (35)] and at the same time lie between 0 and 1 for $0 \leq x \leq x_1$. Any such polynomial would then have $(n + 1)$ points in common with $\frac{1}{2}[1 + T_{2n}(x)]$, double points being counted as such. The symmetry then shows that the two polynomials, each of degree $2n$, have $2n + 2$ points of intersection and so must coincide. If $s \geq \lambda/2$, there are real values of θ corresponding to any x in the interval $0 \leq x \leq 1$. Hence if the side-lobe level of the array is $1/r$, the requirement that $0 \leq G(x) \leq 1$ for $0 \leq x \leq x_1$ must be met, and the only array possible is the Tchebyscheff array.

RADIATING ELEMENTS

9·8. Dipole Radiators.—The various forms of coaxial line-fed dipoles discussed in Chap. 8 can be adapted for use as a linear-array element to be mounted on either coaxial line or waveguide. Design and performance are discussed here in terms of a rectangular guide; however, the fundamental ideas apply to all types of lines. The general properties desired of a dipole element are (1) a balanced excitation of the wings to give a symmetrical pattern, (2) a resistive load presented by the dipole because a reactive component means large reflections in the line, (3) an easily adjustable resistance with minimum frequency dependence, and (4) high power capacity.

The requirement for balanced excitation of the wings favors the use of the slot-fed dipole (*cf.* Sec. 8·4). The open-end termination has been used almost exclusively; the stub-terminated units are more frequency sensitive and are also limited in power capacity by the standing waves in the stub section. The general arrangement of a slot-fed dipole adapted to a rectangular guide is illustrated in Fig. 9·14. The inner conductor of the coaxial line serves as a coupling probe to the waveguide; it is evident that the probe should be parallel to the electric field in the guide for efficient coupling.

The important parameters of the dipole are slot depth, wing length, and outer-conductor diameter. The properties of the element are complicated functions of these parameters, and little is available in the form of systematic data. Breakdown tends to occur between the conductors of the coaxial section. The breakdown potential can be increased by increasing the slot width and the outer-conductor diameter; the extent to which this can be pursued is limited, however, by the unbalancing of the wing excitation. The unbalancing is due to higher modes becoming prominent and producing an asymmetrical field across the line; the simple mode picture drawn in Fig. 8·6 is applicable only for slot widths and

SEC. 9·8] DIPOLE RADIATORS 285

coaxial-line dimensions that suppress the higher modes. The element illustrated in Fig. 9·14 designed for use in the 10.7-cm region has a high power capacity.[1] With the values of the parameters indicated in the figure the balanced condition is maintained, as evidenced by a symmetrical radiation pattern; furthermore, studies of the phase fronts indicate that the unit has a center of feed located in the inner conductor. The waveguide serves as a reflector so that the unit mounted in guide forms essentially a dipole-plate system.

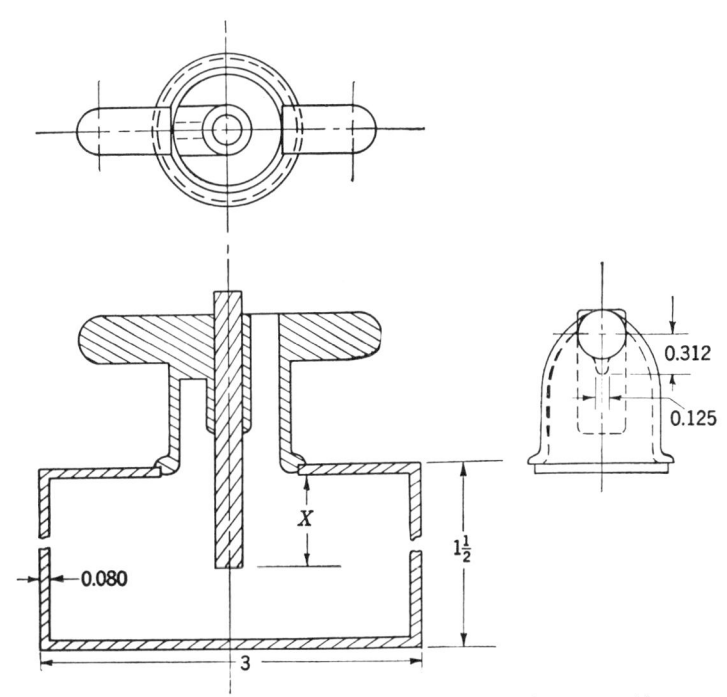

FIG. 9·14.—Cross section of a dipole on rectangular waveguide.

In the arrangement shown in the figure, the dipole behaves like a load shunted across the line. This is proved experimentally by measuring the input admittance of the dipole when it is followed by a variable reactance, which is provided by a movable plunger in the end of the guide; it is found that the conductance of the system is independent of the terminating reactance. The admittance of the dipole is a function of probe depth. With no probe the element presents an inductive susceptance component; the probe, like a tuning screw, is a capacitative susceptance (except for extreme depths of insertion); accordingly it is possible to find a probe depth at which the susceptance of the element as a whole

[1] J. Whelpton, "Admittance Characteristics of Some S-band Waveguide Fed Dipoles," RL Report No. 1082, January, 1946.

vanishes. These relations are illustrated in Fig. 9·15, a plot of the dipole admittance as a function of probe depth. The depth to which the probe may be inserted is limited by breakdown, which can occur between the end of the probe and the bottom of the guide. This difficulty can be obviated in some measure by terminating the probe in a small sphere.

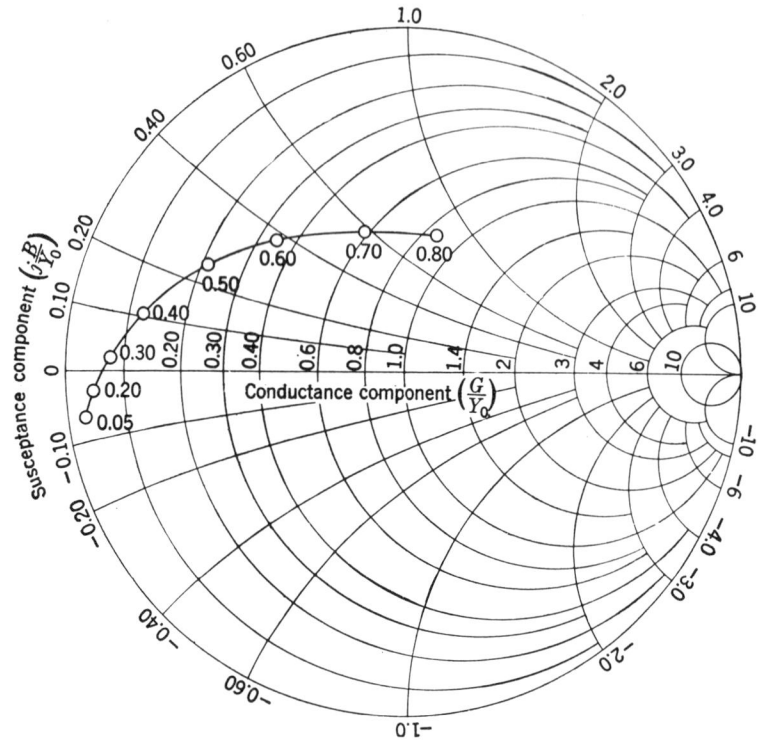

FIG. 9·15.—Dipole admittance as a function of probe depth in inches ($\lambda = 10.7$ cm).

For a given depth of insertion, the sphere causes a slight increase in the capacitative effect of the probe.

The impedances of these dipoles as single elements are practically independent of the orientation with respect to the axis of the line. In an assemblage of elements there are mutual interactions which are decided functions of orientation.

For assemblages of elements the question of reproducibility of an element in production is of considerable importance: it has been found that characteristics can be reproduced quite accurately by centrifugal or die-casting production methods.

9·9. Slots in Waveguide Walls.—It was noted in Chap. 7 that the electromagnetic field in the interior of a waveguide has associated with it a distribution of current over the boundary surfaces of the guide. This current sheet may be regarded properly as that required to prevent penetration of the field into the region exterior to the boundaries; it is indeed true that the metallic structure can be removed, providing the current sheet is maintained, without leakage of energy across the boundaries. If a narrow slot is cut in the wall of a waveguide such that the long dimension of the slot runs along a current line or along the region of the wall where the current is zero, it produces only a minor perturbation of the current distribution and correspondingly very little coupling of the internal field to space. Examples of such slots are elements cut in a coaxial line with the long dimension parallel to the axis of the line or elements of the type c and e cut in a rectangular guide as illustrated in Fig. 9·16; the slot c lying along the central line of the guide is in a region of zero current density. Nonradiating slots offer a means of entry into the guide for studying the internal field and are used for this purpose in impedance measurements (*cf.* Chap. 15).

On the other hand, a slot cut in a guide wall in a direction transverse to the current lines produces a significant perturbation of the current sheet, with the result that the internal field is coupled to space. A slot of this type constitutes a radiating element. The degree of coupling depends on the current density intercepted by the slot and the component of the length of the slot transverse to the current lines. Thus the coupling at a given position on the guide can be adjusted by the orientation of the slot as is indicated for the elements d and f

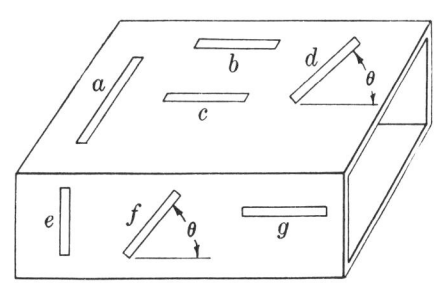

FIG. 9·16.—Slots in the wall of rectangular waveguide.

in Fig. 9·16, or the coupling can be adjusted by position like the radiating slot b and non-radiating slot c in the figure. The type of circuit element that the radiating slot presents to the transmission-line representation of the wave-guide is again a function of position and orientation. Under certain conditions the slot is in effect a shunt element; in others a series element; under very general conditions the slot can be represented adequately only by a T- or Π-section inserted in the line. The general circuit relations and the fundamental properties of slots will be developed in the following section.

9·10. Theory of Slot Radiators.—Let us consider a cylindrical waveguide of arbitrary cross section with its axis the z-axis. It was found in Chap. 7 that the normal modes of such a guide fall into two classes: TE-modes having an H_z- but no E_z-component and TM-modes having an E_z- but no H_z-component. Each mode is characterized by its characteristic admittance $Y_{mn}^{(0)}$ and propagation constant β_{mn}; the latter is real for a freely propagated mode but is to be taken equal to $-j\gamma_{mn}$ for a mode beyond cutoff. From the general discussion in Sec. 7·3 it is seen that the field components of a TE-mode of order $a = mn$ can be written

$$\left.\begin{array}{l} H_z = jH_{az} \exp{(\mp j\beta_a z)}, \\ \mathbf{E}_t = \mathbf{E}_{at} \exp{(\mp j\beta_a z)}, \\ \mathbf{H}_t = \pm \mathbf{H}_{at} \exp{(\mp j\beta_a z)}, \end{array}\right\} \quad (36)$$

where $\mathbf{E}_t$ and $\mathbf{H}_t$ represent the transverse electric and magnetic field vectors and the upper or lower signs are taken according as the wave is going in the positive or the negative z-direction. The general form of the TM-mode field components is the same as in Eq. (36) with H_z replaced by

$$E_z = jE_{az} \exp{(\mp j\beta_a z)}. \quad (37)$$

If β_a is real, the functions E_{az}, H_{az}, $\mathbf{E}_{at}$, and $\mathbf{H}_{at}$ are all real and depend only on a, x, and y. We have also seen (cf. Sec. 7.3) that the component vector functions $\mathbf{E}_{at}$ and $\mathbf{H}_{at}$ have the orthogonality property

$$\left.\begin{array}{ll} \int (\mathbf{E}_{at} \times \mathbf{H}_{bt}) \cdot \mathbf{i}_z \, dS = 0, & a \neq b, \\ \phantom{\int (\mathbf{E}_{at} \times \mathbf{H}_{bt}) \cdot \mathbf{i}_z \, dS} = S_a, & a = b, \end{array}\right\} \quad (38)$$

where S_a is twice the Poynting energy flux for a freely propagated mode and $\mathbf{i}_z$ is a unit vector in the direction Oz. The normal modes of the guide form a complete set in terms of which an arbitrary field distribution over the wall of the guide can be expressed in the form of a Fourier expansion.

Now consider a slot from z_1 to z_2 in the wall of the infinite guide. We assume that the guide is to be excited by a known field distribution along the slot. Then the field in the guide, which is denoted by subscript 1, will consist of outgoing waves on either side of the slot; that is, it will contain only waves going to the right for $z > z_2$ and only waves going to the left for $z < z_1$,

$$\left.\begin{array}{ll} \mathbf{E}_{1t} = \displaystyle\sum_a A_a \mathbf{E}_{at} \exp{(-j\beta_a z)}, & z > z_2, \\ \mathbf{E}_{1t} = \displaystyle\sum_a B_a \mathbf{E}_{at} \exp{(j\beta_a z)}, & z < z_1, \\ \mathbf{H}_{1t} = \displaystyle\sum_a A_a \mathbf{H}_{at} \exp{(-j\beta_a z)}, & z > z_2, \\ \mathbf{H}_{1t} = -\displaystyle\sum_a B_a \mathbf{H}_{at} \exp{(j\beta_a z)}, & z < z_1. \end{array}\right\} \quad (39)$$

The amplitudes of waves going to the right and left are not necessarily equal and are denoted by A_a and B_a respectively; they must be such that on superposing the two sets of waves a field is produced which matches the field over the slot according to the general boundary conditions formulated in Chap. 3.

In order to evaluate the amplitudes A_a and B_a an auxiliary relation must first be derived. Consider two fields $\mathbf{E}_1, \mathbf{H}_1$ and $\mathbf{E}_2, \mathbf{H}_2$ of the same frequency and both satisfying the homogeneous field equations. By virtue of these equations we find

$$\nabla \cdot (\mathbf{E}_1 \times \mathbf{H}_2) = \nabla \cdot (\mathbf{E}_2 \times \mathbf{H}_1) = -j\omega(\epsilon \mathbf{E}_1 \cdot \mathbf{E}_2 + \mu \mathbf{H}_1 \cdot \mathbf{H}_2)_z$$

Hence

$$\nabla \cdot [(\mathbf{E}_1 \times \mathbf{H}_2) - (\mathbf{E}_2 \times \mathbf{H}_1)] = 0.$$

If V is any closed region bounded by a surface S, it follows by the divergence theorem that

$$\int_S (\mathbf{E}_1 \times \mathbf{H}_2 - \mathbf{E}_2 \times \mathbf{H}_1) \cdot \mathbf{n}' \, dS = 0, \tag{40}$$

where $\mathbf{n}'$ is the unit vector normal to dS and directed outward from V. First, we shall evaluate B_a. Let the field $\mathbf{E}_1, \mathbf{H}_1$ be the field set up in the guide by the slot as formulated in Eqs. (39). For the field $\mathbf{E}_2, \mathbf{H}_2$ let us take a normal mode, free propagation of which is supported by the guide, traveling toward the right, and let a be the index of this mode. Furthermore, take as the region V the section of the guide containing the slot, bounded on the left by the plane $z = z_3 < z_1$ and on the right by the plane $z = z_4 > z_2$. The surface S to which Eq. (40) is to be applied consists then of these two planes and the wall of the guide. Over the plane $z = z_4$ the fields 1 and 2 consist of systems of waves traveling in the same direction. When the indicated substitutions are made and the orthogonality property of Eq. 38 is used, the integral vanishes. On the plane $z = z_3$ the fields 1 and 2 are composed of waves traveling in opposite directions. Making use of the orthogonality relation again and noting that for this surface $\mathbf{n}' = -\mathbf{i}_z$, the integral over this surface is $-2B_a S_a$. Considering the integral over the wall, the second term in the integrand is zero everywhere, for it can be written as $\mathbf{H}_1 \cdot (\mathbf{n}' \times \mathbf{E}_2)$, and $\mathbf{n}' \times \mathbf{E}_2$ is zero over the wall, since $\mathbf{E}_2$ is a normal mode. Similarly the field $\mathbf{E}_1$ must satisfy the condition $\mathbf{n}' \times \mathbf{E}_1 = 0$ over the metal wall boundary. The only nonvanishing contribution from the wall area arises from the first term of the integrand over the region of the slot. One thus finds

$$2B_a S_a = -\int_{\text{slot}} (\mathbf{E}_1 \times \mathbf{H}_2) \cdot \mathbf{n} \, dS, \tag{41}$$

where $\mathbf{n}$ is a unit vector normal to the wall and directed into the interior of the guide. If $\boldsymbol{\tau}$ is a unit vector perpendicular to the axis of the guide and tangent to the surface of the guide, then

$$\mathbf{n} = \boldsymbol{\tau} \times \mathbf{i}_z. \qquad (42)$$

Substituting this last relation into Eq. (41), we obtain finally

$$2B_a S_a = \int_{\text{slot}} (-jE_{1\tau}H_{az} + E_{1z}H_{a\tau}) \exp(-j\beta_a z)\, dS. \qquad (43)$$

Second, we shall evaluate A_a. The field $\mathbf{E}_1, \mathbf{H}_1$ is again taken to be that set up in the guide by the slot, and the field $\mathbf{E}_2, \mathbf{H}_2$ is taken to be the normal mode of index a traveling to the left. In this case the plane $z = z_3$ does not contribute to the integral in Eq. (40), and the plane $z = z_4$ contributes $-2A_a S_a$; over the wall of the guide the only nonvanishing contribution arises again from the first term of the integrand over the area of the slot. It is thus found that

$$\begin{aligned} 2A_a S_a &= -\int_{\text{slot}} (\mathbf{E}_1 \times \mathbf{H}_2) \cdot \mathbf{n}\, dS \\ &= -\int_{\text{slot}} (jE_{1\tau}H_{az} + E_{1z}H_{a\tau}) \exp(j\beta_a z)\, dS. \end{aligned} \qquad (44)$$

The interpretation of Eqs. (43) and (44) for the amplitudes becomes clearer if the magnetic field components $H_{a\tau}$ and H_{az} are replaced by surface current densities K_{az} and $-K_{a\tau}$ respectively. These are the components of the surface current, in the direction of the axis of the guide and in the direction transverse to it, that exists over the area of the slot in the nonslotted guide supporting the ath mode. In terms of these currents the amplitudes become

$$2B_a S_a = \int_{\text{slot}} (jE_{1\tau}K_{a\tau} + E_{1z}K_{az}) \exp(-j\beta_a z)\, dS, \qquad (45)$$

$$2A_a S_a = \int_{\text{slot}} (jE_{1\tau}K_{a\tau} - E_{1z}K_{az}) \exp(j\beta_a z)\, dS. \qquad (46)$$

It is evident from these equations that in general the slot does not radiate equally in both directions within the guide. The formulas also show that the slot will couple the ath mode to space only if it cuts across current lines corresponding to that mode. There are various special conditions under which a small slot is symmetrical with respect to the ath mode. If all the dimensions of a slot are small compared with the wavelength, the variation of a phase factor $\exp(\mp j\beta_a z)$ across the slot can be neglected; without loss of generality the slot can be located at $z = 0$, in which case the phase factors are replaced by unity. We then observe that

1. $A_a = B_a$ if E_{1z} or K_{az} is zero. Reference to Eqs. (39) shows that as far as the ath-mode contribution is concerned, $\mathbf{E}_{1t}$ is continuous at the plane $z = 0$ while the magnetic field is discontinuous; in fact, $\mathbf{H}_{1t}^+$ is in opposite phase to $\mathbf{H}_{1t}^-$. With respect to the ath mode the slot acts like a *shunt* element in a transmission line.

2. $A_a = -B_a$ if $E_{1\tau}$ or $K_{a\tau}$ is zero. In this case $\mathbf{E}_{1t}$ is discontinuous and $\mathbf{H}_{1t}$ is continuous at the plane $z = 0$ as far as the ath mode is concerned; the slot behaves like a *series* element in the ath-mode transmission line.

The slots of more general interest are narrow ones having a length of about $\lambda/2$ and a width small compared with the length. The electric-field distribution in such a slot is nearly sinusoidal along the length and independent of the feeding system; the direction of the field is transverse to the long dimension. There are also special conditions under which such slots reduce to series or shunt elements:

1. Axis of the slot perpendicular to the guide axis. In this case the phase factor exp $(\pm j\beta_a z)$ can again be replaced by unity. Furthermore $E_{1\tau} = 0$; hence if $K_{az} \neq 0$, $A_a = -B_a$ and the slot behaves like a *series* element in the ath-mode transmission line.
2. Axis of the slot parallel to the guide axis. In this case $E_{1z} = 0$ and the second members of the integrands of Eqs. (45) and (46) vanish. The variation of the phase factors exp $(\mp j\beta_a z)$ cannot be neglected; however, $K_{a\tau}$ is constant, and $E_{1\tau}$ is an even function along the slot; therefore only the real parts of the phase factors contribute to nonvanishing integrals, and one has $A_a = B_a$. The slot oriented in this manner behaves like a *shunt* element.

Except when special conditions of symmetry are imposed on the field and on the currents in the slot, for orientations more general than (1) and (2) above, $B_a \neq \pm A_a$, and the slot behaves like a more complicated combination of shunt and series element. In this case the slot is represented by a *T*- or Π-section equivalent in the ath-mode transmission line.

9·11. Slots in Rectangular Waveguide; TE_{10}-mode.—The theory of slots in rectangular guide that supports only the TE_{10}-mode will be developed in detail. The discussion will be based on the following assumptions:

1. The slot is narrow; i.e., $2 \log_{10}$ (length/width) $\gg 1$.
2. The slot is cut so that it is to be near the first resonance (length of the slot $\approx \lambda/2$).
3. The field in the slot is transverse to the long dimension and varies sinusoidally along the slot, independent of the exciting system.
4. The guide walls are perfectly conducting and infinitely thin.
5. The field in the region behind the face containing the slot is negligible with respect to the field outside the guide; this is tantamount to extending the face containing the slot into an infinite perfectly conducting plane.

The third assumption concerning the field distribution is closely in accord with experimental conditions. The fifth assumption is probably the most radical in its departure from the actual conditions.

First the equivalent circuits are given for the common types of slot, and then the method is given for calculating the values of the elements by means of Eqs. (45) and (46) and the electromagnetic formulation of Babinet's principle (Sec. 5·15), provided the reactive field of the slot is zero.[1] The rectangular guide has the dimensions shown in Fig. 9·17. The shunt conductance of a slot normalized to the characteristic admittance of the TE_{10}-mode line is g, and the series resistance normalized with respect to the line characteristic impedance is r. We have then (1) for a longitudinal slot in the broad face (shunt element b in Fig. 9·17)

$$g = g_1 \sin^2\left(\frac{\pi x_1}{a}\right), \quad (47a)$$

where

$$g_1 = 2.09 \frac{\lambda_g}{\lambda} \frac{a}{b} \cos^2\left(\frac{\pi\lambda}{2\lambda_g}\right); \quad (47b)$$

(2) for a transverse slot in the broad face (series element c in Fig. 9·17)

$$r = r_0 \cos^2\left(\frac{\pi x_1}{a}\right), \quad (48a)$$

where

$$r_0 = 0.523 \left(\frac{\lambda_g}{\lambda}\right)^3 \frac{\lambda^2}{ab} \cos^2\left(\frac{\pi\lambda}{4a}\right); \quad (48b)$$

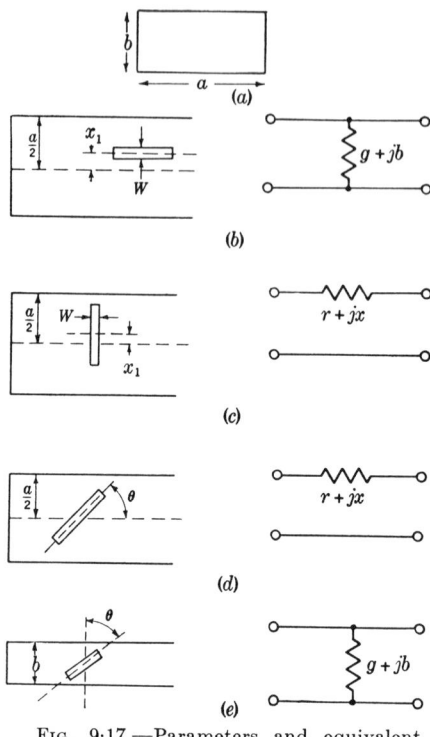

Fig. 9·17.—Parameters and equivalent circuits of slots in rectangular waveguide (reference point for circuit elements is the center of the slot). (a) waveguide dimensions; (b) longitudinal slot in broad face, shunt element; (c) transverse slot in broad face, series element; (d) centered inclined slot in broad face, series element; (e) inclined slot in narrow face, shunt element.

(3) for a centered inclined slot in the broad face (series element d in Fig. 9·17)

$$r = 0.131 \left(\frac{\lambda}{\lambda_g}\right) \frac{\lambda^2}{ab} \left[I(\theta) \sin\theta + \frac{\lambda_g}{2a} J(\theta) \cos\theta \right]^2, \quad (49a)$$

where

$$\left.\begin{array}{c} I(\theta) \\ J(\theta) \end{array}\right\} = \frac{\cos\left(\frac{\pi\xi}{2}\right)}{1-\xi^2} \pm \frac{\cos\left(\frac{\pi\eta}{2}\right)}{1-\eta^2} \quad (49b)$$

[1] The results to be quoted are due to A. F. Stevenson, "Series of Slots in Rectangular Waveguides," Parts I and II, Special Committee on Applied Mathematics, National Research Council of Canada, Radio Reports 12 and 13, 1944.

SEC. 9·11]   SLOTS IN RECTANGULAR WAVEGUIDE; TE_{10}-MODE   293

$$\left.\begin{array}{c}\xi\\\eta\end{array}\right\} = \frac{\lambda}{\lambda_g}\cos\theta \mp \frac{\lambda}{2a}\sin\theta; \qquad (49c)$$

and (4) for an inclined slot in the narrow face (shunt element e in Fig. 9·17)

$$g = \frac{30}{73\pi}\left(\frac{\lambda_g}{\lambda}\right)\frac{\lambda^4}{a^3b}\left[\frac{\sin\theta\cos\left(\frac{\pi\lambda}{2\lambda_g}\sin\theta\right)}{1 - \left(\frac{\lambda}{\lambda_g}\right)^2\sin^2\theta}\right]^2. \qquad (50)$$

As an illustration of the method of deriving the above relations we shall conclude this section with a summary of the procedure for the longitudinal slot in the broad face of the guide, Case (1) above. Choose dimensions as indicated in Fig. 9·17b. Suppose a TE_{10}-wave of amplitude unity to be incident on the slot from the left; this field induces a field across the slot so that the slot radiates waves in both directions in the guide and into space outside the guide. The amplitudes B_{10} and A_{10} (the mode index a is here replaced by 10) of the waves radiated in the interior are given by Eqs. (45) and (46) in terms of the field in the slot; the field, according to the third of our initial assumptions, is

$$\begin{aligned}E_{1r} &= E_0 \cos(kz),\\E_{1z} &= 0,\end{aligned} \qquad (51a)$$

where E_0 is the field at the center of the slot. We have also for the other quantities entering into Eqs. (45) and (46)

$$\begin{aligned}(K_{10})_r &= -Y_{10}^{(0)}\frac{\pi}{ka}\sin\left(\frac{\pi x_1}{a}\right),\\S_{10} &= Y_{10}^{(0)}\frac{ab}{2}\left(\frac{\beta_{10}}{\kappa}\right)^2\\\beta_{10}^2 &= k^2 - \left(\frac{\pi}{a}\right)^2,\end{aligned}\right\} \qquad (51b)$$

where $Y_{10}^{(0)}$ is the characteristic wave admittance of the TE_{10}-mode.[1] On inserting these quantities into the expressions for the amplitudes it is seen at once that $A_{10} = B_{10}$; that is, the slot is a shunt element, in agreement with the previous conclusions relative to slots parallel to the guide axis. The amplitudes are given explicitly by

$$A_{10} = B_{10} = -jwE_0\frac{2}{\pi b}\left(\frac{k}{\beta_{10}}\right)^2\sin\left(\frac{\pi x_1}{a}\right)\cos\left(\frac{\beta_{10}\lambda}{4}\right), \qquad (52)$$

[1] The constants of $(K_{10})_r$ and S_{10} correspond to the mode being so normalized that the electric field across the guide is given by $\frac{\beta_{10}}{k}\sin\frac{\pi x}{a}$.

where w is the width of the slot. It is useful to express the slot excitation in terms of a "voltage" transformation ratio. The "voltage" across the slot is defined to be the line integral of the field across the slot at its center, i.e.,

$$V_0 = wE_0,$$

while the voltage in the guide corresponding to any one of the dominant-mode waves is defined as the line integral of the field across the center of the guide, i.e.,

$$V_1 = bA_{10} = bB_{10}.$$

The voltage transformation ratio is then

$$\frac{V_1}{V_0} = -j\frac{2}{\pi}\left(\frac{k}{\beta_{10}}\right)^2 \sin\left(\frac{\pi x_1}{a}\right) \cos\left(\frac{\beta_{10}\lambda}{4}\right). \tag{53}$$

It is recognized further that the amplitude A_{10} measures directly the reflection coefficient Γ (at $z = 0$) in the transmission-line equivalent of the dominant-mode wave. If the slot is resonant, the value of Γ at $z = 0$ must be real, because the impedance looking to the right is real at that point at resonance. Then if the slot is a shunt element of normalized conductance g, the total admittance at $z = 0$ is $1 + g$; while if the slot is a series element of resistance r, the input impedance at $z = 0$ is $1 + r$. From Eqs. (2·30) and (2·36) g and r may be expressed in terms of Γ by

$$g = -\frac{2}{1+\dfrac{1}{\Gamma}}, \qquad r = -\frac{2}{1-\dfrac{1}{\Gamma}}. \tag{54}$$

The value of Γ can be evaluated for a resonant slot by energy-balance relations. The total energy incident on the slot is equal to the sum of the reflected, transmitted, and radiated energy. The incident power is $S_a/2$ for an incident wave of unit amplitude; the reflected power is $(A_{10})^2\dfrac{S_a}{2}$. The total amplitude of the dominant-mode wave to the right of the slot is $1 + B_{10}$; hence, the transmitted power is

$$\left(\frac{S_a}{2}\right) \operatorname{Re} (1 + B_{10})^2.$$

In computing the power radiated by the slot use is made of a result obtained, by means of an electromagnetic Babinet's principle,[1] for the radiation resistance of a center-driven narrow slot in an infinite perfectly conducting plane sheet of zero thickness. In this case the input resistance is

---

[1] H. Booker, "Babinet's Principle and the Theory of Resonant Slots," TRE Report No. T-1028.

$$R_r = \frac{1}{4 \times 73} \frac{\mu_0}{\epsilon_0}.$$

In the infinite sheet problem the slot radiates to both sides of the sheet; in our case the slot radiates to one side so that the radiation resistance is assumed to be simply twice the above value. The power radiated by the slot is then given by

$$\frac{1}{2} \frac{V_0^2}{2R_r} = 73 V_0^2 \frac{\epsilon_0}{\mu_0} \quad \text{watts}$$

Writing the energy balance equation and remembering that $A_{10} = B_{10}$, we have

$$\frac{S_a}{2} = \frac{S_a}{2} |A_{10}|^2 + \frac{S_a}{2} [1 + |A_{10}|^2 + 2\text{Re}(A_{10})] + 73 V_0^2 \frac{\epsilon_0}{\mu_0}.$$

Finally, since $A_{10} = \Gamma$ is real, we obtain from the above

$$1 + \frac{1}{\Gamma} = -73 \frac{\epsilon_0}{\mu_0} \frac{V_0^2}{S_a(A_{10})^2} \tag{55}$$

Making use of Eqs. (51a) and (51b) and substituting this last result into Eq. (54), the conductance of the resonant shunt slot is

$$g = \frac{120\pi}{73} \frac{a}{b} \left(\frac{\beta_{10}}{k}\right)^3 \left|\frac{V_1}{V_0}\right|^2. \tag{56}$$

We already have the voltage transformation ratio in Eq. (53); substituting this into Eq. (56) gives the final expression for the normalized shunt conductance,

$$g = \frac{480}{73\pi} \frac{a}{b} \frac{\lambda_g}{\lambda} \cos^2\left(\frac{\pi\lambda}{2\lambda_g}\right) \sin^2\left(\frac{\pi x_1}{a}\right). \tag{57}$$

9·12. Experimental Data on Slot Radiators.—Confirmation of the theory developed in the last section has been obtained by experiment for the longitudinal slot in the broad face of the guide (Case b, Fig. 9·17) and for the inclined slot in the narrow face (Case e, Fig. 9·17).[1] The resistance of a longitudinal slot as a function of its position with respect to the center of the guide is shown in Fig. 9·18; the points are in good agreement with the formula

[1] A. L. Cullen, "The Characteristics of Some Slot Radiators in Rectangular Waveguides," Royal Aircraft Establishment, Great Britain, Tech. Note No. Rad. 200; Dodds and Watson, "Frequency Characteristics of Slots," McGill University, PRA-108; Dodds, Guptill, and Watson, "Further Data on Resonant Slots," McGill University, PRA-109; E. W. Guptill and W. H. Watson, "Longitudinally Polarized Arrays of Slots," McGill University, PRA-104.

$$g = g_1 \sin^2\left(\frac{\pi x_1}{a}\right), \tag{9.47a}$$

but the numerical constant g_1 is 1.73 whereas the theoretical value given by Eq. (47b) is 1.63. The discrepancy is probably due to the assumptions underlying the theory. The frequency characteristics of longitudinal

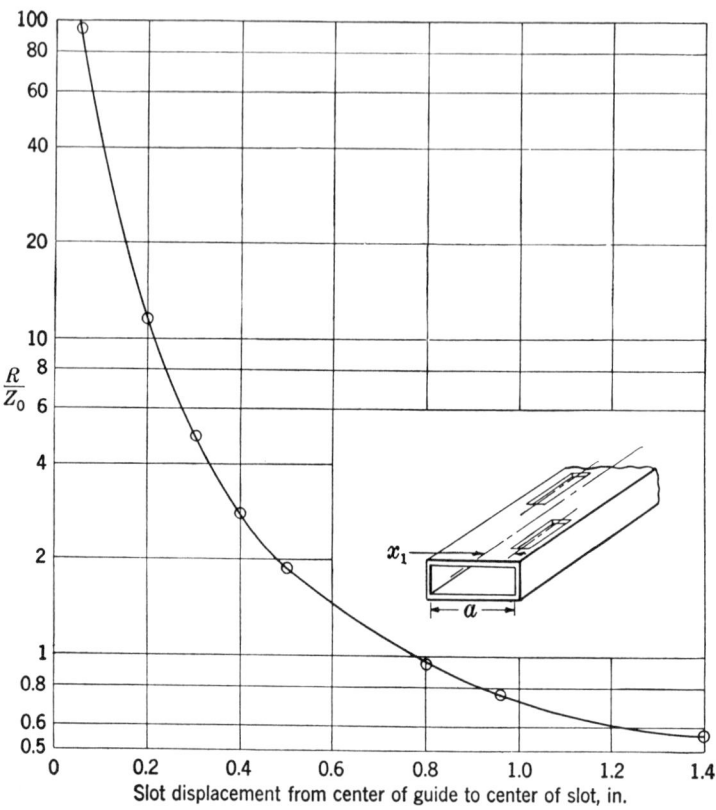

FIG. 9·18.—Resistance offered by a longitudinal slot as a function of its displacement from the center. The slot dimensions are $\tfrac{3}{16}$ by 2 in., the waveguide is $1\tfrac{1}{2}$ by 3 in., λ = 10.7 cm. The data fit the relation $G = Z_0/R = 1.73 \sin^2 [(\pi x/a) 1]$. (From J. W. Dodds, E W. Guptill, and W. H. Watson by permission of the National Research Council of Canada.)

slots as a function of slot width are presented in Fig. 9·19, which shows that the wider the slot the flatter the frequency response. The maximum of conductance does not coincide with the vanishing of susceptance.

For practical convenience dumbbell-shaped slots such as the one illustrated in Fig. 9·20 have been used in arrays in place of rectangular slots. The perimeter of a resonant slot is generally equal to a wavelength. The length of a resonant dumbbell slot is therefore less than that of rectangular ones; they can be used with less sacrifice of mechanical strength, since less guide is cut away. The dumbbell slot is also simpler to machine

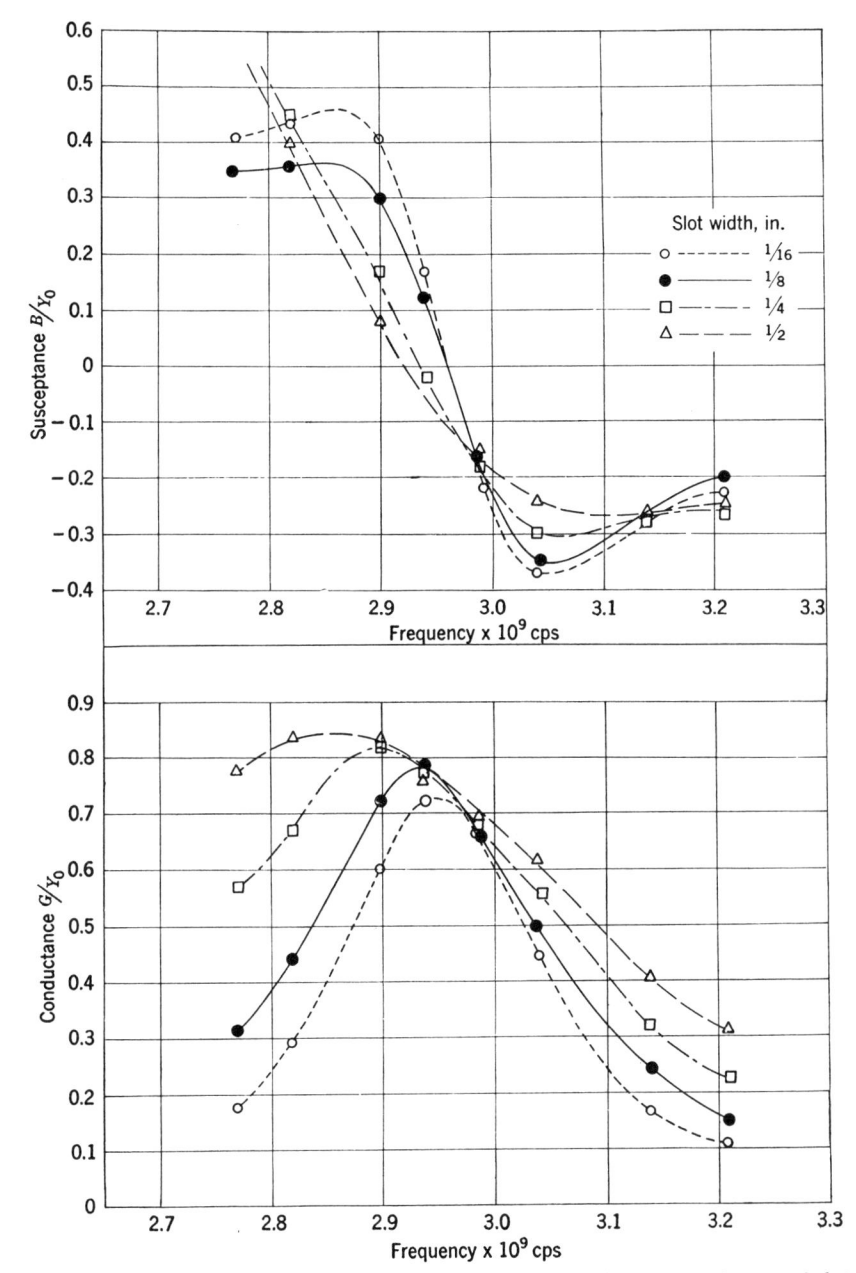

FIG 9·19.—Admittance of longitudinal slot as a function of frequency (center of slot is 1.98 cm from the center of the waveguide). (*From the work of J. W. Dodds and W. H. Watson by permission of the National Research Council of Canada.*)

because the dumbbell areas are drilled rather than cut by a milling machine. Another technique for shortening the resonant length is to place a thin sheet of dielectric over the slot; a sheet of polystyrene of 0.007-in. thickness reduces the resonant length by 1.13 per cent at 10.7 cm. The dielectric sheet also serves as a pressurizing device.

If the conductance and frequency characteristics of each element of a slotted linear array are known, it is possible to place a given number of longitudinal slots $\lambda_g/2$ apart so that they are effectively in parallel and to short-circuit the far end of the guide $\lambda_g/4$ from the last slot so that the admittance in parallel with the last slot is zero. Then if there are n elements, the relative conductance of each slot must (by suitably choosing x_1) be made to equal to $1/n$ in order to provide a good match. Because the slots are placed in the same way as a set of dipoles, end to end, the mutual impedance of the slots is negligible.

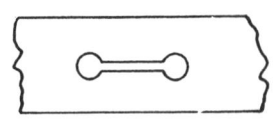

Fig. 9·20.—Dumbbell-shaped slot.

The conductance of a longitudinal slot cut in the broad face of the guide can be readily determined by measuring the input impedance of n slots in parallel because the mutual impedance between slots is negligible. This is not so when the slots are cut in the narrow face. The effective conductance of this slot may be found by measuring the additional conductance produced when one slot is added to an array. In practice a number of slots, for example 10, are cut and the input admittance determined. The input admittance is then again determined when additional slots are cut in sets of, say, 3. Eventually the total susceptance becomes constant and the conductance linearly proportional to n (if the susceptance is also proportional to n, the slot depth is adjusted for resonance). The incremental and ordinary conductances are plotted in Fig. 9·21 as functions of the angle θ. Both obey very well the law

$$g = g_0 \sin^2 \theta$$

over the measured range. This is in good agreement with Eq. (50) for small angles θ.

Slots cut in the narrow face have the very useful feature that the variation of susceptance with frequency is very small compared with that for slots in other positions in the guide. The variation of admittance with slot depth is also small as is shown by Fig. 9·22. Thus a change of ± 1 mm in depth from the resonant point produces a change of only 4 per cent in conductance and only a small change in susceptance. Because the depth of cut can always be accurately controlled in a milling operation, this represents a tolerance which can easily be attained. Since the angle of the slot to the guide axis can also be accurately held, the system represents a satisfactory array from the constructional point of view. A possible objection is that there is an appreciable degree of unwanted polarization in these beams. The field over the slot has a longitudinal component proportional to cos θ; the transverse component of the field does not reverse direction with reversal of the direction of inclination of the slot and gives rise to an unwanted side lobe at about 40° to the main beam. For tilt angles up to 15°, however, the unwanted polarization is less than 1 per cent of the radiated power.[1]

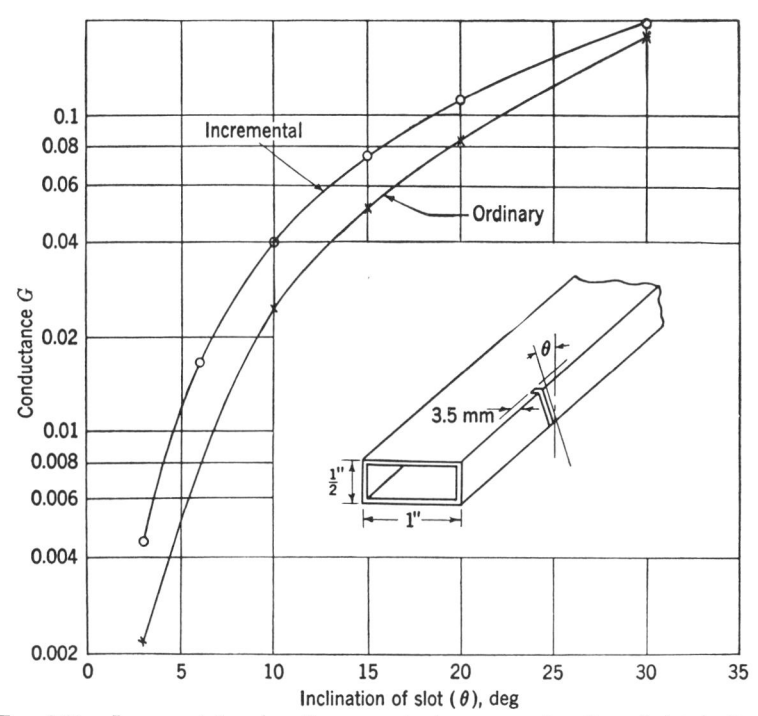

Fig. 9·21.—Incremental and ordinary conductance as a function of slot inclination. (*From the work of E. W. Guptill and W. H. Watson by permission of the National Research Council of Canada.*)

9·13. Probe-fed Slots.—It was pointed out in Sec. 9·9 that there are various positions in a guide and various orientations of the slot axis for which no radiation takes place. It is possible, however, to make any slot of this type radiate by inserting a suitable probe into the guide adjacent to the slot.[2] The probe introduces the necessary asymmetry in the field and current distributions for excitation of a field across the slot. The probe-fed unit has many advantages. In particular the direction of the field across the slot depends on the side in which the probe is

---

[1] Dodds, Guptill, and Watson, *op. cit.*
[2] R. E. Clapp, "Probe-fed Slots as Radiating Elements in Linear Arrays," RL Report No. 455, Jan. 25, 1944.

inserted; the phase of a given slot can be shifted 180° by switching the probe position. An example of this phase reversal is afforded by the array of slots on rectangular guide illustrated in Fig. 9·23; here the phase

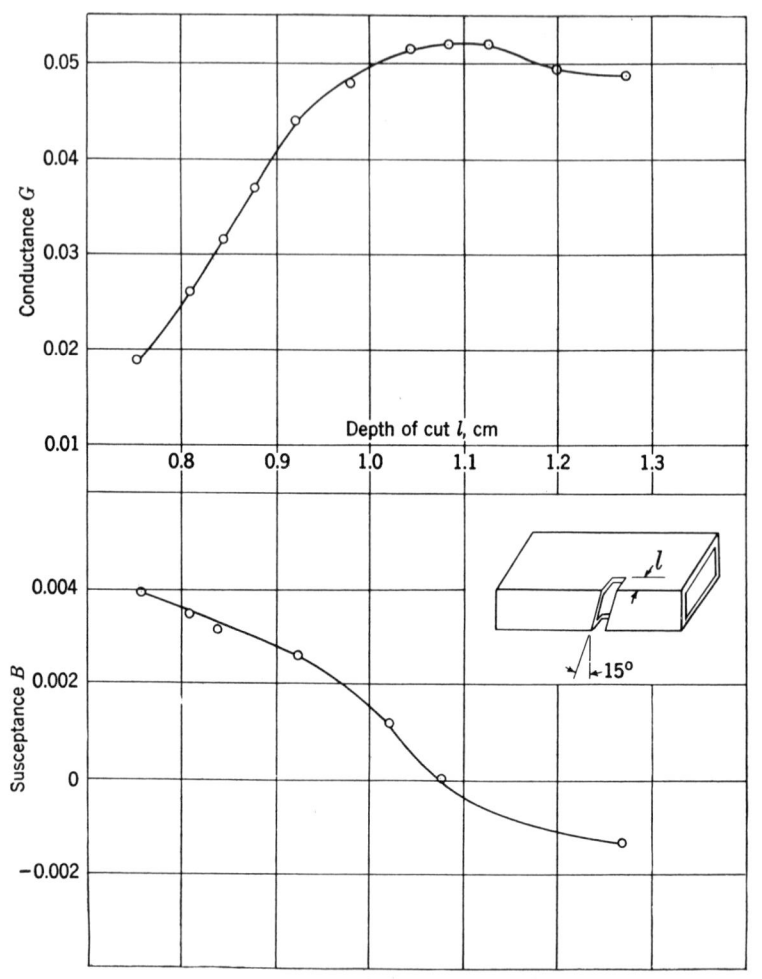

Fig. 9·22.—Admittance of a 15° inclined slot on narrow edge of rectangular waveguide. The waveguide dimensions are $1\frac{3}{8}$ by $2\frac{7}{8}$ in., $\lambda = 10.7$ cm., and the width of the slot is $\frac{1}{4}$ in. (*From the work of J. W. Dodds, E. W. Guptill, and W. H. Watson by permission of the National Research Council of Canada.*)

reversal of the probe is used to compensate for the 180° phase difference corresponding to the $\lambda_g/2$ spacing of the slots; the result is an array of equiphased slot radiators.

Another advantage of the probe-fed unit is that the amount of energy radiated by the slot is controlled by the probe insertion. For the case illustrated in Fig. 9·23 where the probe is parallel to the field, the coupling is adjusted by the probe depth. To excite a slot in the narrow side of a rectangular guide a bent probe is used, as shown in Fig. 9·24; here the coupling can be varied by the angle between the hook of the probe and the electric field. In some cases the screw head of the probe introduces undesirable impedance characteristics; the head of the screw can be

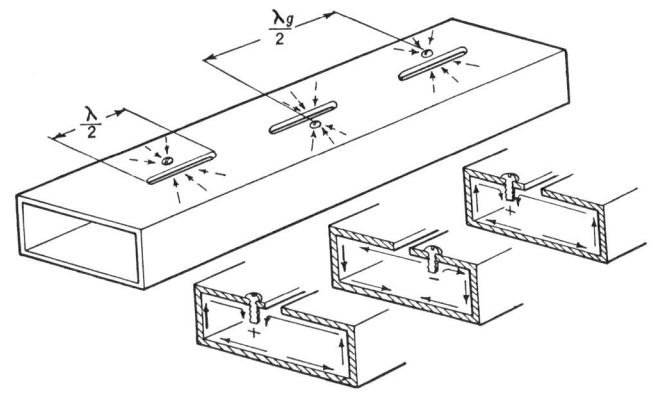

FIG. 9·23.—Probe-fed slots on rectangular waveguide. The arrows show lines of current flow.

ground off after the desired coupling has been obtained, or the unit can be balanced externally by a dummy screw head. Many variants of the probe can be developed for various types of guides and modes; the reader is referred to Clapp's report for details.

9·14. Waveguide Radiators.—The impedance of a radiating element has been seen to consist in general of a resistive and a reactive component. The reactive component is generally undesirable, since it enhances the frequency sensitivity. The reactance vanishes under special conditions, but these are not always optimum operating conditions; for example, in the case of the dipole element discussed in Sec. 9·8 resonance occurs at a probe depth that is generally too small to meet power-extraction requirements. Slots and dipoles suffer another severe disadvantage at short wavelengths as in the 1-cm region where they become so small that they have an insufficient power-handling capacity and the tolerances on the dimensions become impractically restrictive.

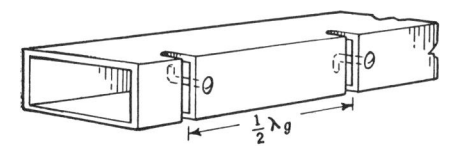

FIG. 9·24.—Probe-fed transverse slots on the narrow face of rectangular waveguide.

The waveguide radiators illustrated in Fig. 9·25 are less subject to the above limitations.[1] The element consists of a waveguide coupled to the main guide by a T-junction. As shown in the figure, two arrange-

[1] W. Sichak and E. M. Purcell, "Cosec² Antennas with a Line Source and Shaped Cylindrical Reflector," RL Report No. 624, Nov. 3, 1944, pp. 7–13.

ments are possible corresponding to longitudinal and transverse polarizations. The longitudinally polarized element, just like a slot with axis transverse to the guide axis, presents an impedance in series with the main line; the transversely polarized element, like a slot with axis parallel to the guide axis, is equivalent to a shunt element across the transmission line, inserted in the plane of symmetry of the radiator that is perpendicular to the guide axis. It has been found experimentally (cf. Sec. 10·11) that the open end of a waveguide can be represented by a load admittance consisting of the radiation resistance in parallel with a capacitive react-

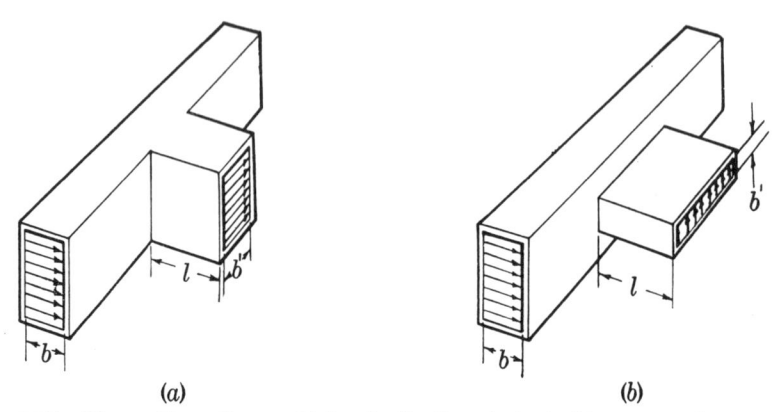

Fig. 9·25.—Waveguide radiators: (a) longitudinally polarized; (b) transversely polarized.

ance. The network equivalent of the T-junction consists in a similar manner of a capacitative reactance in parallel with the input impedance of the branched guide. Both the input and termination capacities are junction effects and may be expected to be of the same order of magnitude. If the length of the branch guide is $\lambda_g'/4$ where λ_g' is the guide wavelength in the branch, the terminal capacitative reactance is transformed into an inductive component at the input end; and since the inductive component is in parallel with the T-junction capacitance, a near-resonant condition should result. In actual practice, however, the length of the section is different from $\lambda_g'/4$. The correct length has been found to be given closely by the result of an analysis of a branched waveguide which takes the junction effects into account, namely,

$$l = \frac{\lambda_g'}{4} - \frac{2b'}{\pi}\left(1 + \ln \frac{b}{2b'}\right). \qquad (58)$$

The dimensions b and b' are defined in Fig. 9·25. With the above length the element has been found to be very closely a pure resistance.

The coupling of the element to the line, i.e., its resistance or conductance, is a function only of the relative dimensions of the branch guide and the main guide. It is the particular advantage of the waveguide element that the coupling factor can be adjusted independently of the resonance condition. On the assumption that the impedance presented to the main guide when the branched guide radiates into free space is not very different from that when the branched guide couples to a second guide parallel to the main guide, the resistance of the longitudinally polarized element has been calculated to be

$$R = \frac{1}{2}\left(\frac{b'}{b}\right)^2.$$

The quadratic dependence on b'/b is in accord with the experimental results; these results indicate, however, that the numerical factor is not $\frac{1}{2}$. No systematic study of a single element has been made as yet. The coupling factor for the transversely polarized element has meaning only in terms of a complete array because with these elements mutual interactions become very significant. The results will be given later in the discussion of nonresonant arrays which make use of these elements.

The length of the element given in Eq. (58) can be increased by any integral multiple of $\lambda'_g/2$ without affecting either the resonance or the coupling factor. This is advantageous in that it provides a method for shifting the phase of the radiator 180°. For the same power extraction the b' dimension of the transversely polarized element must be larger than that of the longitudinally polarized element because the former cuts across transverse currents that are smaller than the longitudinal currents on the broad face of the guide. Consequently, the tolerances are less restrictive for the transversely polarized element, and it therefore is preferred if all other considerations are equal. At short wavelengths, e.g., at 1 cm, the length of the radiator is so small that the wall of the main guide can be constructed of that thickness, and the radiating element then takes the form of a slot in this wall. This produces a sturdy mechanical system.

9·15. Axially Symmetrical Radiators.—For general communication purposes it is desired to have a stationary antenna with an axially symmetrical pattern covering a large region of space. The simplest antenna of this type is a half-wave dipole. The gain of the dipole, however, is too low to meet the usual requirements on range, and it is therefore necessary to design an antenna having the axial symmetry of the dipole but with a more directive meridional pattern. The latter can be achieved by means of a linear array of axially symmetrical radiating elements, an example of which is illustrated schematically in Fig. 9·26. The elements to be discussed fall into two groups distinguished by the polarization of the field: (1) transversely polarized radiators producing a field in which the electric vector lies in planes normal to the axis of the array, (2) longitudinally polarized radiators producing a field in which the electric vectors lie in meridional planes having the axis of the array as a common line of intersection; the transverse element is analogous to a magnetic dipole, and the longitudinal element to an electric dipole. The transverse element in its ideal form should consist of a circular ring of uniform current, while the idealized longitudinal radiator should consist of a short circular cylindrical current sheet of uniform density running parallel to the axis of the cylinder. In practice these elements can best be approximated by an array of elements located at points disposed symmetrically about the array axis in a plane normal to it. Thus the antenna as a whole is, in fact, a three-dimensional array; however, design problems for the azimuthal and meridional patterns are completely separable. The meridional pattern is a straightforward linear-array problem.

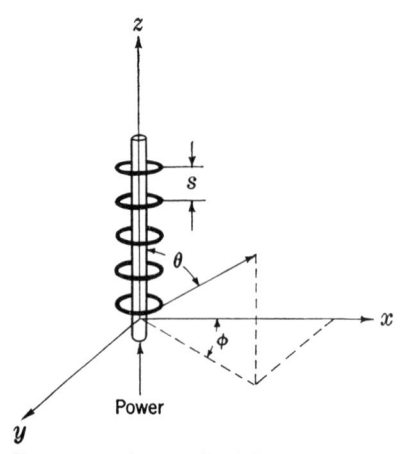

Fig. 9·26.—Array of axially symmetrical radiators.

The azimuth pattern reflects the symmetry of the arrangement of the radiators about the array axis and consequently deviates from a uniform pattern, showing maxima and minima. The ratio of maximum power to minimum is referred to as the *azimuth ratio;* it is generally required that this ratio be less than 2.

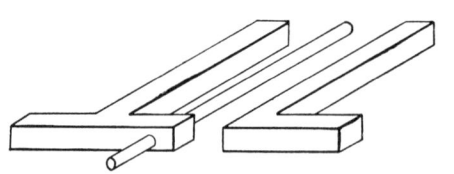

Fig. 9·27.—Dipole fed by three-wire line.

The Tridipole Transverse Element.—First the elements designed for transverse polarization will be considered. A simple approximation to the circular current ring is obtained by arranging three half-wave dipoles on the circumference of a circle. The basic unit illustrated in Fig. 9·27 is a three-wire–line-fed dipole analogous to the slotted dipole discussed earlier. The central line serves as a probe to couple the dipole to the interior of a waveguide. The axially symmetrical tridipole array shown in Fig. 9·28 is designed for use with a coaxial line. The element is made so as to slide over the outer conductor and is soldered to the latter at the appropriate location. In order to maintain the azimuth pattern symmetry it is essential that the three probes be inserted to equal depths.

The line coupling can be achieved either by inserting the probes so as to make contact with the inner conductor of the line or by capacitive coupling in which the probes do not make contact with the inner conductor. In the former case the probes are soldered to the inner line; a more reliable procedure is to have threaded holes in the inner conductor into which the probes can be screwed and then soldered to ensure good contact. For capacitative coupling, probe settings can be made by slipping a shim of suitable thickness over the inner conductor; the shim is subsequently removed.

Satisfactory results have been obtained with tridipole elements over the 10-cm band[1] and at various longer wavelengths. The impedance characteristics of a single unit can be adjusted in the course of design by the choice of the dimensions of the dipole wings; the impedance characteristics of an array of units are adjustable by means of the probe depth. Figure 9·29 shows the frequency sensitivity of the pattern of a tridipole unit designed for the 10-cm band, the unit being fed from a 50-ohm line with a $\frac{7}{8}$ in. OD. The pattern exhibits a

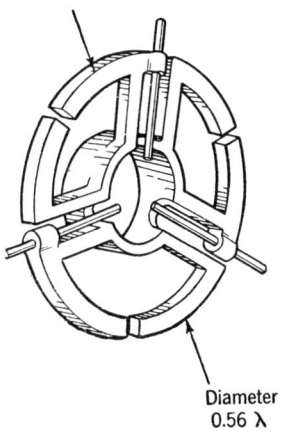

FIG. 9·28.—Tridipole radiator. Diameter 0.56 λ

high degree of stability. The same element with its probes 0.5 mm from the inner conductor handles 10-kw peak power without breakdown. It was found that at 10 cm the dimensions of the unit are not critical and the elements can be produced in quantity by die-casting techniques with good reproduction of performance. A 3-cm version, however, requires manufacturing tolerances too close for practical use.

Axially Symmetrical Slot Array.—Another type of unit for transverse polarization is provided by an array of slots along the circumference in the wall of a circular guide or coaxial line, the long dimension of the slots being parallel to the axis of the guide. A number of factors enter into the design of the unit.

The most important is that the line must carry a radially symmetrical mode so that the slots are excited equally. This condition is fulfilled by a coaxial line supporting only the *TEM*-mode and by a circular waveguide propagating the TM_{01}-mode as indicated in Fig. 9·30. However, for both cases a slot cut parallel to the guide axis does not radiate. It is therefore necessary to excite the slots by means of probes as shown in the figure, and to ensure symmetrical excitation the probe depths must be uniform.

[1] H. Riblet, "Horizontally Polarized Nondirectional Antennas," RL Report No. 517, Feb. 14, 1944.

306   LINEAR-ARRAY ANTENNAS AND FEEDS   [SEC. 9·15

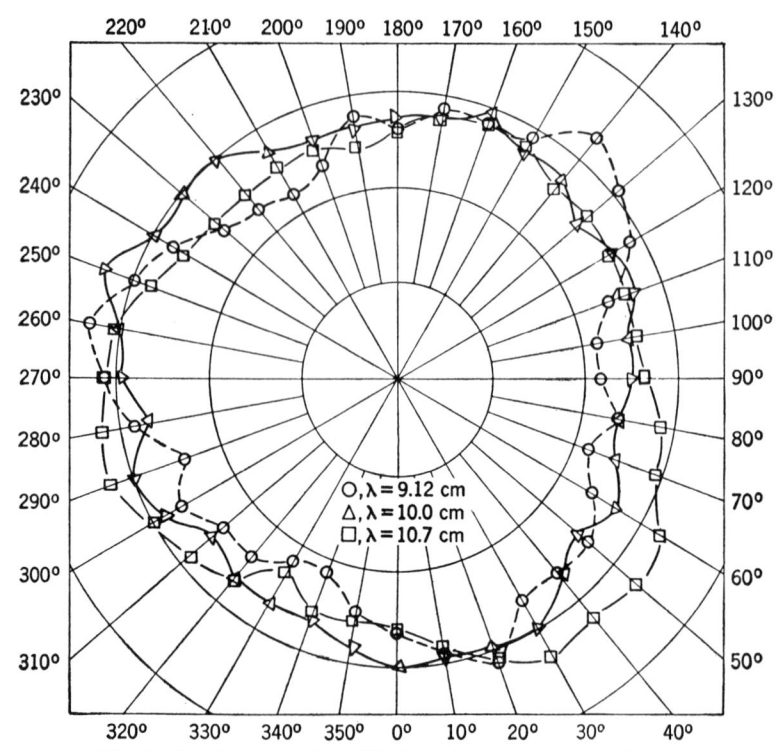

FIG. 9·29.—Patterns of a tridipole unit in the plane of the unit.

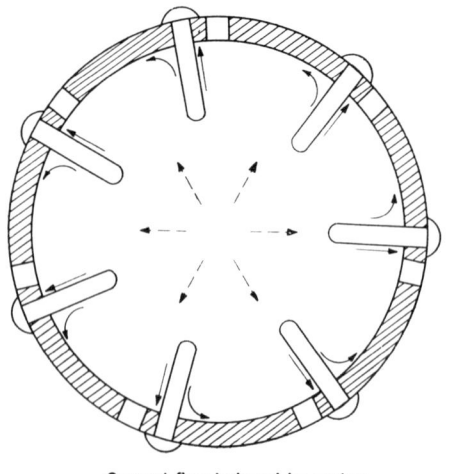

FIG. 9·30.—Axially symmetrical radiating unit formed by a circular array of slots.

Sec. 9·15]   AXIALLY SYMMETRICAL RADIATORS   307

A second factor is the minimum number of slots required to produce a pattern having a satisfactory azimuth ratio. This is found to depend on the size of the line; the larger the guide diameter the greater the number of elements. For a 1-in. OD coaxial line operating in the 3-cm band

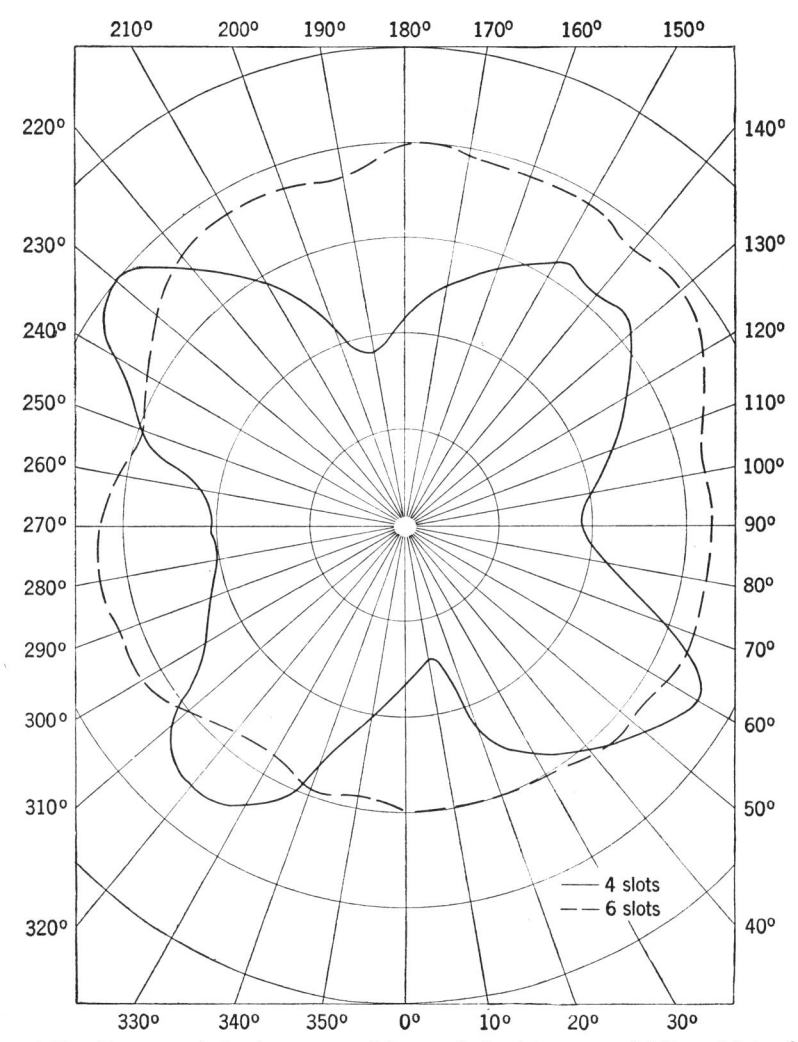

Fig. 9·31.—Patterns of circular arrays of four and six slots on coaxial line of 1 in. OD.

the minimum number is six; Fig. 9·31 shows the patterns obtained from four and six elements; the former reflects strongly the fourfold symmetry of the array. Figure 9·32 illustrates the pattern resulting from a seven-element array on 1¼-in. circular guide, again for the 3-cm band.

The minimum number of elements is also related to a problem of mode control. The conventional coaxial line in the 3-cm band, for which all modes other than the TEM-mode are beyond cutoff, is too limited in its breakdown properties and mechanical strength, the latter being an important factor in long arrays. Line of a larger size is therefore used which can support other modes. Considerable care must be taken at the input end of the line to ensure radially symmetrical excitation. The probe inserts for exciting the slots likewise excite higher modes. No mode will be excited, however, if its planes of symmetry do not contain the symmetries of the geometrical configuration. There is thus a minimum number of probes for which the higher modes excited will attenuate.

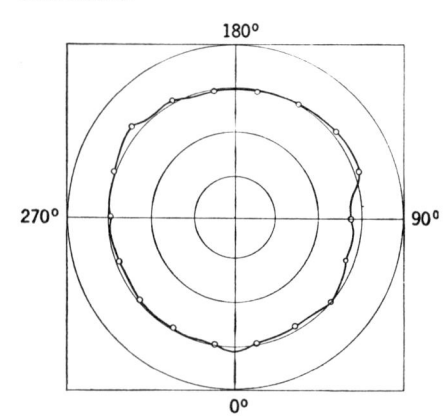

Fig. 9·32.—Pattern of circular array of seven slots on circular waveguide with $1\frac{1}{4}$ in. OD.

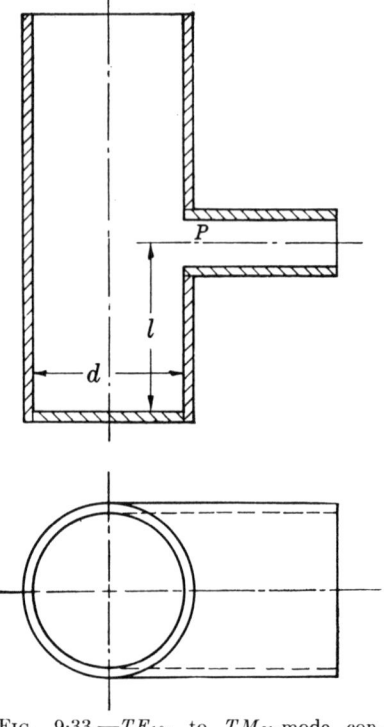

Fig. 9·33.—TE_{10}- to TM_{01}-mode converter.

A similar mode-control problem exists in the circular guide, for a circular guide that can support propagation of the TM_{01}-mode necessarily supports the TE_{11}-mode. It is therefore necessary to feed the guide in such a manner that the TE_{11}-mode is not excited, and again there is a minimum number of slots required. The proper feeding of the circular guide is achieved by transition from the TE_{10}-mode in rectangular guide through a TE_{10}- to TM_{01}-mode converter, which is illustrated in Fig. 9·33. Briefly the principle of its operation is as follows: The distance l is equal to $\lambda_{11}/4$ or $3\lambda_{11}/4$ where λ_{11} is the guide wavelength for the TE_{11}-mode; this puts a large series reactance for this mode at P between the rectangular and circular guide so that the mode is not fed into the latter guide. The diameter d is chosen to be $\lambda_{01}/2$ where λ_{01} is the TM_{01}-

mode guide wavelength; for $l = 3\lambda_{11}/4$, this gives a good match for the TM_{01}-mode.

Longitudinally Polarized Elements.—Satisfactory elements of this type in the microwave region have thus far been developed only for the 10-cm band. A longitudinal element analogous to the tridipole unit can be produced by a circular array of dipoles with axes parallel to the guide axis. It is found, however, that a longitudinally polarized tridipole array gives rise to a pattern having a decided threefold symmetry while a larger number of dipoles results in a unit whose design is very critical. A cylindrical element with three-point excitation provides a simple solution; the element is shown in Fig. 9·34. It can be thought of as being derived from a system of three longitudinal dipoles of the type illustrated in Fig. 9·27 in which the wings have been extended laterally and joined into a cylinder. The currents tend to spread out uniformly over the surface giving a uniform azimuth pattern.

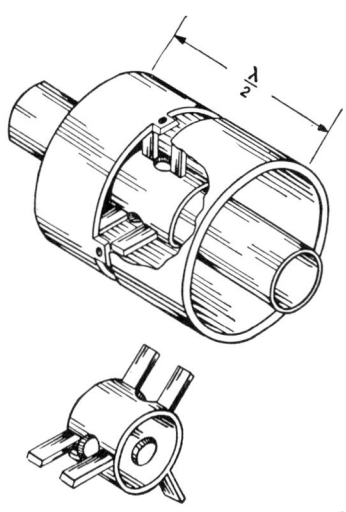

FIG. 9·34.—Longitudinally polarized axially symmetrical radiating unit.

The unit is made in two parts, one consisting of a die-cast spider carrying the two outer lines of the three-wire line-feeding system and the other the pair of cylinders that correspond to the dipole wings. As with the transverse unit the system is fed by probes which couple the cylinders to the line; the general remarks made previously concerning the insertion and alignment of the probes likewise apply here.

Attention should be called to a longitudinally polarized slot radiator which can be designed with a coaxial line. The element illustrated in Fig. 9·35 consists of a slot running completely around the wall. Mechanical support is provided by filling the line with dielectric. The element obviously gives a uniform pattern but suffers from a number of disadvantages. It is very frequency-sensitive; mechanical properties, particularly of long arrays, are poor; satisfactory contact between the dielectric and metal is difficult to maintain particularly under mechanical and thermal stresses with the result that the system becomes susceptible to electrical breakdown. Development of arrays with these units was finally given up because of these limitations and difficulties.

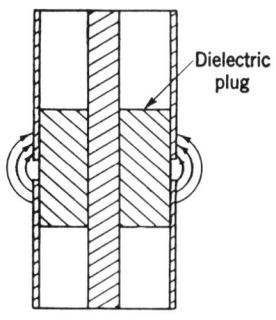

FIG. 9·35.—Longitudinally polarized slot radiator for coaxial line.

9·16. Streamlined Radiators.

Arrays of axially symmetrical radiators have been developed for airborne, ground, and ship installations. With the development of high-speed planes, however, aerodynamic considerations have become increasingly significant in antenna installations. Arrays of elements of the types already discussed produce sufficient aerodynamic drag to present a serious installation problem. It has therefore been necessary to make some compromise between pattern and aerodynamic requirements and to design elements whose geometry has a less deleterious effect on the aircraft. For this purpose various types of streamlined elements have been developed which, though lacking the uniform coverage of the axially symmetrical units, still produce patterns with not too large an azimuth ratio.

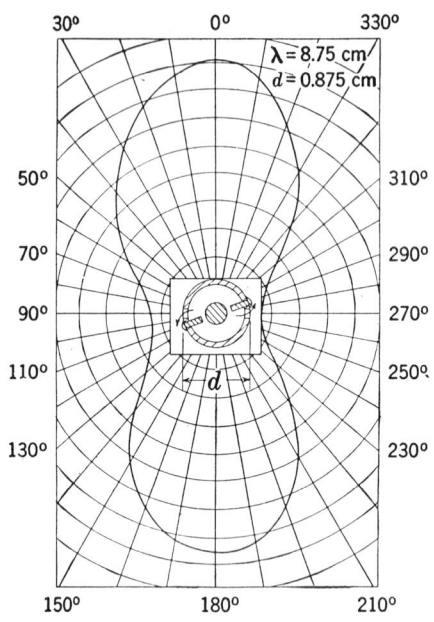

Fig. 9·36.—Pattern produced by a pair of antiphased slots.

Two types have been developed, one for transverse, the other for longitudinal polarization. Let us consider the transverse radiator first. It has been found that two slots cut opposite each other on a coaxial line and excited 180° out of phase produce a pattern with an azimuth ratio not exceeding 5 or 6; this is shown

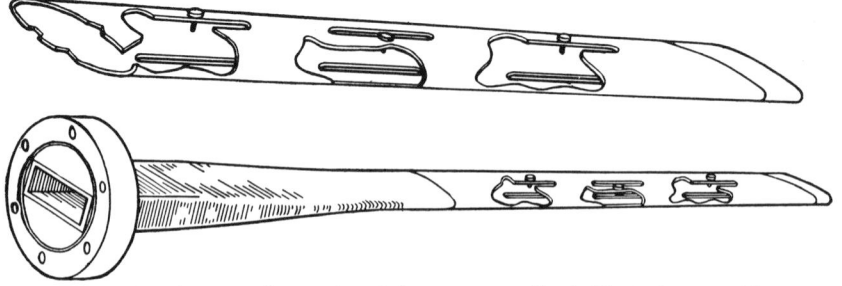

Fig. 9·37.—Array of three pairs of slots on streamlined elliptical waveguide.

in Fig. 9·36. The currents tend to run completely around the cylinder, giving a continuous, if not completely symmetrical, current distribution. Starting from this observation, one can proceed in several directions to the design of streamlined elements. First, the outer conductor instead of

SEC. 9·16]   STREAMLINED RADIATORS   311

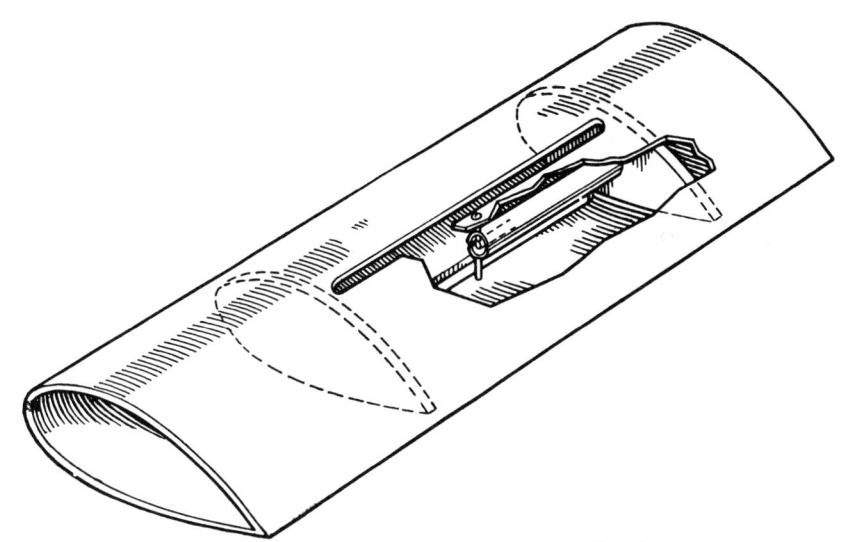

FIG. 9·38.—Transversely polarized streamlined radiator.

being cylindrical can be made elliptical or streamlined, or the inner conductor can be omitted entirely and a streamlined section of sufficient size used instead as a waveguide. Figure 9·37 for example, shows an array of three slots on streamlined elliptical guide for the 3-cm band. Transition to the elliptical guide from rectangular guide is effected by a tapered section. It has been found that if the ratio of major to the minor axis of the ellipse is at least 4 and if the minor axis is approximately $\lambda/4$, the azimuth ratio is in the neighborhood of 2; this figure has been obtained with the three-unit array referred to above.

A method of introducing the r-f that provides a good impedance match is shown in Fig. 9·38 for a 10-cm band system. Here the two slots are cut at the point of maximum width of the guide. A slotted dipole on the end of small coaxial line is used to excite the slots. The wings of the dipole are cut to fit, and each wing acts as an exciting antenna for one slot. The VSWR obtained with a single element is less than 1.2

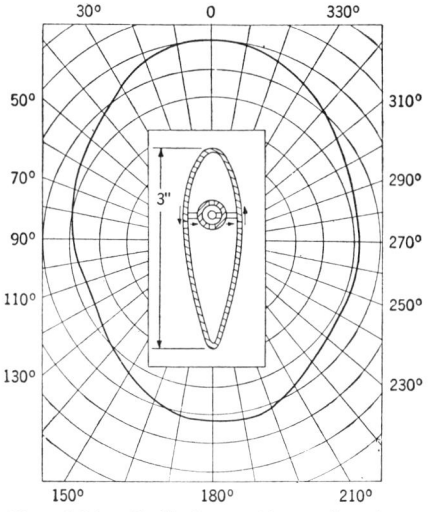

FIG. 9·39.—Radiation pattern of a transversely polarized streamlined radiator.

over a 16 per cent band. The pattern shown in Fig. 9·39 is likewise satisfactory. The only longitudinally polarized unit that has been built is for the 10-cm region. It consists essentially of two vertical dipoles about a quarter wavelength apart supported on opposite sides of coaxial line. Two such dipoles in free space would have an oval pattern which, however, does not have too large an azimuth ratio. This, of course, is modified by the coaxial line; but if the line is small enough (ordinarily $\frac{1}{4}$ in.

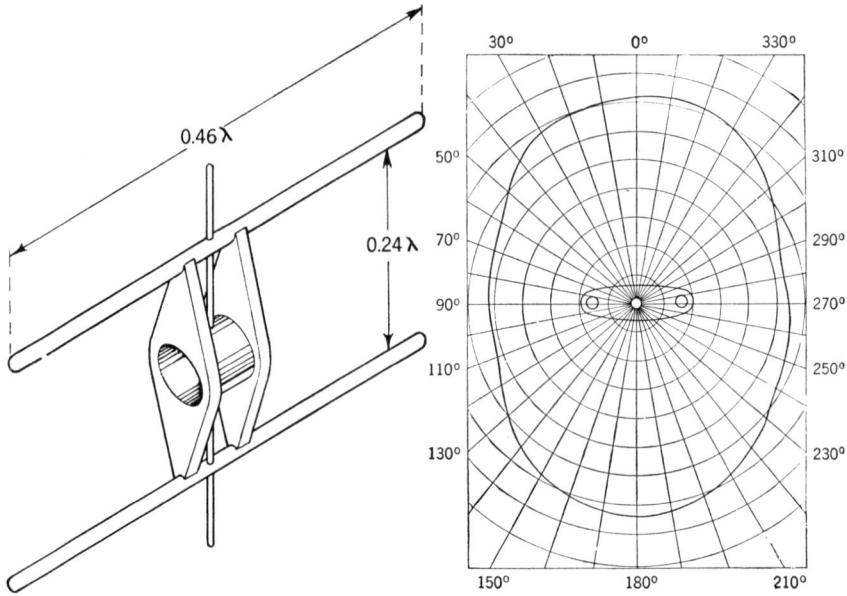

Fig. 9·40.—Longitudinally polarized streamlined radiator (*H*-element).

Fig. 9·41.—Radiation pattern of a longitudinally polarized streamlined radiator (*H*-element).

OD), the effect is small and does not seriously impair the pattern. The unit, generally referred to as an *H*-element, is shown in Fig. 9·40. The two dipole wings are supported by elliptical straps, and the whole unit is placed over the coaxial line and excited by the probes projecting into it. The elliptical straps serve also as a wave trap, to prevent currents running along the coaxial line. The pattern produced by the unit is shown in Fig. 9·41. An array of such elements is ordinarily enclosed in a close fitting elliptical housing.

ARRAYS

It is shown in the sections on general pattern theory that the pattern of a linear array is determined essentially by three factors: (1) the relative amplitude and phase of the current distributions on the elements of the array, (2) the spacing of elements along the axis, and (3) the form factor of the pattern of a single element. In practice these factors are not independent variables; the amplitude and phase of the elements are determined in part by interactions between the elements, which in turn are functions of their spacing. At longer wavelengths feeding techniques are available whereby the amplitude and phase, except for external field coupling between the radiators, are independent of spacing. In microwave antennas the elements must be fed in cascade from a transmission line; the phase of the radiator thus depends on the phase velocity in the line and the position of the element along the line; phase and spacing are thereby most intimately related. The relation becomes complicated further because the feeding arrangement results in a loaded transmission line with propagation constant and characteristic impedance different from those of the unloaded line. Finally mutual interactions between the elements because of their external fields must be considered. The result of these interrelations is that the transition from the properties of a single element to a composite array is not a calculable design procedure but must be determined to a large extent on an empirical basis.

9·17. Loaded-line Analysis.—The relation between the parameters of loaded and unloaded lines will be investigated first. Consider a line, whose unloaded parameters are the characteristic impedance Z_0 and the complex propagation constant $\gamma = \alpha + j\beta$, loaded at regular intervals l with identical radiating elements. Taking a fixed reference point in a radiator, the radiator in general can be regarded as a bilateral passive four-terminal network inserted at the reference point between two segments of line. It was shown in Sec. 2·2 that such a network can be replaced by a T- or Π-section equivalent; in the notation of Sec. 2·2 the three impedance elements of the T-section will be designated by Z_1, Z_2, and Z_3 and the elements of the Π-section by Z_A, Z_B, and Z_C. The relation between the T- and Π-section elements is given in Eq. (2·10). The radiating elements that have been discussed in the earlier sections all have at least one plane of symmetry; if the reference point is taken in this plane, the T- or Π-section equivalent of the radiator is symmetrical; i.e., $Z_1 = Z_2$ and $Z_A = Z_C$. It was shown further in Sec. 2·9 that a section of homogeneous transmission line of length l has a symmetrical T- and Π-section equivalent; from Eqs. (2·8), (2·56a), and (2·56b) the elements of the T-equivalent are found to be

$$\mathfrak{Z}_1 = \mathfrak{Z}_2 = Z_0 \tanh\left(\frac{\gamma l}{2}\right); \qquad \mathfrak{Z}_3 = \frac{Z_0}{\sinh(\gamma l)}. \tag{59}$$

By means of Eqs. (2·7), (2·8), and (2·10) the elements of the equivalent Π-sections are obtained from these. The Π-elements are

$$\mathfrak{Z}_A = \mathfrak{Z}_C = Z_0 \coth\left(\frac{\gamma l}{2}\right); \qquad \mathfrak{Z}_B = Z_0 \sinh(\gamma l). \tag{60}$$

On replacing both the radiators and the line segments by their equivalent T-sections, the loaded line is reduced to a cascade of networks as shown in Fig. 9·42a; the points A and A' are the reference points in the radiating elements. By splitting the shunt element Z_3 into a pair of impedances $2Z_3$ in parallel, the line is further reduced to a chain of symmetrical networks, a single unit of which is shown in Fig. 9·42b.

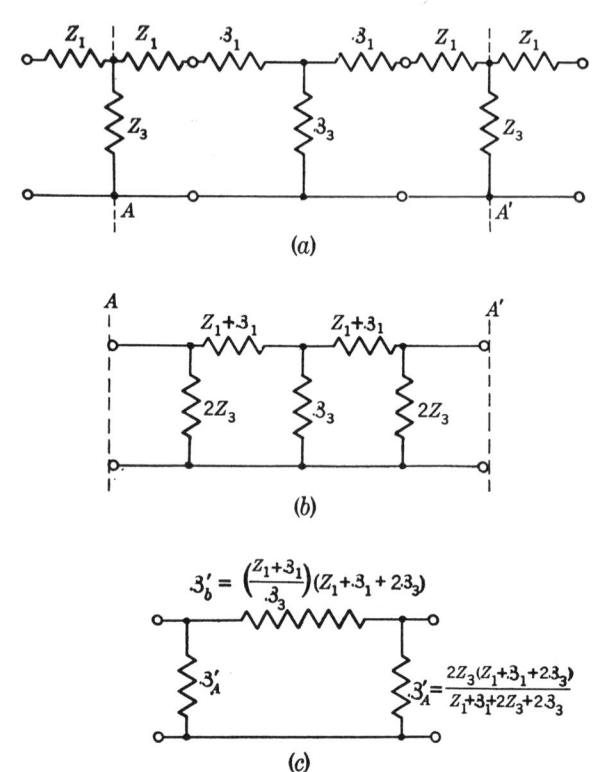

Fig. 9·42.—Network system equivalent to a loaded transmission line: (a) T-section replacements of radiators and line segments; (b) reduction to symmetrical networks; (c) Π-section equivalent of the network in (b).

The characteristic impedance Z'_0 and propagation constant γ' of the loaded line are obtained by reduction of the network in Fig. 9·42b to its equivalent T- or Π-section and subsequently determining the parameters of a homogeneous line having a length l for which the above T-section (or Π-section) constitutes an equivalent representation. In the present case the simplest procedure is to reduce the network to a Π-section by replacing the T-network of elements $Z_1 + \mathfrak{Z}_1$ and $\mathfrak{Z}_3$ by its Π-equivalent. The completely reduced network is shown in Fig. 9·42c. If $\mathfrak{Z}'_A$ and $\mathfrak{Z}'_B$ are the elements of the reduced network, the loaded-line parameters are given by

$$Z'_0 \coth\left(\frac{\gamma' l}{2}\right) = \mathfrak{Z}'_A; \qquad Z'_0 \sinh(\gamma' l) = \mathfrak{Z}'_B. \tag{61}$$

If the values of $\mathfrak{Z}'_A$ and $\mathfrak{Z}'_B$ given in Fig. 9·42c together with the values of $\mathfrak{Z}_1$ and $\mathfrak{Z}_3$ given in Eq. (59) are inserted in Eq. (61), the half-argument identities for hyperbolic functions may be used to obtain Campbell's formulas:

$$\cosh(\gamma' l) = \left(1 + \frac{Z_1}{Z_3}\right)\cosh(\gamma l)$$
$$+ \left(\frac{Z_0}{2Z_3} + \frac{Z_1}{Z_0} + \frac{Z_1^2}{2Z_0 Z_3}\right)\sinh(\gamma l); \tag{62}$$

$$Z'_0 \sinh(\gamma' l) = \frac{\left[Z_1 + Z_0 \tanh\left(\frac{\gamma l}{2}\right)\right]^2}{Z_0}\sinh(\gamma l)$$
$$+ 2\left[Z_1 + Z_0 \tanh\left(\frac{\gamma l}{2}\right)\right]. \tag{63}$$

For the present purposes the attenuation in unloaded waveguides due to conduction losses in the walls may be neglected; under these conditions $\gamma = j\beta$ and the propagation constant of the loaded line is given by

$$\cosh(\gamma' l) = \left(1 + \frac{Z_1}{Z_3}\right)\cos\beta l + j\left(\frac{Z_0}{2Z_3} + \frac{Z_1}{Z_0} + \frac{Z_1^2}{2Z_0 Z_3}\right)\sin\beta l. \tag{64}$$

It is seen at once that the loaded line has a complex propagation constant $\gamma' = \alpha' + j\beta'$ in which both the attenuation and phase constants are functions of the loading and the spacing of the elements. Equation (64) shows, however, that if the spacing is equal to half the wavelength in the unloaded line, the relation reduces to

$$\cosh(\gamma' l) = -\left(1 + \frac{Z_1}{Z_3}\right); \qquad l = \frac{\lambda_g}{2}. \tag{65}$$

If the radiating element is a pure shunt element so that $Z_1 = 0$, it is found directly from Eq. (65) that

$$\gamma' = j\frac{2\pi}{\lambda_g} = \gamma.$$

Similarly if the element is a pure series element, in which case $Z_3 = \infty$, it is found that $\gamma' = \gamma$. *Thus there is no attenuation in a line loaded with pure series or pure shunt elements at half-wavelength intervals.* The same is true of a line loaded at wavelength intervals. For arbitrary spacings the propagation constant of the shunt-loaded line is given by

$$\cosh(\gamma' l) = \cosh(\gamma l) + \frac{Z_0}{2Z_4}\sinh(\gamma l) \tag{66}$$

and for the series-loaded line

$$\cosh(\gamma'l) = \cosh(\gamma l) + \frac{Z_4}{2Z_0}\sinh(\gamma l), \qquad (67)$$

where Z_4 is the impedance of a single element.

The pure series- and shunt-loaded lines with half-wavelength spacing have the additional property of producing a uniformly illuminated array when the line is suitably terminated in a short circuit. For the series-loaded line the short circuit is made an integral number of half wavelengths beyond the final element; by virtue of the half-wavelength spacing the array is equivalent to a simple series circuit of equal impedances; all the elements therefore dissipate equal amounts of power. The shunt-loaded line is terminated $\lambda_g/4 + n\lambda_g/2$ beyond the final element, n being an integer; this array is equivalent to a system of equal impedances all in parallel, and again all the elements dissipate equal amounts of power.

The loaded-line analysis takes no account of coupling between the elements by means of the external fields. Campbell's formulas do show, however, the interrelation between the amplitude and phase of the elements and the spacing and also the relation between the amplitude of the element and its phase, for the phase velocity is a function of the coupling between the radiator and the guide.

9·18. End-fire Array.—The only important examples of end-fire arrays for the microwave region were two very similar antennas for operation at wavelengths of 10.7 and 11.7 cm.[1] They consist of 18 individual radiators a quarter wavelength apart and fed from a coaxial line. The antenna shown in Fig. 9·43 is a 14-element experimental model. An antenna of this type must be terminated in a dummy load to absorb the unradiated power in the line thereby eliminating a reflected wave; otherwise the reflected wave would give rise to an end-fire pattern in its direction of propagation, that is, in a direction 180° away from the principal beam.

The elements are built up from the fundamental dipole shown in Fig. 9·27. Each consists of two such dipoles having their wings bent into arcs of circles and joined to form a unit. Like the axially symmetrical tridipole units these elements are simply slipped over the outer conductor of a coaxial line, and they also can be represented by a shunt impedance.

Two conditions must be satisfied if an end-fire array of this type is to

---

[1] H. J. Riblet and B. L. Birchard, "End-fire Array Antenna," RL Report No. 577, July 11, 1944. Dielectric-rod antennas may be designed to have end-fire patterns with gain, beamwidth, and side-lobe properties as good as those of linear arrays. *Cf.* C. E. Mueller, "The Dielectric Antenna or Polyrod," BTL Report No. 251, Jan. 26, 1942; J. E. Eaton, "Dielectric Rod End-fire Antennas Close to Metal Surfaces," RL Report No. 969, Jan. 23, 1946; R. E. Dillon and L. J. Eyges, "Compact Horns Intermediate between Polyrods and Reflectors," RL Report No. 961, Jan. 31, 1946.

have maximum gain. (1) There is an optimum value for the wavelength λ_g of the coaxial line. It was shown in Sec. 9·9 that maximum gain for quarter-wavelength-spaced end-fire arrays with a constant phase delay occurred when the phase delay between adjacent elements was

$$\frac{\pi}{2} + \frac{2.94}{n}.$$

The total phase delay between the first and last elements is then approximately

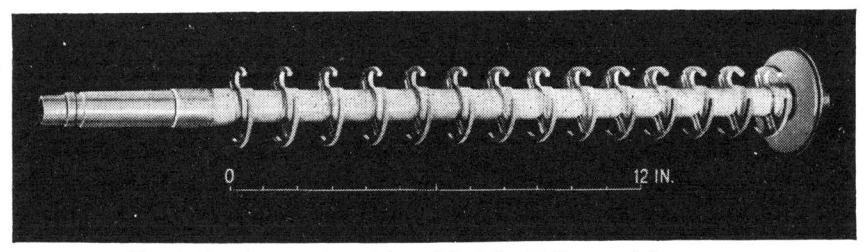

$$\psi = \pi\left(\frac{n}{2} + 1\right). \tag{68}$$

Fig. 9·43.—An experimental model of the 11.7-cm end-fire array.

If $L = n\lambda/4$ is taken as the length of the array, Eq. (68) becomes

$$\frac{2\pi L}{\lambda_g} = \pi\left(\frac{n}{2} + 1\right)$$

or

$$\frac{L}{\lambda_g} = \frac{L}{\lambda} + \frac{1}{2}. \tag{69}$$

(2) The attenuation has a definite optimum value; it must be neither so large that most of the power is radiated from the first few elements nor so small that an excessive amount of power is lost in the dummy load. This optimum attenuation is ordinarily assumed to be that which allows from 5 to 10 per cent of the total power to be absorbed in the dummy load.

The desired attenuation and phase shift can be obtained in principle in a very simple way. From Sec. 9·17 we have seen that periodic loading of a transmission line changes the propagation constant of the line. Hence it should be possible to choose the impedance of individual radiators so that they cause just the right change in attenuation and phase velocity. In fact, if the impedances of the elements are known as a function of several parameters, the propagation constant can be calculated from Eq. (62) as a function of the parameters and the best value chosen.

In the design of the particular arrays described above, the impedances of the individual elements were not known in enough detail to allow this. Hence a different approach was used. The gain was measured as a function of probe depth for various lengths of the dipole wings. This gave the two parameters necessary to adjust for the correct phase velocity and attenuation. The gain of the 11.7-cm array finally obtained in this manner was around 15.4 db, slightly greater than the theoretical value of 15.2 db for such an array; the theoretical value, however, is based on isotropic radiators. The gain of the 10.7-cm array, which used

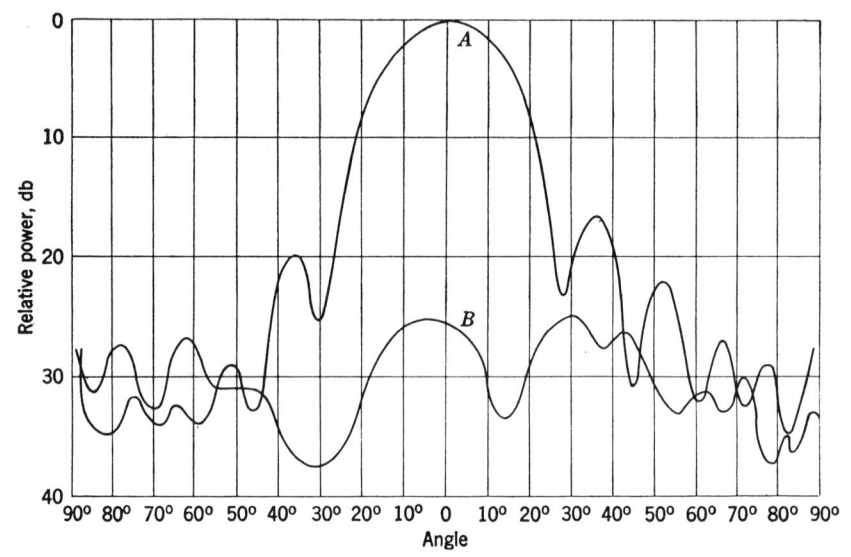

Fig. 9·44.—E-plane pattern of an 18-element end-fire array, $\lambda = 11.7$. A is the pattern in the direction of the main lobe, and B is the pattern in the direction of the back lobe.

11.7-cm elements, was 14.8 db. The E-plane pattern of the 11.7-cm antenna is shown in Fig. 9·44; the H-plane pattern differs from it only in minor details.

BROADSIDE ARRAYS

9·19. Suppression of Extraneous Major Lobes.—The majority of the applications of microwave arrays have called for a beam having the principal maximum in a direction normal to or nearly normal to the axis of the array. Arrays of this type will be referred to as broadside arrays with the arbitrary limit on the classification that the principal maximum lies within 25° of the normal to the array. In general there must be no principal maximum other than that of the broadside lobe, that is, all other maxima must be in the form of side lobes at considerably lower levels. This requirement gives rise to a spacing and phase problem

SEC. 9·19]   SUPPRESSION OF EXTRANEOUS MAJOR LOBES   319

common to all arrays of this type. It was seen in Sec. 9·5 that the elements of a uniform array must all be in phase for an accurately normal main lobe while to produce an off-normal lobe [cf. Eq. (23)] there must be a small progressive phase delay. If there are to be no other major lobes, the spacing between isotropic radiators must be somewhat less than λ, the free-space wavelength. The exact amount depends on n and the acceptable side-lobe level; no portion of an accurately normal main beam will be repeated in the direction $\theta = 0$ if $s = (1 - 1/n)\lambda$. To produce uniform phase, the radiators must be spaced at intervals of λ_g, the guide wavelength. However, for all the air-filled microwave lines discussed in Chap. 7, it was found that $\lambda_g \geqq \lambda$ with the result that the spacing exceeds the limit stated above.

There are various techniques for circumventing the difficulty. The less-than-wavelength spacing limit applies strictly to an array of isotropic radiators. However, in Sec. 9·2 it was shown that the pattern of an array is a product of an array factor corresponding to the pattern of an array of isotropic radiators and the pattern of an individual radiator. If the latter pattern is made sufficiently directive with a maximum in the direction normal to the array, a principal maximum will occur only in the region where the array factor and the radiator pattern simultaneously have appreciable values. In this case the spacing can exceed λ without the appearance of extraneous major lobes. Illustrative of such a directive device is a horn fed by a slot; an array of this type is shown schematically in Fig. 9·45.

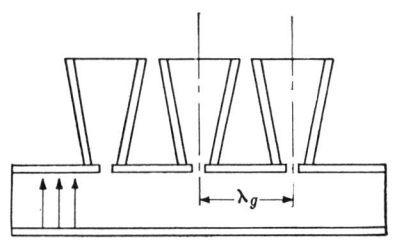

FIG. 9·45.—Array of transverse slots with horns to eliminate end-fire lobes.

A procedure that suggests itself immediately is to shorten the guide wavelength to a value below the allowed spacing limit. The methods that have been used to do this are described here because they have been generally unsatisfactory. The simplest technique is to fill the guide with dielectric and thus reduce the guide wavelength. However, the use of dielectrics gives rise to a number of problems: the loss, particularly in long arrays, results in diminution of the gain; it is difficult to maintain proper contact between the guide walls and the dielectric, with the result that electrical breakdown tends to occur and with it reduction in the power-handling capacity of the array; and also of no small significance is the increase in the weight of the antenna. Another method that has been tried is that of using a corrugated line. With coaxial line the inner conductor is corrugated as shown in Fig. 9·46a, while with rectangular guide one of the broad faces is replaced by a corrugated wall as shown in Fig. 9·46b. The systems can be thought of as a transmission line loaded periodically with reactances. The wavelength in the loaded line[1] has been found to be given approximately for coaxial line by

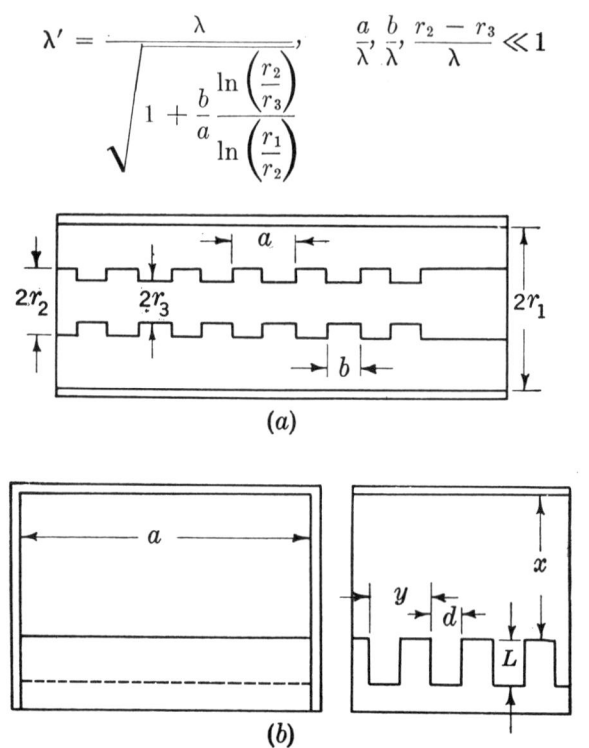

$$\lambda' = \frac{\lambda}{\sqrt{1 + \frac{b}{a}\frac{\ln\left(\frac{r_2}{r_3}\right)}{\ln\left(\frac{r_1}{r_2}\right)}}}, \qquad \frac{a}{\lambda}, \frac{b}{\lambda}, \frac{r_2 - r_3}{\lambda} \ll 1$$

FIG. 9·46.—Corrugated lines for shortening λ_g: (a) coaxial line; (b) waveguide.

and for rectangular guide by

$$\frac{1}{(\lambda'_g)^2} = \frac{1}{\lambda_g^2} + \frac{1}{\lambda_x^2},$$

where λ_x is a solution of

$$\frac{d}{y}\frac{\lambda_x}{\lambda_g}\frac{\tan\frac{2\pi L}{\lambda_g}}{\tanh\frac{2\pi x}{\lambda_x}} = 1.$$

These lines have proved impractical for the same general reasons as the dielectric-filled line: There is a significant increase in weight and great reduction in power-handling capacity, and in addition the corrugated sections are difficult to manufacture. Some of the difficulties, however, are due to the high percentage reduction in wavelength that is being

[1] H. Goldstein, "The Theory of Corrugated Transmission Lines and Waveguides," RL Report No. 494, Apr. 3, 1944.

effected. The corrugated line has been used with more success in other antenna designs where only a small wavelength reduction was attempted. The most successful technique that has been developed is in the design of radiators whose phase can be shifted 180° by simple structural changes; the elements can then be spaced at intervals of $\lambda_g/2$ and brought into phase by the structural phase reversal. Since the guide wavelength is generally in the range $\lambda \leq \lambda_g \leq 1.5\lambda$, this spacing is acceptable. The procedure is also satisfactory from the point of view of the loaded-line analysis. If the elements are pure series or shunt elements, the propagation constant is unaffected by the loading and a uniformly illuminated array results. The phase reversal does not alter the impedance presented by the radiator to the line. A brief summary of the phase-reversal techniques for the various types of elements discussed previously is given below:

1. *Slotted dipole*, Fig. 9·47a, the dipole is rotated through 180° about the coupling probe.
2. *Tridipole radiator*, same as for the slotted dipole.
3. *Shunt slots in broad face of rectangular guide*, Fig. 9·47b, the slots are placed on alternate sides of the axis of the guide.
4. *Shunt inclined slots on the narrow face*, Fig. 9·47c, the inclination of alternate slots is reversed.
5. *Probe-fed slots*, coupling probe is placed on opposite sides in alternate slots, or the orientation of the probe in the guide is reversed (see Figs. 9·23 and 9·24).
6. *Longitudinally polarized waveguide radiator*, the length of alternate slots differs by $\lambda'_g/2$.
7. *Transversely polarized waveguide radiator*, elements are staggered with respect to the guide axis just like the shunt slots in Fig. 9·47b.

9·20. Resonant Arrays.—Broadside arrays can be divided into two general classes: resonant and nonresonant arrays. The resonant type yields an accurately normal beam and is well matched at the design frequency; the impedance match, however, deteriorates rapidly with departure from the design frequency, and the array can be used only over a very narrow frequency band. An array of this type consists of a number of single series or shunt elements, spaced a guide half-wavelength apart on waveguide or coaxial line, with successive elements mechanically reversed in their feeding to give the phase reversal discussed in the preceding section. The resonant array is uniformly illuminated, since, as Eqs. (66) and (67) show, there is no attenuation in a line loaded with half-wavelength-spaced single series or shunt elements. The fact that uniform illumination is produced has been verified experimentally by measurements of the radiation directly in front of the array with a small exploring horn. Furthermore, the secondary patterns of these arrays are in agreement with the patterns of an array of uniformly excited elements. The uniform illumination is an advantageous feature where the

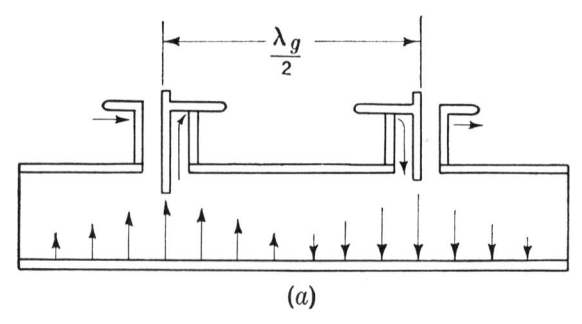

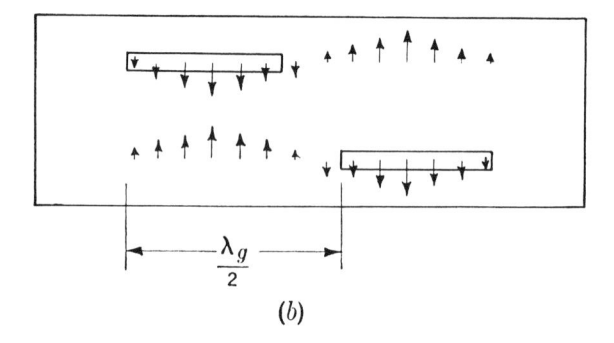

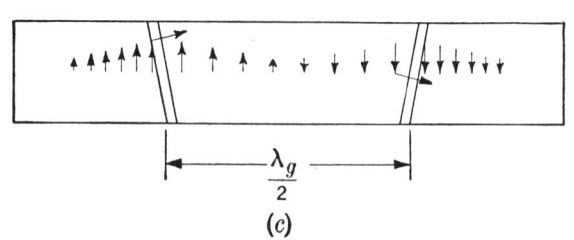

Fig. 9·47.—Phase-reversal technique: (a) phase reversal of dipoles; (b) phase reversal of longitudinal slots in the broad face of rectangular waveguide; (c) phase reversal of inclined slots in the narrow face of rectangular waveguide.

prime requirement is high gain; on the other hand the array is unsatisfactory when side lobes are the major consideration, since the first side lobe is over 4 per cent of the peak intensity.

The impedance match of the array is obtained by choosing the impedances of the elements properly and by adjusting a short-circuiting plunger at the end of the array. The short-circuit termination is a characteristic feature of broadside arrays; the reflected wave causes no difficulties such as would arise in end-fire arrays; for since the elements are half wavelengths apart, the radiation pattern due to the reflected wave is again a normal beam. The well-matched condition on the design frequency and the narrow bandwidth property of the array will be discussed for n series elements; the argument, phrased in terms of admittances, is similar for shunt elements. We assume that the impedance of each element has been adjusted to Z_0/n, where Z_0 is the characteristic impedance of the line. The line is terminated in a short circuit at a distance $\lambda_g/2$ from the last element. Since the spacing is $\lambda_g/2$, the entire array is equivalent to n elements in series. The input impedance is therefore $n(Z_0/n) = Z_0$; that is, the array is matched.

When the exciting frequency is not the design frequency, the elements are no longer exactly a half wavelength apart. Then the impedances do not add up to Z_0, and the array is not matched. The mismatch for frequencies off resonance cannot be calculated unless the frequency variation of the impedances of the elements is known. Their variation can often be neglected over the bands in which one is interested. A simple graphical analysis can then be carried out on an impedance chart. Let us take for example a 10-element $\lambda_g/2$-spaced array of series elements, each of resistance $0.1Z_0$, and plot on an impedance chart, starting from the terminal short circuit, the input impedance seen looking to the right from a point just to the left of each successive element. At the design frequency these points fall along the R/Z_0-axis as indicated on the line S in Fig. 9·48.

Suppose, for example, that the wavelength decreases by 1 per cent. The spacing between elements is now greater than $\lambda_g/2$. The short circuit now presents a small positive reactance in series with the tenth element. As one proceeds from element 10 to 9, the path traversed is greater than $\lambda_g/2$ so that the reactive component increases more than for the resonant wavelength. With each transformation to the next element there is an increase in reactive component due to the excess of the path over $\lambda_g/2$, with the result that the input impedance to the array as a whole has an appreciable reactive element. The transformation is shown as line S' in Fig. 9·48. The frequency sensitivity is evidently greater the longer the line. Common practice has been to limit the length of the array to 15 wavelengths, because longer arrays have been found to be too frequency-sensitive.

There are additional frequency-sensitive characteristics that should be noted. (1) Because the spacing is no longer equal to $\lambda_g/2$ for frequencies off resonance, attenuation sets in [cf. Eq. (67)] and the array is not uniformly illuminated. (2) The beam is no longer accurately normal to the array. These effects are generally less important than the impedance sensitivity because they are relatively insignificant for the narrow band over which the impedance match is acceptable.

The impedance characteristics of a resonant array can be improved by a process of "overloading" the line, i.e., using elements with impedances greater than Z_0/n. Of course, the array is then not matched, and

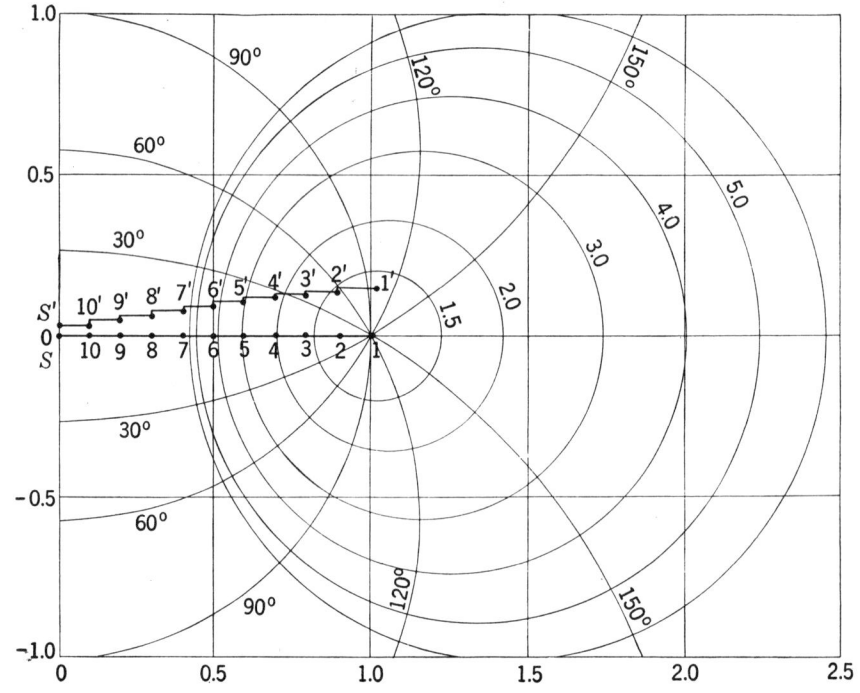

FIG. 9·48.—Input impedance of 10-element resonant array.

a matching transformer must be used. However, the combination of overloaded elements and transformer will generally have a broader band than the array matched by itself. The theory is best shown by example. Let us take again a 10-element $\lambda_g/2$-spaced array with series elements. Suppose now that the resistances of the elements are $0.2Z_0$ and, for definiteness, that the array is matched by a tuning screw. As before, the line 10, 9, 8, . . . in Fig. 9·49 represents the input impedances to successive elements for the frequency at which the spacing is $\lambda_g/2$, and 1 represents the input impedance to the array as a whole.

The array is matched at the design frequency by traveling clockwise on a constant VSWR circle to point P on the unity R/Z_0 line and by inserting a tuning screw there to transform to the point Q where $Z = Z_0$. If the wavelength is again assumed to decrease by 1 per cent, the impedance to the array is given by $1'$. This is quite close to point 1. If this impedance is transformed to the screw, it falls on the point P' and

Sec. 9·20]　　　　RESONANT ARRAYS　　　　325

the input impedance to the array is then Q', which is quite close to Z_0. Thus the array is still fairly well matched. In practice this method is quite successful, sometimes to the extent of doubling or tripling the bandwidth.

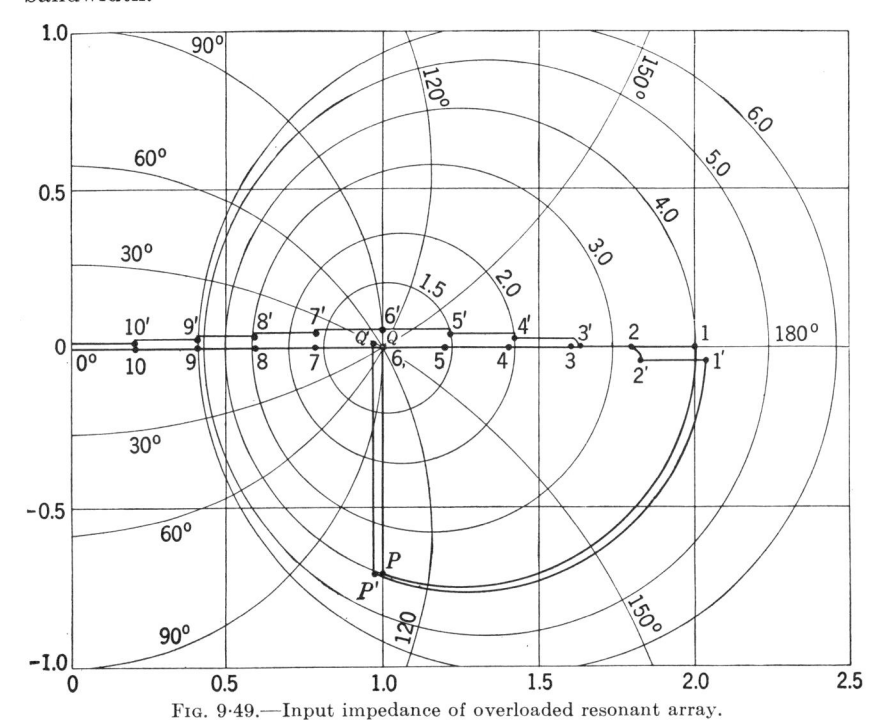

Fig. 9·49.—Input impedance of overloaded resonant array.

As an example of the performance of resonant arrays one antenna of this type will be discussed in detail.[1] Figure 9·50 shows an axially symmetrical array for transverse polarization designed for the 3-cm band.

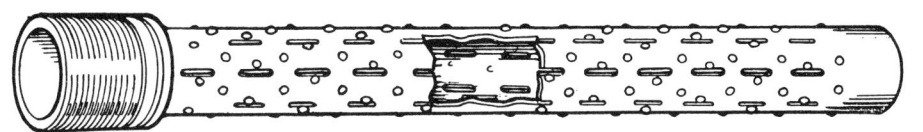

Fig. 9·50.—An axially symmetrical array for transverse polarization.

The elements consist of the axially symmetrical units of slot radiators shown in Fig. 9·30. The line is a circular waveguide having an outer diameter of 1¼ in. and supporting the TM_{01}-mode. It is fed by the converter shown in Fig. 9·33. The distance from slot to slot along the axis of the guide is $\lambda_g/2$. Phase reversal of the slots is achieved by putting

[1] H. J. Riblet, "Horizontally Polarized Non-directional Antennas," RL Report No. 489, Apr. 22, 1944.

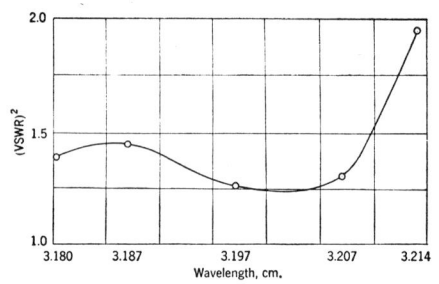

Fig. 9·51.—Frequency sensitivity of axially symmetrical transversely polarized array.

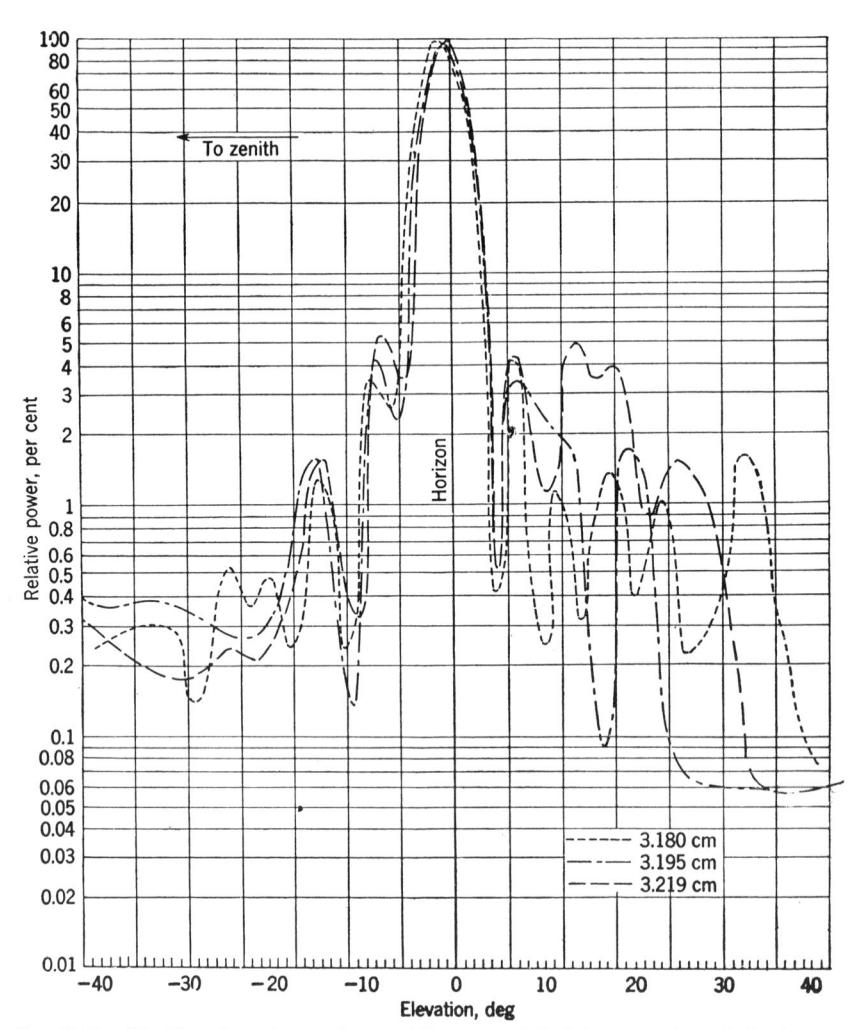

Fig. 9·52.—Meridional patterns of an axially symmetrical transversely polarized array.

the exciting screws on one side of a given slot and on the opposite side of the next slot. Between each bay of slots there is shown another set of screws. It was found that an array without these screws was excessively frequency-sensitive. The screws partially cancel the reflected waves from each bay of slots and hence increase the bandwidth of the array. The input VSWR to this array is shown in Figs. 9·51 and 9·52 shows the meridional pattern. The beamwidth is about 4.5°; the theoretical width calculated from Eq. (14) with $s = .870\lambda$ is 4.9°. The first side lobes are about $4\frac{1}{2}$ per cent, a value expected for uniform illumination. The asymmetry in the pattern is due to spurious reflections from objects surrounding the pattern-measuring equipment.

9·21. Beacon Antenna Systems. In beacon systems the responder (receiver) and transponder (transmitter) are ordinarily on two different frequencies. This necessitates two different antennas, one for transmitting and one for receiving. These two antennas must be so arranged that there is no "crosstalk" between them; i.e., very little energy from the transmitter is picked up directly by the receiver. Actually a little is always picked up, but in satisfactory antennas it is at least 40 db down.

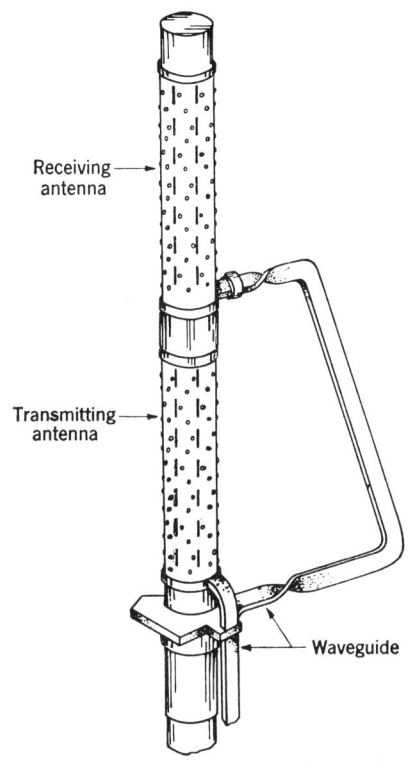

FIG. 9·53.—Beacon antenna with an external feed line.

The ordinary way of arranging a transmitter and receiver is to place one directly above and on the same axis as the other. The major problem is then to feed the upper antenna. This has been solved in two different ways. First, an external feed line can be used. Such an arrangement is shown in Fig. 9·53. The transmitter and receiver of this beacon antenna are resonant arrays of slot-type axially symmetrical radiators. The external feed naturally has an effect on the azimuth pattern of the bottom antenna. This effect is relatively small and not intolerable. It usually takes the form of superimposing a series of sharp maxima and minima on the ordinary azimuth pattern.

For some uses, particularly for airborne beacons, an external feed is so bulky and clumsy that an alternative design is used. It is applicable only when the antennas are built on coaxial line. In this design the inner conductor for the bottom antenna is made hollow and another conductor runs inside it, forming a coaxial feed line for the upper antenna. This "inner" inner conductor is then tapered to normal size as it enters the upper antenna.

Such double antenna systems have been built at both 3 and 10 cm, and almost all the coaxially-fed axially symmetrical radiators previously discussed have been used. Figure 9·54 shows such an antenna for 3

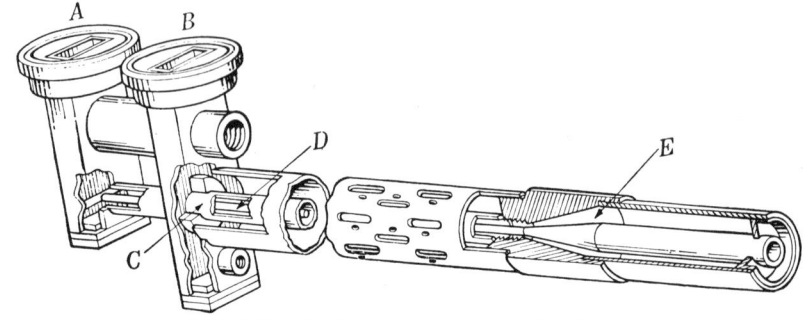

Fig. 9·54.—Double antenna system, $\lambda = 3$ cm.

cm.[1] The elements are axially symmetrical radiators consisting of slots on a coaxial line having a 1 in. OD. The rectangular waveguides A and B feed the coaxial lines for the top and bottom antennas respectively. C is the hollow tube that serves as inner conductor for the bottom antenna and as the outer conductor for the coaxial line feeding the upper antenna. D is the inner conductor of the latter coaxial line. E is a tapered section of coaxial line.

9·22. Nonresonant Arrays.—The nonresonant broadside array may consist of a number of elements spaced a little more or a little less than $\lambda_g/2$ apart. Consequently the beam is not normal to the array but at an angle given by Eq. (23). This may be a disadvantage in some applications. The advantage of this type of array is that its impedance match is generally good. Because the elements are not $\lambda_g/2$ apart, reflections from later elements tend to cancel reflections from earlier ones so that the array remains matched over a much wider band than the resonant array of the same length.

Although the nonresonant array eliminates the matching problem inherent in the resonant array, it presents an illumination problem that the resonant array does not have. The elements of the nonresonant array are not equally excited as in the resonant array; less power reaches the later elements; and if the elements are all alike, an exponential illumination results. Such illumination is undesirable because it reduces

[1] L. J. Eyges, "Omnidirectional Antennas for BUPX," RL Report No. 996, Jan. 17, 1946.

the gain. There are a number of ways to control the illumination and in particular to make the last elements radiate as much as the first. One method is to vary the elements themselves so that the first elements take out small fractions of the power incident on them while the later elements couple out larger and larger fractions. Thus if the elements are dipoles, successive dipoles can have deeper probes; if they are inclined slots on the narrow side of the guide, the inclination can be increased with distance along the array. Another useful and advantageous method permits the elements to be all identical. This consists in taper-

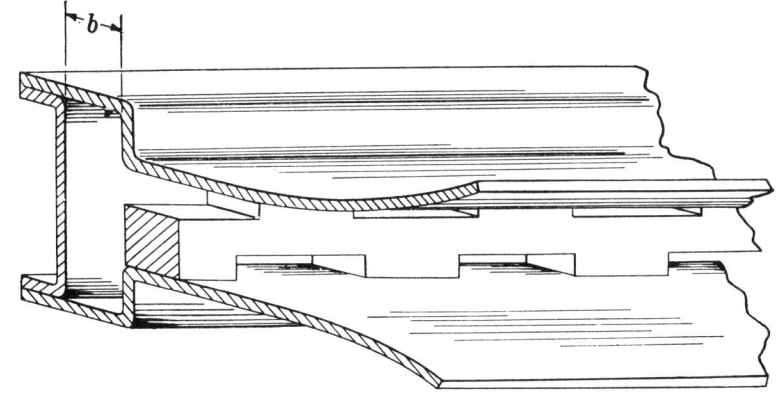

FIG. 9·55.—Transversely polarized array of waveguide radiators.

ing the guide in its narrow dimension so that it is smaller toward the end of the array. If there were no radiating elements, this would mean that the energy density would become larger toward the end of the array because a given amount of energy would be flowing through a smaller and smaller area. When there are radiating elements, the taper can be made to compensate for the loss of energy, thus maintaining a constant energy density in the guide.

In such an array there must be no appreciable wave reflected from the end. If the original wave radiates a beam at an angle θ to the normal, the reflected wave will radiate an undesirable lobe at an angle $-\theta$. To avoid this lobe the array is usually terminated in a matched load. This may be a dissipative load, and ordinarily arrays are designed so that about 5 per cent of the total power gets beyond the last element and is dissipated as heat. To avoid this waste of power, a matched load can be made of one of the radiators backed by a short circuit and matched with an iris. With this on the end of the array there is no reflected wave and all the energy is radiated.

Shown in Fig. 9·55 is a section of a nonresonant array, built for the 1-cm region. The elements are the transversely polarized waveguide radiators shown in Fig. 9·25b. The wall of the guide in which they are cut is a quarter wavelength thick. The elements are spaced nearly $\frac{\lambda_g}{2}$ apart; and for phase reversal, alternate elements are staggered with respect to the center line. The whole array feeds into parallel plates that flare out to the proper size for beam shaping in the other plane. The guide is tapered for uniform illumination. The coupling formula for the fraction of incident power abstracted by each element[1] is

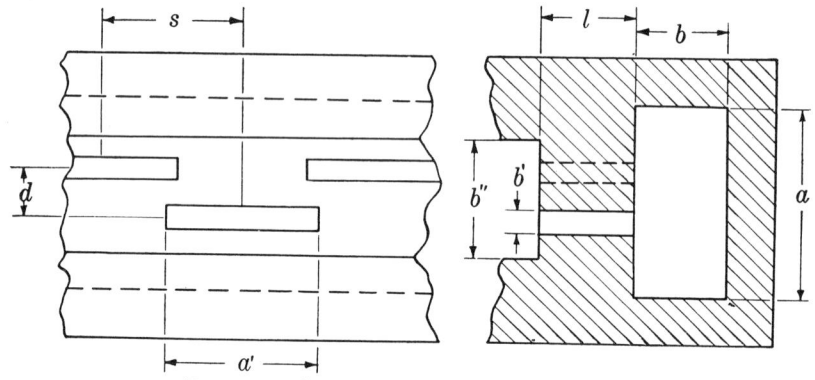

FIG. 9·56.—Geometrical parameters in Eq. (70).

$$\frac{P}{P_0} = \frac{128}{\pi^4} \frac{(a'b')^2}{sabb''} \left(\frac{\lambda_g'}{\lambda_0}\right)^2 \frac{\lambda_0}{\lambda_g} \left(\frac{\lambda_g}{2a}\right)^2 \left[\frac{\cos\left(\frac{\pi a'}{\lambda_g}\right)}{1 - \left(\frac{2a}{\lambda_g}\right)^2}\right]^2 \sin^2\left(\frac{\pi d}{2a}\right), \quad (70)$$

where P/P_0 is the average fraction of power abstracted per radiator and λ_g, λ_g', and λ_0 are respectively the guide wavelengths in the main guide, the branching guide, and the parallel plates. The geometrical parameters are defined in Fig. 9·56. The physical length l of the branching guides must be chosen so that its effective electrical length is $\lambda_g'/4$. An approximate formula for l is

$$l = \frac{\lambda_g'}{4} - \frac{2b'}{\pi}\left(1 + \ln\frac{b}{2b'}\right).$$

Equation (70) has not been checked directly, but arrays based on it have been built, and their performance was almost that expected.

Another type of nonresonant array has been designed that has a normal or closely normal beam, like the resonant array, but is much more broadband in impedance. Like the resonant array it has its element spaced at half-wavelength intervals. In order that the array be matched,

[1] For a derivation of this formula see W. Sichak and E. M. Purcell, "Cosec² Antennas with a Line Source and Shaped Cylindrical Reflector," RL Report No. 624, Nov. 3, 1944, pp. 7–13.

each element is matched to the guide; i.e., it is nonreflecting when terminated in the characteristic impedance of the guide.

There are a number of different ways to realize such a matched element. One obvious method is to match each element individually by a tuning screw or iris in front of it. This is always theoretically possible, but in practice it may be difficult; because the transmission-line equations are not valid close to the radiating elements, it is not always easy to find the proper size and position of the iris or screw to match the elements. This difficulty is avoided by the use of inclined displaced slots. It is possible to choose the length, displacement, and inclination of these slots so that they present an input conductance of unity, shunted by a susceptance.

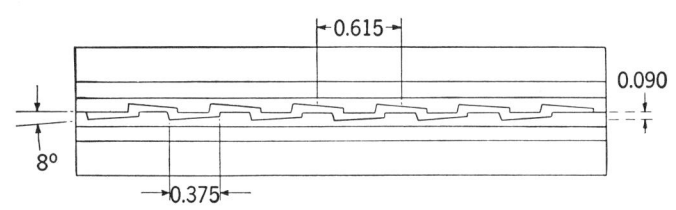

FIG. 9·57.—Section of nonresonant array with spacing $(\lambda_g/2)(\lambda = 1.25$ cm).

A tuning screw placed at the center of the slot will match it. Another type of element that is matched without tuning screws or irises has been built for the 1-cm band. It combines features of the waveguide radiators and inclined displaced slots in that it consists of asymmetrical inclined slots cut through the quarter-wavelength thickness of the broad wall of the waveguide. Successive slots are set on opposite sides of the center of the guide, and succeeding slots run together. The exact dimensions of these slots had to be determined experimentally. Figure 9·57 shows a sketch of these slots.[1]

It is obvious that an element which has an input impedance of Z_0 when terminated in Z_0 cannot be either a simple series or a simple shunt element; it must be represented by some T- or Π-network. Thus the waveguide is equivalent to a line loaded with T- or Π-networks, and (see Sec. 9·17) there is attenuation in such a line. For uniform illumination some device must be used to enable the later elements to abstract as much power as the first. This can be done by increasing the coupling of later elements or by tapering the guide.

9·23. Broadband Systems with Normal Beams.—The various arrays we have discussed thus far have one feature in common: The direction of the beam is a function of frequency. Whether the beam is normal for the design frequency as in the resonant and second type of nonresonant arrays or is not normal as in the first type of nonresonant array, the beam

---

[1] J. Steinberger and E. B. Chisholm, "Linear Array," RL Report No. 771, Jan. 31, 1946.

angle shifts when the frequency changes. This feature is disadvantageous for many applications. This section treats systems of arrays that have the two properties of constant beam angle and broadband impedance match.

The one feature common to all the array systems discussed in this section that causes the beam to remain normal over a band is that they are excited in the center. Such a system can be considered as two end-fed component arrays. These two arrays are arranged so that at the design frequency their component patterns add up exactly to give a resultant normal pattern. When the frequency changes, the beams from the individual arrays move in opposite directions; the resultant beam is still normal to the array. Of course, the resultant beam broadens somewhat and, if the frequency changes excessively, begins to split, but it remains normal to the array system. Three different arrays of this type have been built; the differences among them lie in the methods of obtaining a broadband impedance match.

One consists of two nonresonant arrays each with a beam at an angle θ to the array.[1] The two arrays are arranged in a V of angle $180° - 2\theta$, and the power is applied at the vertex of the V. Generally a parallel-plate waveguide is placed in front of the array to give a satisfactory pattern in the other plane. The impedance properties of such an array are very similar to those for a single nonresonant array, and it remains matched over a broad frequency band. The main disadvantage of this array is that it is not linear. The V-shape and the flaps on the parallel-plate section make it clumsy and heavy.

The disadvantages of size and weight are eliminated in the second example of broadband array with a normal beam. This array consists also of two component arrays excited in the center, but these are of the second type of nonresonant array.[2] Since the beam of each component array is normal to it, the two arrays can be placed in a straight line. Thus, the major disadvantage of the clumsiness of the V-shape is overcome, but there is a new disadvantage in that the component arrays are more complicated.

The third example of array, like the other two, consists of two components excited in the center. In this array, each component is a resonant array.[3] The broadband impedance match is obtained by displacing one array with respect to the other until the reflections from the components cancel each other. Such a system is illustrated in Fig. 9·58 and is made of two arrays, I and II, with identical spacing and phasing;

---

[1] J. R. Risser et al., "Linear Array for Use in the AN/APS-23 Antenna," RL Report No. 973, Mar. 19, 1946, pp. 1–7.

[2] J. Steinberger and E. B. Chisholm, "Linear Array," RL Report No. 771, Jan. 31, 1946.

[3] Risser et al., op. cit., pp. 7–13.

but array I is placed a distance Δ ahead of the other, and the distance x to the first element of array I is different from d, the corresponding distance for array II. There are two requirements for a satisfactory pattern and a broadband impedance match. First, for complete cancellation of the reflected waves at any point P in the main guide, the length from P to the first element of array I must be $\lambda_g/4$ longer than the corresponding distance for array II; then the reflected waves from the two

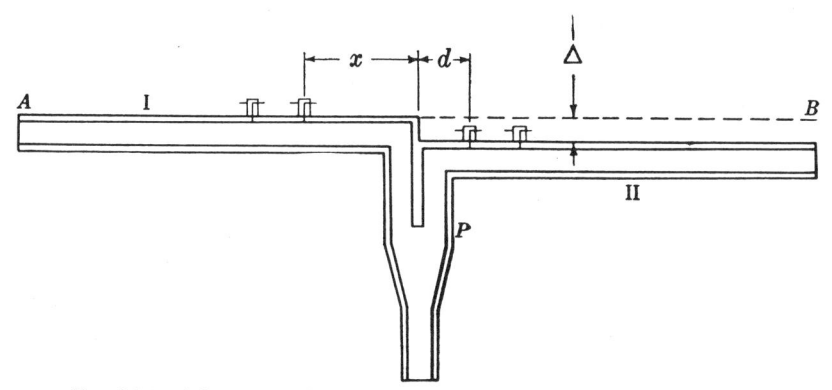

FIG. 9·58.—Schematic of broadband normal-firing resonant array system.

arrays will be just a half wavelength out of phase and will cancel. The condition for this $\lambda_g/4$ path difference is

$$x - d + \Delta = \frac{\lambda_g}{4}. \tag{71}$$

For a satisfactory radiation pattern there is another condition. The line AB must be a line of constant phase. For generality suppose that array II feeds into some sort of parallel-plate system in which the wavelength λ' is not necessarily the free-space wavelength. The condition for equiphase along AB is then

$$\frac{x + \Delta}{\lambda_g} = \frac{d}{\lambda_g} + \frac{\Delta}{\lambda'}. \tag{72}$$

A simultaneous solution of Eqs. (71) and (72) is $\Delta = \lambda'/4$ and

$$x - d = \frac{(\lambda_g - \lambda')}{4}.$$

If $\lambda' = \lambda_g$, then $x = d$.

CHAPTER 10

WAVEGUIDE AND HORN FEEDS

By J. R. Risser

10·1. Radiation from Waveguide of Arbitrary Cross Section.—The problem of radiation from the open end of a waveguide could be discussed in principle from several points of view. Rigorously, the radiation can be considered to arise from the current distribution on the inside walls of the guide, which is just the current distribution associated with the fields propagated in the interior of the guide, together with the currents flowing from the open end out upon the exterior guide surface. Were it not for difficulties in the analysis, this current distribution and the radiation field at an external point could be calculated. This has, however, not yet been accomplished. On the other hand, the approximate methods of diffraction theory developed in Secs. 5·11 and 5·12 have been applied to the problem with some degree of success.[1] The guide opening is presumed to act like a hole or aperture in an infinite screen, the transverse fields in the aperture being assumed to be identical with those in a parallel cross section inside the guide. The vector Huygens principle is applied to obtain the radiation field from the aperture field distribution as discussed in Secs. 5·11 and 5·12.

In all important practical cases the guide allows propagation of only one mode, called the dominant mode. Over a cross section inside the guide sufficiently far from the aperture, any component of the field is the vector sum of the components associated with incident and reflected waves of the dominant mode. In the aperture, however, additional higher-mode fields exist locally, excited by the discontinuity in the guide. It is not possible to determine the details of the higher-mode field distribution empirically; they can be obtained only from a rigorous solution of the boundary problem. The contribution of the higher-mode fields are neglected in the approximate diffraction theory used in this chapter. This is one source of inaccuracy in the method.

The effects of the reflected dominant mode wave can, however, be taken into account. They are expressible in terms of a reflection coefficient Γ which can be determined empirically by standing-wave measurements in the guide. The reflection coefficient Γ is the ratio $(E_t)_r/(E_t)_i$ of the transverse components of the reflected and incident electric field

[1] L. J. Chu, "Calculation of the Radiation Properties of Hollow Pipes and Horns," *Jour. Applied Phys.*, **11**, 603–610 (1940).

vectors; it varies in phase but not in magnitude along the guide. When extrapolated to the plane of the aperture, Γ can be interpreted in terms of an equivalent circuit admittance η for the aperture by the relation

$$\Gamma = \frac{1-\eta}{1+\eta}; \tag{1}$$

the admittance η is normalized to the characteristic wave admittance of the guide. The characteristics of η and Γ will be discussed later (Sec. 10·11). It is assumed here that Γ is a known quantity. The total transverse electric field $\mathbf{E}_t$ of the dominant mode in the aperture is then given by

$$\mathbf{E}_t = (1 + \Gamma)(\mathbf{E}_t)_i, \tag{2}$$

where $(\mathbf{E}_t)_i$ is the transverse electric field of the incident dominant-mode wave. The transverse magnetic field $\mathbf{H}_t$ can be obtained from $\mathbf{E}_t$ using Eqs. (7·33b) and (7·33c):

$$\begin{aligned} (\mathbf{H}_t)_i &= t[\mathbf{i}_z \times (\mathbf{E}_t)_i], \\ (\mathbf{H}_t)_r &= -t[\mathbf{i}_z \times (\mathbf{E}_t)_r], \end{aligned} \tag{3}$$

where

$$\begin{aligned} t &= \frac{\beta_{mn}}{\omega\mu} \quad \text{for } TE\text{-modes} \\ &= \frac{\omega\epsilon}{\beta_{mn}} \quad \text{for } TM\text{-modes.} \end{aligned}$$

Therefore $\mathbf{H}_t$ can be written

$$\begin{aligned} \mathbf{H}_t &= t(1 - \Gamma)[\mathbf{i}_z \times (\mathbf{E}_t)_i] \\ &= t\left(\frac{1-\Gamma}{1+\Gamma}\right)[\mathbf{i}_z \times \mathbf{E}_t]. \end{aligned} \tag{4}$$

The relation between the electric and magnetic fields over the aperture is thus of the form of Eq. (5·104) with the constant $\alpha = t(1-\Gamma)/(1+\Gamma)$. It should be kept in mind that the value of Γ is not altered by the insertion of a matching transformer in the guide because the reflected wave still exists in the region between the transformer and the aperture.

To calculate the radiation field at a point P outside the pipe, we surround P by a closed surface containing the aperture. This surface consists of the aperture, the exterior surface of the guide and the sphere at infinity. The vector Huygens principle is applied to this surface. As in other diffraction problems the sphere at infinity contributes nothing. Over the exterior surface of the guide the electric field is necessarily normal to the surface, and therefore $\mathbf{E}_t$ is zero. There is, however, a tangential component of the magnetic field associated with currents originating at the aperture. As in the case of the higher modes in the aperture, inability to solve the boundary problem at the end of the waveguide means that these currents are unknown, and $\mathbf{H}_t$ is assumed to be zero on the guide surface. This is a second source of error in the method.

The effect of neglecting the higher-mode fields in the aperture and the tangential component of the magnetic field on the outside surface of the guide depends on the dimensions of the aperture as measured in wavelengths. It is reasonable to assume that both factors contribute a smaller fraction of the total radiation field as the aperture dimensions increase. It is, in fact, the case that the calculated radiation field is in increasingly better agreement with experiment as the aperture dimensions increase, so that the limitations of the theory are apparent principally for small apertures. Unfortunately the dimensions of waveguide actually used are fractions of a wavelength. A more rigorous treatment of the problem would be desirable.

By neglecting the higher modes and the current distribution over the exterior surface of the waveguide, the problem is reduced to a simple aperture problem. The radiation field is calculated by means of Eqs. (5·110) and (5·110a). The transverse electric field $\mathbf{E}_r$ appearing in the latter is replaced in the present case by the resultant electric field $\mathbf{E}_t$ of the dominant mode over the aperture. The latter, in turn, is expressed in terms of the incident electric field by means of Eq. (2).

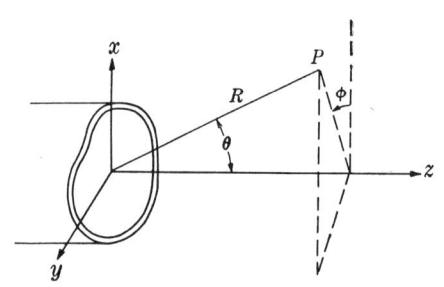

Fig. 10·1.—Coordinate system used in discussing radiation from open waveguide.

The coordinate system is shown in Fig. 10·1. Rectangular coordinates (x,y) are used in the aperture, taken to be the plane $z = 0$, and spherical coordinates R, θ and ϕ are used to locate the point P. From Eqs. (5·111a) and (5·111b) the components of $\mathbf{E}_p$ become

$$\left. \begin{aligned} E_R &= 0, \\ E_\theta &= \frac{jke^{-jkR}}{4\pi R}\left[1 + \iota\left(\frac{1+\Gamma}{1-\Gamma}\right)\left(\frac{\mu}{\epsilon}\right)^{1/2}\cos\theta\right](N_x\cos\phi + N_y\sin\phi), \\ E_\phi &= \frac{-jke^{-jkR}}{4\pi R}\left[\cos\theta + \iota\frac{1-\Gamma}{1+\Gamma}\left(\frac{\mu}{\epsilon}\right)^{1/2}\right](N_x\sin\phi - N_y\cos\phi) \end{aligned} \right\} \quad (5)$$

where $\mathbf{N}$ is the vector

$$\begin{aligned} \mathbf{N} &= \int_A \mathbf{E}_t e^{jk(x\sin\theta\cos\phi + y\sin\theta\sin\phi)}\, dS \\ &= (1 + \Gamma)\int_A (\mathbf{E}_t)_i e^{jk(x\sin\theta\cos\phi + y\sin\theta\sin\phi)}\, dS. \end{aligned} \quad (6)$$

10·2. Radiation from Circular Waveguide.—The radiation vector $\mathbf{N}$ of Eq. (6) can be computed for waveguide of circular cross section using the expression for the transverse field vector of the dominant mode given

Sec. 10·2]   RADIATION FROM CIRCULAR WAVEGUIDE   337

in Sec. 7·13. In computing **N** it is convenient to express $(\mathbf{E}_t)_i$, the incident wave field in the aperture, in rectangular components.

Case 1. *TE-waves.*—In this case the rectangular components of $(\mathbf{E}_t)_i$ are[1]

$$
\left.\begin{aligned}
E_x &= \frac{j\omega\mu\kappa_{mn}}{2}[J_{m-1}(\kappa_{mn}\rho)\sin(m-1)\psi + J_{m+1}(\kappa_{mn}\rho)\sin(m+1)\psi], \\
E_y &= \frac{j\omega\mu\kappa_{mn}}{2}[J_{m-1}(\kappa_{mn}\rho)\cos(m-1)\psi - J_{m+1}(\kappa_{mn}\rho)\cos(m+1)\psi].
\end{aligned}\right\} \quad (7)
$$

Writing $x = \rho\cos\psi$, $y = \rho\sin\psi$, the expressions to be evaluated become

$$
\left.\begin{aligned}
N_x &= \frac{j\omega\mu\kappa_{mn}(1+\Gamma)}{2}\int_0^a\int_0^{2\pi} e^{jk\rho\sin\theta\cos(\phi-\psi)}[J_{m-1}(\kappa_{mn}\rho)\sin(m-1)\psi \\
&\qquad + J_{m+1}(\kappa_{mn}\rho)\sin(m+1)\psi]\rho\,d\psi\,d\rho; \\
N_y &= \frac{j\omega\mu\kappa_{mn}(1+\Gamma)}{2}\int_0^a\int_0^{2\pi} e^{jk\rho\sin\theta\cos(\phi-\psi)}[J_{m-1}(\kappa_{mn}\rho)\cos(m-1)\psi \\
&\qquad - J_{m+1}(\kappa_{mn}\rho)\cos(m+1)\psi]\rho\,d\psi\,d\rho.
\end{aligned}\right\} \quad (8)
$$

These are evaluated with the help of the Bessel-Fourier series

$$e^{j\lambda\rho\cos(\phi-\psi)} = J_0(\lambda\rho) + \sum_{n=1}^{\infty} 2j^n J_n(\lambda\rho)\cos n(\phi-\psi) \quad (9)$$

and the Lommel integral formula

$$\int_0^x xJ_n(\alpha x)J_n(\beta x)\,dx = \frac{x}{\alpha^2-\beta^2}\left[J_n(\alpha x)\frac{d}{dx}J_n(\beta x) - J_n(\beta x)\frac{d}{dx}J_n(\alpha x)\right]. \quad (10)$$

Using these together with the recurrence relations and recalling that $J'_m(\kappa_{mn}a) = 0$, the field components are obtained as follows:

$$
\left.\begin{aligned}
E_\theta &= j^{m+1}\frac{m\omega\mu}{2R}\left[1 + \frac{\beta_{mn}}{k}\cos\theta + \Gamma\left(1 - \frac{\beta_{mn}}{k}\cos\theta\right)\right] \\
&\qquad J_m(\kappa_{mn}a)\,\frac{J_m(ka\sin\theta)}{\sin\theta}\sin m\phi\, e^{-jkR}, \\
E_\phi &= j^{m+1}\frac{ka\omega\mu}{2R}\left[\frac{\beta_{mn}}{k} + \cos\theta - \Gamma\left(\frac{\beta_{mn}}{k} - \cos\theta\right)\right] \\
&\qquad \frac{J_m(\kappa_{mn}a)J'_m(ka\sin\theta)}{1 - \left(\dfrac{k\sin\theta}{\kappa_{mn}}\right)^2}\cos m\phi\, e^{-jkR}.
\end{aligned}\right\} \quad (11)
$$

[1] The following recurrence relations are needed for this section:

$$J'_m(z) = \frac{m}{z}J_m(z) - J_{m+1}(z) = \frac{1}{2}[J_{m-1}(z) - J_{m+1}(z)]$$
$$= -\frac{m}{z}J_m(z) + J_{m-1}(z),$$
$$\frac{m}{z}J_m(z) = \frac{1}{2}[J_{m+1}(z) + J_{m-1}(z)].$$

Case 2. *TM-waves.*—Following the same procedure as above, the integrals to be evaluated are found to be the same. Specifically the integrals in the two cases are related as follows:

$$(N_x)_{TM} = -\frac{\beta_{mn}}{\omega\mu}(N_y)_{TE};$$
$$(N_y)_{TM} = \frac{\beta_{mn}}{\omega\mu}(N_y)_{TE}.$$
(12)

It will be recalled that the characteristic values of κ_{mn} for *TM*-waves are obtained from the roots of $J_m(\kappa_{mn}a) = 0$. On evaluating the field components, it is found that due to this condition, E_ϕ is zero and there is but one component:

$$E_\theta = -j^{m+1}\frac{ka\kappa_{mn}}{2R\sin\theta}\cos m\phi\left[\frac{\beta_{mn}}{k} + \cos\theta + \Gamma\left(\frac{\beta_{mn}}{k} - \cos\theta\right)\right]$$
$$\frac{J_m(ka\sin\theta)J'_m(\kappa_{mn}a)}{1 - \left(\frac{\kappa_{mn}}{k\sin\theta}\right)^2}e^{-jkR}. \quad (13)$$

The TE_{11}-mode which has the lowest cutoff frequency is the one most commonly used in circular-guide antenna feeds. The remainder of the discussion will be confined to this mode. On setting $m = 1$ into Eq. (7) it can be seen that the electric field over the aperture is symmetrical with respect to the yz-plane, which is thus the E-plane of the system. Figure 10·2 taken from Chu's paper shows the calculated E- and H-plane patterns as a function of aperture. The effect of the reflected wave in the pipe on the aperture distribution has been neglected[1] (i.e., Γ has been set equal to zero) in computing these patterns. Figure 10·3 shows a comparison between an observed pattern and the corresponding theoretical pattern. The agreement is quite good considering the factors neglected in the theory.

There are various measures of the sharpness of the beam. One criterion that has been used in the literature is the angle from zero to zero including the main beam. In the E-plane ($\phi = \pi/2$), E_θ is zero when $ka\sin\theta = 3.83$. The beam angles in the E- and H-planes are then

$$\vartheta_E = 2\sin^{-1}\left(\frac{3.83\lambda}{2\pi a}\right),$$
$$\vartheta_H = 2\sin^{-1}\left(\frac{5.33\lambda}{2\pi a}\right).$$
(14)

The beam is thus sharper in the E-plane than in the H-plane. Equation (14) is, of course, meaningless for the E-plane when $2a/\lambda < 1.22$ and for

[1] This is a good approximation for circular guide. For standard Radiation Laboratory waveguide ($2a = 0.75\lambda$) Γ is found to be small.

SEC. 10·2]   RADIATION FROM CIRCULAR WAVEGUIDE   339

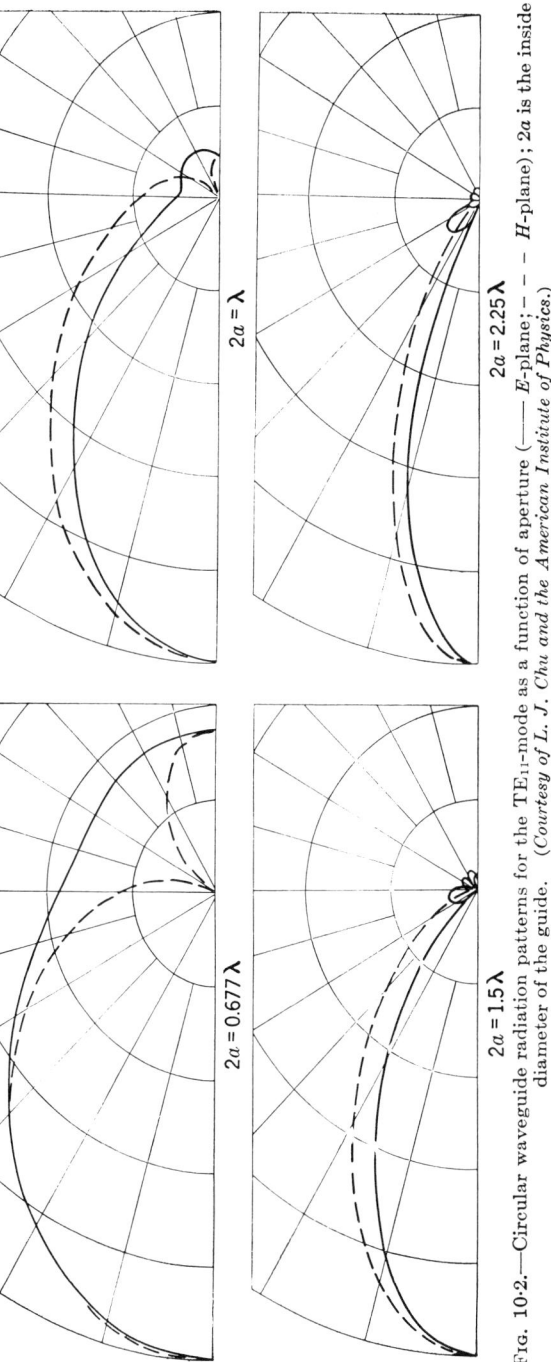

FIG. 10·2.—Circular waveguide radiation patterns for the TE$_{11}$-mode as a function of aperture (——— E-plane; – – – H-plane); $2a$ is the inside diameter of the guide. (*Courtesy of L. J. Chu and the American Institute of Physics.*)

the H-plane when $2a/\lambda < 1.7$. From a practical standpoint, more useful measures of the beam sharpness are the full angular widths between half-power points and tenth-power points. For values of $\lambda/a < 1$, the half-

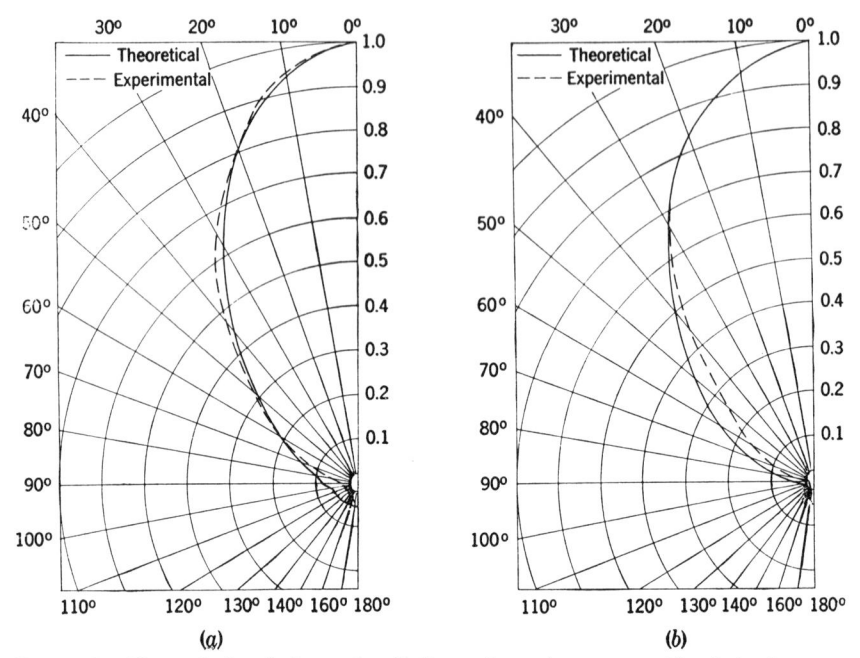

Fig. 10·3.—Theoretical and observed radiation patterns from waveguide of circular cross section; $\lambda = 3.2$ cm. (a) E-plane; (b) H-plane.

power and tenth-power widths in the principal planes are given in degrees by

$$\Theta_E = 14.7° \frac{\lambda}{a}; \qquad \Theta_H = 18.6° \frac{\lambda}{a}; \\ \Theta_E\left(\frac{1}{10}\right) = 25.0° \frac{\lambda}{a}; \qquad \Theta_H\left(\frac{1}{10}\right) = 32.3° \frac{\lambda}{a}. \quad (15)$$

Another characteristic of interest is the gain relative to an isotropic source. It is given by

$$G = 4\pi \frac{P(0,0)}{P_t}$$

where P_t is the total power radiated and $P(0,0)$ is the maximum power radiated per unit solid angle, which is in the direction $\theta = \phi = 0$. This power is

$$P(0,0) = \frac{1}{32}\left(\frac{\epsilon}{\mu}\right)^{1/2} k^2 a^2 \omega^2 \mu^2 \left|1 + \frac{\beta}{k} + \Gamma\left(1 - \frac{\beta}{k}\right)\right|^2 J_1^2(\kappa_{11}a). \quad (16)$$

To find the total power radiated, the Poynting vector $\frac{1}{2}\operatorname{Re}(\mathbf{E}_t \times \mathbf{H}_t^*)$ is integrated over the aperture. This is evaluated as follows:

$$P_t = \frac{\beta(1 - |\Gamma|^2)}{2\omega\mu} \int_0^a \int_0^{2\pi} (|E_x|^2 + |E_y|^2)\rho\, d\psi\, d\rho;$$

inserting the values of E_x and E_y from Eq. (7), we have for the TE_{11}-mode,

$$\begin{aligned}P_t &= \frac{\beta\omega\mu\kappa_{11}^2(1 - |\Gamma|^2)}{8} \int_0^a \int_0^{2\pi} [J_0^2(\kappa_{11}\rho) + J_2^2(\kappa_{11}\rho) \\ &\qquad - 2J_0(\kappa_{11}\rho)J_2(\kappa_{11}\rho)\cos 2\psi]\rho\, d\psi\, d\rho \\ &= \frac{\pi\beta\kappa_{11}^2\omega\mu(1 - |\Gamma|^2)}{4} \int_0^a [J_0^2(\kappa_{11}\rho) + J_2^2(\kappa_{11}\rho)]\rho\, d\rho.\end{aligned}$$

The last integral is evaluated by means of a Lommel formula[1] resulting in

$$P_t = \frac{\pi\beta\kappa_{11}^2 a^2 \omega\mu(1 - |\Gamma|^2)}{8} [J_0^2(\kappa_{11}a) + J_1^2(\kappa_{11}a) + J_2^2(\kappa_{11}a) \\ - J_1(\kappa_{11}a)J_3(\kappa_{11}a)].$$

Making use of the recurrence relations and the boundary condition $J_1'(\kappa_{11}a) = 0$, we obtain finally

$$P_t = \frac{\pi\beta\omega\mu(1 - |\Gamma|^2)}{4} (\kappa_{11}^2 a^2 - 1)J_1^2(\kappa_{11}a). \tag{17}$$

The gain is, therefore,

$$G = \frac{k^3 a^2 \left|1 + \dfrac{\beta}{k} + \Gamma\left(1 - \dfrac{\beta}{k}\right)\right|^2}{4.775\beta(1 - |\Gamma|^2)}, \tag{18}$$

where the value of $\kappa_{11}a = 1.841$ has been inserted. For the region far enough away from cutoff, $\Gamma \approx 0$, $\beta/k \approx 1$, the gain is approximately

$$G \approx 10.5 \left(\frac{\text{area of aperture}}{\lambda^2}\right). \tag{19}$$

10·3. Radiation from Rectangular Guide.[2]—The tangential field components of the dominant mode in the aperture of rectangular guide are obtained from Eq. (7·74) or (7·79) by placing z equal to zero. Then, in the same manner as for circular guide, the radiation vector **N** is calculated from Eq. (6).

---

[1] G. N. Watson, *Bessel Functions*, 2d ed., Macmillan, New York, 1945, p. 135, Eq. (11).

[2] L. J. Chu, *Jour. Applied Phys.*, **11**, 603–610 (1940).

Case 1. *TE-modes.*—The components N_x and N_y are

$$N_x = j\frac{n\pi\omega\mu(1+\Gamma)}{\kappa_{mn}^2 b}\int_0^a \cos\left(\frac{m\pi x}{a}\right) e^{jkx\sin\theta\cos\phi}\,dx$$

$$\int_0^b \sin\frac{n\pi y}{b}\, e^{jky\sin\theta\sin\phi}\,dy = \frac{n^2\pi^2\omega\mu(1+\Gamma)k\sin\theta\cos\phi}{\kappa_{mn}^2 b^2}$$

$$\times \left[\frac{1 - e^{j(ka\sin\theta\cos\phi + m\pi)}}{k^2\sin^2\theta\cos^2\phi - \dfrac{m^2\pi^2}{a^2}}\right]\left[\frac{1 - e^{j(kb\sin\theta\sin\phi + n\pi)}}{k^2\sin^2\theta\sin^2\phi - \dfrac{n^2\pi^2}{b^2}}\right],$$

$$N_y = -j\frac{m\pi\omega\mu(1+\Gamma)}{\kappa_{mn}^2 a}\int_0^a \sin\left(\frac{m\pi x}{a}\right) e^{jkx\sin\theta\cos\phi}\,dx$$

$$\int_0^b \cos\left(\frac{n\pi y}{b}\right) e^{jky\sin\theta\sin\phi}\,dy = -\frac{m^2\pi^2\omega\mu(1+\Gamma)k\sin\theta\sin\phi}{k_{mn}^2 a^2}$$

$$\left[\frac{1 - e^{j(ka\sin\theta\cos\phi + m\pi)}}{k^2\sin^2\theta\cos^2\phi - \dfrac{m^2\pi^2}{a^2}}\right]\left[\frac{1 - e^{j(kb\sin\theta\sin\phi + n\pi)}}{k^2\sin^2\theta\sin^2\phi - \dfrac{n^2\pi^2}{b^2}}\right].$$

The electric-field components of the radiation field are then

$$E_\theta = -\left(\frac{\mu}{\epsilon}\right)^{1/2}\frac{(\pi ab)^2\sin\theta}{2\lambda^3 R k_{mn}^2}\left[1 + \frac{\beta_{mn}}{k}\cos\theta + \Gamma\left(1 - \frac{\beta_{mn}}{k}\cos\theta\right)\right]$$

$$\left[\left(\frac{m\pi}{a}\sin\phi\right)^2 - \left(\frac{n\pi}{b}\cos\phi\right)^2\right]\Psi_{mn}(\theta,\phi),$$

$$E_\phi = -\left(\frac{\mu}{\epsilon}\right)^{1/2}\frac{(\pi ab)^2\sin\theta\sin\phi\cos\phi}{2\lambda^3 R}$$

$$\left[\cos\theta + \frac{\beta_{mn}}{k} + \Gamma\left(\cos\theta - \frac{\beta_{mn}}{k}\right)\right]\Psi_{mn}(\theta,\phi), \quad (20)$$

$$\Psi_{mn}(\theta,\phi)$$

$$= \left[\frac{\sin\left(\dfrac{\pi a}{\lambda}\sin\theta\cos\phi + \dfrac{m\pi}{2}\right)}{\left(\dfrac{\pi a}{\lambda}\sin\theta\cos\phi\right)^2 - \left(\dfrac{m\pi}{2}\right)^2}\right]\left[\frac{\sin\left(\dfrac{\pi b}{\lambda}\sin\theta\sin\phi + \dfrac{n\pi}{2}\right)}{\left(\dfrac{\pi b}{\lambda}\sin\theta\sin\phi\right)^2 - \left(\dfrac{n\pi}{2}\right)^2}\right]$$

$$e^{-j\left[kR - \frac{\pi}{\lambda}\sin\theta(a\cos\phi + b\sin\phi) - (m+n+1)\frac{\pi}{2}\right]}.$$

Case 2. *TM-modes.*—The components N_x and N_y are related to those of the *TE*-modes by

$$(N_x)_{TM} = -\frac{mb\beta_{mn}}{na\mu\omega}(N_x)_{TE},$$

$$(N_y)_{TM} = \frac{na\beta_{mn}}{mb\mu\omega}(N_y)_{TE}.$$

As in the case of the *TM*-modes in a circular guide, the radiation field is found to have only one component:

$$E_\theta = \frac{mn\beta_{mn}\pi^3 ab}{4\lambda^2 R k_{mn}^2} \sin\theta \left[1 + \frac{k}{\beta_{mn}}\cos\theta + \Gamma\left(1 - \frac{k}{\beta_{mn}}\cos\theta\right)\right]\Psi_{mn}(\theta,\phi), \quad (21)$$

while $E_\phi = 0$.

The TE-mode, $m = 1$, $n = 0$, is of special interest. In this case the radiation field reduces to

$$\left. \begin{aligned} E_\theta &= -\left(\frac{\mu}{\epsilon}\right)^{1/2} \frac{\pi a^2 b}{2\lambda^2 R} \sin\phi \left[1 + \frac{\beta_{10}}{k}\cos\theta + \Gamma\left(1 - \frac{\beta_{10}}{k}\cos\theta\right)\right] \\ &\quad \left[\frac{\cos\left(\frac{\pi a}{\lambda}\sin\theta\cos\phi\right)}{\left(\frac{\pi a}{\lambda}\sin\theta\cos\phi\right)^2 - \left(\frac{\pi}{2}\right)^2}\right] \left[\frac{\sin\left(\frac{\pi b}{\lambda}\sin\theta\sin\phi\right)}{\left(\frac{\pi b}{\lambda}\sin\theta\sin\phi\right)}\right] \\ &\quad e^{-j\left[kR - \frac{\pi}{\lambda}\sin\theta(a\cos\phi + b\sin\phi)\right]} \\ E_\phi &= -\left(\frac{\mu}{\epsilon}\right)^{1/2} \frac{\pi a^2 b}{2\lambda^2 R} \cos\phi \left[\cos\phi + \frac{\beta_{10}}{k} + \Gamma\left(\cos\theta - \frac{\beta_{10}}{k}\right)\right] \\ &\quad \left[\frac{\cos\left(\frac{\pi a}{\lambda}\sin\theta\cos\phi\right)}{\left(\frac{\pi a}{\lambda}\sin\theta\cos\phi\right)^2 - \left(\frac{\pi}{2}\right)^2}\right] \left[\frac{\sin\left(\frac{\pi b}{\lambda}\sin\theta\sin\phi\right)}{\frac{\pi b}{\lambda}\sin\theta\sin\phi}\right] \\ &\quad e^{-j\left[kR - \frac{\pi}{\lambda}\sin\theta(a\cos\phi + b\sin\phi)\right]} \end{aligned} \right\} \quad (22)$$

where κ_{10} has been replaced by π/a. The phase factor

$$kR - \left(\frac{\pi}{\lambda}\right)\sin\theta\,(a\cos\phi + b\sin\phi)$$

can be simplified. It will be recalled that in deriving the field expressions the origin was taken at a corner of the guide. It is easily found that if the origin is shifted to the center of the aperture, the phase factor transforms into kR, R now being measured from the new origin. In the case of large apertures $\Gamma \approx 0$, so that the space factor is, therefore, real and the guide is a directive point-source feed, the center of feed being the center of the aperture. In small apertures where Γ is complex, there is no exact center of feed; the guide is only approximately a point source from the point of view of the equiphase surfaces of the radiation pattern.

The electric field over the aperture is polarized in the Y-direction so that the yz-plane is the E-plane of the system while the xz-plane is the H-plane. The patterns in these two principal planes are

a. *E*-plane, $\phi = \pi/2$

$$E_\theta = 2\left(\frac{\mu}{\epsilon}\right)^{1/2} \frac{a^2 b}{\pi \lambda^2 R}\left[1 + \frac{\beta_{10}}{k}\cos\theta\right.$$

$$\left. + \Gamma\left(1 - \frac{\beta_{10}}{k}\cos\theta\right)\right]\frac{\sin\left(\frac{\pi b}{\lambda}\sin\theta\right)}{\frac{\pi b}{\lambda}\sin\theta} e^{-jkR}. \quad (23a)$$

b. *H*-plane, $\phi = 0$

$$E_\phi = -\left(\frac{\mu}{\epsilon}\right)^{1/2} \frac{\pi a^2 b}{2\lambda^2 R}\left[\cos\theta + \frac{\beta_{10}}{k}\right.$$

$$\left. + \Gamma\left(\cos\theta - \frac{\beta_{10}}{k}\right)\right]\frac{\cos\left(\frac{\pi a}{\lambda}\sin\theta\right)}{\left(\frac{\pi a}{\lambda}\sin\theta\right)^2 - \frac{\pi^2}{4}} e^{-jkR}. \quad (23b)$$

It is observed that the predominant factors in the patterns

$$\sin\left(\frac{\pi b}{\lambda}\sin\theta\right)\bigg/\frac{\pi b}{\lambda}\sin\theta \quad \text{and} \quad \cos\left(\frac{\pi a}{\lambda}\sin\theta\right)\bigg/\left[\left(\frac{\pi a}{\lambda}\sin\theta\right)^2 - \frac{\pi^2}{4}\right]$$

are determined by the dimensions of the apertures in the respective planes. It will be further observed that the *E*-plane pattern is essentially the pattern due to uniformly illuminated slit of width *b*. The pattern in the *H*-plane is essentially that due to a slit of width *a* over which the illumination is distributed sinusoidally as it is across the guide in the *x*-direction. This is illustrative of a fairly general characteristic that the patterns in the two principal planes are independent and are determined by the aperture dimension and the distribution of illumination across the aperture in the respective planes. The angular distances between the first zeros on either side of the peak are given by

$$\left.\begin{array}{l}\vartheta_E = 2\sin^{-1}\dfrac{\lambda}{b}, \\ \\ \vartheta_H = 2\sin^{-1}\dfrac{3\lambda}{2a}.\end{array}\right\} \quad (24)$$

Figure 10·4 is a plot of *E*-plane and *H*-plane patterns of 3.2-cm waveguide calculated from Eqs. (23a) and (23b), together with experimentally observed values. Since the guide dimensions are appreciably smaller than a wavelength ($a/\lambda = 0.71$; $b/\lambda = 0.32$), agreement would not be expected to be particularly good in view of the approximations in the theory. Better agreement would be expected with larger aperture dimensions, although from a practical standpoint limitations on size of aperture are imposed by the necessity of suppressing higher modes. The predictions of Eqs. (23a) and (23b) for large apertures are, however,

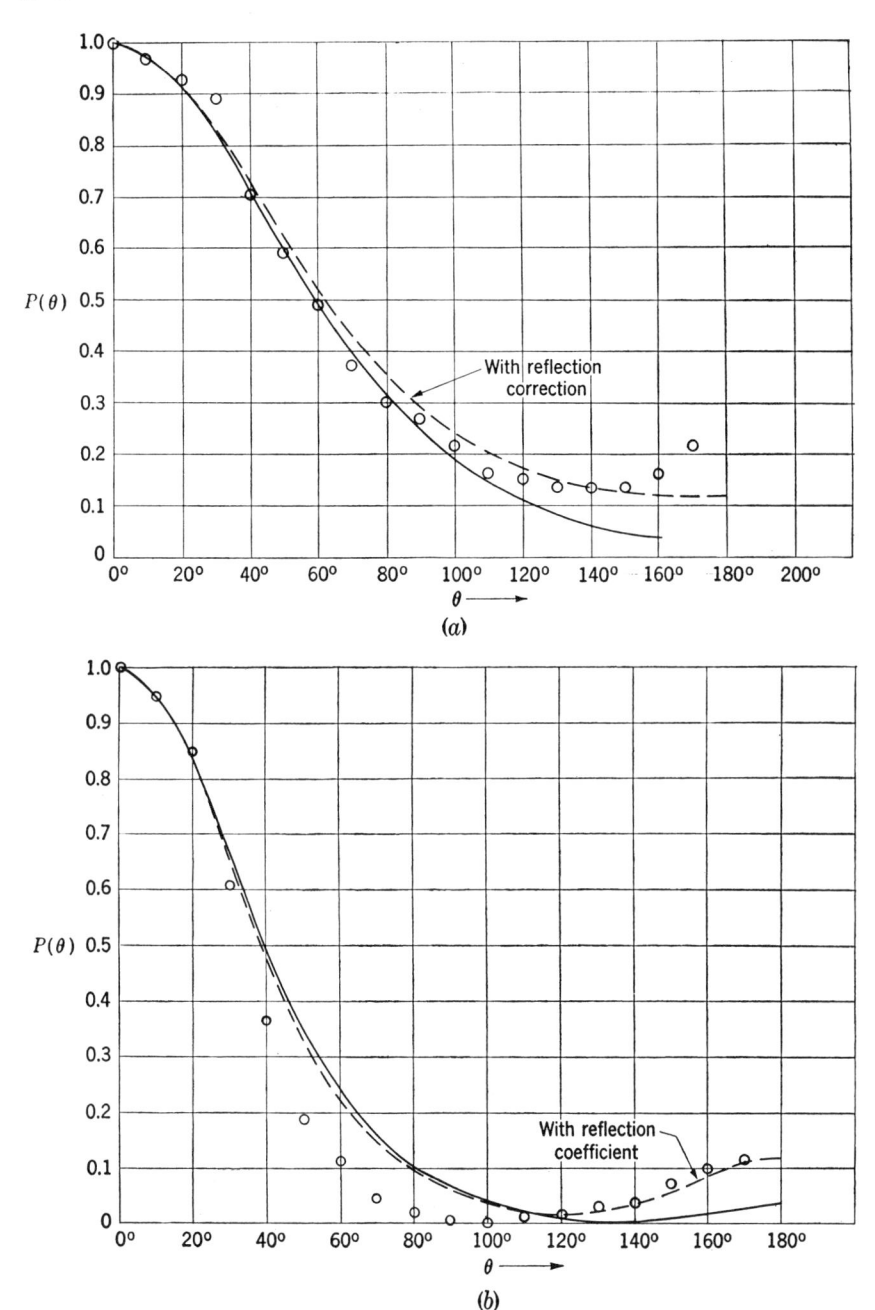

FIG. 10·4.—Theoretical and observed radiation patterns from rectangular waveguide; $\lambda = 3.2$ cm; $\dfrac{a}{\lambda} = 0.71$; $\dfrac{b}{\lambda} = 0.32$; ———— calculated patterns; ○ observed values; (a) E-plane; (b) H-plane.

of interest as a convenient means of predicting the radiation patterns of sectoral horns, with small flare angles. Figure 10·5 is a graph of the 3- and 10-db-widths of the E- and H-plane patterns as a function of the respective aperture dimensions. In computing the latter, Γ was taken equal to zero.

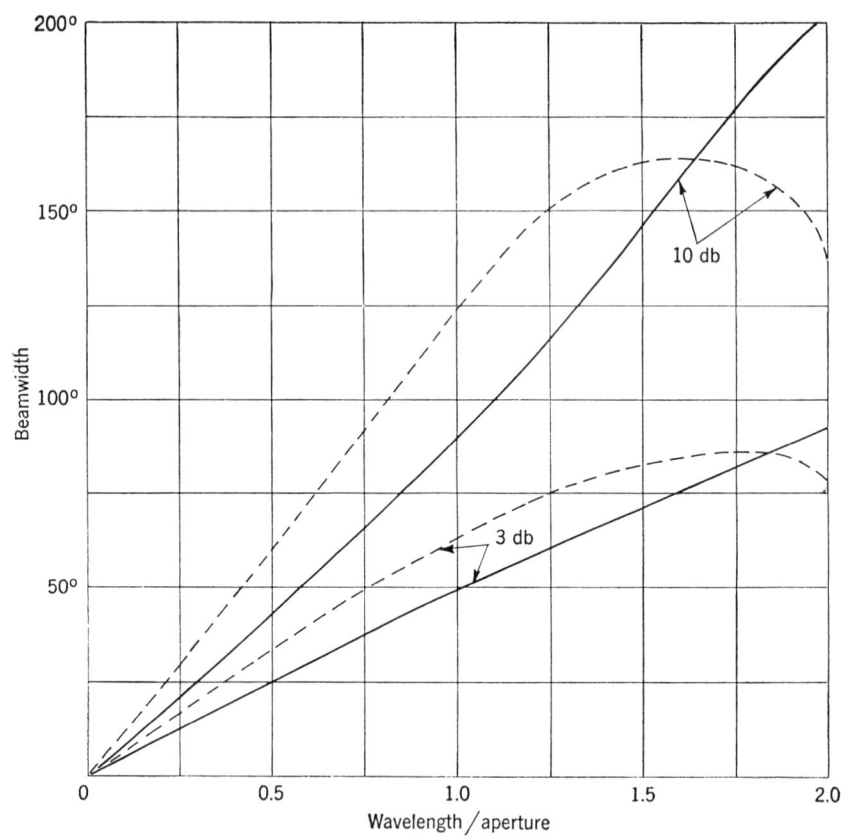

FIG. 10·5.—Relation between the aperture dimension and the 3-db and 10-db widths of the radiation pattern of rectangular waveguide; ——— E-plane; – – – H-plane.

Finally the gain relative to an isotropic point source can be calculated. The power radiated per unit solid angle in the peak direction, $\theta = \phi = 0$, is

$$P(0,0) = 2\left(\frac{\mu}{\epsilon}\right)^{\frac{1}{2}} \left(\frac{a^2 b}{\pi \lambda^2}\right)^2 \left|1 + \frac{\beta_{10}}{k} + \Gamma\left(1 - \frac{\beta_{10}}{k}\right)\right|^2.$$

The total power radiated is obtained as in the case of circular guide by integrating the Poynting vector $\frac{1}{2}$ Re $(\mathbf{E}_t \times \mathbf{H}_t^*)$ over the aperture. This integration is easy to carry through in the present case. We obtain

$$P_t = \frac{(1 - |\Gamma|^2)\omega\mu a^3 b \beta_{10}}{4\pi^2}. \tag{25}$$

The gain is, therefore,

$$G = 4\pi \frac{P(0,0)}{P_t} = \frac{8}{\pi(1 - |\Gamma|^2)} \frac{k}{\beta_{10}} \left|1 + \frac{\beta_{10}}{k} + \Gamma\left(1 - \frac{\beta_{10}}{k}\right)\right|^2 \frac{ab}{\lambda^2}. \quad (26)$$

For large apertures, $\beta/k \approx 1$, $\Gamma \approx 0$, the gain is approximately

$$G \approx 10.2 \left(\frac{\text{area of the aperture}}{\lambda^2}\right). \quad (27)$$

10·4. Waveguide Antenna Feeds.—Waveguide can be used satisfactorily as an antenna feed, but only for very restricted applications. It will be shown in a later chapter that the power radiated by the feed should be down approximately 10 db in the direction of the reflector edge (*cf.* Chap. 12). This requirement determines the reflector shape that can be used efficiently with waveguide feeds. For rectangular waveguide with $b/\lambda = 0.32$ and $a/\lambda = 0.71$, the reflector aperture should subtend an angle at the feed of approximately 180° in the electric plane and 120° in the magnetic plane (see Fig. 10·4). For circular waveguide with $2a/\lambda = 0.75$, these angles should be approximately 150° for the electric plane and 140° for the magnetic plane. While these figures are necessarily approximate because the 10-db specification has some arbitrariness, reflectors of markedly different shape cannot be used without sacrifice in gain or side lobes. In general, the reflector dimensions are determined by the application, and the feed aperture dimensions must be selected accordingly. Flaring the terminal region of the guide to form a simple rectangular or conical horn and placing beam-shaping obstacles in the aperture of the feed constitute the usual solutions to the problem.

Circular waveguide has found a more restricted application than rectangular guide as an antenna feed; in fact its use has been confined to conically scanning antennas. Since long lengths of circular guide are found unsatisfactory, a circular-guide feed is generally excited from rectangular guide through an intermediate tapered section. The feature of circular guide that makes it suitable for conically scanning antennas is that the terminal section can be rotated without distortion of the mode of propagation or rotation of the polarization of the radiated beam. If a circular-guide feed is placed a small distance from the axis of a paraboloidal reflector and is rotated about this axis, the peak of the beam from the paraboloid will describe a cone whose axis coincides with the paraboloid axis. The direction of polarization remains fixed in the course of the rotation. The greatest care must be taken not to deform the guide in bending, because deformations act as transformers converting plane to elliptical polarization.

A number of structures have been developed to enable waveguide to be used in rear feed systems. A "rear feed" is one that enters the paraboloid at or near the vertex from behind and provides a means of deflecting the direction of propagation of the radiation so as to illuminate the paraboloid. With circular guide this is done by a reflecting disk (Fig. 10·6). However, a study of the equiphase surfaces shows that such a feed does not have a point center of feed but behaves rather like a ring source. It is, therefore, not suited for use with a reflector having a point focus.

In the case of rectangular guide, a rear feed system can be obtained by making a U-shaped bend in the guide; this is feasible at short wavelengths where bending the guide is mechanically feasible and the added weight and feed shadow are not important factors. When a more compact rear feed on rectangular guide was needed, modifications such as the two- and four-dipole feeds (Secs. 8·10 and 8·11) or the double-slot feed were been used. The latter feed will be described in the next section.

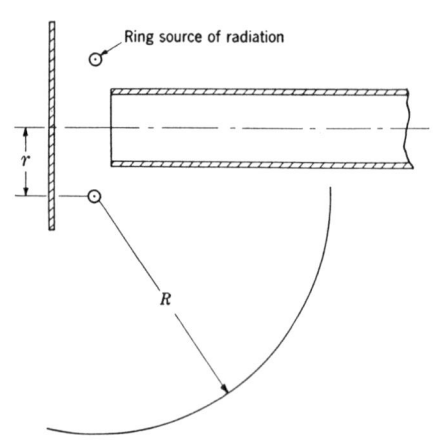

FIG. 10·6.—Circular waveguide and reflecting disk showing position of ring source of radiation.

10·5. The Double-slot Feed.— One form of this type of feed is shown in Fig. 10·7. Essentially the waveguide splits into two waveguide-like branches which turn back and have their open ends directed toward the paraboloid. One opening is above and the other is below the input waveguide, which is tapered to less than normal height to decrease the separation of the slots and consequently the directivity of the feed in the electric plane. As shown in the figure, the two branch paths are contained in a compact cylindrical head designed for ease of manufacture. Each branch consists of half the cylindrical cavity C and the waveguide-like slot S. The slots are pressurized by mica windows. The dimensions of the feed were worked out empirically to obtain good match and pattern over a 6 per cent band ($\Delta\lambda/\lambda_0 = \pm 3$ per cent) centered at 3.2 cm. Over this

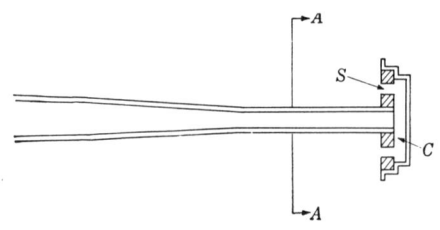

FIG. 10·7.—A double-slot feed.

band the VSWR is less than 1.3. The pattern is somewhat narrower in the electric plane than in the magnetic plane. This feed is useful where a compact straight rear feed is needed.

10·6. Electromagnetic Horns.—It appears at first sight that a radiation pattern of any desired directivity can be obtained from a waveguide by a suitable choice of its dimensions. However, if the dimensions are sufficiently large to allow free propagation of more than one mode, the serious problem of controlling the modes arises. It is difficult to excite a large-sized waveguide so that only a single mode is generated; if several modes are present, their relative phases at the aperture and hence the resultant field over the latter are a function of the length of the guide. The required large aperture with a single-mode-field excitation can be achieved by a gradual transition produced by flaring the terminal section of the waveguide to form an electromagnetic horn. Of course, a number of modes are excited in the throat of the horn at the junction between the latter and the waveguide. However, the throat serves as a filter device, allowing only a single mode to be propagated freely to the aperture. Each mode in the horn can be set into correspondence with a mode in the waveguide into which it passes as the flare angle of the horn is reduced to zero. The horn will not support free propagation of a particular mode until roughly the transverse dimensions of the horn exceed those of a waveguide which would support the given mode. Thus, unless the flare angle is too large, all but the dominant mode will be attenuated to a negligible amplitude in the throat region before free propagation in the horn space is possible.

The discussion in the following sections will be restricted to horns that are derived from a rectangular waveguide. Comparatively little is known about conical horns[1] derived from a circular waveguide, and they have found comparatively few applications in microwave antennas. Rectangular horns are treated in considerable detail in the literature.[2] The reader is referred to the original papers for a complete treatment of the modes in a rectangular horn and the analysis of the filter properties of the throat. If the horn is to serve as the terminal antenna element, there exist optimum relations between the horn length and flare angle for achieving maximum directivity; these relations are given in the sources referred to previously. Horns are used in microwave antennas primarily as a feed to illuminate a reflector or lens. In this case the important design considerations are the impedance characteristics and the efficient

---

[1] G. C. Southworth and A. P. King, *Proc. IRE*, **27**, 95 (1939); A. P. King, *Bell Laboratories Record*, **18**, 247 (1940).

[2] W. L. Barrow and L. J. Chu, *Proc. IRE*, **27**, 51 (1939); W. L. Barrow and F. D. Lewis, *Proc. IRE*, **27**, 41 (1939); L. J. Chu and W. L. Barrow, *Trans. AIEE*, **58**, 333 (1939). The design data are summarized by F. E. Terman, *Radio Engineers' Handbook*, McGraw-Hill, New York, 1943, pp. 824–837.

illumination of the optical system rather than maximum gain from the horn. Only the material not readily available in the literature will be presented in the following sections, with attention being centered on the microwave design problems.

Various types of horns are illustrated in Fig. 10·8. The horns shown in Fig. 10·8a and b are known as sectoral horns; they are flared in one plane only. The fields in the sectoral horns consist of cylindrical waves the axes of which coincide with the line of intersection of the planes containing the flared sides. The compound horn (Fig. 10·8c), allows variation of both aperture dimensions. An alternative procedure to that shown in the figure is to flare both sides of the horn directly from the junction with the waveguide to form a quasi-pyramidal structure. From the point of view of the impedance characteristics the former procedure is preferable.

10·7. Modes in E-plane Sectoral Horns.—The sectoral horns to be considered first are those in which the flare increases the aperture in the direction of the electric vector (Fig. 10·8a). They will be referred to as E-plane sectoral horns. The sectoral character of the space inside the flare and the cylindrical coordinate system (x,r,θ) appropriate to this space can be seen in Fig. 10·9a. The x-axis coincides with the line of intersection of the planes containing the flared sides; the planes of constant x are thus parallel to the unflared sides of the horn. The polar coordinates r and θ locate points in these planes. The unflared sides of the horns are in the planes $x = \pm a/2$. Propagation in the flare is along the radius vector, the wavefronts being coaxial cylindrical surfaces of constant r. The portion of the flare included between any two of these surfaces can be thought of as a length of sectoral guide. In particular, the horn flare is a section of sectoral guide whose length is $(r_2 - r_1)$, where the surfaces $r = r_1$ and $r = r_2$ locate the "throat" and "mouth" of the horn respectively. Maxwell's equations for the sectoral guide space may be written

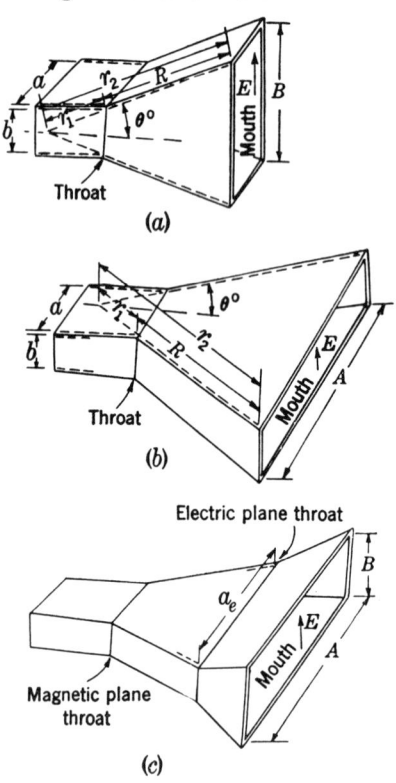

FIG. 10·8.—Horn feed types: (a) electric plane horn; (b) magnetic plane horn; (c) compound horn.

SEC. 10·7]   MODES IN E-PLANE SECTORAL HORNS   351

$$j\omega\epsilon E_r = \frac{1}{r}\frac{\partial H_x}{\partial \theta} - \frac{\partial H_\theta}{\partial x}, \tag{28a}$$

$$j\omega\epsilon E_\theta = \frac{\partial H_r}{\partial x} - \frac{\partial H_x}{\partial r}, \tag{28b}$$

$$j\omega\epsilon E_x = \frac{1}{r}\frac{\partial}{\partial r}(rH_\theta) - \frac{1}{r}\frac{\partial H_r}{\partial \theta}, \tag{28c}$$

$$-j\omega\mu H_r = \frac{1}{r}\frac{\partial E_x}{\partial \theta} - \frac{\partial E_\theta}{\partial x}, \tag{28d}$$

$$-j\omega\mu H_\theta = \frac{\partial E_r}{\partial x} - \frac{\partial E_x}{\partial r}, \tag{28e}$$

$$-j\omega\mu H_x = \frac{1}{r}\frac{\partial}{\partial r}(rE_\theta) - \frac{1}{r}\frac{\partial E_r}{\partial \theta}, \tag{28f}$$

$$\frac{1}{r}\frac{\partial}{\partial r}(rH_r) + \frac{1}{r}\frac{\partial H_\theta}{\partial \theta} + \frac{\partial H_x}{\partial x} = 0, \tag{28g}$$

$$\frac{1}{r}\frac{\partial}{\partial r}(rE_r) + \frac{1}{r}\frac{\partial E_\theta}{\partial \theta} + \frac{\partial E_x}{\partial x} = 0, \tag{28h}$$

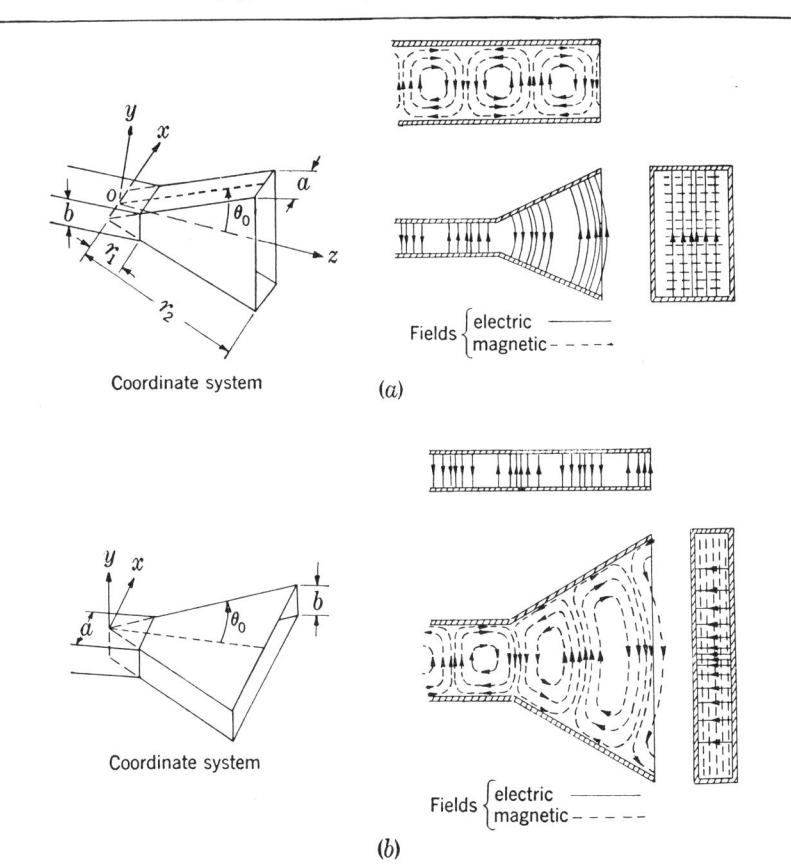

FIG. 10·9.—Coordinate system and lowest-mode field configuration in sectoral guide.

where ϵ and μ are the inductive capacities of the medium filling the sectoral guide.

The Dominant-mode Fields.—The waveguide feeding the horn is assumed to support free propagation of only the TE_{10}-mode which is then the exciting field impressed on the horn. The lowest sectoral guide mode, which is the analogue of the TE_{10}-mode in the uniform guide, will predominate, all the other modes being attenuated in the region of the throat. This mode is characterized by vanishing of all field components except E_θ, H_r, and H_x. Maps of the field lines in the cylindrical wavefronts are qualitatively the same as those in the plane wavefronts of the uniform guide. The electric lines are arcs normal to the flared sides of the guide. The electric field E_θ varies sinusoidally in the x-direction, vanishing at the parallel walls of the guide.

To derive expressions for the dominant-mode field components, the simplifications $E_r = E_x = H_\theta = 0$ are introduced into Maxwell's equations (28a) to (28h), which then become

$$\frac{\partial E_\theta}{\partial \theta} = \frac{\partial H_r}{\partial \theta} = \frac{\partial H_x}{\partial \theta} = 0, \tag{29a}$$

$$j\omega\epsilon E_\theta = \frac{\partial H_r}{\partial x} - \frac{\partial H_x}{\partial r}, \tag{29b}$$

$$j\omega\mu H_r = \frac{\partial E_\theta}{\partial x}, \tag{29c}$$

$$j\omega\mu H_x = -\frac{1}{r}\frac{\partial}{\partial r}(rE_\theta), \tag{29d}$$

$$\frac{1}{r}\frac{\partial}{\partial r}(rH_r) + \frac{\partial H_x}{\partial x} = 0. \tag{29e}$$

Equations (29c) and (29d) serve to express H_r and H_x in terms of the derivatives of E_θ. Substituting the expressions so obtained in Eq. (29b), the following equation is obtained for E_θ:

$$\frac{\partial^2 E_\theta}{\partial r^2} + \frac{1}{r}\frac{\partial E_\theta}{\partial r} + \frac{\partial^2 E_\theta}{\partial x^2} + \left(\omega^2\mu\epsilon - \frac{1}{r^2}\right)E_\theta = 0. \tag{30}$$

As was pointed out previously the electric field E_θ varies sinusoidally along the x-direction as in the case of the TE_{10}-mode of the uniform guide. We have then

$$E_\theta = \cos\left(\frac{\pi x}{a}\right) f(r), \tag{31}$$

where $f(r)$ is a function of r only. The expression for E_θ satisfies the boundary condition that $E_\theta = 0$ at $x = \pm a/2$. Substituting Eq. (31) into Eq. (30), we obtain the following differential equation for $f(r)$:

$$\frac{d^2 f}{d(\beta r)^2} + \frac{1}{\beta r}\frac{df}{d(\beta r)} + \left[1 - \frac{1}{(\beta r)^2}\right] f = 0, \tag{32}$$

where

$$\beta^2 = k^2 - \frac{\pi^2}{a^2} = \left[\frac{2\pi}{\lambda}\sqrt{1 - \left(\frac{\lambda}{2a}\right)^2}\right]^2; \qquad (33)$$

$$k^2 = \omega^2\mu\epsilon.$$

Equation (32) is the Bessel equation of order unity in the argument (βr). The solutions to the equation take a number of different forms; any linearly independent pair of solutions may be taken to construct the general solution. Denoting by $Z_1(\beta r)$ any solution, we have

$$E_\theta = \cos\left(\frac{\pi x}{a}\right) Z_1(\beta r), \qquad (34a)$$

and the corresponding components of the magnetic field are

$$H_r = \frac{j\pi}{\omega\mu a} \sin\left(\frac{\pi x}{a}\right) Z_1(\beta r), \qquad (34b)$$

$$H_x = \frac{j\beta}{\omega\mu} \cos\left(\frac{\pi x}{a}\right) Z_0(\beta r). \qquad (34c)$$

In obtaining H_x use is made of the recurrence relation:[1]

$$\frac{d}{d\rho}[\rho^n Z_n(\rho)] = \rho^n Z_{n-1}(\rho). \qquad (35)$$

The linearly independent solutions to Eq. (32) which are particularly suited to the present problem are the Bessel functions of the second kind—the Hankel functions $H_1^{(1)}(\beta r)$, $H_1^{(2)}(\beta r)$. $Z_1(\beta r)$ in Eqs. (34a) and (34b) is to be taken as representing either one of the two functions; similarly $Z_0(\beta r)$ denotes either of the Hankel functions $H_0^{(1)}(\beta r)$, $H_0^{(2)}(\beta r)$ of order zero. These solutions represent traveling waves as is evident from the asymptotic forms of the functions of order n for large βr:

$$\left. \begin{array}{l} H_n^{(1)}(\beta r) \approx \left(\dfrac{2}{\pi\beta r}\right)^{1/2} e^{j\left(\beta r - \frac{2n+1}{4}\pi\right)}, \\[2mm] H_n^{(2)}(\beta r) \approx \left(\dfrac{2}{\pi\beta r}\right)^{1/2} e^{-j\left(\beta r - \frac{2n+1}{4}\pi\right)}. \end{array} \right\} \qquad (36)$$

It is seen that the first of these represents a wave traveling in the negative r direction, i.e., a wave converging on the cylinder axis $r = 0$, and the second a wave traveling in the direction of increasing r. The solutions correspond to $e^{j\beta z}$ and $e^{-j\beta z}$ in the uniform guide. For large βr the phase fronts are spaced radially in the sectoral guide exactly as they are in the z-direction in the uniform guide. The amplitude is proportional to $r^{-1/2}$ because the energy density associated with a traveling cylinder wave

[1] G. N. Watson, *Bessel Functions*, 2d ed., Macmillan, New York, 1945.

is proportional to r^{-1}, that is, to the reciprocal of the area of the wavefront. It will be noted that the wavelength of propagation $\lambda_g = 2\pi/\beta$ is the same as in the uniform guide. The cutoff condition for the mode is the same as that of the TE_{10}-mode in the uniform guide. In fact, the cutoff conditions for the TE_{no}-modes in the horn are all the same as for corresponding modes in the uniform guide; this explains the attenuation of the higher modes generated at the throat. For small βr values, the interpretation of $H_n^{(1)}(\beta r)$ and $H_n^{(2)}(\beta r)$ is not so simple because there are quadrature terms in the function that represent energy stored in the electromagnetic fields. These terms become more important as βr becomes smaller. For numerical tables of $H_n^{(1)}(\beta r)$ and $H_n^{(2)}(\beta r)$ for small (βr) the reader is referred to Watson.[1]

Having selected the Hankel functions as particular solutions of Eq. (32), the general solution for the dominant-mode field components in the sectoral guide can be written down as follows:

$$E_\theta = A \cos\left(\frac{\pi x}{a}\right)[H_1^{(2)}(\beta r) + \alpha H_1^{(1)}(\beta r)]e^{j\omega t}; \qquad (37a)$$

$$H_r = \frac{j\pi A}{\omega\mu a} \sin\left(\frac{\pi x}{a}\right)[H_1^{(2)}(\beta r) + \alpha H_1^{(1)}(\beta r)]e^{j\omega t}; \qquad (37b)$$

$$H_x = +\frac{j\beta A}{\omega\mu} \cos\left(\frac{\pi x}{a}\right)[H_0^{(2)}(\beta r) + \alpha H_0^{(1)}(\beta r)]e^{j\omega t}, \qquad (37c)$$

where the constants of integration A and α are in general complex. The general field of the dominant mode consists of the incident wave generated at the throat and the wave reflected by the mouth of the horn. The magnitude of α is less than unity because it represents the ratio of the amplitude of the field components in the reflected and incident waves.

Higher-mode Fields.—In addition to the dominant-mode fields which have been considered in detail, fields of other modes exist locally in the sectoral guide. The mouth and throat discontinuities give rise to these modes because the boundary conditions at these points cannot be set up in terms of dominant-mode fields alone. At the throat the amplitudes of the higher-mode fields are small compared with those of the dominant mode unless the flare angle θ_0 is large, and they exist only in the immediate neighborhood of $r = r_1$ because the sectoral guide dimensions are below cutoff. It will be shown (Sec. 10·11) that E-plane sectoral guide admittances can be calculated from the dominant-mode fields alone without appreciable error. At the horn mouth the effect of higher modes is probably not negligible, especially for smal1 apertures. The boundary problem is a difficult one, and no rigorous solution has been obtained. Experimental values of the mouth admittance contain higher-mode contributions but in an unknown proportion. In calculating the radiation field

[1] Watson, *op. cit.*, Table I, Appendix.

from sectoral horns neglect of higher-mode fields in the aperture probably introduces an appreciable error.

10·8. Modes in H-plane Sectoral Horns.—A horn will be referred to as an H-plane sectoral horn when flaring increases the aperture in a plane perpendicular to the electric vector (Fig. 10·8b). The sectoral character of the space in the flare dictates the choice of cylindrical coordinates as in the E-plane case. In this case, the coordinates r, θ, and y will be used (Fig. 10·9b) because the axis of the cylindrical waves in the horn is parallel to the y-axis in the uniform guide. The portion of the flare between any two surfaces of constant r can be considered as a length of H-plane sectoral guide, the flare as a whole being of length $(r_2 - r_1)$, where r_2 and r_1 are chosen as shown in Fig. 10·8. Maxwell's equations for the H-plane sectoral guide are the same as for the E-plane guide [Eqs. (28a) to (28h)] except that x is replaced by y.

The Dominant-mode Fields.—As in the electric-plane case, energy is propagated in only one mode because of the dimensions of the H-plane sectoral guide at the horn throat and the nature of the excitation by TE_{10}-mode in the uniform guide. The dominant mode is characterized by vanishing of all field components except E_y, H_r, and H_θ. The cylindrical character of the space requires that the wavefronts of this mode be surfaces of constant r. To satisfy the boundary conditions E_y must vanish at the walls $\theta = \pm\theta_0$.

The expressions for the dominant-mode field components are derived from Maxwell's equations for the H-plane sectoral guide after introducing the simplifications $E_r = E_\theta = H_y = 0$. Substituting y for x in Eqs. (28a) to (28h) and dropping terms involving E_r, E_θ, and H_y, Maxwell's equations for the dominant mode become

$$\frac{\partial E_y}{\partial y} = \frac{\partial H_r}{\partial y} = \frac{\partial H_\theta}{\partial y} = 0, \tag{38a}$$

$$j\omega\epsilon E_y = \frac{1}{r}\frac{\partial}{\partial r}(rH_\theta) - \frac{1}{r}\frac{\partial H_r}{\partial \theta}, \tag{38b}$$

$$j\omega\mu H_r = -\frac{1}{r}\frac{\partial E_y}{\partial \theta}, \tag{38c}$$

$$j\omega\mu H_\theta = \frac{\partial E_y}{\partial r}, \tag{38d}$$

$$\frac{\partial}{\partial r}(rH_r) + \frac{\partial H_\theta}{\partial \theta} = 0. \tag{38e}$$

Equations (38c) and (38d) serve to express H_r and H_θ in terms of the derivatives of E_y. Substituting the expressions so obtained in Eq. (38b), the following equation for E_y is obtained:

$$\frac{\partial^2 E_y}{\partial r^2} + \frac{1}{r}\frac{\partial E_y}{\partial r} + \frac{1}{r^2}\frac{\partial^2 E}{\partial \theta^2} + \omega^2\mu\epsilon E_y = 0. \tag{39}$$

Since the boundary conditions require that E_y vanish on the walls $\theta = \pm\theta_0$, E_y is of the form

$$E_y = \cos p\theta \, F(r), \tag{40}$$

where

$$p = \frac{\pi}{2\theta_0} \tag{41}$$

and $F(r)$ is a function of r only. Substituting in Eq. (39), the following equation for $F(r)$ is obtained:

$$\frac{\partial^2 F}{\partial (kr)^2} + \frac{1}{kr}\frac{\partial F}{\partial (kr)} + \left[1 - \frac{p^2}{(kr)^2}\right] F = 0, \tag{42}$$

where

$$k^2 = \omega^2 \mu \epsilon = \left(\frac{2\pi}{\lambda}\right)^2.$$

Equation (42) is the form of Bessel's differential equation whose solutions are cylinder functions of order p. The Hankel functions $H_p^{(1)}(kr)$ and $H_p^{(2)}(kr)$ are chosen as particular solutions of this equation again because they represent traveling waves (Sec. 10·7). The general solution for the electric field E_y is therefore

$$E_y = A \cos p\theta [H_p^{(2)}(kr) + \alpha H_p^{(1)}(kr)]. \tag{43a}$$

From Eqs. (38c) and (38d)

$$H_r = \frac{pA}{j\omega\mu}\frac{\sin p\theta}{r}[H_p^{(2)}(kr) + \alpha H_p^{(1)}(kr)], \tag{43b}$$

$$H_\theta = \frac{kA}{j\omega\mu}\cos p\theta \, [H_p^{(2)\prime}(kr) + \alpha H_p^{(1)\prime}(kr)], \tag{43c}$$

where the primes indicate differentiation with respect to kr.

The solutions for the field components in the H-plane sectoral guide [Eqs. (43)] are of the same form as those for the E-plane sectoral guide, but they differ in two noteworthy respects. In the H-plane guide the order p of the functions depends on flare angle θ_0. It is high for small flare angles and is in general not an integer. Thus, for a flare angle of 20°, which is often used in practice, the order is $\frac{9}{2}$. In the H-plane guide the argument of the Hankel functions is $kr(=2\pi r/\lambda)$. From the asymptotic expressions [Eqs. (36)] it is seen that at large kr the equiphase surfaces are separated by a free-space wavelength in contrast to the guide wavelength of the E-plane horn. This is reasonable because the H-plane flare increases the separation of the walls that determine the guide wavelength in the uniform guide. For small θ_0 and high-order p, the asymptotic expressions of the Hankel functions are good approximations only at very large kr, corresponding to the fact that this wall separation becomes large only at very large kr.

Higher-mode Fields.—As in E-plane sectoral guide higher-mode fields are necessarily present at throat and mouth discontinuities. The effect on impedance and radiation patterns of the higher-mode fields at the mouth is not negligible, although it is at present impossible to take them into account in sectoral guide theory.

10·9. Vector Diffraction Theory Applied to Sectoral Horns.—The same considerations discussed in connection with radiation from open waveguide (Secs. 10·1 to 10·3) apply to radiation from horns. If the current distribution on the inside and outside walls of the horn were known, it would be possible to calculate the radiation field at a point outside the guide from this distribution. In the absence of this knowledge, the aperture diffraction method is used as in the case of the waveguides.[1] In the present case the aperture surface is taken to coincide with the cylindrical wavefront of the dominant mode at the mouth of the horn. The aperture field is assumed to be that of the incident wave, the effect of the reflected wave being neglected. The radiation field is computed by means of Eq. (5·103). As usual the radiation field is expressed in terms of spherical coordinates, the origin of which is here taken to coincide with that of the coordinate systems shown in Fig. 10·9. The z-axis of the latter forms the polar axis of the spherical coordinate system, azimuth being measured with respect to the xz-plane in each case. The results are written down in the following paragraphs for the cases in which the medium is the same inside and outside the horn.

Radiation from E-plane Sectoral Horns.—For the E-plane the radiation field at an external point P, as derived from Eq. (5·103), can be shown to be

$$\mathbf{E}_P = \frac{-jkr_2 e^{-jkR}}{4\pi R} \int_A e^{jk(x \sin\Theta \cos\Phi + r_2 \sin\theta \sin\Theta \sin\Phi + r_2 \cos\theta \cos\Theta)}$$
$$\left\{ (\mathbf{i}_x \times \mathbf{R}_0) E_\theta + \left(\frac{\mu}{\epsilon}\right)^{\frac{1}{2}} [\mathbf{i}_\theta - \mathbf{R}_0(\mathbf{i}_\theta \cdot \mathbf{R}_0)] H_x \right\} d\theta \, dx. \quad (44)$$

The quantities R, Θ, and Φ are the spherical coordinates of the point P; r_2, θ, and x are the coordinates on the surface of integration, which is taken to coincide with a wavefront $(r = r_2)$ at the mouth of the horn; $\mathbf{i}_x$, $\mathbf{i}_\theta$, and $\mathbf{R}_0$ are unit vectors in the directions x, θ, and R increasing.

For the plane $\Phi = 90°$ (electric plane):

$$\mathbf{E}_P = -\mathbf{i}_\Theta \frac{jkr_2 e^{-jkR}}{4\pi R} \int_0^a dx \int_{-\theta_0}^{\theta_0} e^{jkr_2 \cos(\Theta - \theta)}$$
$$\left[E_\theta - \left(\frac{\mu}{\epsilon}\right)^{\frac{1}{2}} H_x \cos(\Theta - \theta) \right] d\theta. \quad (45a)$$

[1] L. J. Chu, *Jour. Applied Phys.*, **11**, 603 (1940).

For the plane $\Phi = 0$ (magnetic plane):

$$\mathbf{E}_P = \mathbf{i}_\Phi \frac{jkr_2 e^{-jkR}}{4\pi R} \int_0^a dx \int_{-\theta_0}^{\theta_0} e^{jk(x\sin\Theta + r_2\cos\theta\cos\Theta)}$$
$$\left[E_\theta \cos\Theta - \left(\frac{\mu}{\epsilon}\right)^{\frac{1}{2}} H_x \cos\theta \right] d\theta. \quad (45b)$$

Expressions for E_θ and H_x over the aperture are given by Eq. (37), Sec. 10·8, when r is replaced by r_2.

Radiation from H-plane Sectoral Horns.—For the H-plane sectoral horn

$$\mathbf{E}_P = \frac{e^{-jkR}}{4\pi R} \int_A e^{jk(r_2\sin\theta\sin\Theta\cos\Phi + y\sin\Theta\sin\Phi + r_2\cos\theta\cos\Theta)}$$
$$[jk(\mathbf{R}_0 \times \mathbf{i}_\theta)E_y + j\omega\mu(\mathbf{R}_0(\mathbf{i}_y \cdot \mathbf{R}_0) - \mathbf{i}_y)H_\theta]r_2\, d\theta\, dy. \quad (46)$$

For the plane $\Phi = 90°$ (electric plane):

$$\mathbf{E}_P = \mathbf{i}_\Theta \frac{jkr_2 e^{-jkR}}{4\pi R} \int_0^b dy \int_{-\theta_0}^{\theta_0} e^{jk(y\sin\Theta + r_2\cos\theta\cos\Theta)}$$
$$\left[E_y \cos\theta - \left(\frac{\mu}{\epsilon}\right)^{\frac{1}{2}} H_\theta \cos\Theta \right] d\theta, \quad (47a)$$

For the plane $\Phi = 0$ (magnetic plane):

$$\mathbf{E}_P = \mathbf{i}_\Phi \frac{jkr_2 e^{-jkR}}{4\pi R} \int_0^b dy \int_{-\theta_0}^{\theta_0} e^{jkr_2(\sin\theta\sin\Theta + \cos\theta\cos\Theta)}$$
$$\left[E_y \cos(\Theta - \theta) - \left(\frac{\mu}{\epsilon}\right)^{\frac{1}{2}} H_\theta \right] d\theta. \quad (47b)$$

Expressions for E_y and H_θ are given by Eq. (43) when r is replaced by r_2.

10·10. Characteristics of Observed Radiation Patterns from Horns of Rectangular Cross Section.—When radiation patterns from *sectoral* horns are observed and compared with the patterns obtained from Eqs. (45) and (47) by numerical integration, in general it is found that they do not agree in detail. In view of the fact that the theory neglects the current on the outside walls of the sectoral guide and the higher-mode fields in the aperture, this is not particularly surprising. Only a brief summary of the experimental data will be attempted here.

Figures 10·10 to 10·13 are compilations of patterns from a number of E- and H-plane sectoral horns of large aperture; the apertures were plane surfaces perpendicular to the axis of the guide. They are classified by flare angle and radial length measured in wavelengths (r_2/λ of Fig. 10·9). It will be observed that for a horn of constant flare angle the main lobe undergoes wide changes in width and structure as the horn length increases. This can be correlated qualitatively with the changes in the field over the mouth of the horn. For a given flare angle the aperture area increases directly with the horn length; this alone would tend to

SEC. 10·10] RADIATION PATTERNS FROM HORNS 359

narrow the beam as the length increases. However, the effect of increasing aperture is overshadowed by the phase-error effects. Let δ represent the maximum departure of the wavefront r_2 from the aperture plane (Fig. 10·14). Then $2\pi\delta/\lambda_g$, where λ_g is the wavelength in the sectoral guide at the mouth, is the phase difference between the center of

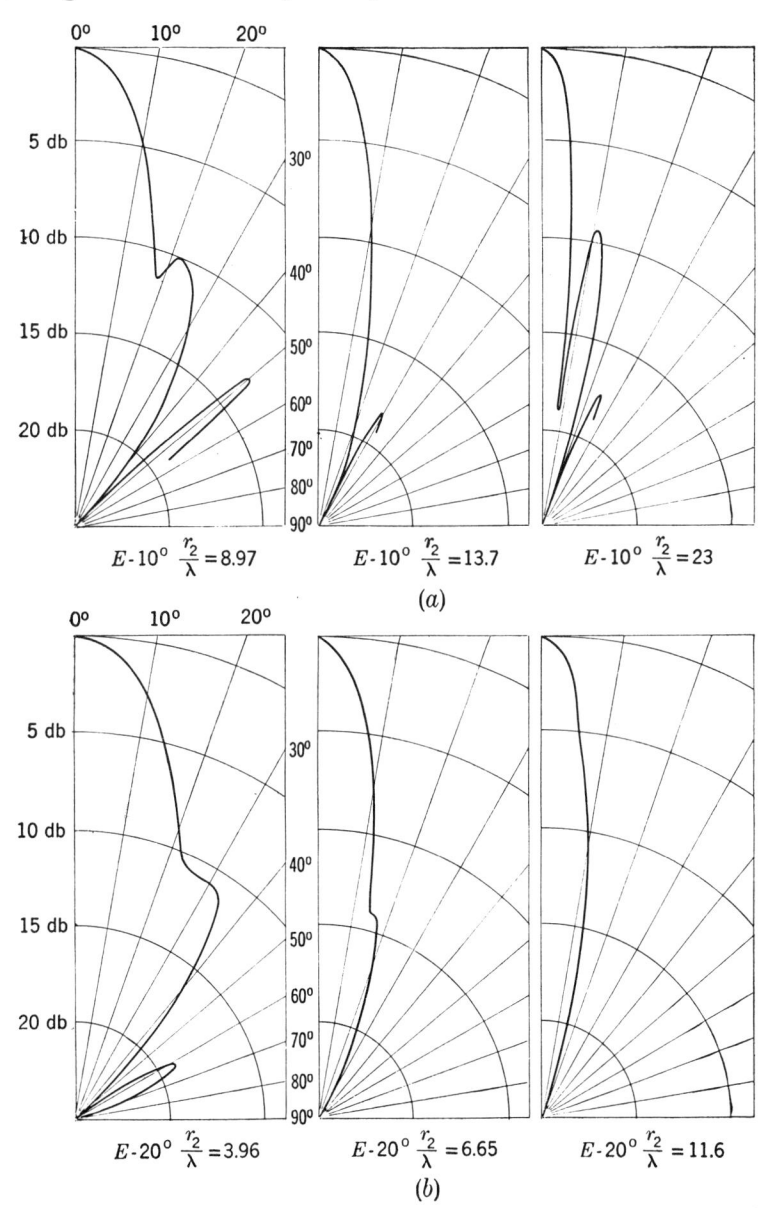

FIG. 10·10.—Radiation patterns of E-plane sectoral horns of various lengths and flare angles: (a) flare angle of $10°$; (b) flare angle of $20°$.

the aperture and the edge. It can easily be seen that

$$\frac{\delta}{\lambda_g} = r_2 \frac{(1 - \cos \theta_0)}{\lambda_g}. \tag{48}$$

Under certain conditions of length and flare angle a phase error is pro-

$E-30°\ \frac{r_2}{\lambda}=1.77$ $E-30°\ \frac{r_2}{\lambda}=2.7$ $E-30°\ \frac{r_2}{\lambda}=3.78$ $E-30°\ \frac{r_2}{\lambda}=5.24$

(a)

$E-40°\ \frac{r_2}{\lambda}=1.21$ $E-40°\ \frac{r_2}{\lambda}=1.53$ $E-40°\ \frac{r_2}{\lambda}=2.4$ $E-40°\ \frac{r_2}{\lambda}=3$

(b)

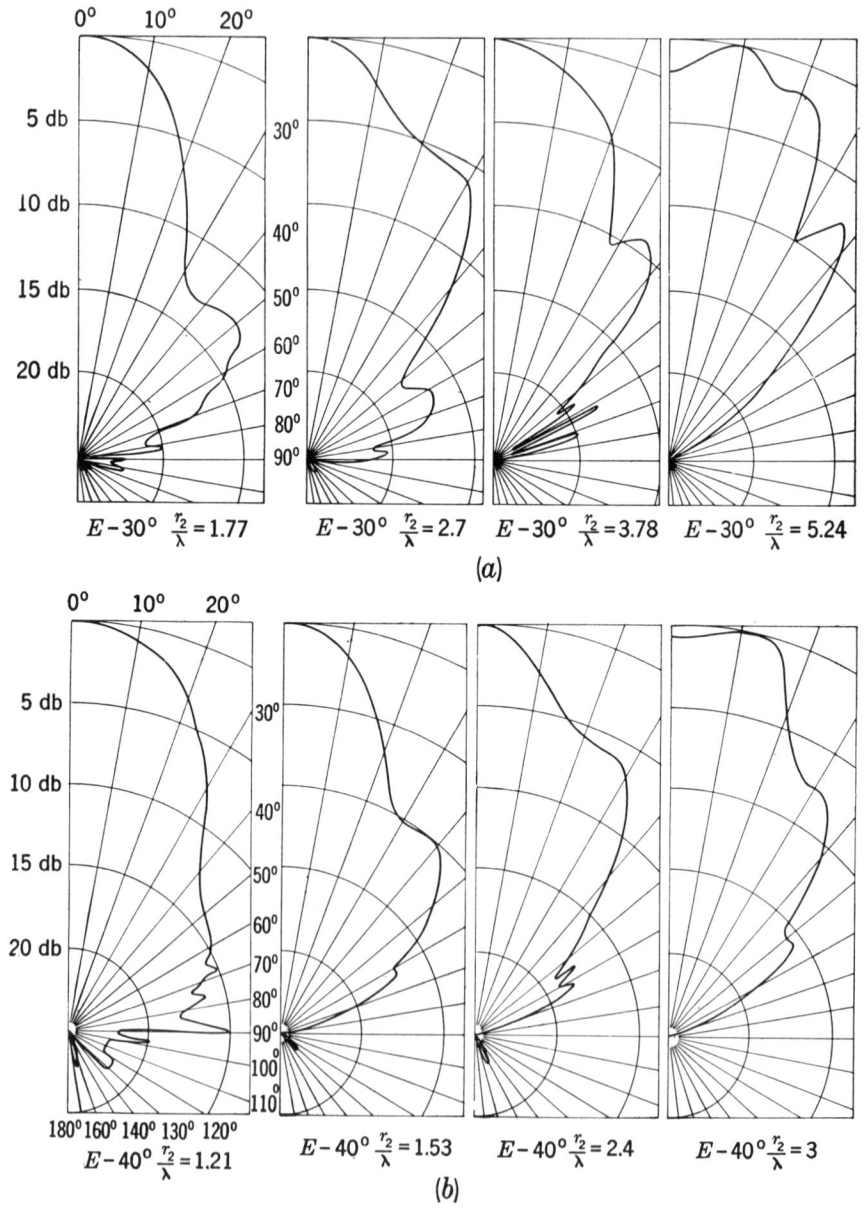

FIG. 10·11.—Radiation patterns of E-plane sectoral horns: (a) flare angle of 30°; (b) flare angle of 40°.

duced over the aperture that leads to a minimum in the main lobe in the forward direction such as may be seen in Fig. 10·11a.

When the aperture or flare angle of a horn is small, δ/λ_g is small and the wavefront at the aperture approximates a plane. Horns are charac-

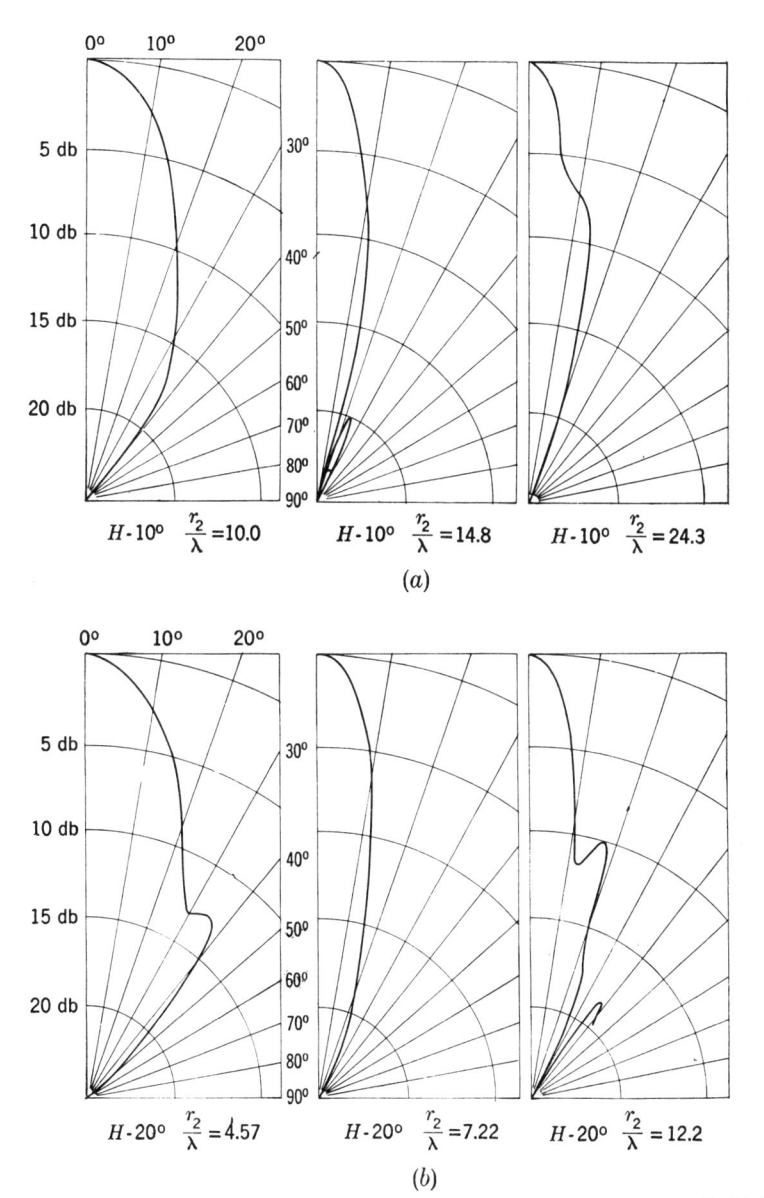

FIG. 10·12.—Radiation patterns of H-plane sectoral horns: (a) flare angle of 10°; (b) flare angle of 20°.

362   WAVEGUIDE AND HORN FEEDS   [SEC. 10·10

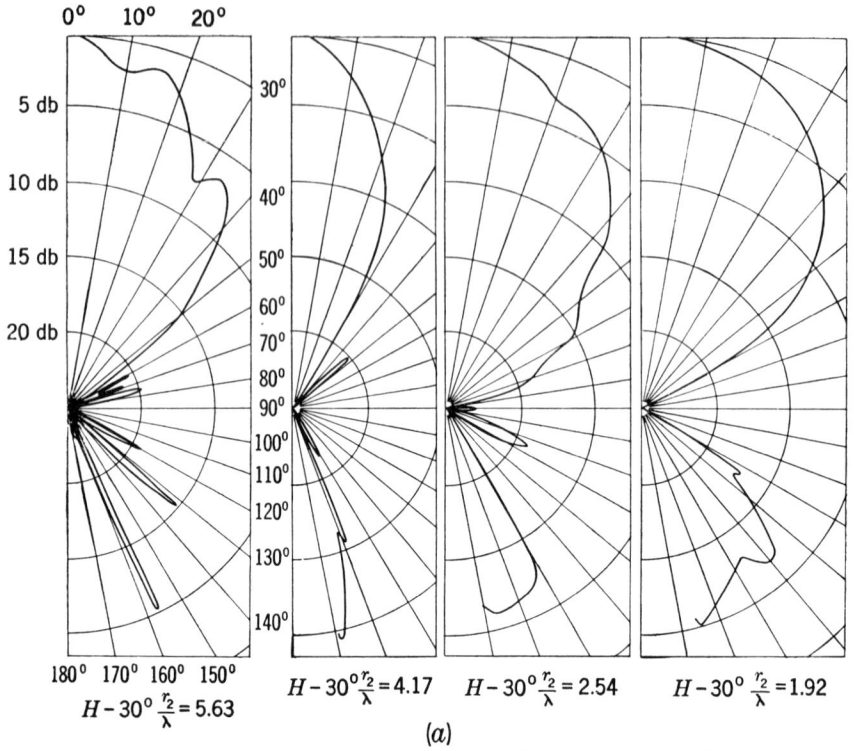

FIG. 10·13.—Radiation patterns of H-plane sectoral

$H - 30° \frac{r_2}{\lambda} = 5.63$ $H - 30° \frac{r_2}{\lambda} = 4.17$ $H - 30° \frac{r_2}{\lambda} = 2.54$ $H - 30° \frac{r_2}{\lambda} = 1.92$

(a)

terized by uniform amplitude distribution across the aperture in the E-plane and sinusoidal distribution in the H-plane [Eqs. (37a) and (43a) and Fig. 10·9]. In the case of approximately uniform phase, therefore, the gain and main lobe width should be functions of aperture corresponding to uniform illumination in the E-plane and sinusoidal illumination in the H-plane. At present these functions can be determined only from experimental data. In Fig. 10·15 the observed 10-db widths of a number of horn patterns are graphed against the reciprocal of the aperture in wavelengths. For all the horns δ/λ_g was less than $\frac{1}{8}$. Results have been obtained for both the sectoral horns and compound horns of the type illustrated in Fig. 10·8; in the latter case the flare again was such that the phase over the aperture was substantially uniform. The E-plane 10-db width (tenth-power width) for all horns lies on the same curve, showing that the E-plane pattern is a function only of the E-plane aperture. The H-plane patterns, on the other hand, depend on both aperture dimensions. Thus, the values obtained from H-plane sectoral horns on standard guide (E-plane aperture approximately $\lambda/3$) fall on Curve II; whereas in the case of the compound horns with an E-plane aperture of a

SEC. 10·10] RADIATION PATTERNS FROM HORNS 363

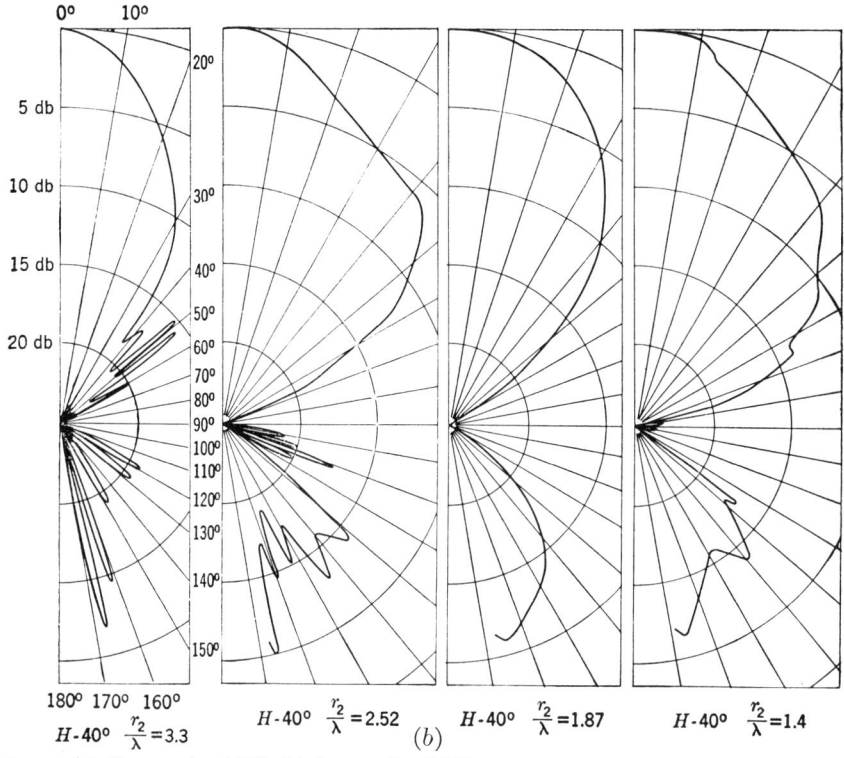

H-40° $\frac{r_2}{\lambda}=3.3$ H-40° $\frac{r_2}{\lambda}=2.52$ (b)   H-40° $\frac{r_2}{\lambda}=1.87$ H-40° $\frac{r_2}{\lambda}=1.4$

horns; (a) flare angle of 30°; (b) flare angle of 40°.

wavelength or greater, the H-plane 10-db widths fall on Curve I. Points for intermediate E-plane apertures which fall between Curves I and II are not shown. At first glance this is somewhat surprising if one assumes that the patterns depend only on amplitude and phase distribution of dominant-mode fields in the aperture. It means, however, that the other factors, namely, higher-mode fields in the aperture and currents on the outside walls of the horn, contribute in the case of small E-plane apertures and are relatively unimportant for E-plane apertures greater than a wavelength. These factors are apparently not dependent on H-plane aperture, at least when this aperture is 0.7λ or more, as in horns on standard rectangular guide, because the observed E-plane widths do not depend on H-plane aperture.

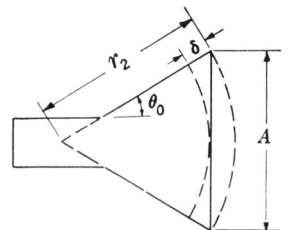

Fig. 10·14.—Origin of phase variation across the aperture of a sectoral horn ($\Delta\phi = 2\pi\delta/\lambda_g$).

The phase variation across the aperture of a horn is small, for a given aperture dimension A in the plane of the flare, only if the flare angle is less than a maximum value (or the length r_2 greater than a minimum value) which depends on aperture and can be obtained from the condi-

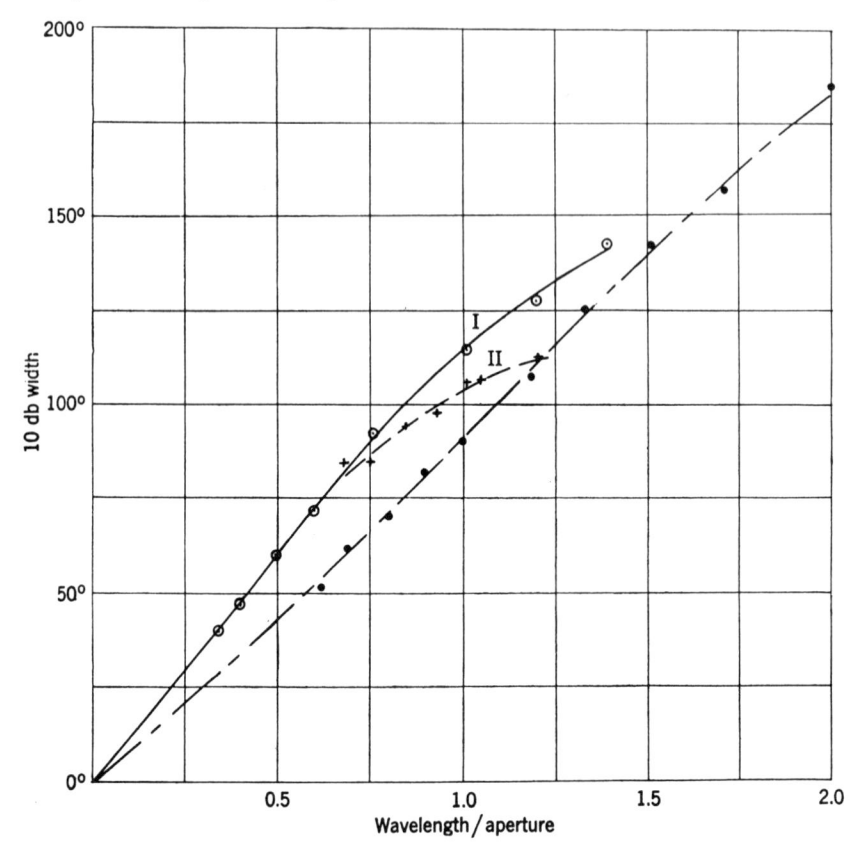

Fig. 10·15.—Experimental 10-db widths of horns having small phase variations over the aperture $\left(\dfrac{\delta}{\lambda_g} < \dfrac{1}{8}\right)$. - - - - - E-plane; - - - - - H-plane sectoral horns; ——— H-plane of compound horns with E-plane aperture equal to or greater than a wavelength.

tion that δ/λ_g shall be small. Using the relation for the separation of an arc and its chord, it is easily shown that

$$\frac{\delta}{\lambda_g} = \frac{A^2}{8r_2\lambda_g} \tag{49}$$

or

$$\frac{\delta}{\lambda_g} = \frac{A \sin \theta_0}{4\lambda_g}. \tag{50}$$

Using $\frac{1}{8}$ as the allowable upper limit for δ/λ_g,

$$(r_2)_{\min} = \frac{A^2}{\lambda_g},$$

$$(\theta_0)_{\max} = \sin^{-1}\left(\frac{\lambda_g}{2A}\right). \tag{51}$$

For many applications the aperture is small and Eq. (51) is satisfied by convenient values of θ_0 and r_2. For large apertures a horn satisfying condition (51) is long and possibly too bulky or heavy for practical applications. In this case the horn designer is forced to compromise on flare angle and aperture. If he is to use a horn feed, he must increase the flare angle and allow for broadening due to phase variation in the aperture by choosing a larger aperture than that predicted by the curves of Fig. 10·15.

In horns of small flare angle (or large r_2/λ_g) the dominant-mode fields near the aperture are described by the asymptotic forms of the Hankel functions [Eqs. (36)], which are exponential functions with slowly varying amplitude. Moreover the departure of the wavefronts from plane surfaces is small. Consequently the dominant-mode fields in the horn closely resemble those in uniform guide. The problem of radiation from horns of small flare angle is therefore approximately the same as that from uniform waveguide, and the predictions [Eqs. (23)] of the vector diffraction theory for waveguide can be applied without serious error to horns satisfying condition (51). It is therefore interesting to compare the curves of Fig. 10·15 with the corresponding theoretical curves for waveguide in Fig. 10·5. Agreement is good for apertures greater than about $2\lambda/3$ in the electric plane and $5\lambda/4$ in the magnetic plane, indicating the probable lower limits at which the factors neglected in the theory are really negligible. It is believed that the predictions of Eq. (23) for waveguide patterns can be useful when properly applied to horns because gain, main-lobe widths at various power levels, side-lobe amplitudes, etc., can be determined for different apertures with relative ease.

Several empirical formulas have been worked out for the 10-db width as a function of aperture for the average horn feed.

1. For the electric plane:

$$\Theta_E\left(\frac{1}{10}\right) = \frac{88\lambda}{B} \text{ (degrees)}, \qquad \frac{B}{\lambda} < 2.5. \tag{52}$$

2. For the magnetic plane:

$$\Theta_H\left(\frac{1}{10}\right) = 31 + 79\frac{\lambda}{A}, \qquad \frac{A}{\lambda} < 3. \tag{53}$$

The symbol $\frac{1}{10}$ represents the 10-db width, and B and A are the apertures in the electric and magnetic planes respectively. These formulas were obtained from a large number of 10-db widths measured at the Radiation Laboratory over a period of several years. The flare angle of the *average* horn is probably about 20°. Since phase variation is not taken into account, the formulas cannot be expected to predict the 10-db widths of individual horns accurately, but they have proved very useful as a first approximation in designing horns.

10·11. Admittance of Waveguide and Horns. *Admittance of Open Waveguide.*—It is observed experimentally that rectangular waveguide of ordinary dimensions when open to space is terminated at the plane of the opening by a capacitive admittance. This type of admittance is to be expected in view of the close spacing between the waveguide walls that are perpendicular to the electric vector (about $\lambda/3$). It is of interest to note that a rigorous treatment of the radiation from the open end of a semi-infinite parallel-plate line carrying the TEM-mode leads to the result that the line is terminated by a capacitive admittance.[1] This property of the waveguide will be useful in analyzing the admittance characteristics of horns.

Admittance of Sectoral Horns.—From the transmission-line point of view a sectoral horn consists of a length of sectoral guide terminated by a mouth admittance at one end and joined to uniform guide at the other. The discussion of its admittance characteristics will be based on sectoral guide transmission-line arguments. The input horn admittance observed in the uniform guide depends on the aperture admittance terminating the sectoral guide, the guide length, and the transformation associated with the junction to uniform guide at the horn throat.[2] In the following discussion sectoral guide characteristics will be summarized. The summary will be followed by a discussion of mouth admittances, junction effects at the throat, and the influence of both factors on horn admittances. Particular attention will be given to E-plane sectoral horns. In what follows, when the term "horn admittance" is used, it will be understood to refer to the admittance measured in the uniform guide and referred to the plane of the junction between the guide and the horn.

Characteristics of E-plane Sectoral Guide.—In Sec. 10·7 expressions were developed for the lowest-mode field components in E-plane sectoral guides [Eqs. (37)]. As in uniform guide, one can define and use a wave admittance, consisting of the ratio of the transverse magnetic to transverse electric fields. If the admittance is expressed in units that make the characteristic admittance of the TE_{10}-mode in the uniform guide equal to unity, the admittance for the E-plane guide becomes

$$Y = \left(\frac{k(\mu)^{\frac{1}{2}}H_x}{\beta(\epsilon)^{\frac{1}{2}}E_\theta}\right).$$

By inspection of the expressions for the field components [Eqs. (37)] it can be seen that this ratio is a function of βr only:

$$Y(\beta r) = j\frac{H_0^{(2)}(\beta r) + \alpha H_0^{(1)}(\beta r)}{H_1^{(2)}(\beta r) + \alpha H_1^{(1)}(\beta r)}. \tag{54}$$

[1] The results of the analysis are given in "Waveguide Handbook Supplement," RL Group Report No. 41, Jan. 23, 1945, Sec. 60c.

[2] J. R. Risser, "Characteristics of Horn Feeds on Rectangular Waveguide," RL Report No. 656, December 1945.

The complex constant α is determined in magnitude and phase from the ratio of incident and reflected waves in the sectoral guide. It can be expressed in terms of the output admittance Y_2 evaluated at the aperture end of the guide where $r = r_2$:

$$\alpha = -\frac{jY_2 H_1^{(2)}(\beta r_2) + H_0^{(2)}(\beta r_2)}{jY_2 H_1^{(1)}(\beta r_2) + H_0^{(1)}(\beta r_2)}. \tag{55}$$

Substituting this value of α in Eq. (54) an expression is obtained for the admittance at a general point r in terms of Y_2 and line parameters.

$$Y(\beta r) = j\frac{\mathcal{K}_{00} + jY_2 \mathcal{K}_{10}}{\mathcal{K}_{01} + jY_2 \mathcal{K}_{11}}, \tag{56}$$

where the symbol $\mathcal{K}_{ij}$ is used to represent combinations of Hankel functions as follows:

$$\mathcal{K}_{ij} = H_i^{(1)}(\beta r_2) H_j^{(2)}(\beta r) - H_i^{(2)}(\beta r_2) H_j^{(1)}(\beta r) \tag{57}$$

A degree of simplification of Eq. (56) is obtained by expressing the Hankel functions in terms of amplitude and phase, using the property that for real values of βr, $H_n^{(2)}(\beta r)$ is the complex conjugate of $H_n^{(1)}(\beta r)$. Let

$$\left.\begin{array}{l} H_0^{(1)}(\beta r) = F e^{j\psi}, \\ H_1^{(1)}(\beta r) = G e^{j\phi}, \end{array}\right\} \tag{58}$$

where F, G, ψ, and ϕ are real functions of βr. Numerical values of these functions are listed in tables of Bessel functions[1] for small values of βr; for large βr the asymptotic values can be used.[2] Substituting in Eq. (56) and using subscripts 2 for the functions evaluated at the aperture where $r = r_2$,

$$Y(\beta r) = -\frac{F}{G}\frac{\sin(\psi_2 - \psi) + j\dfrac{Y_2 G_2}{F_2}\sin(\phi_2 - \psi)}{j\sin(\psi_2 - \phi) - \dfrac{Y_2 G_2}{F_2}\sin(\phi_2 - \phi)}. \tag{59}$$

The characteristic admittance $Y_c(\beta r)$ of E-plane sectoral guide can be written down from Eq. (54) by making α equal to zero,

$$\begin{aligned} Y_c(\beta r) &= j\frac{H_0^{(2)}(\beta r)}{H_1^{(2)}(\beta r)} \\ &= j\frac{F}{G}e^{j(\phi - \psi)}. \end{aligned} \tag{60}$$

---

[1] Watson, "A Treatise on the Theory of Bessel Functions," Cambridge, London, 1944, Table I. The relation to Watson's notation is as follows: $F(\beta r) = |H_0^{(1)}(x)|$; $G(\beta r) = |H_1^{(1)}(x)|$; $\psi(\beta r) = \arg H_0^{(1)}(x)$; $\phi(\beta r) = \arg H_1^{(1)}(x)$; $\beta r = x$.

[2] $F(\beta r) = G(\beta r) = \left(\dfrac{2}{\pi \beta r}\right)^{1/2}$; $\psi(\beta r) = \beta r - \dfrac{\pi}{4}$; $\phi(\beta r) = \beta r - \dfrac{3\pi}{4}$.

For large βr, Y_C approaches unity, since $F/G \to 1$ and $(\phi - \psi) \to -\pi/2$. For small βr, $Y_C(\beta r)$ is complex. It is graphed in Fig. 10·16. The complex character of Y_C for small βr is due to the fact that the fields in the region of the horn apex store as well as transmit energy.

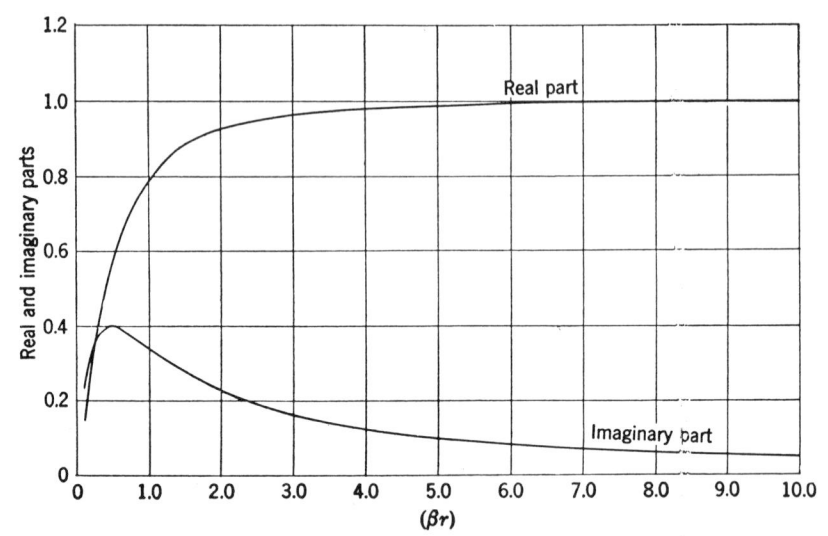

Fig. 10·16.—Characteristic wave admittance $Y_c(\beta r)$ of a sectoral guide.

It is often useful to speak in terms of a reflection coefficient Γ_s in the sectoral guide and to use its transformation properties along the guide; it is defined as the ratio of the electric vector in the reflected wave to that in the incident wave. Then

$$\begin{aligned}\Gamma_s &= \Gamma_s(\beta r) \\ &= \alpha \frac{H_1^{(1)}(\beta r)}{H_1^{(2)}(\beta r)} \\ &= \alpha e^{j2\phi}.\end{aligned} \quad (61)$$

Since $\Gamma_s e^{-j2\phi}$ is equal to the complex constant α, Γ_s transforms down the sectoral guide according to the relation

$$\Gamma_s' e^{-j2\phi'} = \Gamma_s e^{-j2\phi}, \quad (62)$$

where Γ_s, ϕ are evaluated at the point r and Γ_s', ϕ' at r'. It can then be shown that

$$\Gamma_s = \frac{1 - \dfrac{Y}{Y_c}}{\dfrac{Y}{Y_c} - e^{j2(\psi-\phi)}}. \quad (63)$$

Conversely

$$\frac{Y}{Y_c} = \frac{1 + \Gamma_s e^{j2(\psi-\phi)}}{1 + \Gamma_s}. \quad (64)$$

For large βr, $2(\psi - \phi)$ is equal to π, so that Eq. (63) becomes identical with the corresponding expression for Γ in uniform guide.

Mouth Admittance of E-plane Sectoral Horns.—The admittance of an open sectoral guide has not been successfully treated theoretically. Qualitative arguments can be applied from the theory of open parallel-plate transmission lines. As long as the electric plane aperture is small, the mouth admittance should be capacitive as in the case of the waveguide. As the electric plane aperture is increased, the capacitive term in the admittance should decrease. In the limit of large electric-plane aperture the admittance is probably determined by the separation of the guide walls in the magnetic plane.

Experimentally, the determination of the mouth admittance of an E-plane sectoral horn is comparatively easy. It has been obtained in a number of cases by using Eq. (59) to extrapolate down the sectoral guide from the throat where the admittance is determined from measurements in the uniform guide.[1] The capacitive susceptance term, which decreases with increasing aperture, is seen to be present. For large apertures the admittance is independent of aperture to a first approximation. The magnitude of Γ_{s2}, the reflection coefficient in the sectoral guide referred to the aperture,

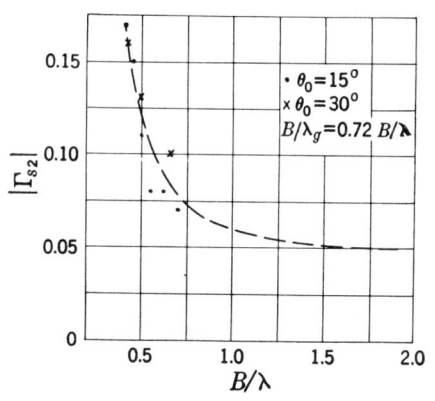

Fig. 10·17.—Reflection coefficient in sectoral guide referred to the aperture B, where B is the aperture dimension in the E-plane.

decreases rapidly with increasing aperture, being small and approximately independent of aperture for apertures above $3\lambda/4$ (Fig. 10·17).

Except in the region near cutoff the mouth admittance and reflection coefficient Γ_{s2} are not sensitive to wavelength changes of the order of 10 per cent. This is due to the fact that at large apertures for which the aperture-to-wavelength ratio changes rapidly with wavelength, the admittance is practically independent of B, where B is the aperture dimension in the E-plane.

10·12. Transformation of the E-plane Horn Admittance from the Throat to the Uniform Guide. *The E-plane Throat Transition.*—The effect on admittance of the transition from sectoral to uniform guide at the horn throat depends primarily on βr_1, where r_1 is the inner radius of the sectoral guide (Fig. 10·8). It is informative to consider first the case where there is no reflected wave in the sectoral guide, so that the char-

---
[1] Risser, *op. cit.*, p. 19.

acteristic admittance of sectoral guide at $r = r_1$ is $Y_c(\beta r_1)$. As can be seen by inspection of Fig. 10·16 for values of $\beta r > 5$, $Y_c(\beta r_1)$ approaches unity, that is, it becomes equal to the characteristic admittance of the waveguide. The throat mismatch becomes negligible as βr_1 increases or as the flare angle decreases; βr_1 depends on waveguide height b, flare angle θ_0, and guide wavelength, as follows:

$$\beta r_1 = \frac{\pi b}{\lambda_g \sin \theta_0}. \tag{65}$$

To show the effect of θ_0 on the throat admittance, a plot of calculated admittances for a series of θ_0 values at 10.0 cm for horns built on stand-

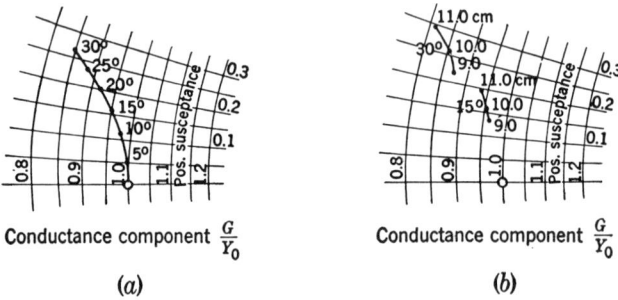

Fig. 10·18.—Throat admittance as a function of flare angle and frequency: (a) $\lambda = 10$ cm, θ_0 varied; (b) λ varied for flare angles of 15° and 30°.

ard guide ($b = 0.34\lambda$; $\lambda_g/\lambda = 1.39$) is shown in Fig. 10·18. The throat mismatch is small for flare angles less than 10° and increases with increasing θ_0. For the 15° and 30° cases the admittances are plotted in Fig. 10·18b for wavelengths from 9 to 11 cm. From this the mismatch can be seen to increase in the direction of the long wavelength end of the band, becoming very large when the wavelength approaches cutoff. In choosing b values for applications involving nonstandard guide, it is necessary to be careful because small values of b are equivalent to large values of θ_0 or λ_g.

When the sectoral guide is not matched, the admittance $Y(\beta r_1)$ in the sectoral guide at the throat is given by Eq. (59) with $r = r_1$. In either case, when computing the admittance in the uniform guide at the throat, the admittance in the sectoral guide at $r = r_1$ must be multiplied by a factor that ensures continuity of voltage and current at the junction. The continuity of current is ensured by continuity in H_x. However, voltage is proportional to the product of the length of the electric-field lines and the field strength. Thus, in the uniform guide the voltage is $bE_y(= 2r_1 \sin \theta_0 E_y)$ and in the sectoral guide $2r_1\theta_0 E_\theta$. Therefore, the current-voltage ratio is proportional to $H_x/(2r_1\theta_0 E_\theta)$ in the sectoral guide and to $H_x/(2r_1 \sin \theta_0 E_y)$ in the uniform guide, so that

Sec. 10·12]   TRANSFORMATION OF THE E-PLANE HORN   371

$$Y_H = \frac{\sin \theta_0}{\theta_0} Y(\beta r_1), \tag{66}$$

where Y_H is the horn admittance in the uniform guide referred to the plane of the junction with the throat of the horn. For values of θ_0 up to 30° the ratio $(\sin \theta_0)/\theta_0$ differs only slightly from unity, and the correction is not important. In general, when guides of different geometry are joined, account must be taken of the contribution of higher-mode fields to the admittance. However, in sectoral horns with values of θ_0 up to 30° this effect can be neglected.

Admittances of E-plane Sectoral Horns.—A discussion of the relation between the parameters of an E-plane horn and its admittance can best be carried out in terms of reflection coefficients. If Γ_H is the reflection coefficient in the *uniform guide* referred to the junction with horn throat, where according to the usual definition of reflection coefficient

$$\Gamma_H = \frac{1 - Y_H}{1 - Y_H},$$

it can be shown from Eqs. (54), (60), (61), and (66) that if the reflection coefficients are small, in particular $|\Gamma_1 \Gamma_{s1}| \ll 1$,

$$\Gamma_H = \Gamma_1 + \Gamma_2, \tag{67}$$

where

$$\Gamma_1 = \frac{1 - \dfrac{\sin \theta_0}{\theta_0} Y_c(\beta r_1)}{1 + \dfrac{\sin \theta_0}{\theta_0} Y_c(\beta r_1)} \tag{68}$$

$$\Gamma_2 = \frac{4 \dfrac{\sin \theta_0}{\theta_0} \dfrac{F_1(\beta r_1)}{G_1(\beta r_1)}}{1 + \left(\dfrac{\sin \theta_0}{\theta_0} \dfrac{F_1}{G_1}\right)^2 + 2 \dfrac{\sin \theta_0}{\theta_0} \dfrac{F_1}{G_1} \sin(\psi_1 - \phi_1)} \Gamma_{s1} e^{-j\epsilon}. \tag{69}$$

Γ_{s1} is the reflection coefficient in the sectoral guide at the horn throat, [Eq. (61)], and ϵ is a small angle given by

$$\epsilon = 2 \tan^{-1} \frac{\dfrac{\sin \theta_0}{\theta_0} \dfrac{F_1(\beta r_1)}{G_1(\beta r_1)} \cos(\psi_1 - \phi_1)}{1 + \dfrac{\sin \theta_0}{\theta_0} \dfrac{F_1}{G_1} \sin(\psi_1 - \phi_1)}. \tag{70}$$

Equation (67) states that Γ_H is the sum of two components Γ_1 and Γ_2; Γ_1 is the reflection coefficient in the uniform guide at the throat when the sectoral guide is matched, i.e., when $\Gamma_{s1} = 0$, and the admittance of the sectoral guide at $r = r_1$ is its characteristic admittance. For any of the 10-cm horns whose throat admittances $(\sin \theta_0/\theta_0) Y_c(\beta r_1)$ are plotted in Fig. 10·18, Γ_1 is the vector drawn from $Y/Y_0 = 1$ on the chart to the admittance point. From Eqs. (62) and (69)

$$\Gamma_2 = \frac{4 \dfrac{\sin \theta_0}{\theta_0} \dfrac{F_1(\beta r_1)}{G_1(\beta r_1)}}{1 + \left(\dfrac{\sin \theta_0}{\theta_0} \dfrac{F_1}{G_1}\right)^2 + 2 \dfrac{\sin \theta_0}{\theta_0} \dfrac{F_1}{G_1} \sin(\psi_1 - \phi_1)} \Gamma_{s2} e^{-j\Delta}, \qquad (71)$$

where

$$\Delta = \epsilon + 2[\phi(\beta r_2) - \phi(\beta r_1)].$$

For reasonable flare angles $(\sin \theta_0)/\theta_0 \approx 1$ and $F_1(\beta r_1)/G_1(\beta r_1) \approx 1$ so that Γ_2 differs essentially only in phase from Γ_{s2}, the reflection coefficient at the horn mouth. The phase angle Δ consists of the sum of a term

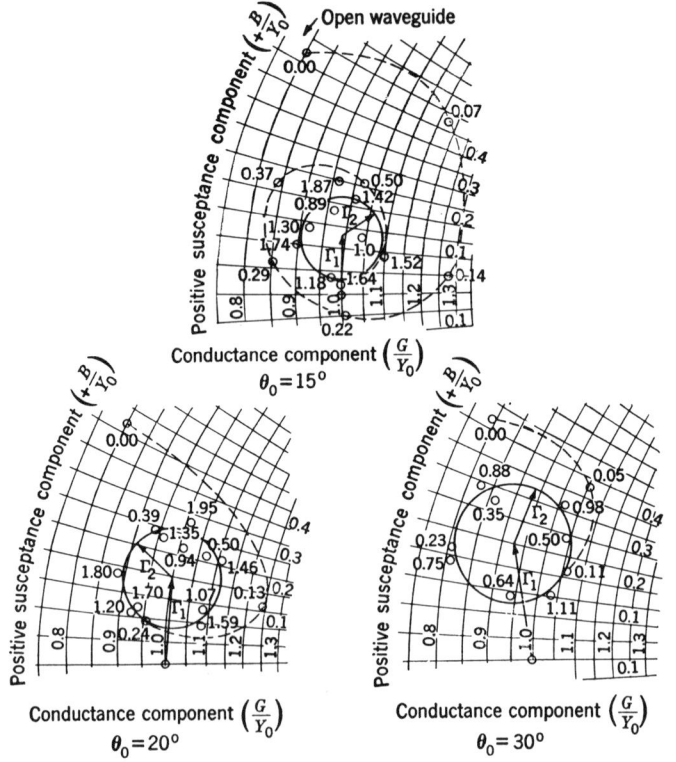

Fig. 10·19.—Admittances of 15°, 20°, and 30° electric plane horns for different flare lengths $(r_2 - r_1)/\lambda_g$.

depending on the horn length, $2[\phi(\beta r_2) - \phi(\beta r_1)]$, and ϵ, a small phase shift at the throat. For the 10-cm horns of Figs. 10·18, ϵ varies from 3° when $\theta_0 = 5°$ to 15° when $\theta_0 = 30°$.

The manner in which the admittance Y_H of a sectoral horn and its reflection coefficient Γ_H depend on the vector sum of Γ_1 and Γ_2 is illustrated in Fig. 10·19, where the admittances at 10 cm of a series of horns

Sec. 10·12]   TRANSFORMATION OF THE E-PLANE HORN   373

with different lengths are plotted on a reflection coefficient chart for three different values of θ_0. These admittances were determined experimentally. The admittance of all horns with the same θ_0 lie on a spiral whose center is determined by Γ_1 and whose periphery is determined by $\Gamma_1 + \Gamma_2$. The decrease in the radius of the spiral with increasing horn length (and aperture) is due to the decrease in the magnitude of Γ_{s2} with

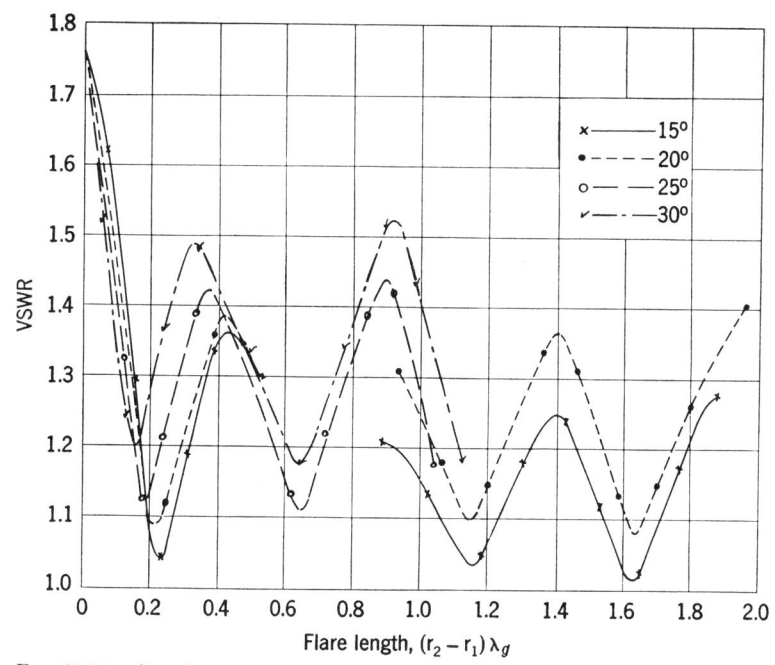

Fig. 10·20.—Standing-wave ratio vs. flare length for typical electric plane horns.

increasing aperture (Fig. 10·17). Moreover, since the aperture height B increases more rapidly with increasing length for larger flare angles, the inner portion of the spiral is reached for smaller values of $\beta(r_2 - r_1)$ for larger flare angles. The relation between B and $(r_2 - r_1)$ is

$$B = 2(r_2 - r_1) \sin \theta_0 + b. \tag{72}$$

Since the inner radii of the spirals are independent of aperture for long horns of large aperture, Γ_{s2} must have a small constant value independent of B/λ for large values of B.

For a given θ_0 a series of horn lengths exist for which Γ_1 and Γ_2 are 180° out of phase and the match is optimum. An empirical formula for these optimum lengths is

$$(r_2 - r_1)_0 = 0.17\lambda_g + N\frac{\lambda_g}{2} \quad \begin{array}{l} N = 0,1,2, \cdots \text{ for } \theta = 25°, 30° \\ N = 1,2,3 \cdots \text{ for } \theta = 5°, 10°, 15°, 20°. \end{array} \tag{73}$$

From Eqs. (72) and (73) horns can be designed to be matched at any aperture and wavelength, since the value of θ_0 is not critical. The degree of match attainable is indicated in Fig. 10·20.

The mouth and throat admittances and consequently Γ_{s2} and Γ_1 are not sensitive to wavelength changes of the order of 10 per cent (Fig. 10·18b). The principal frequency sensitivity of an E-plane sectoral horn therefore arises from variation of the effective sectoral guide length $2[\phi(\beta r_2) - \phi(\beta r_1)]$. This is clearly shown by the admittance-frequency curves for several typical horns shown in Fig. 10·21. For the long horns

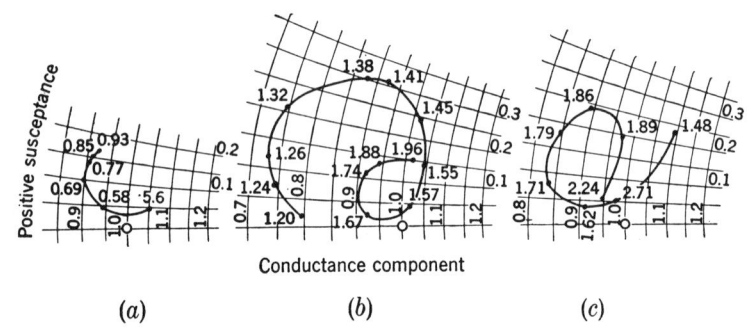

Fig. 10·21.—Admittances of three 15° electric plane horns of different lengths as a function of wavelength: (a) $r_2 - r_1 = 10.6$ cm, λ varied from 9.0 to 11.5 cm with best match at 11.4 cm; (b) $r_2 - r_1 = 22.8$ cm, λ varied from 9.1 to 11.5 cm with best match at 10.1 cm; (c) $r_2 - r_1 = 26.3$ cm, λ varied from 8.1 to 11.5 cm with best match at 10.9 cm.

the frequency variation causes the admittance to traverse more than a complete loop, corresponding to a change in Δ [Eq. (71)] of more than 360°. As discussed previously in this section (see Fig. 10·18) increasing wavelength results in increasing mismatch at the mouth and throat, so that Γ_{s2} and Γ_1 both increase in magnitude as the long-wavelength end of the band is approached. To obtain a low SWR over a very wide band, it is advisable to choose the horn length for optimum match at the long-wavelength end of the band; a comparison of curves b and c of Fig. 10·21 indicates that improvement is obtained by so doing.

10·13. Admittance Characteristics of H-plane Sectoral Horns.—The wave admittance for H-plane sectoral guide is $(k/\beta)(\mu/\epsilon)^{1/2}(H_\theta/E_y)$, using units in which the characteristic admittance of the TE_{10}-mode in uniform guide is unity. Then

$$Y(kr) = -j\frac{k}{\beta}\left[\frac{H_p^{(2)'}(kr) + \alpha H_p^{(1)'}(kr)}{H_p^{(2)}(kr) + \alpha H_p^{(1)}(kr)}\right], \tag{74}$$

where the prime indicates the derivative of the function with respect to (kr). Noting that Eq. (74) for the H-plane differs from Eq. (54) for the E-plane in having $H_p^{(n)'}(kr)$ in place of $H_0^{(n)}(\beta r)$, $H_p^{(n)}(kr)$ in place of $H_1^{(n)}(\beta r)$, and $(-jk/\beta)$ in place of j, equations for the H-plane analogous to each of E-plane equations [Eqs. (54) to (64)] can be written down. There are practical limitations to the usefulness of the H-plane sectoral

Sec. 10·13] ADMITTANCE CHARACTERISTICS OF H-PLANE

guide equations, since the order $p\left(=\dfrac{90°}{\theta_0}\right)$ is not the same for all horns, as it is in the E-plane case. The order is in general high, horns for most applications having small flare angles.

While mouth admittances of H-plane sectoral horns are somewhat difficult to calculate because of the high p, it is not difficult to predict them approximately from E-plane horn measurements and parallel-plate theory. Since the aperture B in the electric plane is small compared with the magnetic plane aperture and is equal to the height b of the uniform guide on which the horn is built, the mouth admittances of all H-plane sectoral horns would be expected to be approximately equal to the mouth admittance of open uniform waveguide and to exhibit the same degree of frequency sensitivity.

The throat transition in magnetic plane horns of small angle has a minor effect on the horn admittance, and the throat reflection is very small compared with the reflection at the mouth. This has been shown experimentally by making standing-wave measurement on magnetic plane horns over a band of frequencies and again at a constant frequency while the horn length was cut

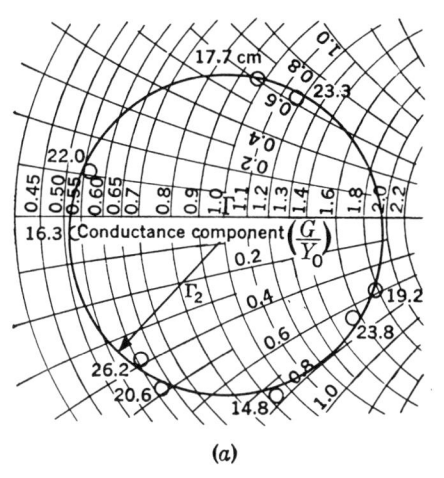

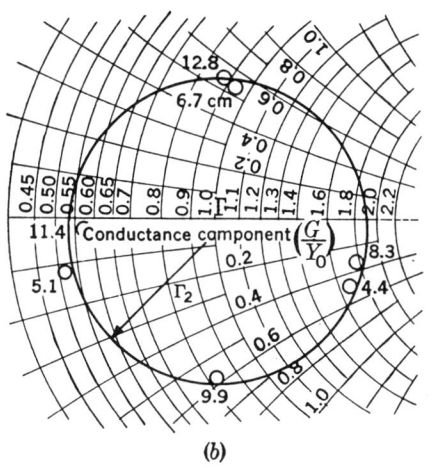

Fig. 10·22.—Admittances of 15° and 30° magnetic plane horns for different flare lengths: (a) $\theta = 5°$; (b) $\theta = 30°$.

down (decreasing the aperture simultaneously with the length). In both cases the shift in phase in the standing-wave pattern indicated that the large reflection occurred at the mouth.

The admittances measured at 10.0 cm of a series of H-plane sectoral horns are shown in Fig. 10·22. The admittance values correspond to different lengths (and apertures). The form of the plot can be explained by postulating two components of the reflection coefficient: a small fixed component representing the throat reflection and a large component due to the mouth which has a magnitude independent of aperture and a phase dependent on sectoral guide length. Experiments indicate that the frequency sensitivity of the horn admittance arises primarily because the effective sectoral guide length varies.

10·14. Compound Horns.—Doubly flared horns must be used to enlarge both the E- and H-plane dimensions. The admittance characteristics of the sectoral horns provide the basis for the design of a broadband compound horn of the type illustrated in Fig. 10·8c. The principle is to use the frequency characteristics of an E-plane flare to compensate those of an H-plane flare. Consider, for example, the problem of producing an aperture having an H-plane dimension larger than the E-plane dimension. The desired aperture is obtained by first flaring in the H-plane to the required dimension and subsequently flaring in the E-plane. Since the H-plane flare introduces a negligible mismatch at the throat, the mismatch arises entirely in the E-plane horn, one component at the junction with the H-plane sectoral guide and the second component at the mouth. The E-plane flare in this case does not differ much from that on uniform guide, and the data obtained for E-plane sectoral horns can be used to determine the E-plane flare angle and horn length so that the mouth reflection cancels the junction reflection (reflection at the throat of the E-plane flare). The shortest possible length is chosen for the E-plane flare consistent with the matching conditions. This is to eliminate the "long-line effect" which would cause the phase of the mouth reflection transformed to the throat to vary rapidly with respect to the throat reflection as the frequency changed. In actual practice, the mouth reflection is chosen to cancel only partially the throat reflection, because it is necessary to close the mouth of the horn by a pressurizing device which likewise gives rise to a reflection. The reflection coefficient of the pressurizing device can be designed to be equal to the residual mismatch of the mouth and throat and phased properly relative thereto by positioning the device with respect to the mouth to give an over-all reflection-free system (see Sec. 10·17).

The technique is essentially the same in the case where the E-plane aperture dimension is larger than the H-plane dimension. The E-plane flare is introduced first followed by the H-plane flare. In this case, the major sources of the mismatch are widely separated—one at the junction between the uniform guide and the E-plane sectoral guide and the second at the mouth. Although the latter reflection can be chosen to counteract the first at some one frequency, the bandwidth is small because of the long-line effect. It is preferable to cancel the mouth reflection by means of the pressurizing device and to cancel the throat reflection by matching window in the uniform guide. In this way the sources between any pair of compensating reflections can be put close together so that rapid phase variations due to long electrical paths are eliminated.

10·15. The Box Horn.—The box horn is a special horn type devised to have greater directivity in the H-plane than a flared horn of the same aperture. It is so constructed as to introduce a third harmonic 180° out of phase with the fundamental mode in the aperture plane. This alters the amplitude distribution across the aperture from the cosine type associated with the fundamental mode to one more nearly uniform.[1]

The box horn is not a true horn in that there is no throat that is used to filter out higher modes. Its essential features are sketched in Fig. 10·23a. The horn consists primarily of a piece of waveguide of length L, frequently referred to as a "box," whose magnetic plane dimension A is large enough to support $TE_{n,0}$-modes with values of n up to 4. It is open to space at one end and fed at the other by a waveguide or H-plane sectoral horn of aperture A' located centrally so as to excite only the modes having nonzero amplitude at the center, i.e., the TE_{10}- and TE_{30}-modes. The ratio of the amplitudes of the TE_{30}- and TE_{10}-modes depends on the ratio A'/A. Since the velocity of propagation of the two modes is not the same, the length L of the box determines their relative phase at the aperture. The horn may be made as directive as desired in the E-plane by introducing an E-plane flare.

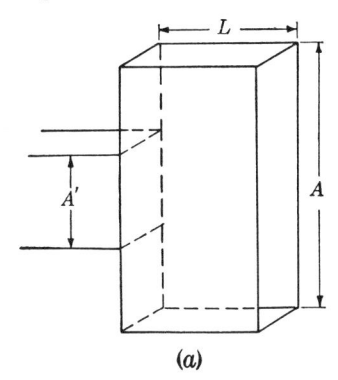

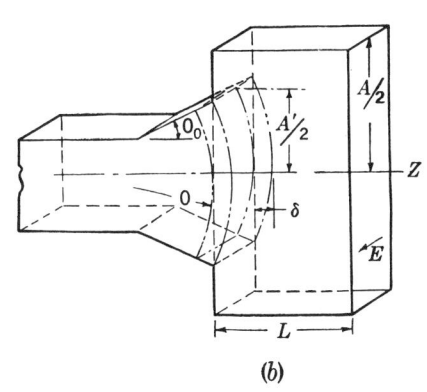

FIG. 10·23.—Box horn: (a) direct junction type; (b) sectoral guide junction type.

It is easy to show approximately how the ratio of the amplitudes of the TE_{30}- and TE_{10}-modes depends on the ratio A'/A of the dimensions of the two guides. The fields $E_y(x,z)$ in the box can be represented as a superposition of the modes excited at the junction. Neglecting the effect of the nonpropagating modes, we have

$$E_y(x,z) \approx a_1 \cos\left(\frac{\pi x}{A}\right) e^{-j\beta_{10}z} + a_3 \cos\left(\frac{3\pi x}{A}\right) e^{-j\beta_{30}z}. \tag{75}$$

The junction between the two guides is taken to be in the plane $z = 0$.

[1] S. J. Mason, "Flared Box Horn," RL Report No. 653, July 1945.

378 WAVEGUIDE AND HORN FEEDS [SEC. 10·15

It is assumed that the field over the common area between the guides is that of the dominant mode in the smaller guide; that is,

$$E_y(x,0) = \cos\frac{\pi x}{A'}, \quad |x| \leq \frac{A'}{2};$$
$$E_y(x,0) = 0, \quad \frac{A'}{2} < |x| < \frac{A}{2}. \quad (76)$$

On setting $z = 0$ in Eq. (75) and making use of Eq. (76) for the field, we can obtain the coefficients a_1 and a by the usual procedure for determining Fourier coefficients. The ratio of the harmonic components is found to be

$$\frac{a_3}{a_1} = \frac{\int_{-A'/2}^{A'/2} \cos\left(\frac{\pi x}{A'}\right) \cos\left(\frac{3\pi x}{A}\right) dx}{\int_{-A'/2}^{A'/2} \cos\left(\frac{\pi x}{A'}\right) \cos\left(\frac{\pi x}{A}\right) dx} \quad (77)$$

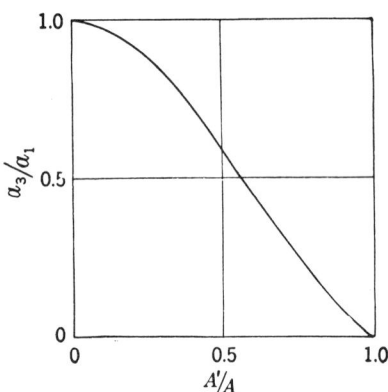

Fig. 10·24.—Ratio of the amplitudes of the TE_{30}- and TE_{10}-modes in a box horn vs. A'/A.

Figure 10·24 is a plot of a_3/a_1 vs. A'/A. The ratio a_3/a_1 decreases with increasing A'/A, reaching zero when A'/A is unity. If it were possible to make A'/A very small, values of a_3/a_1 approximately equal to 1 could be obtained. The lower limit of A'/A is 0.20, corresponding to

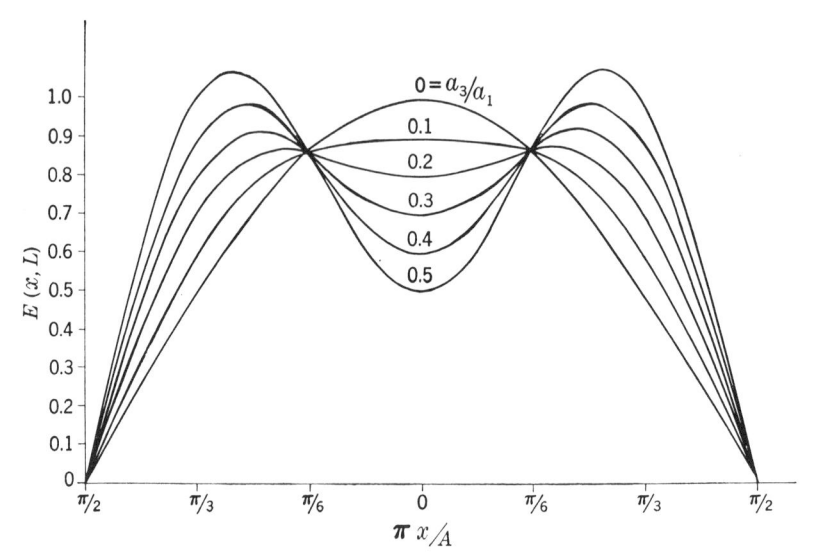

Fig. 10·25.—Aperture illumination (amplitude) of a box horn for a series of values of the ratio a_3/a_1.

SEC. 10·15]   THE BOX HORN   379

$A' = 0.5\lambda$ the cutoff in the input guide, so that the maximum a_3/a_1 is 0.93. Actually much smaller values of a_3/a_1 are desirable. Figure 10·25 is a plot of amplitude illumination across the aperture for a_3/a_1 values from 0 to 0.5, calculated on the assumption that the horn length is correctly chosen to

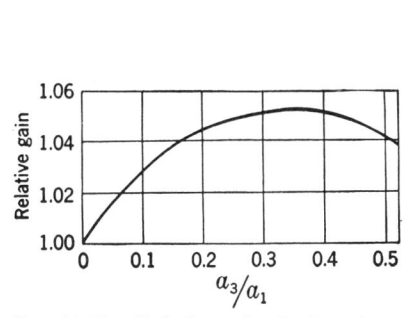

FIG. 10·26.—Relative gain of a box horn vs. a_3/a_1.

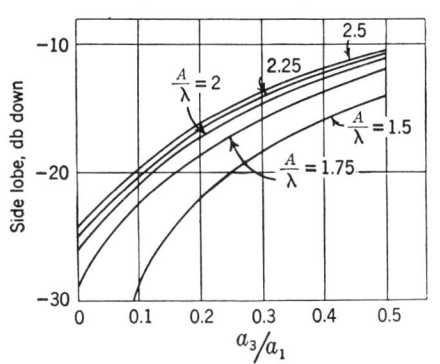

FIG. 10·27.—Magnitude of the first side lobe in the power pattern of a box horn vs. a_3/a_1.

make the two propagating modes 180° out of phase at $z = L$ with respect to their relative phase at $z = 0$. The curve for a_3/a_1 equal to 0.3 is a fairly good approximation to uniform illumination.

The H-plane radiation patterns in the Fraunhofer region have been calculated from scalar diffraction theory (Chap. 6) for the amplitude distributions of Fig. 10·25. The results are summarized in Figs. 10·26 to 10·28, where Fig. 10·26 shows relative gain, Fig. 10·27 the magnitude of the first side lobe in the power pattern in decibels down from peak, and Fig. 10·28 the full angular width of the pattern at tenth power. Gain is seen to be maximum in the neighborhood of a_3/a_1 equal to 0.35, where the amplitude distribution across the aperture approximates uniformity. In this region also the first side lobes are approximately 13 db down, the theoretical value for uniform illumination. Illuminating the edges of the aperture more strongly increases side lobes and cuts down gain, although it somewhat increases the directivity.

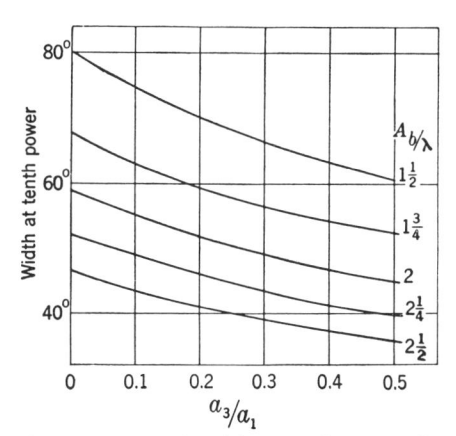

FIG. 10·28.—Full width at tenth power of a box horn pattern vs. a_3/a_1.

The length L of the box is obtained from the relation

$$\pi = (\beta_1 - \beta_3)L, \tag{78}$$

where

$$\begin{aligned}\beta_1 &= \frac{2\pi}{\lambda}\left[1 - \left(\frac{\lambda}{2A}\right)^2\right]^{1/2}, \\ \beta_3 &= \frac{2\pi}{\lambda}\left[1 - \left(\frac{3\lambda}{2A}\right)^2\right]^{1/2}.\end{aligned} \qquad (79)$$

The ratio A'/A can be varied over wider limits than that set by the uniform guides by means of a sectoral guide transition as shown in Fig. 10.23b. In this case, however, because of the cylindrical waves in the sectoral guide, a phase error is introduced into the field over the plane of the junction with the box horn. This error is not taken into account in Eq. (77); it is usually neglected in designing the horn. From existing measurements[1] it is difficult to evaluate the influence of this factor on box horn performance. It is found true in general that the effect of replacing a plane wave front by a cylindrical one is barely detectable experimentally where $2\pi\delta/\lambda_g$ is $\pi/4$ or less. The same criterion would be expected to be valid for box horns. In terms of A' and θ_0, it is written [Eq. (49)]

$$A' \sin \theta_0 \leq \frac{\lambda_g}{2}. \qquad (80)$$

The use of a box horn is, of course, limited to applications requiring H-plane tenth-power widths from about 36° to 70°. Its principal advantage is compactness. The contrast in size between a box horn and flared horn of the same tenth-power width is greatest in the region of small apertures. Here, near cutoff for the T_{30}-mode, L is small because $(\beta_1 - \beta_3)$ is approximately equal to β_1. A box horn with $A = 1.6\lambda$, $a_3/a_1 = 0.5$, and $\Theta(\frac{1}{10}) = 67°$ is 1.3λ long, while the flared horn of the same tenth-power width is twice as long. However, for A from 2.0λ to 2.5λ, a box horn is only about 20 per cent shorter than the corresponding flared horn. The box horn is especially useful therefore for applications requiring tenth-power widths from about 55° to 65°, with apertures from 1.6λ to 1.7λ. A flared horn has an advantage over the box horn in having side lobes 5 to 10 db lower, although this is of no concern in many applications.

10·16. Beam Shaping by Means of Obstacles in Horn and Waveguide Apertures.—There are antenna applications requiring very broad or very narrow feed patterns for which waveguide and horn feeds are not strictly suitable. Thus, to obtain primary patterns in the H-plane with 10-db widths greater than about 120°, the 10-db width obtained from ordinary waveguide, special beam-shaping techniques are required. At the other extreme of very narrow 10-db widths, horns become too bulky for many applications. To solve these problems beam shaping by means

---

[1] Mason, *op. cit.*, p. 18, Fig. 13.

of obstacles in horn and waveguide apertures has been investigated and techniques developed.[1]

a. Beam Broadening: H-plane.—To obtain a broader pattern than that of ordinary waveguide, the device shown in Fig. 10·29 is used. A metallic post is inserted across the waveguide just inside the aperture at the center, and the waveguide corners are removed by symmetrical cuts AA. Figure 10·30 shows the 10-db widths observed with this arrangement on 3.2 cm guide for a series of values of C and ϕ, where C is the distance from the waveguide edge at which the diagonal cut begins and ϕ is the angle of cut. Figure 10·31 shows a comparison of a waveguide pattern with a broadened pattern.

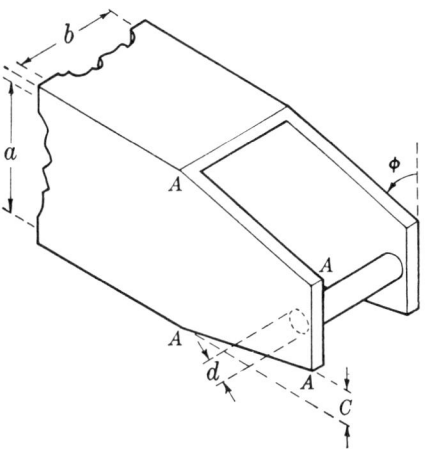

Fig. 10·29.—Device for broadening the pattern in the H-plane.

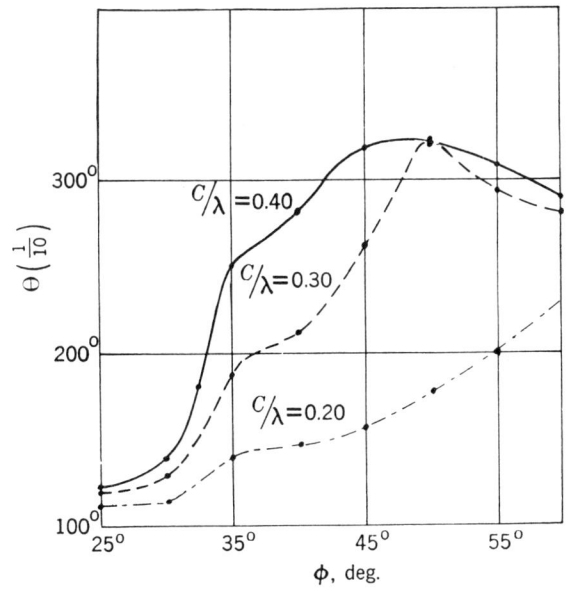

Fig. 10·30.—10-db widths of the pattern obtained with the device of Fig. 10·29 as a function of C/λ and ϕ (λ = 3.2 cm, a = 0.900 cm, b = 0.400 cm, d = 0.063 cm).

The wavefronts, or surfaces of constant phase, from any device yielding a very broad pattern must receive critical examination by the antenna

[1] C. S. Pao, "Shaping the Primary Pattern of a Horn Feed," RL Report No. 655, January 1945.

designer. The gain obtainable from a properly illuminated secondary reflector may not be realized if phase irregularities exist over its aperture.

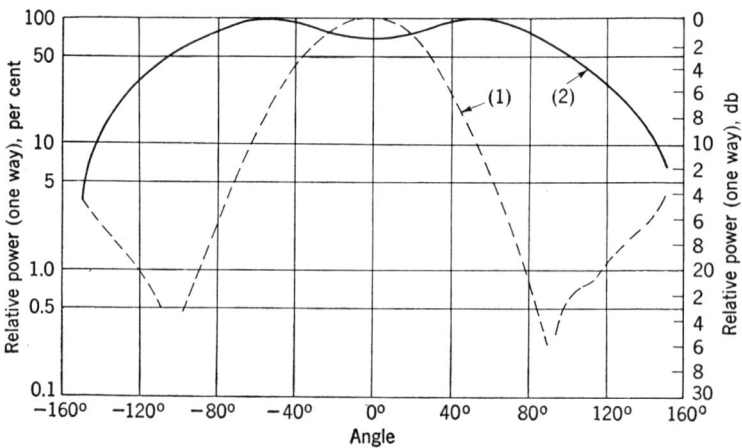

Fig. 10·31.—Primary patterns of (1) an open waveguide and (2) a cut corner waveguide ($\phi = 63.4°$ and $C = \frac{1}{4}$).

The tendency to irregularities on the phase fronts of wide-angle devices arises from the fact that they usually consist of several radiating elements with noncoincident directive patterns located an appreciable fraction of a wavelength apart. An analysis of a simple case (Fig. 10·32) will illustrate this. Consider two identical radiating elements, 1 and 2, separated by a distance $2d$ which have pattern maxima in the Fraunhofer region at angles θ_1 and θ_2 with the forward z direction, respectively. Then letting $f_1 = (1/r)f(\theta - \theta_1)$ and $f_2 = (1/r)f(\theta - \theta_2)$ equal the amplitude functions of the two sources, the amplitude and phase on a circle of radius R about O is given by

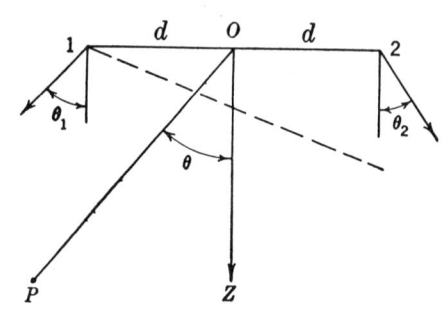

Fig. 10·32.—Phase distortion in the fields of primary feeds of low directivity.

$$E = f_1 e^{j(\omega t - kR + kd \sin \theta)} + f_2 e^{j(\omega t - kR - kd \sin \theta)}$$
$$= [f_1^2 + f_2^2 + 2f_1 f_2 \cos (2kd \sin \theta)]^{\frac{1}{2}} e^{j(\omega t - kR)} e^{j\psi}, \quad (81)$$

where

$$\tan \psi = \frac{f_1 - f_2}{f_1 + f_2} \tan (kd \sin \theta). \quad (81a)$$

If $f_1 = f_2$, the phase is constant on a circle of radius R. However, if $f_1 \neq f_2$, the phase at angle θ departs from its value at $\theta = 0$ by an angle Δ where

$$\Delta = \tan^{-1} \left[\frac{f_1 - f_2}{f_1 + f_2} \tan (kd \sin \theta) \right]. \quad (82)$$

For large values of kd, Δ may exceed the allowable deviation from constant phase, particularly if f_1 is appreciably greater or smaller than f_2. Usually the beam shaping need not be carried to the point where the phase departure Δ has to be reckoned with.

b. Narrowing the Primary Pattern: H-plane.—The pattern in the H-plane is narrowed when several metallic pins are placed in the aperture of an H-plane horn. For example, the 10-db width of a 3.2-cm horn with a flare angle 30° and magnetic plane aperture 2λ decreases from 78° to 56° when two $\frac{1}{16}$-in.-diameter pins are placed just inside the aperture at a distance 0.44λ on either side of the center. The impedance match is improved rather than impaired by the presence of the pins. A number of other arrangements also have been found to be effective.[1]

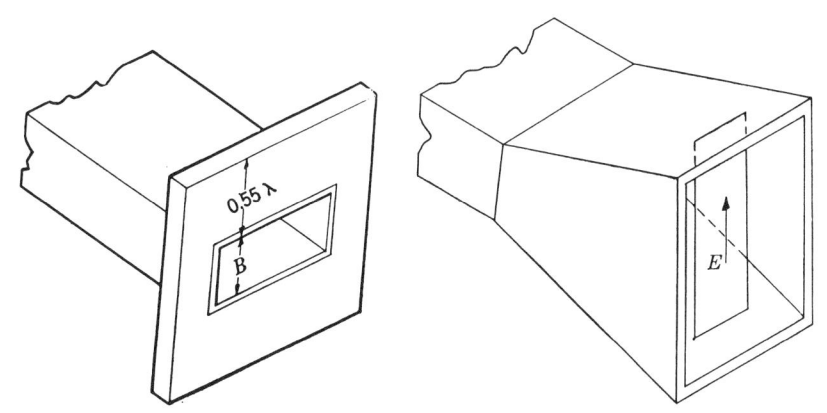

FIG. 10·33.—Flange for E-plane beam shaping.

FIG. 10·34.—Strip for narrowing of the beam in the E-plane.

Beam Shaping: E-plane.—Since ordinary waveguide has an E-plane aperture of about $\lambda/3$, there is rarely a need for special techniques for beam broadening. However, the form of the pattern can be improved by adding a flange approximately 0.55λ wide in the E-plane as shown in Fig. 10·33. With the flange, the power drops off more rapidly at small angles from the forward direction and less rapidly at large angles, thereby yielding a more nearly uniform illumination across the secondary aperture.

Narrowing of the E-plane pattern has been accomplished by inserting a metallic strip with a considerable H-plane width, as shown in Fig. 10·34. This strip, however, causes a troublesome mismatch,[1] and it seems questionable from the data if it is more effective than straightening the phase fronts in the horn by making the horn longer (Sec. 10·10).

10·17. Pressurizing and Matching.—In most applications waveguide and horn feeds must be at least weatherproof and preferably capable of holding pressure. Several techniques have been evolved to utilize the

[1] Pao, *op. cit.*

pressurizing or weatherizing enclosure as a matching device. One of the most successful is illustrated in Fig. 10·35. The enclosure, roughly cylindrical in form, is placed over the end of the waveguide or horn with the axis of the cylinder approximately in the aperture plane. Matching is accomplished by adjusting the thickness t and position d so that the reflections from enclosure and feed are equal in magnitude and 180° out of phase in the direction of the generator. While t and d can be

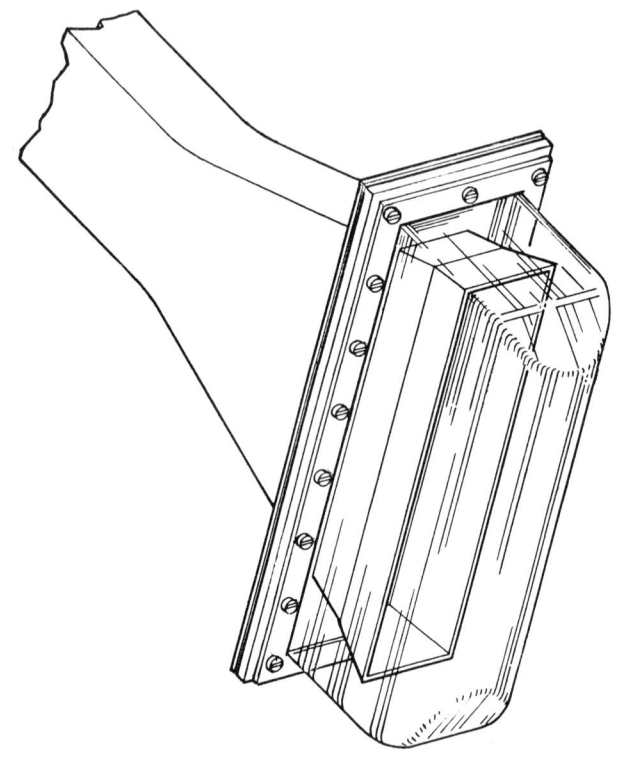

FIG. 10·35.—Pressurizing and matching enclosure.

calculated to a rough approximation for a feed with a given mismatch, the matching procedure is actually empirical. Cylinders are formed with thicknesses and radii ranging in value about the calculated t and d. Each cylinder is placed over the mouth of the feed, and impedance measurements are made for a series of d values. From these measurements the final choice of enclosure parameters is made. Figure 10·36 is a typical impedance plot obtained during this procedure. It will be observed that the best impedance match for this example is obtained with an enclosure of thickness 0.030 in. and mounted at a distance $d = 1.2$ cm (Curve III). It is desirable, although not essential, that the enclosure radius be approximately equal to d. The mounting flange is positioned behind the aperture plane and a position can be found that improves the feed SWR; it should be in place during the determination of enclosure parameters and the pattern of the horn.

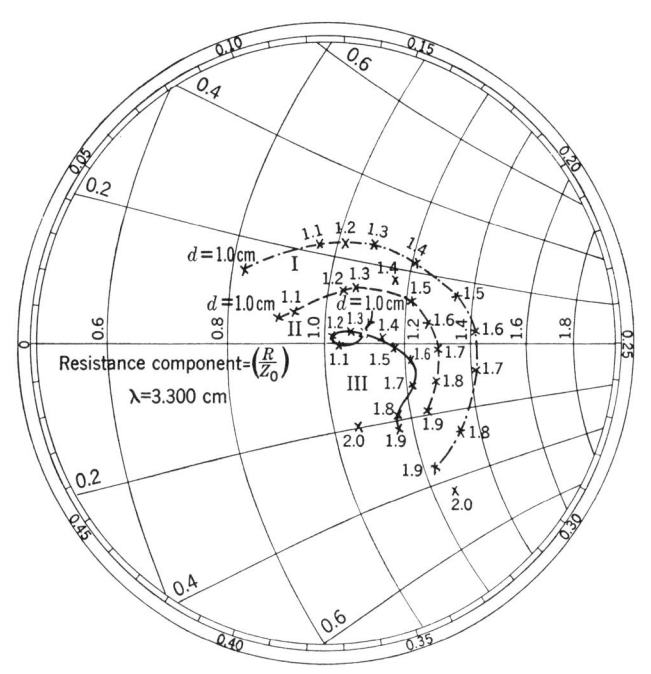

FIG. 10·36.—Impedance of a typical horn as a function of the position of the pressurizing enclosure for different wall thicknesses.

The materials used for pressurizing enclosures have been low-loss or glass thermoplastics: plexiglas or lucite

$$\left[\text{index of refraction } n = \left(\frac{\epsilon}{\epsilon_0}\right)^{1/2} = 1.60\right]$$

for wavelengths above about 6 cm, polystyrene ($n = 1.60$) and styraloy ($n = 1.60$) at 3 cm and above. Laminates can be used, but because of the high dielectric constants of these materials the enclosure walls are thin. At 1.25 cm Corning 707 glass has been used for two-dimensional pressurizing enclosures, but it presents obvious, although possibly not insuperable, difficulties in three dimensions. It also necessitates building the feed of metals with low coefficients of thermal expansion, such as invar and covar. Plastic enclosures are sealed by means of gaskets; glass by platinizing or bonding metal to the edges and soft soldering.

For a given material the wall thickness t and position d of an enclosure can be estimated from the measured reflection coefficient

$$\Gamma_F (= |\Gamma_F| e^{j\phi_F}) \tag{83}$$

of the feed referred to the aperture. The assumption is made that the reflection coefficient Γ_E from the enclosure is the same as that of an infinite sheet of the same material for plane waves. The latter is given by

$$\Gamma_E = \frac{n^2 - 1}{2n} \frac{\sin\left(\frac{2\pi nt}{\lambda}\right)}{\left[1 + \frac{(n^2 - 1)^2}{4n^2} \sin^2\left(\frac{2\pi nt}{\lambda}\right)\right]^{\frac{1}{2}}} e^{j\left\{\pi + \tan^{-1}\left[\frac{2n}{n^2+1} \cot\left(\frac{2\pi nt}{\lambda}\right)\right]\right\}}.$$

(84)

On this assumption the feed will be matched when the thickness t is such that

$$|\Gamma_F| = \frac{n^2 - 1}{2n} \frac{\sin\left(\frac{2\pi nt}{\lambda}\right)}{\left[1 + \frac{(n^2 - 1)^2}{4n^2} \sin^2\left(\frac{2\pi nt}{\lambda}\right)\right]^{\frac{1}{2}}} \quad (85)$$

and the distance d satisfies the relation

$$\frac{d}{\lambda} = \frac{1}{4\pi} \left\{ \tan^{-1}\left[\frac{2n}{n^2 + 1} \cot\left(\frac{2\pi nt}{\lambda}\right)\right] - \phi_F - 2m\pi \right\}. \quad (86)$$

where m is an integer. If the enclosure were in the Fraunhofer region and were designed so that its surfaces coincide with equiphase surfaces from the feed, the assumption involved in using Eqs. (84) to (86) would be justified. Because the enclosure is actually close to the feed aperture and cuts across equiphase surfaces, a given thickness t corresponds to smaller $|\Gamma_F|$ than is indicated by Eq. (85). For this reason also, the average path from the aperture to the inner surface of the enclosure is less than d, so that the experimental optimum d is larger than the value indicated by Eq. (86). In fact, for feeds with small E-plane apertures ($B < 3\lambda/4$), where there is a capacitive mismatch localized at the aperture (ϕ_F is approximately 270°),[1] the value of d calculated from Eq. (85) is $\lambda/4$; experimentally d is about 0.35λ for a number of typical cases. Where the feed mismatch is small or the wavelength very short, it may not be practical to use the smallest thickness t_1 calculated from Eq. (85). For these cases, as inspection of Eq. (85) will show, a reflection of the same magnitude can be obtained by using thicknesses $[t_1 + (\lambda/2n)]$; the spacing is unaffected.

The use of an external pressurizing enclosure has a number of advantages to recommend it. The possibility of breakdown at high power is minimized because the pressurized region extends beyond the feed aperture and the dielectric housing is located in a region of low field strengths. The impedance match is reproducible in the sense that it is independent

[1] "Waveguide Handbook Supplement," RL Group Report No. 41, Jan. 23, 1945.

of small fluctuations in enclosure dimensions and positioning. This is in distinct contrast to the properties of pressurizing diaphragms with flanges in the plane of the aperture, where the geometry of the flange and even the tightness of the retaining bolts have an effect on pattern and match. As a matching device the pressurizing enclosure is especially effective when the mismatch of the feed arises at the aperture. This is true for open waveguide and properly designed horns whose electric plane aperture is less than about $3\lambda/4$. Then, since enclosure and aperture are closely spaced ($d \sim \lambda/3$), the impedance match is insensitive to fre-

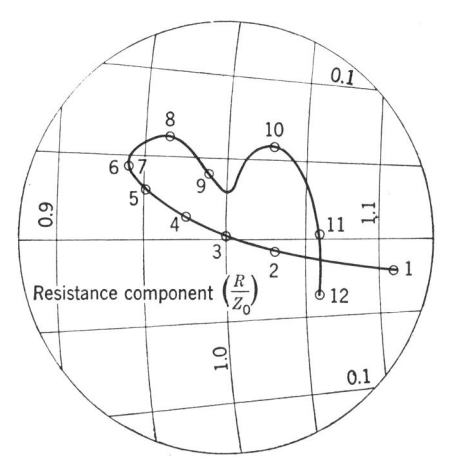

Fig. 10·37.—Impedance vs. frequency for a typical horn with properly designed pressurizing enclosure.

quency changes in which $\Delta\lambda/\lambda$ is of the order of 10 per cent. The impedance-wavelength curve of a properly designed pressurized horn is shown in Fig. 10·37 to illustrate the bandwidth of the device.

Another successful pressurization technique which has received only preliminary trial[1] consists in soldering a diaphragm of Corning 707 glass with a bonded metal rim inside the feed near the aperture. The glass window is somewhat smaller than the inside dimensions of the feed. The thickness of the glass and the dimensions of the opening can be chosen to make the effect of the diaphragm resonant, inductive, or capacitive so that any feed can be matched by this technique.

[1] M. D. Fiske, "Resonant Windows for Vacuum Seals in Rectangular Waveguides," *Rev. Sci. Instruments*, **17**, 478 (1946).

CHAPTER 11

DIELECTRIC AND METAL-PLATE LENSES

By J. R. Risser

11·1. Use of Lenses in Microwave Antennas.—The utilization of optical methods is an outstanding feature of microwave antenna design. It is natural, therefore, to consider a much-used optical device, the lens. Dielectric lenses of conventional optical design are, in general, too cumbersome for use in microwave antenna systems, but when they are zoned so that the dielectric is nowhere more than several wavelengths thick, their use is a distinct possibility. At wavelengths in the microwave region, a practical lens can also be constructed using parallel metal plates spaced a fraction of a wavelength apart, because for radiation with the electric vector parallel to the plate surfaces, the space between the plates is characterized by a longer wavelength than the free-space wavelength and consequently has the properties of a refracting medium with an index of refraction less than unity.

Lenses and reflectors are interchangeable in microwave antennas, because both perform the same basic function—modification of phase. Thus, for example, lenses can be substituted for reflectors to produce pencil beams; cylindrical lenses of suitable contour can be used with line sources to obtain asymmetrically flared beams; and line sources can be formed by the use of two-dimensional lenses between parallel plates. Reflectors have many advantageous features: mechanical simplicity, lightness, and freedom from chromatic aberration. Lenses, in turn, have characteristics that render them invaluable for many applications. Thus, in a lens system, the feed is out of the path of the main beam, a consideration of particular importance in parallel-plate line sources. Lenses are also particularly suited for insertion into optical systems to perform special corrective functions; for example, correcting lenses of the Schmidt type can be used in conjunction with a reflector to obtain a wide field in antennas for rapid scanning, and metal-plate lenses are used for phase-front correction in sectoral horns.

Because of the difference in wavelengths, microwave lens techniques are free of certain restrictions which obtain in optics. Surface tolerances are large. Dielectric lenses can be made of relatively soft thermoplastics, such as lucite or polystyrene, instead of glass, and the lens surfaces can be turned on a lathe or molded. Consequently, the surfaces need not be spherical but can be cut to contours appropriate to the function of the lens. Metal-plate lenses, likewise, can be produced by ordinary machine-shop methods. Since the fundamental function of a lens is to modify the phase fronts from a radiating source, the lens surfaces are designed using the laws of geometrical optics. However, the radiation pattern of the antenna as a whole—lens and primary source—must be considered from the standpoint of diffraction theory. The far zone field produced by the lens is obtained from the amplitude and phase distribution over its aperture by the methods of Chaps. 5 and 6.

It is the aim of this chapter to point out the methods of design, types of structure, and general problems involved in the use of lenses. Correcting lenses and other lenses designed for special purposes will not be considered. Discussion will be confined to those lenses whose function is to convert the spherical (or cylindrical) phase front from a point (or line) source at the focus of the lens into a plane phase front across the aperture. This is the most frequently recurrent problem in microwave antenna design, because, by diffraction theory, a plane phase front results in the most directive pattern for an aperture of a given size with a given amplitude distribution across it.

DIELECTRIC LENSES

11·2. Principles of Design.—The general principles of geometrical optics were formulated in Chap. 4. Lens design is based on two of these principles (Secs. 4·8 and 4·9): (1) the principle of equality of optical paths along rays between pairs of wavefronts and (2) Snell's law of refraction. The procedure of lens design is commonly referred to as "ray tracing" because it deals exclusively with the optical paths or rays, that is, the normals to the equiphase surfaces or wavefronts. In a homogeneous medium the rays are straight-line segments. In empty space the optical path length is just the length of the ray segment; in a dielectric medium, it is the length times the index of refraction n (equal to $\sqrt{k_e}$). The rays are refracted at the lens surfaces in the way described by Snell's law. The ray-tracing method consists in determining the lens surfaces, so that the combined optical length $(l_1 + nl_2 + l_3)$ (see Fig. 11·1) along any one ray between two equiphase surfaces S_1 and S_2 on opposite sides of the lens is the same as the length $(l'_1 + nl'_2 + l'_3)$ along any other ray between S_1 and S_2. The reciprocity theorem can be invoked to show that it is immaterial whether the direction of propagation is from S_1 toward S_2 or the reverse; a lens designed to convert a spherical equiphase surface from a point source F, located to the left of the lens, into a plane

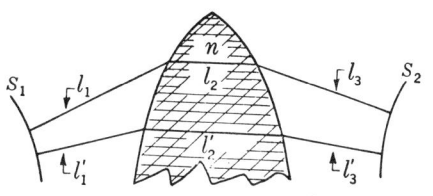

FIG. 11·1.—Optical paths and equiphase surfaces.

equiphase surface to the right, will function equally well in bringing to a focus at F the energy in a plane wave incident from the right.

11·3. Simple Lenses without Zoning.—Simple lenses will be divided into two categories according to the number of refracting surfaces. If one lens surface is an equiphase surface of the incident or emergent wave with the result that the rays are normal to the surface and pass through undeviated, the term "one-surface lens" will be applied. A "two-surface lens" is one in which refraction occurs at both lens surfaces. The design of a one-surface lens is a relatively simple problem and will accordingly be treated first.

The first to be considered is shown in Fig. 11·2. The boundary TT' between air on the left and the dielectric of index of refraction n on the right is to be determined so that the phase front S_1 from a source at F is converted into the plane phase front S_2 in the dielectric. The dielectric is terminated on the right in a plane parallel to S_2. The problem is essentially two-dimensional, whether the lens is cylindrical and F is a line source—in which case the line source and cylinder axis are perpendicular to the plane of the diagram—or the source at F is a point source and the lens is spherical,[1] that is, has rotational symmetry about the optical axis FQ_2, the line through F perpendicular to S_2. The equation of the lens surface is obtained from the condition that the optical path length $[(P_1P) + n(PP_2)]$ through an arbitrary point P shall be equal to the optical path length $[(Q_1Q) + n(QQ_2)]$ on the axis or, more simply, that

$$(FP) = (FQ) + n(QQ_2'). \tag{1}$$

In terms of $FQ \ (= f)$ and the polar coordinates (r,θ) of the point P, Eq. (1) can be written

$$r = f + n(r \cos \theta - f). \tag{2}$$

Solving for r, the equation for TT', the generating curve of the lens surface, is given by

$$r = \frac{(n-1)f}{n \cos \theta - 1}. \tag{3}$$

Since $n > 1$, this is the equation of a hyperbola of eccentricity n with the origin at one focus. The asymptotes make an angle θ_a with the lens axis given by

$$\theta_a = \cos^{-1}\frac{1}{n}. \tag{4}$$

The angle θ_a is 51° for polystyrene and plexiglas for each of which $n = 1.6$. It may be noted that the law of refraction is not used in deriving Eq.

[1] For convenience, a lens will be designated as "spherical" if its surfaces are generated by a rotation about the axis and "cylindrical" if the generating motion is a translation parallel to the line source F.

SEC. 11·3]   SIMPLE LENSES WITHOUT ZONING   391

(3); it is easy to show, however, that the law of refraction is satisfied at this boundary for the ray construction of Fig. 11·2, and it is, in fact, known from general considerations (Sec. 4·9) that it must be satisfied. It is likewise easily seen that the right-hand boundary of the dielectric has no effect on the optical paths because it coincides with an equiphase surface and, consequently, is normal to the rays PP_2, QQ_2, etc.

In evaluating the usefulness of the lens, its effect on the amplitude distribution over the aperture must be ascertained, because this as well as the phase determines the diffraction pattern. The effect differs in spherical and cylindrical lenses. For a spherical lens with a point source at F that has an axially symmetric pattern, if $P(\theta)$ is the power radiated per unit solid angle by the point source in the θ-direction and $P(\rho)$ the corresponding power per unit area in the aperture at a distance from the axis $\rho(= r \sin \theta)$, then from geometrical considerations,

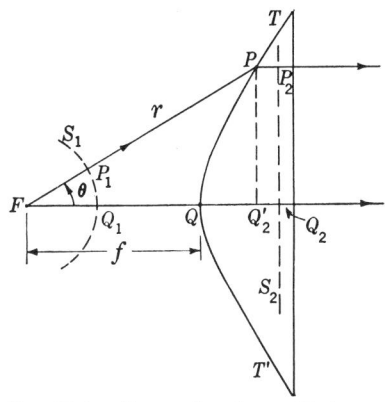

FIG. 11·2.— One-surface lens with hyperbolic contour.

$$\frac{P(\rho)}{P(\theta)} = \frac{\sin \theta \, d\theta}{\rho \, d\rho}. \tag{5}$$

Reflection at the lens surface is being neglected. For the hyperbolic surface generated by the curve of Eq. (3)

$$\frac{P(\rho)}{P(\theta)} = \frac{(n \cos \theta - 1)^3}{f^2(n - 1)^2(n - \cos \theta)}, \tag{6}$$

whereas the corresponding amplitude ratio is

$$\frac{A(\rho)}{A(\theta)} = \sqrt{\frac{(n \cos \theta - 1)^3}{f^2(n - 1)^2(n - \cos \theta)}}. \tag{7}$$

For a cylindrical lens with a line source at F, $P(\theta) \, d\theta$ is the power radiated per unit length by the line source between the angles θ and $\theta + d\theta$. Then if $P(y) \, dy$ is the power per unit length in the corresponding aperture interval between y and $y + dy$, where y is again the distance $r \sin \theta$ from the axis,

$$\frac{P(y)}{P(\theta)} = \frac{d\theta}{dy}. \tag{8}$$

For the hyperbolic surface,

$$\frac{P(y)}{P(\theta)} = \frac{(n \cos \theta - 1)^2}{(n - 1)f(n - \cos \theta)} \tag{9}$$

and
$$\frac{A(y)}{A(\theta)} = \sqrt{\frac{(n\cos\theta - 1)^2}{(n-1)f(n-\cos\theta)}}. \tag{10}$$

The amplitude ratios for spherical and cylindrical lenses normalized at $\theta = 0$ are plotted in Fig. 11·3. They drop off rapidly with increasing θ, an effect that impairs the usefulness of the lens. For the spherical lens of polystyrene or plexiglas, the aperture amplitude has dropped off 50 per cent relative to the feed amplitude at an angle of 30° with the axis, with the result that it is scarcely feasible to use a lens aperture extending beyond this point because of the high degree of taper in the illumination. This results in a serious reduction in the gain and an increased width of the main lobe of the antenna pattern (cf. Sec. 6·6).

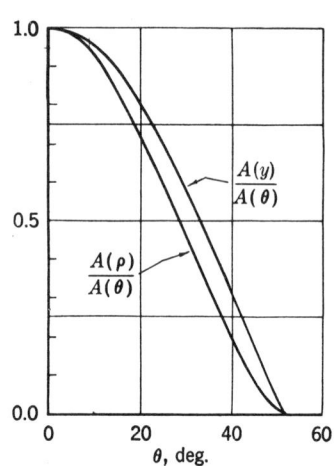

Fig. 11·3.—$A(\rho)/A(\theta)$ for a spherical lens and $A(y)/A(\theta)$ for a cylindrical lens with hyperbolic contours; $n = 1.6$.

Another design for a one-surface lens can be obtained by considering the source F (see Fig. 11·4) to be immersed in a dielectric medium of index of refraction n bounded on the right by a dielectric-air surface. In this case the equation for the generating curve T_1T_2 is found to be

$$r = \frac{f(n-1)}{n - \cos\theta}, \tag{11}$$

where f is the distance along the axis from the focus to T_1T_2. This is the equation of an ellipse of eccentricity $1/n$ with the origin at the focus farther from Q. An actual lens would be constructed, as shown in Fig. 11·4, where the source F is outside the dielectric and the incident dielectric surface $T_1'T_2'$ is spherical or cylindrical, as the case may be, and normal to the rays FP and FQ. For a given focal length, the aperture of the lens cannot be larger than $2b$, where b is the semiminor axis of the ellipse:

$$b = \sqrt{\frac{n-1}{n+1}}f. \tag{12}$$

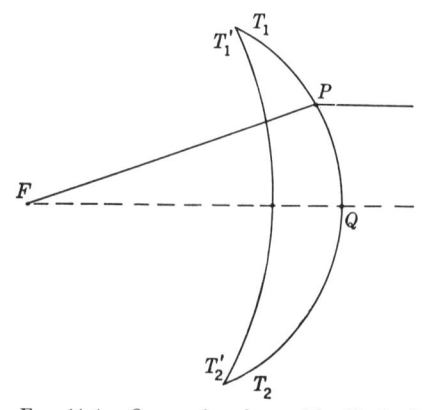

Fig. 11·4.—One-surface lens with elliptical contour.

The semiminor axis b subtends an angle θ_m at the feed given by

$$\theta_m = \cos^{-1}\left(\frac{1}{n}\right). \tag{13}$$

The minimum ratio of the focal length to the diameter is

$$\frac{f}{2b} = \frac{1}{2}\sqrt{\frac{n+1}{n-1}}, \tag{14}$$

which is 1.04 for plexiglas or polystyrene. For a spherical lens with an elliptical contour, $A(\rho)/A(\theta)$ is given by

$$\frac{A(\rho)}{A(\theta)} = \sqrt{\frac{(n-\cos\theta)^3}{(n-1)^2 f^2 (n\cos\theta - 1)}} \tag{15}$$

and for a cylindrical lens by

$$\frac{A(y)}{(A\theta)} = \sqrt{\frac{(n-\cos\theta)^2}{(n-1)f(n\cos\theta - 1)}}. \tag{16}$$

Plots of these amplitude ratios normalized to unity at $\theta = 0$ are given in Fig. 11·5. The amplitude in the aperture increases relative to the feed amplitude with increasing θ. This property of the elliptical contours is desirable for microwave work, because the lens compensates for the directivity of the feed pattern, producing a more efficient illumination over the aperture from the standpoint of antenna gain. On the other hand, the more uniform illumination enhances the side-lobe structure of the pattern, and the lens is not suited for use with an antenna feed of too low a directivity.

Practical considerations of bulk and efficiency place a limit on the useful aperture of this lens. For polystyrene or plexiglas, the maximum diameter $5b$ is $0.96f$, and the half angle θ_m subtended at the feed F is 51°. For a 30° half angle, on the other hand, the lens diameter is 85 per cent of maximum; and in the case of the spherical lens, the volume of dielectric is approximately one-fourth of the volume with maximum diameter. For a spherical lens cut at $\theta = 30°$, $A(\rho)/A(\theta)$ is 1.7, with the result that the feed pattern must be down 14.5 db from $\theta = 0$ in order to make the power at the edge of the aperture 10 db down from center.

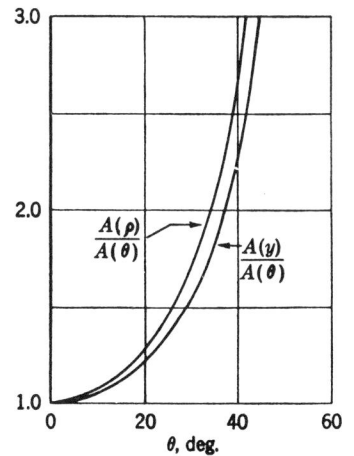

Fig. 11·5.—$A(\rho)/A(\theta)$ for a spherical lens and $A(y)/A(\theta)$ for a cylindrical lens with elliptical contours; $n = 1.6$.

A "two-surface lens" is one in which neither of the lens surfaces coincides with a surface of equal phase and the rays undergo refraction upon entering and leaving the lens. The use of a second refracting surface is one way of increasing the versatility of the lens. The optics of a two-surface lens will be outlined briefly, although there will be no detailed discussion of an example. Figure 11·6 is a sketch of the geometry. As before, F is a point or line source to the left of the first surface whose contour is denoted by $T_1 T_1'$. Since refraction occurs at both surfaces, there is no unique form for the equiphase surfaces in the dielectric. It is convenient in this case to discuss the problem in terms of the angle θ' (Fig. 11·6) which a ray entering the dielectric at the point (r, θ) makes with the axis of the lens. The angle θ' is determined from the form of the contour $T_1 T_1'$, or vice versa, in accordance with Snell's law. In terms of θ' the differential equation for the contour $T_1 T_1'$ can be shown to be

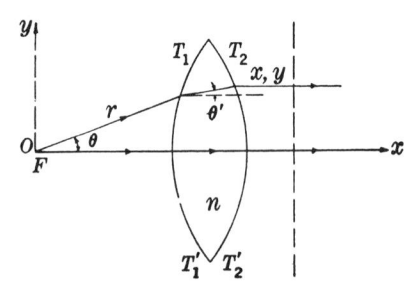

Fig. 11·6.—Coordinate system for a two-surface lens.

$$\frac{1}{r}\frac{dr}{d\theta} = \frac{n \sin(\theta - \theta')}{n \cos(\theta - \theta') - 1}. \qquad (17)$$

The coordinates x, y of the point where the ray from (r, θ) intersects the second surface $T_2 T_2'$ are determined by the geometrical relationship

$$\frac{y - r \sin \theta}{x - r \cos \theta} = \tan \theta' \qquad (18)$$

and by the condition for the equality of optical paths. Since the equiphase surfaces to the right of $T_2 T_2'$ are required to be planes, the condition on the optical paths is given by

$$r + n \sqrt{(y - r \sin \theta)^2 + (x - r \cos \theta)^2} - x = \text{constant.} \qquad (19)$$

As is easily seen, Eqs. (17) to (19) are not sufficient to determine uniquely the coordinates of both surfaces. Another condition, essentially equivalent to a condition on θ', may be imposed. For example, it may be required that the lens be free of coma (to render it suitable for use in a scanning antenna) or that the amplitude ratio $A(y)/A(\theta)$ of Eq. (5) or (8) be specified as a function of y or θ. For a general-purpose microwave antenna it would be desirable that the amplitude ratio be constant or at least a slowly varying function of y or θ, with the result that the taper in the angular pattern of the feed is reproduced in the aperture. As far as is known, a practical solution of this problem has not yet been obtained.

SEC. 11·4]   ZONED DIELECTRIC LENSES   395

When the contour $T_1 T_1'$ is determined by arbitrarily setting θ' equal to $\theta/2$, it is found that $A(y)/A(\theta)$ is very nearly constant for a cylindrical lens of moderate aperture and only very slowly varying for a spherical lens. This lens is thick, however, for reasonable apertures. In general, with the large apertures used in microwave applications, considerations of weight and bulk make zoned construction a practical necessity. Consequently, the design of lenses without zoning is of somewhat academic interest, and the conditions on the lens surfaces are not complete without provision for the zone steps. This will be treated in the next section.

Before leaving the subject of unzoned lenses, however, we may take note of an expression for the thickness of a simple converging lens on the axis. This expression is useful in estimating lens proportions. It can easily be seen from the principle of equality of optical paths that if R and θ_0 are the coordinates of the apex of the lens where the dielectric reaches zero thickness, the thickness t on the axis is given by

$$R(1 - \cos \theta_0) = (n - 1)t. \qquad (20)$$

For example, a polystyrene lens with R equal to 20 wavelengths and θ_0 equal to 30° is 4.5 wavelengths (free-space) thick.

11·4. Zoned Dielectric Lenses.—A simple lucite or polystyrene lens of the type described in the preceding section is many wavelengths thick if its focal length and aperture are large compared with a wavelength. For a simple lens the optical path length along the axis is the same as the length by way of the edge. This condition is unnecessarily restrictive, however, at microwave frequencies where the wavelength is large compared with ordinary manufacturing tolerances. The surfaces of microwave lenses can be divided into zones with the optical paths differing by integral multiples of a wavelength from one zone to another. A lens may accordingly be designed with its cross section similar to those shown in Fig. 11·7. Starting with zero thickness at the edge of the lens the thickness of dielectric may be progressively increased toward the lens axis, as required by the phase condition, until the path difference introduced by the presence of the dielectric is equal to a wavelength. At this point the path in the dielectric can be reduced to zero without altering the wavefronts from the lens. This is then the outer boundary of another zone, through which the optical path lengths are one wavelength less than those through the outermost zone. This zone likewise increases

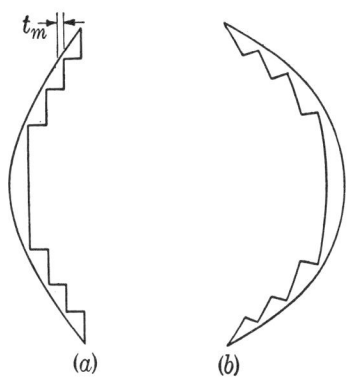

FIG. 11·7.—Zoning of nonrefracting lens surfaces.

in thickness in the direction of the axis until the point is reached where reduction of the dielectric thickness to zero results in an optical path length smaller by another wavelength, and so on. The resulting lens is similar to a conventional zone plate except that the path difference between zones is equal to a single wavelength. When the lens has K zones, the optical paths through the outermost zone are $(K - 1)\lambda$ longer than those passing through the zone on the lens axis.

The maximum thickness of a zoned lens is approximately equal to $\lambda/(n - 1)$, because the maximum path difference $(n - 1)t$ introduced in a path of length t by the presence of the dielectric is approximately a wavelength. In actual practice a small thickness t_m (Figs. 11·7 and 11·8) must be left at the thinnest points for reasons of mechanical strength, so that the maximum thickness is greater than $\lambda/(n - 1)$ by this amount.

A good example of a zoned two-surface lens which has been tested and used[1] is shown in Fig. 11·8. This is a plexiglas lens, 13.5 in. in diameter, for use at 3.3 cm. The surface toward the feed is chosen somewhat arbitrarily to be a plane. This choice has much to recommend it, however, because a plane surface should have somewhat less back reflection to the feed than a concave surface and better illumination characteristics than a convex surface. Except for the inclusion of zones, the lens is designed in the manner described for two-surface lenses in Sec. 11·3. Once a plane for the first surface and the distance from this plane to the focus (6 in. in this case) are chosen, the lens structure is completely determined by the requirement that the equiphase surfaces to the right of the lens shall be planes: Snell's law determines the directions of the rays in the dielectric of the lens; then the second surface is determined from the principle of equality of optical paths with the provision that the paths differ by a wavelength from one zone to another. The following equations describe the zoned surface:

$$d = \left[\frac{(K - 1)\lambda + (n - 1)D + f - \sqrt{f^2 + r^2}}{n - \sqrt{1 - \frac{r^2}{n^2(r^2 + f^2)}}} \right] \sqrt{1 - \frac{r^2}{n^2(f^2 + r^2)}} \quad (21)$$

and

$$R = r \left[1 + \frac{d}{\sqrt{n^2(f^2 + r^2) - r^2}} \right]. \quad (22)$$

The notation is that of Fig. 11·8. The zone number K is unity for the central portion of the lens. This zone is carried out from the axis to a point where the thickness of material t_m is considered a minimum for mechanical strength. There the step is made to the surface determined by $K = 2$, and so on to the edge of the lens.

[1] A. M. Skellett, "Plexiglas Lens Antenna for Microwaves," BTL Report MM-43-170-15, September 1943.

SEC. 11·4]   ZONED DIELECTRIC LENSES   397

In designing a zoned lens care must be taken to avoid excessive shadow area between zones in the lens aperture. Shadow invariably occurs when zone steps are cut in a refracting surface of the lens because the ray

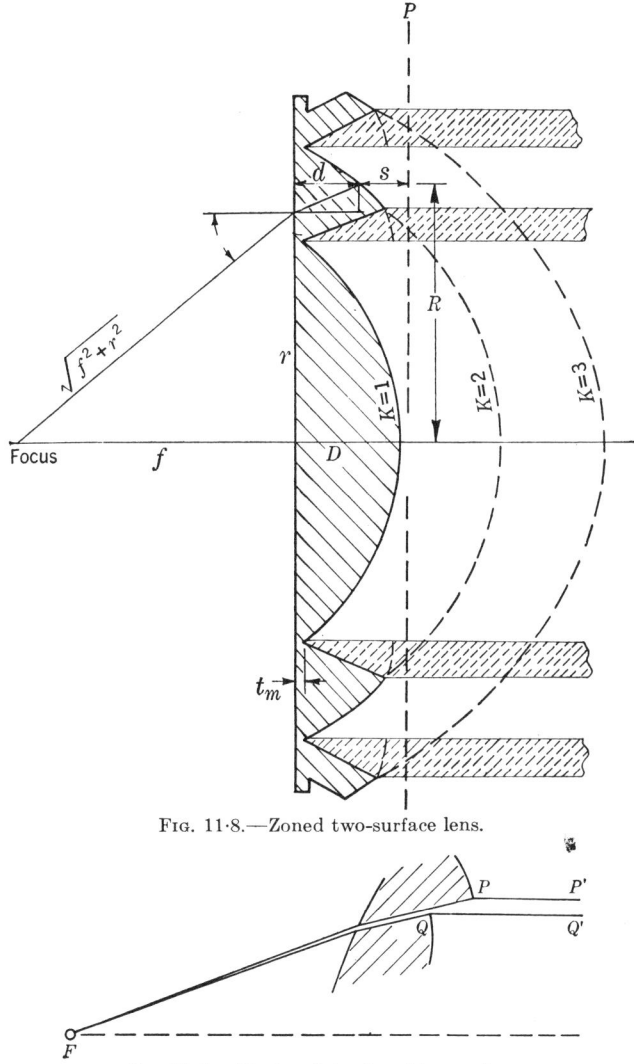

FIG. 11·8.—Zoned two-surface lens.

FIG. 11·9.—Shadow introduced by zoning.

adjacent to a step inside the dielectric has a different direction from that of the adjacent ray outside. This is shown in Fig. 11·9 where the rays FPP' and FQQ', which are together on entering the lens, have undergone considerable separation at the step. In transmission the aperture illumination is zero between P and Q; in reception the energy incident between P and Q does not reach F and is therefore lost. Shadow does not occur where the step is cut in an equiphase lens surface because the directions of the rays are not altered at the surface and hence undergo no separation. The effect of shadow regions, of course, shows up in the secondary pattern of the lens as a decrease in gain and increase in side lobes. Data are lacking at present on the relation between these effects and the size and position of the shadow regions.

Frequency Sensitivity.—Since dielectric constants are independent of frequency in the microwave region, an unzoned dielectric lens performs its function regardless of frequency. With zoning, however, frequency sensitivity is introduced. Let P_1 be the length of optical path from the focus F along the axis to any plane perpendicular to the axis on the far side of the lens. If K is the total number of zones, the optical path P_2 through the Kth zone is $P_1 + (K - 1)\lambda_0$ where λ_0 is the design wavelength. At wavelength λ_0, the radiation from the first and Kth zones will be in phase in any plane perpendicular to the optical axis. However, at a near-by wavelength $\lambda_0 + \Delta\lambda$, the wavefront from the Kth zone will be displaced a distance δ along the axis relative to the wavefront from the first zone. When the changes in wavelength are small, δ in wavelengths is given by

$$\frac{\delta}{\lambda} = \left[\frac{d}{d\lambda}\left(\frac{P_2 - P_1}{\lambda}\right)\right]_{\lambda=\lambda_0} \Delta\lambda, \qquad (23)$$

whence

$$\frac{\delta}{\lambda} = -(K - 1)\frac{\Delta\lambda}{\lambda}.$$

The usual criterion for microwave work is that the displacement δ shall not exceed 0.125λ.[1] Using this criterion and defining the bandwidth as twice the maximum allowable fractional change in wavelength expressed in per cent,

$$\text{Bandwidth} \approx \frac{25}{K - 1}. \qquad \text{per cent.} \qquad (24)$$

A lens of 4 per cent bandwidth can have seven zones with λ_0 steps between zones. The formula is approximate because $\Delta\lambda$ is assumed small in the derivation. It should be noted that $(K - 1)$ is actually the number of wavelength steps introduced by zoning; if there are steps of two or more wavelengths, this must be taken into account. It is believed that Eq. (24) gives a conservative estimate of bandwidth, because with tapered aperture illumination, higher values of δ/λ might be tolerated for certain applications.

11·5. Use of Materials with High Refractive Indexes.—Recently materials with high refractive indexes and low losses have been developed. The use of these materials would greatly reduce the bulk of microwave

---

[1] The relation between the gain of a pencil beam antenna and the phase error over the aperture is discussed in Sec. 12.5.

lenses. Two distinct types exist. Polyglas[1] mixtures with titanium dioxide or titanate fillers have refractive indexes from 1.7 to 4.9, depending on composition. Power factors are 0.002 to 0.003 at 10^{10} cps. These materials have the advantage of possessing coefficients of thermal expansion near those of copper and brass. Very high refractive indexes (about 10) are obtained from titanium dioxide and titanate ceramics.[2] The titanium dioxide ceramics have power factors below 10^{-3} at 3×10^9 cps. There is no fundamental obstacle preventing the use of these materials for microwave lenses. At present lack of development of manufacturing techniques for heat treating and molding of large samples is the principal difficulty. The degree of control necessary for successful manufacture is indicated in the reports on the materials. Because tolerances against warpage and twisting are large for lens surfaces (Sec. 11·6), a lens could be made as an assemblage of small sections, in order to decrease the size of furnaces and molds.

In addition to techniques of manufacture, materials with high refractive indexes present several problems to the lens designer. The tolerances on lens thickness become important (Sec. 11·6), and reflections from lens surfaces result in prohibitively high transmission losses (Sec. 11·7) unless surface-matching sections are added.

11·6. Dielectric Losses and Tolerances on Lens Parameters.—In evaluating the usefulness of microwave lenses, it is necessary to consider the practical problems arising from properties of lens materials and possible limitations in methods of manufacture. Attenuation in the dielectric must be reasonably small. It must be possible to fabricate the lens to satisfactory tolerances on the contours of both surfaces and on the thickness. Many materials with suitable refractive indexes are lossy at microwave frequencies. The attenuation in decibels per (free-space) wavelength in an unbounded dielectric medium is given by

$$A = 27.3n \frac{\epsilon''}{\epsilon'}, \qquad (25)$$

where n is the index of refraction and ϵ' and ϵ'' are the real and imaginary parts of the complex dielectric constant $(\epsilon' - j\epsilon'')$ characteristic of lossy materials. (The index of refraction is the square root of ϵ'/ϵ_0, which is the inductive capacity k_e ordinarily quoted.) The ratio ϵ''/ϵ' is equal to the power factor of the material when it is small compared with unity. Since the maximum thickness times $(n - 1)$ is about a wavelength for a zoned microwave lens, the upper limit to the attenuation in a zoned lens

---

[1] A. von Hippel, S. M. Kingsbury, and L. G. Wesson, "Low Thermal Expansion Plastics," NDRC 14-539, October 1945.

[2] A. von Hippel, R. G. Breckenridge, A. P. de Bretteville, Jr., J. M. Brownlow, F. G. Chesley, G. Oster, L. Tisza, and W. B. Westphal, "High Dielectric Constant Ceramics," NDRC 14-300, August 1944.

is approximated by

$$A_M \approx 27.3 \left(\frac{\epsilon''}{\epsilon'}\right) \frac{n}{n-1}, \qquad (26)$$

which is about $70\epsilon''/\epsilon'$ for polystyrene or lucite and about $35\epsilon''/\epsilon'$ for a dielectric with n equal to 4.5. Thus, power factors up to 0.003 can be used without introducing more than a few tenths of a decibel attenuation. For polystyrene ($\epsilon''/\epsilon' \approx 0.0003$) the attenuation is negligible.

The tolerance to be placed on any lens parameter is proportional to the maximum allowable irregularity in the wavefronts, or equiphase surfaces, formed by the lens. This is again taken to be $\lambda/8$. The tolerances on thickness and index of refraction are interrelated because the compensation in optical path introduced by the presence of dielectric of thickness t is $(n-1)t$. Setting $\lambda/16$ as the upper limit on wavefront irregularities arising from variation in either t or n to allow for variations in both quantities, we have, approximately,

$$\Delta t \leq \frac{\lambda}{16(n-1)} \qquad (27)$$

and

$$\Delta n \leq \frac{\lambda}{16t}. \qquad (28)$$

Since $(n-1)t$ is of the order of a wavelength for zoned microwave lenses, Eq. (28) becomes

$$\frac{\Delta n}{n-1} < \frac{1}{16}. \qquad (29)$$

The tolerance on thickness becomes important only for materials with high index of refraction. For polystyrene at a wavelength of 3.2 cm, Eq. (27) gives a thickness tolerance of $\frac{1}{8}$ in., whereas for a substance with a refractive index of 4.5, it is 0.020 in. at the same wavelength. As regards the dielectric constant, a variation greater than 3 per cent is not likely to occur when reasonable care is taken in manufacture, even in materials of high dielectric constants. This variation is well within the limits prescribed by Eq. (29). If variation in thickness alone or dielectric constant alone is considered, the tolerances given by Eqs. (27) and (28) may be increased by a factor of 2.

Some restriction on the surface contours of a lens arises from the tolerance on the thickness t. The two surfaces can be deformed simultaneously by warping, however, without affecting the thickness appreciably. From Fermat's principle, the length of optical path through any portion of a lens has an extremum value and consequently, small displacements of any section of the lens result in changes in optical path that are small compared with the displacements. The tolerance on a surface contour, except when it affects thickness, is therefore at least as large as the maximum allowable irregularity in the wavefronts.

11·7. Reflections from Dielectric Surfaces.—Reflections from dielectric surfaces can cause feed mismatch and power loss. Feed mismatch is most likely to occur when a lens surface coincides with an equiphase surface because the reflection from the entire surface is then in phase at the feed. In this case the surface reflection coefficient is given by the well-known expression for normal incidence

$$R = -\frac{n-1}{n+1}. \quad (30)$$

Equation (30) yields a value of 0.23 for plexiglas or polystyrene and, of course, larger values for higher dielectric constants. A reflection of this magnitude is too large if picked up by the feed, and so the use of an equiphase surface as a lens surface is to be avoided whenever possible.

For high indexes of refraction and large angles of incidence, power loss itself becomes important. For a refractive index of 4.5 the loss is 40 per cent at normal incidence, so that surface-matching sections are necessary. The reflection coefficient R depends not only on n but also on the angle of incidence. This is shown in Fig. 11·10, where $|R|$ is plotted for a plane wave incident on the plane surface of an infinite dielectric slab ($n = 1.6$) at angles up to 90°. The curve for polarization with the electric vector perpendicular to the plane of incidence is a plot of the well-known relation

$$|R| = \frac{\sin(i-r)}{\sin(i+r)}$$

with
$$r = \sin^{-1}\left(\frac{1}{n}\sin i\right). \quad (31)$$

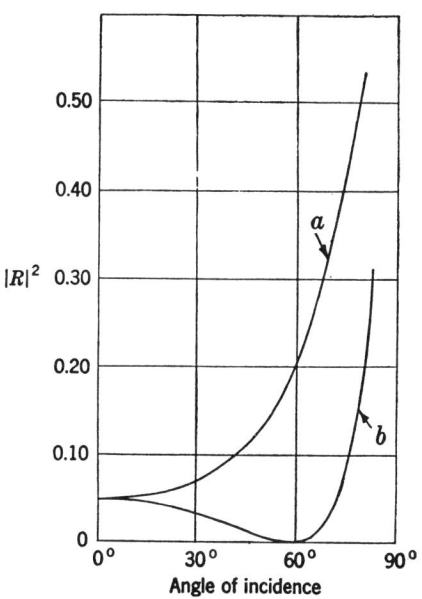

FIG. 11·10.—Fraction of incident power reflected from the surface of an infinite dielectric slab vs. angle of incidence: (a) electric vector perpendicular to the plane of incidence; (b) electric vector in the plane of incidence; $n = 1.6$.

In this case $|R|$ increases with angle of incidence from its value at normal incidence, slowly at small angles and rapidly in the neighborhood of 90°. The power loss reaches a value of 10 per cent at about 40° for $n = 1.6$. Account must be taken of this effect in the design of lenses. For polarization with the electric vector in the plane of incidence,

$$R = \frac{\tan(i-r)}{\tan(i+r)}. \tag{32}$$

In this case $|R|$ decreases with angle of incidence until it reaches zero at the Brewster angle, $\tan^{-1} n$; beyond this angle it again increases. Matching devices can be used to cut down surface reflections, as, for example, a quarter-wavelength-thick surface layer of material whose refractive index is the geometric mean of the refractive index of the lens dielectric and that of air. Such a matching section is, of course, an additional complication in the design and manufacture of the lens.

METAL-PLATE LENSES

11·8. Parallel-plate Lenses.—A lens structure using spaced conducting planes instead of a dielectric has been developed[1] for use at microwave frequencies. A common form consists of parallel strips of sheet metal held apart by accurate nonreflecting spacers. Where the electric vector is parallel to the plate surfaces and the plate spacing a is less than λ, but greater than $\frac{\lambda}{2}$ (cf. Sec. 7·15), the wavelength between the plates is given by

$$\lambda_P = \frac{\lambda}{\sqrt{1 - \left(\dfrac{\lambda}{2a}\right)^2}}. \tag{33}$$

Since this wavelength is greater than the free-space wavelength, the parallel-plate space has the properties of a refracting medium with index of refraction less than unity. When a thickness of this medium is introduced into the paths of the rays, optical path lengths are reduced from their free-space values instead of increased as in the case of dielectrics. Thus a converging parallel-plate lens is thinner on the axis than at the edge of the lens, and a diverging lens is thicker. This is in contrast to dielectric-lens structure. A parallel-plate lens can be designed to have variable thickness, like a dielectric lens, or it can have a uniform thickness and variable plate spacing. The former, a more common design, will be considered here.

The refractive index of the parallel-plate space is given by

$$n = \frac{\lambda}{\lambda_P} = \sqrt{1 - \left(\frac{\lambda}{2a}\right)^2} \tag{34}$$

[1] W. E. Kock, "Experiments with Metal Plate Lenses for Microwaves," BTL Report MM-44-160-67, March 1944; "Wire Lens Antennas," MM-44-160-100, April 1944; "Metal Plate Lens Design Considerations," MM-44-160-195, August 1944; "Metal Plate Lenses for Microwaves," MM-45-160-23, March 1945; "Metal-Lens Antennas," *Proc. IRE*, **34**, 828–836; November 1946.

Sec. 11·8]   PARALLEL-PLATE LENSES   403

with $\lambda/2 < a < \lambda$. Values of n lie between 0 and about 0.86. Naturally the smallest practical value of n should be used in order to minimize the thickness of the lens. Owing to difficulties arising from the use of plate separations near cutoff, namely, frequency sensitivity, mismatch at the lens surface, and difficulty in maintaining the tolerances on plate spacing, 0.5 is generally considered the minimum practical value of n. One important difference between dielectric and parallel-plate lenses consists in the constraints placed on the rays in the lens by the presence of the parallel plates. The direction of propagation in the lens must be parallel to the plate surfaces.

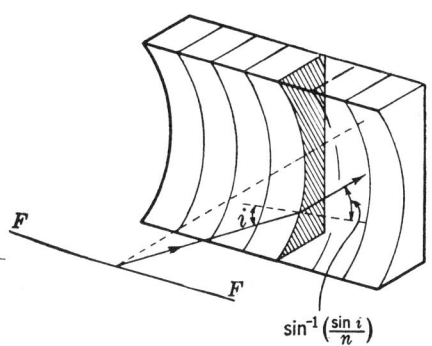

FIG. 11·11.—Cylindrical parallel-plate lens in which change of angle at the surface is determined by Snell's law.

Thus Snell's law does not, in general, describe the change in direction of a ray at the lens surface. In the case of cylindrical lenses this fact does not essentially change the lens design. Here two cases may be distinguished. When the electric vector and lens plates are perpendicular to the cylinder axis, Snell's law is valid and the constraints do not enter (Fig. 11·11). When the electric vector and plate edges lie in planes parallel to the cylinder axis, Snell's law is replaced by the condition of the constraint (Fig. 11·12). However, the design of a spherical lens consisting only of plates parallel to the E-plane of the feed becomes a three-dimensional problem instead of the two-dimensional one discussed for dielectrics. The constraint exists in the magnetic but not in the electric plane, and the lens surface is not symmetrical with respect to a rotation about the axis. The design can, of course, be reduced to a two-dimensional problem by use of a cellular construction to introduce identical constraint in the electric plane.

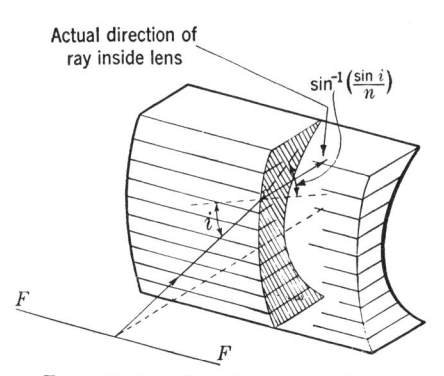

FIG. 11·12.—Cylindrical parallel-plate lens in which change of angle at the lens surface is determined by the condition that the ray is constrained to pass between the plates.

There is one useful example in which the constraint does not enter explicitly, and most spherical parallel-plate lenses used hitherto have been of this type. This is the one-surface lens in which the refracting surface is on the side of the feed with the second surface a plane perpendicular to the lens axis. The rays are then parallel to the axis, and the constraints are automatically satisfied. As discussed in Sec. 11·3, the contour of this lens may be designed solely from the condition of equality of optical paths. Referring to Sec. 11·3, to the case where the refracting medium is to the right of the contour TT' (Fig. 11·2), the coordinates of TT' are given again by Eq. (3):

$$r = \frac{(1-n)f}{1 - n \cos \theta}. \quad (3)$$

In the present case, however, with $n < 1$, this is the equation of an ellipse; the hyperbolic face of the dielectric lens is replaced by an elliptical face in the metal-plate lens. Since little use is made of unzoned lenses, it is desirable to rewrite this equation to apply to a zoned lens. If the zone on the lens axis is taken as the first, the equation for the surface of the Kth zone is given by

$$r_K = \frac{(1-n)f_K}{1 - n \cos \theta} \quad (35)$$

with

$$f_K = f_1 + \frac{(K-1)\lambda}{1-n}. \quad (36)$$

In this equation the assumption is made that there are steps of one wavelength between adjacent zones; otherwise $(K-1)$ in Eq. (36)

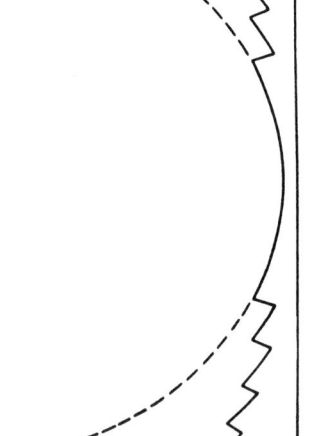

Fig. 11·13.—Elliptical contour with five zones; $n = 0.5$; $f = 40\lambda$.

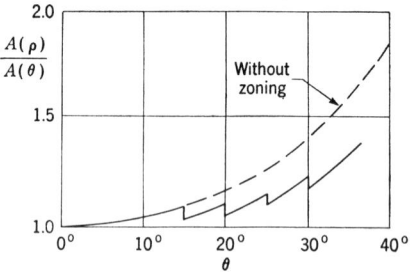

Fig. 11·14.—Ratio $A(\rho)/A(\theta)$ for zoned contour of Fig. 11·13.

must be replaced by the total number of wavelengths in the steps between the axis and the Kth zone. Figure 11·13 shows a five-zone arrangement of this contour using 0.5 for n and 40λ for f_1. The ratio of f_1 to aperture is 0.86. The unzoned contour is shown for comparison. The amplitude-illumination ratio (Sec. 11·3) which relates the amplitude across the aperture to the amplitude pattern of the feed is given for a "spherical" lens by

$$\frac{A(\rho)}{A(\theta)} = \sqrt{\frac{(1 - n \cos \theta)^3}{(1-n)^2 f_K^2 (\cos \theta - n)}} \quad (37)$$

and for a cylindrical lens by

$$\frac{A(y)}{A(\theta)} = \sqrt{\frac{(1 - n \cos \theta)^2}{(1 - n)f_K(\cos \theta - n)}}. \tag{38}$$

The ratio $A(\rho)/A(\theta)$ of Eq. (37) for the spherical case is plotted in Fig. 11·14 for the five-zoned elliptical contour of Fig. 11·13. The ratio exhibits a slow stepwise increase from the center to the edge of the lens

Fig. 11·15.—Rear view of a 1.25-cm parallel-plate lens. (*Courtesy of the Bell Telephone Laboratory.*)

in a fashion favorable for microwave use. The improvement introduced by zoning is seen by comparing with the dotted curve for an unzoned contour. Figure 11·15 is a photograph of the rear view of a spherical lens;[1] it is 48 in. in diameter and designed for use in the 1-cm range. It is a zoned single-surface lens with elliptical contours. This lens is constructed of thin, equally spaced metal plates parallel to the E-plane of the feed. The time involved in cutting the plates for lenses of this type is a factor, because the plates differ in contour except for corresponding pairs on opposite sides of the optical axis. Since surface tolerances are large (Sec. 11·10), the contours can be cut by sawing or filing to a scribe line or by stacking the plates with temporary wooden spacers and cutting on a lathe.

[1] W. E. Kock, "Metal Plate Lenses for Microwaves," BTL Report MM-44-160-100, April 1944.

11·9. Other Metal-lens Structures.—Several other methods of lens construction have been tried with success.[1] One procedure that leads to good structural characteristics makes use of polystyrene foam ($\epsilon = 1.018$ and weight 1 to 2 lb/ft^3) as a dielectric medium between the plates. The plates are actually sheets of metal foil bonded to the polystyrene. The slabs of polystyrene foam with the metal foil sidings may be molded into a block, and the lens contour can then be cut out on a lathe.

Another method is to replace the lens plates by a system of parallel wires. Lenses of this type are interesting principally because they point to the possibility of using lenses at long wavelengths where metal plates are out of question but curtains of wire suspended from poles are feasible. The lens structure is based on the fact that slots in the wall of a waveguide which are parallel to the electric-field vector do not radiate. The system of parallel wires may be thought of as a limiting condition arrived at by cutting slots in the parallel plates constituting the lens. The primary problem in the design is the

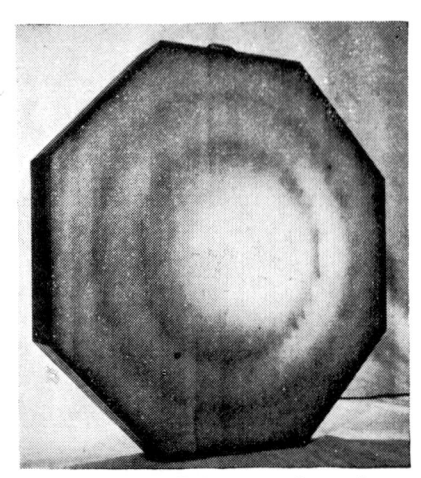

FIG. 11·16.—Polystyrene-foam lens. (*Courtesy of the Bell Telephone Laboratory.*)

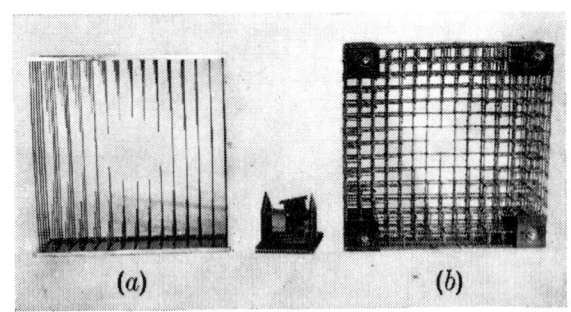

FIG. 11·17.—Wire lenses: (*a*) parallel-wire lens; (*b*) wire mesh lens. (*Courtesy of the Bell Telephone Laboratory.*)

determination of the practical ratio of slot width to conductor width. It has been found[2] that at a wavelength of 3.2 cm wires of diameter 0.049 in. and spaced 0.3 in. center to center have a loss of 0.1 db per inch or about $\frac{1}{8}$ db per wavelength. Since the lens is only a few wavelengths

[1] W. E. Kock, "Metal Plate Lenses for Microwaves," BTL Report MM-45-160-23, March 1945.
[2] W. E. Kock, "Wire Lens Antennas," BTL Report MM-44-160-100, April 1944.

thick, this is not prohibitive. The over-all performance of a lens of this type has been found to compare favorably with an equivalent lens making use of solid plates.

Several of these types of lenses are illustrated in Figs. 11·16 and 11·17. The former shows a lens making use of a polystyrene-foam base. Figure 11·17a shows a parallel wire lens, and Fig. 11·17b shows a wire mesh lens. The latter takes as its starting point the use of rectangular waveguides as the lens medium. The waveguide walls are replaced by wire mesh, again making use of the fact that slots in a waveguide wall, when suitably oriented, do not radiate.

11·10. Metal-plate Lens Tolerances.—Tolerances on the lens surface are large with respect to deformation by warping and twisting, as they are for dielectric lenses. Extreme rigidity in the cellular structure is consequently not necessary. Tolerances on the lens thickness and plate spacing are interrelated. Using the same criteria as for the dielectric lenses (allowing an error either in thickness or refractive index alone to cause more than half the allowable phase error $\lambda/8$),

$$\Delta t \leq \frac{\lambda}{16(1-n)} \tag{39}$$

and

$$\Delta n \leq \frac{\lambda}{16 t}, \tag{40}$$

where Δn depends both on plate spacing and wavelength. Reserving discussion of changes with wavelength until later (Sec. 11·11), the variation of n with plate spacing a is given by

$$\Delta n = \frac{(1-n^2)}{n} \frac{\Delta a}{a}. \tag{41}$$

If in addition $(1-n)t$ is assumed to be approximately a wavelength, as it is for zoned lenses, Eq. (40) becomes

$$\frac{\Delta a}{a} \leq \frac{n}{16(n+1)}. \tag{42}$$

When n is equal to 0.5, Δa must be less than 0.024λ, which is 0.030 in. at 3.2 cm. At this wavelength the tolerance on a would place a lower practical limit of about 0.3 on n, because this would lead to a tolerance of about 0.014 in. From Eq. (39) a value of 0.5 for n leads to an exceedingly liberal tolerance on t, i.e., $\Delta t \leq \lambda/8$. This is about twice the tolerance on a reflector contour for the same over-all phase error of $\lambda/8$; a discrepancy of $\lambda/16$ in a reflector contour leads to a phase error of approximately $\lambda/8$. The tolerances given here are conservative, based on the assumption that both thickness and plate spacing are in error. If the process of fabrication is such as to hold one or the other to better than the stated values, the alternate tolerance may be increased.

11·11. Bandwidth of Metal-plate Lenses; Achromatic Doublets.—Since the index of refraction of a metal-plate medium depends on the ratio of plate spacing to wavelength, metal-plate lenses are frequency-sensitive devices. The relation between small changes in λ and corresponding changes in n is obtained by differentiating Eq. (34):

$$\Delta n \approx -\frac{1-n^2}{n}\frac{\Delta\lambda}{\lambda}. \qquad (43)$$

The effect on the wavefronts from an unzoned lens will be considered first. Let L_2 represent the total length of the line segments FP and PP' (Fig. 11·18) passing through the edge of the lens and L_1 the length FQQ' on the axis. If P_2 and P_1 are the optical lengths of L_2 and L_1 respectively and t is the difference between the thickness of the lens at the edge and at the center, the optical path difference $(P_2 - P_1)$ is given by

$$P_2 - P_1 = (L_2 - t + nt) - L_1. \qquad (44)$$

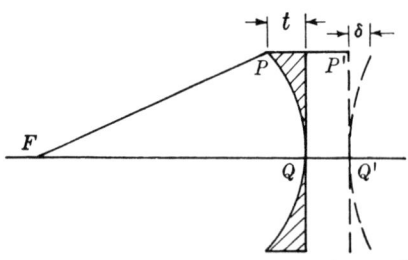

Fig. 11·18.—Effect on a wavefront of change in wavelength from the design wavelength.

The path difference $(P_2 - P_1)$ is a function of wavelength because n depends on wavelength. At the design wavelength λ_0, $(P_2 - P_1)$ must be zero in order that $P'Q'$ may represent a wavefront. Hence

$$L_2 - (1 - n_0)t = L_1. \qquad (45)$$

At a near-by wavelength $\lambda_0 + \Delta\lambda$, the wavefront at P' is displaced a distance δ along the axis with respect to the wavefront at Q', where δ/λ is given by

$$\frac{\delta}{\lambda} \approx \left[\frac{d}{d\lambda}\left(\frac{P_2 - P_1}{\lambda}\right)\right]_{\lambda=\lambda_0} \Delta\lambda. \qquad (23)$$

The approximation is good only for very small $\Delta\lambda$. Substituting from Eqs. (43) and (45),

$$\frac{\delta}{\lambda} \approx -\frac{1+n_0}{n_0}\frac{(1-n_0)t}{\lambda_0}\frac{\Delta\lambda}{\lambda_0}. \qquad (46)$$

If the limits of bandwidth are defined by $|\delta|/\lambda$ equal in magnitude to 0.125, the bandwidth is given by

$$\text{Bandwidth} \approx \frac{25n_0}{1+n_0}\frac{\lambda_0}{(1-n_0)t} \quad \text{per cent.} \qquad (47)$$

In the special case n_0 equal to 0.5,

$$\text{Bandwidth} \approx 8.3 \frac{\lambda_0}{(1 - n_0)t} \quad \text{per cent.} \tag{48}$$

Since $(1 - n_0)t$ is at least several times λ_0, bandwidths are of the order of a few per cent. It is believed that Eq. (48) gives a conservative estimate of the bandwidth, because, with considerable taper in the aperture illumination, values of $(|\delta|/\lambda)_{\text{max}}$ up to 0.25 might be tolerated for certain applications.

In determining the bandwidth of zoned lenses, the same type of procedure is followed. The frequency sensitivity of zoned lenses is due not only to the variation in n but also to the steps. As for unzoned lenses the maximum deviation in the wavefront can be calculated by comparing the optical path along the lens axis with the path by way of the edge. Using the same notation as before, we have

$$P_2 - P_1 = (L_2 - t + nt) - L_1, \tag{44}$$

with the somewhat different condition at λ_0 introduced by the zoning:

$$P_2 - P_1 - (K - 1) \lambda_0 = 0. \tag{49}$$

Here K is the number of zones, counting the zone on the axis as the first. In this case

$$\left[\frac{d}{d\lambda} \left(\frac{P_2 - P_1}{\lambda} \right) \right]_{\lambda = \lambda_0} \Delta\lambda = -(K - 1) \frac{\Delta\lambda}{\lambda_0} - \frac{(1 + n_0)}{n_0} \frac{(1 - n_0)t}{\lambda_0} \frac{\Delta\lambda}{\lambda_0} \tag{50}$$

and the bandwidth is approximately given by

$$\text{Bandwidth} \approx 25 \frac{1}{(K - 1) + \frac{1 + n_0}{n_0} \frac{(1 - n_0)t}{\lambda_0}} \quad \text{per cent.} \tag{51}$$

For zoned lenses $(1 - n_0)t$ is approximately one wavelength at the thickest portions. For practical purposes, therefore,

$$\text{Bandwidth} \approx 25 \frac{n_0}{1 + Kn_0} \quad \text{per cent.} \tag{52}$$

For the special case $n_0 = 0.5$,

$$\text{Bandwidth} \approx \frac{25}{2 + K} \quad \text{per cent.} \tag{53}$$

Zoning increases the bandwidth of a lens. For example, the use of Eqs. (48) and (53) to compare equivalent zoned and unzoned lenses which introduce compensation of five wavelengths in the longest optical path [$(1 - n_0)t = 5\lambda_0$ for the unzoned lens; $(1 - n_0)t = \lambda_0$, $K = 5$, for the zoned lens] shows that the zoned lens has slightly more than twice the bandwidth of the unzoned lens, 3.57 per cent as compared with 1.67

per cent. For large values of K the zoned lens with n_0 equal to 0.5 has approximately three times the bandwidth of the equivalent unzoned lens since a lens with K zones is equivalent to an unzoned lens with $(1 - n_0)t$ equal to $K\lambda_0$. This indicates the advisability of zoning on the basis of bandwidth alone. Moreover it suggests the possibility[1] of using a doublet consisting of zoned and unzoned lenses with opposite frequency characteristics to obtain increased bandwidth. A zoned converging lens, for example, is stronger than an unzoned diverging lens of opposite frequency sensitivity, and the combination is a converging lens. It must be remembered, however, that the focal length of the doublet is much longer than that of the uncorrected converging lens unless the compensat-

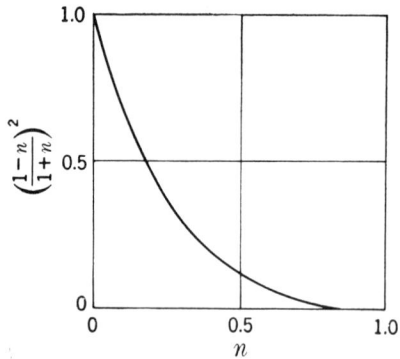

Fig. 11·19.—Power reflection at normal incidence as a function of n.

ing lens can be made optically thin and given the requisite frequency sensitivity by spacing the plates closely to yield small n.

For any lens the first-order effect of change in frequency is to alter the effective focal length of the lens. For frequencies close to the design frequency the deformed wavefronts are so nearly spherical that moving the feed along the axis effectively removes the deformation. The effective bandwidth of a lens is consequently increased by a provision in the antenna system for feed motion.

11·12. Reflections from Surfaces of Parallel-plate Lenses.—While the general problem of reflection from the surfaces of a parallel-plate lens has not been solved, some indication of magnitude can be obtained from a study of the reflection of a plane wave from the edges of an array of parallel, equally spaced plates when the edges lie in a plane. This problem has received rigorous theoretical treatment.[2] It seems reasonable to expect that the values of R derived for this case at various angles of incidence should be a good approximation to local values of R on a

[1] W. E. Kock, "Experiments with Metal Plate Lenses for Microwaves," BTL Report MM-160-67, March 1944.

[2] J. F. Carlson and Albert E. Heins, "The Reflection of an Electromagnetic Plane Wave by an Infinite Set of Plates, I," *Quart. Applied Math.*, **4**, 313–329, January 1947.

lens surface, providing that lens surfaces and wavefronts do not appreciably depart from a plane over distances comparable to a wavelength. For normal incidence R is given by

$$R = \frac{1-n}{1+n} e^{j\Phi}. \qquad (54)$$

This expression differs from the expression for normal incidence on a dielectric surface only by the presence of the phase angle Φ. Because n is less than unity, the magnitude of R is larger, however, than that from a dielectric surface with the same value of $|1 - n|$. The magnitude

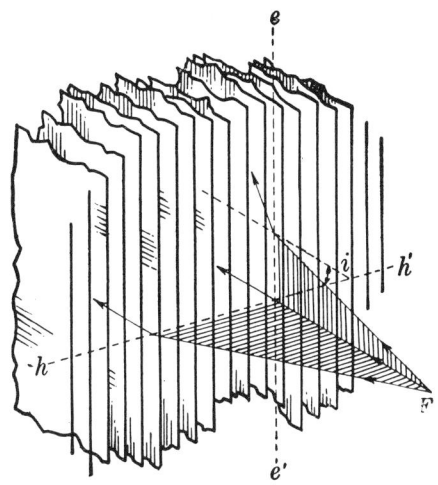

FIG. 11·20.—Planes of incidence at the plane face of an infinite set of parallel plates.

of $|R|^2$ at normal incidence is plotted in Fig. 11·19. Its value for small n obviously puts a lower practical limit on n, especially for equiphase lens surfaces where the reflected wave is in phase at the feed. For n equal to 0.5, the reflection is already quite large (11 per cent power reflection). Surface-matching devices are, of course, a possibility.

The average reflection over a lens surface is probably less than the value derived for normal incidence because the magnitude of the reflected wave probably decreases with the angle of incidence in both planes. For the magnetic plane ($hh'F$ in Fig. 11·20) where the change in angle at the surface is determined by the constraint and not by Snell's law, an expression has been derived for $|R|$ as follows:

$$|R| = \left|\frac{\cos i - n}{\cos i + n}\right|. \qquad (55)$$

This expression is valid for a restricted range of angles:

$$0 \leqq i \leqq \sin^{-1}\left(\frac{\lambda}{a} - 1\right)$$

with

$$1 < \frac{\lambda}{a} < 2.$$

The restriction on i arises from the fact that grating lobes become possible for angles of incidence larger than $\sin^{-1}\left(\dfrac{\lambda}{a} - 1\right)$. Inspection of Fig. 11·20 and a plot of $|R|^2$ in Fig. 11·21 shows that $|R|$ decreases with increasing i, reaches zero for i equal to $\cos^{-1} n$, and increases beyond that angle.

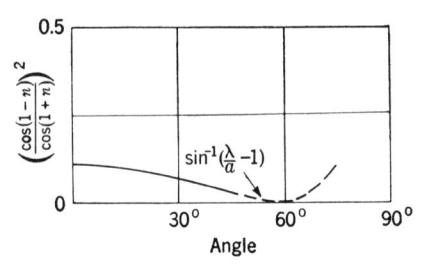

Fig. 11·21.—Power reflection at the plane face of an infinite set of parallel plates as a function of angle of incidence.

For the electric plane ($ee'F$ of Fig. 11·20) the expression for the variation of R with angle of incidence has not been derived. In this plane the change of angle upon refraction is described by Snell's law. One might expect some similarity to the dielectric case where the magnitude of the reflected wave decreases from its value at normal incidence with increasing i until it reaches zero at the Brewster angle, $\tan^{-1} n$, and then increases. It is hoped that a solution for this plane will become available in the future.

CHAPTER 12

PENCIL-BEAM AND SIMPLE FANNED-BEAM ANTENNAS

By S. Silver

PENCIL-BEAM ANTENNAS

12·1. Pencil-beam Requirements and Techniques.—The term "pencil beam" is applied to a highly directive antenna pattern consisting of a single major lobe contained within a cone of small solid angle and almost circularly symmetrical about the direction of peak intensity. As used here, it will apply to beams with half-power width less than 15°. These beams are analogous to searchlight beams, and, as with an optical searchlight, the elevation and azimuth coordinates of a target in space can be simply correlated with the similar coordinates that define the orientations of the antenna. In connection with the technique of using radar echoes for obtaining range information, the pencil-beam antenna serves to define the position of a target completely.

There are several possible techniques for producing pencil beams. The simplest in conception and from the point of view of practical design is that of placing a point source at the focus of an "optical" system, such as a reflector or lens, to produce a beam of parallel rays. It is evident that to produce a circularly symmetrical beam, the optcial system should have rotational symmetry with the feed located on the axis of rotation (optical axis). This presupposes that the primary feed pattern likewise has rotational symmetry about the same axis; in practice this is approximated by a feed pattern having a pair of orthogonal principal planes (symmetry planes) that intersect along the optical axis, with nearly equal half-power widths in the two planes. In many calculations this actual feed pattern can be replaced by an equivalent circularly symmetric pattern that is the average of the patterns in the two principal planes. If the simple geometrical picture—that the beam produced by the optical system consists of a family of parallel rays— were strictly valid, the beam would have "zero" width as plotted in a polar diagram. However, this simple picture is markedly modified by diffraction phenomena due to the limited aperture of the optical system. The aperture is the projected area of the reflector or lens on a plane normal to the optical axis, and for a rotationally symmetrical system it is circular in shape. As a result of diffraction the antenna pattern has a major lobe of finite width and characteristic side-lobe structure.

The general theory of apertures and diffraction has been developed in Chap. 6. It was shown there that of all the phase and amplitude distributions over a plane aperture that give rise to a beam with maximum intensity in the direction normal to the aperture, a uniform amplitude and phase distribution gives rise to maximum gain; in general, minimum beamwidth is concomitant with maximum gain.

The relation between antenna gain and range in radar systems has been noted in Sec. 1·2 [Eq. (1·12)]. The beamwidth is also an important factor in the precision with which target location can be effected. The considerations here are partly optical (of exactly the same nature as those which determine the resolving power of a telescope[1]) and in part involve system factors such as pulse width; a rather complete discussion of resolving power of a radar set and its bearing on beamwidth requirements is given in Vol. 1 of this series.[2] On the basis of gain and beamwidth considerations *a fundamental design requirement for radar antennas is that the phase distribution over the aperture be uniform—in terms of geometrical optics, that the optical system produce a beam of parallel rays.* It should be noted that in addition to the gain and beamwidth requirements, the greatest possible suppression of all secondary lobes is desirable; for if a target is sufficiently close to be detected by the side lobe, it becomes indistinguishable from a target detected by the main lobe at the same range. However, as was found in the treatment of general diffraction theory, requirements of maximum gain and minimum sidelobe level are generally incompatible. The necessary compromise between them in antenna design is made in optical systems by adjusting the illumination, that is, the amplitude distribution, over the aperture.

The advantages of microwaves become strongly evident in the design of the pencil-beam antennas. Within reasonable limits on the over-all size of the antenna, the distance from the reflector (or lens) to the antenna feed can be made so large that the optical device is in the radiation zone of the feed. Thus the difficulties associated with the phase quadrature of the induction field are avoided; that is, it is possible to operate in that region of the feed pattern where the feed is essentially a point source. Because the dimensions of the reflectors and lenses are fairly large compared with the wavelength, it is possible to simplify the theoretical considerations by suitable approximations. As a result the design of a pencil-beam antenna becomes to a large extent a calculable procedure.

Inasmuch as lenses have been discussed in detail in Chap. 11, the treatment of design problems in the present chapter will be confined almost entirely to reflectors. Many problems are common to both: The secondary pattern is determined essentially by the field over the aperture, and the requirements to be imposed on the latter, which will be

---

[1] M. Born, *Optik*, Edwards Bros., Inc., Ann Arbor, Mich., Chap. 4.

[2] Ridenour, *Radar System Engineering*, Vol. 1, RL Technical Series, Chap. 14.

arrived at from the discussion of reflectors, can be transferred directly to lenses. The latter just began to commond serious attention at the close of the war period, and their study and use are still in the initial stages.

PARABOLOIDAL REFLECTORS[1]

12·2. Geometrical Parameters.—The nature of a reflector that transforms a spherical wave, arising from a point source, into a plane wavefront was discussed in Sec. 4·9, where it was found to be a paraboloid of revolution with the source at the focus. In discussing these systems it is convenient to use several different coordinate systems simultaneously; these are defined in Fig. 12·1. A rectangular coordinate system x, y, z will be used, with the origin at the vertex of the paraboloid and the z-axis the axis of revolution. In these coordinates the equation of the paraboloidal surface is

$$x^2 + y^2 = 4fz, \qquad (1)$$

where $f = OF$ is the focal length. We shall also use cylindrical coordinates r, ξ, z, where r and ξ are polar coordinates in the planes z = constant, ξ being measured from the xz-plane. In these coordinates the equation of the surface is

$$r^2 = 4fz. \qquad (2)$$

In expressing the relation of the primary feed pattern to the reflector, there is employed a spherical coordinate system ρ, ψ, ξ, with the origin at the focus F and the polar axis directed in the negative z-direction; the aximuth angle ξ is the same as that defined in the cylindrical system, and ψ is the polar angle. The equation of the surface referred to these spherical coordinates was obtained in Sec. 4·9 [Eq. (4·69)]; it is

$$\rho = \frac{2f}{1 + \cos \psi} = f \sec^2 \left(\frac{\psi}{2}\right). \qquad (3)$$

Lastly, to discuss the final antenna pattern we use a spherical coordinate system with polar axis in the *positive* z-direction and origin again at the

---

[1] The material to be presented in the following sections represents a summary of British and American work done during the war period; the following is a partial bibliography of reports on paraboloidal reflectors: L. J. Chu, "Theory of Radiation from Paraboloidal Reflectors," RL Report No. V-18, Feb. 12, 1941; E. U. Condon, "Theory of Radiation from Paraboloid Reflectors," Westinghouse Report No. 15, Sept. 24, 1941; G. F. Hull, Jr., "Application of Principles of Physical Optics to Design of UHF Paraboloid Antennas," BTL Report MM-43-110-2, Feb. 8, 1943; and F. R. N. Nabarro, "Theoretical Work on the Paraboloid Mirror," British Report, Ministry of Supply, A. C. 1435, RDF 103, Com. 72, Nov. 27, 1941. References to earlier French and German work are given by F. E. Terman, *Radio Engineers' Handbook*, McGraw-Hill, New York, 1943, p. 837.

focus; the coordinates are R, θ, ϕ, with θ the polar angle and ϕ the azimuth angle, the latter being measured from the xz-plane.

The reflector is cut off by the "aperture plane" A at $z = z_0$. The diameter of the aperture will be designated by D, and its area by A. The "shape" of the reflector is specified by the ratio of focal length to

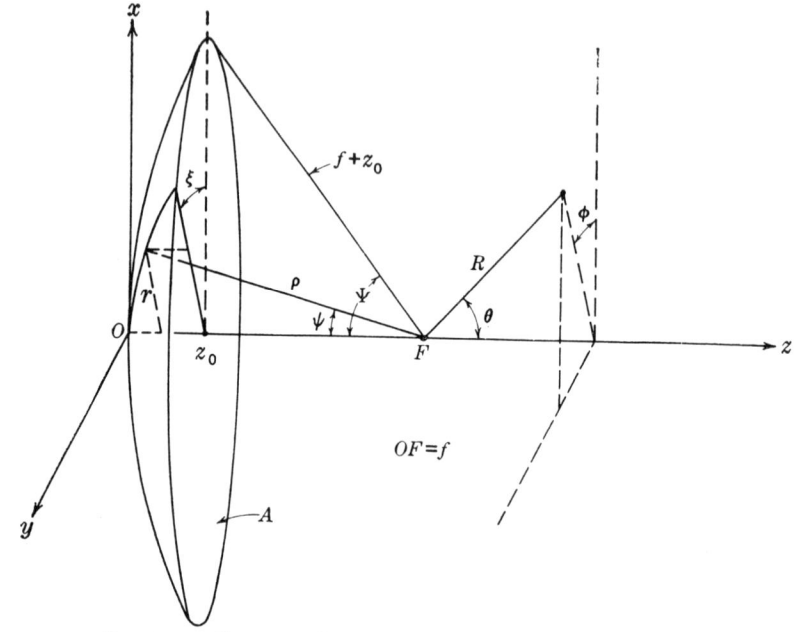

FIG. 12·1.—Geometrical parameters for the paraboloidal reflector.

diameter, f/D, or alternatively by the angular aperture Ψ, that is, the angle subtended at the focus by a radius of the aperture. The relation between the f/D ratio and the angular aperture is given by

$$\sin \Psi = \frac{1}{2}\left(\frac{\dfrac{D}{f}}{1 + \dfrac{D^2}{16f^2}}\right), \tag{4a}$$

$$\tan \Psi = \frac{1}{2}\left(\frac{\dfrac{D}{f}}{1 - \dfrac{D^2}{16f^2}}\right). \tag{4b}$$

One of the most important design problems is the determination of the shape that gives maximum antenna gain for a given aperture diameter and a given primary feed pattern.

The geometrical properties of paraboloids are well known. Any section of the surface containing the z-axis is, of course, a parabola with focus at F. In addition, however, the curve of intersection of the surface with any plane parallel to the z-axis (normal to the xy-plane) is also a parabola of the same focal length f as the paraboloid. As a consequence of the property, only a single parabolic template is needed in the construction of the reflector to test the accuracy of all parts of the surface. The normal to the surface at a point ρ, ξ, ψ lies in the plane containing this point and the z-axis and makes an angle $\psi/2$ with the incident ray from F.

12·3. The Surface-current and Aperture-field Distributions.—In the treatment of the general theory of reflectors developed in Chap. 5 it is shown that the over-all pattern of the antenna, that is, the secondary pattern, arises by the superposition upon the radiation field of the antenna feed of the radiation field of the distribution of current generated on the surface of the reflector in the presence of the feed. It was shown further that the reflector field can be determined either from the surface-current distribution directly or in the form of a diffraction pattern from the field distribution over the aperture of the mirror. Before proceeding with the calculation of the surface-current and aperture-field distributions, some fundamental ideas and assumptions which underlie all of the subsequent discussions should be noted. The feed pattern in the presence of a reflector, in general, differs from its free-space pattern, because the reaction of the reflector on the antenna feed modifies its current system. If, however, the focal length of the paraboloid is at least several wavelengths in magnitude and the mirror is in the radiation zone of the free-space pattern of the feed, the interaction between the mirror and the antenna feed is a second-order effect as far as the primary pattern is concerned. These conditions are usually realized in microwave antennas; and subject to their realization, it will be assumed that the feed pattern in the presence of the reflector is the same as under free-space conditions.

To avoid the complex problem of interference between the fields of the reflector and the antenna feed in the formation of the main structure of the antenna pattern [cf Eqs. (5·75) and (5·76)], the directivity of the feed pattern should be such that the major portion of the energy lies within the cone defined by the feed and the reflector. Referring to Fig. 12·1, if—taking an ideal case—the primary pattern is zero for angles $\psi > 90°$, the main structure of the beam is determined by the reflector currents alone. The directive feeds discussed in Chaps. 8 and 10 approximate this condition rather closely; their back lobes, however, are not completely negligible and have significant effects not only on the wide-angle side lobes where the back-lobe field is comparable to the weak reflector field but also on the peak intensity, that is, on the antenna gain. The effect of the back lobe on gain will be investigated in Sec. 12·5.

The general approximation procedure based on geometrical optics and plane-wave boundary conditions, which is discussed in Chap. 5, will be used to evaluate the surface-current and aperture-field distributions. Let the principal E- and H-planes of the primary feed pattern coincide with the xz- and yz-planes, respectively, in Fig. 12·1. If P_T is the total power radiated by the feed and $G_f(\xi,\psi)$ its gain function, the power $P(\xi,\psi)$ radiated per unit solid angle in the direction ξ, ψ is

$$P(\xi,\psi) = \frac{P_T}{4\pi} G_f(\xi,\psi).$$

The electric-field-intensity primary pattern, reduced to the unit sphere about the center of feed, is then given by

$$[\mathbf{E}(\xi,\psi)]_{\rho=1} = \left\{ 2\left(\frac{\mu}{\epsilon}\right)^{\frac{1}{2}} \left[\frac{P_T}{4\pi} G_f(\xi,\psi)\right] \right\}^{\frac{1}{2}} \mathbf{e}_0(\xi,\psi), \tag{5}$$

where $\mathbf{e}_0$ is a unit vector defining the polarization in the primary pattern.[1] The field intensity in the incident wave at a point ρ, ξ, ψ on the reflector is therefore given by

$$\mathbf{E}_0 = \left[2\left(\frac{\mu}{\epsilon}\right)^{\frac{1}{2}} \frac{P_T}{4\pi} \right]^{\frac{1}{2}} \frac{[G_f(\xi,\psi)]^{\frac{1}{2}}}{\rho} e^{-jk\rho} \mathbf{e}_0. \tag{6}$$

The field intensity $\mathbf{E}_1$ in the reflected wave at the same point is

$$\mathbf{E}_1 = \left[2\left(\frac{\mu}{\epsilon}\right)^{\frac{1}{2}} \frac{P_T}{4\pi} \right]^{\frac{1}{2}} \frac{[G_f(\xi,\psi)]^{\frac{1}{2}}}{\rho} \bar{e}^{jk\rho} \mathbf{e}_1, \tag{7}$$

where $\mathbf{e}_1$ defines the polarization in the reflected wave; according to the plane-wave boundary condition [Eq. (5·25)] the vectors $\mathbf{e}_0$ and $\mathbf{e}_1$ are connected by the relation

$$\mathbf{n} \times (\mathbf{e}_0 + \mathbf{e}_1) = 0, \tag{8}$$

in which $\mathbf{n}$ is the unit vector normal to the reflector at the point of incidence. The vector $\mathbf{n}$ will be taken to be directed outward from the reflector into free space. Following Eq. (5·57a), the surface-current density $\mathbf{K}$ is given in terms of the incident wave by

$$\mathbf{K} = \left[8\left(\frac{\epsilon}{\mu}\right)^{\frac{1}{2}} \frac{P_T}{4\pi} \right]^{\frac{1}{2}} \frac{[G_f(\xi,\psi)]^{\frac{1}{2}}}{\rho} e^{-jk\rho} [\mathbf{n} \times (\boldsymbol{\varrho}_0 \times \mathbf{e}_0)], \tag{9}$$

where $\boldsymbol{\varrho}_0$ is a unit vector in the direction of the incident ray. Expanding the vector product, we obtain

$$\mathbf{K} = \left[8\left(\frac{\epsilon}{\mu}\right)^{\frac{1}{2}} \frac{P_T}{4\pi} \right]^{\frac{1}{2}} \frac{[G_f(\xi,\psi)]^{\frac{1}{2}}}{\rho} e^{-jk\rho} \left[\mathbf{e}_0 \cos\frac{\psi}{2} + (\mathbf{e}_0 \cdot \mathbf{n})\boldsymbol{\varrho}_0 \right]. \tag{9a}$$

The current can be expressed in a similar manner in terms of the reflected

[1] It is being assumed that the radiation field of the primary feed is linearly polarized at every point but that $\mathbf{e}_0$ is a function of ξ, ψ.

field by making use of Eq. (5·57b), noting that the reflected ray is parallel to the z-axis; we have then

$$\mathbf{K} = \left[8 \left(\frac{\epsilon}{\mu} \right)^{1/2} \frac{P_T}{4\pi} \right]^{1/2} \frac{[G_f(\xi,\psi)]^{1/2}}{\rho} e^{-jk\rho} [\mathbf{n} \times (\mathbf{i}_z \times \mathbf{e}_1)] \tag{10}$$

or

$$\mathbf{K} = \left[8 \left(\frac{\epsilon}{\mu} \right)^{1/2} \frac{P_T}{4\pi} \right]^{1/2} \frac{[G_f(\xi,\psi)]^{1/2}}{\rho} e^{-jk\rho} \left[-\mathbf{e}_1 \cos \frac{\psi}{2} + \mathbf{i}_z(\mathbf{n} \cdot \mathbf{e}_1) \right]. \tag{10a}$$

To obtain the field over the aperture we note that since the reflected rays are all parallel, the field intensity remains constant in magnitude

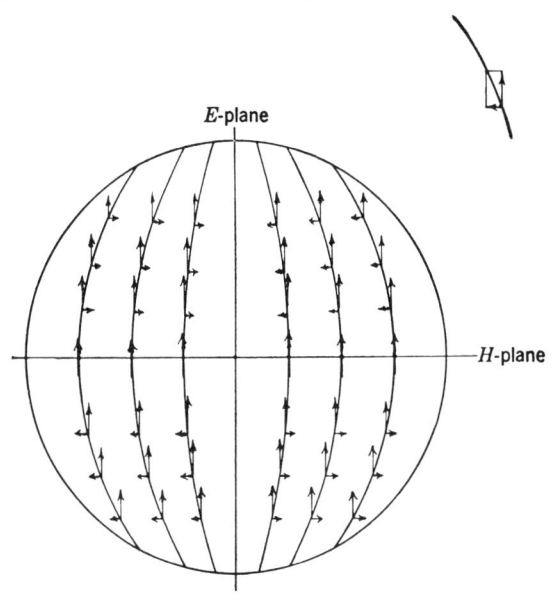

Fig. 12·2.—Typical aperture-field distribution; the field is resolved into principal and cross-polarization components.

along the reflected ray (*cf.* Sec. 4·4). The electric-field intensity $\mathbf{E}(r,\xi)$ at a point (r,ξ) on the aperture is thus given directly by $\mathbf{E}_1$ at the corresponding point (ρ,ξ,ψ), except for the phase retardation corresponding to the path from the reflector to the aperture plane. The relation is

$$\mathbf{E}(r,\xi) = \mathbf{E}_1(\rho,\xi,\psi) e^{-jk(z_0-z)}$$
$$= \left[2 \left(\frac{\mu}{\epsilon} \right)^{1/2} \frac{P_T}{4\pi} \right]^{1/2} \frac{[G_f(\xi,\psi)]^{1/2}}{\rho} e^{-jk(\rho+z_0-z_1)} \mathbf{e}_1. \tag{11}$$

The distance $\rho + z_0 - z$ is the total optical path from F to the aperture plane; it is therefore independent of the point (r,ξ), and more specifically it is equal to $f + z_0$. Comparing the surface-current distribution as given by Eq. (10a) with the aperture-field distribution [Eq. (11)], it is seen that except for constants, the aperture field is the projection of the surface-current distribution into the aperture plane. In this connection it should be noted that the longitudinal component of the current given by the term $(\mathbf{e}_1 \cdot \mathbf{n})\mathbf{i}_z$ in Eq. (10) finds no counterpart in the aperture distribution because the field over the aperture is wholly transverse to the z-axis. This longitudinal component of the current has generally been neglected in paraboloid theory. It contributes nothing to the field in the forward direction and therefore does not enter into the computation of the peak intensity, but it does modify the side-lobe structure of the beam. The general character of the aperture distribution (and the transverse component of the current distribution) is illustrated in Fig. 12·2. It is seen that the polarization reflects the symmetry of the primary feed pattern. The component e_{1x} of the aperture polarization which is parallel to the principal E-plane of the feed is known as the principal polarization component, and the component e_{1y}, which is transverse thereto, is known as the cross-polarization component. By virtue of the symmetry conditions the cross-polarization components at any pair of points that are symmetrical with respect to the principal planes are effectively 180° out of phase with one another.

12·4. The Radiation Field of the Reflector.—The secondary pattern produced by the reflector may now be calculated from either the current distribution or the aperture field, using the methods of Secs. 5·9 and 5·12, respectively. The two calculations do not lead to completely concordant results; the differences between them vanish, however, in the limit of zero wavelength. The discrepancy lies in the fact that the aperture field which would be calculated as produced by the surface currents is equal to that calculated on the basis of the reflected rays only under the limiting condition of zero wavelength. To exhibit the relationships we shall set up the expressions for the radiation field as obtained from the current distribution. Letting $\mathbf{R}_1$, $\mathbf{i}_\theta$, $\mathbf{i}_\phi$ be unit vectors associated with the spherical coordinates R, θ, ϕ (Fig. 12·1), we have, by Eqs. (5·74a) to (5·74c), that the radiation field of the reflector is

$$\left.\begin{array}{c}E_\theta \\ E_\phi\end{array}\right\} = -\frac{j\omega\mu}{2\pi R}e^{-jkR}\left[\left(\frac{\epsilon}{\mu}\right)^{\frac{1}{2}}\frac{P_T}{2\pi}\right]^{\frac{1}{2}}\left\{\begin{array}{c}\mathbf{i}_\theta \cdot \mathbf{I} \\ \mathbf{i}_\phi \cdot \mathbf{I}\end{array}\right. \quad (12)$$

where the vector $\mathbf{I}$, expressed in terms of the incident field on the reflector is

$$\mathbf{I} = \int_0^{2\pi}\int_0^{\Psi}\frac{[G_f(\xi,\psi)]^{\frac{1}{2}}}{\rho}\mathbf{n} \times (\varrho_0 \times \mathbf{e}_0)e^{-jk\rho[1+\cos\psi\cos\theta-\sin\psi\sin\theta\cos(\xi-\phi)]}$$

$$\times \rho^2 \sin\psi \sec\frac{\psi}{2}\,d\psi\,d\xi \quad (12a)$$

and, in terms of the reflected field,

$$\mathbf{I} = \int_0^{2\pi}\int_0^{\Psi}\frac{[G_f(\xi,\psi)]^{\frac{1}{2}}}{\rho}[\mathbf{n} \times (\mathbf{i}_z \times \mathbf{e}_1)]e^{-jk\rho[1+\cos\psi\cos\theta-\sin\psi\sin\theta\cos(\xi-\phi)]}$$

$$\times \rho^2 \sin\psi \sec\frac{\psi}{2}\,d\psi\,d\xi. \quad (12b)$$

Comparing with Eqs. (10) and (10a) it is seen that in the form of Eq. (12b), the vector **I** is resolved into a transverse component parallel to the xy-plane,

$$\mathbf{I}_t = \int_0^{2\pi} \int_0^{\Psi} \frac{[G_f(\xi,\psi)]^{1/2}}{\rho} \left(-\mathbf{e}_1 \cos \frac{\psi}{2} \right) e^{-jk\rho(1+\rho^2)} \sin \psi \sec \frac{\psi}{2} d\psi \, d\xi, \quad (13a)$$

and a longitudinal component

$$\mathbf{I}_z = \mathbf{i}_z \int_0^{2\pi} \int_0^{\Psi} \frac{[G_f(\xi,\psi)]^{1/2}}{\rho} (\mathbf{n} \cdot \mathbf{e}_1) e^{-jk\rho(1+\rho^2)} \sin \psi \sec \frac{\psi}{2} d\psi \, d\xi. \quad (13b)$$

As regards the longitudinal component it is observed that $\mathbf{I}_z$ makes no contribution to the E_ϕ-component of the field because $\mathbf{i}_\phi$ is always in a plane normal to $\mathbf{i}_z$. Furthermore, since $\mathbf{i}_z \cdot \mathbf{i}_\theta = \sin \theta$, the longitudinal component makes no contribution to the field in the direction $\theta = 0$—the physical basis for this being that a current element is equivalent to a dipole and does not radiate in the direction along its axis. The contribution of $\mathbf{I}_z$ is significant only at wide angles. For the systems with which we are concerned that produce narrow beams, the contribution of $\mathbf{I}_z$ is a second-order effect; it vanishes in the limit of zero wavelength. There is no counterpart of the $\mathbf{I}_z$ contribution in the calculation of the pattern from the aperture-field distribution.

Considering the transverse component $\mathbf{I}_t$, it will be observed that if the radiation field is confined to a small angular region about the $\theta = 0$ axis, the variation of $\cos \theta$ in the phase term of Eq. (13a) can be neglected; we have then $\rho(1 + \cos \psi \cos \theta) \approx 2f$. Also it will be noted on comparing with Eq. (11) that except for a multiplicative constant—which is contained in the field expressions [Eq. (12)]—the factor $[G_f(\xi,\psi)]^{1/2} \mathbf{e}_1/\rho$ is the field in the aperture plane at the point (r,ξ) which corresponds to the point (ρ,ξ,ψ) on the reflector. Equation (13a) is, therefore, given approximately by the integral

$$\mathbf{I}_t \approx -e^{j2kf} \int_0^a \int_0^{2\pi} \frac{[G_f(\xi,\psi)]^{1/2}}{\rho} \mathbf{e}_1 e^{jkr \sin\theta \cos(\xi-\phi)} r \, d\xi \, dr \quad (14)$$

over the aperture plane. On setting up the radiation field on the basis of the aperture field by the methods of Sec. 5·12 it will be found that the same result is obtained for the pattern as that from the use of Eq. (14) in conjunction with Eqs. (12). Thus, the current-distribution method passes into the aperture-field method as the angular spread of the pattern decreases, that is, as the ratio of the wavelength to aperture diameter, λ/D, approaches zero. The significant difference between the results of the two methods is the dependence of the pattern on the ratio λ/D. It was shown in Sec. 6·8 that on the basis of the aperture-field calculation, the angular distribution of the secondary pattern is proportional to λ/D for a given relative distribution over the aperture and the side-lobe intensities are independent of λ/D. On the other hand, it has been found in a study of special forms of $G_f(\xi,\psi)$ by the current-distribution method, using the complete expressions Eqs. (12a) and (12b), that the side-lobe intensity is also a function of λ/D, which asymptotically approaches the value given by the aperture method as λ/D approaches zero.

Principal E- and H-plane Patterns.—In an arbitrary direction the field has both the E_θ- and the E_ϕ-component. They are generally out of phase with the result that the field is elliptically polarized (*cf.* Sec. 3·12). However, in the principal planes—the planes $\phi = 0$ and $\phi = \pi/2$—the field is linearly polarized in the direction determined essentially by the principal polarization component of the aperture field. Considering the E-plane, $\phi = 0$, we see that the y-component of $\mathbf{I}_t$ which arises from the cross-polarization component of the current distribution (or aperture field) vanishes because contributions from points in the reflector that are symmetrically located with respect to the xz-plane are 180° out of phase. The field is produced by the I_z- and I_{tx}-components and, therefore, has only an E_θ-component which lies in the E-plane. Similarly it is found that in the H-plane the field has only an E_ϕ-component and is, therefore, everywhere normal to the H-plane and parallel to the principal component of the aperture field. Again, since the cross-polarization components of the current at a pair of points on the reflector that are symmetrically located with respect to the yz-plane are 180° out of phase, their resultant contribution to the H-plane vanishes. It was noted earlier that the longitudinal current element contributes nothing to the E_ϕ-component; therefore the H-plane field is produced entirely by the principal component of the aperture field. Using the aperture-field approximation [Eq. (14)], we find that the principal plane patterns are

a. *E-plane*:

$$E_\theta = \frac{j\omega\mu}{2\pi R} e^{-jkR} \left[\left(\frac{\epsilon}{\mu}\right)^{\frac{1}{2}} \frac{P_T}{2\pi}\right]^{\frac{1}{2}} \cos\theta I_{tx}, \qquad (15a)$$

with

$$I_{tx} = e^{-j(2kf)} \int_0^a \int_0^{2\pi} e_{1x} \frac{[G_f(\xi,\psi)]^{\frac{1}{2}}}{\rho} e^{jkr\sin\theta\cos\xi} r\, d\xi\, dr; \qquad (15b)$$

b. *H-plane*:

$$E_\phi = \frac{j\omega\mu}{2\pi R} e^{-jkR} \left[\left(\frac{\epsilon}{\mu}\right)^{\frac{1}{2}} \frac{P_T}{2\pi}\right]^{\frac{1}{2}} I'_{tx}, \qquad (16a)$$

with

$$I'_{tx} = e^{-j(2kf)} \int_0^a \int_0^{2\pi} e_{1x} \frac{[G_f(\xi,\psi)]^{\frac{1}{2}}}{\rho} e^{jkr\sin\theta\sin\xi} r\, d\xi\, dr. \qquad (16b)$$

The two patterns have the same value, of course, along the axis (in the direction $\theta = 0$).

Cross Polarization.—The polarization of the field in a pencil beam is generally expressed with reference to the x- and y-axes rather than the spherical coordinate directions as we have done above. The use of the cartesian components has associated with it an error in that the field is transverse to the radial direction from the origin and not to the z-axis; but if the beam is narrow, the error is small. The latter mode of description has the advantage that the E_x-component is associated directly with principal polarization component of the aperture field and the E_y-component with the cross-polarization component. The E_x- and E_y-components are designated correspondingly as the principal polarization and cross-polarization components of the secondary pattern. They are given by relations equivalent to Eqs. (12) with $\mathbf{i}_\theta \cdot \mathbf{I}$ and $\mathbf{i}_\phi \cdot \mathbf{I}$ replaced by I_{tx} and I_{ty}, respectively.

By using the cartesian components, the secondary pattern is resolved into a principal polarization pattern and a cross-polarization pattern. The E- and H-plane patterns given by Eqs. (15a) and (16a) belong to the former. It is obvious that the symmetry properties of the aperture field with respect to the principal planes, which lead to zero cross polarization in those planes, do not hold for other directions in space. The cross-polarization pattern must, therefore, have maxima in the four quadrants between the principal planes. A detailed analysis[1] shows that the cross-polarization pattern takes the form of four lobes whose maxima lie in the 45° planes between the principal planes. Any two lobes related by reflection in a given principal plane are out of phase by 180°. The maxima of one set of lobes occur at angular distances from the paraboloid axis equal to the position of the first minimum of the principal polarization pattern, which is very closely equal to the half-power width of the main lobe. A second set of cross-polarization lobes appears at much wider angles; the peaks are quite low, but the lobes are very broad and therefore represent a not completely negligible fraction of the total energy.

Cross-polarization studies should be made on all antennas on which the side-lobe specifications are very stringent. Although the principal polarization lobes may meet the operational requirements, the cross-polarization lobes may not. Furthermore, since they lie close into the main beam, they effectively increase its width.

12·5. The Antenna Gain.—The gain is generally the primary consideration in the design of the antenna. The factors affecting the gain are treated conveniently in three parts: (1) the dependence of the optimum angular aperture Ψ on the feed pattern, for a fixed diameter D in the aperture plane, assuming that $G_f(\xi,\psi) = 0$ for $\psi \geqq 90°$; (2) the backlobe interference effect; and (3) phase-error considerations. In this

[1] E. U. Condon, "Theory of Radiation from Paraboloid Reflectors," Westinghouse Report No. 15, Sept. 24, 1941.

discussion the primary pattern will be taken to be circularly symmetrical, independent of ξ; as was noted in Sec. 12·1 this means that in practice the feed pattern is replaced by the arithmetic mean of its principal E- and H-plane patterns. It is immaterial for the calculation of gain whether the surface-current or aperture-field distribution is taken as the starting point, because as was pointed out above, the longitudinal component of the current is ineffective in determining the peak intensity.

Optimum Angular Aperture Relations.—The field intensity in the secondary pattern on the axis at a distance R_0 from the focus is given by either Eq. (15a) or (16a) for $\theta = 0$. For the present purpose, it is more convenient to express I_{tx} in the form of Eq. (13a) as an integral over the surface of the reflector rather than in the form of Eq. (15b); we have then

$$E(R,0,0) = i_x \frac{j\omega\mu}{2\pi R_0} e^{-jkR_0} \left[\left(\frac{\epsilon}{\mu}\right)^{1/2} \frac{P_T}{2\pi} \right]^{1/2}$$
$$\int_0^{2\pi} \int_0^{\Psi} e_{1x} [G_f(\psi)]^{1/2} e^{-jk\rho(1+\cos\psi)} \rho \sin\psi \, d\psi \, d\xi. \quad (17)$$

The polarization component e_{1x} is in general a function of ξ and ψ because of the presence of cross polarization. However, in most cases of interest the cross polarization e_{1y} is a very small fraction of the total field and the variation of e_{1x} over the aperture may then be neglected. Introducing the equation of the paraboloid [Eq. (3)] and performing the integration over ξ, we get

$$E = \frac{j\omega\mu f}{R_0} \left[8 \left(\frac{\epsilon}{\mu}\right)^{1/2} \frac{P_T}{4\pi} \right]^{1/2} e^{-jk(R_0+2f)} \int_0^{\Psi} [G_f(\psi)]^{1/2} \tan\frac{\psi}{2} \, d\psi. \quad (18)$$

The power per unit solid angle $P(0,0)$ radiated in the forward direction is given by

$$P(0,0) = \frac{1}{2} R_0^2 \left(\frac{\epsilon}{\mu}\right)^{1/2} |E(R_0,0,0)|^2, \quad (19)$$

and the antenna gain is obtained from it as

$$G = \frac{P(0,0)}{\frac{P_T}{4\pi}} \quad (20)$$

because the total power radiated by the antenna as a whole equals that radiated by the feed. The gain is thus found to be

$$G = \frac{16\pi^2 f^2}{\lambda^2} \left| \int_0^{\Psi} [G_f(\psi)]^{1/2} \tan\frac{\psi}{2} \, d\psi \right|^2. \quad (21)$$

The focal length is related to the angular aperture and the aperture diameter D by

$$f = \frac{D}{4} \cot\frac{\Psi}{2}. \quad (22)$$

Substituting into the preceding relation, we obtain finally the working formula

$$G = \left(\frac{\pi D}{\lambda}\right)^2 \cot^2 \frac{\Psi}{2} \left| \int_0^\Psi [G_f(\psi)]^{\frac{1}{2}} \tan \frac{\psi}{2} d\psi \right|^2. \qquad (23)$$

The factor $(\pi D/\lambda)^2$ is the gain for a uniformly illuminated constant-phase aperture; the rest is the gain factor or efficiency

$$\mathcal{G} = \cot^2 \frac{\Psi}{2} \left| \int_0^\Psi [G_f(\psi)]^{\frac{1}{2}} \tan \frac{\psi}{2} d\psi \right|^2. \qquad (23a)$$

Thus the efficiency is a function only of the feed pattern and the angular aperture; that is, for a given feed pattern, the efficiency is the same for all paraboloids having the same f/D ratio.

It is instructive to consider the class of feed patterns defined by

$$G_f(\psi) = G_0^{(n)} \cos^n \psi, \qquad 0 \leq \psi \leq \frac{\pi}{2},$$
$$= 0 \qquad\qquad\qquad \psi \geq \frac{\pi}{2}. \qquad (24)$$

Many feed patterns can be represented by some one member of this class over a sizable portion of the main lobe. The gain $G_0^{(n)}$ is determined by the condition that

$$\int G_f(\psi) \, d\Omega = 4\pi,$$

$d\Omega$ being the element of solid angle; this gives

$$G_0^{(n)} = 2(n + 1). \qquad (24a)$$

Substituting Eqs. (24) and (24a) into Eq. (23a), we obtain

$$\mathcal{G} = 2(n + 1) \left[\cot \frac{\Psi}{2} \int_0^\Psi \cos^{n/2} \psi \tan \left(\frac{\psi}{2}\right) d\psi \right]^2, \qquad (25)$$

with the following explicit expressions for the even values of n between $n = 2$ and $n = 8$:

$$\mathcal{G}_2 = 24 \left(\sin^2 \frac{\Psi}{2} + \ln \cos \frac{\Psi}{2}\right)^2 \cot^2 \frac{\Psi}{2};$$

$$\mathcal{G}_4 = 40 \left(\sin^4 \frac{\Psi}{2} + \ln \cos \frac{\Psi}{2}\right)^2 \cot^2 \frac{\Psi}{2};$$

$$\mathcal{G}_6 = 14 \left[2 \ln \cos \frac{\Psi}{2} + \frac{(1 - \cos \Psi)^3}{3} + \frac{1}{2} \sin^2 \Psi \right]^2 \cot^2 \frac{\Psi}{2};$$

$$\mathcal{G}_8 = 18 \left[\frac{1 - \cos^4 \Psi}{4} - 2 \ln \cos \frac{\Psi}{2} - \frac{(1 - \cos \Psi)^3}{3} - \frac{1}{2} \sin^2 \Psi \right]^2 \cot^2 \frac{\Psi}{2}.$$

These results are shown graphically in Fig. 12·3, where $\mathcal{G}_n$ is plotted as

426 PENCIL-BEAM AND SIMPLE FANNED-BEAM ANTENNAS [SEC. 12·5

a function of the angular aperture Ψ. For each primary pattern there is an optimum aperture for which the maximum gain factor is attained. The more directive the feed pattern the smaller is the optimum aperture and, since the diameter of the aperture plane is constant, the longer is the optimum focal length. The general course of the curves and the

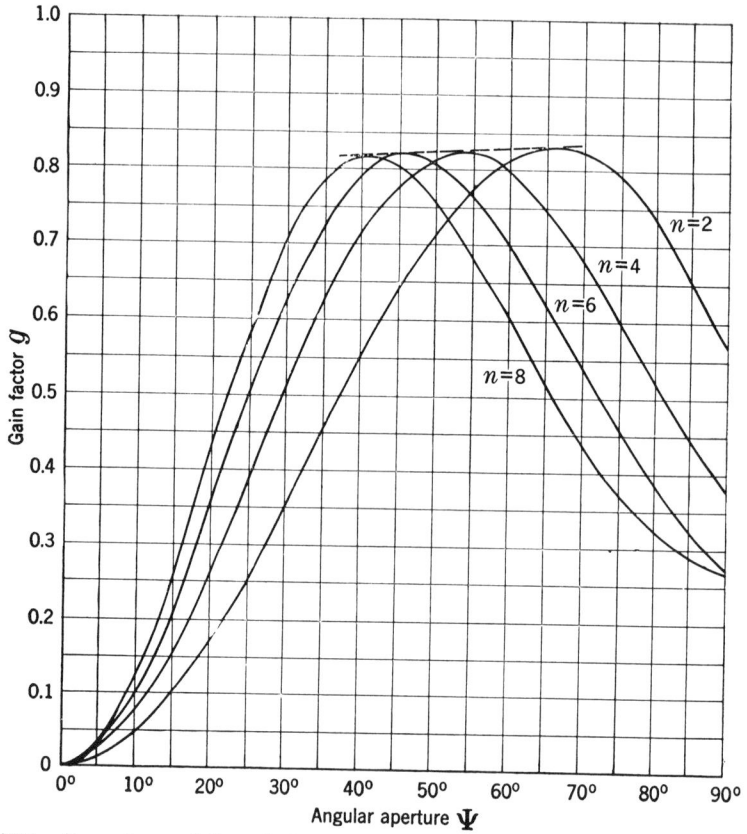

FIG. 12·3.—Dependence of the gain factor on angular aperture and primary feed pattern.

$$G_f(\psi) = 2(n+1)\cos^n \psi, \quad 0 \leq \psi \leq \frac{\pi}{2}$$
$$= 0, \quad \psi > \frac{\pi}{2}.$$

existence of a maximum are readily understood when it is recognized that the gain factor arises essentially as a product of two factors: (1) the fraction of the total power radiated by the antenna feed that is intercepted by the reflector and is thus made available to its aperture for the main beam and (2) the efficiency with which the aperture concentrates the available energy in the forward direction. The first factor obviously increases with increasing angular aperture. The second factor, determined by the field distribution over the aperture, decreases with increasing Ψ; for as Ψ increases, the illumination over the aperture becomes more and more tapered toward the edge relative to the center. This tapering is accentuated by the superposition of the space attenuation factor $1/\rho$ on the already directive feed pattern $[G_f(\psi)]^{1/2}$; as was shown in Chap. 6, such tapering of the illumination results in a decrease in aperture efficiency. The optimum angular aperture represents the proper compromise between spillover of the feed energy and aperture efficiency. For an arbitrary $G_f(\psi)$, the optimum angular aperture is obtained as a solution of

$$\left(\sin^2 \frac{\Psi}{2}\right)[G_f(\Psi)]^{1/2} = \frac{1}{2}\int_0^\Psi [G_f(\psi)]^{1/2} \tan \frac{\psi}{2} d\psi, \qquad (26)$$

a relation obtained by setting the derivative of Eq. (23a), $d\mathcal{G}/d\Psi$, equal to zero.

The values of the gain factor at the maxima in Fig. 12·3 are considerably higher than the values realized in practice. This is because idealized feed patterns have been assumed in which no feed energy is radiated beyond $\psi = 90°$. As a result the gain $G_0^{(n)}$ of the idealized pattern is much greater than the gain G_{f0} of an actual feed whose main lobe can be represented closely by $G_{f0} \cos^n \psi$ but which in addition radiates beyond 90°. The gain factor $\mathcal{G}$ realized with the actual feed is related to $\mathcal{G}_n$ by

$$\mathcal{G} = \frac{G_{f0}}{2(n+1)} \mathcal{G}_n. \qquad (27)$$

The value of the optimum angular aperture is unaffected by this scaling in the primary feed gain. It will be observed that the value of the maximum varies but slowly with the illumination function. The broader the primary feed pattern the broader is the maximum in the $\mathcal{G}$-curve and the less critical is the choice of angular aperture. It is convenient to designate the optimum angular aperture in terms of the decibel level of the primary pattern at the edge of the aperture relative to its maximum. Thus for a cosine-squared pattern the optimum value of Ψ corresponds to that angle in the primary feed pattern at which the power is 8 db down from the peak intensity. The decibel-cutoff point in the primary pattern is plotted as a function of the directivity in Fig. 12·4. The decibel-cutoff point again is not a sensitive function of the directivity. For most feeds the average optimum figure is from 9 to 10 db.

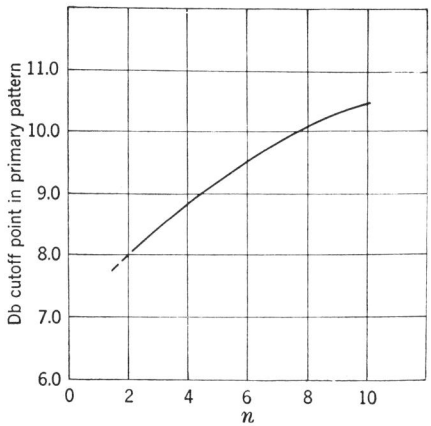

FIG. 12.4—Cutoff point in primary feed pattern for maximum gain as a function of the sharpness of the feed pattern.

The optimum angular aperture can also be expressed in terms of the intensity of illumination at the edge of the reflector relative to that at the vertex. This is obtained by multiplying the ratio of the primary pattern intensities $G_{f0}/G_f(\psi)$ by the ratio of the space attenuation factors ρ^2/f^2. It is found in the case of each of the distributions studied above that the optimum angular aperture corresponds to an edge illumination 11 db below the vertex illumination.

Back-lobe Interference.—The above results may be modified significantly by the effect of interference between the back lobe of the primary feed pattern and the reflector field. Let G_π be the gain of the antenna feed in the direction $\psi = 180°$. The back-lobe field intensity at the field point $(R_0,0,0)$ along the axis is then

$$\mathbf{E}_\pi = \pm \frac{1}{R_0} \left[2 \left(\frac{\mu}{\epsilon}\right)^{1/2} \frac{P_T}{4\pi} G_\pi \right]^{1/2} e^{-jkR_0} \mathbf{i}_x. \tag{28}$$

The choice of positive or negative sign is made according to whether the field of the feed in the direction $\psi = \pi$ is parallel (in phase) or antiparallel (180° out of phase) to that in the direction $\psi = 0$. Superposition of the back-lobe field on the reflector field [Eq. (18)] yields the total field intensity

$$\mathbf{E} = \mathbf{i}_x \frac{1}{R_0} \left[2 \left(\frac{\mu}{\epsilon}\right)^{1/2} \frac{P_T}{4\pi} \right]^{1/2}$$
$$e^{-jkR_0} \left[\frac{4\pi f}{\lambda} e^{-j\left(\frac{4\pi f}{\lambda} - \frac{\pi}{2}\right)} \int_0^\Psi [G_f(\psi)]^{1/2} \tan\frac{\psi}{2} d\psi \pm G_\pi^{1/2} \right]. \tag{29}$$

By the procedure followed previously the gain factor is found to be

$$\mathcal{G} = \left(U_0 \cot\frac{\Psi}{2}\right)^2 \left[1 \pm \frac{2\lambda G_\pi^{1/2}}{\pi D U_0} \tan\frac{\Psi}{2} \sin\left(\frac{\pi D}{\lambda} \cot\frac{\Psi}{2}\right) + \frac{\lambda^2 G_\pi}{(\pi D U_0)^2} \tan^2\frac{\Psi}{2} \right], \tag{30}$$

where

$$U_0 = \int_0^\Psi [G_f(\psi)]^{1/2} \tan\frac{\psi}{2} d\psi. \tag{30a}$$

In most cases of interest G_π is so small that the last term in Eq. (30) is negligible; under this condition the gain factor becomes

$$\mathcal{G} = \left(U_0 \cot\frac{\Psi}{2}\right)^2 \left[1 \pm \frac{2\lambda G_\pi^{1/2}}{\pi D U_0} \tan\frac{\Psi}{2} \sin\left(\frac{\pi D}{\lambda} \cot\frac{\Psi}{2}\right) \right]. \tag{31}$$

The term in brackets is the modification of the previous result introduced by the back-lobe interference. This modification introduces an additional λ/D dependence; the interference effect depends on the ratio of the back-lobe field intensity to the reflector field intensity, and the latter is proportional to D/λ. For a given primary pattern $\mathcal{G}$ is no longer a function of the paraboloid shape alone.

SEC. 12·5] THE ANTENNA GAIN 429

The back-lobe effect is illustrated graphically in Fig. 12·5. The curves pertain to an actual feed—the ⅞-in. stub-supported dipole-disk feed of Sec. 8·8—and a reflector with a 30-in. aperture diameter; the wavelength is 10 cm. The main lobe of this particular feed is fitted closely by the function

$$G(\psi) = 7.0 \cos^4 \psi.$$

It will be noted that the gain is 7, as compared with $G_0^{(4)} = 10$ for the idealized $\cos^4 \psi$ pattern used previously. The back-lobe gain G_π is 0.142. Curve A is the relation between the gain factor and aperture,

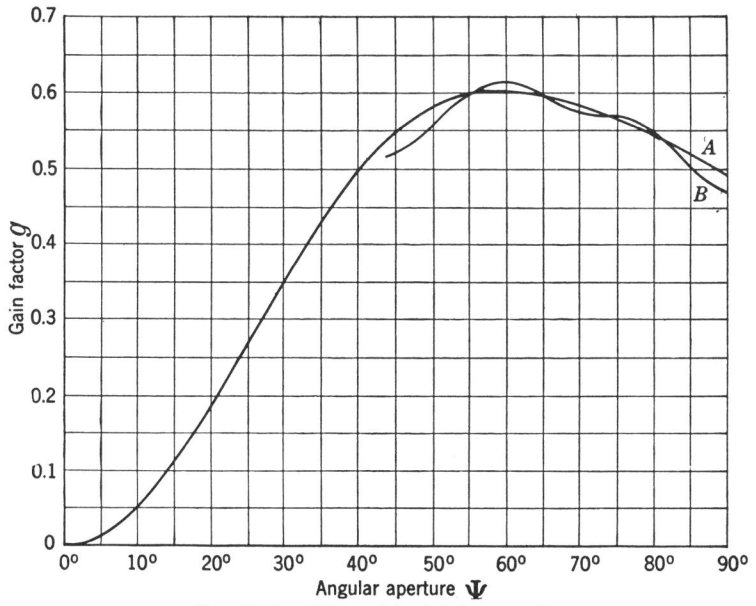

FIG. 12·5.—Effect of back lobe on gain.

neglecting the back lobe, while Curve B includes the interference effect. The gain falls above or below Curve A according to whether the back lobe is in phase or out of phase with the reflector field. The two fields add when the focal distance is such that, together with the 180° phase change at the reflector, the field of the latter is brought into phase with the back lobe. The points of maximum deviation from Curve A correspond to differences in focal length very nearly equal to $\lambda/2$. The optimum aperture is not altered noticeably, but the maximum realizable gain factor increases by 2.5 per cent. The effect is small for this particular feed because the back-lobe level is so low relative to the main lobe. With feeds such as the 3-cm-band double-dipole feed discussed in Sec. (8·9), having a comparatively high back-lobe level, the back-lobe interference effect is much more significant.

Phase-error Effects.—It was pointed out earlier that a diminution of gain results from any departure from uniform phase over the aperture that, however, leaves the peak intensity on the axis of the paraboloid. The direction of peak intensity remains unchanged if the phase-error distribution over the aperture is independent of ξ; this discussion is confined to such distributions. The phase deviation can arise from a number of sources: (1) deviation of the reflector from a paraboloidal shape, (2) defocusing (displacement of the feed center from the focus), or (3) deviation of the antenna-feed wavefronts from spherical wavefronts. From the point of view of the aperture it is immaterial which of the three factors is operative. To tie in with the preceding discussion of the relation between the reflector and the feed pattern, the phase-error source will be taken to be the third of the above, that is, the absence of a true center of feed. Back-lobe interference will be neglected. The final results can easily be interpreted in terms of equivalent errors arising from surface distortion or defocusing. Let us then assume that the field-intensity pattern of the antenna feed has the form

$$\mathbf{E} = \left[2 \left(\frac{\mu}{\epsilon} \right)^{1/2} \frac{P_T}{4\pi} G_f(\psi) \right]^{1/2} e^{-jk\delta(\psi)} \mathbf{e}_0(\xi,\psi), \qquad (32)$$

where $\delta(\psi)$ represents the phase error in the feed pattern. A review of the steps leading to the field intensity $E(R_0,0,0)$ of the secondary pattern on the axis, given in Eq. (18), will show that the only change introduced in Eq. (18) is the replacement of $[G_f(\psi)]^{1/2}$ by $[G_f(\psi)]^{1/2} e^{-jk\delta(\psi)}$. By precisely the same development as before, the gain factor is given by

$$\mathcal{G} = \cot^2 \frac{\Psi}{2} \left(\left\{ \int_0^\Psi [G_f(\psi)]^{1/2} \cos \left[\frac{2\pi\delta(\psi)}{\lambda} \right] \tan \frac{\psi}{2} d\psi \right\}^2 + \left\{ \int_0^\Psi G_f(\psi)^{1/2} \sin \left(\frac{2\pi\delta(\psi)}{\lambda} \right) \tan \frac{\psi}{2} d\psi \right\}^2 \right). \qquad (33)$$

By way of illustration, the effect of a quadratic phase error has been computed for the primary pattern of the dipole-disk feed considered above in connection with back-lobe interference. The phase function is taken to be

$$\frac{\delta(\psi)}{\lambda} = \alpha\psi^2.$$

The optimum angular aperture in the absence of phase error is taken as a base for comparison and α is adjusted to produce a preassigned phase error at the edge of the aperture for that case. The curves given in Fig. 12·6 are for values of α that result in phase errors of $\lambda/24$, $\lambda/16$, $\lambda/8$, and $\lambda/4$ at the edge of an aperture of angle $\Psi = 61°$. The loss in gain is 2 per cent for an error of $\lambda/16$ at the edge, 6 per cent for $\lambda/8$, and 20 per cent for $\lambda/4$. The effect of a highly tapered illumination is shown in Fig. 12·6 by the values of the gain factor for the aperture of angle $\Psi = 90°$. For this value of Ψ the α_1 curve represents a phase error of approximately $\lambda/11$ at the edge, while the α_2 curve represents an error of approximately $\lambda/6$; the corresponding losses in gain relative to the $\alpha = 0$ curve are 2.7 and 3.6 per cent respectively. Since the gain curves are not very sensitive to the illumination, the results obtained here for the $\cos^4 \psi$ distribution may be taken as characteristic; a conservative evaluation sets $\lambda/8$ as the maximum allowable phase deviation over the aperture.

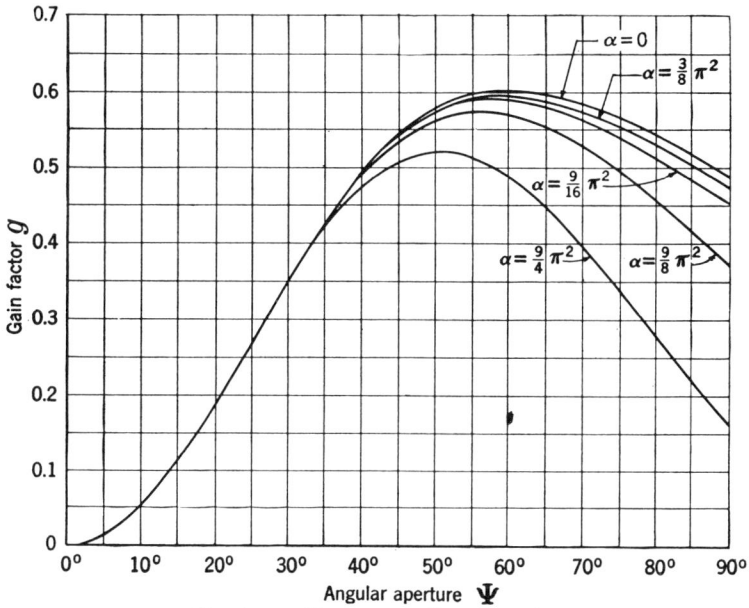

FIG. 12·6.—Phase-error effects on gain.

There is another aspect of the feed pattern that should be noted in connection with phase-error effects. The discussion above is based on the assumption that the reflector is illuminated by the main lobe of the feed. For some purposes it may be desirable to accept the loss in gain associated with a large angular aperture in order to suppress the side lobes. The angular aperture, however, must not extend beyond the first minimum of the feed pattern. Generally, in passing through a minimum (more exactly a null) in the feed pattern there is a discontinuity of 180° in the phase. Inclusion of any portion of the pattern beyond the minimum thus introduces completely out-of-phase illumination at the periphery of the aperture, with a very serious reduction in gain.

Results similar to those obtained above are obtained when the phase error arises from defocusing. As shown in Fig. 12·7, if the center of feed is displaced a distance δ_0 from the focus along the axis, the phase-error function is

$$\frac{\delta}{\lambda} = \frac{\delta_0}{\lambda} \cos \psi. \qquad (34)$$

The $\lambda/8$ criterion indicates a focusing tolerance related to the angular aperture by

$$\delta_0 = \frac{\lambda}{8} \sec \Psi.$$

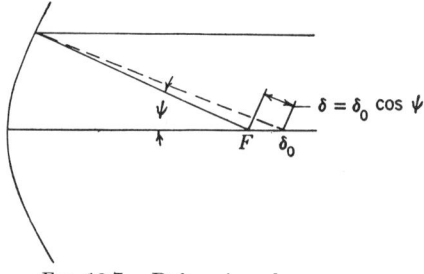

FIG. 12·7.—Defocusing phase errors.

In practice, the focusing condition is not adhered to rigidly. It is not practicable to tailor every reflector to the feed, because frequently it is necessary to interchange feed systems. In these cases the back-lobe interference effect may be a decided asset; by defocusing to bring the back lobe in phase with the main beam it may be possible to achieve an increase in gain that far exceeds the loss due to defocusing phase errors. This is particularly true with feeds such as the 3-cm-band double-dipole feed which has a very large back lobe.

Design Procedures.—The theoretical analysis may be summed up in terms of design procedures for realizing a maximum gain factor:

1. The shape factor f/D is to be chosen so that the full angle subtended by the reflector at the feed is in the range between the 9- and 10-db widths of the primary feed pattern. A more exact value for a given primary pattern is obtained by solving Eq. (26).
2. The focal length of the paraboloid should be an integral number of half wavelengths $f_m = m\lambda/2$ if the back lobe of the primary pattern is 180° out of phase with the main lobe; if the back lobe and main lobe of the primary pattern are in phase, the focal length should be $f_m = (2m + 1)\lambda/4$ where m again is an integer. Under these conditions the back lobe will be in phase with the paraboloid beam and add to the gain. If it is not possible to satisfy these requirements exactly, the feed should be placed at the point nearest the focus, at which the distance to the vertex satisfies the half- or quarter-wavelength requirement.
3. Deviations from constant phase of the aperture should be kept within $\lambda/8$ and certainly should not exceed $\lambda/4$. Two factors contribute to phase error: distortion of the paraboloid surface and deviation of the primary wavefronts from spherical waves. With reference to the first of these the phase-error criterion can readily be converted to tolerances that may be allowed in constructing the reflector. As concerns the feed, the phase-error criterion serves

to define the point-source cone (*cf.* Sec. 8·1). The angular aperture of the paraboloid should lie within the point-source cone.

12·6. Primary Pattern Designs for Maximizing Gain.—Mention should be made of the technique of shaping the primary feed pattern so as to produce uniform illumination over the aperture and thereby to maximize the gain. The required primary pattern is obtained directly from the expression for the aperture field in Eq. (11). For $E(r,\xi)$ to be constant, the primary pattern must be such that within the cone subtended by the reflector at the feed

$$\frac{[G_f(\xi,\psi)]^{1/2}}{\rho} = \text{const}, \qquad (35)$$

or

$$G_f(\xi,\psi) = G_{f0} \sec^4 \frac{\psi}{2}. \qquad (35a)$$

In addition, the feed must radiate no energy outside the angular aperture Ψ in order to realize the gain of $4\pi A/\lambda^2$. The value of G_{f0} is obtained from the condition

$$\int_0^{2\pi} \int_0^{\pi} G_f(\xi,\psi) \sin \psi \, d\psi \, d\xi = 4\pi,$$

whence

$$G_0 = \frac{1}{\displaystyle\int_0^{\Psi} \sec^4 \frac{\psi}{2} \sin \psi \, d\psi} = \cot^2\left(\frac{\Psi}{2}\right). \qquad (36)$$

It is, of course, impossible to produce a pattern having a sharp cutoff, but the required pattern can be approximated quite closely. Techniques of shaping the primary patterns of horn feeds are discussed in Sec. 10·16. It will be noted that the pattern [Eq. (35a)] has a minimum in the direction $\psi = 0$. In order to produce such a minimum considerable phase distortion must be introduced over the mouth of the horn. Such feeds must be used with caution, for a concomitant effect of the phase distortion to that of producing the desired intensity distribution may be that of eliminating the center of feed. This will result in phase errors in the field over the aperture of the reflector that may well cancel the gains which might have been made by the uniform illumination.

12·7. Experimental Results on Secondary Patterns.—The relation between the secondary pattern and the aperture-field distribution can be studied by evaluating the expressions in Sec. 12·4 for the secondary pattern for a number of different types of gain functions $G_f(\xi,\psi)$. The essential results of such calculations have been summarized in Sec. 6.8. In this section the relation between the principal-plane patterns and the aperture will be discussed by reference to experimental data. The material will also serve as a presentation of the performance of several of the more important types of feeds described in earlier chapters.

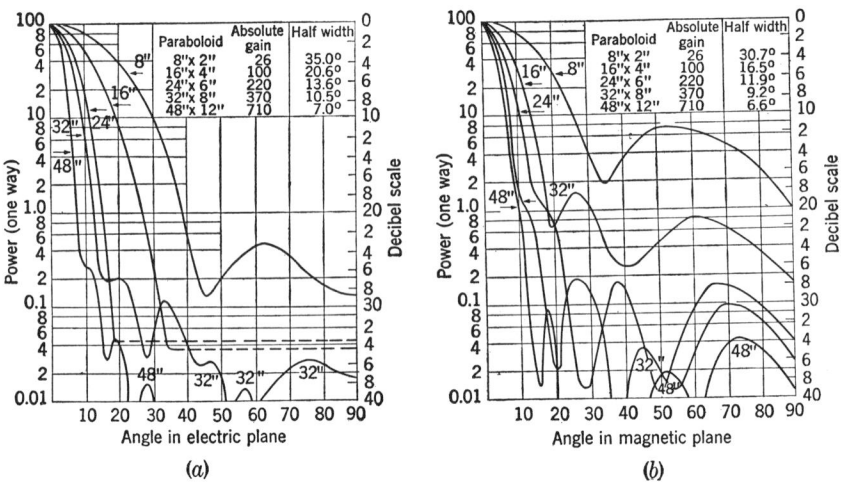

FIG. 12·8.—Principal plane patterns as a function of diameter for a series of paraboloids of $f/D = 0.25$; $\lambda = 4.00$ in.: (a) E-plane; (b) H-plane.

The dependence of the pattern on the diameter for a given relative distribution over the aperture is exemplified by the series of patterns,[1] shown in Fig. 12·8, for a set of paraboloidal antennas all of the same shape, $f/D = 0.25$, and illuminated by the same antenna feed. The latter is a coaxial-line-fed double-dipole feed of the same general type as was discussed in Sec. 8·9. The installations are the rear-feed type in which the feed line lies along the axis of the reflector, passing through its vertex (cf. Sec. 12·11). The focal length in each case is an integral multiple of a half wavelength so that the interaction between the aperture beam and back lobe of the feed along the axis is the same for all members of the series.

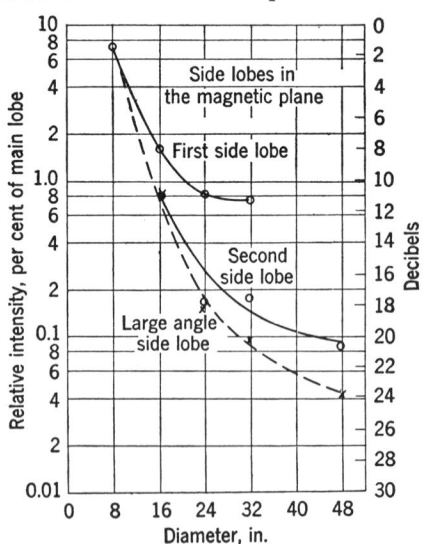

FIG. 12·9.—Dependence of relative height of side lobes on aperture diameter.

It is observed that with increasing diameter the beamwidth decreases

[1] L. C. Van Atta, "Effect of Paraboloid Size and Shape on Beam Patterns," RL Report No. 54-9, Aug. 5, 1942.

and the side lobes move in toward the axis. The intensities of the side lobes are diameter dependent, contrary to the results of the aperture theory. However, as was noted in Sec. 12·4, such deviations are to be expected for large values of λ/D. Figure 12·9 shows the variation of side-lobe intensity with diameter. The intensity approaches an asymptotic value as λ/D decreases, becoming independent of the diameter as the latter becomes large compared with the wavelength; the asymptotic limit agrees with aperture theory predictions. The diameter dependence of the side lobes may be accounted for only in part by the corrections to the aperture theory that are contained in the current-distribution method for calculating the pattern (*cf.* Sec. 12·4). Another significant factor is the overlapping between the primary feed pattern and the aperture pattern. The overlapping also has the effect of filling in the minima. It is seen that in some cases the side lobes have been fused into the main lobe and appear only as shoulders. The same effect is produced by phase errors in the aperture field (*cf.* Sec. 6·7).

The beamwidth also shows an anomalous behavior from the point of view of aperture theory. According to the latter the product of the beamwidth and D/λ is a constant for a given distribution over the aperture. The products for each of the principal plane patterns of

TABLE 12·1.—BEAMWIDTH AND GAIN FACTOR AS A FUNCTION OF DIAMETERS.
Θ is in radians

Diam., in.	$\dfrac{D}{\lambda}\Theta_E$	$\dfrac{D}{\lambda}\Theta_H$	$\mathcal{G}$
8	1.22	1.07	0.66
16	1.44	1.15	0.63
24	1.42	1.25	0.62
32	1.46	1.28	0.59
48	1.47	1.38	0.50

Fig. 12·8 are listed in Table 12·1; it is seen that the product, for each of the principal planes, varies with the diameter. The E-plane half width appears to be approaching an asymptotic value that is proportional to λ/D. The difference between the E- and H-plane beamwidths can be correlated with the directivity of the feed. Because of the directivity of a single dipole in the E-plane, the pattern of the double-dipole system is likewise more directive in the E-plane than in the H-plane. Consequently, the aperture field is more tapered in the E-plane than the H-plane, and the former has a broader secondary pattern.

The variation of the gain factor $\mathcal{G}$ with diameter, as shown by Table 12·1, arises from the back-lobe interference effect. Along the axis in each case, the back lobe of the feed adds to the field produced by the reflector. Since the latter is proportional to D/λ, the addition of the back-lobe intensity produces a greater fractional increase in the total intensity and peak power for smaller diameters than for large diameters and correspondingly larger gain factors.

It will be of interest to record the data on the performance of the waveguide double-dipole feed shown in Fig. 8·14 and of the stub-terminated dipole-disk feed shown in Fig. 8·10 because of their extensive use. The beam characteristics obtained with the double-dipole feed[1]

TABLE 12·2.—PERFORMANCE OF THE DOUBLE-DIPOLE FEED IN VARIOUS PARABOLOIDS.
V is the distance from the vertex of the paraboloid to the front edge of the waveguide

Paraboloid		V, cm	$\dfrac{g}{(\pm 3\%)}$	Beamwidths		Side lobes, db down			
Diam, in.	Focal length, in.			$\dfrac{D}{\lambda}\Theta_E$	$\dfrac{D}{\lambda}\Theta_H$	H_1	H_2	E_1	E_2
18	4.5	10.8	0.61	1.2	1.27	27	30	25	30
18	5.67	13.9	0.61	1.25	1.2	24	29	25	29
18	6.0	14.2	0.64	1.15	1.15	26	27	26	30
24	8.0	19.5	0.63	1.13	1.20	22	28	23	28
30	10.0	24.0	0.60	1.25	1.16	22	28	25	28

at a wavelength of 3.2 cm are summarized in Table 12·2. H_1 and H_2 are the first and second side lobes in the H-plane; E_1 and E_2 designate the corresponding lobes in the E-plane.

The data for the three 18-in. diameter paraboloids can be compared for the effect of tapered illumination; the longer the focal length the less tapered is the aperture illumination with a given primary pattern. The effects are quite evident in the decrease in the H-plane beamwidth and the rise in the H-plane side-lobe intensity levels; the H_1 lobe of the paraboloid of 5.67-in. focal length is an exception to the general behavior. The E-plane characteristics are also anomalous. The discrepancies are caused by the peculiar properties of the feed. As was pointed out in Sec. 8·9, the centers of feed are different in the E- and H-planes; this gives rise to small defocusing phase errors. In addition, the back-lobe intensity is large, and the position of the feed on the axis is determined primarily by the optimum interaction between the back lobe and main lobe rather than by the focal point of the reflector. The last three rows form a sequence of paraboloids of the same shape; here too it is seen that the behavior is not in accord with the more systematic characteristics observed in the set of patterns considered in Fig. 12·8. While the characteristics of the feed leave much to be desired from the standpoint of theoretical analysis of the patterns, the pattern characteristics given in Table 12·2 are highly satisfactory for operational purposes.

[1] W. Sichak, "Double-dipole Rectangular Wave Guide Antennas," RL Report No. 54-25, June 26, 1943.

The stub-supported dipole-disk feed is the one to which the gain factor curves in Fig. 12·5 apply. The patterns obtained[1] with a reflector of 30-in. diameter and 10.6-in. focal length, at a wavelength of 10 cm, are shown in Fig. 12·10. The angular aperture of the paraboloid is

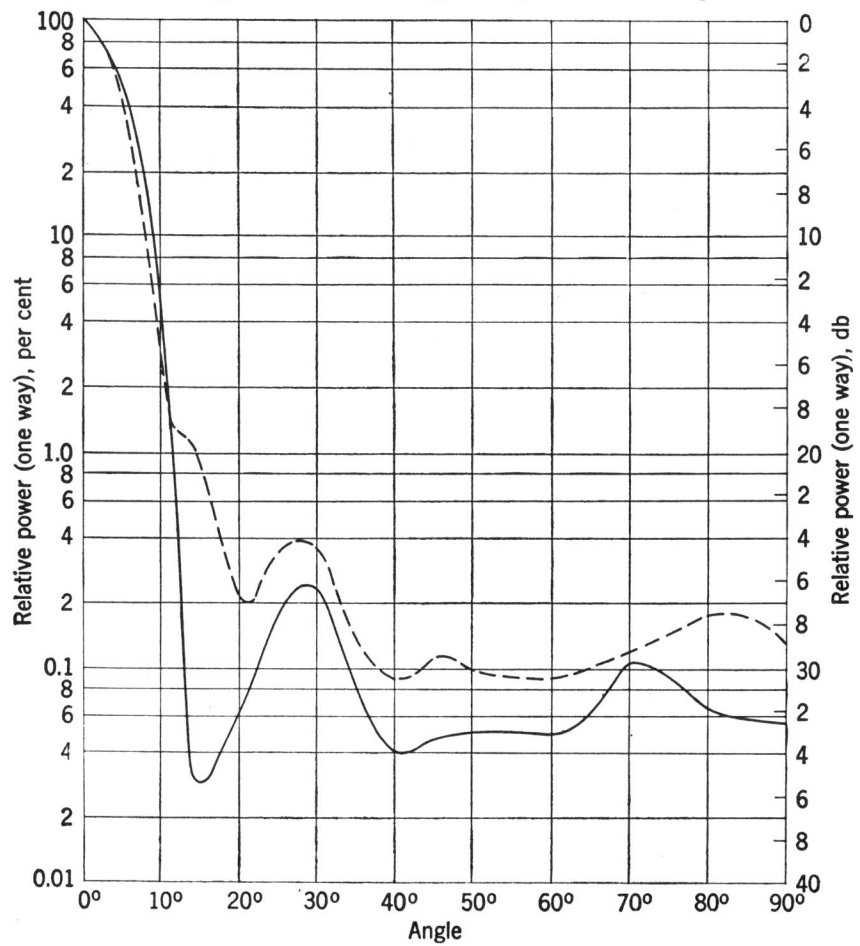

Fig. 12·10.—Principal plane patterns of 30-in.-diameter paraboloid ($f/D = 0.354$) illuminated by the dipole-disk feed; $\lambda = 10$ cm; —— E-plane; - - - H-plane; $\Theta_E = 10.2°$; $\Theta_H = 9.5°$.

$\Psi = 70.5°$, larger than the theoretical value of $60°$ for a maximum gain factor. Whereas this represents a small loss in gain, the larger angular aperture results in a more tapered illumination over the aperture plane and better side-lobe characteristics. The half-power widths (in radians) are

$$\Theta_E = 1.40 \frac{\lambda}{D}; \qquad \Theta_H = 1.26 \frac{\lambda}{D}.$$

[1] S. Breen and R. Hiatt, "Antenna Feeds for ⅞-in. Stub-supported Coaxial Line," RL Report No. 54-23, June 21, 1943.

Attention was called in Sec. 8·3 to the fact that with a feed of this type the axis of the beam does not coincide with the axis of the reflector. The deviation is not shown in Fig. 12·10 because of its small magnitude; it is less than half a degree. The squint phenomenon has great operational value; by rotating the feed about its axis, the antenna beam is made to describe a cone, thus creating an effective cusp-shaped mini-

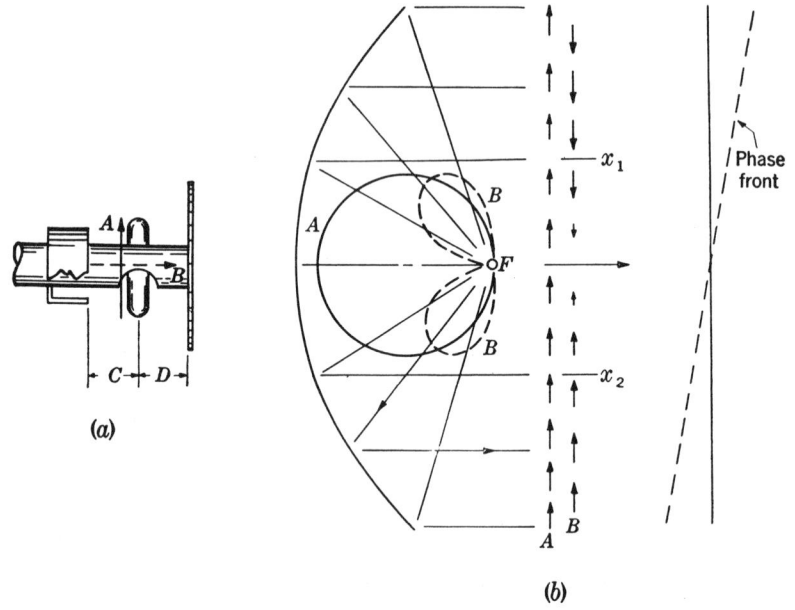

FIG. 12·11.—Production of squint by the asymmetric dipole: (a) current on the feed; (b) distortion of the phase front.

mum along the axis of the paraboloid. The intensity differentiation in the cusp is more sensitive than on the peak of the beam, and by this technique the accuracy of pointing the antenna at a target is increased.

The production of the squint may be understood by reference to Fig. 12·11. It will be recalled (Sec. 8·3) that the asymmetric dipole termination gives rise to currents along the outer conductor of the coaxial line, and the effect of the choke is to confine the line current to the terminal region as shown in Fig. 12·11a. The feed can be regarded as two radiating elements: A the transverse dipole current and B the axial current. The relative magnitudes and phases are determined by the position of the choke. The primary pattern of A is the normal type of pattern shown in Fig. 12·11b and gives rise to the field distribution in the aperture that we have discussed previously (cf. Fig. 12·2). The pattern of the element B has a null along the axis; it produces a field distribution over the aperture in which the electric vector along any diameter undergoes a reversal in direction through the center, which is equivalent to a 180° reversal in phase. The space relationship between the aperture fields of A and B in the E-plane are shown in Fig. 12·11b. If the current elements A and B are in phase, the fields at two diametrically opposite points in the E-plane, such as x_1 and x_2 in the figure, are $a - b$ and $a + b$, respectively; there is no distortion of the phase front, providing b is always less than a. If, however, there is a phase difference Φ between the currents, the resultant fields at the same two points are $a - be^{-j\Phi}$ and $a + be^{-j\Phi}$, respectively; the resultant phases are

$$x_1;\ \tan^{-1}\frac{b\sin\Phi}{a - b\cos\Phi},$$

$$x_2;\ -\tan^{-1}\frac{b\sin\Phi}{a + b\cos\Phi}.$$

The aperture is no longer an equiphase surface; the phase front is tipped with respect to the aperture as shown in the figure.

For a given position of the choke, the beam deviation varies with frequency. In the case of the antenna whose patterns are shown in Fig. 12·10, the observed variation is as follows: $\lambda = 9.7$ cm, deviation $= 0.3°$; $\lambda = 10.0$ cm, deviation $= 0.19°$; $\lambda = 10.3$ cm, deviation $= 0.38°$.

Finally, it should be mentioned that the current element B produces cross polarization in the H-plane. This, however, does not affect the accuracy of pointing, since the cross polarization is zero along the axis.

12·8. Impedance Characteristics.—Another consideration of major importance in the design of an antenna is the impedance bandwidth. The impedance characteristics are the resultant effects of the impedance characteristics of the antenna feed in free space and the mismatch produced by the interaction between the reflector and the antenna feed. The latter problem was treated quite generally in Sec. 5·10. It was shown that if the feed in free space is itself matched to the line, the reflector gives rise to a reflection coefficient

$$\Gamma_r = e^{-jk\delta_1}\int_{S_0}\frac{G_f(\psi,\xi)}{4\pi\rho^2}\cos i e^{-j2k\rho}\,dS \tag{5.97}$$

in the transmission line. $G_f(\psi,\xi)$ is the gain function of the feed; i is the angle of incidence at the point (ρ,ψ,ξ) on the reflector. If the feed in free space is mismatched, with a reflection coefficient Γ_f, measured at the same point in the line to which Γ_r is referred, the total mismatch of the antenna is to a good approximation the sum

$$\Gamma = \Gamma_f + \Gamma_r; \tag{37}$$

that is, the reflection coefficients add vectorially on the reflection coefficient chart (Sec. 2·8).

If the wavelength is small compared with the focal length and aperture diameter, the asymptotic value of Eq. (5·97), given by Eq. (5·98), may be used. For the present case of the feed at the focus of the paraboloid, the radii of curvature R_ξ and R_η at the point of normal incidence, which is the vertex, are both equal to $-2f$, and ρ_n is equal to f. We have then

$$\Gamma_r = \frac{G_{f0}\lambda}{4\pi f} e^{-j(2kf+\delta)}. \tag{38}$$

More generally, if the feed is on the axis near the focus, but at a distance p from the vertex, the reflection coefficient of the reflector is

$$\Gamma_r \approx \frac{G_{f0}\lambda}{4\pi f} e^{-j(2kp+\delta)}. \tag{38a}$$

The magnitude of the reflection coefficient

$$|\Gamma_r| = \frac{G_{f0}\lambda}{4\pi f} \tag{39}$$

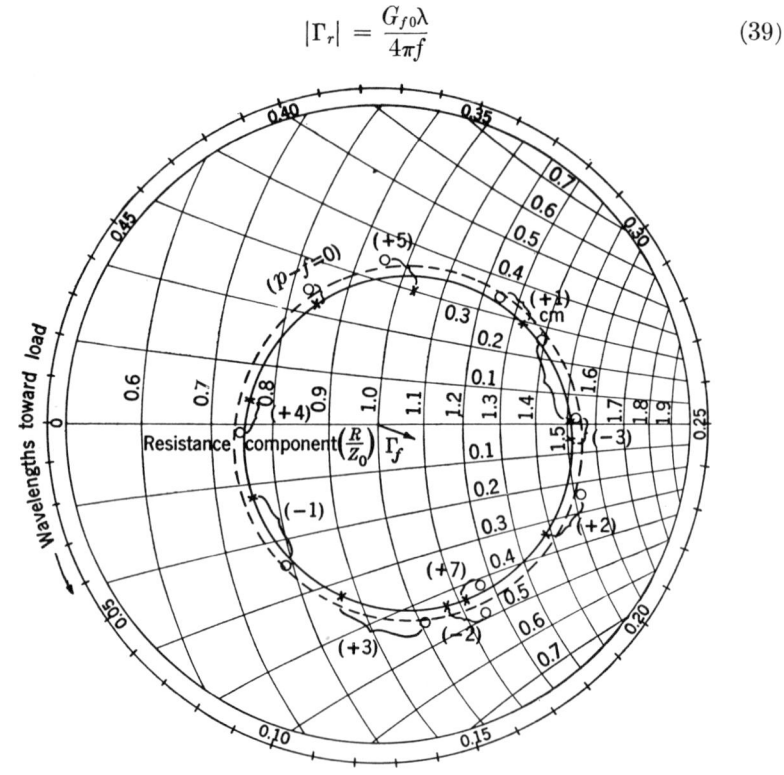

FIG. 12·12.—Variation of the reflection coefficient with position of the feed along the axis; – – – experimental, ——— theoretical curve as obtained from Eq. (38a).

can be determined by measuring the total reflection coefficient Γ as a function of position of the feed along the axis. The feed reflection coefficient Γ_f remains fixed, whereas Γ_r undergoes a cyclic variation by virtue of the changing distance to the reflector. Over small distances about the focal point $|\Gamma_r|$ is essentially constant. As the feed is moved along the axis, the total reflection coefficient therefore describes a circle in the reflection coefficient plane corresponding to the rotation of Γ_r about the terminal point of the vector Γ_f; this is illustrated in Fig. 12·12. The magnitude of Γ_r is determined directly from the radius of the circle.[1] The measurements can, in fact, be used to obtain the gain G_{0f} of the feed pattern by use of Eq. (39).

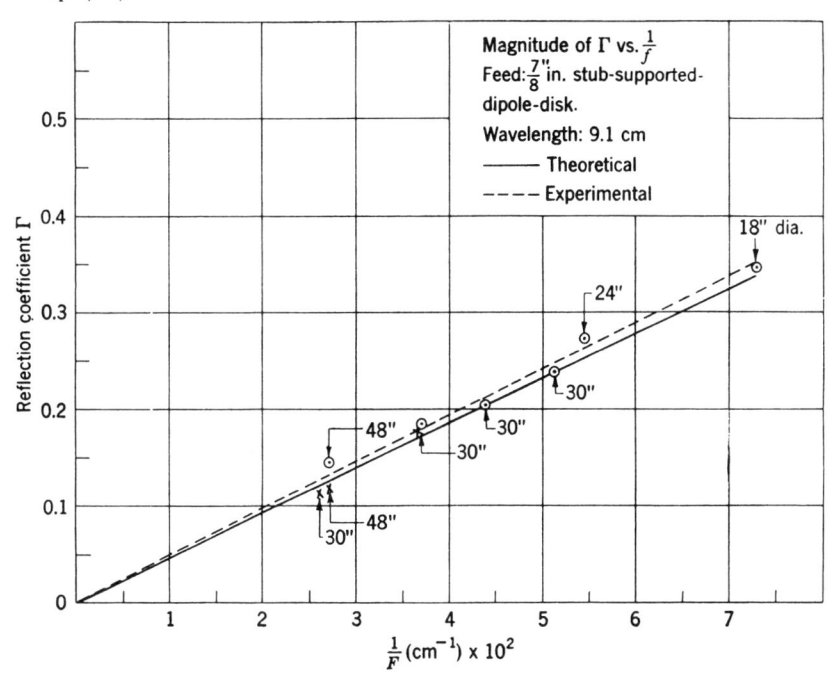

FIG. 12·13.—Contribution of the paraboloid to the reflection coefficient as a function of focal length.

The data presented in Fig. 12·12 were obtained with the stub-supported dipole-disk feed shown in Fig. 8·10 and a paraboloidal reflector of 10.6 in focal length and having an aperture diameter of 30 in. The gain of the feed was evaluated by graphical integration of its primary pattern, and the theoretical curve of Fig. 12·12 was then obtained from Eq. (38a), the constant δ being adjusted to make the theoretical and experimental values agree at the focal point. Similar studies with the same feed in a series of reflectors of different focal lengths gave the results shown in Fig. 12·13, demonstrating the applicability of Eq. (39).[2]

It is seen from Eq. (37) that the process of matching the antenna—

[1] S. Silver, "Contribution of the Dish to the Impedance Mismatch of an Antenna," RL Report No. 442, Sept. 17, 1943.
[2] *Ibid.*

reducing Γ to zero—by means of a transformer in the line can be regarded as that of transforming the mismatch Γ_f of the feed in free space into $-\Gamma_r$. Therefore, if an antenna is matched with the feed at a position p_1, it will also be matched with the feed at positions $p_1 \pm n\lambda/2$, where n is an integer, for Γ_r has the same value at all these points [cf. Eq. (38a)]. Furthermore, the feed can be placed in an entirely different paraboloid; and providing the distance from the vertex is $p_2 = p_1 \pm n\lambda/2$, a good impedance match will be obtained. A small difference will be observed from the value obtained with the original reflector because of the different magnitude of Γ_r, but the phase relations between Γ_f and Γ_r in each case are the optimum for minimizing the total reflection coefficient. If the distance p_1 is chosen to be the closest to the focal length that is equal to an integral number of half wavelengths, the feed may be placed at the corresponding half-wave points in other paraboloids with both the proper conditions for impedance match and the constructive superposition of the back lobe and main lobe being maintained.[1]

The seriousness of the mismatch caused by the reflector lies in its frequency sensitivity. Since the focal length is large compared with λ, a small change in the latter produces a large change in the phase of Γ_r. The antenna can easily be matched at one wavelength λ_0 by a conventional type of matching transformer (cf. Chap. 7). However, the characteristics of the transformer do not vary rapidly enough with frequency to follow the rapid change in the phase of Γ_r and in any case do not necessarily vary in the proper direction. The total reflection coefficient, therefore, varies rapidly with frequency. For this reason it is necessary to eliminate the mismatch caused by the reflector by other methods in order to realize satisfactory impedance characteristics over a wide frequency band.

There are two obvious solutions to the problem. One is to reduce the reflection coefficient of the reflector to zero.[2] For this purpose we must return to Eq. (5·97), which formulates the reflection coefficient as a superposition of contributions from the entire reflector surface. The matching technique that suggests itself immediately is to divide the reflector into two areas, which give integrated effects of equal magnitude, and then by a small displacement of one of the areas with respect to the other to make their contributions 180° out of phase. Since only a small displacement of one area with respect to the other is

[1] H. Krutter, R. Hiatt, J. Bohnert, "Some Matching Properties of Antenna Feeds," RL Report No. 54-13, Nov. 17, 1942.

[2] N. Elson and A. B. Pippard, "Wide Band Matching of Waveguide Radiators and Paraboloids," ADRDE (British) Report No. 220; W. Kock, "Method for Reducing Reflection Effects in Antenna Feeds," BTL Report MM-42-160-92; S. Silver, "Analysis and Correction of the Impedance Mismatch Due to a Reflector," RL Report No. 810.

involved, the matching process is not very frequency sensitive. The bandwidth of the antenna is then largely determined by the impedance characteristics of the feed in free space. A second solution is to render the feed insensitive to the reflected radiation. This will be accomplished if the polarization of the latter is rotated through 90° by the reflecting surface. Such a rotation can be effected by introducing a suitable grating over the surface of the reflector. The details of the two methods will be developed in the following two sections.

12·9. The Vertex-plate Matching Technique.—A complete evaluation of Eq. (5·97) involves considerable numerical work. For the present purposes the computation can be simplified by replacing $G(\psi,\xi)$ by a circularly symmetrical function $G(\psi)$ which is the mean value of the gain functions in the E- and H-planes of the feed pattern. By virtue of the symmetry we can take as the element of area dS the circular zone subtending the angle $d\psi$ at the focus. It is more convenient to base the integral upon the projection of dS on the aperture plane:

$$dS = \frac{2\pi r\, dr}{\cos\left(\frac{\psi}{2}\right)}. \tag{40}$$

The gain function $G(\psi)$ can be expressed as a function of r through the relation

$$\sin\psi = \frac{\dfrac{r}{f}}{1 + \dfrac{r^2}{4f^2}}. \tag{41}$$

Since $i = \psi/2$, we have for the reflection coefficient contributed by the portion of the reflector of aperture radius r,

$$\Gamma(r) = \frac{1}{2}\int_0^r \frac{G\left(\dfrac{r}{f}\right)}{f^2\left(1 + \dfrac{r^2}{4f^2}\right)^2} e^{-j\frac{\pi f}{\lambda}\left(\frac{r^2}{f^2}\right)} r\, dr, \tag{42}$$

constant terms in the phase being discarded. Changing variables to

$$v = \frac{r^2}{f^2}, \tag{43}$$

we get

$$\Gamma(v) = \frac{1}{4}\int_0^v \frac{G(v)}{\left(1 + \dfrac{v}{4}\right)^2} e^{-j\frac{\pi f}{\lambda}v}\, dv. \tag{44}$$

If we take the real and imaginary parts of $\Gamma(v)$,

$$R(v) = \frac{1}{4}\int_0^v \frac{G(v)}{\left(1+\dfrac{v}{4}\right)^2}\cos\left(\frac{\pi f}{\lambda}v\right)dv,$$
$$I(v) = -\frac{1}{4}\int_0^v \frac{G(v)}{\left(1+\dfrac{v}{4}\right)^2}\sin\left(\frac{\pi f}{\lambda}v\right)dv,\qquad(45)$$

and plot $I(v)$ against $R(v)$, we obtain a reflection coefficient spiral for the paraboloid as shown in Fig. 12·14a. The vector Γ_v from the origin to any point on the spiral is the reflection coefficient due to the portion of the

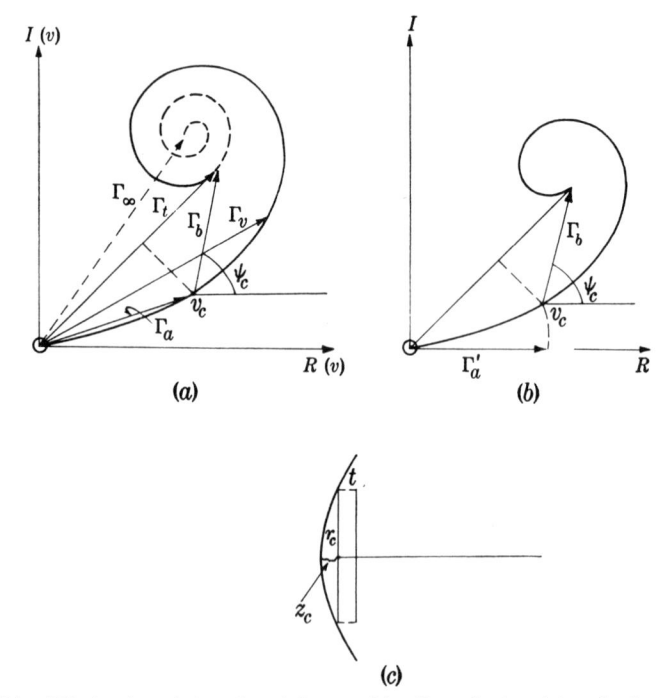

Fig. 12·14.—Elimination of the mismatch caused by the reflector: (a) reflection coefficient spiral; (b) effect of infinitely thin zone plate; (c) final position of zone plate.

paraboloid whose aperture radius r corresponds to that point v on the spiral. The reflection coefficient due to the entire paraboloid is given by the vector to the terminal point corresponding to

$$v_t = \frac{D^2}{4f^2}. \qquad (46)$$

For any particular case $R(v)$ and $I(v)$ can be evaluated numerically once the gain function of the feed has been measured, and the spiral constructed accordingly. It has been found that in many cases the function

$$\frac{G\left(\dfrac{r}{f}\right)}{\left(1+\dfrac{r^2}{4f^2}\right)^2}$$

can be fitted satisfactorily by an exponential

$$G_0 e^{-\frac{\alpha r^2}{f^2}} \to G_0 e^{-\alpha v}, \tag{47}$$

G_0 being the gain of the feed. Assuming this form we can evaluate $R(v)$, $I(v)$ analytically:

$$\begin{aligned}R(v) &= \frac{G_0}{4\left(\alpha^2 + \dfrac{\pi^2 f^2}{\lambda^2}\right)} \left\{\alpha + e^{-\alpha v}\left[\frac{\pi f}{\lambda}\sin\left(\frac{\pi f}{\lambda}v\right) - \alpha\cos\left(\frac{\pi f}{\lambda}v\right)\right]\right\}; \\ I(v) &= \frac{G_0}{4\left(\alpha^2 + \dfrac{\pi^2 f^2}{\lambda^2}\right)} \left\{\frac{\pi f}{\lambda} - e^{-\alpha v}\left[\alpha\sin\left(\frac{\pi f}{\lambda}v\right) + \frac{\pi f}{\lambda}\cos\left(\frac{\pi f}{\lambda}v\right)\right]\right\}.\end{aligned} \tag{48}$$

The limit point of the spiral, corresponding to $v = \infty$, comes at

$$R_\infty = \frac{\alpha G_0}{4\left(\alpha^2 + \dfrac{\pi^2 f^2}{\lambda^2}\right)}; \qquad I_\infty = \frac{\pi f G_0}{4\lambda\left(\alpha^2 + \dfrac{\pi^2 f^2}{\lambda^2}\right)}. \tag{49}$$

This is to be contrasted on the one hand with the Cornu spiral in which $R_\infty = I_\infty$ and on the other hand with the circular aperture diffraction spiral in which $R_\infty = 0$ and I_∞ is the radius of curvature of the initial portion of the spiral. The magnitude of the vector to the limit point is

$$|\Gamma_\infty| = \frac{G_0 \lambda}{4\pi f}\left(1 + \frac{\lambda^2 \alpha^2}{\pi^2 f^2}\right)^{-\frac{1}{2}}. \tag{50}$$

If the aperture of the paraboloid is large, the difference between $|\Gamma_t|$, the reflection coefficient due to the entire paraboloid, and $|\Gamma_\infty|$ is small. We observe further that if $f \gg \lambda$, $|\Gamma_\infty|$ becomes $G_0\lambda/4\pi f$, independent of the illumination function. This is the result obtained previously from Eq. (38).

It is further of interest to note that the radius of curvature of the spiral is

$$\rho = \frac{G_0 \lambda}{4\pi f} e^{-\alpha v}. \tag{51}$$

In the limit of a very large aperture and $f \gg \lambda$, the center of curvature of the spiral in the neighborhood of $v = 0$ coincides with the limit point. Under these conditions we obtain the result that the spiral has the form of the diffraction spiral for a point on the axis of a circular aperture, independent of the feed illumination function.

The method of impedance correction is as follows: The perpendicular bisector of Γ_t, the reflection coefficient due to the reflector, is erected, and its intersection point v_c on the spiral is determined (refer back to Fig. 12·14a). This divides the surface into two zones, one within the radius $r_c = f(v_c)^{1/2}$ contributing the vector Γ_a, the other the region outside r_c contributing Γ_b. The magnitudes of Γ_a and Γ_b are equal. Now suppose that an infinitely thin plate of radius r_c is placed against the surface (Fig. 12·14c). The path from F to the edge is $f + z_c$; the path to the center is $f - z_c$; the average path length is f. The plate thus rotates Γ_a onto the R-axis to coincide in phase with the contribution from the vertex area of the paraboloid (Fig. 12·14b). It is desired to bring Γ_a 180° out of phase with Γ_b. This is achieved by rotating Γ_a by moving the plate forward a distance t (or making the plate of that thickness),

$$t = (2n + 1)\frac{\lambda}{4} - \frac{\lambda \psi_c}{4\pi}, \qquad n = 0, 1, 2, \cdots, \tag{52}$$

ψ_c being the angle between Γ_b and the R-axis. It is evident that only a small portion of the spiral in the neighborhood of $v = 0$ and the terminal vector Γ_t are required to determine the parameters of the correction plate. In most cases Γ_t can be replaced for this purpose by Γ_∞, and the final position of the plate is adjusted empirically to compensate for the error.

In the limit $\lambda/f = 0$ and large apertures, the parameters of the correction plate become practically independent of the aperture and primary feed illumination, providing the latter is not too sharply peaked. We have noted above that in the limit indicated the resultant Γ_t differs negligibly from Γ_∞ which (in this case) lies on the I-axis. Also from Eq. (51) it is seen that if the primary feed illumination is not too sharp, i.e., magnitude of α is not too large, the initial portion of the spiral can be regarded with small error as a circle of constant radius $|\Gamma_\infty|$ and center on the I-axis. When the procedure outlined in the preceding paragraph is applied to this case, it is found that the diameter of the correction plate is

$$2r_\infty = \left(\frac{4f\lambda}{3}\right)^{1/2} \tag{53}$$

and its thickness

$$t = (2n + 1)\frac{\lambda}{4} - \frac{5\lambda}{24}. \tag{54}$$

We have assumed that the current distribution over the correction plate is, except for phase, the same as that over the corresponding area of the paraboloid. In general the area of the correction plate is small, and when a small obstacle is irradiated, there is an appreciable current distribution over the shadow area of the obstacle as well as on the illuminated region. To eliminate the former it is preferable to use a plate of the thickness specified by Eq. (52), making good electrical contact with the paraboloid, rather than a thin plate set at the specified distance.

The one major objection to the vertex plate is the deleterious effect on the secondary pattern. The displacement of the vertex area produces a phase error in the field over a corresponding area of the aperture, with a resulting loss in gain, increase in beamwidth and side-lobe intensity. If the specifications on the side lobes are very stringent, the vertex-plate technique cannot be used.

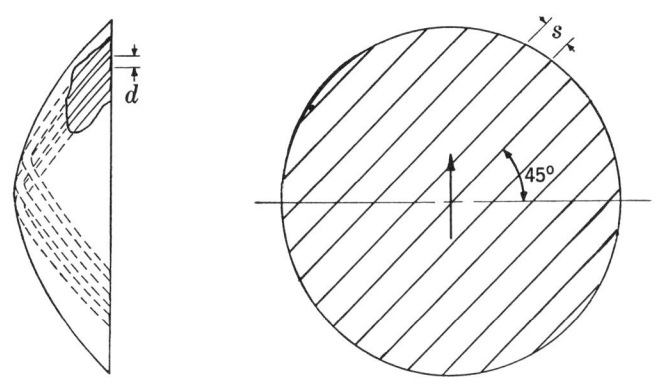

FIG. 12·15.—Quarter-wave grating to rotate the polarization of the electric vector and eliminate the mismatch.

12·10. Rotation of Polarization Technique.—The electric vector of the wave reflected by the paraboloid can be rotated through 90° with respect to the incident wave by means of a quarter-wave grating. The system is illustrated in Fig. 12·15. The grating is made up of parallel plates cut to the contour of the reflector; the plates are oriented to make an angle of 45° with the E-plane of the feed.

The grating makes use of the property of parallel plates (Sec. 7·15) that they will not support free propagation of a wave having the electric vector parallel to the plates unless the spacing s between them is greater than $\lambda/2$. If $s < \lambda/2$, the wave is attenuated; if $s < \lambda/8$, the parallel plates reflect almost completely an incident wave with the electric vector parallel to the plates. With the grating oriented at angle of 45° with respect to the E-plane, the incident electric vector can be resolved into two equal components, one parallel to the plates and one perpendicular to them. The plate spacing is such that the parallel component is reflected, with a change in phase of 180°. The perpendicular component, on the other hand, propagates between the plates with free-space velocity. If the depth of the plates d is $\lambda/4$, the latter component after reflection from the paraboloid emerges from the plates in the same direction as it had on entry. Combination with the reversed parallel component then results in a resultant vector perpendicular to the E-plane.

448   PENCIL-BEAM AND SIMPLE FANNED-BEAM ANTENNAS   [SEC. 12·11

Preliminary experiments conducted at the Radiation Laboratory to test the effectiveness of the technique gave promising results in so far as the impedance characteristics were concerned. The effects on the secondary pattern were not determined. It is to be expected that the grating does not function properly at the edges because of the oblique incidence of the primary radiation, thus introducing phase distortion. Further study of the subject is needed in order to evaluate the relative values of the grating and vertex-plate techniques.

12·11. Structural Design Problems.—An antenna must generally meet certain mechanical specifications such as a minimum weight/strength factor, low wind resistance, and visual transparency in addition to fulfilling the requirements on the secondary pattern.

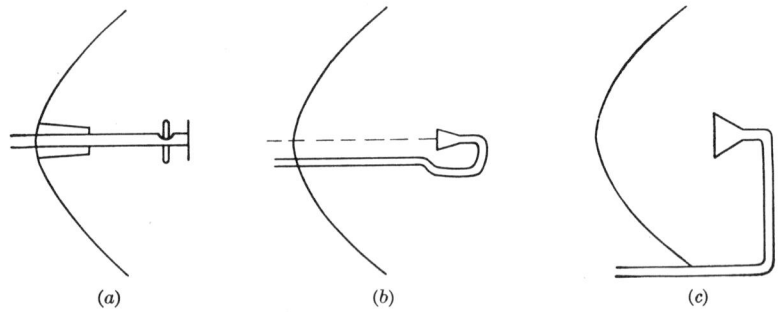

FIG. 12·16.—Rear-feed and front-feed installations: (a) rear-feed technique for a dipole-disk feed; (b) rear-feed technique for a horn; (c) front-feed technique for a horn.

Rear-feed and Front-feed Systems.—The first factor to be considered is the type of feed installation. Two general methods—rear-feed and front-feed installation—are illustrated in Fig. 12·16. The rear-feed installations (Fig. 12·16a and b) have the advantages of compactness and requiring a minimum length of transmission line. The latter has important bearing on the impedance presented by the system at the generator terminals. If the focal length is short, a simple flange connection between the transmission line and reflector is sufficient to support the feed system. If the focal length is large, a more extended collar such as is shown diagrammatically in Fig. 12·16a is necessary to prevent free play of the feed. The rear-feed installation of a horn, such as illustrated in Fig. 12·16b, is feasible only at short wavelengths (3 cm or less). Even for the latter it is not to be recommended because of the asymmetry and possible phase distortion introduced into the primary pattern.

The front-feed installation (Fig. 12·16c) is recommended for all horn feeds. It suffers from one serious defect of obstructing too much of the aperture. The interference is reduced somewhat if the waveguide is placed in the H-plane. This may make it necessary to put a twist in the waveguide in order to orient the horn properly with respect to the horizontal plane.

Grating and Screen Reflectors.—The weight and wind resistance of the paraboloid can be reduced considerably by replacing the continuous reflector surface by a perforated surface or a grating structure. An example of an antenna using a perforated paraboloid is shown in Fig. 1·5; examples of grating reflectors will be found in Fig. 12·23 and in several photographs in Chap. 13.

The reflectivity of the perforated surface is insensitive to polarization. The perforations can be regarded as short waveguides designed to be far beyond cutoff for the frequency band over which the antenna is to be used. For example, if the reflector is a wire screen with square openings, the edge length a of the openings must be such that

$$a < \frac{\lambda}{\sqrt{2}}. \qquad (55)$$

This is the condition for cutoff in a square waveguide.

The gratings are sensitive to polarization. The space between the grating elements may be thought of as waveguides beyond cutoff for the electric vector parallel to grating element. The grating elements

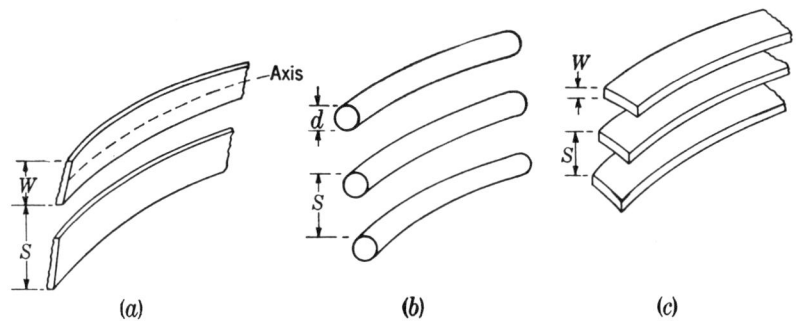

Fig. 12·17.—Grating reflectors: (*a*) broadside strips; (*b*) round bars; (*c*) edgewise strips.

may be divided into three groups: (1) broadside strips, (2) bars, and (3) edgewise strips; these are illustrated in Fig. 12·17. The various types of gratings have been studied experimentally[1] to determine the relationships between the grating dimensions and wind resistance and transmissivity. There are two major restrictions that apply to all gratings:

1. The electric vector of the incident wave must be in the plane determined by the incident ray and the axis of the grating element.
2. The center to center spacing of the elements must be less than $\lambda/(1 + \sin \theta)$, where θ is the angle between the incident ray and

[1] W. D. Hayes, "Grating and Screens as Microwave Reflectors," RL Report No. 54-20, Apr. 1, 1943.

450 PENCIL-BEAM AND SIMPLE FANNED-BEAM ANTENNAS [SEC. 12·12

the normal to the axis of the grating element. Larger spacings cause the appearance of undesirable higher-order lobes in the secondary pattern.

The edgewise strips are generally to be preferred. Their transmission characteristics are summarized in Fig. 12·18, which gives the relation between the strip depth and the spacing for fixed values of transmissivity. The properties vary, of course, with the width of the strips; the reader is referred to the report by Hayes for more extensive data. The variation of the depth of the strips to control the r-f transmission has a negligible effect on wind resistance; both can be made quite low. Mechanical

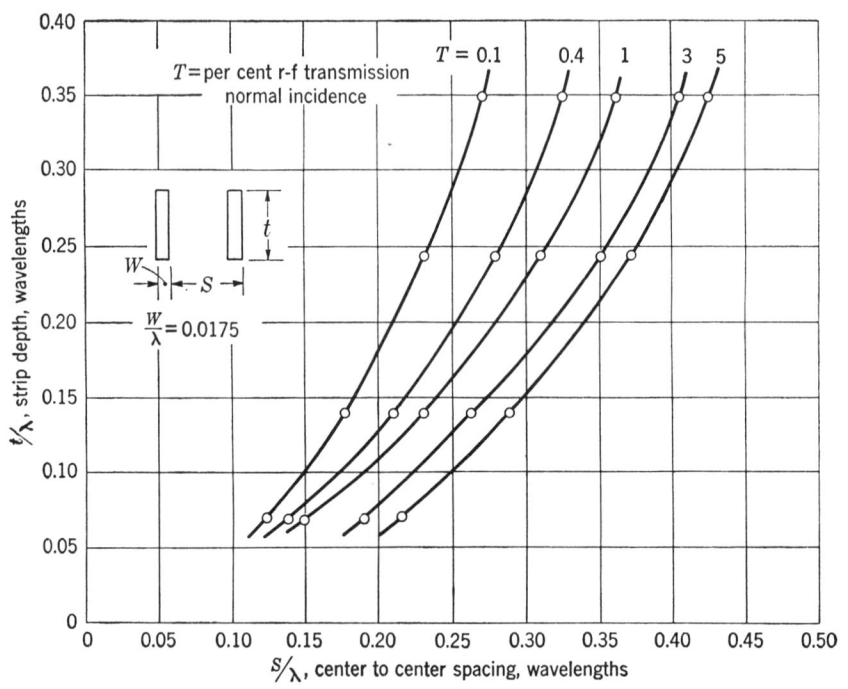

Fig. 12·18.—Grating of edgewise strips: Relation between strip depth and spacing for constant transmission.

rigidity can be obtained by proper bracing. The strips also have the advantage that the reflector shape can be obtained by a cutting operation; in making up a paraboloid all the strips can be identical punchings of flat sheet metal.

SIMPLE FANNED-BEAM ANTENNAS

12·12. Applications of Fanned Beams and Methods of Production.— The singular advantage of a pencil beam for locating a target with accuracy is offset by the difficulty of intercepting a target in the course of a random search because the beam covers only a narrow cone of space at a given instant. Further difficulties are encountered in the case of antennas on ships; in the course of the roll and pitch of the ship the beam swings into the water where it serves no purpose or up above the horizon losing its effectiveness in locating surface vessels. To counter these various difficulties it is necessary to sacrifice the directivity by flaring the beam in one of its principal planes—generally the vertical plane. By retaining the narrow width in azimuth, resolution is maintained in this aspect and the radar echo technique supplies information on range.

The present chapter concerns itself only with simple fanned beams which may be thought of as being developed by distorting the almost circularly symmetrical beam into a symmetrical elliptical beam. The more complex fanned beams which are designed for highly specialized operational functions will be treated in the next chapter. From the general relations developed in Chap. 6, between the symmetry of the aperture and aperture field and the symmetry of the beam, two basic techniques suggest themselves: (1) to use an aperture with two highly different dimensions in the principal planes, the beamwidths in the principal planes being inversely proportional to the aperture dimensions, and (2) to taper the illumination differently in the two principal planes. The second of these may be dismissed as an isolated technique because the beamwidth is not sufficiently sensitive to the illumination. The only practical technique, therefore, is the first, of using an aperture with suitable dimensions in the principal planes. The illumination technique may be used as auxiliary to the other method.

The fanned-beam antennas take the following forms: (1) an ovoid section of a paraboloidal reflector with a point-source feed at the focus, (2) a parabolic cylinder with a line source producing a rectangular aperture, and (3) a parallel-plate antenna consisting of a parabolic cylindrical reflector illuminated by a simple feed at the focus and located between parallel plates that are perpendicular to the generator of the cylinder; this likewise produces a rectangular aperture. Design techniques will be presented for each of these types of antennas.

12·13. Symmetrically Cut Paraboloids.—The simplest procedure is to cut a paraboloid symmetrically by a pair of parallel planes as shown in Fig. 12·19a. The long dimension will be denoted by d_1, and the narrow dimension by d_2. The results obtained from a circular aperture with many types of feeds and paraboloid shapes show that the beamwidth is in the range $(1.2 \pm 0.2)\lambda/D$. These results have been extrapolated to the cut paraboloid, and the relations between the dimensions of the latter and the principal-plane beam widths are generally taken to be

$$\Theta_1 = 1.2 \frac{\lambda}{d_1}; \qquad \Theta_2 = 1.2 \frac{\lambda}{d_2}. \tag{56}$$

It is quite evident that a circularly symmetrical primary feed pattern is unsuited to illuminate the reflector; a large fraction of the energy would be wasted in spillover. The primary feed pattern must be shaped to the same symmetry as the reflector. Taking the results for optimum performance of a circular aperture again as a criterion, we may require that the 10-db width of the feed pattern in a given principal plane be equal to the angle subtended by the reflector at the feed in the given principal plane.

Horns with rectangular apertures lend themselves particularly well to the design of suitable feeds, since the beamwidths in the two principal

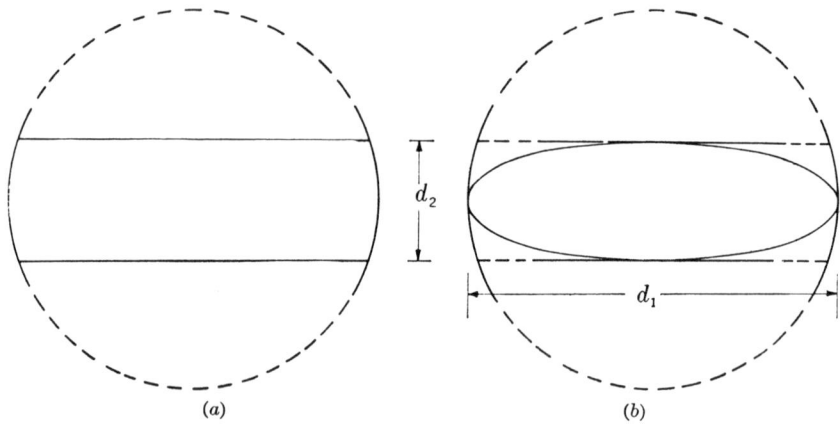

Fig. 12·19.—Symmetrically cut paraboloids; (a) simple line cut; (b) equi-intensity contour cut.

planes can be controlled virtually independently of one another by choice of the principal plane dimensions. The relation betweens the primary pattern 10-db beamwidth and the horn dimensions are given by Eqs. (10.52) and (10.53). The primary pattern beamwidths that are required are determined by the dimensions d_1 and d_2 of the reflector aperture and the focal length. The latter should be chosen as small as possible to keep the primary pattern 10-db width large; otherwise the design of a practical horn becomes very difficult. Difficulties are encountered if the ratio d_1/d_2 is too large; since the dimension of the horn in the d_1-plane must be so much smaller than that in the d_2-plane that the resulting horn has widely different centers of feed in the two planes. This will give rise to serious phase errors and loss in gain.

The primary pattern of the horn designed to meet the principal plane requirements has an elliptical cross section. Consequently, the equi-intensity illumination contours on the reflector are also elliptical in shape. There are several reasons for cutting the reflector along such a contour as shown in Fig. 12·19b. It is found in general that the gain factor increases and the general features of the pattern are improved by a reduction in side lobes in the principal planes. The basis for this lies in the fact that the effective illumination for say the d_2-principal plane at a given point on the d_2-axis is the integrated intensity across the aperture parallel to the d_1-plane. With the aperture of the type shown in Fig. 12·19b the integrated area tapers along the d_2-axis, and the effective illumination is, therefore, more tapered than in the corresponding case of Fig. 12·19a, hence the improved side-lobe characteristics. The ovoid shape of Fig. 12·19b also has advantages of low wind resistance and smaller moments of area and inertia which are of considerable importance in connection with the mechanical problems of support and rotation of the antenna.

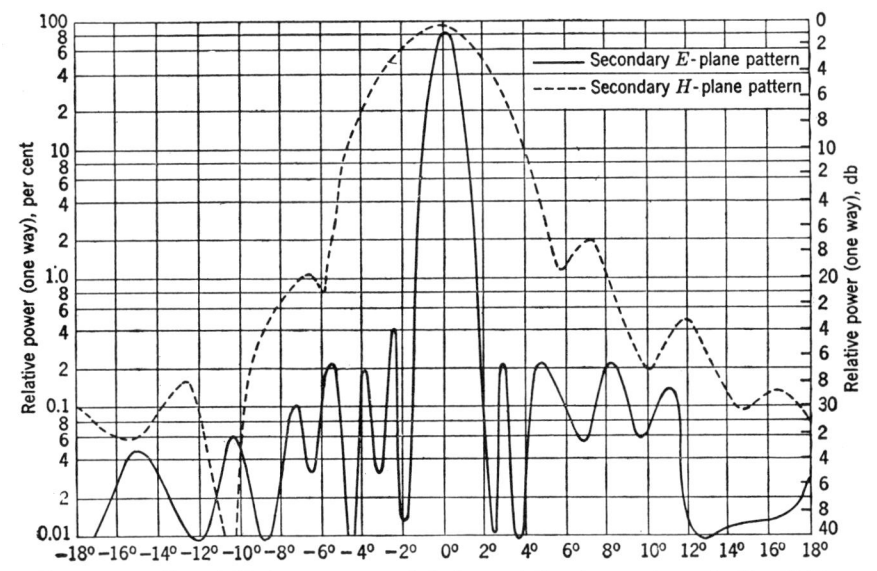

Fig. 12·20.—E- and H-plane patterns of the beavertail antenna, shown in Fig. 13·12a.

An antenna using the symmetrical ovoid-shaped reflector[1] is shown in the following chapter in Fig. 13·12a. The dimensions of the reflector are $d_1 = 20$ ft, $d_2 = 5$ ft, and the focal length $f = 5$ ft. The feed that was finally adopted for this antenna is a flared box horn[2] designed to meet the illumination requirements in the principal planes. The secondary patterns of the antenna are shown in Fig. 12·20. The ratio of the half-power widths Θ_E/Θ_H is 0.29, and the ratio of the aperture dimensions $d_2/d_1 = 0.25$. The H-plane side-lobe levels are all down below 17 db with no prominent wide-angles lobes; the E-plane side lobes are all below 23 db, showing that the illumination is properly distributed over the reflector.

12·14. Feed Offset and Contour Cutting of Reflectors.—The symmetrically cut paraboloids have the drawback that the feed must be located

[1] C. S. Pao, "The Beavertail Antenna," RL Report No. 1027, Apr. 9, 1946.
[2] S. J. Mason, "Flared Box Horn," RL Report No. 653, July 9, 1945; see also Chap. 10.

in the center of the aperture. In this position it is in the path of the reflected rays from the most intensely illuminated area, and hence the mismatch introduced by the reflector is quite significant. Furthermore, the use of a horn feed introduces a large section of waveguide, which in large reflectors necessitates additional supporting structures; these together with the feed block out aperture area, causing a loss in gain and increase in side lobes (*cf.* Sec. 6·7).

Both of these defects are eliminated by the offset feeding technique which is illustrated schematically in Fig. 12·21. The center of feed is placed at focus of the paraboloid as in the previous case, but the horn is tipped so that the peak of the primary pattern makes some angle ψ_0 with the paraboloid axis. The major portion of the lower section of the paraboloid is discarded. The dimension d_2 is again determined from the secondary pattern beamwidth by Eq. (56). The offset feeding removes the horn and its supporting structure out of the way of the most intensely illuminated area of the aperture with resulting improvement in gain and in side-lobe characteristics.

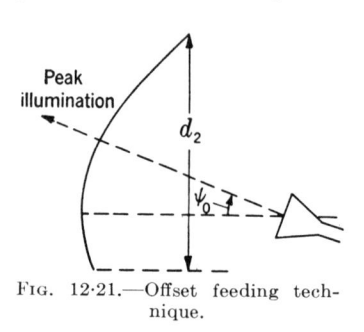

Fig. 12·21.—Offset feeding technique.

The reduction of the mismatch can be understood in terms of the geometrical-optics picture that the radiation returning to the feed comes from the area around the vertex of the paraboloid. The magnitude of the mismatch is given by a relation equivalent to Eq. (39):

$$|\Gamma_r| = \frac{G_f(\psi_0)\lambda}{4\pi f}, \qquad (57)$$

where $G_f(\psi_0)$ is now the gain of the feed in the direction along the axis. By offsetting the feed the reflection coefficient is reduced by the ratio $G_f(\psi_0)/G_{f_0}$, where G_{f_0} is the peak gain.

The design procedure is essentially the following: The dimensions d_1 and d_2 are chosen in accordance with the beamwidth relations [Eq. (56)]. The focal length and the dimensions of the horn aperture are chosen as though the reflector is to be cut symmetrically; the angle subtended by d_1 at the focus should not exceed 160°. The horn is constructed, pressurized, and matched by the methods discussed in Chap. 10. Let Γ_f be the residual mismatch of the feed and Γ the allowable total mismatch of the antenna; the allowable reflector mismatch is then

$$|\Gamma_r| = |\Gamma| - |\Gamma_f|. \qquad (58)$$

Using a circular paraboloid of the focal length of the final antenna, the paraboloid reflection coefficient is determined as a function of the feed

SEC. 12·14] CONTOUR CUTTING OF REFLECTORS

offset ψ_0 by the circle diagram method referred to previously (cf. Fig. 12·12), or the peak gain G_{f0} is determined from the mismatch at the angle $\psi_0 = 0$, and the mismatch at any other angle is computed by means of Eq. (57) from the knowledge of the primary pattern.

For the chosen value of ψ_0 the primary feed pattern is transformed into equi-intensity illumination contours on the surface of the paraboloid by taking into account space attenuation according to the inverse square

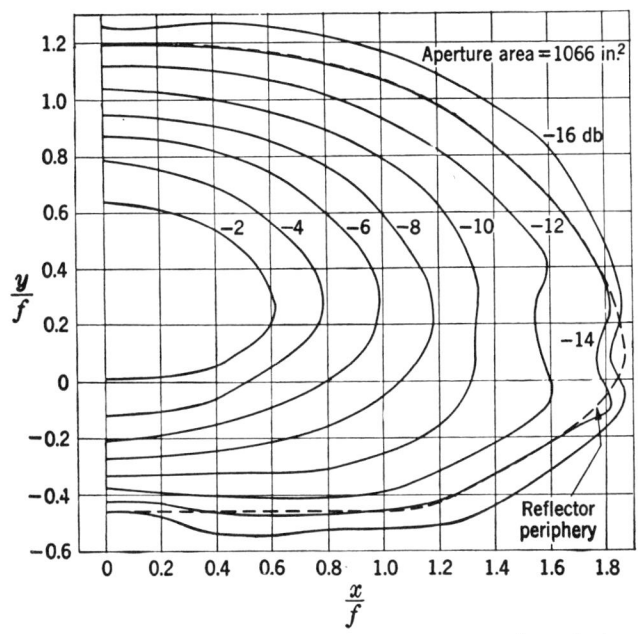

FIG. 12·22.—Constant intensity contours in paraboloid aperture (horn feed axis tilted 20° relative to paraboloid axis).

law. An example of such an equi-intensity contour plot is shown in Fig. 12·22. The paraboloid is then cut to follow an equi-intensity contour, generally chosen as the 14-db contour.

A number of antennas have been designed according to this procedure with very successful results.[1] Figure 12·23 is a photograph of the antenna[2] to which the constant intensity contours (Fig. 12·22) apply. The reflector dimensions are $d_1 = 54$ in., $d_2 = 24$ in., $f = 14.5$ in. The horn aperture dimensions are 2 cm in the E-plane, 6.0 cm in the H-plane with flare angles of 10° and 40° in the respective planes. The offset angle is 20°; this was chosen so that $\Gamma_r < 0.04$ in order that the resultant mismatch of the feed and paraboloid over the entire band of 8600 to

[1] T. J. Keary and J. I. Bohnert, RL Report No. 659, Mar. 7, 1945; RL Report No. 660, Feb. 19, 1945; RL Report No. 779, Aug. 30, 1945; J. I. Bohnert and H. Krutter, RL Report No. 665, Feb. 7, 1945.

[2] T. J. Keary and J. I. Bohnert, RL Report No. 659, Mar. 7, 1945.

456   PENCIL-BEAM AND SIMPLE FANNED-BEAM ANTENNAS   [SEC. 12·14

FIG. 12·23.—Fanned-beam antenna using the offset feed technique.

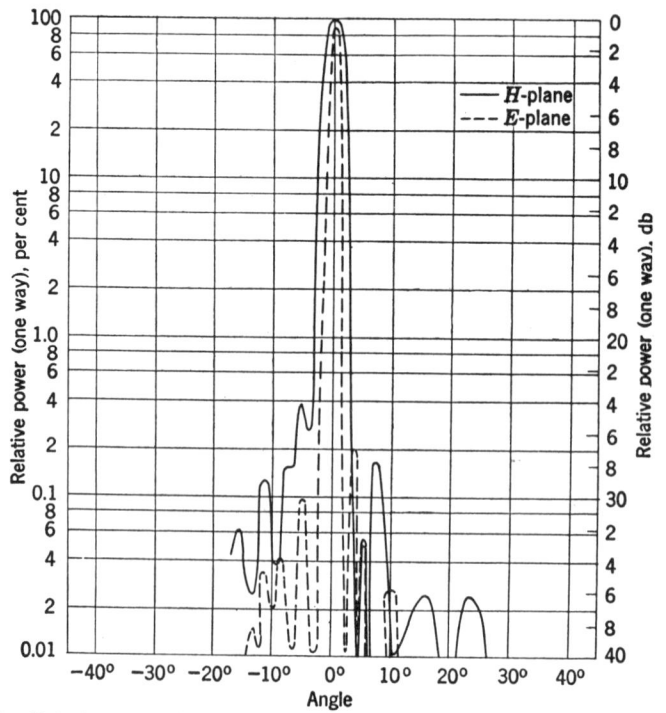

FIG. 12·24.—Principal E- and H-plane polar diagrams of the antenna shown in Fig. 12·23.

9700 Mc per sec should represent a reflection coefficient less than 0.091. The performance of the antenna is demonstrated by the E- and H-plane patterns shown in Fig. 12·24. The ratio of the beamwidths Θ_E/Θ_H again is very closely equal to the ratio d_2/d_1. The low level of the side lobes attest further to the validity of the design procedure.

The elimination of one of the planes of symmetry by the offset feeding technique produces one serious effect. The process destroys the symmetry of the cross-polarization component of the aperture field leading to cross polarization in the plane of the large dimension of the aperture. The cross-polarization pattern has lobes on either side of the main lobe in the plane of the narrower beamwidth, which may seriously affect the

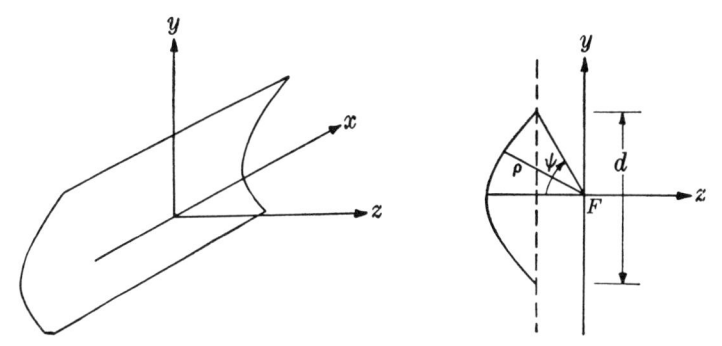

FIG. 12·25.—Parabolic cylinder and line source.

performance of the system. Cross-polarization studies should be made in the narrow-width plane for all antennas of this type.

12·15. The Parabolic Cylinder and Line Source.—In principle a simple fanned beam is most easily produced by using a rectangular aperture with a separable type of aperture field such as was discussed in Secs. 6·5 and 6·6. The principal plane patterns are then determined completely by the aperture dimension in the given plane and the field distribution in that aspect. There is no interaction between the distributions in the principal planes. A second advantage is the reduction of cross polarization.

The required aperture configuration and field distribution are readily obtained by illuminating a parabolic cylinder by a line source located along its focal line. An antenna of this type is shown in Chap. 1, Fig. 1·6. The general theory of such systems has been developed from the standpoint of the reflector currents in Sec. 5·9 and from the aperture field standpoint in Secs. 6·8 and 6·9. We shall here simply state the results which are particular to the parabolic cylinder. In Fig. 12·25, the line source is taken along the x-axis which is also the focal line of the parabolic cylinder. Let l be the length of the source and Ψ be the angular aperture of the reflector. The performance of the system depends on the fact that the reflector is in the *cylindrical wave cone* of the source. It is, therefore, necessary that

$$l \gg \lambda, \qquad \rho_{\max} < \frac{l^2}{\lambda}, \tag{59}$$

where $\rho_{\max}$ is the maximum radial distance from the source to the reflector. For wavelengths greater than about 10 cm conditions (59) imply that the length of the cylinder is greater than the height of the aperture.

It is clear that all rays from the source incident on the reflector in a plane parallel to the yz-plane are reflected in that plane into a family of rays parallel to the z-axis. The reflector thus produces a uniform phase distribution over the aperture. Also, since the reflected rays are parallel, the field intensity at a given point on the aperture is the same in magnitude as that of the reflected field (or incident field) at the corresponding point on the reflector. The intensity distribution, $F(x)$, in the x-direction over the aperture is, therefore, the same as that of the line source, and the aperture distribution in the transverse direction is determined entirely by the two-dimensional gain function $G(\psi)$ of the cylindrical wave zone of the line source (*cf.* Sec. 5·9). Evaluating the field in the forward direction by means of Eqs. (5·86) and (5·88), we find that for both the longitudinally and transversely polarized systems the gain is given by

$$G_M = \frac{A}{2\lambda^2} \cot \frac{\Psi}{2} \left[\frac{1}{l} \int_{-l/2}^{l/2} [F(x)]^{\frac{1}{2}} dx \right]^2 \left[\int_{-\Psi}^{\Psi} [G(\psi)]^{\frac{1}{2}} \sec \frac{\psi}{2} d\psi \right]^2. \tag{60}$$

The gain factor $\mathcal{G} = G_M \lambda^2 / 4\pi A$ is, therefore,

$$\mathcal{G} = \frac{1}{8\pi} \cot \frac{\Psi}{2} \left[\frac{1}{l} \int_{-l/2}^{l/2} [F(x)]^{\frac{1}{2}} dx \right]^2 \left[\int_{-\Psi}^{\Psi} [G(\psi)]^{\frac{1}{2}} \sec \frac{\psi}{2} d\psi \right]^2. \tag{61}$$

The term involving $F(x)$ gives the effect of the deviation from uniform illumination along the x-direction. The second term gives the dependence on the angular distribution of the primary pattern. As in the case of the paraboloid of revolution there is an optimum angular aperture for a given feed pattern that represents the compromise between spillover and tapered illumination over the aperture in the y-direction. The optimum angular aperture can be found by graphical methods as was done in Sec. 12·5 for the paraboloid of revolution.

For maximum gain, the distribution $F(x)$ should be equal to unity. This, however, gives maximum side lobes in the longitudinal pattern—that is, in the planes containing the line source—as compared with tapered distributions. The longitudinal pattern can be studied as a two-dimensional problem, independently of the transverse pattern. All of the results of Sec. 6·6, other than the actual values of the gain, can be applied here without modification. The gain is affected by the transverse distribution as is shown by Eq. (60).

The essential difficulty with antennas of this type is in producing an efficient line source. Linear arrays such as are discussed in Chap. 9 are frequently used. The impedance characteristics are generally poor due to strong interaction between the reflector and the source.

12·16. Parallel-plate Systems. *Cheese and Pillbox Antennas.*—The limitations imposed on the antenna design by conditions (59) can be eliminated by placing the parabolic cylinder between parallel plates as shown in Fig. 12·26. The feed may then be a waveguide or a horn with one of its aperture dimensions equal to the distance h between the plates.

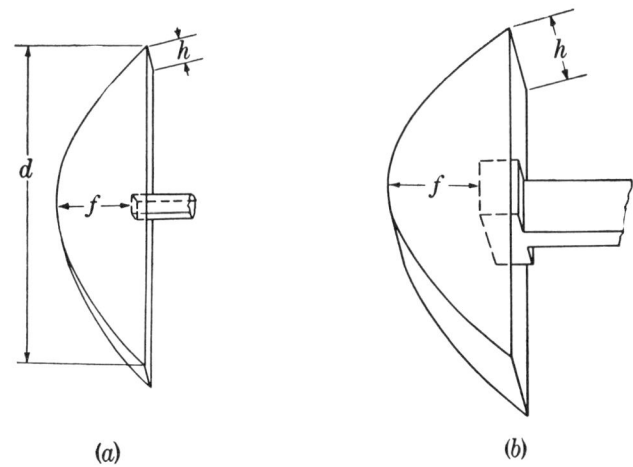

FIG. 12·26.—Parallel-plate systems: (b) pillbox antenna; (a) cheese antenna.

From the point of view of the rays between the plates, the system is equivalent to a segment of an extended line source and parabolic cylinder.

The antennas differ from the open system of the preceding section in that propagation can take place between the plates in various modes (*cf.* Sec. 7·15). The parallel plates support free propagation of a principal wave—the TEM-mode—in which the electric vector is normal to the plates; the velocity of propagation and the wavelength is the same as in free space. TE- and TM-modes are also possible, which are equivalent to the modes in a rectangular waveguide. We need concern ourselves only with the lowest TE-mode in which the electric vector is parallel to the plates and varies in magnitude along the line normal to the plates according to $\sin(x/h)$, where h is the distance between them. The plates will support propagation in this mode only for free-space wavelengths that satisfy the condition

$$\lambda < 2h. \qquad (62)$$

The wavelength of propagation is

$$\lambda_g = \frac{\lambda}{\left[1 - \left(\frac{\lambda}{2h}\right)^2\right]^{1/2}}. \qquad (63)$$

The cutoff condition for the next higher mode is

$$h < \lambda. \qquad (64)$$

The parallel-plate systems may be classed into two groups: (1) those with spacing $h < \lambda$ which support free propagation in the TEM-mode and possibly the TE_1-mode if $h > \lambda/2$, (2) those with spacing $h > \lambda$ which support additional modes. The two groups are labeled pictorially, the former being called the pillbox antennas, the latter cheese antennas.

The cheese antennas can be designed to meet any length-to-height ratio desired. If only the TEM-mode is desired, the feed must be designed with great care in order to avoid the excitation of other modes. The difficulty of eliminating other modes is the major objection to the TEM-cheese antenna. On the other hand, the feed can be designed to excite various TE- and TM-modes purposefully. Each mode travels with a characteristic phase velocity, and the superposition of the modes is used to synthesize various types of phase distributions over the aperture.[1]

The limitations imposed on the secondary pattern of a pillbox in the plane containing the h-dimension, because of the restrictions on the latter, can be obviated to some extent by flaring the mouth of the pillbox into a two-dimensional horn. This has a further advantage of reducing the reflection by the aperture of the wave between the plates. Another method of controlling the pattern is by means of flaps such as are shown in the half-beacon antenna[2] in Fig. 12·27. Half a pillbox is used in this particular case; it is fed by an H-plane sectoral horn. The h-dimension is equal to $\lambda/3$ so that the plates can support only the TEM-mode. Attention should be called to the curled edge of the flap; the curl follows an exponential spiral in order to reduce the impedance mismatch arising from the discontinuity at the edge of the flap. The pattern obtained in the plane of the h-dimension is shown in Fig. 12·28.

12·17. Pillbox Design Problems.—There are three major problems to be considered in the design of the pillbox: (1) the f/d ratio for maximum gain factor, (2) impedance mismatch, (3) structural problems.

---

[1] The cheese antenna received more attention in Britain than in the United States. Information pertaining to British reports may be obtained from the British Scientific Commission office in Washington, D.C. or the British Central Radio Bureau in London. Much of the British work is appearing in the new section Part IIIa, "Radiolocation," of the journal of the Institute of Electrical Engineers.

[2] A. Braunlich, "Half Beacon Antenna," RL Report No. 419, Sept. 6, 1943.

PILLBOX DESIGN PROBLEMS

Optimum Shape.—The analysis of the gain factor[1] proceeds along the same lines as for the parabolic cylinder in Sec. 12·15. The result is essentially that of Eq. (61) except for multiplicative constants. The

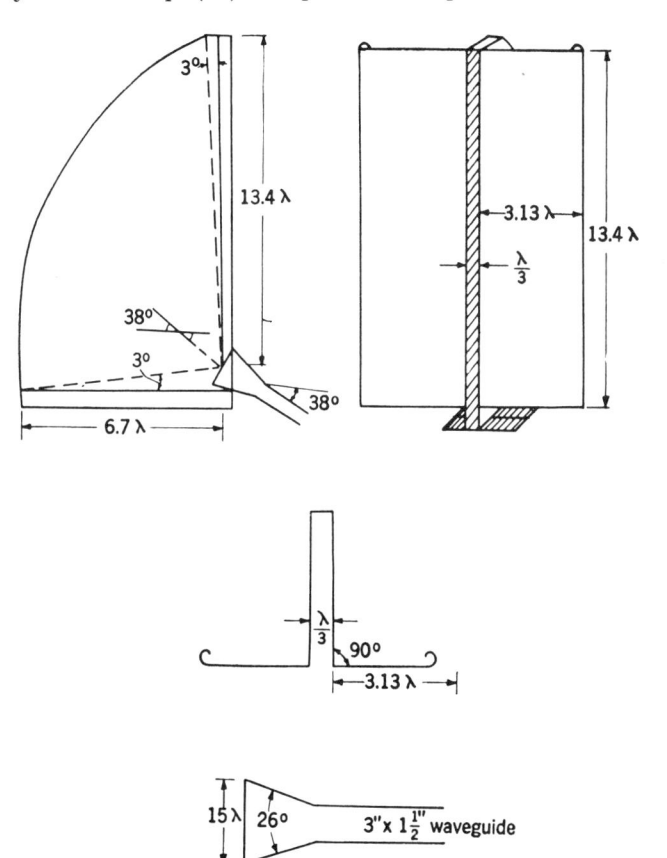

Fig. 12·27.—Half-beacon antenna.

optimum angular aperture Ψ (*cf.* Fig. 12·29) is the value for which the expression

$$\cot\left(\frac{\Psi}{2}\right)\left[\int_0^\Psi [G_f(\psi)]^{1/2} \sec \frac{\psi}{2}\, d\psi\right]^2 \tag{65}$$

has its maximum value. The gain function $G_f(\psi)$ is that of the feed radiating between parallel plates, not in free space; like that of the cylindrical wave zone of a line source it is two-dimensional.

The optimum angular aperture is generally less than 90°. The pillbox is then constructed as shown in Fig. 12·29 with the parallel plates

[1] T. J. Keary, A. R. Poole, J. R. Risser, H. Wolfe, "Airborne Navigational Radar Antennas," RL Report No. 808, Mar. 15, 1946.

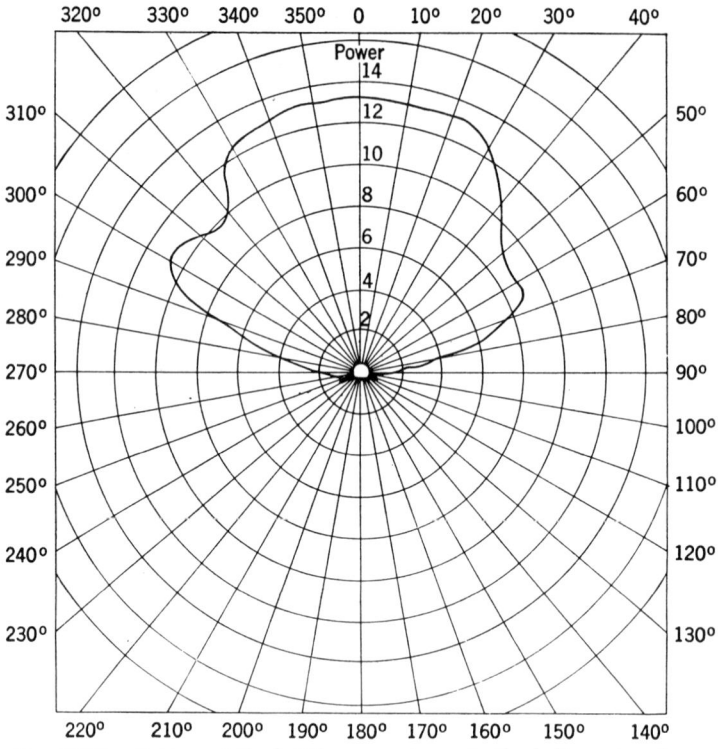

462 PENCIL-BEAM AND SIMPLE FANNED-BEAM ANTENNAS [SEC. 12·17

FIG. 12·28.—Pattern of the half-beacon antenna in the h-dimension plane.

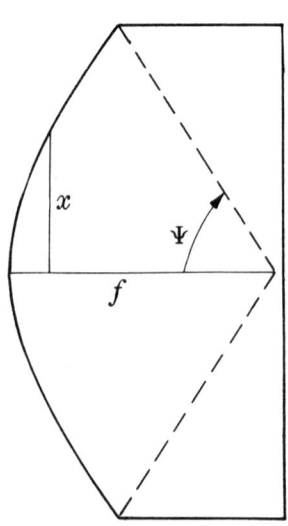

FIG. 12·29.—A pillbox of angular aperture Ψ.

extending a little beyond the focal point. The gain factors realized by pillboxes are considerably higher than that of paraboloidal antennas, ranging in value upward from 0.8.

Impedance Mismatch.—The mismatch produced by the parallel-plate system arises from the parabolic strip and from reflection at the aperture. The latter can be reduced, as was noted before, by flaring the mouth of the pillbox into a two-dimensional horn. The reflection coefficient produced by the parabolic strip can be developed along lines similar to that followed in the case of the paraboloid.[1] The essential difference is that the field between the parallel plates is in the form of a cylindrical wave rather than a spherical wave. The reflection coefficient is found to be

$$\Gamma_r = \frac{1}{\pi} \int_0^\Psi G(\psi) \frac{e^{-jk\rho}}{\rho} \cos\frac{\psi}{2} ds, \tag{66}$$

where ds is the element of length along the reflecting strip. Let x measure position along a line parallel to the aperture; on introducing the variable

$$v = \frac{x}{f} = 2\tan\frac{\psi}{2} \tag{67}$$

the reflection coefficient becomes

$$\Gamma_r(v) = \frac{1}{\pi} \int_0^v \frac{G(v)}{1 + \frac{v^2}{4}} e^{-j\frac{\pi f}{\lambda}v^2} dv, \tag{68}$$

disregarding all constant-phase terms. If $f/\lambda \gg 1$, the integral of Eq. (68) is very closely equal to

$$\Gamma_\infty = \frac{G_0}{2\pi}\left(\frac{\lambda}{f}\right)^{\frac{1}{2}}, \tag{69}$$

where G_0 is the peak gain of the feed. This is the two-dimensional analogue of Eq. (39).

The mismatch can be eliminated by means of a vertex plate as in the case of the circular paraboloid. The technique of determining the dimensions of the plate is the same as that described in Sec. 12·9. It should be noted that the intersection point v_c on the spiral gives x_c which is only half the length of the plate. The vertex plate has the same undesirable effects on the secondary pattern as in the paraboloid: reduction in gain, increase in side-lobe intensity.

Structural Problems.—Special attention must be paid to the structure and assembly of the pillbox. The feed must make good electrical contact with the parallel plates. The contact can be established by solder-

[1] Details are given by S. Silver, "Analysis and Correction of the Impedance Mismatch Due to a Reflector," RL Report No. 810, Sept. 25, 1945.

ing; a better technique, however, is to make a choke joint between the feed and the pillbox.[1] Good electrical contact must also be maintained between the reflecting strip and the parallel plates. This is a more important consideration for the TEM-mode, in which the electric-field vector is normal to the plates, than the TE-mode, in which it is parallel to the plates, and zero at the surface of the plates. The space between the reflecting strip and the parallel plate, in the case of poor contact, is too small to propagate a TE-mode.

It is recommended that the parabolic strip be cut out of a metal plate and cut to a sizable thickness so that the plates can be bolted to the strip without warping the parabolic curve.

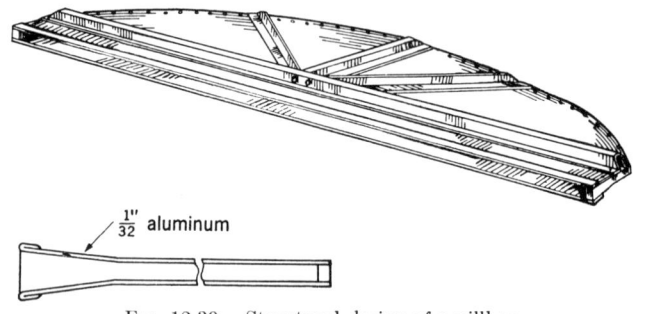

Fig. 12·30.—Structural design of a pillbox.

Maintaining a uniform spacing between the plates poses a number of difficult problems. The spacing problem is not too serious for the TEM-mode. If the spacing h is well below $\lambda/2$, there is no significant mode control problem. In this case, proper reinforcement of the parallel plates[2] as shown in Fig. 12·30 together with the spacing support provided by the feed is sufficient. Additional support is necessary only in extremely large structures.

The tolerances on the spacing are more restrictive in the case of the TE-mode. The phase velocity varies with the spacing [Eq. (63)]; a nonuniform spacing produces phase distortion over the aperture. The spacing can be maintained by a distribution of metal or dielectric posts. These scatter the energy, however, producing both a mismatch and distortion of the field over the aperture. The pins should be kept out of the high intensity region of the primary pattern of the feed.

[1] T. J. Keary et al., "Airborne Navigational Radar Antennas," RL Report No. 808, Mar. 15, 1946.

[2] W. Sichak and E. Purcell, "Cosec² Antennas with a Line Source and Shaped Cylindrical Reflector," RL Report No. 624, Nov. 3, 1944.

CHAPTER 13

SHAPED-BEAM ANTENNAS

By L. C. Van Atta and T. J. Keary

The highly directive beams attainable with microwave antennas have been utilized to achieve large antenna gain, precision direction finding, and a high degree of resolution of complex targets. The exploration of a wide angular region with such sharp beams requires an involved scanning operation in which the scanning time becomes a limiting factor. This problem is much simplified if the required scanning can be reduced to only one direction, the coverage of the angular region being completed by fanning the beam broadly. The characteristics of simple fanned-beam antennas have been discussed in Chap. 12. For many applications, however, the characteristic shape of the fanned beam obtained

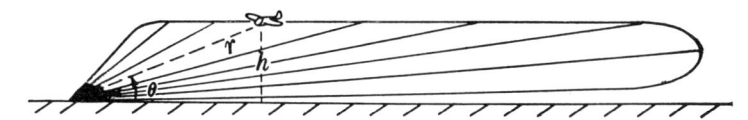

FIG. 13·1.—Beam from ground-based or shipborne antenna providing coverage on aircraft.

by simply reducing the corresponding dimension of the aperture is unsatisfactory; it may be wasteful of the limited microwave power, or it may result in a very unequal illumination of targets in different directions. To overcome these limitations it is necessary to impose on the beam by special design techniques some shape not characteristic of the normal diffraction lobe. These beams are referred to as *shaped beams*, and the antennas that produce them as *shaped-beam antennas*.

The purpose of this chapter is to describe several applications for shaped beams, to discuss requirements imposed on the beam by these applications, and to present a number of design techniques for producing shaped-beam antennas.

13·1. Shaped-beam Applications and Requirements.—There are a number of radar applications for microwave systems that impose more or less severe beam-shaping requirements upon the antenna. The applications and requirements will be considered here; the means for realizing the shaped beams will be deferred to later sections.

Surface Antenna for Air Search.—For use in search for aircraft, an antenna on the ground or on a ship is required to produce a beam sharp in azimuth but shaped in elevation; the azimuth coverage is obtained by scanning. The elevation shape of the beam must provide coverage on aircraft up to a certain altitude and angle of elevation and out to the maximum range of the system. This is to be accomplished without wasteful use of available power. Figure 13·1 indicates the general shape of the coverage required in the vertical plane. The antenna beam need not meet the coverage requirement very accurately, since conservation of power and a relatively constant signal on a plane at a fixed altitude are the only objectives.

In order to maintain a fixed minimum of illumination on the aircraft at various points along the upper contour of the coverage diagram, it is necessary that the amplitude of the antenna pattern be proportional to

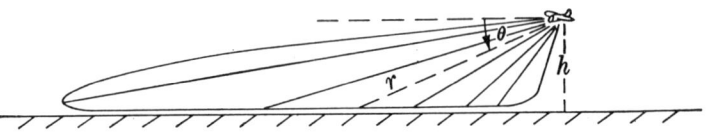

Fig. 13·2.—Beam from antenna of airborne radar for surface search.

the distance r from the antenna to the aircraft on that contour. In other words, the coverage contour of Fig. 13·1 can be taken to be the amplitude pattern of the antenna (cf. Sec. 1·2). Since $r = h \csc \theta$, the amplitude pattern must be proportional to $\csc \theta$, or the power pattern must be proportional to $\csc^2 \theta$. The proportionality must hold over the region from a minimum angle arc sin h/r_{max}, to the maximum elevation angle for which coverage is required.

Airborne Antenna for Surface Search.—An airborne antenna is required to produce a beam sharp in azimuth but so shaped in elevation as to provide uniform illumination on the ground; azimuthal coverage again is achieved by scanning. Figure 13·2 illustrates the vertical coverage requirement; this was shown in the previous paragraph to be identical with the vertical amplitude pattern. Both this and the previous pattern assume isotropic scattering by target objects. Deviations from this assumption for various target objects will be discussed in the next section.

When an airborne antenna is used primarily for surface search over sea against such point targets as ships and buoys, the purpose of beam shaping is to conserve power, to maintain a relatively constant signal as the target is approached, and to avoid overloading the indicator scope with sea return. None of these objectives impose exacting requirements on the beam shape. However, for successful surface search over land it is necessary to illuminate the ground very uniformly in order to obtain "solid painting" on the indicator scope and a fully intelligible picture. The results from the operator's viewpoint of satisfactory and unsatisfactory elevation patterns are described in Sec. 14·4 and its accompanying figures.

SEC. 13·1]   *SHAPED-BEAM APPLICATIONS AND REQUIREMENTS*   467

Shipborne Antenna for Surface Search.—A shipborne antenna for use in surface search must scan in azimuth with a sharp azimuth pattern. To accommodate roll and pitch the beam of an unstabilized antenna must be broad in elevation. This broadening will be more conservative of power and will provide a more constant illumination of the target if it is accomplished with a shaped beam (Fig. 13·3) rather than a simple fanned

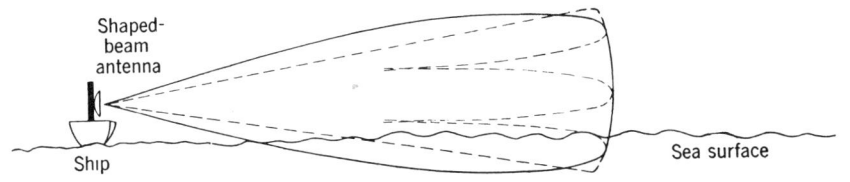

FIG. 13·3.—Sector shaped beam for surface search by shipborne antenna.

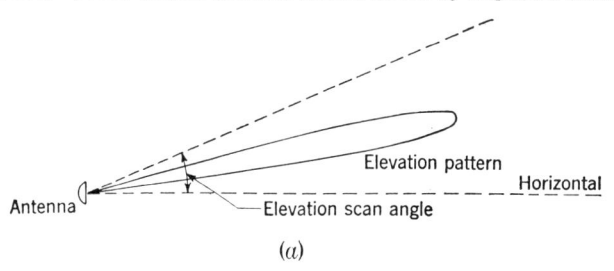

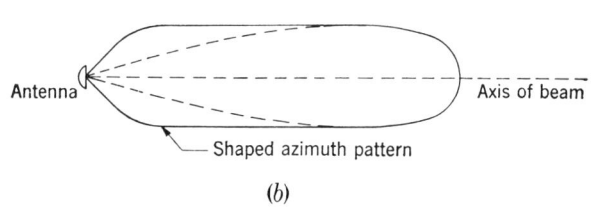

FIG. 13·4.—Beavertail beam for height-finding antenna: (*a*) elevation pattern; (*b*) shaped azimuth pattern.

beam. The ideal beam shape for this purpose would be given by $I = I_{max}$ for angles in the region $+\theta_1$ to $-\theta_1$ and $I = 0$ for angles outside that region. This sector shape can be approximated more closely as the vertical aperture of the antenna is increased, but a close approximation is not justified.

Surface Antenna for Height Finding.—A ground or ship antenna designed for height finding must have a sharp elevation beam for obtaining precise elevation information and a rapid elevation scan. Provision must also be made for scanning the antenna slowly in azimuth or for turning the antenna to an assigned azimuth. The beam must be relatively broad in azimuth in order that the target will be held in the beam long enough to obtain height information. If the beam is assumed to be stationary in azimuth, an airplane flying across the beam will be illuminated for a period proportional to its distance away. To increase the time of illumination on near-by crossing targets, a low-intensity broadening of the azimuth beam is required. If a fixed minimum of illumination is to be achieved at a given linear distance on both sides of the center line of the azimuth beam, the amplitude pattern must have the so-called "double csc θ" or "beavertail" shape illustrated in Fig. 13·4.

13·2. Effect of a Directional Target Response.—In the previous section it was assumed that the target response is isotropic. The effect of a directional target response is to alter the beam shape required of the antenna from that predicted by simple "inverse-square" considerations. The power received by a *radar system* from a target in a given direction is proportional to the "radar cross section" of the target; the radar cross section[1] may be defined as the interception cross section multiplied by the scattering gain for that direction. Since the radar cross section of some targets varies widely with direction, this effect must be taken into consideration in establishing the required beam shape for the antenna. The power transmitted in a *communication system* from a shaped-beam antenna to a receiving antenna in a given direction is proportional to the product of the gains of the two antennas along the line joining them. A directional receiving antenna will modify, therefore, the beam shape required of the transmitting antenna.

Let us consider in greater detail the case of a radar antenna located a perpendicular distance h from a plane (Fig. 13·2) and required to obtain equal signals from identical targets located arbitrarily in that plane. Let us assume that the antenna is to scan with a sharp beam in azimuth, as in the case of the first two shaped-beam applications described in the previous section. Then the specifications for the vertical polar diagram may be derived if quantities are defined as follows:

P = power emitted by the antenna
$G(\theta)$ = power gain of the antenna at depression angle θ
r = slant range to the target
$\sigma(\theta)$ = interception cross section of the target for a plane wave from the direction of the antenna[2]

---

[1] Also known as the back-scattering coefficient.

[2] The definition of the scattering cross section is being set up here in more detailed form than was done in Sec. 1·2. It is based on the physical picture that the target presents an interception area such that it removes from the plane wave all the energy incident thereon and redistributes it in space in a scattering pattern. Both the interception area and the scattering pattern vary with the aspect presented by the target to the incident wave. The back-scattering cross section is the product of the interception area and the gain of the scattering pattern in the direction of the transmitter. It is the cross section of the equivalent sphere that would produce the same

SEC. 13·2] EFFECT OF A DIRECTIONAL TARGET RESPONSE 469

$\gamma(\theta)$ = power gain of the scattering object in the direction of the antenna
p = power received by the antenna from the target

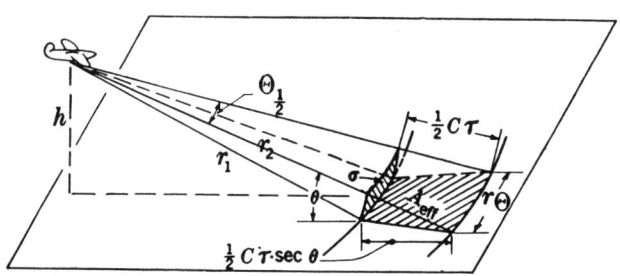

FIG. 13·5.—Effective ground target area and its interception cross section for airborne pulsed radar.

The fraction of the transmitted power received by the antenna is given by

$$\frac{p}{P} = \frac{G(\theta)}{4\pi r^2} \frac{\sigma(\theta)\gamma(\theta)}{4\pi} \cdot \frac{G(\theta)\lambda^2}{4\pi r^2}$$
$$= \frac{[G(\theta)]^2 \lambda^2}{(4\pi)^3 r^4} \sigma(\theta)\gamma(\theta). \tag{1}$$

To impose the condition that equal signals be received from identical targets in the plane, let us note that $r = h \csc \theta$ and write $p/P = C^2$, a constant. Then

$$G(\theta) = C(4\pi)^{3/2} \frac{h^2}{\lambda} \frac{\csc^2 \theta}{[\sigma(\theta)\gamma(\theta)]^{1/2}}. \tag{2}$$

If $[\sigma(\theta)\gamma(\theta)]$ is independent of angle, we obtain the earlier result that $G(\theta)$ for the shaped-beam antenna is proportional to $\csc^2 \theta$. This should be recognized as only a crude approximation in the majority of cases of actual interest.

In the particular case of reflections from ground targets, serious consideration has been given to the angular dependence of the quantity $[\sigma(\theta)\gamma(\theta)]$.[1] The effective area of the target on the ground depends upon beamwidth $(\Theta_{1/2})$, range (r), pulse length (τ), and depression angle (θ). By reference to Fig. 13·5, it is evident that the effective target area A_{eff} on the ground as determined by the pulse length is related to depression angle by the proportionality

$$A_{\text{eff}} \propto r \sec \theta \propto \csc \theta \sec \theta$$

---

return signal at the transmitter as does the target; thus, the product $\sigma(\theta)\gamma(\theta)$ used here is equal to the scattering cross section σ of Sec. 1·2.

[1] R. E. Clapp, "A Theoretical and Experimental Study of Radar Ground Return," RL Report No. 1024, Apr. 10, 1946.

and that the projection of this area in the direction of incident radiation is

$$\sigma(\theta) = A_{\text{eff}} \sin \theta \propto \sec \theta. \tag{3}$$

The angular distribution $\gamma(\theta)$ of the radiation scattered by the area $\sigma(\theta)$ will depend upon the nature of the target or terrain. A mathematical expression derived for $\gamma(\theta)$ will depend upon the simplified target model assumed. Best agreement with experience is obtained by assuming a flat plane made up of closely spaced components which scatter isotropically. The radiation will then be equal in all directions for any given condition of illumination; i.e.,

$$\gamma(\theta) = 1. \tag{4}$$

Combining Eqs. (3) and (4) gives the angular dependence

$$[\sigma(\theta)\gamma(\theta)] \propto \sec \theta. \tag{5}$$

Introducing this dependence into Eq. 2 gives the proportion

$$G(\theta) \propto \csc^2 \theta \sqrt{\cos \theta} \tag{6}$$

for the assumed ground target model.

An antenna with a vertical pattern shaped according to Eq. (6) would produce, within the limits of the assumptions, a range trace of uniform brightness on an indicator scope for any given azimuth setting of the antenna. A succession of range traces from an antenna scanning in azimuth would still be displayed with uniform brightness on a *B*-scope presentation which makes a rectangular plot of range vs. azimuth. On a plan position indicator (PPI), however, the range traces are presented radially and the azimuth angle, as polar angle. The spacing between range traces therefore varies in direct proportion to the range with the result that the scope is brightened toward the center. This effect can be compensated if the vertical pattern of the antenna is used to modify the received power by a factor of $1/r$; i.e., the gain function $G(\theta)$ of the antenna should be modified by a factor $r \propto \csc \theta$. For the case of PPI presentation then, Eq. (6) becomes

$$G(\theta) \propto \csc^{2.5} \theta \sqrt{\cos \theta} = \csc^2 \theta \sqrt{\cot \theta}. \tag{7}$$

The several "ideal" curves for $G(\theta)$ discussed above are presented in Fig. 13·6:

Curve *A*. $\csc^2 \theta$ dependence for a uniform range trace with isolated isotropic targets.

Curve *B*. $\csc^2 \theta \sqrt{\cos \theta}$ dependence for a uniform range trace with closely packed isotropic targets.

Curve *C*. $\csc^2 \theta \sqrt{\cot \theta}$ dependence for uniform PPI presentation with closely packed isotropic targets.

Curve D. $\csc^2 \theta \cdot \cos \theta$ dependence which approximates the experimental optimum pattern shape obtained from considerable flight experience with a number of antenna designs at wavelengths between 10 and 1.0 cm.

These curves are all plotted for a minimum depression angle of 10°; this corresponds to the case of an airborne radar system with a maximum

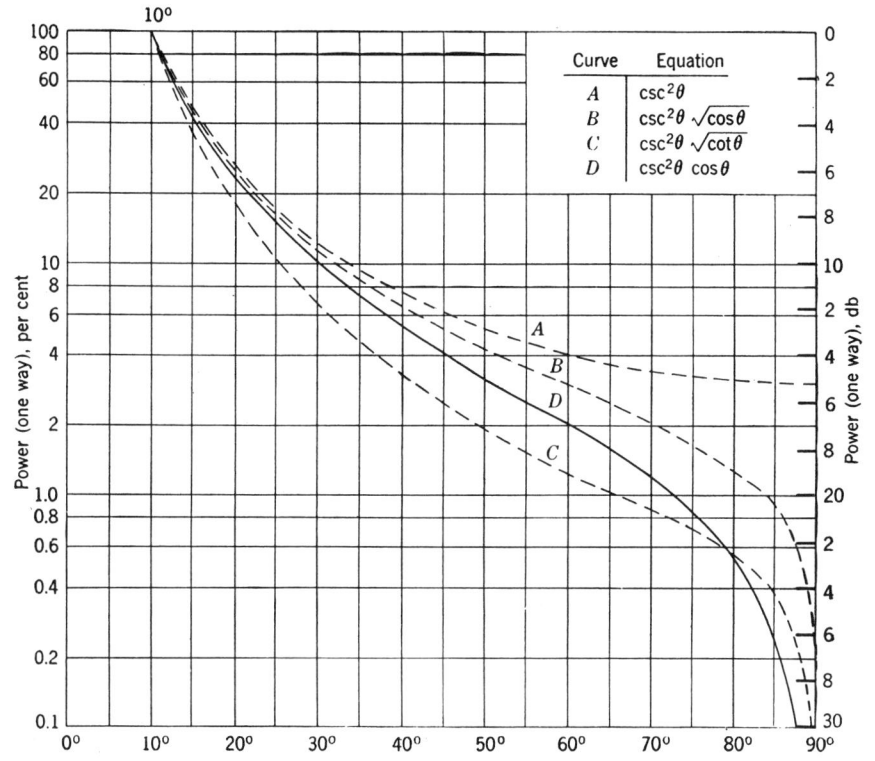

Curve	Equation
A	$\csc^2\theta$
B	$\csc^2\theta \sqrt{\cos\theta}$
C	$\csc^2\theta \sqrt{\cot\theta}$
D	$\csc^2\theta \cos\theta$

Fig. 13·6.—Ideal curves for dependence of vertical pattern on depression angle.

range about six times the altitude of the aircraft. Several curves of $\csc^2 \theta \cdot \cos \theta$ for different values of minimum depression angle are presented in Fig. 13·7.

13·3. Survey of Beam-shaping Techniques.—In the preceding sections we considered various applications for shaped-beam antennas and the requirements that they impose on the beam shape. In this section we will discuss the physical principles involved in various beam-shaping techniques and survey a number of antenna designs that have been used for producing beams of various shapes.

Physical Principles.—In Chap. 12, the characteristics of pencil beams and simple fanned beams were described. Such beams were shown to have a common shape, characteristic of the main lobe in the diffraction

472 SHAPED-BEAM ANTENNAS [Sec. 13·3

pattern of a constant-phase aperture. This is true independent of the shape or size of the aperture—for apertures larger than about 2λ—and independent of the intensity of illumination across the aperture. The effect of these variables is to change only the scale factor for the angular coordinate of the pattern. The only means available for altering this characteristic shape is to vary the *phase* of the illumination across the aperture.

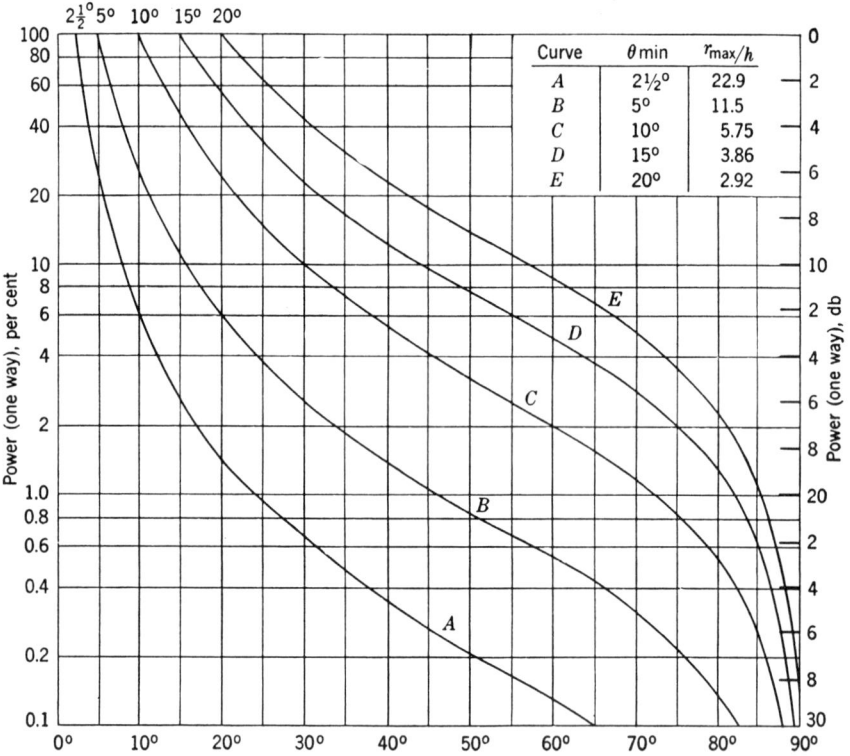

Curve	θ min	r_{max}/h
A	$2\frac{1}{2}°$	22.9
B	$5°$	11.5
C	$10°$	5.75
D	$15°$	3.86
E	$20°$	2.92

Fig. 13·7.—A family of curves, $\csc^2 \theta \cos \theta$, for different values of minimum depression angle.

The elementary principles of beam shaping can be understood in terms of geometrical optics. From Huygens' principle of propagation normal to the phase front, it is evident that a curved phase front will produce a more dispersed beam than a plane phase front. For a fixed aperture size, beam shaping can be accomplished only at the expense of antenna gain, since the curved portion of the phase front subtracts from the total aperture available for contribution in the direction of maximum gain. The radiation intensity for a given direction in a shaped beam will depend upon the radius of curvature of the phase front normal to the given direction and upon the intensity of illumination in that region; the exact relations involved will be derived in Sec. 13·6. Any shaped-beam

SEC. 13·3] SURVEY OF BEAM-SHAPING TECHNIQUES 473

antenna can be considered to be a device for obtaining the proper phase and intensity of illumination across an aperture to realize a specified beam shape.[1]

In the language of ray optics, a constant phase front across an aperture produces a collimated beam of rays from that aperture. This collimated beam is obtained usually by focusing the diverging rays from an antenna feed either with a parabolic reflector or with a lens. The process of forming a shaped beam can then be visualized as a defocusing

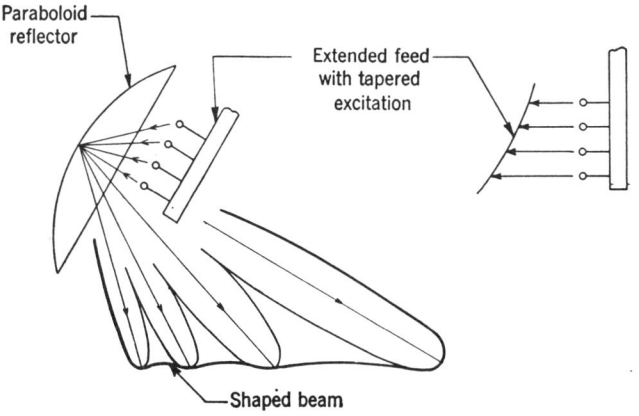

FIG. 13·8.—Formation of a shaped beam by means of a feed array in a paraboloid reflector.

process; the rays emerging from the aperture will not all be collimated but will be distributed through a range of angles with a variable density dependent upon the pattern required. Defocusing in one plane can be accomplished either by extending the point source into a line source in that plane or by modifying the reflector or lens in that plane.

The extension of a point source into a line source can be accomplished by disposing an array of dipoles or horn feeds in a line in or near the focal plane and by exciting them in the proper intensity and phase. The formation of a shaped beam by such a feed array in a paraboloid reflector is illustrated in Fig. 13·8. Each of the elements in the array can be visualized as a point source that forms its own sharp beam in the paraboloid. The intensity of this beam will depend upon the intensity of excitation of its feed; the angular displacement of the beam from the axis will be proportional to the angular displacement of the feed point about the vertex on the opposite side of the axis. The overlapping beams formed by an array of point sources will synthesize by amplitude addition into a shaped beam, as illustrated in Fig. 13·8. The resulting beam can be quite smooth if the component beams are properly spaced and phased.

[1] R. C. Spencer, "Synthesis of Microwave Diffraction Patterns with Application to $Csc^2 \theta$ Patterns," RL Report No. 54-24, June 23, 1943.

It has been assumed above that a feed moved off axis from the focal point will form in a paraboloid a sharp beam on the other side of the axis. This is true for small displacements from the axis; for large displacements of a point source, its individual beam is broadened in the plane of displacement, which is not serious, since the beam is being broadened intentionally in that plane, but it is also broadened in the perpendicular plane, which *is* serious, since it reduces resolution and gain (Sec. 13·4). The extended feed method of beam shaping is therefore not recommended as a means for forming wide-angle patterns. It *is* recommended for forming shaped beams confined to small angles, since it accomplishes the beam shaping by increasing the size of the small feed rather than by increasing—for equal gains—the size of the relatively large reflector or lens.

A sharp beam formed by a point source and paraboloid reflector or by a line source and parabolic cylinder can be dispersed in a controlled way by modifying the shape of the reflector. The process can be thought of as one of dispersing collimated rays into new directions dictated by the shaped-beam pattern, or it can be visualized as one of controlling the phase and intensity of illumination across the aperture. In the latter case, the next step to the far-field pattern can be made by use of Huygens' principle in some cases and by a Fourier transform process in other cases (*cf.* Chap. 6). Figure 13·9 shows two reflector modifications for obtaining an asymmetrical flared beam. It has been shown that one aperture illumination which gives an asymmetrical beam consists of a sharply peaked amplitude distribution with a sudden 180° phase reversal in the region of maximum amplitude. One means of realizing this is to use a point source feed with two paraboloid reflectors of focal lengths f_1 and $f_2 = f_1 + \lambda/4$ for the top and bottom halves of the aperture respectively. Other methods

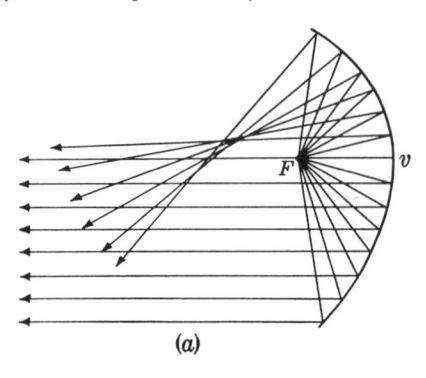

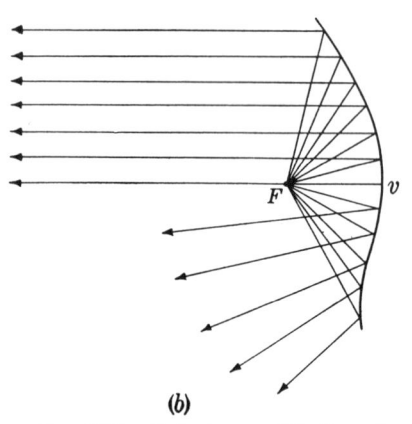

Fig. 13·9.—Reflector modifications for producing an asymmetrical flared beam: (*a*) by shaping the reflector on the opposite side from the flare; (*b*) by shaping the reflector on the adjacent side to the flare.

SEC. 13·3] SURVEY OF BEAM-SHAPING TECHNIQUES 475

of beam shaping will be described in connection with specific beam-shaping problems.

Symmetrical Shaped Beams.—A *sector shaped* beam with sharp sides and a square end, approximating the requirement illustrated in Fig. 13·3,

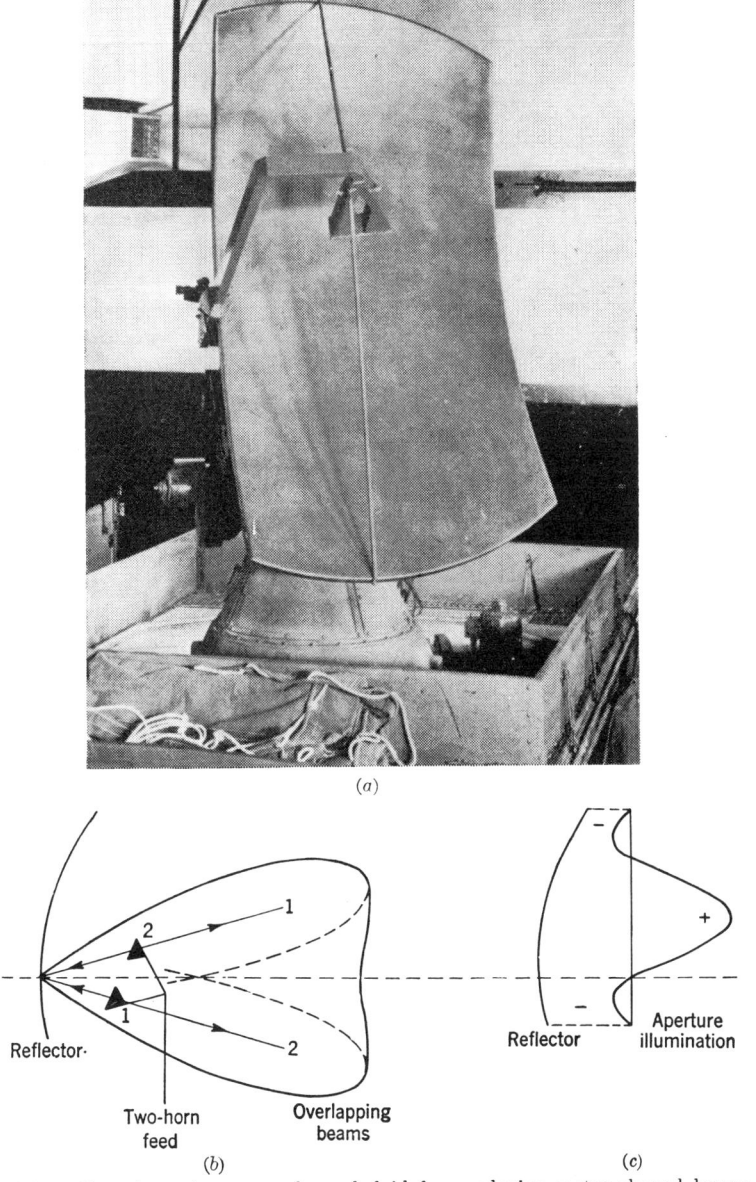

FIG. 13·10.—Two-element array and paraboloid for producing sector shaped beam: (a) antenna; (b) overlapping beams; (c) aperture illumination.

476 SHAPED-BEAM ANTENNAS [SEC. 13·3

can be produced by means of either an extended feed or a shaped reflector. If an array of radiating elements in the focal plane are equally excited, so spaced that their individual patterns cross over at the half-amplitude point and so phased that the patterns add in amplitude, the result will be a beam with sharp sides—determined by the size of the aperture—and a square end. Figure 13·10a shows two elements combined in this way. The design procedure can be interpreted as above and illustrated in Fig. 13·10b, or the following. The two feeds excited equally and in phase will form a symmetrical interference pattern. If each feed is so dimensioned that its individual pattern properly illuminates the reflector,

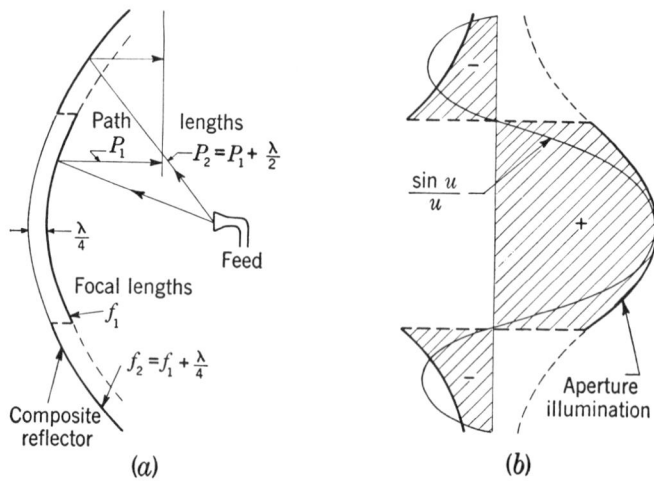

Fig. 13·11.—Cut paraboloid method for obtaining sector shaped beam: (a) antenna; (b) aperture illumination.

and if the two feeds are correctly spaced, their interference pattern will result in the $(\sin u)/u$ *aperture illumination* shown in Fig. 13·10c. The Fourier transform of this illumination curve will be approximately the sector shaped pattern required (Chap. 6). The antenna shown in Fig. 13·10 produces its sector shaped beam in the horizontal plane. The reflector is cut with a slight asymmetry to bring the null in the illumination pattern opposite the feed for improved impedance performance.

The sector shaped beam can also be obtained with a point-source and modified paraboloid reflector. Let the aperture be divided into three parts along the lines corresponding to the nulls of Fig. 13·10c. Let the two outer portions of the aperture be illuminated with segments of a paraboloid having a focal length $\lambda/4$ longer than that of the central paraboloid. This situation is illustrated in Fig. 13·11a. Then the phase of the illumination over the outer portions of the aperture will be delayed by $\lambda/2$. If the antenna feed provides a normal primary pattern, the aperture illumination shown in Fig. 13·11b will then be obtained. This will be a crude approximation to the $(\sin u)/u$ aperture illumination required for a sector shaped beam.

The beavertail beam illustrated in Fig. 13·4 is not obtained conveniently by means of an extended feed; the angle of coverage in actual applications is too large, and the taper in two directions from the center complicates the feed. It is obtained very easily, however, with a reflector modification. The simplest arrangement is a narrow vertical strip down the middle of the reflector set out from the surface of the main reflector by a fraction of a wavelength. The factors affecting the width and offset of the strip can be appreciated by reference to Fig. 13·12 which illustrates the design and the mechanism of beam shaping for an actual case. The width of the strip affects the total power that it intercepts and the directivity of its pattern. Its offset from the main reflector establishes the phase relationship for amplitude addition which is important in the region where the two amplitudes are of the same order of magnitude. In the actual case[1] at $\lambda = 10$ cm, a reflector with a 20- by 5-ft aperture and a 5-ft focal length was fitted with a strip running the long way of the reflector. The optimum width of the strip proved to be 8 in., and the offset, $\frac{1}{4}$ in. It is evident that the presence of a strip of this width will introduce interference side lobes in the pattern of the remainder of the reflector, which will impair the quality of the final beavertail pattern. An improved pattern would be obtained if the flaring of the beam were accomplished by modification at the two edges of the reflector rather than at the center, in which case the interference lobes would be less prominent.

Asymmetrical Shaped Beams.—Extended feed and modified reflector designs have both been used successfully to obtain asymmetrical shaped beams. Extended feed designs have been used in general for ground and ship antennas for which the required elevation coverage was limited usually to small angles and for which the reflectors were too large for convenient modification. Extended feeds are readily adapted also to the use of multiple transmitters when the prescribed coverage requires high power. Modified reflector designs have been used almost exclusively in airborne antennas which are required to provide wide-angle elevation coverage and to possess smooth and stable pattern characteristics.

Both dipoles and horns have been used in linear-array feeds for shaped-beam antennas; in some cases both have been used in the same array. The choice between a dipole and a horn as the radiating element in a given situation depends upon the power to be handled, the required impedance characteristics, and convenience in construction. The design and performance of three extended feed antennas will be described in order of increasing complexity. A cut paraboloid reflector with a two-

---

[1] C. S. Pao, "The Beavertail Antenna," RL Report No. 1027, Apr. 9, 1946.

478   SHAPED-BEAM ANTENNAS   [SEC. 13·3

(a)

FIG. 13·12.—Central strip in reflector for producing beavertail beam: (a) antenna with-

dipole array feed[1] for operation at $\lambda = 10$ cm is illustrated in Fig. 13·13. The aperture dimensions are 8 by 4 ft, and the focal length, 27.5 in. The elevation pattern is shown in Fig. 13·16a. This pattern and the impedance match (VSWR < 1.2) were maintained satisfactorily over a 2 per cent bandwidth. A cut paraboloid reflector with the long dimension horizontal and with a four-horn array feed[2] for $\lambda = 10$ cm is shown in Fig. 13·14. The aperture dimensions are 5 by 14 ft, and the focal length, 60 in. The feed design will be discussed in the next section. The elevation pattern is shown in Fig. 13·16b. In impedance match, the feed showed a VSWR < 1.12 for a 6 per cent band when tested in free space and a VSWR < 1.25 for a 3 per cent band when tested in the

[1] C. F. Porterfield and L. J. Chu, "A Simplified Search Antenna," RL Report No. 486, Jan. 1, 1945.
[2] W. J. West, "A Four-Horn Feed to Give $Csc^2 \theta$ Antenna Patterns," RL Report No. 896, Mar. 15, 1946.

SEC. 13·3] SURVEY OF BEAM-SHAPING TECHNIQUES

(b)

(c)

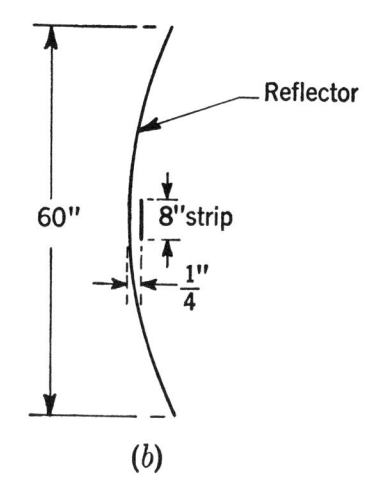

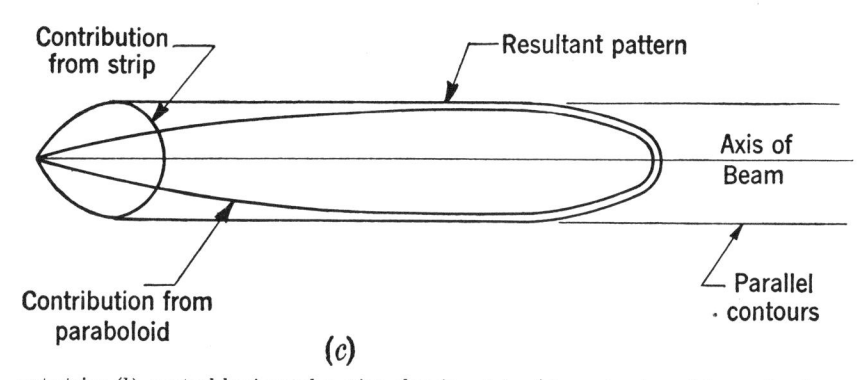

out strip; (b) central horizontal section showing strip; (c) mechanism of beam shaping.

reflector. A large and complicated antenna[1] for $\lambda = 10$ cm is shown in Fig. 13·15. The reflector is a cut paraboloid 10 by 25 ft with a 78-in. focal length. Of the eleven elements in the feed array, one element is fed from a first transmitter, two elements from a second transmitter, and eight elements from a third transmitter. Horn radiators are used with the first two transmitters because of the concentration of power, and dipoles with the third transmitter. The three-lobe pattern is shown in Fig. 13·16c. This elevation coverage represents the limit practicable with this antenna, since the azimuth beamwidth of a point feed 30 in. off axis (or 21° referred to the vertex) is increased by about 70 per cent. The impedance match presented to all three transmitters was VSWR < 1.12 over at least a 4 per cent band.

A variety of reflector shapes have been used for beam shaping in airborne antenna designs. Of these the most common and successful

[1] C. G. Stergiopoulos, RL Report No. 951, Feb. 12, 1946.

was the so-called *barrel reflector* antenna.[1] The shape of this reflector is obtained by replacing the top half of a paraboloid reflector with a figure of revolution produced by rotating the generating parabola of the parab-

(a)

(b)

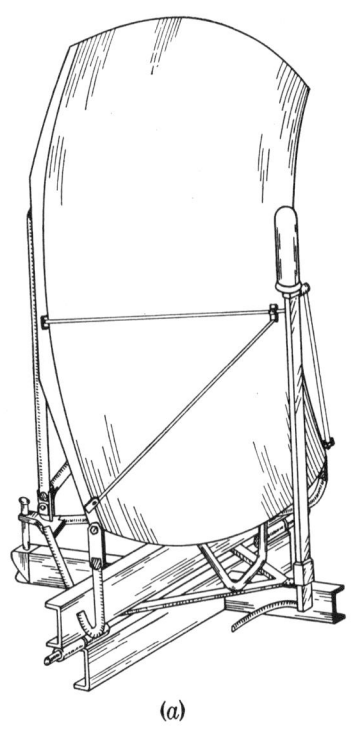

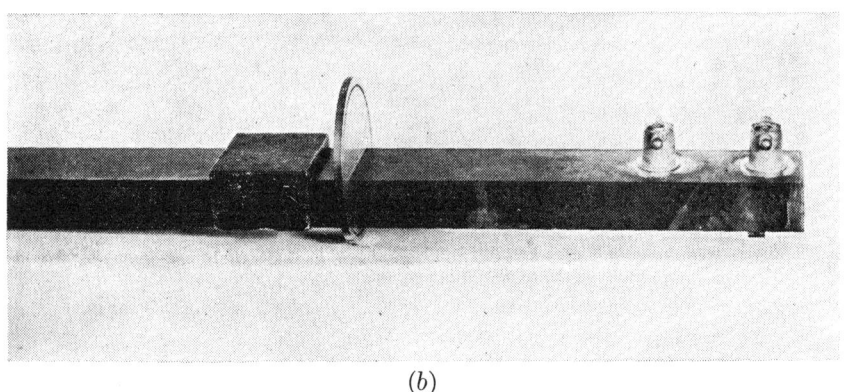

Fig. 13·13.—Cut paraboloid reflector with two-dipole array feed: (a) assembly; (b) feed.

oloid about a horizontal line through the focal point. The central vertical section through this reflector is of the type shown in Fig. 13·17b. The complete antenna is illustrated in Fig. 13·17 and will be discussed

[1] A. S. Dunbar, RL Report No. 411, Aug. 3, 1943.

SEC. 13·3]   SURVEY OF BEAM-SHAPING TECHNIQUES   481

further in Sec. 13·8. Its elevation pattern, shown in Fig. 13·20a, is best suited to high-altitude use. Elevation patterns suitable for low altitude can be obtained with several reflector shapes. The *shovel reflector*[1] belongs to the family defined in Fig. 13·9b. It is obtained by

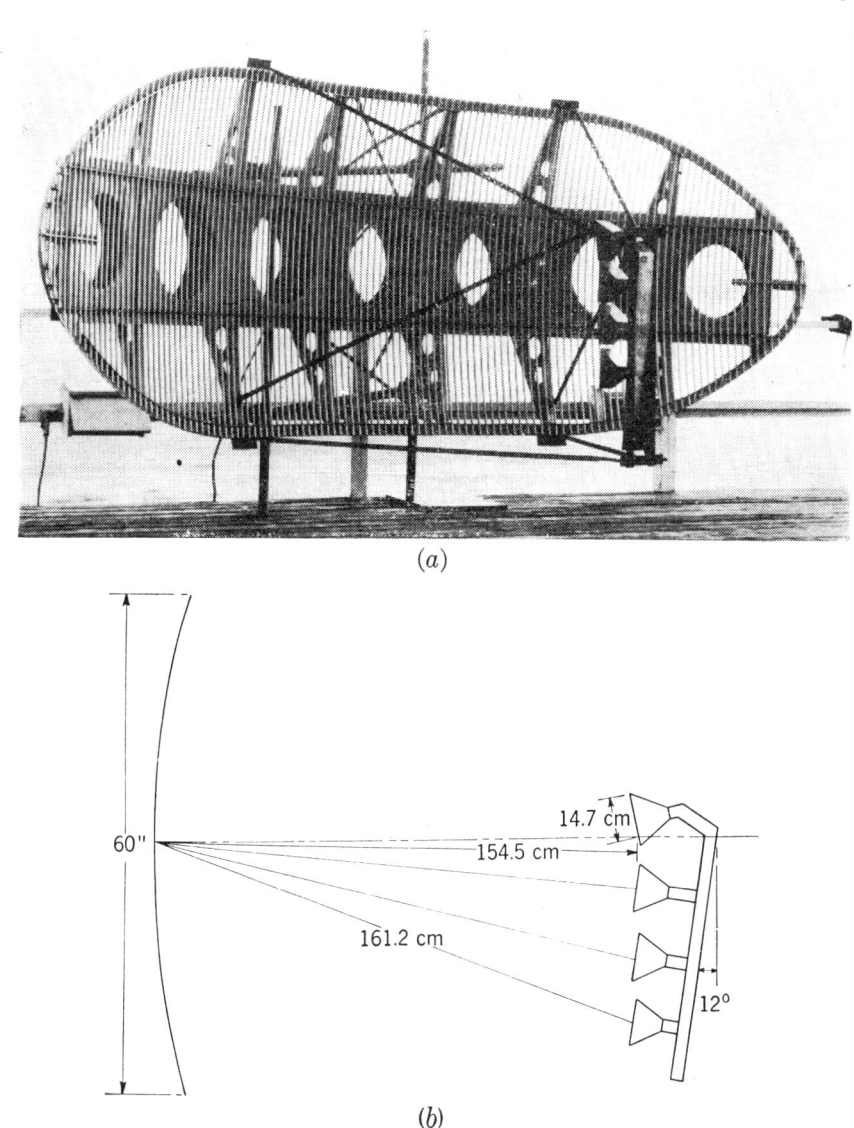

FIG. 13·14.—Cut paraboloid reflector with four-horn array feed: (a) antenna; (b) central vertical section.

[1] J. H. Gardner, "Low Altitude Navigation Antennas," RL Report No. 615, Oct. 3, 1944.

482   SHAPED-BEAM ANTENNAS   [SEC. 13·3

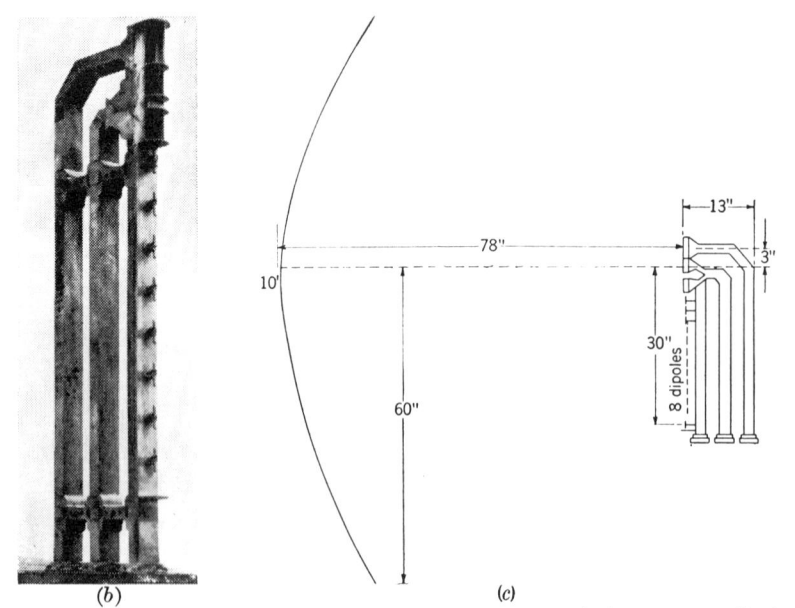

FIG. 13·15.—Cut paraboloid reflector with 11-element array feed: (a) antenna; (b) feed; (c) central vertical section.

SEC. 13·3] SURVEY OF BEAM-SHAPING TECHNIQUES   483

replacing the lower third or so of a paraboloid reflector with a parabolic cylinder. The shovel-reflector antenna is illustrated in Fig. 13·18; its elevation pattern is given in Fig. 13·20b. The general shape and smoothness of the pattern depend upon the point of attachment of the shovel,

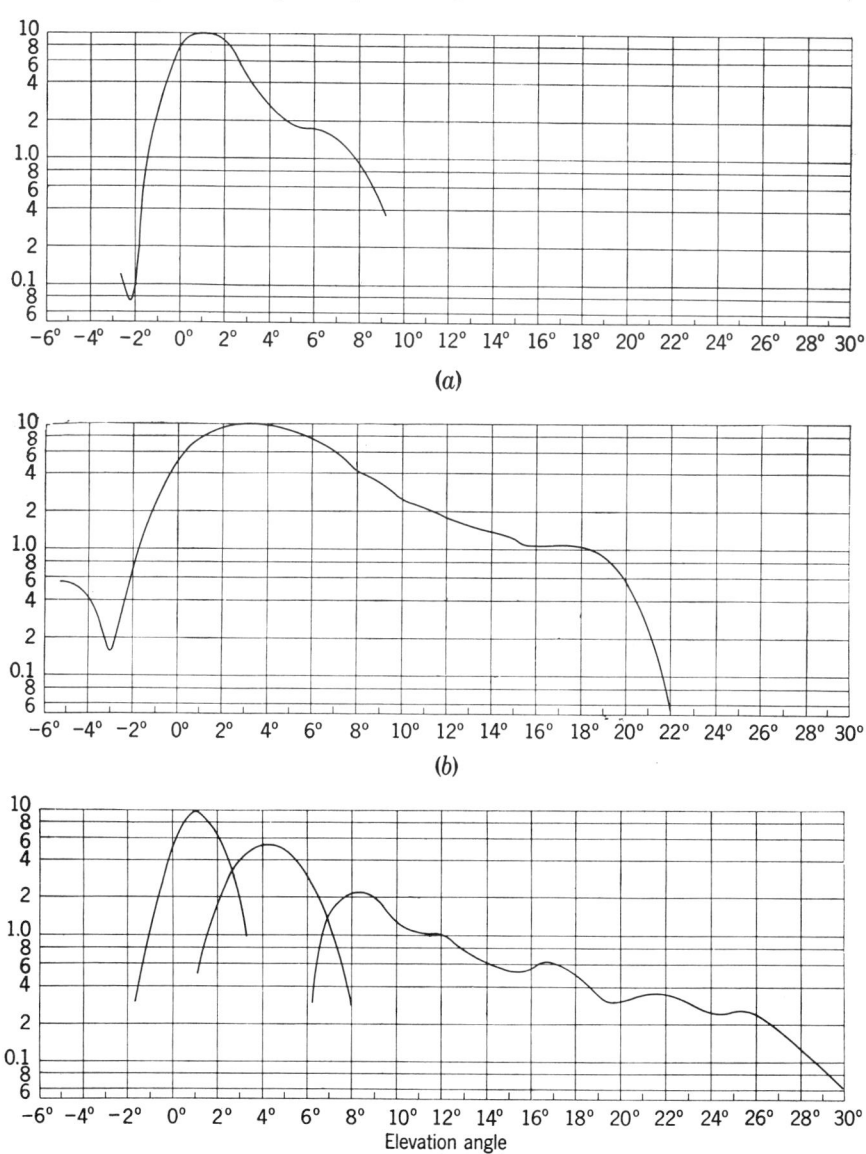

FIG. 13·16.—Elevation patterns obtained with cut paraboloid reflectors and linear-array feeds: (a) two-element array of Fig. 13·13; (b) four-element array of Fig. 13·14; (c) eleven-element, three-transmitter array of Fig. 13·15.

its displacement normal to the surface of attachment, and its tilt with respect to the tangential direction. A low-altitude beam can be obtained also by inserting a narrow horizontal strip just above the center of a

(a)

(b)

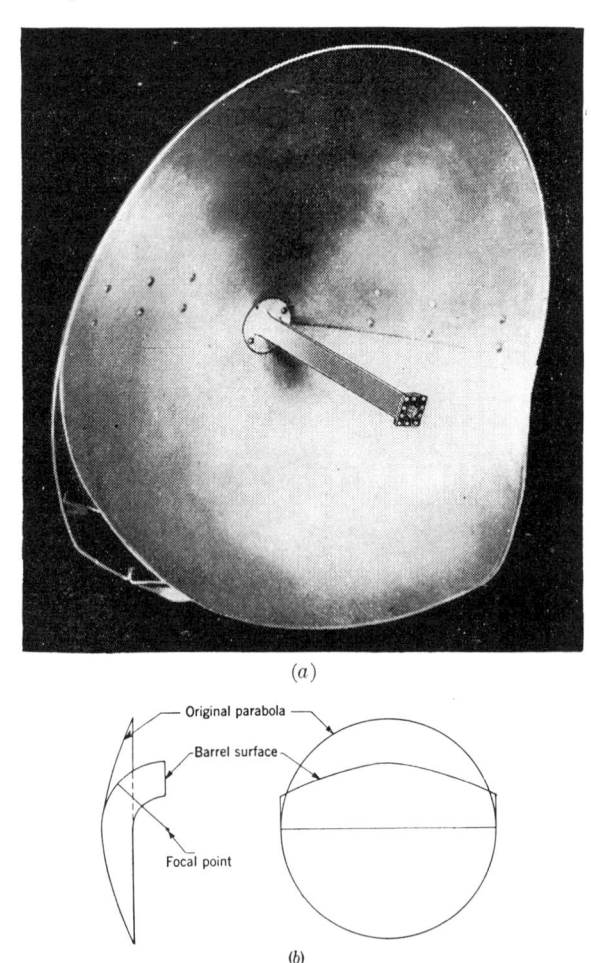

Fig. 13·17.—Barrel-reflector antenna for high-altitude beam shaping: (a) photograph; (b) drawing.

paraboloid reflector with the proper width, offset, and tilt.[1] A *strip-reflector* antenna is illustrated in Fig. 13·19, and its elevation pattern is given in Fig. 13·20c. The basic limitation with this strip-reflector design, as pointed out earlier in connection with the beavertail-shaped beam, is that the strip divides the aperture into two parts between which inter-

---

[1] C. C. Cutler, "Notes on the Design of Asymmetrical (Cosecant) Antennas," BTL Report MM-43-160-192, Nov. 12, 1943; J. H. Gardner, "Low Altitude Navigation Antennas," RL Report No. 615, Oct. 3, 1944.

ference occurs. Even with the relatively narrow strips required for low-altitude beams, the interference lobe that appears on the opposite side of the peak from the flared portion is only 10 db down from the peak gain. In many aircraft installations this upward-directed interference lobe would

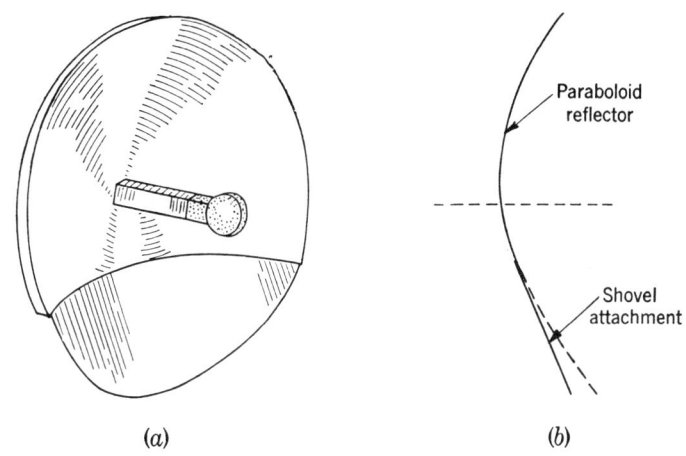

FIG. 13·18.—Shovel-reflector antenna for low-altitude beam: (*a*) antenna; (*b*) central vertical section of reflector.

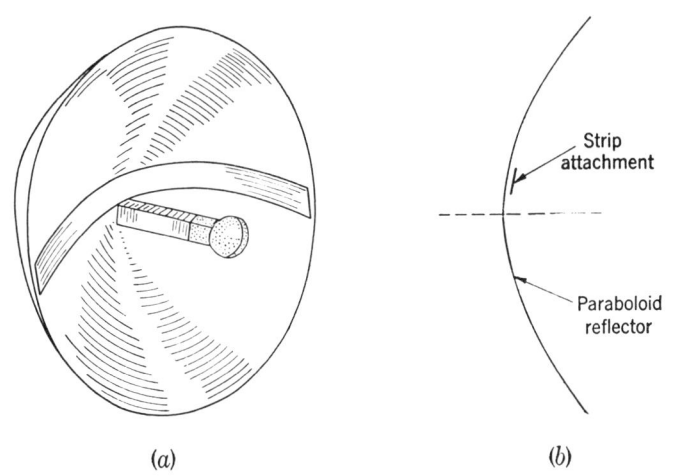

FIG. 13·19.—Strip-reflector antenna for low-altitude beam: (*a*) antenna; (*b*) central vertical section of reflector.

be reflected from the under side of the fuselage and wings and produce an interference ripple in the pattern on the ground. The best simple design for a low-altitude shaped-beam antenna has been one that uses a cut-down barrel reflector. This antenna[1] consists of a 29-in. diameter by 10.6-in. focal length paraboloid with a barrel insert cut down in

[1] J. H. Gardner, "Low Altitude Csc² θ Antenna," RL Report No. 1073, Feb. 21, 1946.

the vertical to 19 in. (7½ in. above the axis, 11½ in. below) in order to eliminate that part of the barrel which contributes to the steep angle portion of the pattern. This reflector is then fed with a horn so directed

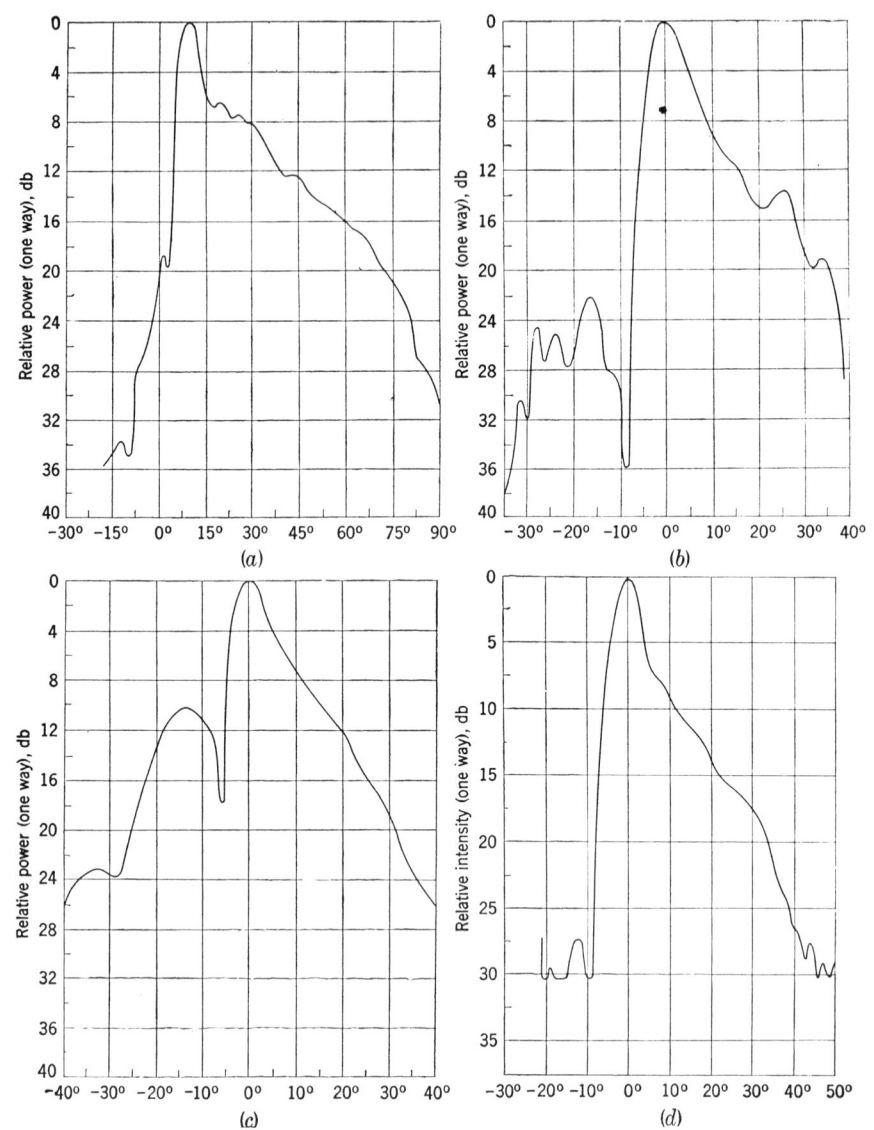

FIG. 13·20.—Elevation patterns obtained with airborne shaped-reflector antennas: (a) barrel-reflector antenna of Fig. 13·17; (b) shovel-reflector antenna of Fig. 13·18; (c) strip-reflector antenna of Fig. 13·19; (d) cut-down barrel-reflector antenna.

as to obtain proper illumination. The antenna can be visualized by reference to the uncut antenna shown in Fig. 13·17. Its vertical pattern is shown in Fig. 13·20d.

A number of shaped-beam antennas obtained by reflector modifications were described in the previous paragraph. These were truly reflector modifications in the sense that the design started with an existing paraboloid reflector and proceeded on the basis of obtaining the required beam shape with a minimum of remodeling. Time-consuming cut-and-try processes were involved, and the ultimate result was frequently less than satisfactory. A satisfactory design is obtainable more rapidly and reliably if the reflector is visualized as a device for transforming the primary pattern of the feed into the required secondary pattern. In this sense the reflector shape is dictated by specifying the two patterns; the problem is one for computation rather than cut-and-try. The computation of shaped cylindrical reflectors is a straightforward problem and has received experimental confirmation in a number of antenna designs. These results will be described in detail in Sec. 13·6. The line source used with a cylindrical reflector could be used equally well with a shaped cylindrical lens[1] to obtain an antenna with certain advantages over the reflector antenna. The calculation of the complete shape of a *double curvature* reflector for beam shaping is more involved and subject to further study, but experimental confirmation of present design procedures has been obtained in a limited number of cases.

13·4. Design of Extended Feeds.—In the previous section, several shaped-beam antennas utilizing extended feeds were described as to over-all design and performance. In actual fact the principal r-f design problem is concentrated in the extended feed itself. It has been pointed out that both dipoles and horns have been used as the radiating elements in linear-array extended feeds and that the choice between them depends upon power-handling, impedance, and mechanical considerations. These questions and others involved in the design of these arrays will be considered in this section. We will consider first, however, some optical focusing problems common to all extended feeds used in paraboloids (*cf.* Sec. 6·7).

Optical Focusing Properties of Paraboloid Reflectors.—Let us begin with a single radiating element at the focal point of a circular (uncut) paraboloid reflector. The resulting diffraction pattern will depend upon the directivity of the feed and upon the shape (focal length to diameter ratio, F/D) of the paraboloid. As the feed is moved off axis[2] by a rotation about the vertex, the beam will move off axis on the side opposite the feed and in direct proportion to the feed displacement. This proportionality factor would be unity for a flat plate according to Snell's law; it is slightly less than unity for paraboloids in the useful range of shapes,

[1] A. S. Dunbar, "Metal Plate Lens for Csc² Antenna," RL Report No. 1070, Feb. 15, 1946.

[2] S. Silver and C. S. Pao, "Paraboloid Antenna Characteristics as a Function of Feed Tilt," RL Report No. 479, Feb. 16, 1944.

as shown in Fig. 13·21. As the beam moves off axis, it deteriorates, at first slowly and then more rapidly as the angle increases. The gain decreases; the beamwidths increase; and a series of side lobes (the so-called coma lobes) appear on the axis side of the displaced beam. These effects can be described uniquely in terms of the angular displacement expressed in beamwidths, but only if the reflector shape and feed directivity are held constant. The variation of gain with feed tilt for reflectors of different shapes with a relatively directive feed is plotted in Fig. 13·22. In addition to these pattern changes, an antenna initially matched in impedance with the feed on axis will undergo a series of impedance changes with feed tilt. These effects are illustrated by the measured data plotted in Fig. 13·23, which are susceptible of quite accurate theoretical verification. The curves have been rotated into separate quadrants of the Smith Chart for clarity in presentation.

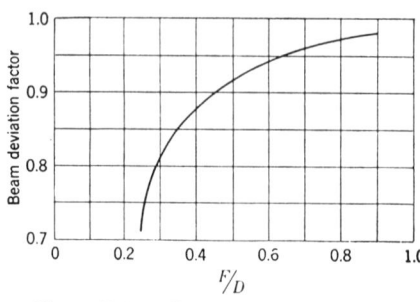

Fig. 13·21.—Beam deviation proportionality factor as a function of reflector shape.

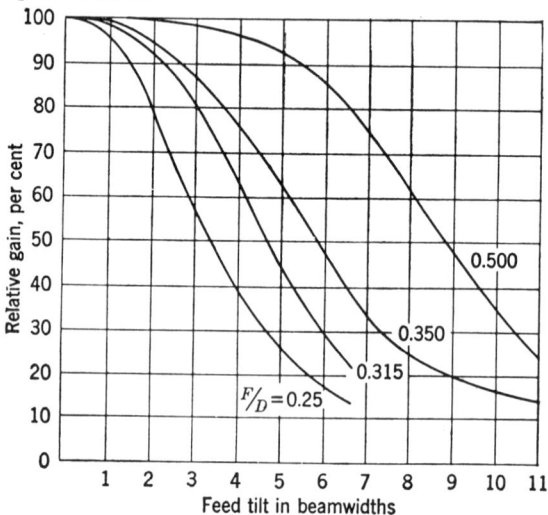

Fig. 13·22.—Dependence of antenna gain on feed tilt and paraboloid shape. F/D is the focal length to diameter ratio.

When the paraboloid is cut to reduce the aperture in one dimension or the other, its off-axis focusing properties are modified considerably. These effects can be understood qualitatively by reference to Fig. 13·24. The circle represents the aperture of a paraboloid reflector. When the feed is moved off axis in the vertical plane to some position in the rec-

SEC. 13·4]   *DESIGN OF EXTENDED FEEDS*   489

tangle shown at the center of the circle, rays reflected from various portions of the paraboloid will deviate from parallelism with the ray reflected from the vertex. The equal deviation contours plotted in the figure are purely schematic, since actual contours would depend upon reflector shape and feed displacement. However, they serve to illustrate the fact that the outer portions of the reflector lying between the principal planes are responsible for the most serious deviations. From Fig. 13·24a

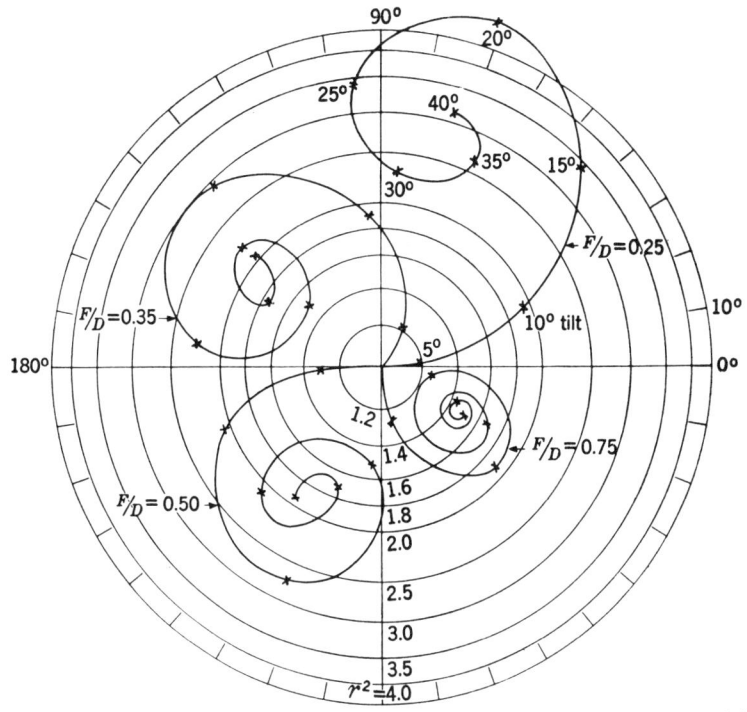

FIG. 13·23.—Impedance changes at $\lambda = 10$ cm in 30-in. paraboloid reflectors of different shapes (F/D) as the feed is tilted off axis.

it is evident that cutting down the aperture along lines BB' will eliminate regions of high deviation and increase the allowable angle of feed displacement. It is misleading now to express this angle in beamwidths, since there are two different beamwidths in the two planes.

In actual practice the reflector of a carefully designed antenna is not cut along the straight lines shown in Fig. 13·24a, but rather along one of the *equal illumination* curves of the feed (*cf.* Chap. 12), e.g., the equal illumination curve 14 db down from the point of maximum illumination. Three such curves are shown in Fig. 13·24b superimposed on the equal deviation contours: Curve A is appropriate to a feed on axis and pointed at the vertex of the paraboloid; Curve B, to a feed on axis but pointed into the top half of the reflector; and Curve C, to a feed below the axis but pointed at the vertex. Whereas a reflector shape that follows Curve B is satisfactory for a point-source feed on axis, it would introduce serious defocusing when used with an extended feed because of the high deviation contours that it includes. For this reason symmetrically cut reflectors are always used with extended feeds. In the case of a feed array extending downward from the axis, the top element would favor the reflector shape given by Curve A while the bottom element would favor the shape given by Curve C. The actual reflector shape will therefore be a compromise between these.

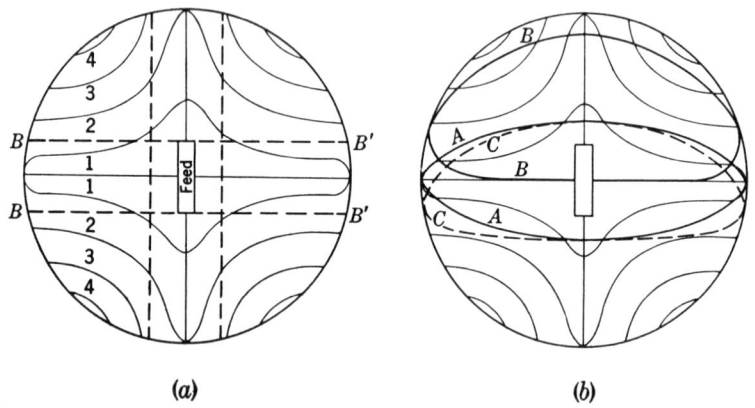

FIG. 13·24.—Aperture of paraboloid reflector showing contours of equal deviation of rays from direction of central ray as feed is moved off focus in vertical plane: (a) Lines BB' represent straight cuts to narrow the aperture in one plane; (b) Curves A, B, and C represent equal illumination contours for different feed positions and orientations.

In the preceding discussion the feed displacement has been described as a simple rotation about the vertex. It is not necessarily true that this places the feed at the distance from the vertex corresponding to maximum gain. Measurements made under a variety of conditions have shown that the feed must then be moved away from the vertex to or slightly beyond the vertical plane through the focal point. This optimum distance for a given displacement angle will depend upon how the reflector has been cut.

To obtain optimum performance of a cut paraboloid with an extended feed, it is necessary to carry out an experimental design procedure. This procedure with its results will be described for the 10 by 25-ft reflector shown in Fig. 13·15 at $\lambda = 10$ cm.[1] A dipole feed was moved transversely over a range of 30 in. off axis (21° referred to the vertex). At each displacement the feed was then moved parallel with the axis to find the point of maximum gain. The optimum feed point proved to be 5 in. farther out along the axis for the maximum feed displacement than for the feed on axis. For these optimum feed points, the variation with

[1] C. G. Stergiopoulos, RL Report No. 951, Feb. 12, 1946.

displacement of gain and azimuth beamwidth (perpendicular to the displacement plane) were measured with the results shown in Fig. 13·25. The reduction in gain is not so serious as it appears, since a major part of the reduction is due to beam spreading in the vertical plane in which a flared beam is required. The increase in azimuth beamwidth *is* serious and sets the limit for this reflector on the angular range over which it is practicable to flare the beam.

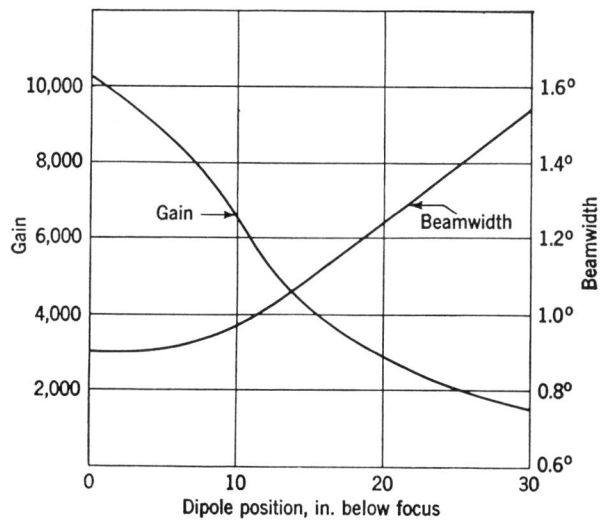

FIG. 13·25.—Variation of gain and azimuth beamwidth with dipole feed displacement for the reflector of Fig. 13.15.

Dipole-array Extended Feeds.—Several problems arise in the design of a dipole array to be placed near the focal plane of a paraboloid for the purpose of obtaining a flared beam. First the array should be located with reference to the focal plane so that the individual overlapping lobes comprising the flared beam have maximum gain. The dipoles along the waveguide must be so spaced that the individual lobes are all in phase with each other. The input impedance of the dipole array must be such as to terminate the transmission line properly. Finally, the available power must be divided among the several dipoles in such proportions as to obtain the desired beam shape.

The array is oriented with respect to the paraboloid axis and focal plane on the basis of gain and beamwidth information of the type presented earlier in this section. The individual lobes will combine in phase if the radiations from successive dipoles arrive at the vertex in phase. With the waveguide oriented for maximum gain of the individual lobes, the relative phases are controlled by the spacing between dipoles along the waveguide. In terms of the quantities shown in Fig. 13·26, the spacing for reversed dipoles is given by the relation

$$\frac{d}{\lambda_g} + \frac{r_n}{\lambda} - \frac{r_{n+1}}{\lambda} = \frac{1}{2}.$$

The dipole array is usually fed from a traveling wave which is realized by terminating the array in a dummy load or preferably by using the last dipole in the array as a load. If the latter technique is used, the last dipole in the array must have its impedance such that it absorbs all the power transmitted down the waveguide to it. This condition is attained by proper adjustments of the depth of the probe feeding the

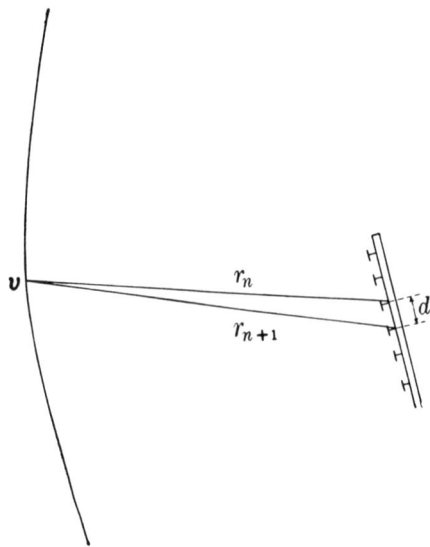

FIG. 13·26.—Array of dipoles on waveguide in paraboloid reflector, showing dipole spacing and orientation of array with respect to vertex.

last dipole and of the distance from the probe to the shorting plug at the end of the waveguide. The probes of the other dipoles are next inserted into the waveguide in succession starting from the last dipole in the array and proceeding to the first dipole. The depths to which the successive dipole probes are inserted are determined by the power division among the dipoles necessary to produce the required flared beam. Once the desired antenna pattern is obtained, the final impedance match of the array may be accomplished by inserting an inductive iris of appropriate dimensions at the proper location in the waveguide.

The principal advantage of the dipole array is its simplicity in design and construction. The disadvantage of this array is the interdependence of spacing and phasing of the individual elements. Each element of the array independently should provide proper illumination of the reflector. The dipole elements suffer from the disadvantage that their radiation patterns are dependent upon the polarization required and are relatively inflexible.

Horn-array Extended Feeds.—The problems in the design of a horn array are similar to those arising in the design of a dipole array. Optimum gain locations for the horns near the focal plane of the paraboloid are determined in the same manner as for the dipole elements. The radiation from the several horns of the array is made to arrive at the

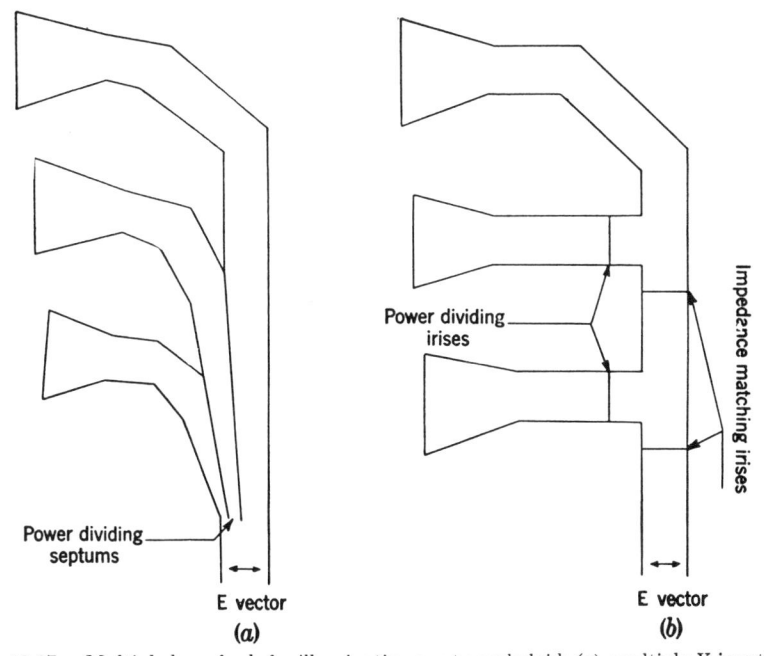

FIG. 13·27.—Multiple horn feeds for illuminating a cut paraboloid: (*a*) multiple *Y*-junction; (*b*) successive *T*-junctions.

paraboloid vertex in phase by a proper choice of the lengths of waveguide extending to the individual horns.

Various methods of dividing the power in multiple horn feeds exist; two methods are illustrated in Fig. 13·27. In the multiple Y-junction, the power is proportioned among the several horns by septums extending into the main waveguide. The impedance match of this multiple feed may be accomplished by a single iris in the main waveguide. In the array employing successive T-junctions (Fig. 13·27*b*), the division of power between the upper horn and the middle horn is determined by an iris in the section of waveguide feeding the middle horn. The impedance match of the combination of upper and middle horns to the waveguide is accomplished by an iris just below the junction. The division of power between the combination of top and middle horns and the lower horn is determined by the iris in the section of waveguide feeding the lower horn. The over-all impedance match of the multiple feed is accomplished by the iris in the main waveguide. A four-horn array of the latter type is used in the antenna shown in Fig. 13·14.

The advantage of the horn array lies in the fact that the phasing between successive elements and the radiation pattern of each element are completely at the disposal of the designer and are susceptible of calculation. The disadvantages of this type of array are the extreme complexity of the design and the bulk and weight of the resulting feed.

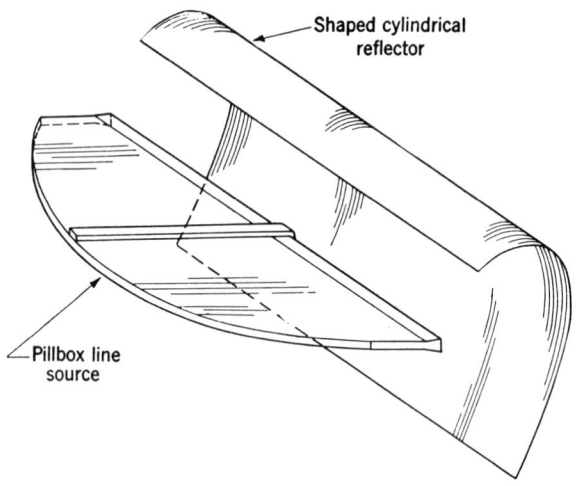

FIG. 13·28.—Shaped cylindrical reflector illuminated by a pillbox line source.

13·5. Cylindrical Reflector Antennas.—The technique of obtaining an asymmetrical beam by a shaped cylindrical reflector and line source has been used extensively in airborne navigational radar antenna design where sharp azimuth beams and wide-angle vertical coverages are required. Figure 13·28 illustrates a shaped cylindrical reflector illuminated by a pillbox which serves as a line source parallel to the generating line of the reflector.

The general theory of cylindrical reflectors and line sources is treated in Sec. 5·9. It was shown there that the pattern in the plane perpendicular to the generator of the cylinder—the vertical plane in the present discussion—is determined by the energy distribution of the source in that plane and the cross-section contour of the reflector, and that the pattern in the transverse planes is determined by the energy distribution along the axis of the source. The cross-section contour of the reflector is so shaped and oriented with respect to the line source that its lower section concentrates the rays from the source into approximately parallel directions, thereby concentrating the energy into the peak of the beam. The upper part of the reflector is bent forward with increasing curvature to disperse its rays into a broad flare. The net result is an asymmetrical beam, the exact vertical diffraction pattern of which depends upon the shape of the cylinder and the distribution of illumination from the line source on the reflector surface.

General Requirements.—The specifications on the vertical polar diagram are determined by the operational requirements. For example, if a navigational radar is to be designed for an aircraft flying at an altitude of about 1 mile and is to be capable of covering a radius of about 20 miles, then the peak of the antenna beam should be depressed about 3° from the horizontal and the beam should be asymmetrical so as to produce approximately constant illumination of the terrain. Whereas the beam specifications depend on operational requirements, the overall size of the reflector usually depends on the available installation space. If the antenna is part of an airborne radar, the allowable over-all height of the reflector is limited; the antenna is installed with the sharply curved upper portion of the reflector retracted into the fuselage, so that the protuberance below the fuselage for housing the antenna need not be large (*cf.* Sec. 14·3).

In order that the power reflected from the cylinder back into the line source be negligible and that the line source have no destructive effect on the vertical polar diagram of the reflector, it is necessary that the reflector be so shaped that the complete system of reflected rays clears the line source. To achieve this condition the radiation from the latter is directed down into the reflector and the major part of the reflector is below the line source. It is also important that the radiation reflected optically from the top of the reflector should not strike the bottom of the reflector. A profile view of a typical reflector and line source is shown in Fig. 13·29. The orientation of the top of the reflector is such that the radiation reflected there passes clear of the line source and the bottom of the reflector.

Line Sources.—With the reflector height limited, it is necessary so to design the reflector contour and feed aperture that the angle subtended by the reflector at the feed includes most of its radiation; the illumination should taper to a low value at the top and bottom of the reflector. This precaution is necessary if the amplitude of radiation from the line source going past the edge of the reflector is to be maintained low and if the side lobes in the vertical diffraction pattern of the reflected radiation are to be kept down. In addition, the feed must be designed and oriented in the reflector with a view to minimizing the amount of feed back-lobe radiation in the angular region of the flared beam. The azimuth diffraction pattern of the antenna is determined by the design of the line source. Since the beamwidth should be as narrow as possible in azimuth to secure good resolution on objects, and since relatively high side lobes in the azimuth pattern are allowable, the line source is required to have as sharp an azimuth beam as can be obtained with the length of line source used. This means that the intensity of the source should be uniform along its axis.

The various designs of line-source feeds used in this connection can be classified according to two general types: (1) arrays of radiating elements arranged in a line on a waveguide and radiating approximately broadside and (2) parallel-plate linear focusing systems that have rec-

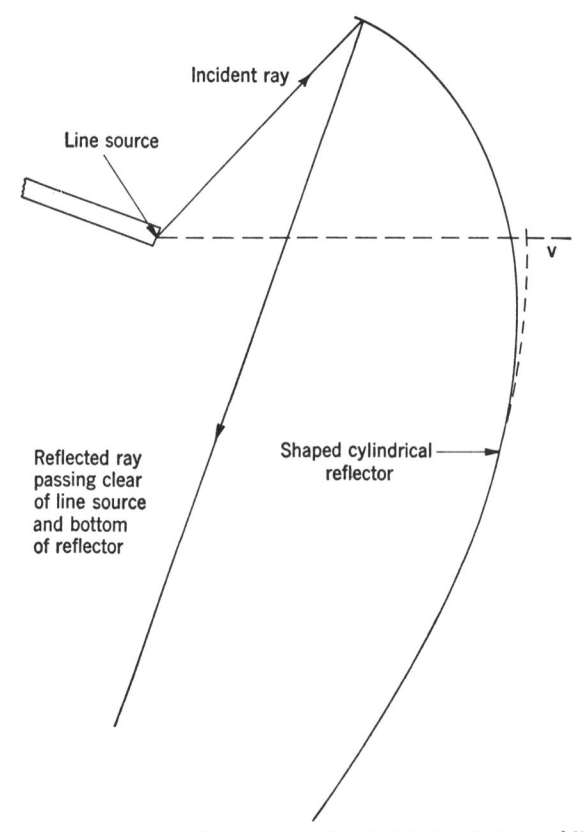

Fig. 13·29.—Vertical section through shaped cylindrical reflector and line source.

tangular apertures with a large length-to-breadth ratio. In the terms used in the treatment of linear arrays in Chap. 9, either resonant or nonresonant arrays may be used;[1] design procedures and performance characteristics are treated in detail in Chap. 9. The advantages of linear arrays for this application are their compactness and light weight. The disadvantages of the nonresonant array are (1) that the beam scans through a small conical angle as the result of frequency fluctuations and

[1] J. R. Risser, A. M. Steenland, J. Steinberger, L. J. Eyges, RL Report No. 973, Mar. 19, 1946.

(2) that the beam must be kept off normal to avoid cumulative impedance mismatch; this results in a conical beam that illuminates a hyperbolic trace on the ground. The resonant array suffers from the disadvantages of frequency sensitivity in impedance match and pattern.

A variety of parallel-plate focusing systems have been examined for their usefulness in this connection. One type consists of a rapidly flaring sectoral horn with either a dielectric[1] or metallic lens[2] in the horn so designed that radiation across the horn aperture is all in phase.

Another type of parallel-plate line source is the pillbox antenna discussed in Sec. 12·16. The pillbox line source is simple to design and relatively simple to construct but suffers from certain basic disadvantages. The center of the aperture is necessarily obstructed by the feed at the focus of the parabola; this results not only in side lobes in the azimuth pattern but also in an impedance frequency sensitivity as the result of radiation reflected back down the feed line. Efforts to correct one or both of these effects, have taken various forms: the design of a matching plate to be put at the vertex of the parabola,[3] the design of new pillbox feeds,[4] and the design of double pillboxes to obtain canceling reflections.[5]

13·6. Reflector Design on the Basis of Ray Theory.—A successful procedure has been developed for designing the shape of the reflector that is required to produce a specified vertical-plane pattern. The latter is usually specified in idealized form: the power distribution is to be a prescribed function $P(\theta)$ between the depression (or elevation) angles θ_1 and θ_2, and zero for all other angles. It must be recognized, however, that it is impossible to realize a discontinuous power distribution of this type accurately with a reflector of finite extent. Diffraction phenomena are unavoidable, and the best that one can hope to achieve is to approach the idealized pattern within acceptable limits of deviation from the prescribed $P(\theta)$ in the range (θ_1,θ_2) and with an acceptable low level of intensity outside the given range.

A good first approximation to the cross-section curve can be arrived at on the basis of geometrical optics.[6] If it then proves to be necessary,

---

[1] C. C. Cutler, "Line Sources of Microwave Energy for Feeding Cylindrical Reflectors," BTL Report MM-45-160-3, Jan. 5, 1945.

[2] M. A. Taggart, "Horn with Metal Lens," RL Report No. 863, Nov. 13, 1945.

[3] S. Silver, "Analysis and Correction of the Impedance Mismatch Due to a Reflector," RL Report No. 810, Sept. 25, 1945.

[4] M. A. Taggart, "A New Pillbox Feed," RL Report No. 862, Nov. 7, 1945; L. J. Eyges, "Lens Feed for Pillboxes," RL Report No. 869, Jan. 23, 1946.

[5] W. O. Smith, "A Broad-band *TEM* Pillbox," RL Report No. 901, Jan. 11, 1946.

[6] R. C. Spencer, "Synthesis of Microwave Diffraction Patterns with Application to Csc² θ Patterns," RL Report No. 54-24, June 23, 1943, describes L. J. Chu's method for calculating the reflector shape. See also C. C. Cutler, BTL Reports MM-44-160-37, Feb. 14, 1944; MM-45-160-4, Jan. 5, 1945.

498   SHAPED-BEAM ANTENNAS   [Sec. 13·6

the curve is modified by means of the more exact analysis given in the following section. We shall consider here the geometrical-optics technique. The cross-section configuration of the reflector, the line source, and the rays is shown in Fig. 13·30. The z-axis is taken in the horizontal direction; F is the trace of the line source. Positive angles are measured in the clockwise sense as shown. The reflector subtends a total angle $\psi_2 - \psi_1$ at the source F. Let ρ be the radius vector from F to an arbi-

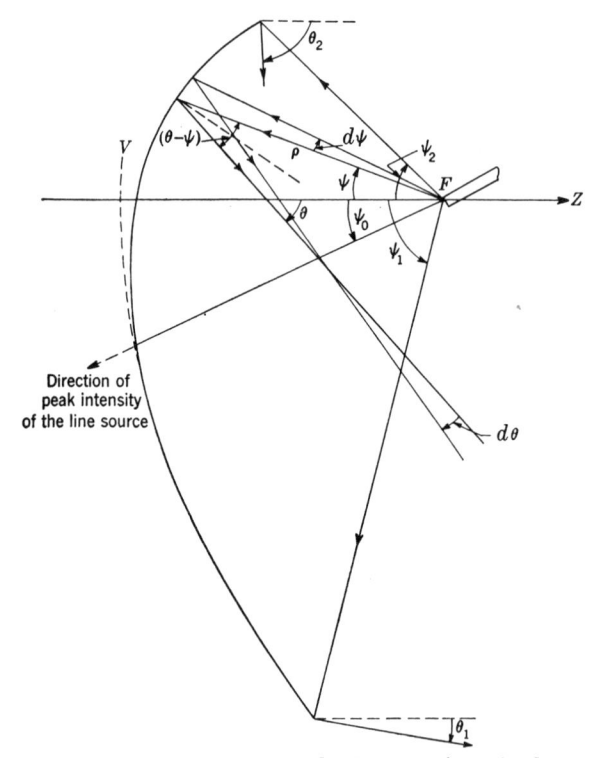

Fig. 13·30.—Geometry of ray reflection at surface of reflector.

trary point on the curve. From the law of reflection it follows that the angle between the normal to the curve and the incident or reflected ray is $(\theta - \psi)/2$; the differential equation of the curve is then readily found to be

$$\frac{1}{\rho}\frac{d\rho}{d\psi} = \tan\frac{\psi - \theta}{2}. \tag{8}$$

The functional relation between θ and ψ must be determined to produce the proper dispersion of the primary feed energy into the secondary pattern distribution. To this end consider the wedge of incident rays between ψ and $\psi + d\psi$; on reflection this becomes a wedge of rays between θ and $\theta + d\theta$. By the energy balance principle of geometrical optics the power in the incident wedge is equal to the power in the reflected wedge. Let $I(\psi)$ be the distribution in the primary pattern about the line source F. The incident power is then proportional to $I(\psi)\,d\psi$. Similarly the reflected power is proportional to $P(\theta)\,d\theta$. We have then

$$I(\psi)\,d\psi = KP(\theta)\,d\theta. \tag{9}$$

The constant K is determined from the condition that the total primary feed energy intercepted by the reflector must appear in the secondary pattern in the required range $(\theta_1 \theta_2)$, whence

$$K = \frac{\int_{\psi_1}^{\psi_2} I(\psi)\,d\psi}{\int_{\theta_1}^{\theta_2} P(\theta)\,d\theta}. \tag{10}$$

Similarly the total primary feed energy in an arbitrary range (ψ_1, ψ) must appear in a corresponding range (θ_1, θ) of the secondary pattern. This leads to the integral relation

$$\int_{\theta_1}^{\theta} P(\theta)\,d\theta = \frac{1}{K}\int_{\psi_1}^{\psi} I(\psi)\,d\psi \tag{11}$$

which serves to determine θ as a function of ψ for prescribed distributions $I(\psi)$ and $P(\theta)$. For example, if $P(\theta)$ is $\csc^2 \theta$, Eq. (11) gives

$$\cot \theta = \cot \theta_1 + \frac{\cot \theta_2 - \cot \theta_1}{\int_{\psi_1}^{\psi_2} I(\psi)\,d\psi} \int_{\psi_1}^{\psi} I(\psi)\,d\psi. \tag{12}$$

The primary pattern function $I(\psi)$ is generally known only in numerical form from experimental data. The integrations over ψ must therefore be carried out numerically or graphically.

The functional relation $\theta(\psi)$ determined from Eq. (11) is then substituted into Eq. (8) to obtain the equation of the curve. The integral of Eq. (8) is

$$\ln \frac{\rho}{\rho_0} = \int_0^{\psi} \tan\left(\frac{\psi - \theta(\psi)}{2}\right) d\psi. \tag{13}$$

This integration must also be performed graphically or numerically. It is seen that this leads only to the shape and not the absolute scale of the reflector. The distance ρ_0 from F to the reflector along the axis is determined so that the required height of the reflector conforms to the total angle $\psi_1 - \psi_2$ subtended at F.

There are a number of arbitrary variables in the procedure, the choice of which can be determined only by experience. It will be observed that the line source is oriented so that its peak intensity is in a direction ψ_0. This takes cognizance of the fact that the lower portion of the reflector is required to produce the high-intensity region of the secondary pattern. The choice is not critical, and ψ_0 may be as low as 0° and as much as $-25°$. The angle $\psi_2 - \psi_1$ subtended by the reflector should be sufficiently large to make efficient use of the primary feed energy. As a general guide $\psi_2 - \psi_1$ should correspond closely to the 10-db width of the primary pattern $I(\psi)$.

It is interesting to observe that the lower section of the reflector in general turns out to be very nearly a parabola with focus at F and axis FV parallel to the reflected ray in the direction θ_1. The physical basis for this property is readily evident, for this portion of the reflector must converge the divergent rays from the feed into a narrow beam to produce the required peak intensity. The total flared pattern may be regarded as a superposition of a narrow beam produced by the parabolic segment and a broad beam produced by the dispersive section of the reflector.

13·7. Radiation Pattern Analysis.—The secondary pattern that will actually be obtained with the reflector determined on the basis of geometrical optics can be calculated to a high degree of accuracy by the methods of Sec. 5·9. Such a calculation serves a several-fold purpose. The extent of the deviations from the idealized pattern can be determined prior to construction and test of an experimental model. The diffraction effects are quite sensitive to the scale factor ρ_0 in Eq. (13). If there is some latitude allowed in choice of the reflector dimensions, the radiation pattern can be calculated for several values of ρ_0 for a given shape to determine the best dimensions within the allowed range. For a given choice of dimensions, the effect of small alterations in the cross-section contour from the geometrical-optics curve can be studied to arrive at a curve that yields an acceptable pattern. These results also serve as a basis for setting the tolerances that are to be required on the reflector shape in the production of the antenna.

The reader is referred to Secs. 5·7 to 5·9 for the theoretical details. The calculation is based on the assumptions that the reflector is in the cylindrical wave zone of the source, that the minimum value of ρ is large compared with the wavelength, and that the maximum value of ρ is less than $\frac{l^2}{\lambda}$, where l is the length of the source. The primary pattern of the source is specified by the distribution function $F(x)$ along the length of the source and the angular distribution $G(\psi)$ around the source. Thus, if P is the total power radiated, the power radiated in a segment of angular width $d\psi$ and length dx is

$$dP = \frac{P}{2\pi l} F(x) G(\psi) \, dx \, d\psi. \tag{14}$$

The properties of the distribution functions are given in more detail in Eqs. (5·81) and (5·82).

The procedure is to calculate the current distribution induced on the reflector by the primary field and then to calculate the radiation pattern of the surface-current distribution. Two cases are to be distinguished: (1) longitudinal polarization—horizontal polarization in the present discussion—in which the electric vector of the primary field is parallel to the line source and (2) transverse—vertical—polarization in which the electric vector is parallel to the planes perpendicular to the line source. The coordinate system is described in detail in Fig. 5·9 and will not be reproduced here. Only one change need be noted: In the present section the positive direction of the angle Θ is reversed with respect to that chosen in Sec. 5·9. The complete expressions for the patterns are given for the two types of polarization by Eqs. (5·86) and (5·87), respectively. For the vertical plane pattern, that is, the plane $\Phi = 0$ in Fig. 5·9, these reduce to

1. Horizontal polarization:

$$E(\Theta) = -jA \int_{\psi_1}^{\psi_2} [\rho G(\psi)]^{1/2} \cos i \left[1 + \frac{1}{\rho}\left(\frac{d\rho}{d\psi}\right)^2\right]^{1/2} e^{-jk\rho[1+\cos(\psi-\Theta)]} d\psi. \quad (15)$$

2. Vertical polarization:

$$E(\Theta) = -jA \int_{\psi_1}^{\psi_2} [\rho G(\psi)]^{1/2} \cos (\mathbf{n},\mathbf{R}_1) \left[1 + \frac{1}{\rho}\left(\frac{d\rho}{d\psi}\right)^2\right]^{1/2} e^{-jk\rho[1+\cos(\psi-\Theta)]} d\psi, \quad (16)$$

where

$$A = \frac{\omega\mu}{2\pi R} e^{-jkR} \left[\left(\frac{\epsilon}{\mu}\right)^{1/2} \frac{P}{\pi l}\right]^{1/2} \int_{-l/2}^{l/2} [F(x)]^{1/2} dx. \quad (17)$$

The angle i is the angle between ρ and the normal $\mathbf{n}$ to the surface; the angle $(\mathbf{n},\mathbf{R}_1)$ is between the normal and the unit vector $\mathbf{R}_1$ in the given direction of observation in the secondary pattern. We note also that from the preceding section

$$\left[1 + \left(\frac{1}{\rho}\frac{d\rho}{d\psi}\right)^2\right]^{1/2} = \sec i. \quad (18)$$

The reflector curve being given, ρ, $\cos i$, and $\cos (\mathbf{n},\mathbf{R}_1)$ are known functions of position on the reflector. The integrals are then evaluated numerically for successive values of Θ. The power pattern is then obtained by the usual methods which are discussed in Chap. 5:

$$P(\Theta) = \frac{1}{2}\left(\frac{\epsilon}{\mu}\right)^{1/2} R^2 |E|^2, \quad (19a)$$

and the gain in a given direction is

$$G(\Theta) = \frac{4\pi P(\Theta)}{P}. \quad (19b)$$

The calculations are laborious but straightforward. The reliability of the method may be judged by the comparison between the calculated and measured patterns shown in Fig. 13·31.[1] The positions of the maxima and minima are predicted accurately, and the maximum deviation from the measured values is 2 db. The performance of the reflector could have been improved by the procedure, outlined at the beginning

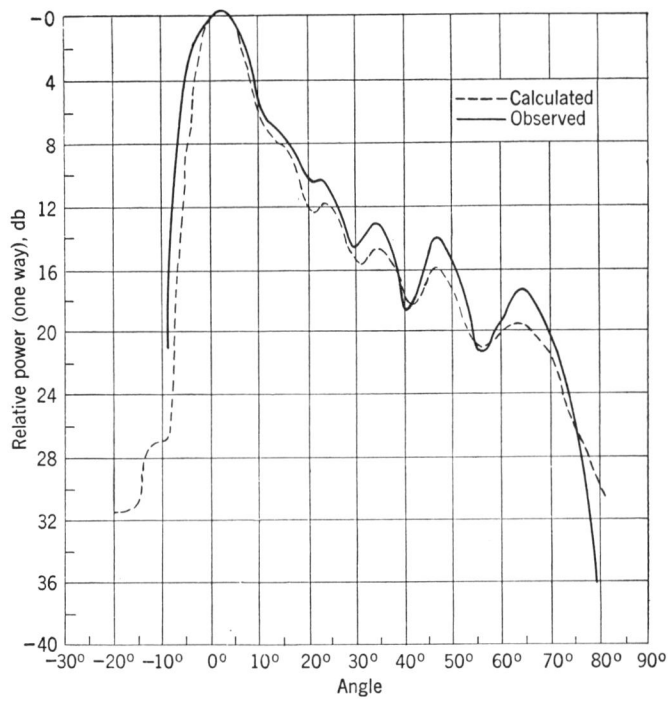

FIG. 13·31.—Calculated and observed vertical patterns of cylindrical reflector antenna with horizontal polarization.

of the section, of modifying the reflector curve until the amplitude fluctuations in the calculated pattern are reduced to an acceptable value.

13·8. Double-curvature Reflector Antennas.—The line-source feeds required to illuminate the cylindrical reflectors discussed in the preceding section have a number of disadvantages as compared with point-source feeds. They are in general bulkier, heavier, less satisfactory as to impedance properties, and more complicated to design, build, and pressurize. When a reflector is used with a point-source feed to form a shaped beam, it is required to provide a pattern of the specified shape in one plane and to focus in transverse planes. A number of cut-and-try improvisations for accomplishing this end were described in Sec. 13·3 under Asymmetrical Shaped Beams.

[1] Taken from T. J. Keary, "Calculation of Vertical Polar Diagrams and Power Gains of Antennas for Airborne Navigational Radars," RL Report No. 750, Sept. 10, 1945.

Barrel Reflector Antenna.—The barrel reflector antenna shown in Fig. 13·17 was the most successful of these improvisations but is subject to some serious limitations. A consideration of these limitations will provide helpful guide lines in the design of double-curvature reflectors. In the first place, the shape of the central vertical section of the reflector was determined, not as the proper transformation from primary to secondary pattern, but by the simplest geometry which provides a focus in transverse planes. The bottom half of the reflector focuses half of the radiation from the feed into a pencil beam. The barrel-shaped top half of the reflector focuses the primary pattern in transverse planes but reflects it unchanged in the vertical plane. The superposition of the two portions of the secondary pattern, shown schematically in Fig. 13·32,

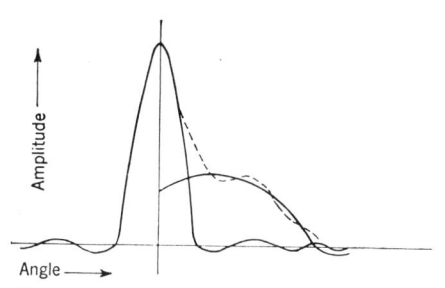

FIG. 13·32.—Superposition of barrel and paraboloidal amplitude patterns.

inevitably results in a depression in the pattern near the peak of the beam, as can be seen in the pattern of Fig. 13·20a. Second, since the feed is at the center of curvature of the barrel section in the plane of symmetry, all the reflected rays in that plane pass through the feed. This causes an excessive amount of power to be returned to the transmission line resulting in a frequency-sensitive impedance mismatch. A third difficulty is that the right and left upper portions of the barrel reflector are set at such angles as to give rise to strong cross-polarized components in the wide-angle portion of the flared beam. These cancel in the median plane but add to the sides of the normal polarization lobe to give a considerably broader effective transverse pattern. These several difficulties with the barrel reflector antenna can be avoided by proper design[1] of the generalized surfaces to be discussed below.

General Considerations.—The primary feed pattern and the reflector are to have as a common plane of symmetry the plane in which the beam is flared—plane XFZ of Fig. 13·33. Two considerations enter the design: The central vertical section of the reflector is to be adjusted on the basis of the central vertical pattern of the feed to give the specified secondary pattern; the remainder of the surface is to be so shaped as to obtain pencil-beam characteristics in the transverse planes. The second consideration requires that all rays from the point source after reflection

[1] R. C. Spencer, "Synthesis of Microwave Diffraction Patterns with Applications to $Csc^2 \theta$ Patterns," RL Report No. 54-24, June 23, 1943; J. F. Hill, G. G. Macfarlane, W. Walkinshaw, TRE Report No. 1878, May 17, 1945; S. Silver, "Double Curvature Surfaces for Beam Shaping with Point Source Feeds," RL Report No. 691, June 15, 1945.

from the surface emerge parallel to the central plane; the reflected wavefront is thus a cylinder whose generator is normal to the central plane.

Transverse Sections of the Reflector.—The transverse sections of the surface are determined by the requirement that the reflector is to convert a spherical wave into a cylindrical wave. The condition is easier to formulate from the point of view of reception. Referring to Fig. 13·34, consider the sheet of rays, all parallel to the central plane, incident on the reflector in the plane $OANP$. The latter is perpendicular to the central plane. We require all these rays to be brought to a focus at F. If then we take any line in the plane $OANP$ perpendicular to the rays, the optical path from that line to F is the same for all rays in the sheet. Let ρ be the radius vector from F to the central section curve, ϕ its angle

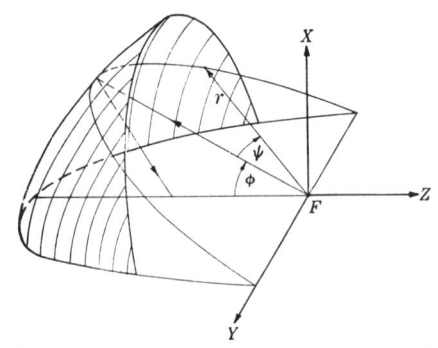

Fig. 13·33.—Double curvature surface and spherical coordinate system.

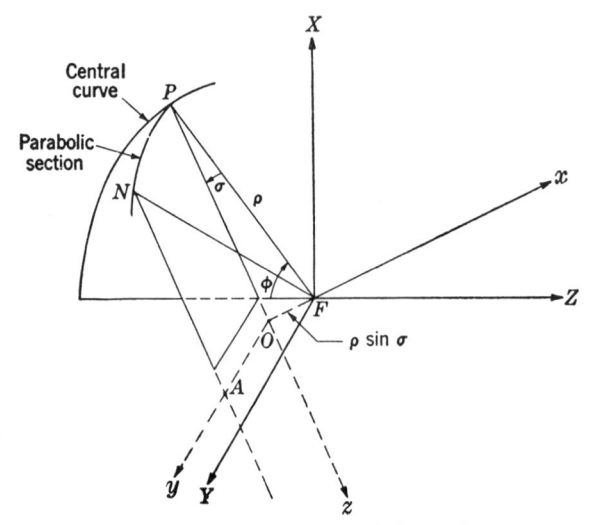

Fig. 13·34.—Surfaces of parabolic section.

of elevation, and σ the angle between the incident and reflected ray in the central section. Through F draw Ox normal to the plane $OANP$. In the plane $OANP$ set up the orthogonal axes Oy, Oz, with Oy normal to the central plane. Writing the condition of the optical path (*cf.* Sec. 4·9) we have

$$AN + NF = OP + \rho$$

or

$$(\rho^2 \sin^2 \sigma + y^2 + z^2)^{1/2} - z = \rho(1 + \cos \sigma). \qquad (20)$$

SEC. 13·8] DOUBLE-CURVATURE REFLECTOR ANTENNAS 505

This gives
$$y^2 = 4\rho \cos^2 \frac{\sigma}{2} (z + \rho \cos \sigma). \tag{21}$$

The section of the surface *in the plane OANP* is, therefore, a parabola with vertex at P and focal length
$$f(\phi) = \rho(\phi) \cos^2 \frac{\sigma(\phi)}{2}. \tag{22}$$

Given the central section curve, $\rho(\phi)$, $f(\phi)$ and the associated reflection plane are determined. It is readily seen that the whole family of reflection planes corresponds to a cylindrical wavefront. The barrel section reflector is obtained by setting ρ = constant, $\sigma(\phi) = 0$.

Central Section Curve.—The central section curve is to be determined so that the system will radiate a desired Fraunhofer pattern. There is no simple decisive procedure for relating the central curve to this pattern, which takes diffraction effects into account properly. The method discussed here is based on geometrical concepts of energy balance between the primary feed and Fraunhofer patterns. The latter is a three dimensional pattern for whose specification we shall use spherical coordinates such as are shown in Fig. 13·33. The angle of elevation with respect to the XY-plane will be denoted by θ (ϕ being used for the primary feed pattern).

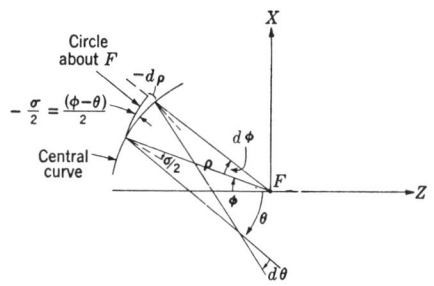

FIG. 13·35.—Geometry of central curve.

ψ is the azimuth angle in the planes θ = constant. The pattern is specified in the form $P(\theta, \psi)$, the power radiated per unit solid angle in direction (θ, ψ). The central plane pattern $P(\theta, 0)$, hereafter designated as $P(\theta)$, is assumed to be specified.

Referring to Fig. 13·35 it is seen that the differential equation of the central curve is given by
$$\frac{1}{\rho} \frac{d\rho}{d\phi} = -\tan \frac{\sigma}{2},$$

σ being the angle between the incident and reflected ray. Or since $\sigma = \theta - \phi$,
$$\frac{1}{\rho} \frac{d\rho}{d\phi} = \tan \left(\frac{\phi - \theta}{2} \right). \tag{23}$$

The positive directions of θ and ϕ are shown in the figure. The relation between θ and ϕ, which is necessary for integrating the above equation, is obtained from energy balance considerations.

This method is based strictly on geometrical optics. The assumptions involved are that the beam is narrow in the transverse (ψ) directions and that the transverse aspect of the beam is virtually independent of θ. The effect of diffraction is assumed to be the same in all such planes. Accordingly we assume that we have a cylindrical wavefront in the far field corresponding to geometrical optics and that the energy flow through the cylindrical wavefront between the planes θ and $\theta + d\theta$ per unit length along the generating element is of the form

$$P(\theta)\, d\theta\, F(y),$$

where $F(y)$ represents the distribution in the transverse aspect of the beam.

Let us take a small cone of rays from the source F in the central section defined by the planes ϕ and $\phi + d\phi$ and azimuth extent $d\psi$. The energy in this cone is

$$I(\phi)\, d\phi\, d\psi,$$

where $I(\phi)$ is the power radiated per unit solid angle in direction $(\phi,0)$.[1] On reflection this energy appears in a wedge defined by the planes θ and $\theta + d\theta$, since the reflected rays are parallel.[2] The width of the wedge is $\rho\, d\psi$, so that in terms of $P(\theta)$ the energy contained therein is

$$P(\theta)\, d\theta\rho\, d\psi.$$

Equating this to the incident energy gives

$$\rho P(\theta)\, d\theta = I(\phi)\, d\phi \tag{24}$$

or

$$P(\theta) \frac{d\theta}{d\phi} = \frac{I(\phi)}{\rho} \tag{24a}$$

Because of the factor ρ occurring in Eq. (24), the latter cannot be integrated as in the case of the cylindrical reflector to give the relation between θ and ϕ. Taking logarithmic derivatives with respect to ϕ of Eq. (24a) and substituting from Eq. (23), we get instead the differential equation

$$\frac{d^2\theta}{d\phi^2} + \left[\tan\left(\frac{\phi - \theta}{2}\right) - \frac{I'(\phi)}{I(\phi)}\right]\frac{d\theta}{d\phi} + \frac{P'(\theta)}{P(\theta)}\left(\frac{d\theta}{d\phi}\right)^2 = 0, \tag{25}$$

which is to be integrated numerically. Here

$$I'(\phi) = dI/d\phi, \qquad P'(\theta) = \frac{dP}{d\theta}.$$

The general arrangement of the central curve and feed is shown in

---

[1] The solid angle in the spherical coordinates used here is $\cos\psi\, d\phi\, d\psi$. For the central section $\cos\psi = 1$.

[2] The correspondence is not strictly true, since rays in an incident sheet do not lie in a plane on reflection. We are assuming that the error is negligible in the neighborhood of the central section.

Fig. 13·36. The angular limits ϕ_1, ϕ_2 of the reflector are arbitrary. A satisfactory choice is to take these to correspond to the 10-db points in the primary pattern. The angular aperture $\phi_2 - \phi_1$ must be taken fairly large, or the surface will be found to curl in too rapidly in the transverse aspects. By setting up the reflected rays as shown in the figure the feed is kept out of their paths and the mismatch due to the reflector is kept at a low level. Since the region $\phi_1 \leq \phi \leq 0$ contributes to the high energy region of the Fraunhofer pattern, the primary feed

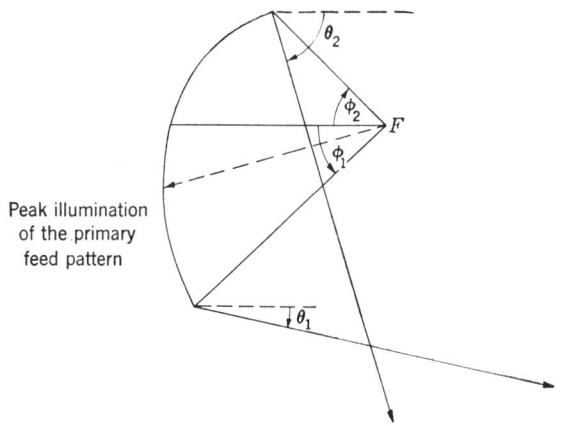

FIG. 13·36.—General arrangement of central curve and feed.

pattern is tipped to illuminate that area more strongly. The angle of tip is arbitrary. For cylindrical reflectors with a line source this is usually taken in the neighborhood of 15°. This should be satisfactory here. With the choice of these various factors decided upon, the integration problem is defined.

Integration for the $\csc^2 \theta$ *Pattern.*—We shall discuss the integration of the central curve equations with particular reference to the $\csc^2 \theta$ pattern used in airborne navigational antennas. $P(\theta)$ is required to be a $\csc^2 \theta$ distribution between the angular limits θ_1 and θ_2. That is,

$$P(\theta) = K^2 \csc^2 \theta, \quad \theta_1 \leq \theta \leq \theta_2, \quad \theta_1 > 0$$
$$= 0$$

outside this range. Using Eq. (25), we first find $\theta(\phi)$ so that

$$\theta = \theta_1 \quad \text{when} \quad \phi = \phi_1,$$
$$\theta = \theta_2 \quad \text{when} \quad \phi = \phi_2.$$

For the $\csc^2 \theta$ pattern we get

$$\frac{d^2\theta}{d\phi^2} + \left[\tan\left(\frac{\phi - \theta}{2}\right) - \frac{I'(\phi)}{I(\phi)}\right]\frac{d\theta}{d\phi} - 2 \cot \theta \left(\frac{d\theta^2}{d\phi}\right)^2 = 0. \quad (26)$$

The numerical integration of this equation subject to the end-point conditions at ϕ_1 and ϕ_2 requires a guess as to the value of the slope $d\theta/d\phi$ at the starting point. The integration is carried through for the trial value of initial slope, and the resulting end value θ_2 is determined. On the basis of this result a second assumption is made for the initial slope, and the integration is repeated. The second or, at most, a third guess is sufficient, especially since the end value θ_2 is not critical. The initial slope may be expected to be small, since this region of the reflector contributes to the peak of the beam and the reflected rays are very nearly parallel. Throughout the course of the curve $d\theta/d\phi$ must be positive.

13·9. Variable Beam Shape.—It is frequently specified for an airborne navigational radar antenna that the shaped elevation pattern be controllable, either continuously or in a limited number of steps. This requirement is imposed because of the need to operate the aircraft at a range of altitudes, whereas the $\csc^2 \theta$ pattern is designed for a single altitude. Several means have been employed to exercise control over the elevation beam shape. In one cylindrical reflector antenna the directivity of the feed was altered by means of a flap[1] to direct more of the radiation onto the sharply curved portion of the reflector and thus put more power into the wide-angle part of the elevation pattern for high-altitude operation. In the case of the cylindrical reflector—pillbox antenna, measurements have been made on the change in shape of the secondary pattern of the antenna as the pillbox is rotated about the long axis of its aperture. Similar measurements have been made with encouraging results by rotating a directive horn feed in the barrel reflector.[2] In all such cases, after the altered pattern shape has been obtained by a change in feed, it is then necessary to rotate the entire antenna to obtain the correct minimum depression angle. Rotation of the reflector about a horizontal axis near the vertex changes the illumination and depression angle together in such a way as to give satisfactory patterns in an actual case[3] for depression angles from 3° to 13°.

It is sometimes necessary to compensate for roll and pitch of the aircraft by line-of-sight stabilization which maintains a fixed elevation angle of the beam with respect to the horizon. For simplified mechanical control and to eliminate mechanical interferences, there are advantages in obtaining stabilization in the beam by rotating the reflector alone in the vertical plane. In this case, it is desirable to change the position of the beam without changing its shape. This is difficult because displacement of the feed causes changes in the illumination over the reflector.

Both of these problems would be much simplified if the reflector were

---

[1] C. C. Cutler, BTL Report MM-45-160-4.

[2] J. H. Gardner, "Low Altitude $\csc^2 \theta$ Antenna," RL Report No. 1073, Feb. 21, 1946.

[3] T. J. Keary, A. R. Poole, J. R. Risser, H. Wolfe, "Airborne Navigational Radar Antennas," RL Report No. 808, Mar. 15, 1946.

not asked to perform the dual function of focusing and shaping the beam. In one antenna design[1] a point source and cut paraboloid produced a collimated beam which was then shaped in the vertical plane by a second reflector. In this case an excellent elevation pattern was obtained. By control of the second reflector alone it was possible to change the shape and direction of the beam for depression angles ranging from 1° to 9° and to accomplish line-of-sight stabilization over a range of $\pm 15°$. The serious disadvantage of the double-reflector antenna is the space that it requires.

[1] E. B. Chisholm and B. R. Vogel, "Double Reflector Antenna," RL Report No. 775, July 16, 1945.

CHAPTER 14

ANTENNA INSTALLATION PROBLEMS

BY L. C. VAN ATTA AND R. M. REDHEFFER

GENERAL SURVEY OF INSTALLATION PROBLEMS

The customary procedure in microwave radar antenna development has been to design the antenna and to carry out the early experimental tests on the assumption of free-space conditions surrounding the antenna. Whereas this represents a good approximation in general, it is necessary eventually to consider the effect upon the antenna performance of the supporting structure on which it must be mounted. Also, it is generally necessary to place the antenna in a dielectric housing—the radome—which likewise affects its performance. It is sometimes possible so to choose the antenna location on the structure and to design the radome that the original performance of the antenna is unimpaired. The final result can be predicted with greater certainty, however, if the electrical design of the antenna is considered from the beginning in conjunction with that of the radome and with a view to the structure and location that the antenna-radome system must occupy.

The purpose of this chapter is to present the problems imposed by installation requirements and the practices that have been adopted for dealing with them. These considerations are intended to serve merely as background for one engaged in antenna design.

14·1. Ground Antennas.—The mechanical design of an antenna for ground use must effect a compromise between the factors of electrical reliability, mechanical ruggedness, and portability. An antenna in a permanent site has moderate weight and portability requirements, is not required to survive repeated rough handling, but should be ideally sited and should be designed to operate reliably over an extended period in the face of local conditions of weather and wind.

The antenna of a high-performance radar set which must be moved quickly from time to time over considerable distance to a new site is still not seriously restricted in weight but must disassemble into relatively lightweight components of limited dimensions for transportation by air or truck. Its construction must permit precision reassembly after repeated handling. Components must be susceptible to repair or replacement under field conditions.

The antenna of a somewhat lighter radar set may be truck-mounted with the rest of the set. This eliminates the necessity for disassembly and facilitates ground transport when relocation of the set is required. At the same time it rules out the desirable features of air transportability. Very lightweight sets have been designed for transport by mule or pack for use in inaccessible regions. So much is sacrificed in reliability and performance, however, by this severe portability requirement that such sets have not been demonstrated to be practicable.

Some ground antennas, particularly truck-mounted units, are provided with radomes. More often the antenna is exposed. It proves to be more practicable in these cases to design the antenna for operation in the open than to provide a satisfactory housing. The antenna feed must then be weatherized by means of a dielectric cup or plate. The reflector is generally of mesh, perforated, or grating construction to reduce windage effects. Mechanical and electrical components must be enclosed in a metal housing.

14·2. Ship Antennas.—The distinctive features of a ship as an antenna location are the small number of suitable sites and the large amount of other gear that invariably interferes with the placement and performance of an antenna. The top of a mast is obviously the ideal location either for an antenna with an omnidirectional pattern or for one with a complete azimuth scan. Any other location will involve obstruction of the beam by the mast or by other parts of the superstructure. Such obstruction will result in blind regions, false signals, or transmitter pulling. In addition to competing for favorable sites, antennas obstruct each other's view and jam each other when frequency relations permit. Furthermore, mechanical and electrical considerations are frequently at cross purposes. A consideration of each antenna's performance argues for placing it at the top of the mast, whereas a consideration of ship stability would place it below deck. An antenna mounted alongside a mast on a bracket should be far out from the mast for electrical reasons but close in for mechanical reasons. The electrical performance is seldom improved by the modifications introduced to provide resistance to shock and vibration.

A structure of limited extent equipped with a large number of antennas is referred to as an antenna system. The antenna system problem then is to obtain satisfactory performance from the several antennas by relocating them, by combining their functions or otherwise reducing their number, by redesigning them, or by redesigning the supporting structure. The ship antenna system problem is still far from solution, and each new ship type and new equipment serve to increase the over-all problem. Extensive measurements, including model measurements, are required to assess the performance of existing antennas in present locations and to predict their performance in other locations. Ship superstructures and antennas must be more closely integrated in the design stages. One approach to the problem that is commanding considerable attention is the design of series or parallel coupling circuits to permit the multiple use of antennas of certain types. Rejection filters can be designed to eliminate jamming between antennas and to reduce background noise.

Special arrangements have been employed for improving the performance of individual equipments in the presence of screening structures. Radars with scanning antennas may be duplicated fore and aft so that their regions of effectiveness supplement one another. An antenna with an omnidirectional pattern may be replaced by two antennas on opposite sides of the superstructure with 180° azimuth patterns.

Another coverage problem is introduced by interference nulls in the elevation pattern when sufficient radiation from the antenna is reflected from the deck or sea. One solution to this problem which has been employed in the case of nonscanning antennas is to replace the antenna by several properly distributed in height. Such a set of nonscanning antennas may be connected to different receivers or may be connected to a single receiver with a "diversity" hookup which leaves the antenna receiving the strongest signal actually connected to the receiver.

In designing an antenna for shipboard use, the effects of dampness, salt spray, condensation, temperature extremes, high wind velocities, and icing conditions must be considered. Some small nondirectional antennas and scanning antennas are enclosed in radomes. The larger scanning antennas are weatherized at the feed and are provided with openwork reflectors of perforated plate or grating construction to reduce windage effects.

14·3. Aircraft Antennas.—An aircraft with its many antennas for communication, navigation, instrument landing, radar, identification, and radar countermeasures provides an antenna system problem of great complexity. To the problems of siting, avoiding interference between antennas, and obtaining a proper pattern with the antenna on the structure is added the problem of meeting serious aerodynamic requirements. In the faster aircraft it is desirable to have the antenna totally contained within the airframe. When this is not possible, the extension should present minimum frontal area and should be streamlined with a housing that must have a greater elongation in the direction of motion for a higher design speed of the aircraft. Any changes in airframe imposed by antenna requirements must be incorporated in the very early stages of the aircraft design. To reduce drag, to protect the antenna from wind forces and weather, and in some cases to provide for pressurization, every scanning antenna must be provided with a radome. This is true whether the antenna is totally included within the airframe or is exposed in a streamlined housing, often referred to as a blister or nacelle. The electrical and mechanical design requirements imposed on such radomes have become increasingly severe because of the trend toward shorter wavelengths, larger antennas, and more complete streamlining; the satisfactory solution of the radome design problem has required increasingly more sophisticated engineering.

Improper siting of the antenna on the aircraft or unsatisfactory design of the radome may result in a variety of performance defects in the system. The radiation pattern of the antenna may be seriously altered by near-by conducting edges or surfaces. Excessive absorption or reflection by the dielectric wall of the radome may introduce a number of undesirable pattern and impedance effects. Such defects have been tolerated to a certain extent in the recent past as unavoidable in the face of the rapid development in aircraft design and in the variety of antenna installations. However, with a better understanding of the design problems involved and with the possibility of accommodating antennas more satisfactorily in new aircraft designs, performance defects in aircraft radio systems can be drastically reduced. It will be necessary, however, to accomplish this improvement in electrical performance with antenna designs that are at the same time more satisfactory aerodynamically.

14.4. Scanning Antennas on Aircraft.—A scanning antenna employed in air-to-air search is required to have a narrow pencil beam and to scan a forward angular region only. Such an antenna can be located in the nose of a multiengine plane or in a wing nacelle in a single-engine plane. The wing nacelle can be located in the leading edge or at the tip of the wing without introducing serious drag. The performance of a narrow pencil-beam antenna is not appreciably affected by metal parts of the aircraft. It need not be affected seriously by the radome wall except in cases of poor radome design. The effect of the latter will be discussed later and is illustrated in Fig. 14·8. Difficulty is also encountered with streamlined radomes designed to meet the aerodynamic requirements of very high mach numbers, since this necessitates near-grazing angles of incidence of radiation upon the walls.

An antenna scanning in azimuth for air-to-ground search is required to have a beam that is sharp in azimuth and achieves with high accuracy a prescribed shape in elevation such as was described in Chap. 13. If an intelligible picture of the ground is to be presented on the cathode-ray screen of the radar set, the elevation pattern must follow the $\csc^2 \theta$ shape over a wide range of angles with an accuracy of 1 db for closely spaced variations. If a maximum range of 50 miles is to be covered from an altitude of 5 miles, the steep-angle portion of the pattern is at least 20 db down from the near-horizontal portion. Surfaces or edges near the antenna can reflect or diffract a small amount of power from near-horizontal portions of the beam into directions corresponding to steep angle portions of the beam. If such an unwanted contribution at steep angles is present even in power intensities 40 db down from the peak or 20 db down from steep-angle portions of the beam, the resulting interference effect will produce a 1-db ripple in this portion of the elevation pattern.

The most serious installation problem is encountered when an antenna that is employed in air-to-ground search is required to scan through 360° in azimuth. The ideal location for this antenna is at the lowest point in the fuselage. Since obstructions on this portion of the airframe introduce serious drag, the antenna and radome must be so designed as to minimize the transverse area exposed and to realize streamlined flow. The problem is aggravated by the fact that these antennas are generally large. Certain design features may be incorporated to meet this situation. The antenna may be so shaped (Sec. 13·4) that the upper portions of the reflector contribute to the steep-angle pattern and only the lower portion of the reflector protrudes from the fuselage. The well into which the antenna is recessed may be made large so that metal edges are further removed from the antenna and are more completely cleared by the slightly depressed beam. However, we still have a situation in which an antenna that is required to produce very accurately a prescribed pattern is closely surrounded by a metal surface into which it is partially recessed. Furthermore the beam is required to pass through a radome wall at near-grazing incidence. This combination of conditions makes these antennas particularly subject to the performance defects mentioned in the previous section. Experience with scanning antennas will serve therefore to illustrate some of the major defects encountered in airborne microwave radar performance due to faulty design or siting of the antenna-radome component.

Performance defects from the operator's viewpoint are best illustrated by actual photographs of the cathode-ray tube that supplies him with visual information. A PPI photograph substantially free of defects is first presented for the sake of comparison. PPI stands for plan-position-indicator in which slant range is displayed radially and azimuth angle is represented by polar angle. Figure 14·1 is a PPI photograph of fairly flat wooded terrain taken with 3.2-cm radiation. The antenna employed a 29-in. paraboloid reflector with a barrel-shaped insert in the upper half to obtain a shaped beam. The uniform illumination of the ground and especially the absence of lobes in the elevation pattern allow such details as the small lakes and the river with its islands and bridges to be clearly recognized. The black disk in the center is a measure of the distance of the aircraft above the ground; the bright circle surrounding the disk is caused by perpendicular reflection from the earth and is called the altitude circle. This photograph is to be compared with some less satisfactory ones which follow.

There are certain defects in performance that can be traced to the presence of conducting edges or surfaces near the antenna and can be interpreted in terms of such physical phenomena as reflection, refraction, and interference. Microwave radiation shows a sufficiently optical type of behavior that antenna sites which would involve total obstruction

SEC. 14·4] SCANNING ANTENNAS ON AIRCRAFT 515

of the beam by portions of the aircraft in important directions are recognized and generally avoided. Two examples of excusable obstruction are shown, however, to illustrate the effect. Figure 14·2 is a PPI photograph taken with 1.25-cm radiation in a shaped beam produced by a 42-in. shaped cylindrical reflector. In addition to hills, a river, and three

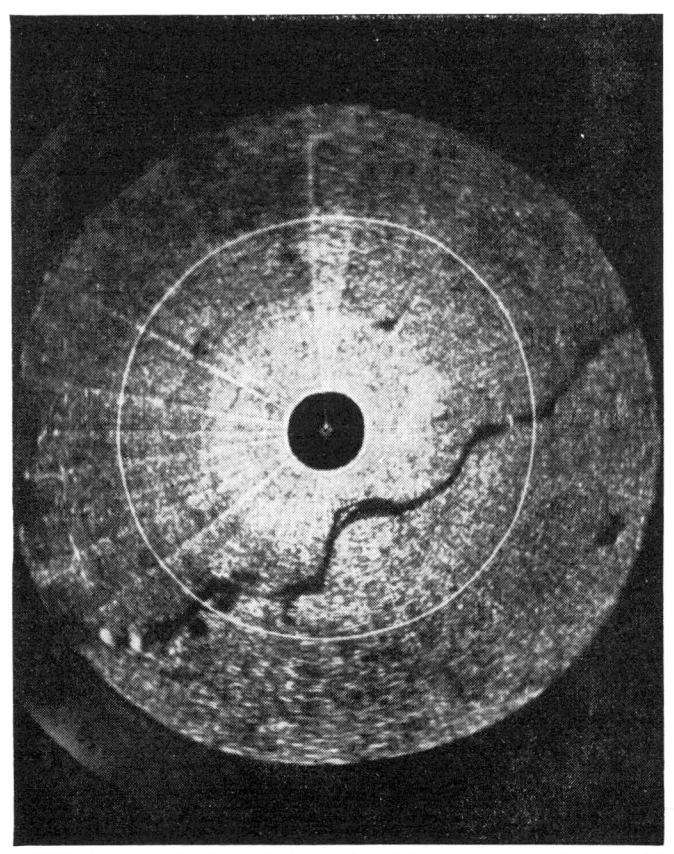

FIG. 14·1.—PPI photograph of wooded terrain. This photograph is essentially free of defects due to antenna pattern.

airfield runways, the photograph shows two black sectors extending in to steep angles and caused by the lowered landing wheels. In this PPI the altitude circle had been subtracted out. In Fig. 14·3 the shaped beam of 3.2-cm radiation from a 30-in. paraboloidal reflector with a barrel insert has been obstructed by a second aircraft below the antenna. The effect of this obstruction is to block off a portion of the radiation and to create a radar "shadow" against the illuminated background.

The two elevation patterns shown in Fig. 14·4 illustrate an effect that can result from mounting the antenna on an aircraft. This antenna, designed to give a shaped beam of 3.2-cm radiation, employs a pillbox line-source feed and a 60-in. shaped cylindrical reflector. The reflector is 12-in. high but extends only 4 in. below the center line of the fuselage for the pattern shown in Fig. 14·4a. The beam is directed forward along the line of flight with the peak tilted down 6° below horizontal; as the tilt angle is decreased from 6°, the pattern becomes rapidly worse. The interference effect which is evident in this pattern results from a combination of the direct radiation from the antenna with a small amount of radiation scattered by the straight rim of the rectangular hole into which

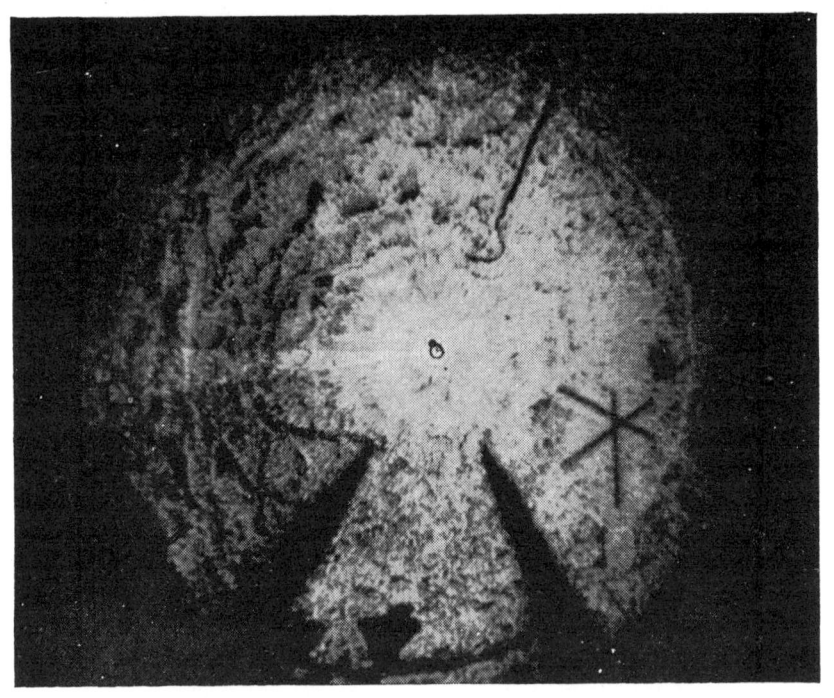

Fig. 14·2.—PPI photograph showing blank sectors produced by landing wheels.

the antenna is recessed. This scattered radiation is spread broadly in elevation but is essentially confined in azimuth to the sector of the original beam when the antenna is pointed forward. Reflection from the underside of the fuselage can be shown by geometrical arguments not to be a contributing factor. Figure 14·5 shows the appearance of this interference effect in the central portion of a PPI photograph. This photograph was taken from an altitude of 25,700 ft over Lake Okeechobee, Florida, with a 50-mile maximum range setting.

Whereas flat surface reflections did not contribute to the effects described above, there were a number of cases observed in which interference effects were due to reflections from the undersurfaces of fuselage or wings. Some PPI photographs show small regions of interference fringes at right angles to the direction of flight. These fringes occur with very close spacing in angle and can be shown to result from interference between the direct beam and radiation reflected from the undersurfaces of the wings. The fringe pattern observable in Fig. 14·6, on the other hand, is quite extensive fore and aft with a wider angular spacing between fringes which also varies with azimuth angle. Calculations confirm the assumption that this interference pattern is due to reflections from the curved undersurface of the fuselage.

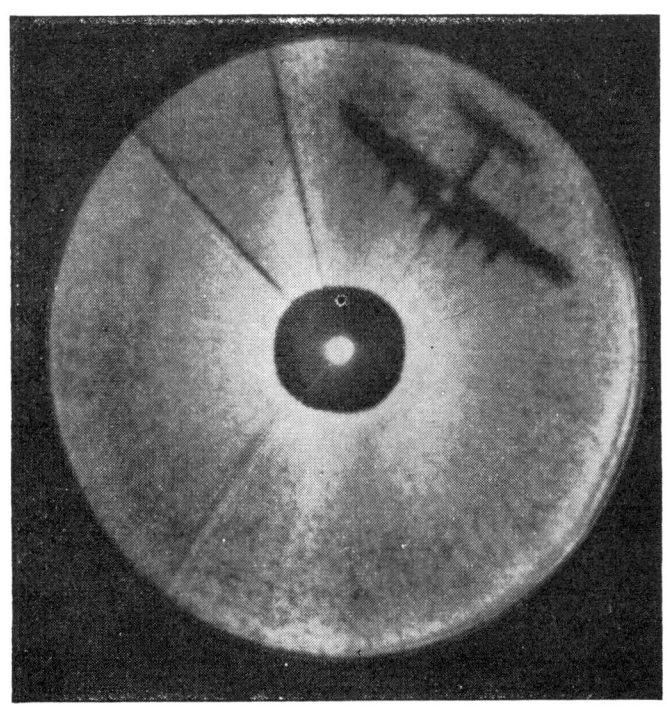

Fig. 14·3.—PPI photograph showing blank area due to obstruction by another aircraft.

There are other defects in the performance of a radar set which can be traced to electrical effects upon the beam caused by the dielectric walls of the radome housing the antenna. These defects may be listed as complete blanking out of all signals in certain azimuth sectors, reduction in range, obscuring of the screen, false signals, and displacement of the target. Severe blanking in certain sectors, as shown in Fig. 14·7, is due to reflections at the radome wall. These reflections direct radiation back down the r-f line to the transmitter which is thereby pulled in frequency off the pass band of the receiver. A smaller reduction in range which is at the same time more uniform in azimuth results from excessive absorption of radiation in the wall of the radome.

518   ANTENNA INSTALLATION PROBLEMS   [SEC. 14·4

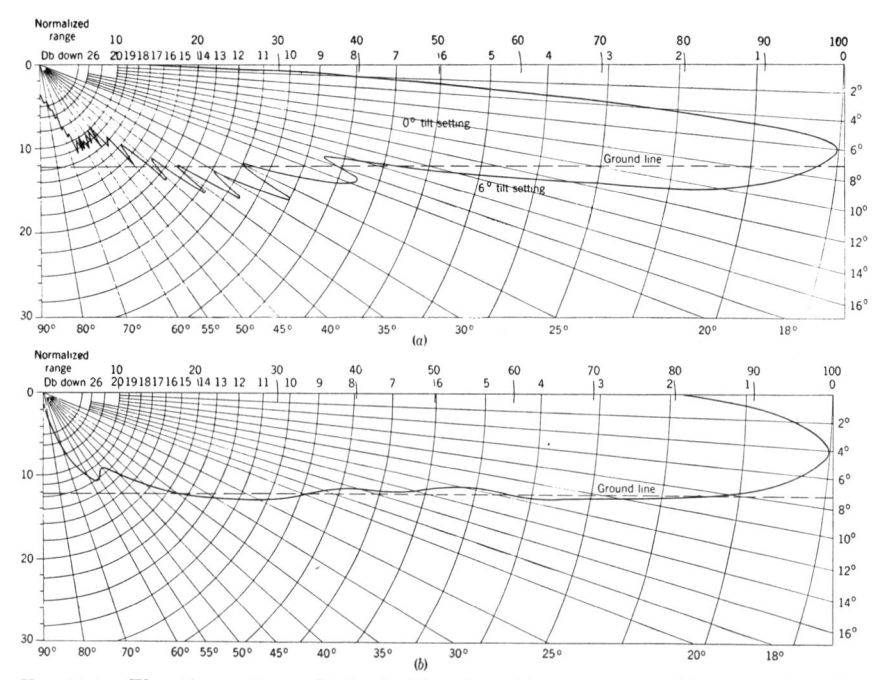

Fig. 14·4.—Elevation patterns obtained with a shaped-beam antenna: (a) mounted on the underside of a fuselage; (b) mounted in free space.

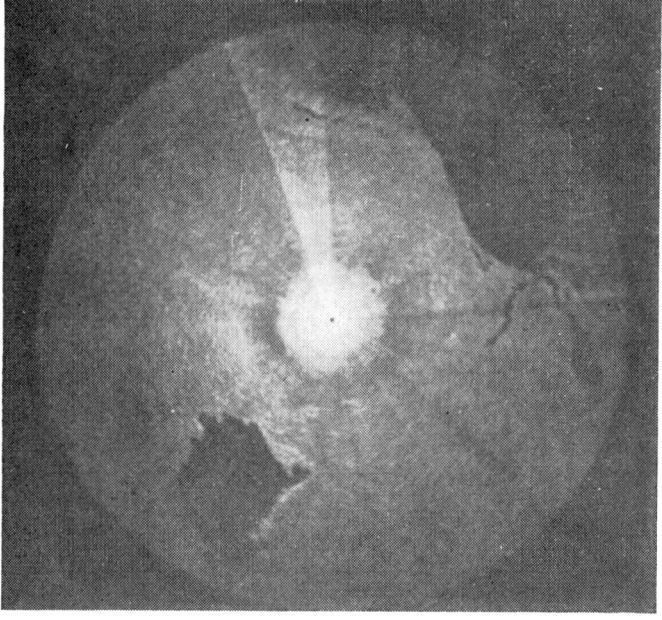

Fig. 14·5.—PPI photograph showing interference fringe due to edge reflection.

SEC. 14·4]   SCANNING ANTENNAS ON AIRCRAFT   519

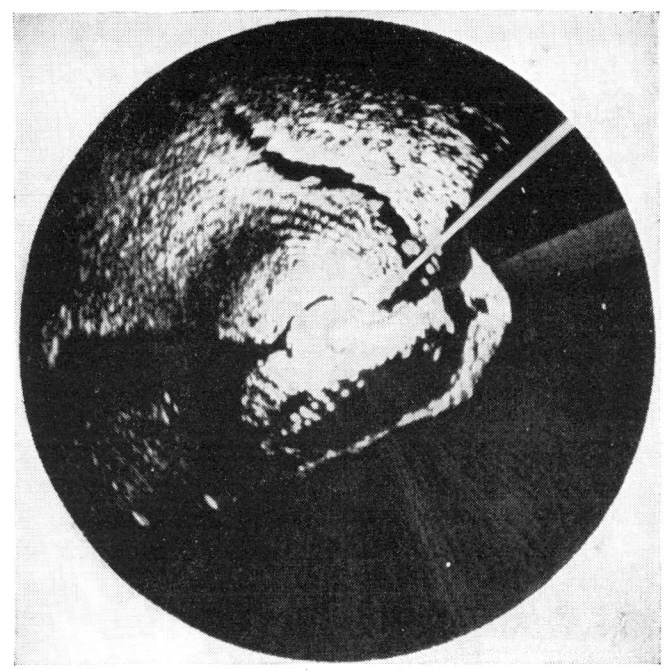

FIG. 14·6.—PPI photograph showing interference due to reflections from fuselage surface.

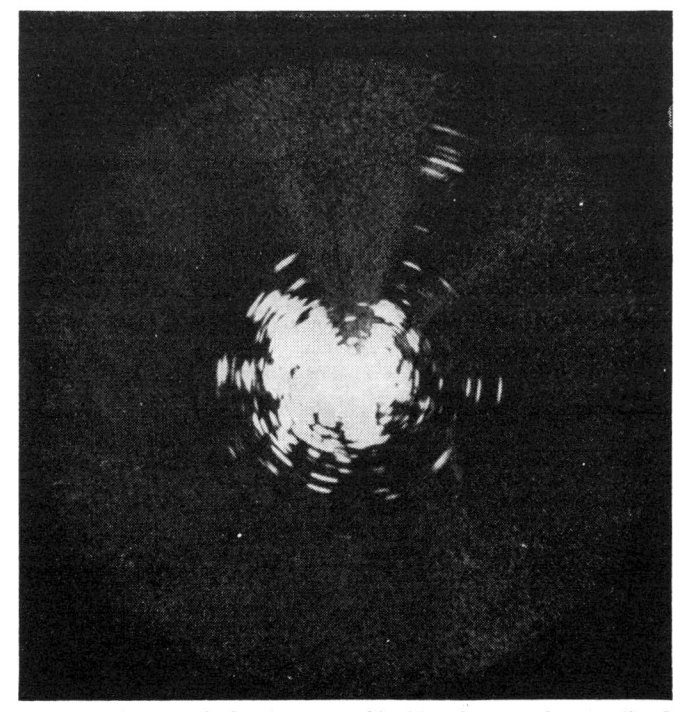

FIG. 14·7.—PPI photograph showing sector blanking due to radome wall reflections.

Other effects than transmitter pulling can result from radome wall reflections. Figure 14·8 presents four photographs of a B-scope (range vs. azimuth angle) taken with 10-cm radiation and a 29-in. paraboloid antenna located in the streamlined dielectric nose of a two-engine air-

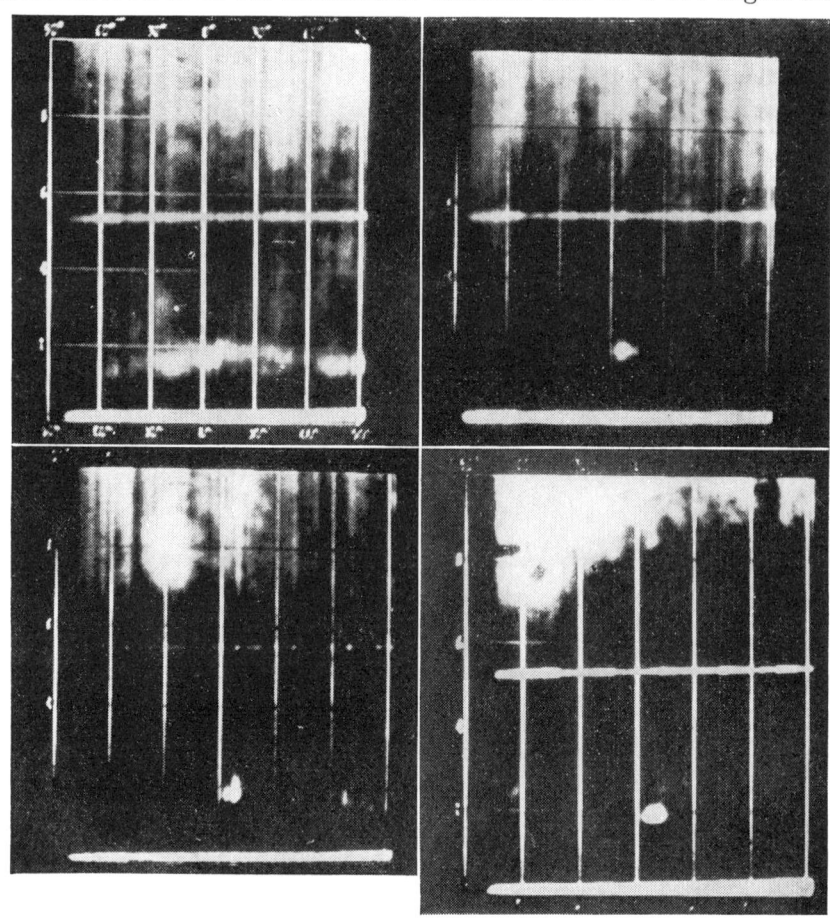

Fig. 14·8.—B-scope photographs illustrating various degrees of obscuration of target signal by altitude signal: (a) thick radome and horizontal polarization; (b) thick radome and vertical polarization; (c) thin radome and horizontal polarization; (d) thin radome and vertical polarization.

craft. In each case the altitude of the aircraft is 3000 ft, the range sweep is 15,000 ft, the electronic range marker is at 10,000 ft, and the target aircraft is at 3000 ft so that its indication coincides with the altitude signal. Trouble experienced in following the target aircraft through the altitude signal was traced to reflections from the upper half of the radome. These reflections directed a small fraction of the radiation downward toward the ground which then returned a signal by the same path. Two methods of reducing this ground signal were proposed: the use of vertical polarization and the use of a radome with a thinner wall. Figure 14·8a shows the partial obscuration of the target signal by the altitude signal even with optimum gain setting, when the thick radome and horizontal polarization are used. The reduction in altitude signal accomplished by changing to vertical polarization is shown in Fig. 14·8b. The improvement resulting from the use of a radome with the wall thickness reduced by one-third is shown in Fig. 14·8c. The improvement realized when both vertical polarization and the thin radome are employed is evident in Fig. 14·8d, since it has been possible to bring the target signal out quite strongly without bringing out the altitude signal. Effects of the radome on system performance will be discussed more fully later in the chapter.

14·5. Beacon Antennas on Aircraft.—The majority of long-wave antennas on aircraft are required to have omnidirectional patterns. Because of the strong and unavoidable influence of the aircraft on the antenna pattern at these wavelengths, the omnidirectional requirement is usually not well satisfied. In the case of microwave beacon antennas, which are also required to have omnidirectional patterns, the influence of the aircraft on the pattern is still strong but is more predictable and also more nearly avoidable. There are a limited number of sites on an aircraft suitable for a microwave beacon antenna; the top of the vertical stabilizer, the highest point on the upper side of the fuselage, and the lowest point on the underside of the fuselage are three favored locations. In selecting such a site the influence of the supporting structure on the pattern must be considered, not only with the aircraft in level flight but also under conditions of roll and pitch. The ideal elevation pattern would have uniform intensity for 10° about the horizontal plane to allow for roll and pitch of the aircraft and would have an approximately cosecant-squared decrease in intensity on the lower side of the beam to provide uniform illumination of the ground in to steep angles. The lowest point on the underside of the fuselage is the only location from which an unobstructed view of the ground at steep angles can be obtained. Even with this location the view in near-horizontal directions may be obstructed by roll of the aircraft if the bottom of the fuselage is flat.

Let us consider the elevation patterns that are obtained with a beacon antenna mounted on the underside of a fuselage. The various factors that must be considered are the polarization of the radiation, the vertical directivity of the antenna, the distance between the radiating elements and the fuselage, and the extent and curvature of the fuselage. A radiating element located below the fuselage will send some radiation directly toward the ground and some indirectly toward the ground by reflection in the fuselage. The over-all effect can be simply described by postulating an image of the radiating element in the fuselage. Because of the nature of the reflection process, the image of a vertically polarized element will be in phase, while the image of a horizontally polarized element will be 180° out of phase with the radiating element. If for the moment we consider the underside of the fuselage to be an infinite horizontal plane, it is evident that the over-all elevation pattern from a radiating element and its image will have a maximum in the horizontal plane in the case of vertical polarization and a minimum in the horizontal plane in the case of horizontal polarization.

Let us now assume a requirement of continuous elevation coverage for an angular range from 2° to 30° below horizontal. For vertical polarization the coverage requirement in the horizontal plane is automatically met except for the effects of roll and pitch. The steep-angle requirement can be met by limiting the combined length of the array and its image in accordance with the relation $d = \lambda/\sin \theta$ where θ is chosen somewhat greater than 30°. For horizontal polarization it is more difficult to obtain coverage in directions near the horizontal. Under the assumption of an infinite plane fuselage it would be necessary to make the linear array long and to accept an interference pattern in the region to be covered. Since the fuselage is actually curved and of limited extent, it is possible to obtain the required coverage by the use of a relatively directive array either located near the fuselage or set off at some distance from it, depending upon the shape of the fuselage. It is evident that this amounts to solving the electrical problem by creating an aerodynamic problem.

Microwave beacon antennas have regularly been enclosed in radomes. Many of the smaller arrays originally were provided with cylindrical radomes, but later it was recognized that no antenna protrusion is small enough to justify the omission of streamlining. In the case of vertical polarization the streamlined radome can have serious effects on the azimuth pattern. This situation favors large reflections, since it involves wide angles of incidence for radiation polarized perpendicular to the plane of incidence. A solid radome of low-density material affects the azimuth pattern also because of the focusing effect due to path-length variation in the material with azimuth direction. Vertical wires properly spaced in the dielectric offer a possible means to maintain unity dielectric constant.

RADOME DESIGN PROBLEMS AND PROCEDURES

The relation of the dielectric housing, or radome, to the general installation problem has been considered in the previous sections. The principal purpose of this part of the chapter is to analyze the problems and to describe the procedures associated with radome design.[1,2] It

---

[1] This subject has been treated more fully in *Radar Scanners and Radomes*, Vol. 26, Radiation Laboratory Series.

[2] This material is a severe condensation of the subject matter in a series of RL

will be necessary to consider several wall designs and two general radome types in terms of the mechanical and aerodynamic, as well as the r-f requirements that are involved. These considerations will be assisted by a preliminary examination of the nature and magnitude of the effects that radomes can have on system performance. An appreciation of these effects establishes the need for further investigation and indicates the difficulties that must be avoided in order to achieve good radome design.

14·6. Relation of the Radome to System Performance.—In Sec. 14·4 radar system performance defects caused by faulty radome design were presented from the operator's viewpoint. In the present section various reactions of the radome on the radar system will be presented from the radome designer's viewpoint.

---

reports on radome subjects. These reports are listed here to avoid repeated multiple references:

Y. Dowker, "Dielectric Constant and Loss Tangent Computation," No. 483-19, Aug. 7, 1945.
———, "Transmission of Lossy Sandwiches," No. 483-22, Jan. 23, 1946.
Elizabeth Everhart, "Recent Dielectric Constant and Loss Tangent Measurements." No. 483-5, July 14, 1944.
———, "Sandwich Walls at Variable Angles of Incidence," No. 483-8, Dec. 19, 1944.
———"Radome Wall Reflections at Variable Angles of Incidence," No. 483-20, Jan. 4. 1946.
M. Hegarty, Y. Dowker, R. M. Redheffer, E. D. Winkler, "Current Progress on R-f Research," No. 483-17, May 10, 1945.
E. B. McMillan, "Outline of Radome Development Problems," No. 483-1, Dec. 2, 1943.
H. A. Perry, "Electrical Test Methods for Radomes," No. 483-26, Jan. 11, 1946.
R. M. Redheffer, "An Outline of the Electrical Properties of Radomes," No. 483-2, Dec. 20, 1943.
———, "Transmission and Reflection of Single Plane Sheets, No. 483-4, July 12, 1944.
———, "Radomes and System Performance," No. 483-6, Nov. 17, 1944.
———, "Transmission and Reflection of Parallel Plane Sheets," No. 483-12, Jan. 26, 1945.
———, "Electrical Properties of Double Wall and Sandwich Radomes," No. 483-11, Feb. 1, 1945.
———, "Elliptical Polarization Produced by Streamlined Radomes," No. 483-13, Feb. 12, 1945.
——— and E. D. Winkler, "The Measurement of Dielectric Constants in the One-centimeter Band," No. 483-15, May 11, 1945.
———, "The Interaction of Microwave Antennas with Dielectric Sheets," No. 483-18, Mar. 1, 1946.
E. R. Steele, "Some Electrical Aspects of Microwave Sandwich Radome Design," No. 483-16, May 9, 1945.
T. J. Suen, Elizabeth Everhart, "Dielectric Constants and Loss Tangents of Radome Materials," No. 483-25, Jan. 11, 1946.
J. S. White, "Ice Formation on Shipborne Radomes," No. 483-3, Feb. 15, 1944.

Range.—A radome does not transmit all the r-f energy incident upon it but reflects and absorbs certain fractions of that energy. If sufficient reflected r-f energy finds its way back down the feed line to the transmitter, frequency pulling of the transmitter can result in total blanking of the receiver or in the very severe range reduction illustrated in Fig. 14·7. Independent of this effect, an appreciable reduction in range can result from attenuation of the transmitted signal, especially since attenuation in the outgoing signal is repeated on the return signal. Radar range is proportional to the fourth root of the transmitter power [*cf.* Eq. (1·17)], so that range reduction is related to radome transmission by Eq. (1):

$$\text{Range reduction } (\%) = \frac{\text{one-way transmission loss } (\%)}{2}. \qquad (1)$$

To pick large but not unusual values, if one-way attenuation due to radome wall reflection is 12 per cent and due to absorption is 8 per cent, then the range will be reduced by 10 per cent.

Pattern.—Reflection and absorption at the radome wall can reduce the absolute value of the radiation pattern by the processes described above. Due to a variation in effectiveness of these processes with direction and to related mechanisms, the shape of the pattern as well as its magnitude can be altered. Several mechanisms that have been guilty of pattern distortion will be described because of the seriousness of this effect in certain cases. Pattern distortion is especially objectionable in the case of a pencil beam that is being used in a conical scan for precision direction finding and in the case of a shaped beam that is being used for uniform illumination of the ground.

Pattern distortions have been traced to a variety of causes. In some cases the effects were barely appreciable, but in most cases they were serious enough to require study and elimination. Some of the causative mechanisms are listed below with brief descriptions.

1. In a streamlined radome the reduction in transmission due to absorption and reflection varies considerably with azimuth angle and can therefore produce a minor change in the shape of the main lobe.
2. An antenna transmitting through a spherical portion of a radome can experience trouble owing to focusing of reflections from the radome wall. The focal point of these reflections may lie near the focal point of the antenna reflector and act as a secondary source. The beam produced by the focused rays from this secondary source can combine in various ways with the original beam to produce distorted patterns. This process is illustrated in Fig. 14·9a.

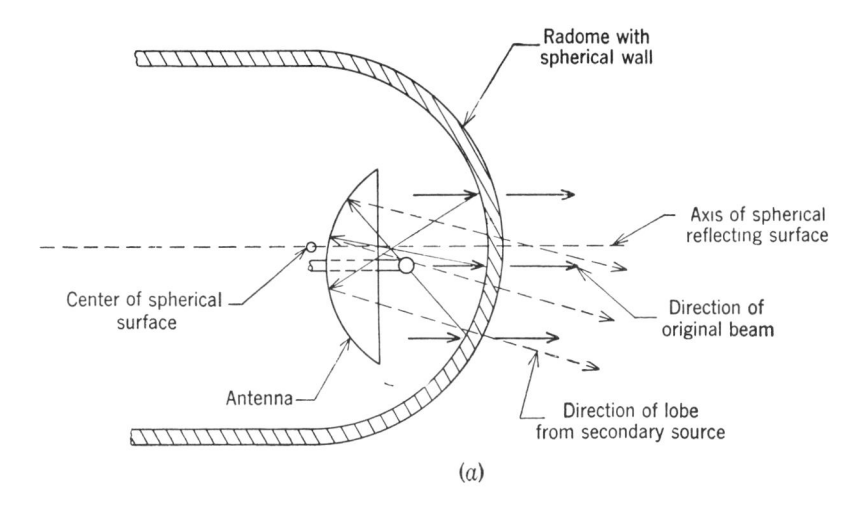

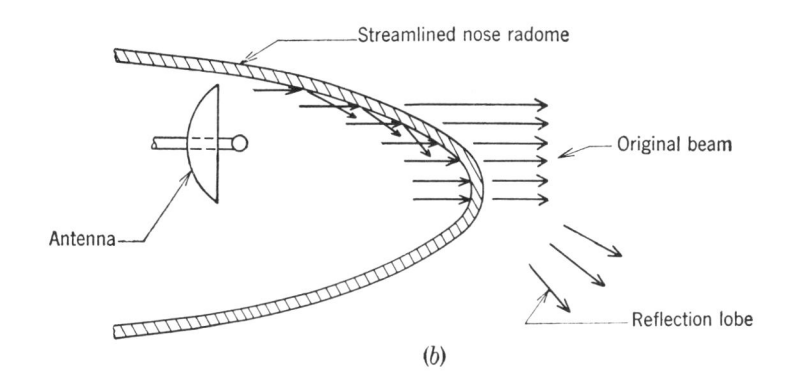

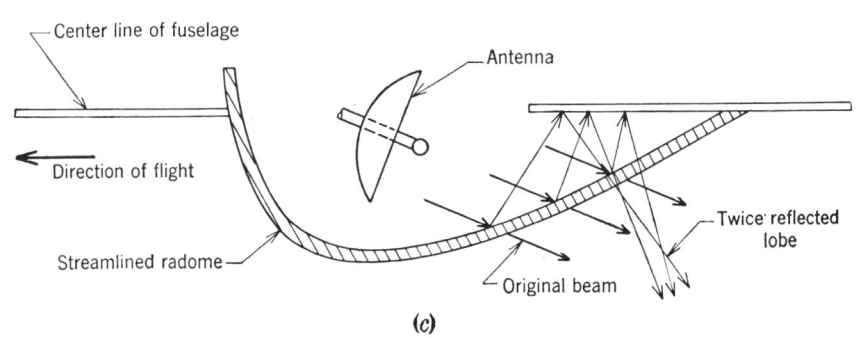

SEC. 14·6] *RELATION OF THE RADOME TO SYSTEM PERFORMANCE* 525

FIG. 14·9.—Mechanisms by which radome wall reflections can distort patterns: (*a*) secondary source produced by focused reflections; (*b*) wide-angle lobe produced by reflection from the upper portion of a radome; (*c*) wide-angle lobe produced by a double reflection involving the undersurface of the fuselage.

3. With the geometry illustrated in Fig. 14·9b, excessive reflections from the upper portion of the radome can produce a broad downward lobe at a wide angle from the original beam. This lobe can produce a strong ground return capable of obscuring aircraft targets at the same range, as illustrated in Fig. 14·8. In case the antenna beam is shaped for surface search, a spurious downward lobe will produce a very objectionable interference ripple in the elevation pattern when combined with the original beam.

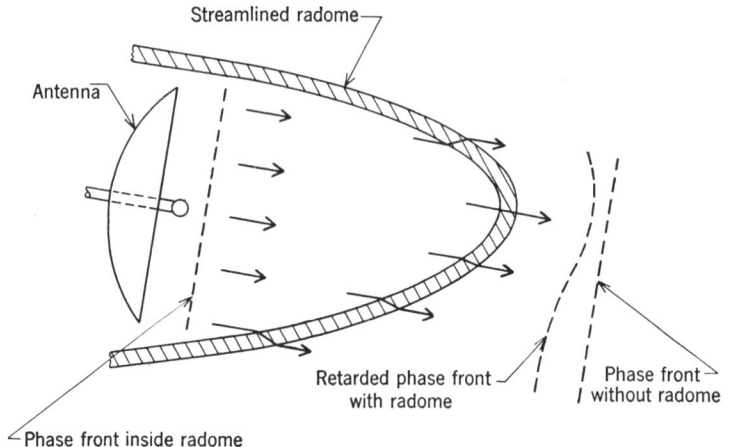

Fig. 14·10.—Distortion of the phase front of the radiation from an antenna by variable phase delays in sharply curved portions of the radome.

4. A streamlined radome mounted on the undersurface of an aircraft can produce a downward lobe by the process illustrated in Fig. 14·9c. This lobe will have the objectionable features described in the preceding paragraph.
5. Phase delays in the transmitted radiation caused by the dielectric of the radome wall vary appreciably over sharply curved portions of the wall. When the antenna aperture spans such a portion of the radome, the constant phase fronts in the near zone of the antenna are distorted upon transmission through the wall. The result upon the pattern is a distortion of the main lobe. This process is illustrated in Fig. 14·10.

Apparent Pattern.—The discussion of the previous paragraphs referred to the radiation pattern which can be defined as the variation of radiation intensity or receiving sensitivity with angle for a fixed position of the antenna. In distinction to the radiation pattern, the apparent pattern from the radar observer's viewpoint can be defined in terms of the return signal from an effective point-source target as presented on his radar scope while the antenna is scanning. In particular, the apparent pattern can be defined as the extent along an angular coordinate in the radar scope of the return from a point target. Because of the limited range in spot intensity on the scope, the description of an apparent pattern consists simply of a statement of the angular region on the scope coordinate that is occupied by the target signal.

The apparent pattern depends upon the beamwidth of the radiation pattern, but it depends on other factors as well. An important modification of apparent pattern is introduced by a transmitter frequency pulling which varies with the scan angle of the antenna. This variation in pulling arises from a variation in the phase or magnitude of the reflection back in the r-f line toward the transmitter.

The modification in apparent pattern most pertinent to the subject matter of this section is that which arises as the result of an angular variation in the reflection from the radome wall. This pulling effect can be serious even in the case of a radome that has not appreciably attenuated or distorted the transmitted beam. An antenna mounted slightly off axis in a cylindrical radome and scanning in azimuth will be subject to a reflection of constant magnitude but slowly changing phase. A streamlined radome can introduce large variations of magnitude as well as more abrupt changes in phase of the reflected wave. The most abrupt changes in both magnitude and phase are encountered in the case of a relatively thin walled radome with internal reenforcing ribs.

Serious operational effects can be traced to a variable frequency pulling by radome reflections. The resulting change in the apparent pattern, even though not accompanied by a noticeable reduction in range, can affect the apparent direction of the target in sufficient amount to ruin the performance of a radar employed in precision direction finding. This effect is especially serious, since it is not revealed by careful measurements of the radiation pattern.

Automatic Frequency Control.—No discussion of radomes and system performance is complete without mention of automatic frequency control (AFC), since its use profoundly modifies radome design requirements. AFC causes the local oscillator frequency to follow the transmitter frequency in such a way as to maintain between them a constant difference, the i-f frequency. AFC is strong enough in the case of a relatively stable magnetron to follow larger frequency changes than those ordinarily produced by radomes. However, the AFC circuit involves a time constant that may be considerably longer than the interval required for the change in the radome reflection. An extreme example of a rapid change in radome reflection is that which occurs as the antenna sweeps by an internal supporting rib in the radome wall. The AFC will not follow rapidly enough to correct for the changing reflection and, in addition, will not return to the original frequency setting until long after the radome reflection has returned to its original value. In this case the frequency shift has been extended in time rather than reduced by AFC. This is not to be interpreted as an argument against the use of AFC but rather against the use of localized supporting ribs in radome design. In general, AFC changes the radome design problem from one of reducing reflection to one of reducing rate of change of reflection.

Classification of Radomes.—The system performance effects described in this section are not all present simultaneously for a given radome, but some are emphasized and some minimized according to the type of radome considered. For this reason it is convenient to classify radomes under several headings and to follow a different design procedure in each case. A natural classification on the basis of use provides three general groups: pressurizing seals for antenna feeds, beacon housings, and housings for scanning antennas. The first group, pressurizing seals, will not be discussed here. For the other two it is convenient to make a further subdivision into cylindrical radomes and streamlined radomes, a distinction that is of considerable importance in practical work. Not only are the equations for radome wall design different in the two cases, but the underlying objectives are altered. The objective in the case of cylindrical radomes (normal incidence) is to reduce the reflection, but in the case of streamlined radomes it is primarily to improve transmission. After presenting the quantitative considerations of the next section, it will be possible to state this distinction more precisely.

14·7. Radome Wall Design.—An ideal radome wall would completely transmit an incident electromagnetic wave with neither reflection nor absorption, and for such a radome wall the deleterious effects associated with one or the other of these causes would be eliminated. Although this ideal situation cannot be attained in practice, the transmission can be maximized or the reflection minimized, depending upon the type of radome under consideration.

The calculation of transmission and reflection by the radome wall is most conveniently carried out in terms of transmission and reflection coefficients. Whereas such coefficients cannot be defined suitably for a curved surface, a curved surface can be replaced by a plane sheet for approximate analysis if the radius of curvature is large. The complex field distribution of the antenna that provides the incident wave is also replaced by a plane wave of uniform amplitude. The investigation is thus based on the coefficients that exist and are easily defined for the simpler situation of plane sheets and uniform plane waves.

Physical Principles.—Reflection of an electromagnetic wave occurs only at a discontinuity, that is, at the transition from one medium to another. Every radome wall without exception may be regarded as a set of pairs of such discontinuities. The over-all reflection will result from superposition of the individual reflections; its magnitude will be determined by their magnitudes and relative phases. Reflections can be reduced or transmission increased either by reducing each of the individual reflections or by adjusting the spacing between the discontinuities to obtain partial or complete phase cancellation.

In Fig. 14·11 a generalized radome wall is represented as a pair of discontinuities, or sheets. Individual transmission and reflection coefficients for sheets 1 and 2 and over-all transmission and reflection coefficients for the wall are defined in the figure. In terms of these coefficients, the advantage to be gained from minimizing the individual reflections and adjusting their phases for cancellation can be expressed by the relations

$$|R|_{\max} = |r_1| + |r_2|, \quad (2)$$

and

$$|R|_{\min} = |r_1| - |r_2|; \quad (3)$$

these neglect higher-order interaction terms and absorption.

A precise phase cancellation of the component reflections requires an accurate spacing between their sources. Since deviations from this optimum spacing must be allowed in manufacture, the effect of such deviations on transmission and reflection must be investigated. The investigation can be carried out by dividing the wall configuration into two groups and by expressing the over-all transmission and reflection in terms of the individual transmission and reflection coefficients of the two groups. In the case of a single, uniform sheet, the two groups would be the two air-dielectric interfaces of the sheet. In the case of a double-wall configuration each group would represent the total effect of one of the walls. The method can be extended to more complicated configurations. In any case the nomenclature of Fig. 14·11 applies.

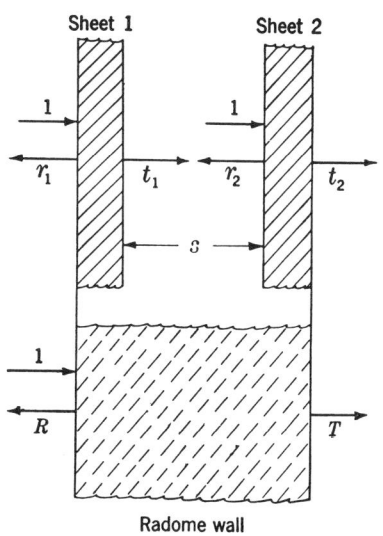

FIG. 14·11.—Generalized radome wall showing division into two groups or sheets which are treated as separate sources of reflection. The figure serves to define symbols used in the text for reflection and transmission coefficients.

As the geometrical spacing s between the sheets of Fig. 14·11 is varied, the over-all transmission coefficient T will reach the maximum value

$$|T|^2_{\max} = \frac{|t_1 t_2|^2}{(1 - |r_1 r_2|)^2} \quad (4)$$

for the optimum spacing s_0 given by

$$s_0 = \frac{n\lambda}{2} - \frac{\lambda}{4\pi}(\phi_1 + \phi_2); \quad n = 0, 1, 2, \cdots, \quad (5)$$

where λ is the wavelength in the region between the sheets and ϕ_1 and ϕ_2 are the phases of r_1 and r_2 respectively. If $\delta = s_0 - s$ is the deviation from optimum spacing, the over-all transmission will be

$$|T|^2 = \frac{|t_1 t_2|^2}{1 - 2|r_1 r_2| \cos\frac{4\pi\delta}{\lambda} + |r_1 r_2|^2}. \qquad (6)$$

The corresponding over-all reflection, for the case of zero absorption, is

$$|R|^2 = \frac{|r_1|^2 - 2|r_1 r_2| \cos\frac{4\pi\delta}{\lambda} + |r_2|^2}{1 - 2|r_1 r_2| \cos\frac{4\pi\delta}{\lambda} + |r_1 r_2|^2}. \qquad (7)$$

From these equations nearly all the results needed for radome wall design may be obtained directly.

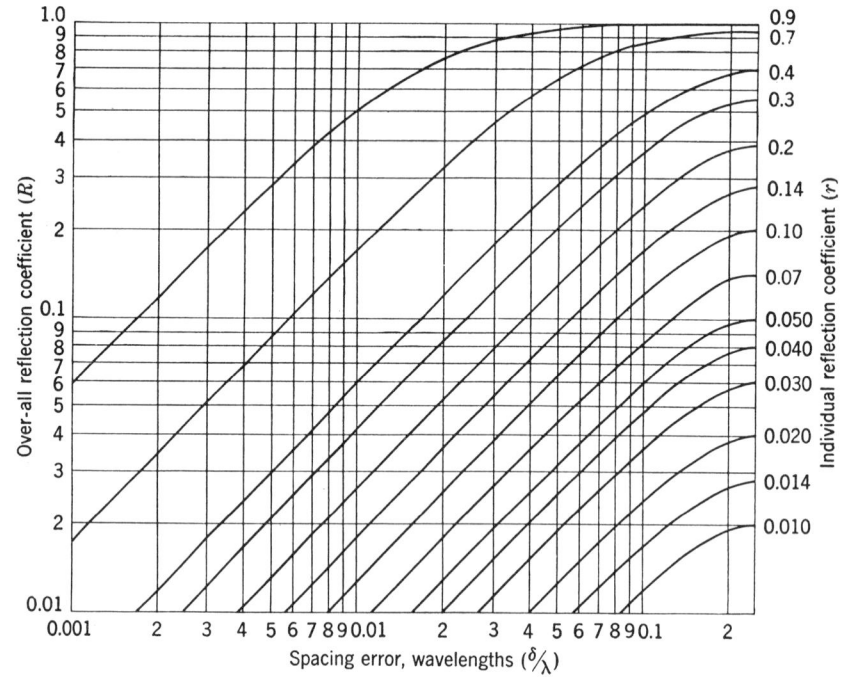

Fig. 14·12.—Over-all reflection coefficient (R) for a symmetrical configuration of zero absorption plotted as a function of the error in spacing (δ) and the individual reflection coefficient (r).

The symmetrical radome wall consisting of identical groups of approximately zero absorption spaced in accordance with Eq. (5) will be encountered almost invariably in practice. For the symmetrical configuration, $t_1 = t_2 = t$, $r_1 = r_2 = r$, $\phi_1 = \phi_2 = \phi$ and Eqs. (4), (5), (6), and (7)

Sec. 14·7]   RADOME WALL DESIGN   531

assume simplified forms which will not be repeated here. The expression for the over-all reflection can be simplified further to

$$|R| \cong \frac{\frac{4\pi |r| \delta}{\lambda}}{1 - |r|} \quad (8)$$

for small δ/λ, or alternatively to

$$|R| \cong 2|r| \sin \frac{2\pi \delta}{\lambda}, \quad (9)$$

when $|r|$ is small enough to be neglected. The exact value of $|R|$ from Eq. (7) is plotted in Fig. 14·12 for the symmetrical configuration.

Single Wall.—The simplest radome wall design is a single, uniform sheet of dielectric material. In this case the two sources of reflection treated above and illustrated in Fig. 14·11 are the two air-dielectric interfaces. If the specific inductive capacity k_e of the material is denoted by β, the over-all reflection is given by Eq. (7) with

$$|r_1| = |r_2| = \frac{\sqrt{\beta} - 1}{\sqrt{\beta} + 1}. \quad (10)$$

Equation (10) is plotted in Fig. 14·13. The optimum spacing is

$$s_0 = \frac{n\lambda}{2} = \frac{n\lambda_0}{2\sqrt{\beta}}; \quad n = 0, 1, 2, \cdots, \quad (11)$$

since $\phi_1 = \phi_2 = 0$; λ_0 is the free-space wavelength.

The two design procedures mentioned above—reduction of r or adjustment of phase—permit of simple interpretation for single-walled radomes. The first method indicates the use of a low-dielectric-constant material to reduce r [Eq. (10)]. The second method requires the use of a thin sheet or in general one that satisfies Eq. (11). If a thin sheet is used, the tolerance δ in the equations above stands for the thickness itself. The single-wall construction becomes increasingly attractive at longer wavelengths where a small value of δ/λ can be realized with a wall thickness sufficient for mechanical strength.

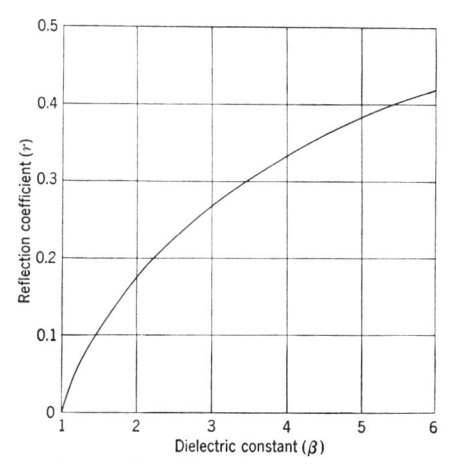

Fig. 14·13.—The reflection coefficient (r) of an air-dielectric interface as a function of the specific inductive capacity (β)

The assumption of zero absorption loss, even in the case of actual radome materials, does not invalidate the results for reflection and optimum spacing, but the results for transmission must be modified appropriately. The general equation for a single sheet is quite complicated except when the thickness satisfies Eq. (11). For a half-wavelength thickness of relatively low-loss dielectric

$$|T|^2 \cong 1 - \frac{\pi}{2}\left(\sqrt{\beta} - \frac{1}{\sqrt{\beta}}\right)L; \qquad (12)$$

L is the loss tangent ϵ''/ϵ' for the material where ϵ' and ϵ'' are the real and imaginary components of the complex dielectric constant (see Sec.

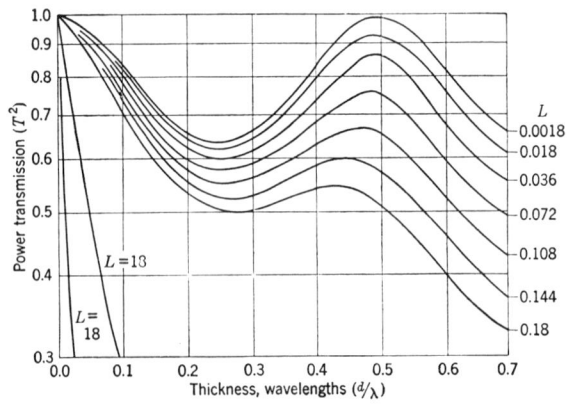

FIG. 14·14.—Power transmission $|T^2|$ as a function of thickness for various values of loss tangent $L = \epsilon''/\epsilon'$.

3·2). The exact values of the power transmission $|T|^2$ as a function of thickness for various values of L is given in Fig. 14·14 for the particular specific inductive capacity $\beta = 4$.

The single-wall design is usually not practicable at microwave frequencies. The half-wavelength wall is frequently too heavy for airborne installations, whereas the wall thin enough to have good electrical properties is not satisfactory mechanically.

Sandwich Wall.—A major improvement both mechanically and electrically is realized by substituting sandwich-wall for single-wall design. The sandwich-wall design consists of a sheet of low-density core material faced on both sides with thin, high-density skins. Sandwich construction has seen considerable use in purely mechanical installations where the strength-to-weight ratio must be high. From the electrical point of view, the skins show low reflection because they are thin, and the core, because it has a low dielectric constant. Furthermore the core provides a means of accurately holding the skin separation to a value favorable for canceling reflections.

A double-wall construction which is obtained by omitting the core material from between the skins loses most of the advantages of the sandwich construction. It has been used, however, since it provides means for deicing by circulating hot air between the skins of the radome wall. The optimum spacing of the skins for reducing reflection when separated by air is given in terms of their thickness d and specific inductive capacity α by the approximate relation

$$s_0 \cong (2n + 1) \frac{\lambda_0}{4} - d \sqrt{\alpha}. \quad (13)$$

The over-all reflection is nearly proportional to the spacing error and to the reflection from a single sheet when these are small and is given by Eqs. (7), (8), and (9). The power transmission is given by Eq. (6) in the general case and by $1 - R^2$ when the loss is zero.

The general sandwich wall is shown in Fig. 14·15 which serves to define some of the symbols to be used in the discussion. The electrical thicknesses are $D = d \sqrt{\alpha}/\lambda_0$ and $S = s \sqrt{\beta}/\lambda_0$. The individual reflection coefficient is now that for the three-medium transition from the air through the skin to the core. It is given by

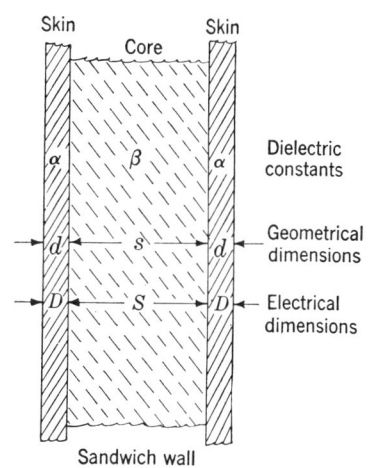

Fig. 14·15.—Sandwich-wall design. The figure serves to define symbols used in the text.

$$|r|^2 = \frac{A - 4\alpha \sqrt{\beta}}{A + 4\alpha \sqrt{\beta}}, \quad (14)$$

where

$$A = (\alpha + \beta)(\alpha + 1) - (\alpha - \beta)(\alpha - 1) \cos (4\pi D). \quad (14a)$$

The value of r obtained from Eq. (14) can be used in Fig. 14·12 together with the appropriate value of the spacing tolerance to determine the over-all reflection coefficient. The optimum electrical spacing is similarly found to be

$$S_0 = \frac{n}{2} - \frac{1}{2\pi} \tan^{-1} \frac{2 \sqrt{\alpha\beta} \, (\alpha - 1) \sin (4\pi D)}{(\beta - \alpha)(\alpha + 1) + (\alpha + \beta)(\alpha - 1) \cos (4\pi D)}. \quad (15)$$

Under certain conditions this equation can be approximated by

$$S_0 \cong \frac{n}{2} - 2D \frac{(\alpha - 1)}{(\beta - 1)} \left(\frac{\beta}{\alpha}\right)^{1/2}. \quad (16)$$

The range of validity of Eq. (16) is indicated in Fig. 14·16 which is a typical family of curves calculated from Eq. (15) for a fixed value of α and a range of values of β.

In certain methods of sandwich-wall fabrication the skins are glued rather than molded to the core. The layer of glue introduced by this process effectively increases the skin thickness and introduces appreciable error into the calculations. This concept of an *effective skin thickness* which includes the effect of the glue is an approximation that ceases to be valid when the angle of incidence is variable or when both tolerance

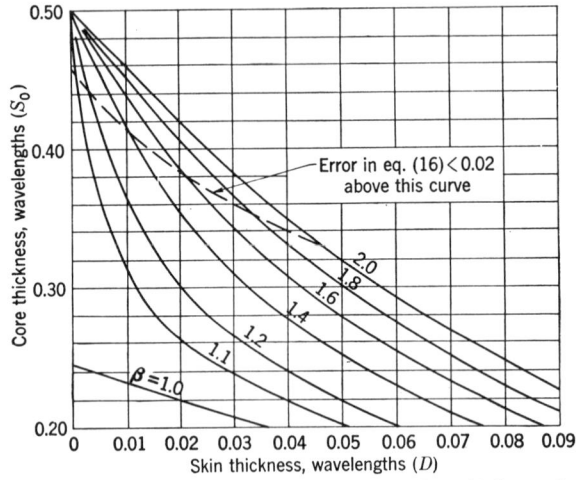

FIG. 14·16.—Optimum core thickness as a function of skin thickness for $\alpha = 4$ and β ranging from 1.0 to 2.0.

and spacing are involved. It has proved very useful and is entirely justified when one is concerned with optimum spacing alone at a fixed angle of incidence.

Arbitrary Incidence.—The results derived hitherto apply to normal incidence only. They may be used for arbitrary incidence at either polarization, however, if the quantities D, S, S_0, α, and β are replaced by appropriate quantities as given in Table 14·1. Similar equivalence relations for arbitrary loss will not be given here. Limiting values for

TABLE 14·1.—EQUIVALENCE RELATIONS FOR CONVERTING FROM NORMAL TO ARBITRARY INCIDENCE VALUES

$p = \sin^2\theta_0$; θ_0 = angle of incidence measured from normal

Normal incidence quantities	D	S	S_0	α	β
Arbitrary incidence values, perpendicular polarization	$\dfrac{d}{\lambda}\sqrt{\alpha - p}$	$\dfrac{s}{\lambda}\sqrt{\beta - p}$	$\dfrac{s_0}{\lambda}\sqrt{\beta - p}$	$\dfrac{\alpha - p}{1 - p}$	$\dfrac{\beta - p}{1 - p}$
Arbitrary incidence values, parallel polarization	$\dfrac{d}{\lambda}\sqrt{\alpha - p}$	$\dfrac{s}{\lambda}\sqrt{\beta - p}$	$\dfrac{s_0}{\lambda}\sqrt{\beta - p}$	$\dfrac{\alpha^2(1-p)}{\alpha - p}$	$\dfrac{\beta^2(1-p)}{\beta - p}$

grazing incidence, simplified forms at Brewster's angle, and approximate relations for thin skins, though useful for computation, illustrate no new principles and will not be included in the present discussion.

Considerations up to this point have been based upon the assumption of a single angle of incidence—whether normal or otherwise—and a single plane of polarization for the incident wave. It will be shown in Sec. 14·9 that a given portion of the wall of a streamlined radome will be required to transmit for a range of angles of incidence and for a range of polarizations. If all portions of the radome wall are considered, the ranges involved will be correspondingly increased. In the interest of simplicity in fabrication it is very desirable to use the same wall structure throughout the radome. The resulting problem, therefore, is to find a wall design that will be satisfactory for a *range* of angles of incidence and for *both* polarizations.

To this end the equations for arbitrary incidence are plotted as shown in Fig. 14·17 where contours of constant reflection for a sandwich wall are presented as a function of core thickness and angle of incidence. Similar charts can be used for single-wall construction. With such diagrams the optimum thickness is readily determined for a specified range of θ and for a specified tolerance in thickness; the thickness referred to here is the total wall thickness of the single-wall design or the core thickness of the sandwich design.

The representation for the single wall requires only three variables; the angle of incidence, the dielectric constant, and the thickness in wavelengths. A single series of charts for various dielectric constants therefore provides complete information for single-wall design work. In the case of the sandwich wall, five variables are required: skin dielectric constant and thickness in wavelengths, core dielectric constant and thickness in wavelengths, and angle of incidence. A very large number of charts are required; in practice percentage reflection contours are plotted against core thickness and angle of incidence, each chart being for specific values of skin thickness, skin ϵ', and core ϵ'.

Mechanical Requirements.—Besides satisfying the electrical requirements that form the main subject of the present chapter, a radome must satisfy certain mechanical requirements as well. An airborne radome must withstand the distributed load produced by windage; it must not deform when a concentrated load is applied as is necessary in installation procedures; and it must often stand the impact of rocks and water in landings. To these mechanical requirements must be added certain requirements in physical properties; e.g., the radome should not be soluble in gasoline or in any other solvent likely to be brought in contact with it; it should withstand high temperatures without softening, low temperatures without becoming brittle; it must not absorb moisture to any appreciable extent; and in certain cases it must be provided with

536   ANTENNA INSTALLATION PROBLEMS   [SEC. 14·7

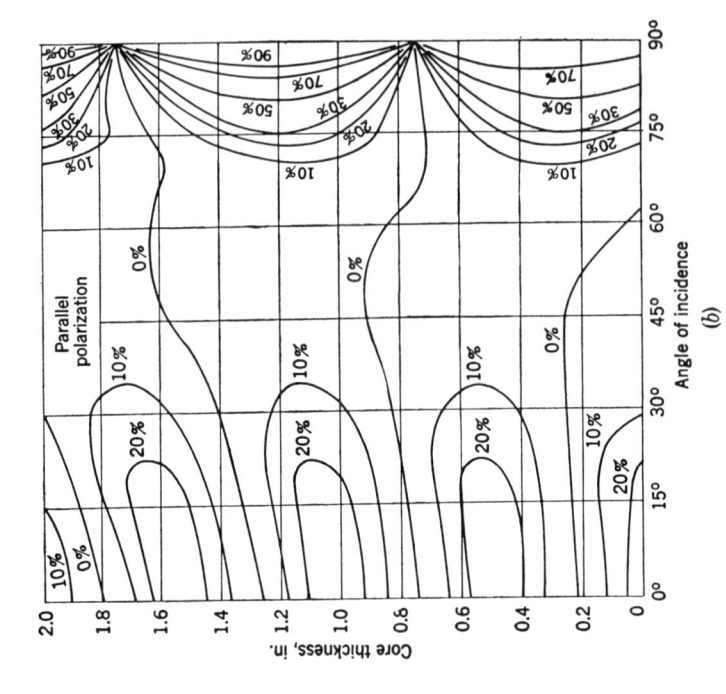

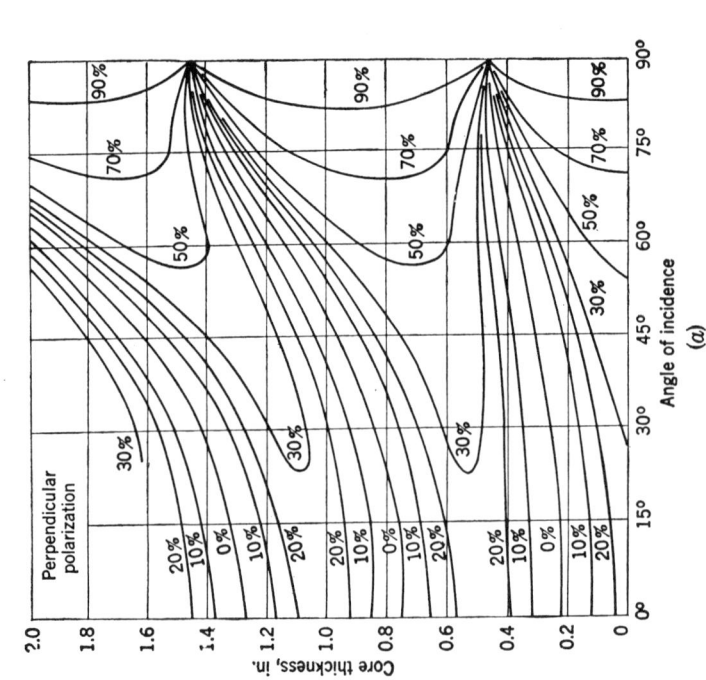

FIG. 14·17.—Typical charts showing contours of constant reflection as a function of core thickness and angle of incidence for the following conditions: $\lambda_0 = 3.2$ cm; core $\epsilon'/\epsilon_0 = 1.4$; skin $\epsilon'/\epsilon_0 = 3.7$; skin thickness = 0.050 in. (a) Polarization perpendicular to the plane of incidence; (b) parallel polarization.

means for removing ice deposits as they are formed. Some of these requirements are clearly of the utmost importance; particularly in airborne systems, failure of the radome can have very serious consequences and should be avoided even at the expense of electrical performance. It is this simultaneous presence of mechanical and electrical requirements, which must be satisfied without excessive complication in manufacture, that constitutes the radome problem. These mechanical requirements are met by methods standard in the aircraft industry; the significant change introduced by electrical requirements is the restriction to suitable materials and to suitable relations, as determined above, for the linear dimensions. Without giving a detailed discussion of fabrication or mechanical design techniques, it is therefore sufficient here to observe their great importance.

14·8. Normal Incidence Radomes.—There have been earlier allusions in Secs. 14·4, 14·5, and 14·6 to the fact that the most serious problem with normal incidence (cylindrical) radomes is the resulting impedance mismatch at the transmitter, while the most serious problem with streamlined radomes is the resulting attenuation and distortion in the transmitted pattern. These problems and procedures for solving them will be considered further in this and the following section. First consideration will be given to the pattern and impedance effects encountered in normal incidence radomes.

Pattern Effects.—The effect of a cylindrical radome on antenna gain can be minimized in a straightforward manner by use of the normal incidence relations of Sec. 14·7. In case the radome wall is uniform, this procedure also minimizes the effect on pattern, since it leads to small reflection as well as to large transmission. There is another source for pattern distortion, however, even in a radome having complete transmission. A radome wall having a structural rib, overlap, or buttstrap could be designed for complete transmission through both the thin part and the thick part of the wall. Whereas this would maintain the amplitude of the transmitted wave unchanged, it would not compensate for the change in phase. A double thickness section of wall would introduce twice the phase delay and so distort the transmitted phase front. This phase distortion can become so serious that the presence of a thick dielectric rib may be more harmful than a metal rib of the same dimensions. A similar effect introduced by sharply curved surfaces is illustrated in Fig. 14·10. If discontinuities are avoided and the reflection is minimized, then a cylindrical radome will not have an appreciable effect upon the antenna pattern.

Impedance Effects.—The problem of antenna mismatch and transmitter pulling is much more serious for normal incidence radomes. For quantitative consideration of the case of a slightly tilted antenna in a cylindrical radome, the amplitude of the reflection back into the line may be written

$$\rho_L = \frac{7R\sqrt{\lambda_0 p}}{19a} e^{-(4\ln 2)(\phi/\Theta)^2}, \qquad (17)$$

where the symbols have the following definitions:
R = over-all reflection coefficient for the radome wall flattened into a plane sheet
λ_0 = free-space wavelength
p = radius of the cylindrical radome
a = radius of the antenna aperture
ϕ = angle between axis of the pencil beam and a line normal to the generator of the cylinder
Θ = half-power width of the antenna beam

Equation (17) is valid only for small values of ϕ, but this is obviously the condition for maximum reflection into the line and therefore the condition of greatest interest.

If the antenna has a sharp vertical beam fixed at zero elevation, ρ_L can be greatly reduced by the use of a truncated cone instead of a cylinder for the radome shape. In general it must be assumed that the scanning range will cause the maximum reflection to be attained; furthermore it must be assumed that the phase of the reflection will vary by at least a half wavelength. Under these conditions the frequency pulling of the transmitter will be

$$\Delta f = 5\rho_L(p.f.), \qquad (18)$$

where the symbols have the following definitions:
Δf = frequency pulling of the transmitter in megacycles
ρ_L = value from Eq. (17) with the exponential factor set equal to unity
$p.f.$ = the pulling figure[1] of the transmitter.

A modified investigation is required if the radar system is equipped with AFC. The problem is then to estimate the rate of change of frequency due to transmitter pulling. This involves the use of the Rieke diagram for the transmitter, results from Eq. (17), similar results for the phase of the reflection, and a time factor introduced by the scan rate.[2] These calculations become important in estimating the apparent shift in the direction of a pencil beam or in the crossover point of a lobe switching beam.

Design Considerations.—The considerations bearing on the design of cylindrical radomes will be summarized in several quasi-chronological steps. For this purpose it is assumed that the designer is equipped with

---

[1] The pulling figure of a magnetron has been defined as the maximum frequency shift in megacycles from the initial frequency, which can be induced by a (VSWR) = 1.5 of arbitrary phase.

[2] *Radar System Engineering*, Vol. 1, and *Radar Scanners and Radomes*, Vol. 26, Radiation Laboratory Series.

considerable information regarding the dielectric constants and loss tangents of materials known to be available commercially and to be suitable for fabrication purposes. It is assumed also that he is equipped with radome wall design charts showing the proper dimensional relationships for different wall designs, dielectric constants, losses, and incident angles and polarizations.

1. A background of pertinent information must be obtained from several sources: tactical application of the system; r-f requirements of the system including allowable maximum mismatch and rate of change of mismatch; information regarding the antenna, such as size, angle and rate of scan, impedance match, far-field pattern, near-field ray diagram; mechanical requirements of the actual installation including method of mounting, windage, shock loading, chemical exposure, weather conditions.
2. The most favorable wall design must be selected in the light of available materials and on the basis of electrical and mechanical requirements from among the several possibilities: thin wall of arbitrary dielectric constant, low-dielectric-constant material of arbitrary thickness, half-wave thick wall, double wall, sandwich construction.
3. The specific materials to be used in the selected wall design must be determined. This involves using the dielectric constants and loss tangents of the several alternative materials and the design charts for the radome wall to calculate the transmission and reflection that would result from the choice of each material. A comparison can then be made between these r-f performance figures and the r-f requirements and between similar mechanical performance figures and mechanical requirements to arrive at the final choice of materials.
4. A number of flat panels must be fabricated using the materials and dimensions selected. These dimensions and the method of fabrication may be varied somewhat to obtain optimum performance. The panels are tested for all properties deemed relevant in the radome under design, e.g., structural strength, moisture absorption, temperature resistance, transmission, and reflection.
5. A complete radome must be fabricated on the basis of results obtained in the panel tests. Final tests are made of those characteristics requiring the entire radome, e.g., strains due to continuous loading and effects on the antenna pattern.

In the sequence of operations outlined above the less expensive and time-consuming operations are carried out first, whereas the major undertaking of building the complete radome is delayed until all preliminary problems have been resolved.

14·9. Streamlined Radomes.—The cylindrical radomes discussed in the previous section are suitable for ground and ship installations and have been used in the past on slow aircraft. For modern airborne installations, however, it is necessary for the radome to be severely streamlined, preferably to the extent of being completely incorporated into the original airframe. These streamlined radomes present problems quite different from those discussed above; antenna gain and pattern require detailed investigation, whereas the problem of transmitter pulling assumes secondary importance. Both favorable and unfavorable aspects stem from the wide angles of incidence usually encountered. A nonscanning antenna, such as a linear-array beacon antenna, is free of transmitter pulling because of the fixed relationship of antenna to radome.

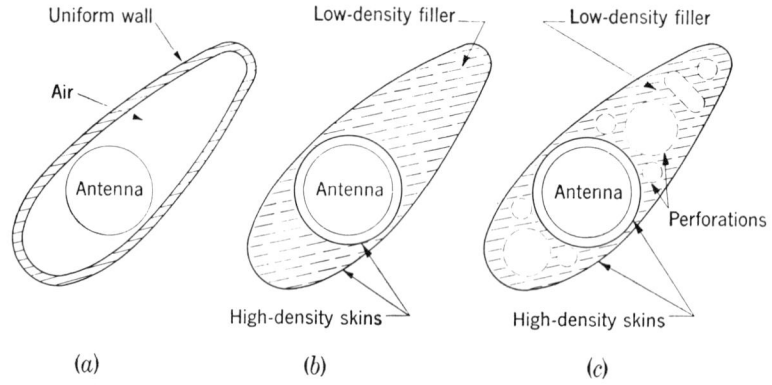

FIG. 14·18.—Beacon radome designs: (a) uniform wall; (b) solid wall; (c) perforated solid construction.

Pattern Effects.—The pattern effects caused by streamlined radomes for beacon antennas can best be discussed in terms of the several beacon radome designs illustrated in Fig. 14·18. In transmission through sharply curved portions of the uniform radome wall shown in Fig. 14·18a, the cylindrical phase front is appreciably distorted by variable phase delays. This mechanism is similar to that illustrated for a plane phase front in Fig. 14·10. The pattern is further distorted in the case of vertical polarization by large reflections from the tapered portion of the radome wall. The use of high-density material in the streamlined shape can be confined to a single thin skin if a low-density filler is used. This design (Fig. 14·18b) reduces the reflections below those resulting from the uniform wall but increases the variations in phase delay. A system of perforations (Fig. 14·18c) determined partly by experiment reduces these variations in phase delay to such a degree that the azimuth pattern is not seriously affected by the radome. Amplitude reduction in the transmitted wave is present in sufficient amount to cause appreciable distortion in the pattern but is not nearly so serious as the other effects just discussed.

The pattern effects to be expected from a proposed streamlined radome design for a scanning antenna can be predicted qualitatively by drawing elevation and plan views of the antenna with its ray diagram and located in the radome. This type of drawing is illustrated schematically in Fig. 14·19 of a streamlined belly radome and an antenna employ-

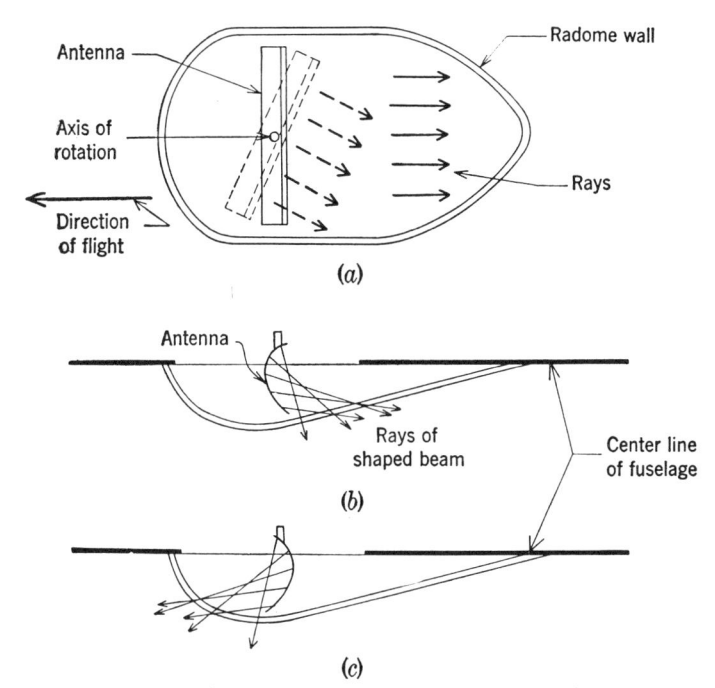

FIG. 14·19.—Schematic drawing of a shaped-beam antenna with its ray diagram and located in a streamlined radome: (a) plan view, antenna looking backward; (b) elevation, antenna looking backward; (c) elevation, antenna looking forward.

ing a shaped cylindrical reflector. The rays from the antenna are incident upon the radome wall under a wide variety of conditions, as can be seen by a study of the figure. The angle of incidence varies from 0° for steeply deflected rays to almost 90° for near-horizontal rays looking backward. One polarization with respect to the plane of incidence will prevail over the bottom of the radome, the opposite over the sides, with intermediate polarizations in intermediate directions. A specific portion of the radome may be required to transmit rays having a wide range of angles of incidence at each of two opposite polarizations; this can be seen by studying parts (b) and (c) of the figure with the antenna stationary as shown or by examining a portion of the side wall of the radome in part (a) of the figure with the antenna scanning. The procedures to be followed in radome wall design to meet situations of this type were discussed in Sec. 14·7 under Arbitrary Incidence.

Design Considerations.—It is possible now to outline in several steps the considerations involved in the procedure of designing a streamlined radome for a scanning antenna. In the early stages this procedure is similar to that described for normal incidence radomes in Sec. 14·8. The following description is based on the antenna-radome system shown in Fig. 14·19:

1. The major dimensions of the radome must be determined. These depend upon the size of the antenna, the size of the hole in the fuselage, the amount of retraction of the antenna. These last two items are interrelated, since a larger fuselage hole will permit more retraction in the case of a slightly depressed beam. Amount of roll and pitch of the airplane and presence or absence of antenna stabilization are also involved.
2. The shape of the radome must be established. The first approximation is based strictly upon aerodynamic considerations. This shape is then used together with the antenna shape and ray diagram to construct an accurately dimensioned figure of the type indicated schematically in Fig. 14·19. The radome shape is modified to keep angles of incidence below some maximum figure, say 70°, but subject to approval from aerodynamic considerations.
3. The radome wall design must be calculated. Further study of the antenna-radome diagram is necessary to fix the ranges of incidence angles at both polarizations. The wall must be designed to meet the system mismatch requirements by keeping down reflections at small angles of incidence and to meet pattern and gain requirements by keeping down reflection and absorption at large angles of incidence.

Further steps, as in the case of cylindrical radomes, are construction and testing of flat panels of the radome wall and construction and testing of the complete radome. Mechanical as well as electrical tests are made on the panels, while tests on the complete radome include aerodynamic performance and effect on antenna patterns.

CHAPTER 15

ANTENNA MEASUREMENTS—TECHNIQUES

By H. Krutter

15·1. Introduction.—The principles and techniques of antenna design were developed in the preceding chapters without consideration of the methods for obtaining design data and for testing the performance of the completed antenna. This and the following chapter will be devoted to a discussion of measurement techniques and a survey of the equipment required for such measurements.

The antenna characteristics to be measured fall into four groups: impedance, primary feed patterns, secondary patterns, and gain. The impedance measurement techniques differ little in detail from those for other r-f components of microwave systems; the problem is complicated to a small degree by the fact that the antenna is a radiating load. The importance of the primary feed pattern increases with the progress that is made in reducing antenna design from an empirical to a calculable procedure. The study of pencil beams and fanned beams has shown the need for a detailed knowledge of both the phase and intensity distribution in the primary feed pattern. The over-all characteristics of the antenna are particularly sensitive to the phase characteristics of the feed. It is evident from Chap. 13 that the design of shaped-beam antennas would be decidedly limited in scope without a complete knowledge of this primary feed pattern.

The pattern of the antenna as a whole, referred to as the secondary pattern, is taken partly in the course of the design and developmental research and finally, of course, as a test of the antenna performance. A particular advantage of the microwave region is that, on the one hand, the secondary pattern can be determined so as to be closely identical with that of the antenna in free space and, on the other, by use of full-scale models, the distortion of the free-space pattern due to the installation and housing can be studied. Secondary patterns of pencil-beam and fanned-beam antennas are generally confined to the principal E- and H-planes; however, the importance of complete space patterns is being recognized particularly in regard to the effect of cross polarization in reducing the resolving power of the antenna beam. Complete space patterns are, of course, always necessary in the design of shaped-beam antennas; here again the polarization of the field must be determined completely in order to arrive at a correct evaluation of the antenna performance.

The range of a system for a given amount of available power is limited by the gain of the antenna (Sec. 1·2). From the point of view of technique, the direct measurement of gain is perhaps more exacting than that of other antenna measurements. A necessary complement to the instrumentation of an antenna laboratory is a set of primary and secondary gain standards.

IMPEDANCE MEASUREMENTS

15·2. Transmission-line Relations.—The subject of two-wire lines and the relation of waveguides thereto were treated in considerable detail in Chaps. 2 and 7, respectively. For the sake of continuity of the discussion in the present chapter the principal transmission-line relations will be reviewed here briefly.

The voltage and current at a position z along the line are

$$V(z) = A_1 e^{-\gamma z} + A_2 e^{\gamma z} \tag{2.18}$$

$$i(z) = \frac{1}{Z_0}(A_1 e^{-\gamma z} - A_2 e^{\gamma z}), \tag{2.19}$$

where γ is the propagation constant, in general a complex number

$$\gamma = \alpha + j\beta. \tag{2.17}$$

The constants α and β are respectively the attenuation and phase constants; the latter is related to the wavelength on the line by

$$\lambda_q = \frac{2\pi}{\beta}.$$

The amplitudes of the component waves A_1 and A_2 are evaluated in terms of the conditions at the two ends of the line. The origin $z = 0$ will be taken at the termination, and the input end of the line will be located at $z = -L$. If Z_L is the terminating (load) impedance, we have by Eq. (2·23)

$$\frac{A_2}{A_1} = \frac{Z_L - Z_0}{Z_L + Z_0} = \Gamma_L; \tag{1}$$

Γ_L is the load reflection coefficient defined by Eq. (2·27) on setting $z = 0$. At the input end of the line $z = -L$, we have

$$V_G = V(-L) + Z_G i(-L), \tag{2}$$

where V_G is the generator emf and Z_G is its internal impedance. A reflection coefficient Γ_G may be defined corresponding to the mismatch between the generator impedance Z_G and the characteristic impedance of the line:

$$\Gamma_G = \frac{Z_G - Z_0}{Z_G + Z_0}. \tag{3}$$

Upon substitution of these relations into Eqs. (2·25a) and (2·25b) we obtain

$$A_1 = \left(\frac{V_G Z_0}{Z_0 + Z_G}\right) \frac{e^{-\gamma L}}{1 - \Gamma_G \Gamma_L e^{-2\gamma L}} \qquad (4)$$

and

$$A_2 = \left(\frac{V_G Z_0}{Z_0 + Z_G}\right) \frac{\Gamma_L e^{-\gamma L}}{1 - \Gamma_G \Gamma_L e^{-2\gamma L}}. \qquad (5)$$

The voltage and current at any point in the line are therefore expressed in terms of the source, the transmission line, and the load by

$$V(z) = \left(\frac{V_G Z_0}{Z_0 + Z_G}\right) \frac{1 + \Gamma_L e^{2\gamma z}}{1 - \Gamma_G \Gamma_L e^{-2\gamma L}} e^{-\gamma(z+L)}, \qquad (6)$$

$$i(z) = \left(\frac{V_G}{Z_0 + Z_G}\right) \frac{1 - \Gamma_L e^{2\gamma z}}{1 - \Gamma_G \Gamma_L e^{-2\gamma L}} e^{-\gamma(z+L)}. \qquad (7)$$

The line has impedance transformation properties that are described by Eq. (2·32), from which we may obtain the impedance at a point $z = -l$ relative to the load impedance at $z = 0$; i.e.,

$$\frac{V(-l)}{i(-l)} = Z(-l) = Z_0 \left[\frac{Z_L + Z_0 \tanh(\gamma l)}{Z_0 + Z_L \tanh(\gamma l)}\right]. \qquad (8)$$

Similarly the reflection coefficient transforms along the line, and we have from Eq. (2·31)

$$\Gamma(-l) = \Gamma_L e^{-2\gamma l}. \qquad (9)$$

It is evident that Z_L or Γ_L can be determined by these relations from measured values of $Z(-l)$ or $\Gamma(-l)$ at any point along the line a distance l from the load.

For a lossless line Eqs. (8) and (9) simplify to

$$Z(-l) = Z_0 \left(\frac{Z_L + jZ_0 \tan \beta l}{Z_0 + jZ_L \tan \beta l}\right) \qquad (10)$$

and

$$\Gamma(-l) = \Gamma_L e^{-2j\beta l}. \qquad (11)$$

The relations for admittance ($Y = 1/Z$) are given by Eqs. (8) and (10) with Z everywhere replaced by Y. Also Γ_L in terms of admittance is given by

$$\Gamma_L = \frac{Y_0 - Y_L}{Y_0 + Y_L}. \qquad (12)$$

15·3. Standing-wave Ratios.—Instruments for voltage measurement used in impedance determinations indicate some function (not necessarily linear) of the time average of the square of the real voltage; this is given in terms of the complex voltage by $\frac{1}{2}VV^*$, where V^* is the complex conjugate of V. Carrying out the indicated operation on Eq. (6), we obtain

$$\frac{1}{2} VV^* = \frac{|A_1|^2}{2} [e^{-2\alpha z} - 2|\Gamma_L| \cos (2\beta z + \delta) + |\Gamma_L|^2 e^{2\alpha z}], \qquad (13)$$

where δ is the phase angle of $\Gamma_L (\Gamma_L = |\Gamma_L| e^{j\delta})$.

The attenuation constant in microwave transmission lines is small and may usually be neglected in impedance measurements of antennas. The simplified expressions that follow are strictly true only for a lossless line. Assuming that $\alpha = 0$, Eq. (13) simplifies to

$$\frac{1}{2} VV^* = \frac{|A_1|^2}{2} [1 + 2|\Gamma_L| \cos (2\beta z + \delta) + |\Gamma_L|^2]. \qquad (14)$$

$\frac{1}{2} VV^*$ is a periodic function of z with maxima occurring at $2\beta z + \delta$ equal to even multiples of π and minima at $2\beta z + \delta$ equal to odd multiples of π. Voltage maxima occur, therefore, every half wavelength, and minima occur every half wavelength, with adjacent maxima and minima separated by a quarter wavelength.

The ratio of maximum voltage squared to minimum voltage squared, obtained from Eq. (14), is called the power standing-wave ratio and is

$$r^2 = \left[\frac{1 + |\Gamma_L|}{1 - |\Gamma_L|} \right]^2. \qquad (15)$$

The square root of r^2 is referred to as the voltage standing-wave ratio and is given by

$$r = \frac{1 + |\Gamma_L|}{1 - |\Gamma_L|}. \qquad (16)$$

Accordingly, the magnitude of the reflection coefficient of the load is given by

$$|\Gamma_L| = \frac{r - 1}{r + 1}. \qquad (2 \cdot 44b)$$

Following the argument of Sec. 2·7 it is observed that the reflection coefficient is equal to $|\Gamma_L|$ at a voltage maximum and to $-|\Gamma_L|$ at a voltage minimum. Correspondingly at a voltage maximum the impedance is real and equal to rZ_0, and at a voltage minimum the impedance is real and equal to Z_0/r. If l is the distance from a voltage minimum to the load terminals, the load impedance can be expressed in terms of r and l by replacing $Z(-l)$ in Eq. (10) by Z_0/r, thus obtaining

$$Z_L = Z_0 \frac{1 - jr \tan \beta l}{r - j \tan \beta l}$$

or, separating Z_L into a resistive component R_L and reactive component X_L,

$$Z_L = R_L + jX_L$$
$$= Z_0 \frac{r \sec^2 \beta l}{r^2 + \tan^2 \beta l} - jZ_0 \frac{(r^2 - 1) \tan \beta l}{r^2 + \tan^2 \beta l}. \qquad (17)$$

The load impedance can thus be calculated by means of Eq. (17) from the measured values of the standing-wave ratio and the distance from a voltage minimum point to the load. The calculation may be performed graphically on the reflection coefficient or bipolar charts as was discussed in Sec. 2·8.

15·4. Measurement of Voltage Standing-wave Ratio.—The most common method of determining VSWR is by means of the apparatus[1]

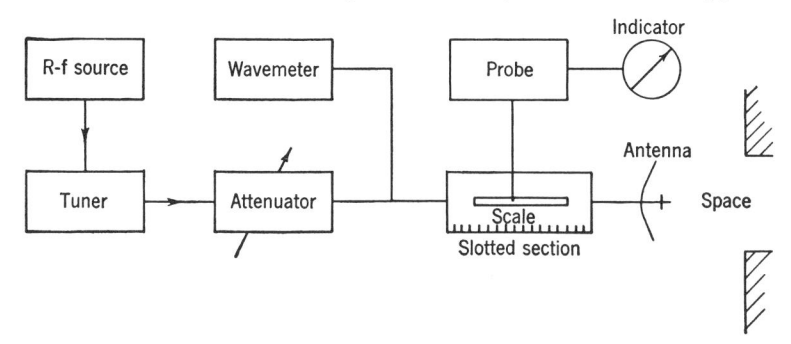

FIG. 15·1.—Block diagram of impedance-measurement apparatus.

shown schematically in Fig. 15·1. The first unit is an r-f power source which at microwaves is usually a velocity-modulated tube capable of being tuned over the wavelength band on which measurements are to be made. Since these tubes are sensitive to the impedance mismatch, an r-f tuner is generally connected close to the source and tuned for maximum stable output. The tuner is then followed by a variable attenuator which is preferably matched. The attenuator serves to control the r-f power level and to reduce the pulling effect of a variable load. A wavemeter should be in the set but should be detuned during standing-wave measurements. The attenuator is followed by a slotted section of line; the slot is narrow and cut in such a way that it does not interrupt the current lines in the waveguide wall appreciably. The latter is necessary in order that the coupling between the line and space be negligible; then the characteristic impedance of the slotted section does not differ significantly from that of the uncut line. The field inside the guide is explored by means of a probe mounted on the slotted section. The microwave instruments generally employ an electric-field probe—a wire or needle entering into the guide parallel to the electric vector of the field. Such a probe measures the voltage standing wave. Mounted in the probe is a detector that supplies direct current or audio frequency to an indicator. The response of the detector-indicator combination is a measure of the field intensity at the probe. The slotted

[1] The details of various r-f components such as tuners, attenuators, wavemeters, slotted sections, and probes together with the techniques of their use are given in Vol. 11 of this series.

section is placed as close to the antenna as possible in order to reduce errors in measurement of electrical length.

The actual measurements consist of moving the probe along the line and determining the maximum response and minimum response which should be separated by a distance of $\lambda_g/4$. From the calibrated response the VSWR is obtained. The distance of a voltage minimum (the reason for choosing a voltage minimum rather than a voltage maximum is given later) from the load is noted. These two quantities together with knowledge of the wavelength in the transmission line suffice to determine the impedance of the load.

Precautions in Standing-wave Measurements.—The procedure described above is exactly the same as that used in measuring any r-f component, except for one very important difference. The antenna is a radiating load and therefore precautions must be taken so that reflections from near-by walls or objects do not affect the measurements of VSWR or phase (position of minimum or maximum voltage). To avoid or diminish the effect of reflections the most intense portions of the radiation should be directed toward an open space with as much open space in all other directions as possible. Measurements may be taken inside the laboratory by directing the main beam at an angle of approximately 45° toward a wall, preferably of low reflection. To ensure that the space in which measurements are being made is satisfactory, the antenna should be moved to several positions (varying the distance to the wall and changing the angle of incidence plus or minus a few degrees from 45°) and the effect on phase as well as VSWR observed. If no changes occur, the site is satisfactory. Particular caution must be exercised when impedance measurements on low gain, nondirective antennas such as beacons are to be made. If such antennas are to be mounted on a metallic sheet in actual use, impedance measurements should be made in such a way as to simulate as closely as possible the actual final conditions.

Several other precautions must be taken to ensure the accuracy of VSWR and phase measurements. The probe should be loosely coupled to the line in order to avoid alteration in the standing wave pattern which will occur if the probe has an appreciable reflection coefficient. The reflection coefficient of a probe is a function of the tuning of the probe as well as the probe insertion. The effects of this will be discussed in more detail in Sec. 15·7. For a matched generator this reflection results in apparent VSWR less than the true VSWR and asymmetry in the standing-wave pattern with the maxima and minima not separated by a quarter wavelength. The position of the minimum is not affected appreciably. If Γ_P is the reflection coefficient of a tuned probe when the line is terminated in its characteristic impedance, the measured VSWR corresponding to a load of reflection coefficient Γ_L is less than the true value by the factor $(1 - |\Gamma_L\Gamma_P|)/(1 + |\Gamma_L\Gamma_P|)$. Thus, for example, a probe presenting a reflection coefficient $\Gamma_P = 0.05$ will measure a standing-wave ratio $r^2 = 4$ as $r^2 = 3.75$. If the internal impedance of the generator is different from the characteristic impedance of the line (mismatched generator), these effects will be even more aggravated.

The probe mount should be tunable so as to obtain maximum output for various frequencies and for a given probe insertion. Tuning should be smooth and not subject to erratic contact; the probe is generally a fairly high Q device, and the response will be affected easily by instability. Furthermore, for a given probe insertion the reflection coefficient of the probe will be real when the probe is tuned; under the latter condition no asymmetry will result in the standing-wave pattern although the measured VSWR will still be less than the true VSWR.

The response of the probe and indicator must be calibrated. It is not safe to assume that the response is proportional to the square of the voltage. A crystal detector[1] is accurately square law only for very low r-f power levels. The law of the crystal varies from crystal to crystal; for a given crystal it is a function of the power level and the load of the indicating system. A current-biased bolometer element, such as a Littelfuse, is very accurately square law over a very large range of power levels except in the neighborhood of burnout. A simple and convenient method for calibrating the detector will be given in Sec. 15·6.

Irregularities in the standing-wave pattern are frequently due to structural defects in the slotted section such as erratic contact between the probe carriage and the line. A common failing of coaxial-line sections is that the inner conductor is not accurately concentric with the outer. The usual effect is a probe depth varying almost linearly along the section resulting in a standing-wave pattern that appears to be superimposed on a monotonic voltage; this effect is known as *slope*. The effect of slope can be compensated for by calculating the VSWR from the ratio of the geometric mean of two maximum values of response separated by $\lambda/2$ to the minimum value between them or by taking a maximum response divided by the geometric mean of the two minimum responses on both sides of the maximum. Both procedures should give the same result. Actually an arithmetic average is satisfactory. The average of two maxima divided by the average of two minima should not be used in correcting for slope. Maximum and minimum values should be taken near the center of the slotted section so as to avoid the edge effect at the ends of the slotted section.

The impedance of the slotted section should be the same as that of the feed line of the antenna being investigated, and good electrical contact between the two lines should occur. For most accurate results the lines should also be geometrically the same. For example, if two 50-ohm

[1] A general discussion of detectors is given in Sec. 16·3.

coaxial lines with an appreciable difference in the radii of their respective conductors are connected together, the two lines will not be matched because of the capacitance introduced by the junction.

15·5. Determination of Electrical Length.—In determining the proper position for insertion of impedance-matching devices, accurate knowledge of the position of a voltage minimum point close to the antenna terminals is required. For good impedance matching, this information frequently must be known to a higher degree of accuracy than the value of VSWR. To reduce the error in determining the position of the voltage minimum the standing-wave section is always put as close to the antenna as is physically possible.

Voltage minima in the slotted section are first determined. Determination of the position of a minimum by adjustment of the probe position so that the response is a minimum is inaccurate, since with normal VSWR the minima are broad and the position is hard to determine exactly. A more accurate method is to determine the position of the probe for equal response on both sides of the minimum; the average of these positions will then be the location of the minimum. For greater accuracy the average position for several different responses may be taken. Having located such a minimum position, it may be transferred up the line an integral number of half wavelengths to a point near the antenna terminals. In actual practice, except for certain simple cases, this is easier said than done.

Perhaps the best method for transferring the position of the minimum is to short-circuit the transmission line at a point close to the antenna terminals and to use this as a reference point. This method, of course, assumes that such a short circuit can be made. This will usually be true for experimental antennas, but not usually with production antennas. Consider as an illustration Fig. 15·2 which shows the experimental setup for determining the voltage minimum near the terminals of a coaxial-line-fed antenna. Let l_1 be the position of a voltage minimum in the slotted section in Fig. 15·2a. The feed is now removed at the fitting, and a shorting plate is placed at the end of the coaxial line as in Fig. 15·2b. A new voltage minimum (zero) is located at l_2 on the load side of position l_1. Transferred up the line, this new minimum is at the short circuit or an integral number of half wavelengths from the short circuit and therefore provides a convenient reference point to which the load voltage minimum may be referred. Measuring a distance $|l_1 - l_2|$ toward the generator from the shorting plate determines the load voltage minimum position relative to the short circuit.

For air-filled coaxial lines, the measured physical lengths and electrical lengths will show good agreement, because the wavelength is independent of variations in line impedance. However, in waveguides the guide wavelength depends on the dimensions of the guide; hence the use

SEC. 15·5] DETERMINATION OF ELECTRICAL LENGTH 551

of an incorrect guide wavelength will lead to an error in the location of a transferred voltage minimum. For example, suppose that λ_g were actually 4.52 cm in the transmission line and a guide wavelength as determined in the slotted section were 4.50 cm. If one wished to find the position of the voltage minimum approximately 90 cm (about $20\lambda_g$) from the minimum in the slotted section, then the error resulting from the use of the value 4.50 cm instead of 4.52 cm would be $20 \times 0.02 = 0.4$ cm. The shorting method discussed in the previous section would almost

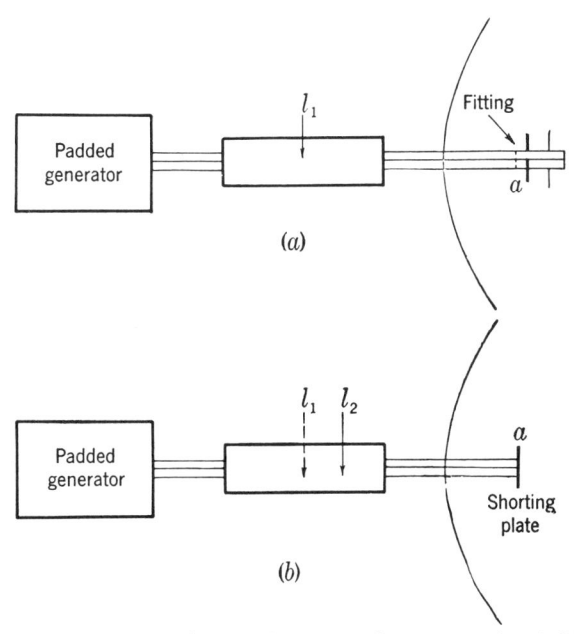

FIG. 15·2.—Short-circuited line technique for determining electrical length.

completely cancel this error. It should be emphasized that this procedure assumes that the connectors, bends, or small variations in impedance in the transmission line are reflectionless and therefore cause no phase shift.

When the shorting method cannot be used, the problem becomes more difficult, and various tricks may be used with more or less accuracy to give the desired information. One method frequently used in waveguide matching problems is to have made a set of experimental inductive irises of various openings, similar to that in Fig. 15·3, so that the outside dimensions of the frame of the experimental iris are such as to ensure a snug fit inside the waveguide with just sufficient clearance to permit sliding. To a given percentage open area of the inductive iris there corresponds a definite mismatch which will be eliminated by locating the iris in the proper position. Knowing the VSWR of the load from measurement, an iris whose open area is most nearly that required to cancel the mismatch is chosen from the set and slid along inside the waveguide to a point in that portion of the waveguide where it is desired that the matching transformer should be. The position of the experimental iris is varied until a matched (or nearly matched) condition is obtained. The position of the iris so determined for a minimum VSWR then determines the proper point for actual insertion of an inductive iris. This method has been found to give the proper position to within a millimeter in 1 by $\frac{1}{2}$-in. waveguide. The size of the actual iris to be inserted is obtained from the VSWR of the load and knowledge of the inductive susceptance introduced by an iris as a function of the iris opening.[1]

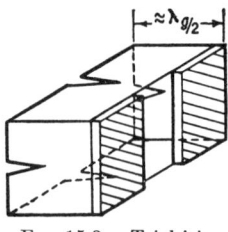

Fig. 15·3.—Trial iris.

Another method that may be used in waveguides is to cut a small hole in the center of the broad side of the waveguide near the antenna terminals and to introduce a capacitive screw. The VSWR and position of the minima are obtained for several screw depths. A voltage minimum position is assumed to be at the screw position. Referring to this point, the admittance of each value of VSWR and phase are plotted on the admittance diagram.[2] It will then be observed that the points do not fall along a vertical straight line corresponding to capacitance being added with increasing screw depth. The points will all be on a straight line when they have all been rotated the same phase angle on their respective r^2 circles. In the example shown in Fig. 15·4, rotating each point 30° on its corresponding VSWR circle the points then fall on a vertical straight line with added screw insertion increasing the capacitance in parallel. This then means for this example that the true voltage minimum is located a distance $l = 30\lambda/720$ toward the generator from the position of the center of the screw. In practice, this method is subject to error because there is, in addition to the capacitance of the screw, a small shunt conductance which increases with increasing diameter of the screw. For small screw diameters and small screw depths this method gives an approximate position for the minimum.

15·6. Calibration of Detection System.—As has been stated previously, the law of response of the detection system must be known for accurate measurement of VSWR. If an accurate attenuator is available and the power level is known, the response as a function of power input can easily be determined. Or a calibration is available if the response system can be compared with a known response system. Both of these methods have been used. However, the method discussed below is a simple one, utilizes the apparatus already available for VSWR measure-

[1] *Waveguide Handbook.* (Vol. 10 of this series.)
[2] See Sec. 2·8.

ments, and can be carried out easily from time to time if a variation in the law of the detection system is suspected (as may happen when a crystal is being used).

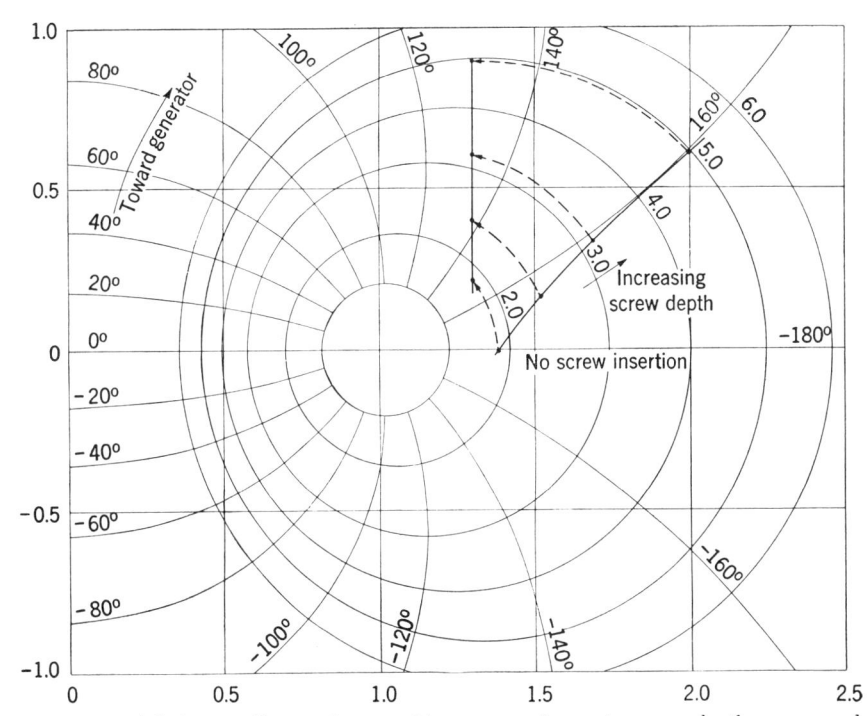

FIG. 15·4.—Admittance diagram for capacitive screw. Increasing screw depths correspond to increasing capacitance in parallel.

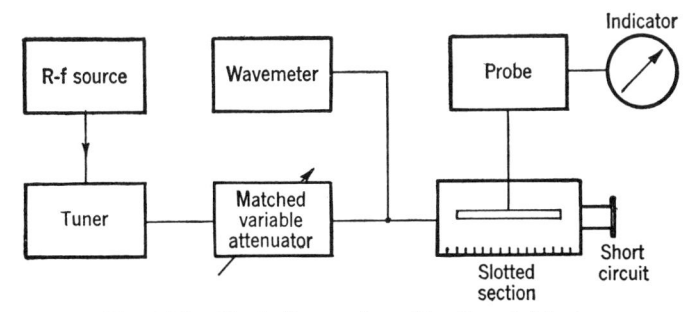

FIG. 15·5.—Block diagram for calibration of detector.

Figure 15·5 shows a schematic setup for this experiment showing a shorted standing-wave section being fed by a generator, the power level of which is adjusted by a variable matched attenuator. Since with a short circuit, the voltage is a sinusoidal function of position [Eq. (14) with $|\Gamma_L| = 1$, $\delta = \pi$] with period $\lambda_g/2$, the response I for the detection system as read on a meter will be given by

$$I = I_0 \cos^n \left(\frac{2\pi l}{\lambda_g}\right), \tag{18}$$

where the exponent n is the so-called "law" of the system and I is the maximum response at $l = 0$ (note that $l = 0$ here refers to the position of maximum response). The law of the detector may be a function of power level, a not uncommon occurrence in the case of crystals. Equation (18) can be written

$$n = \frac{\log_{10} \frac{I}{I_0}}{\log_{10} \cos \frac{2\pi l}{\lambda_g}}. \tag{18a}$$

It is advisable to plot the experimental values of $\log_{10} I/I_0$ vs. $\log_{10} \cos 2\pi l/\lambda_g$ in order to smooth out experimental irregularities. The slope of the curve at any indicated level I is equal to n.

If $n = 2$, the system is a square-law system and the meter readings are therefore proportional to the square of the voltage. The ratio of maximum response to minimum response for a load then gives directly the r^2 of the load. If $n \neq 2$, r^2 is given by

$$\frac{(\text{maximum response})^{2/n_1}}{(\text{minimum response})^{2/n_2}}$$

if the law of the system is n_1 at the maximum reading and n_2 at the minimum reading. For low (VSWR)^2 the ratio of maximum to minimum response may be used uncorrected if n does not differ greatly from 2. For example, suppose that the meter reading were 15 and 10; then (VSWR)^2 would be written as 1.5; but if the law of the system at both levels were 2.2, the true VSWR would be $(1.5)^{2/2.2} = 1.45$.

It is much simpler, when n does not vary greatly for different power levels, to determine n experimentally over 3-db power intervals. If $2l = d$ corresponds to the separation of half response, that is, when $I = \frac{1}{2}I_0$, then Eq. (18a) may be approximated by

$$n = \frac{0.2206}{\frac{d}{\lambda_g} - 0.1397}. \tag{19}$$

The error in this approximation is less than 0.02 for the range

$$1.8 < n < 2.4.$$

Figure 15·6 is a plot of n as given by Eq. (18a) as a function of the full width d/λ_g at half response. It should be noted that d/λ_g must be determined accurately because an error of 0.01 in d/λ_g results in an error in n

SEC. 15·6] CALIBRATION OF DETECTION SYSTEM 555

of 0.2. For small wavelengths a dial indicator calibrated in thousandths of an inch is mounted on the standing-wave section for accurate measurements of d.

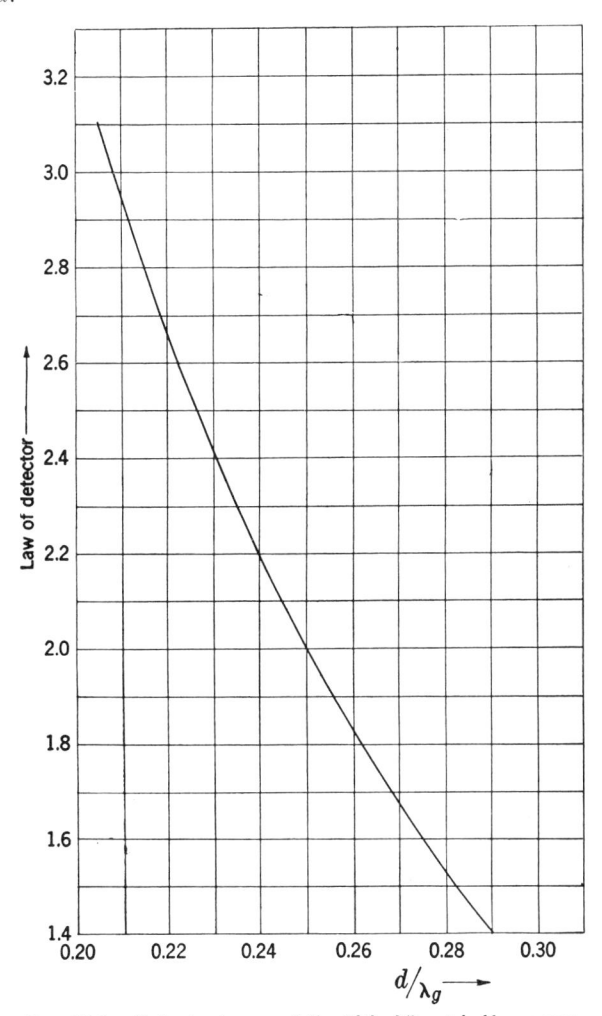

FIG. 15·6.—Detector law vs. full width d/λ_g at half response.

A possible procedure to be used is as follows: (1) Use the same microammeter or audio amplifier that will be used with the crystal or bolometer in practice. This is necessary, since the apparent law of the detector depends on the impedance of the metering arrangement as well as on the r-f power level. (2) Adjust the power level by varying the line power level or by varying the probe insertion. The probe insertion should be so small that the standing-wave pattern is not disturbed as it is moved along. This may be tested by monitoring with a second probe held at a fixed position in a slotted section between the generator and the section carrying the detector being calibrated. If the fixed probe reading is independent of the position of the moving probe, the insertion depth of the latter is acceptable. (3) Measure the separation of the points 3 db down from the maximum for several maxima positions along the standing-wave detector. Average these values of d. (4) The distance between alternate minima will be the wavelength λ_g in the transmission line. In the case of coaxial lines the wavelength is best obtained by means of a precision wavemeter. (5) The average value of n for these conditions is then determined from Fig. 15·6. (6) Determine n at other power levels by a similar procedure.

15·7. Probe Reflections.—The whole of the previous discussion in regard to making standing-wave measurements by means of a probe in a standing-wave detector is based on the assumption that the probe itself does not affect the field being measured. If the probe has a reflection coefficient different from zero, considerable error may result in the determination of the law of a detector or in the measurement of standing-wave ratios, especially if the mismatch being measured is great.[1] The most noticeable effect of probe reflections is a distortion of the observed standing-wave pattern.

For the most part, results that occur in standing-wave measurements may be accounted for by assuming that the probe is a shunt admittance. We may obtain the voltage V at any point of the line by Eq. (6) upon replacing Γ_L by the combined reflection coefficient of the load and of the probe. For simplicity, assume that the generator is matched so that $\Gamma_G = 0$. The voltage will be proportional to $(1 + \Gamma_{\text{eff}})$, where Γ_{eff} is the total reflection coefficient at the probe. The latter can be shown to be

$$\Gamma_{\text{eff}} = \frac{\Gamma'_L + \Gamma_P + 2\Gamma'_L \Gamma_P}{1 - \Gamma'_L \Gamma_P}, \qquad (20)$$

in which Γ_P is the reflection coefficient of the probe, given in terms of the probe admittance Y_P by

$$\Gamma_P = \frac{-Y_P}{Y_P + 2Y_0},$$

and Γ'_L is the reflection coefficient of the load referred to the probe position

$$\Gamma'_L = \Gamma_L e^{-2j\beta z}.$$

The voltage variation, therefore, in terms of Γ_L and Γ_P is given as a function of z by

$$V = \text{const} \, \frac{1 + \Gamma'_L}{1 - \Gamma_P \Gamma'_L} \, e - j\beta z. \qquad (21)$$

[1] Y. Dowker and R. M. Redheffer, "An Investigation of R-F Probes," RL Report No. 483-14; W. Altar, F. B. Marshall, L. P. Hunter, "Probe Errors in Standing-wave Detectors," *Proc. IRE*, **34**, 1, 33–44.

The effect of a finite reflection coefficient as given by Eq. (21) in measuring VSWR and phase shows the following results for a matched generator (for a mismatched generator the state of affairs becomes even more complicated). For a tuned probe the measured VSWR will always be less than the true VSWR; however, the minima and maxima will occur in their proper positions. For an untuned probe the measured VSWR will also be less than the true VSWR; the standing-wave pattern, however, becomes asymmetric with maxima and minima displaced from their correct positions and maxima and minima no longer separated by $\lambda_g/4$, despite the fact that adjacent minima are still separated by $\lambda/2$ as are also adjacent maxima. The position of the minimum is less affected than the position of the maximum; the higher the VSWR the less the minima are affected; and it is for this reason that minimum position has been recommended instead of maximum position for phase determinations. For example, for a load mismatch $|\Gamma_L| = 0.1$ ($r^2 = 1.49$), with $\Gamma_P = -0.1 e^{j\pi/4}$ the minimum position is displaced only $0.0063\lambda_g$ from its true position: and for $\Gamma_L = -1$, the minimum is not displaced at all. In the calibration method of Sec. 15·6 a probe reflection as small as $\Gamma_P = -.005$ will lead to an apparent value of $n = 1.97$ where the correct value is $n = 2$.

PRIMARY FEED PATTERN MEASUREMENTS

15·8. Primary Pattern Apparatus for Point-source Feeds.—The dimensions of microwave radiating systems that are employed as point-source feeds for reflectors and lenses are such that the physical distance from the feed at which the radiation zone sets in is within the dimensions of normal laboratory space. The distance is, of course, large compared with the feed dimensions in conformity with the requirements of the far-zone fields (Sec. 3·11). A complete primary pattern consists of the spatial distribution of radiated energy about the feed, the surfaces of constant phase, and the orientation of the electric vector (polarization) at all points on a sphere centered at the feed. The techniques for determining each of these components of the pattern in the radiation zone will be considered in the following sections.

The technique of measurement is, in general, the choice of the experimenter. It has been found convenient, however, to take transmitting patterns of primary feeds; that is, the particular feed in question is made to transmit microwave energy of the required frequency, and a small horn or pickup is then used to measure the intensity of the radiated energy, phase, and polarization. Later, in discussing the measurement of secondary patterns—the pattern of the composite antenna—it will be convenient to consider receiving patterns.

Transmitting patterns may be obtained in principle by receiving the radiated energy from the transmitting antenna in a polarized pickup antenna at all points on a large sphere centered about the transmitting

558   ANTENNA MEASUREMENTS—TECHNIQUES   [SEC. 15·8

antenna. A suitable form of apparatus to accomplish this for an essentially point-source primary feed is sketched in Fig. 15·7. The r-f components illustrated are typical 9400-Mc equipment. The general layout

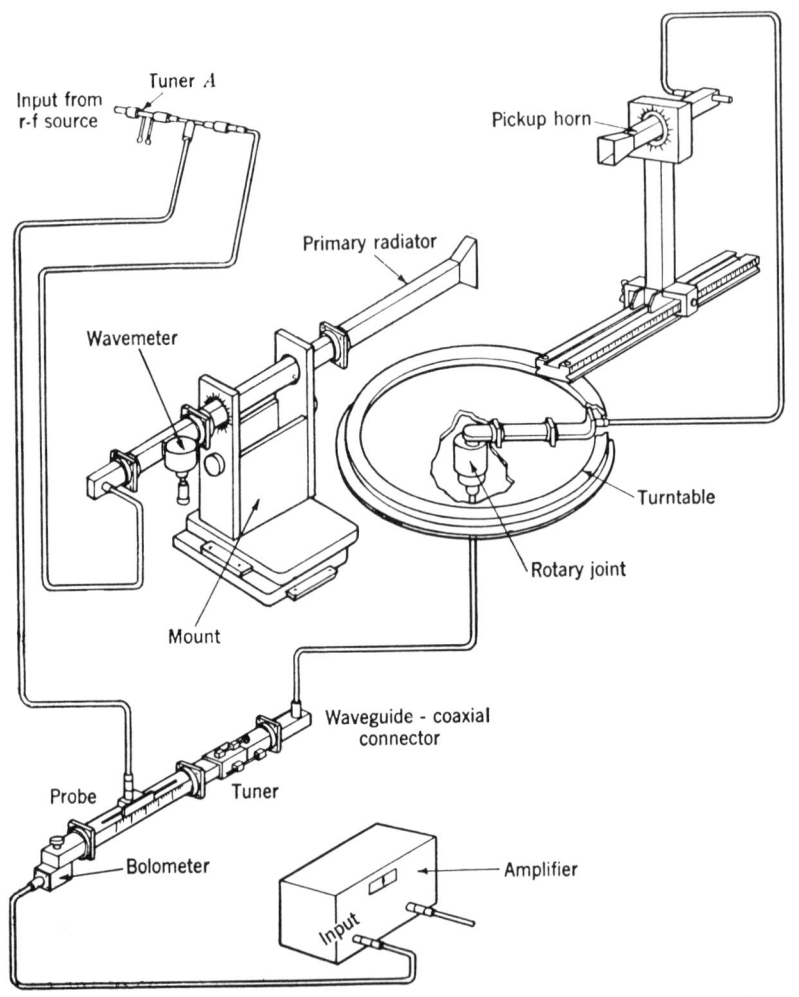

Fig. 15·7.—Apparatus for primary pattern measurements on point-source feeds.

and ideas, however, may be used throughout the microwave region. As an r-f source, a square wave modulated klystron has been utilized to give a steady output. The output from the generator is connected by means of coaxial cable to the waveguide that is terminated by the feed under study, which is shown here as a horn. A cavity wavemeter is included in the r-f line. The line and feed are clamped in an adjustable mount so that the apparatus may be properly aligned. These adjustments include horizontal and vertical displacements as well as means of rotating the feed about its axis (feed axis being here defined as the direction of peak radiation). The feed is mounted so that its approximate center of phase is located on the azimuth axis of rotation of the pickup antenna. The pickup antenna is mounted on a turntable illustrated here as a 30-in. gun mount, the upper ring of which is free to rotate on ball bearings between it and the lower ring. The pickup may be rotated about a horizontal axis, and its distance from the axis of the gun mount is adjustable.

If only intensity and polarization measurements are of interest, the detecting element may be placed behind the pickup and the output delivered directly to the indicating system. Illustrated in Fig. 15·7, however, is a system suitable for the measurement of phase as well as of intensity and polarization. In order to eliminate cable flexing which may produce phase errors, the energy is delivered to the detecting system via a cable, fastened to the pickup mount, to a rotating joint mounted with its axis coinciding with the turntable axis, and from the joint to the detecting system by a fixed cable or a rigid waveguide. The detecting system consists of a tuner, a slotted section on which is mounted a movable tunable probe through which r-f energy from the source may be introduced for phase determinations,[1] and finally the detecting element, crystal or Littelfuse. For intensity measurements the cable from the r-f source to the probe is disconnected.

A detecting element, which has been found to be very satisfactory, is a current-biased Littelfuse or Wollaston wire. This is in series with the transformer input of a low-noise, narrow bandpass, 10,000-gain, linear amplifier. The change in the resistance of the element is proportional to the modulated r-f power input over a very wide range of power level, and the resultant readings on a model 300 Ballantine voltmeter or other suitable electronic voltmeter are very accurately proportional to the square of the field intensity. A crystal may be used as a detector with the above apparatus; but although the sensitivity is of the order of 10 db greater than the Littelfuse, the law of the crystal varies as a function of the r-f power level. Consequently, it is necessary to know the calibration or law of the crystal as a function of the power level before correct field intensities may be obtained. This objection to crystals holds only for intensity measurements and not for phase measurements. In addition, the law of the crystal has been known to change with time and handling. Crystals are preferable for both intensity and phase measurements if a c-w source and superheterodyne detection system are used instead of the modulated source and amplifier-voltmeter system, since the r-f power level required is greatly reduced and the crystal is square law for low power levels.

As far as possible, all r-f components should be well matched to obtain

[1] An alternative method for providing a phase reference is discussed in Sec. 15·12.

maximum-power transfer and to prevent possible interactions. Generators should usually be loosely coupled to avoid a change in power output and frequency due to a variable load. The same techniques and cautions required in other r-f measurements are generally true for antenna measurements. However, in some cases, mismatch in the various elements does not result in incorrect measurements. Wherever good matching is required, this will be emphasized in the text.

There are several requirements that the pickup antenna should satisfy. (1) It should be polarized and arranged so that reflections from its mount will be negligible. (2) The pickup should be mounted so that its feed axis is perpendicular to and intersects the turntable azimuth axis. (3) The pickup should have some directivity in order to minimize the effect of possible reflections. (4) The pickup should be capable of being rotated about its axis in order to determine the polarization of the antenna being studied as well as to measure intensity for various orientations of the transmitting antenna. It is also desirable to be able to vary the distance of the pickup from the turntable axis so that the field may be measured at different distances from the feed when necessary. A simple and suitable pickup antenna consists of a rectangular waveguide horn of mouth dimensions such that the aperture in the electric plane is equal to the width of the waveguide in the magnetic plane. For example, in the 3-cm region a fairly well matched polarized pickup horn is obtained by flaring 1- by $\frac{1}{2}$-in. rectangular waveguide to an aperture 1 by 1 in.

The primary pattern apparatus should be located in as open a space as possible to eliminate undesirable reflections from surrounding objects. It is very difficult to establish a criterion for this because it is a function of the desired accuracy, the size and directivity of the feed as well as of the pickup, and the scattering cross sections of the various objects in the neighborhood of the apparatus. In many cases it is difficult to ascertain the cause of unexpected peculiarities in the pattern. If there are objects near-by suspected of reflecting, the effects of their removal naturally should be observed. If a symmetrically constructed feed is being studied, the presence of a side lobe on one side and not on the other at the same angle denotes either an error in the alignment of apparatus or the presence of a reflecting object on one side. A useful precaution against the effect of reflections, in particular those of near-by walls, is to cover the scattering objects with microwave absorbing material. One must be certain that the material is actually an absorber in the wavelength region being considered, particularly if a lossy material backed up by a metal sheet is used as an absorber.[1] In this case, the angle of incidence as well as the wavelength must be considered.

[1] An effective absorbing screen can be constructed by backing a slab of wood a

In general, the apparatus itself should be as free as possible of reflecting surfaces or objects. Patterns should be taken preferably in azimuth because reflections from a horizontal surface may be made symmetrical and with the usual directivity of feed and pickup they will be negligible. The primary radiator and pickup should be located as far as possible from horizontal surfaces such as the table and ceiling in order to minimize the effect of these reflections. In most cases, if the distance between the table and the feed is of the order of 10 wavelengths or greater, very little trouble will be encountered.

A number of other more elaborate primary pattern setups have been devised, primarily with a view to greater accuracy in phase measurements. One possibility is to use a vertical transmission path, thus eliminating reflections from either the floor or the ceiling which may conceivably cause errors in phase measurements made on the apparatus described above. However, a vertical-path device is considerably more complex mechanically and is warranted only where definitive measurements are required.

15·9. Intensity Measurements.—The intensity patterns in the principal electric and magnetic planes (hereafter to be denoted by E- and H-planes respectively) are usually sufficient for simple antenna designs; sometimes, however, a knowledge of a greater portion of the space pattern is required in more careful antenna design. The method for determining the field intensity as a function of direction will be discussed first for the principal E- and H-planes and then for other planes or cuts.

The r-f components need not be well matched when only relative intensity measurements are required. Mismatch in the various components will affect only the power level at the detector and will not therefore affect the measurement unless the mismatch is variable. The separation of feed and pickup should be of the order of $2d^2/\lambda$ or greater, where d is the maximum aperture of the feed. For small d of the order of a wavelength, this requirement may be open to question, and one should in this case have a feed-to-pickup separation of at least several wavelengths. These criteria ensure that the feed is being examined in its Fraunhofer region. The distances suggested will minimize interaction between the transmitter and receiver. If the feed is to be used in combination with a reflector such as a paraboloid, it is desirable that the dis-

---

quarter wavelength (dielectric) thick with a metal sheet and facing it with *resistive cloth*. The latter is an aquadug-impregnated fabric sold by the U.S. Rubber Company. On the assumption that the thickness of the fabric is negligible the surface resistance of the cloth should be 377 ohms. It has been found, however, that the fabric thickness is not negligible and that a surface resistance in the neighborhood of 600 ohms is more effective. The screen with $\frac{\lambda}{4}$ thickness is most effective at normal incidence.

tance from the pickup to the feed should be roughly equal to the distance from the feed to the contemplated reflector.

A check on whether the Fraunhofer pattern is being measured may be obtained by moving the pickup away from the feed at a fixed angle, say the peak direction, and determining whether the power response, as measured on the electronic voltmeter, varies inversely with the square of the distance. Figure 15·8 illustrates the dependence of the measured

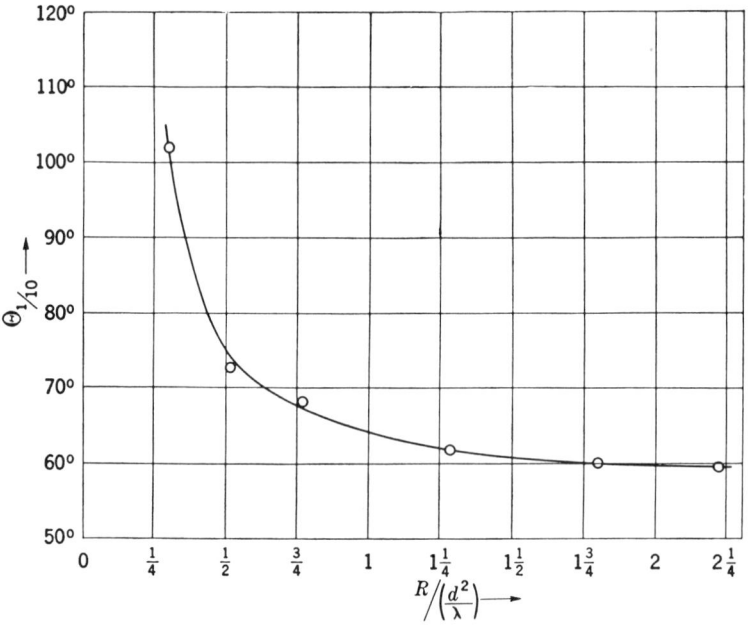

Fig. 15·8.—Tenth-power width $\Theta_{\frac{1}{10}}$ as a function of the path distance R between test horn and pickup horn.

10-db width of the intensity pattern upon the separation between a feed and the pickup. It will be seen that the beamwidth becomes practically constant for path distances greater than $1.5d^2/\lambda$. This is in agreement with the choice of $2d^2/\lambda$ as a minimum separation (*cf.* also Sec. 6·9).

To measure the *H*-plane intensity distribution, the primary feed under study is mounted so that the estimated center of feed[1] is located directly over the center of rotation of the turntable and the feed oriented so that the *E*-vector is vertical. The pickup horn with *E*-vector vertical is aligned with the feed so that the axes of the feed and of the pickup, if extended, become a common horizontal line. The *H*-plane intensity pattern is then determined by moving the pickup feed in azimuth at a fixed distance and taking readings from the voltmeter as a function of

[1] The center of feed or center of phase is that point in the primary feed which is the center of the circle that most nearly approximates the curve of equiphase (see Sec. 15·11).

angle. With a square-law detection system the relative power per unit solid angle is thereby determined as a function of angle for the principal H-plane of the radiation field. By rotating the pickup and feed 90° about their own axes and carrying out the procedure described for the H-plane, the principal E-plane pattern is determined.

Various methods of presenting this data have been used, each with its own merit. All too frequently in the literature and reports there is a failure to label properly the intensity axis as field intensity or square of field intensity (power pattern). A semilogarithmic plot of the power as a function of angle is commonly used and has the advantage of emphasizing the lobe structure. The relative field intensity (or simply the relative field intensity squared) as a function of angle on a polar plot is valuable for obtaining a good picture of the field distribution; in particular for field intensity, this plot gives the relative range as a function of angle at which a given receiver will receive a constant signal from the fixed transmitting antenna (*cf*. Sec. 1·2).

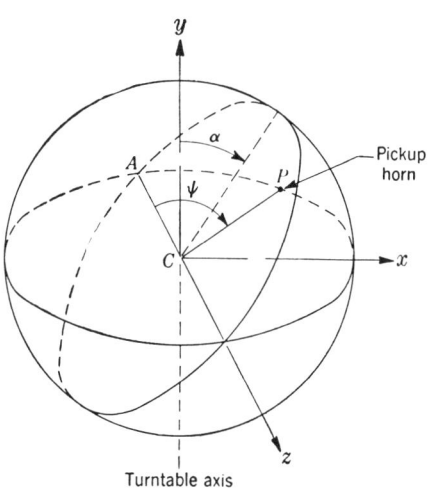

Fig. 15·9.—Coordinate system for primary patterns.

We now have the principal E- and H-plane intensity patterns; let us consider the problem of obtaining the space distribution of intensity and the polarization. We define a coordinate system α and ψ, shown in Fig. 15·9. The z-axis is the axis of the primary feed; ψ represents rotation of the pickup about the axis of the turntable; and α represents the orientation of the E-vector at points on the feed axis relative to the plane defined by the y- and z-axes. The apparatus is first lined up, as described before so as to obtain the principal H-plane pattern. To check the polarization in this plane, at each azimuth position of the pickup, the pickup is rotated about its axis until the maximum power is received. The angle of the pickup then determines the angle of polarization of the radiation at that point. After the polarization and intensity have been determined for the principal H-plane pattern, the feed and pickup are returned to their initial condition of $\alpha = 0$ and $\psi = 0$. The feed is then rotated through an angle α (say 10°) about its axis, and the pickup is rotated through a similar angle so that their polarizations are parallel. The intensity reading should be exactly the same as the initial reading. Note that this method has the desirable feature that as each cut is taken the power level is checked at $\psi = 0$, that is, for peak intensity. The pickup is again moved in azimuth; and at each angle ψ, the maximum intensity and polarization are obtained as above. This procedure is repeated for successive values of α. If the field-intensity-squared data are plotted as a function of α and ψ on suitable coordinate paper, contours of constant intensity can be drawn.

Direct polarization on a paraboloid is that component of the radiated primary field which, after reflection from the paraboloid, is parallel to the dominant polarization of the antenna pattern. Direct- and cross-polarization components exist for each point on the paraboloid. If one is interested only in the direct-polarization component of this feed for use with a paraboloid, the procedure discussed above is considerably simplified. For example, the H-plane pattern is taken without rotating the pickup about its axis for polarization determination. Likewise, if the feed is rotated by an angle α for the α cut, then the pickup is rotated the same angle α and kept there while the intensity as a function of ψ is observed by moving the pickup in azimuth. This procedure is correct as may be observed by considering the properties of a paraboloid of revolution. If for any reason the cross-polarization pattern of the feed is desired, it may be obtained exactly as above with the exception that when the feed is rotated α degrees, the pickup is rotated $\alpha + 90°$.

If we now desire to determine the contours of constant intensity on a paraboloidal reflector, in order to obtain optimum illumination of the reflector area,[1] we may, by a suitable transformation of coordinates and remembering to account for space attenuation, project the contours of constant intensity already drawn on the sphere onto the paraboloid.[2]

15·10. Phase Determinations.—In determining the phase front of a primary radiator it is usually considered sufficient to examine the phase in only the principal E- and H-planes, since, if the curves of constant phase in these two planes are circles with a common center, it is reasonable to expect that the phase front is spherical. However, if other planes are of interest, they may be examined in a manner similar to that employed for intensity measurements. The methods for determining phase involve the comparison of energy from the pickup horn with energy from the r-f source. The apparatus for the method described here is illustrated in Fig. 15·7. The power from the r-f source is divided between the primary radiator and the tunable sliding probe inserted in the waveguide. Energy from the pickup and from the source will then add in or out of phase at the detector depending on the probe position.

If the phase distribution in the H-plane is being investigated, then

[1] See Sec. 12·14.
[2] J. I. Bohnert and T. J. Keary, RL Report No. 659, March 1945; J. I. Bohnert, RL Report No. 665, February 1945; S. J. Mason, "Horn Feeds for Parabolic Reflectors," RL Report No. 690, January 1946, pp. 21–22.

the apparatus is lined up as before for H-plane intensity measurements with the axes of the pickup horn and primary radiator coinciding. Keeping the pickup fixed, the position of the probe is varied until a minimum signal is observed on the electronic voltmeter. For this situation, the monitored power and the pickup power are out of phase. This will be the reference value, and the position of the probe is recorded. It will be noted that a similar minimum will occur if the probe is moved a distance λ_g (λ if the detection components are coaxial), since this represents a 360° change in phase. For this setup, motion of the probe toward the bolometer is equivalent, so far as phase is concerned, to keeping the probe fixed and moving the pickup toward the primary radiator along a radius through the center of feed. Therefore, moving the pickup a distance d toward the radiator is equivalent to moving the probe a distance $p = d(\lambda_g/\lambda)$ toward the bolometer. The reference value for the probe position having been determined, additional points on the phase front are determined in he following manner. The pickup is moved in azimuth through an angle ψ; again the position of the minimum is noted; and the difference p from the reference minimum position is obtained. This is carried out through the required range in angle ψ.

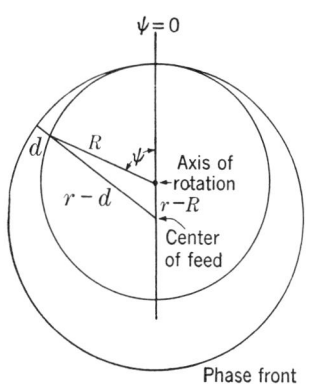

FIG. 15·10.—Center of feed correction. Pickup moves on circle of radius R. True wavefront is circle of radius r. Reference point $\psi = 0$.

If the position of minimum deflection is independent of ψ, then the phase front in that plane is a circle whose center is exactly the point on the feed directly over the turntable axis of rotation. This determines the center of feed in that plane. If the minima positions are different, the estimated center of feed, whose location is known to be on the z-axis as shown in Fig. 15·10, may be corrected by means of the equation

$$r - R = d \frac{1 - \dfrac{d}{2R}}{1 - \cos\psi - \dfrac{d}{R}}, \qquad (22)$$

where $r - R$ is the correction to be added to the assumed center of feed located in this plane and on the axis of the turntable. Values of d at corresponding angles ψ are obtained from $d = \pm p(\lambda/\lambda_g)$ with due regard to the choice of sign, the plus sign corresponding to motion of probe away from bolometer. Since R is large and $r - R$ is small, the correction term may be obtained with a high degree of accuracy. The relation above may be obtained by considering Fig. 15·10. It will be noted that d or $p(\lambda/\lambda_g)$

is measured along the radius r. By the law of cosines,

$$(r - d)^2 = R^2 + (r - R)^2 + 2R(r - R) \cos \psi,$$

or solving for r,

$$r = R \frac{\left[1 - \cos \psi - \dfrac{\left(\dfrac{d}{R}\right)^2}{2}\right]}{\left[1 - \cos \psi - \dfrac{d}{R}\right]};$$

when R is subtracted from both sides, the desired relation is obtained.

Experimental Precautions.—Several practical precautions should be taken in order to ensure proper phase measurements. The power from the probe and from the pickup into the bolometer should be of approximately the same level in order that well-defined minima should occur. It is undesirable to change the power level from the generator or the probe tuning during the phase measurements, since not only is it a nuisance but there is also the possibility of a change in phase. A recommended procedure is to adjust the tunable probe at first so that the power which it delivers to the bolometer is 6 db below the power delivered by the pickup in peak position. This may be accomplished by first disconnecting the probe and then rotating the pickup in azimuth until the power delivered to the bolometer is 6 db below peak value. The probe connection is then made; the probe moved to a minimum position; the probe tuning adjusted to deepen the minimum; and finally, the minimum may be sharpened by adjusting the probe depth. Slightly more tuning may be necessary after the probe depth is changed. The probe depth should not be so great as to have a large reflection coefficient. After the above adjustments have been made the probe tuning is not varied during the phase measurements. The least change between maxima and minima over a 12-db range of pickup power will then be that in which the powers from the pickup and the probe differ by a factor of 4 (6 db), and therefore the amplitudes will differ by a factor of 2. The resulting ratio of the maxima amplitude to the minima will be $(2 + 1)/(2 - 1) = 3$, and the ratio of maximum to minimum voltage as observed on the Ballantine voltmeter will be 9, which is sufficiently large for accurately determining minima.

With the system described above it is necessary that a good match be ensured looking from the bolometer to the pickup, since the power radiated by the probe into the line divides, part going to the bolometer, part going toward the pickup horn. To obtain a good match a tuner is placed between the probe section and the r-f line connected to the pickup horn, as in Fig. 15·7. If this section is mismatched, the power delivered by the probe to the bolometer is a function of its position along the line, with a maximum to minimum ratio proportional to the (VSWR)^2 (looking from bolometer to pickup), and a period of $\lambda_g/2$. As a result false minima may occur with roughly $\lambda_g/2$ spacing instead of the expected λ_g spacing. A good check on the phase-measuring equipment is to make sure that no maxima or minima occur other than those exactly λ_g apart.

The effect of a mismatched pickup may be illustrated by the following example. Suppose that the voltage amplitude at the bolometer from a matched pickup is unity and that the amplitude from the probe at the

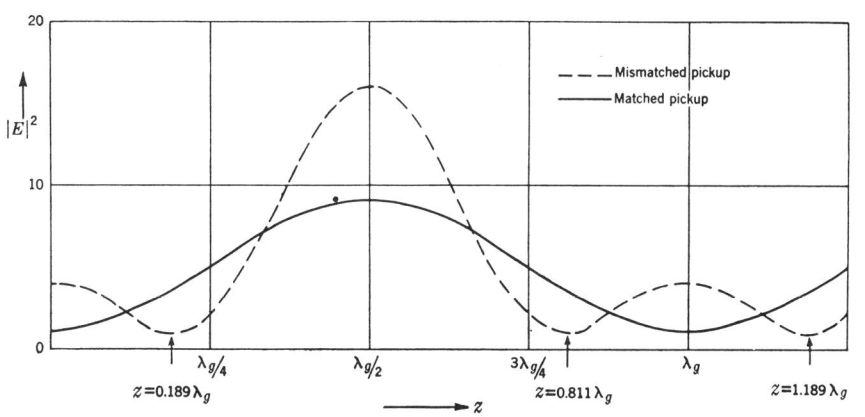

Fig. 15·11.—Power at the bolometer E^2 vs. distance z along the line for both a matched and mismatched pickup; $\phi = \pi$.

bolometer is 2, its phase being dependent on its position; then the resultant field at the bolometer is

$$E = 1 + 2e^{j(\beta z + \phi)}, \tag{23}$$

where z is the distance of the probe from the pickup and ϕ is a phase angle that represents the relative phase between the voltage at the probe and at the pickup. The power into the bolometer is proportional to EE^* which is

$$EE^* = 5 + 4 \cos (2\beta z + \phi). \tag{24}$$

Illustrated in Fig. 15·11 is a plot of EE^* as a function of z for the case $\phi = \pi$. Here $|E|^2$ has a minimum value of 1 at $z = 0$ and a maximum value of 9 at $z = \lambda_g/2$ and other minima at $\lambda_g, 2\lambda_g$ . . . and other maxima at $3\lambda_g/2, 5\lambda_g/2$ . . . . This example shows the proper periodicity that is obtained with a well-matched pickup.

If, however, the pickup is mismatched (for example VSWR = 3), the resultant field at the bolometer depends on the probe position and is given by

$$E = 1 + 2e^{j(\beta z + \phi)} + e^{j(-\beta z + \phi + \alpha)}, \tag{25}$$

where the voltage amplitudes are the same as in the above example;

α depends on the phase of the reflection coefficient of the pickup and will be chosen equal to zero in this example without affecting the qualitative nature of the illustration. The bolometer power, for $\phi = \pi$, will be

$$|E|^2 = 2 - 6 \cos \beta z + 8 \cos^2 \beta z. \tag{26}$$

The dashed curve in Fig. 15·11 shows the variation of $|E|^2$ vs. z. It will be noted that instead of a minimum at $z = 0$, there is actually a maximum, and likewise minima do not occur at λ_g spacing but at spacings of $0.378\lambda_g$ and $0.622\lambda_g$ and are poorly defined. By variation of the amplitudes and phases other undesirable configurations may be obtained.

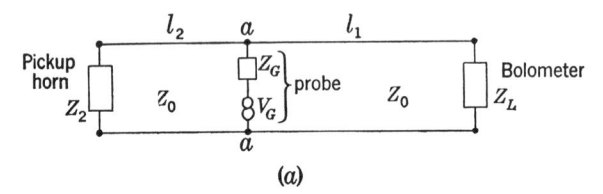

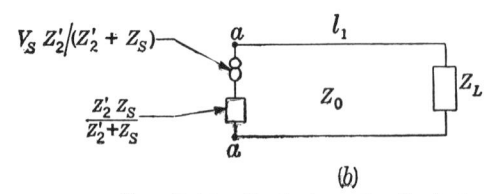

FIG. 15·12.—Equivalent circuit of r-f probe.

That the variation of power as determined by the probe position is a function of the mismatch of the pickup may be verified by the following analysis. If a generator of voltage V_G and internal impedance Z_G is connected to a load Z_L by means of a transmission line whose characteristic impedance is Z_0 and length is L, and if Z_L and Z_G are assumed to be real (this assumption will not affect the general result), the power delivered to the load is $V(0)I^*(0)/2$ as obtained from Eqs. (6) and (7); setting $\gamma = j\beta$, we have

$$\text{Power to the load} = \frac{Z_0 V_G^2}{2(Z_0 + Z_G)^2} \frac{1 - \Gamma_L^2}{\left(1 - 2\Gamma_L\Gamma_G \cos \frac{4\pi L}{\lambda_g} + \Gamma_L^2 \Gamma_G^2\right)} \tag{27}$$

Note that the power delivered is a function of the line length. The interaction between the generator and the load is contained in the expression $(1 - 2\Gamma_L\Gamma_G \cos 4\pi L/\lambda_g + \Gamma_L^2 \Gamma_G^2)$ which varies between $(1 + \Gamma_L\Gamma_G)^2$ and $(1 - \Gamma_L\Gamma_G)^2$, depending upon the line length.

The properties of the probe in the phase-measuring circuit, as represented in the equivalent circuit of Fig. 15·12a, may be determined from Eq. (27). The impedance of the pickup arm is represented by Z_2. The probe is to be considered as a source with very high impedance; motion of the probe will cause l_2 and l_1 to vary, but their sum will be fixed. By means of Thévenin's theorem (Sec. 2·4), the above circuit can be replaced by the circuit in Fig. 15·12b in which Z_2' is the impedance Z_2 transformed to the terminals aa. The replacement of V_G in Eq. (27) by $V_G Z_2'/(Z_2' + Z_G)$ and of Z_G by $Z_2' Z_G/(Z_2' + Z_G)$ gives as the power into the load

$$P = \frac{Z_0 V_G^2}{8 Z_G^2} \frac{(1 - \Gamma_L^2)(1 + 2\Gamma_2 \cos 2\beta l_2 + \Gamma_2^2)}{(1 - 2\Gamma_2 \Gamma_L \cos 2\beta L + \Gamma_2^2 \Gamma_L^2)} \tag{28}$$

for $Z_G \gg Z_L$ or Z_2. The only factor that contains the position of the probe is $(1 + 2\Gamma_2 \cos 2\beta l_2 + \Gamma_2^2)$ which has a maximum value of $(1 + \Gamma_2)^2$ and a minimum value of $(1 - \Gamma_2)^2$. Thus the ratio of maximum to minimum power received by the bolometer by varying the probe position is $(1 + \Gamma_2)^2/(1 - \Gamma_2)^2$ or the (VSWR)^2 of the pickup arm.

To remove this variable effect of position a tuner may be used effectively to make $\Gamma_2 = 0$. Another possibility is to use a well-matched attenuator pad. A 3-db matched pad or attenuator (looking both ways) reduces the reflection coefficient by one-half; a 6-db pad, by one-quarter. Suppose, for example, that the pickup (VSWR)^2 is 9 ($\Gamma_2 = \frac{1}{2}$) and that a 6-db pad is put between the pickup and the probe section; then the square of the resulting mismatch will be

$$r^2 = \left[\frac{1 + \frac{1}{4} \times \frac{1}{2}}{1 - \frac{1}{4} \times \frac{1}{2}} \right]^2 = 1.65, \tag{29}$$

thus appreciably reducing the effect of the pickup mismatch. The latter method can be used only if the loss in power level can be tolerated.

Frequency Sensitivity.—In order to reduce frequency sensitivity in the primary pattern phase measurements, the path lengths from the r-f source to the mixer must be chosen so as to minimize the change in phase of the two paths as a function of the wavelength. Consider the path from tuner A (Fig. 15·7) to the mixer via the feed and pickup, and let there be a length L_a for which the wavelength is the free space wavelength λ_0 and a length L_g for which the wavelength is λ_g. If the path length L_a includes a section of dielectric-filled coaxial line of length l_c, then $L_a = l_a + l_c k_e^{1/2}$, where l_a is the path length in free space or air-filled coaxial line and k_e is the specific inductive capacity of the dielectric section. The corresponding distances from point A to bolometer via the probe will be S_a and S_g. The difference in phase between the two paths is this:

$$\phi = 2\pi \left(\frac{L_a}{\lambda_0} + \frac{L_g}{\lambda_g} \right) - 2\pi \left(\frac{S_a}{\lambda_0} + \frac{S_g}{\lambda_g} \right) \tag{30}$$

$$= 2\pi \left(\frac{L_a - S_a}{\lambda_0} + \frac{L_g - S_g}{\lambda_g} \right). \tag{30a}$$

For waveguide $\lambda_g = \lambda_0/\sqrt{1 - (\lambda_0/2a)^2}$ so that $d\lambda_g/d\lambda_0 = (\lambda_g/\lambda_0)^3$, whence, on differentiating ϕ with respect to λ_0, we have

$$\frac{d\phi}{d\lambda_0} = 2\pi \left[-\frac{L_a - S_a}{\lambda_0} - \frac{L_g - S_g}{(\lambda_G)^2} \left(\frac{\lambda_g}{\lambda_0}\right)^3 \right]. \tag{31}$$

For a minimum phase variation $d\phi/d\lambda_0 = 0$, and we see that the difference in path in waveguide for the two paths should be related to the difference in path lengths in coaxial line or free space by

$$S_g - L_g = \frac{\lambda_0}{\lambda_g} (L_a - S_a). \tag{32}$$

It will be noted that this is not the same result which one would obtain by simply making the electrical path lengths equal, which gives the result $S_g - L_g = (\lambda_g/\lambda_0)(L_a - S_a)$. Of course, if only transmission lines with wavelength λ_0 are used, equal path lengths in the two directions are necessary to minimize the frequency sensitivity.

As an example of the decrease in frequency sensitivity, suppose that $\lambda_g/\lambda_0 = 1.4$ and that the frequency of the source changes by 0.2 per cent. Calculating from Eq. (30a) the change in phase that occurs for this change in frequency when the proper distance given by Eq. (32) is used, one finds it to be $\frac{1}{1000}$ of the phase change obtained by using equal path lengths. If $L_a - S_a = 50\lambda_0$, the change in phase is 0.034° using proper path lengths, whereas with equal electrical path lengths a phase error of the 34.6° results. The former is negligible in phase measurements. In actual practice the requirements of Eq. (32) need be met only approximately if the line lengths in terms of wavelength are not large.

15·11. Line-source Primary Pattern.—The measurement of the intensity and phase distribution along the length of a line source such as a linear array or pillbox is of interest. The measurement of the field close to the line source, which usually has a small effective vertical aperture and wide horizontal aperture, affords a valuable check on the design of the line source. The techniques involved in this measurement are exactly the same as those discussed above in connection with point sources, with the exception that motion of the pickup must be parallel to the line source as the intensity and phase measurements are taken. This does not represent any difficulty if intensity distribution alone is desired. For phase determinations, however, the apparatus becomes more complicated, since several rotating joints must be utilized in order to eliminate the flexing cables which may cause apparent changes in phase. Figure 15·13 is a sketch of a suitable apparatus, containing three rotating joints in order to make possible the required linear motion. These three are necessary if the line source being studied is horizontally

SEC. 15·11]   *LINE-SOURCE PRIMARY PATTERN*   571

polarized. For vertical polarization the rotary joint attached to the optical bench can be eliminated by using a coaxial pickup mounted in a bearing. Care must be taken that the rotary joints are properly designed so as not to show a change in phase with rotation. They should be fastened in such a way that they will not go out of alignment. The

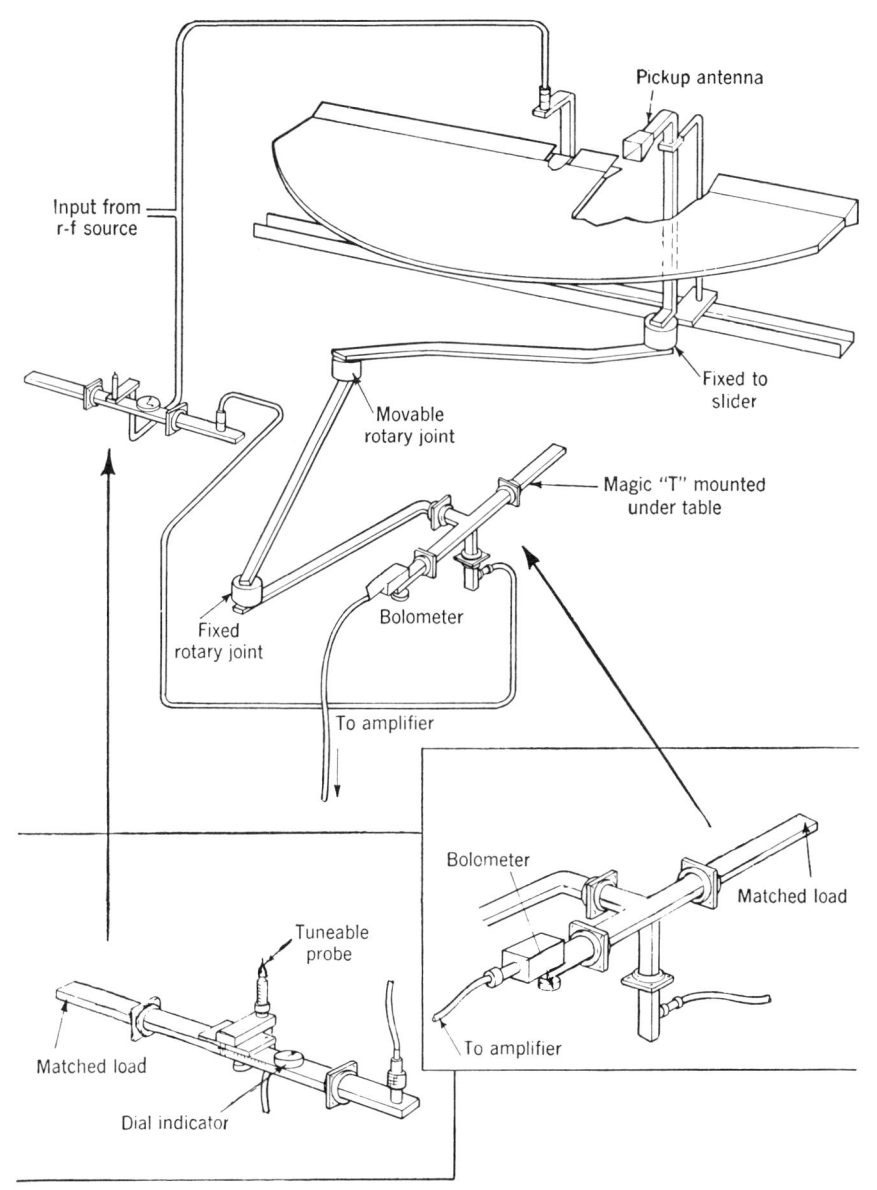

FIG. 15·13.—Apparatus for primary pattern measurements on line-source feeds.

rotary joints are all mounted in a plane with the first joint fastened to the traveling arm of the optical bench, the second movable about a circle whose center is the third fixed rotary joint. The r-f output from the third rotary joint is shown here being fed into the phase detection system. The mixing system illustrated here utilizes the so-called Magic T that will be discussed below. The power coupled from the generator output for a phase reference is fed into a tunable sliding probe which excites the waveguide, one end of which is terminated in a matched load and the other end connected by cable to the Magic T. A dial indicator (calibrated in 0.001 in.) is shown mounted on the probe section so that the motion of the probe may be accurately measured. For intensity measurements the cable to the probe is disconnected. Usually the distance between line source and pickup is only of the order of 2 to 4λ in order to stay within the cylindrical wave zone of the source. The pickup should be as small as feasible in order to prevent any interactions of the pickup in the field of the line source. The sketch also indicates that the movable parts of the apparatus are removed from the intense portion of the radiated field so as to reduce extraneous reflections. The operator should likewise be out of the strong field.

15·12. Magic T.—The "Magic T"[1] may be advantageously used in phase measurements to reduce interaction effects discussed in connection with the apparatus shown in Fig. 15·7. The waveguide form of this device is shown in Fig. 15·14. If power is fed into branch P (parallel arm) and A and B are terminated by matched loads, then the power divides equally between branches A and B, since a symmetric condition exists and there is no component of field available to excite arm S. If power is fed into arm S (series arm), the power again divides equally between A and B but the fields in each arm are 180° out of phase and no power is delivered to branch P. In order to prevent reflected power in arms P and S respectively, it is necessary that a match exist looking into arms P and S respectively when arms A and B are terminated by matched loads. In this way interaction between the two sources delivering power from P and S to arms A and B is made negligible.

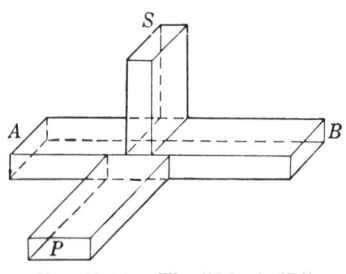

FIG. 15·14.—The "Magic T."

In the Magic T as used in the primary pattern apparatus, arm A is terminated by a matched load and arm B is terminated by a matched bolometer. Power from the pickup is fed to P as shown in Fig. 15·13, and power from the source is fed to arm S (this order can, of course, be

---

[1] See Vol. 11 of this series.

reversed). The relative phase of the fields from the two effective sources is varied by means of the sliding probe. Motion of the probe a distance $p = d(\lambda_g/\lambda)$ toward the bolometer is exactly equivalent to motion of the pickup horn a distance d perpendicular to its path toward the source.

15·13. Beacon Azimuth Patterns.—A microwave beacon antenna may be examined for uniformity of azimuth pattern by utilizing the apparatus described for point sources, with the simplification mentioned where intensity only is measured.[1] The accuracy required for azimuth patterns is not very high because one is usually interested in uniformity to within 1 or 2 db. Because of the essentially uniform azimuth patterns, reflections from many sources may affect the results; therefore particular care must be taken that reflections from surrounding walls and objects are negligible and the apparatus should be located in as open a space as possible, since a 360° pattern is desired. The pickup should be as small as possible if measurements close to the beacon are being made. However, if sufficient r-f power is available so that a larger distance between pickup and beacon can be used, then the pickup may be larger and more directive. Since in the measurement of azimuth patterns only a few decibels variation are observed, greater sensitivity in the apparatus may be obtained by the use of crystal instead of a bolometer as a detecting element. Over a range of several decibels the law of the crystal will not appreciably affect the relative readings. For example, assuming a square-law crystal, the ratio of peak power to minimum power might be 100 to 50 or 2; if the exponent in the law of the crystal were 2.3; the corresponding readings would have a ratio of 2.2 which is within $\frac{1}{2}$ db. For a little more accuracy, the ratio may be corrected to a sufficient degree if the approximate law of the crystal is known.

There may be some question about the distance required for measurement; obviously a distance between pickup and beacon of $2L^2/\lambda$, where L is the vertical length of the beacon, would certainly be safe. However, because the directivity in the azimuth plane is almost nonexistent, this distance could easily be reduced by a factor of 2 or 4 before an appreciable change in azimuth pattern is observed.

Two schemes have been used for beacon measurements. In one of these the transmitting beacon is kept fixed and the pickup is rotated in azimuth about the vertical axis passing through the beacon. The other procedure is to rotate the beacon about its vertical axis, feeding power into the beacon by means of a rotary joint, while the pickup is kept fixed. In lining up the apparatus, the pickup is adjusted vertically so that a cut through the peak of the beam is taken. Usually the axis of the pickup will be in a horizontal plane bisecting the beacon.

[1] Sec. 15·8.

SECONDARY PATTERN MEASUREMENTS

15·14. Siting Considerations.—In general, the term "secondary pattern" is associated with directive antennas that are large, measured in wavelengths, and with a large distance between the transmitter and receiver. A directive antenna will usually consist of a primary radiator together with a reflector or lens or combination of reflectors or lenses. The secondary pattern is the Fraunhofer pattern of the antenna in question. The techniques involved in measurement are fairly simple and closely related to those in the discussion of primary feed patterns. The simplifying consideration here lies in the fact that only the relative field intensity or power per unit solid angle, and not the phase, is of interest.

The discussion will center about the method of taking receiving patterns. In brief, a distant transmitter sends an essentially plane-polarized wave toward the receiving antenna. The power received by the receiving antenna as a function of its orientation with respect to the line of sight between the transmitting and receiving antennas is recorded either manually or by a recording device; the data thus obtained provide a pattern of relative field intensity or relative power per unit solid angle for the antenna under study.

The distance between the transmitting and receiving antennas is dictated by the size of the antenna being investigated. The site should be chosen with the largest antenna to be investigated in mind. It is required by theory that a plane wave be incident on the receiving antenna; actually, this requirement is met within a certain tolerance. A paraboloidal antenna as a transmitter will appear to be a point source when viewed from a large distance. If the distance is sufficiently large, the wavefront over a small portion of the main beam will deviate from a plane wave by only a small amount.

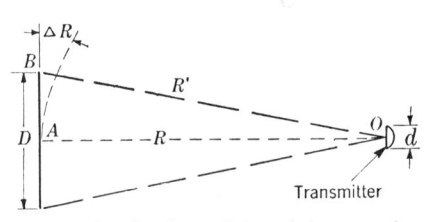

FIG. 15·15.—On the path length for secondary patterns.

Referring to Fig. 15·15, let D be the aperture dimension over which a plane wave is desired for pattern measurements and R be the distance between the transmitter, of aperture d, and the antenna to be investigated; then the difference in path length between the outer edge of D and the center is given by

$$\Delta R = R' - R.$$

Summing the squares of the sides of the right triangle OAB we have

$$R^2 + \left(\frac{D}{2}\right)^2 = (R + \Delta R)^2.$$

If we require that ΔR shall be a small fraction of a wavelength, $(\Delta R)^2$ is negligible, and there results

$$R = \frac{D^2}{8\Delta R}. \tag{33}$$

For a path difference $\Delta R = \lambda/16$ we have

$$R = \frac{2D^2}{\lambda} \tag{34}$$

which is a safe distance to use. The effect of such small deviation from a plane wave only slightly affects any gain determinations[1] and causes very slight changes in the pattern obtained as compared with that which would be obtained if the wave were truly plane. This distance will also minimize any interaction between transmitter and receiver.[2] Actually if space or power limitations do not allow such a great distance, then a distance of D^2/λ may be tolerated. This will lead, in general, to an apparent decrease in measured gain, an apparent increase in the minima of the side-lobe structure, practically no effect on the maxima of the side lobes, and greater possibility of transmitter and receiver interaction. Calculations may be carried out for certain ideal cases illustrating the semiquantitative nature of the above remarks.

In the foregoing discussion the distance R is determined by a consideration of the phase deviation of the incident wave. Another factor to be considered is the uniformity of the power distribution over the aperture D. If we require that the power at the edge of the aperture shall be a certain fraction of the power at the center, another criterion for R results. In the vicinity of the peak of the beam of the transmitter of aperture d, the power in direction θ may be approximated by

$$P = P_0 \left[1 - 2\left(\frac{\theta}{\Theta}\right)^2\right], \tag{35}$$

where $\Theta = 1.2\lambda/d$ is the full width of the transmitting beam at half power. Then if $P = 0.9P_0$ and $\theta \cong D/2R$ radians at the edge of the aperture, there results by substitution

$$R = \frac{2dD}{\lambda}. \tag{36}$$

Accordingly, if the transmitter aperture is equal to the receiving aperture, the criteria of distance for proper phase and for intensity over the receiving aperture are the same. The transmitting antenna is usually smaller than the receiving aperture; and therefore under the previous criterion [Eq. (34)], the power at the edge of the receiving aperture is the same within a few per cent as that at the center of the aperture.

[1] See Sec. 6·9.
[2] See Sec. 15·22.

Other factors determining the separation of transmitter and receiver are the power available in the r-f source and the sensitivity of the receiving system. Given two antennas separated by a distance R, the power received, P_r, by the receiving antenna when power P_t is transmitted with gain G_T is given by

$$P_r = \frac{P_t G_T}{4\pi} \frac{A_r}{R^2} = \frac{P_t G_T G_R \lambda^2}{(4\pi)^2 R^2} \tag{37}$$

with the absorption cross section (A_r) given by $G_R \lambda^2/4\pi$. Here one is really concerned with the smallest antenna that can be investigated, since this will be the limiting factor on P_r.

To illustrate the application of the criteria consider the problem of choosing a site for measurement of antennas at wavelengths varying from 3 to 10 cm. Assume (1) that r-f (magnetron) sources of 50 watts average power are available, (2) that antennas to be studied vary from 1 to 10 ft, and (3) that 1 mw of average power received in the bolometer detecting system corresponds to 100 volts on the electronic voltmeter. The limiting conditions are most stringent at the shortest wavelength; and if satisfactory for this wavelength, they will be more than suitable for the larger wavelengths as far as distances and power are concerned. Also, higher power sources are usually available at the longer wavelengths. For the above assumptions the following conditions result:

1. Specifying that the phase variation over the aperture D, which is taken equal to 10 ft, shall not exceed $\lambda/16$ for a wavelength of 3 cm, the distance R required is found to be 2130 ft by Eq. (34).
2. The maximum diameter d of the transmitting antenna can be as large as 10 ft and still satisfy the 90 per cent power requirement at the edge of D.
3. To read 100 volts on the voltmeter for the smallest antenna being studied, namely, $D = 1$ ft, a conservative estimate for the gains of the receiving and transmitting antennas is given by $G_R = \frac{1}{2}(\pi D/\lambda)^2$ and $G_T = \frac{1}{2}(\pi d/\lambda)^2$. Setting $R = 2000$ ft in accordance with condition 1, we have by Eq. (37) that the minimum aperture of the transmitter is $d = 2.5$ ft. Thus all the conditions of the problem are satisfied.

At a wavelength of 10 cm the transmitter aperture required would be approximately 7 ft in order to receive power of 1 mw at the detector of a 1-ft antenna at the receiver, assuming the transmitter power is 50 watts as before.

The next most important condition for choosing a site, when the power and distance requirements are satisfied, is the absence of reflecting objects, particularly buildings. If 360° patterns are desired, then clear surroundings for 360° must be available. For distances as great as those required for studying large microwave antennas, one usually must choose a site between two high points, such as buildings or hills. The interference between the direct beam from the transmitter and the reflected beam from the ground may result in a poor field distribution over the aperture that is being studied. To eliminate this interference, the transmitting antenna should be as directive as possible so that its first minimum will be in such a direction that even a specular reflection will not affect the field. This is illustrated in Fig. 15·16. Since the first minimum in the transmitting beam will occur at approximately λ/d radians from the peak, then $2h/R \cong \lambda/d$ or

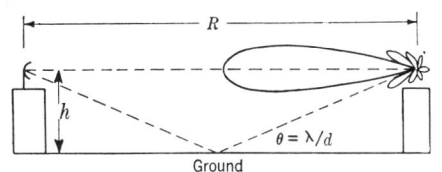

FIG. 15·16.—On the conditions for the height of the transmitter.

$$h = \frac{\lambda R}{2d}; \qquad (38a)$$

the larger the transmitting dish the lower the height required, and the longer the wavelength the greater the height for a given distance R. If $R = 2D^2/\lambda$, then

$$h = \frac{D^2}{d}. \qquad (38b)$$

In the problem discussed before with $D = 10$ ft and $d = 2.5$ ft, Eq. (38b) would mean a required height of $100/2.5 = 40$ ft. However, since the criterion for the 10-cm wavelength is not fulfilled, bad reflections may occur at the longer wavelengths. This is one of the difficulties involved in having one site for a large spread in wavelengths. A possible method to minimize the effect of the reflected beam is to place absorption screens or diffracting edges halfway between the two sites. It might appear that a reflection of 1 per cent in power may be negligible; however, one must remember that if we have two waves of respective powers 100 and 1, and if they interfere constructively or destructively, the resulting variation in power received is not 101 to 99 but rather $(10 + 1)^2$ to $(10 - 1)^2$ or 121 to 81, since it is the amplitudes that add, not the powers.

The various conditions have been stated, and a suitable compromise must be made between the various factors involved, such as heights and separation of sites, spread of wavelengths being considered, power available, sensitivity of the detection system, and the accuracy desired in the radiation pattern. In the example discussed, it might be necessary to resolve the conflict between the various conditions by the use of several sites.

After the site has been chosen, it should be checked for uniformity of field. The procedure to be followed is to direct the transmitter beam so that the center of its peak is on the center of the receiving site. A pickup antenna, either a paraboloid or a horn, is then moved over the aperture of interest, and the field intensity is examined for uniformity. The field should also be examined in depth to be sure that the intensity does not fluctuate seriously for motions toward or away from the transmitter corresponding to the depth of the antenna system. If the distance requirement is satisfied and the field intensity is uniform over the aperture, it will not be necessary to check the phase.

15·15. Pattern Measurements.—The mount on which the receiving antenna is to be placed should have at least two rotation axes: an azimuth axis and an elevation axis, so that complete space patterns may be obtained without too much difficulty. Although other axes may be more convenient for some purposes, the two stated are certainly sufficient. Whenever possible, patterns should be taken in azimuth, since reflections from the ground can usually be minimized. For mechanical reasons the azimuth patterns are likewise desirable, as it is easier to make an accurate mount for azimuth rotations than for elevation.

The transmitting antenna should be on a mount which permits motion through an angle sufficient to direct the peak of the beam at the receiving antenna, and there should be provision for locking the antenna in place. For convenience, the mount should also have the property that turning the antenna 90° for changing the polarization does not require shutting off the transmitter source and does not change the direction of the peak of the beam. If a paraboloid antenna is used as a transmitter, the undesirable cross-polarization component may be reduced by fastening to the aperture of the paraboloid a grating structure with spacing approximately $3\lambda/8$ and depth approximately $\lambda/4$, with polarization of the antenna perpendicular to the grating slats. Such a waveguide-beyond-cutoff grating will decrease the cross polarization of the transmitting antenna about 10 db below its normal value. Magnetron sources have proved satisfactory as a fairly constant, high-level r-f source for the transmitting antenna. Modulated high-power klystrons may be used if the power requirements are satisfied, or an unmodulated c-w transmitter, if a superheterodyne detection system is used.

The antenna under study is mounted, for example, with its dominant polarization vertical so that the H-plane pattern may be studied in azimuth. The transmitter must have vertical polarization. The antenna is adjusted in azimuth and elevation to receive maximum power. The transmitter is then adjusted to be sure that its peak is directly pointed toward the receiving antenna. The antenna is then repeaked for maximum power, and the mechanism controlling elevation is locked. These last two steps should be necessary only if the transmitter has been replaced or moved since the original siting measurements. A monitoring receiving antenna roughly peaked on the transmitter should be available to check the transmitting power level at any time. The vertical axis of the antenna mount should be perpendicular to the line of sight between the transmitter and test antenna to ensure that the proper azimuth cut is taken. This condition is particularly important when narrow-beam antennas are being investigated. The bolometer detection system need not be matched to the transmission line, since the mismatch of the bolometer does not affect relative response. The antenna is then rotated in azimuth, and the power received as a function of angle is recorded either manually or by means of a recorder. The value of a recorder lies mainly in its speed when the effects on the pattern caused by changing variables are being studied and in the continuity of data as well as in the permanent value as a record. For single patterns its value is questionable as far as time saved is concerned, since most of the time required for antenna measurements is used in setting up the antenna and preparing the electrical equipment for measurements.

To obtain the E-plane pattern, the antenna is returned to its peak direction and locked in azimuth. The antenna is then rotated in elevation, and its E-plane pattern taken. Data taken pointing into the ground may be questionable; to get the remaining 180° of the pattern it will be necessary to reverse the mounting of the antenna. A better procedure is to rotate the antenna 90° so that its polarization is horizontal (the transmitter must also be rotated 90°) and its E-plane pattern taken in azimuth.

The simplest procedure for obtaining complete space coverage patterns with a two-axis mount is to take the normal E- or H-plane pattern in azimuth and then to rotate the antenna $\theta°$ in elevation and take an azimuth pattern, thus obtaining the $\theta°$ cut. This is repeated for all angles θ of interest. Space patterns are usually taken only for fanned beams of the type used for navigation purposes. Certain precautions must be observed, however, in the choice of axes in the event that pattern cuts are required for a shaped-beam antenna such as described in Chap. 13. The antenna should be mounted so that the plane containing the flare of the beam is vertical. The angular widths measured in the transverse cut patterns will then be true. If the antenna is mounted with the fanned beam in the horizontal plane, the angular widths, now measured in the vertical plane, are too large by a factor of sec θ, where θ is the cut angle. These conditions are imposed by the mechanical aspects of the mount design.

The cross-polarization pattern for any desired cut is obtained by simply rotating the transmitting antenna 90° and taking the pattern as usual. The grating in front of the transmitting paraboloid ensures that the cross polarization of the receiving antenna is measured and not that of the transmitter. Together with this precaution lies the additional large factor of safety that if the transmitter is a symmetric paraboloidal antenna, the cross polarization along the peak of the beam is negligible. The maximum of the cross polarization occurs roughly at an angle of λ/d radians from the peak and is usually at least 16 db down from peak power.

15·16. Gain Measurements.—Thus far the procedures for determining relative field intensity or relative power per unit solid angle in all directions have been discussed. However, for the calculation of the transmission or reception of radiated energy it is necessary to place the radiation pattern on an absolute basis. To do this a standard uniform radiator is assumed, and the directive gain of an antenna is then defined as the ratio of the peak radiated power per unit solid angle to the radiated power per unit solid angle from an isotropic radiator, assuming the same total radiated power in each case. Knowledge of the gain and the radiation pattern therefore fixes the radiation in any direction.

For practical applications, one would like to have a quantity that expresses the power per unit solid angle in the direction of maximum radiated power in terms of the power delivered to the antenna terminals. Or conversely, if the antenna is used as a receiving antenna, one would like to know the maximum power delivered to a load matched to the antenna transmission line of assumed zero loss when the power per unit solid angle incident on the antenna is known. This effective gain, as defined above, will differ from the definition of directive gain that was used in previous chapters only in so far as it takes into account heating losses in the antenna and the loss of power due to reflection as a result of having a mismatched antenna. It is assumed that the same losses will result whether the antenna is used as a transmitter or a receiver, and therefore the receiving and transmitting gains of an antenna are identical.

Typical procedures for determining directive gain and effective gain will be discussed, and procedures for determining effective gain standards will be outlined.

15·17. Directive Gain.—The directive gain of a transmitting antenna referred to an ideal isotropic radiator is given by

$$\text{Directive gain} = \frac{\text{peak power radiated/unit solid angle}}{\text{total power radiated}/4\pi},$$

or that of a receiving antenna by

$$\text{Directive gain} = \frac{\text{peak power received}}{\text{average power received}}.$$

This definition does not take into account any heating losses or reflection losses.

Experimentally the directive gain is obtained directly from the radiation pattern (either receiving or transmitting). If the relative power per unit solid angle $P(\theta,\phi)$ as a function of orientation θ and ϕ has been determined, then

$$\text{Directive gain} = \frac{4\pi P(0,0)}{\int_0^\pi \int_0^{2\pi} P(\theta,\phi) \sin\theta \, d\theta \, d\phi}, \tag{39}$$

where $\theta = 0$ and $\phi = 0$ is the direction of peak radiation. For accuracy a complete space pattern is required.

For an antenna with essentially a pencil-beam pattern, the assumption is frequently made that $P(\theta,\phi)$ may be replaced by the average of the radiation pattern in the principal E- and H-planes with no dependence on ϕ. If $P_E(\theta)$ and $P_H(\theta)$ are the patterns in the principal E- and H-planes respectively, then

$$\text{Directive gain} = \frac{4P(0,0)}{\int_0^\pi P_E(\theta) \sin\theta \, d\theta + \int_0^\pi P_H(\theta) \sin\theta \, d\theta}. \tag{40}$$

For integration purposes (planimeter or Simpson's rule), it is convenient to write $x = 1\cos\theta$ and plot P_E and P_H as a function of x, leading to the equation

$$\text{Directive gain} = \frac{4P(0,0)}{\int_0^2 [P_E(x) + P_H(x)] \, dx}. \tag{41}$$

This approximate procedure has been found to be fairly accurate (within 10 per cent) for pencil-beam antennas in which not too large a portion of the radiated energy is contained within the side-lobe structure).

The experimental determination of directive gain is very tedious and is subject to many possible errors such as incompleteness of radiation pattern measurements, spurious lobes due to improper setting, inaccuracy of angle determinations, improper evaluation of noise, and errors in graphical integration. It also suffers from the fact that the time required for such a measurement is long and thus rapid gain determinations cannot be made.

15·18. Gain Comparison.—The best method for determining effective gain, which is the quantity of most interest, is by comparison of the antenna under investigation with that of a gain standard, either on reception or on transmission. The procedure for determining such a gain standard is discussed later.

The experimental setup for gain comparisons of a receiving antenna is the same as that utilized for receiving pattern measurements. Uniformity of field across the pattern mount is essential for accurate gain comparisons. The antenna is first peaked in azimuth, and elevation for maximum received power and the received power P_u is noted. The antenna is then replaced by the standard antenna which is also peaked in azimuth and elevation for maximum received power P_s using the same detection apparatus. The effective gain of the antenna is then

$$G = \frac{P_u}{P_s} \times \text{gain of standard}.$$

If the field is uniform, the gain standard may be clamped to the mount near the antenna under study. The reciprocal procedure is utilized for measuring gain on transmission.

Several precautions must be taken to ensure accurate gain comparisons. (1) It is essential that the field distribution be uniform; otherwise the gain comparison will depend on the relative positions of the two antennas. (2) A monitoring antenna at the receiving station peaked on the transmitter should always be available to check the constancy of the transmitter output. If the output has varied during the measurements, the ratio P_u/P_s must be corrected for the change in power level. (3) The same detection system should be used for both the antenna and antenna gain standard. (4) The detection system should be matched to the transmission line. This may be accomplished either by means of a tuner or by a suitable matching transformer. In any case, the matching device should be considered as part of the detection system and not changed during the determination of P_u and P_s. (5) The readings P_u and P_s must be corrected if the response of the detection system is not square law. (6) The matched gain standard should be directive and preferably have a gain comparable (within 10 db) to that of the antenna under study.

Assuming that the gain standard is matched to the transmission line and that the detector is likewise matched to the line, the gain determined in the above procedure measures the efficiency of the test antenna compared with the gain standard. If the mismatch of the test antenna is known, the measured gain under matched conditions may be corrected by multiplying by $1/(1 - |\Gamma|^2)$, where Γ_L is the reflection coefficient of the antenna.[1] In principle the effective gain of the matched antenna can be obtained by inserting a tuner in the transmission line and adjusting the tuner for maximum received power. This serves to emphasize the fact that there is actually no difference between (1) matching both the detector and the antenna to the line and (2) matching the detector to the antenna, but owing to unavoidable losses in most tuning devices it is more satisfactory to match the detector to the transmission line by other methods and then make corrections for the mismatch of the antenna.

15·19. Primary Gain Standard Determination.—Given two identical matched antennas separated by a large distance R with power P_T being delivered to the transmitting antenna, how much power will be received in the terminating load at the receiving antenna? It will be assumed that free-space propagation exists, that the transmission line between the receiving antenna terminals and the load is lossless, and that the load

[1] See Sec. 2.15.

is matched to the transmission line. The two antennas are peaked so that maximum power is received in the receiving antenna.

The maximum power transmitted per unit solid angle is given by $P_T G/4\pi$, where G is the effective gain of the antenna. The solid angle subtended at the transmitter by the effective absorption area A_r of the receiving antenna is given by A_r/R^2. The power P_R received by the latter is therefore

$$P_R = \frac{P_T G}{4\pi} \frac{A_r}{R^2} = \frac{P_T G^2 \lambda^2}{(4\pi)^2 R^2}, \qquad (42)$$

where A_r has been replaced by $G\lambda^2/4\pi$.

It will be noted that losses in the antenna have been combined into the factor G and that the receiving gain and transmission gain have been assumed equal. The factor G determined by means of this equation is what is meant by effective gain. If such an experiment is performed and a number G obtained, all other antennas at the same wavelength may be compared with this antenna and their effective gains may then be obtained. The effective gain defined in this manner may be expressed on transmission by

$$G = \frac{\text{peak power radiated/unit solid angle}}{\text{power delivered to antenna}/4\pi}.$$

The experimental determination of

$$G = \frac{4\pi R}{\lambda} \sqrt{\frac{P_R}{P_T}} \qquad (43)$$

requires the determination of R, λ, P_R and P_T. Wavelength can be very accurately measured by means of a wavemeter. R, the separation between the two antennas, can be determined by measurement with good accuracy. The exact points between which R should be measured is a little doubtful; however, with R large ($R \geqq 2d^2/\lambda$, corresponding to a phase variation of less than $\lambda/16$ over the aperture of width d) the use of the aperture to aperture distance is sufficiently exact. By utilizing a method that involves measuring the ratio P_R/P_T instead of P_R and P_T separately, G can be determined quite accurately. With suitable precautions in experimental technique G can easily be measured to better than 5 per cent.

The procedure for determining G is to first match two practically identical antennas and match the calibrated detection system to the transmission line. The antennas should be separated by a distance R greater than $2d^2/\lambda$ and in a clear space so that reflections from the ground or near-by objects are negligible in comparison with the direct beam between the two antennas. The setup is shown schematically in Fig. 15·17. The electrical apparatus used is the same as that used in pattern measurements. The transmitting and receiving antennas are peaked for maximum received power. The procedure for this is first to line up the antennas roughly and, with the transmitting antenna fixed, to adjust the receiving antenna for maximum received power. The receiver is then fixed and the transmitter adjusted so as to make the received power a maximum. The transmitting antenna is again fixed, and the receiving antenna repeaked. This should be sufficient for accurately lining up the antennas. The r-f transmitter should be loosely coupled to the transmission line by means of a matched attenuator pad so that removing the

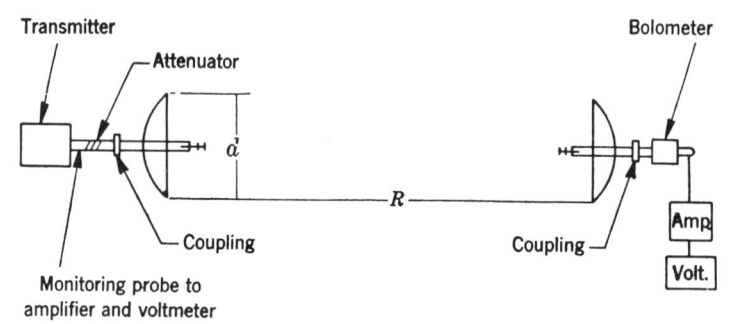

FIG. 15·17.—On the method for the determination of the gain of identical antennas.

antenna does not affect the power output of the r-f transmitter (this can be verified by using a monitoring probe in the line between transmitter and attenuator). Assuming the use of a bolometer-linear-amplifier-detection system the electronic voltmeter reading is recorded at the receiving end and is proportional to P_R. The bolometer system is then disconnected from the receiving antenna, and the transmitting antenna is disconnected. The bolometer detecting system is then brought over and attached to the transmitter, and a reading on the electronic voltmeter proportional to P_T is obtained. The ratio of these readings is then P_R/P_T. The usual experimental procedure of repeating the experiment several times and with several distances R should be followed in order to obtain a good degree of accuracy. There is no technical reason for not being able to perform this experiment to an accuracy within 5 per cent.

Actually in this experiment $G = \sqrt{G_1 G_2}$ is obtained where G_1 and G_2, although supposedly identical, may differ by a few per cent due to inability to make two exactly duplicate antennas. In order to differentiate between the two, a comparison experiment, which has been described for determining the gain of an unknown antenna, is made. Antenna G_1 is placed on the receiving mount for secondary pattern antenna measurements and set up in the usual fashion, and the power received is noted on the voltmeter as P_1. It is then replaced by antenna G_2, and the voltmeter reading P_2 is noted. The gain of antenna G_1 will be $G_1 = P_1G_2/P_2$ so that

$$G_1 = \sqrt{\frac{P_1}{P_2}}G \quad \text{and} \quad G_2 = \sqrt{\frac{P_2}{P_1}}G.$$

This experiment then produces two absolute gain standards at the operating wavelength.

Even if these two gain standards were lossy, the result of the experiment would be the determination of the desired value of G for practical application. With no heating losses the value G obtained in this fashion would be exactly that obtained on the basis of the definition of the directive gain.

15·20. Reflection Method for Gain Determination.—A modification[1] of the preceding method for determining absolute gain utilizes a single

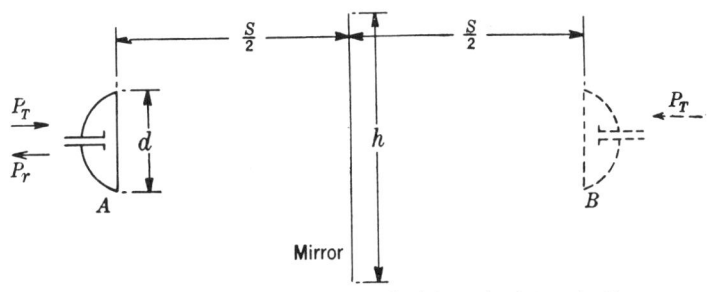

Fig. 15·18.—On the reflection method for gain determination.

matched antenna and a plane metallic reflecting surface as shown in Fig. 15·18. The second antenna used in the method discussed in the preceding section is replaced here by the image in the reflector. Energy incident on the latter is reflected and absorbed by antenna A giving rise to a reflected wave in the transmission line. The ratio P_r/P_t is then found by measuring the standing-wave voltage ratio in the line:

$$\frac{P_r}{P_t} = \left(\frac{r-1}{r+1}\right)^2 = \frac{G^2\lambda^2}{(4\pi)^2 S^2}, \tag{44}$$

where S is the distance from the antenna to its image in the reflector. Solving for G, we obtain

$$G = \frac{4\pi S}{\lambda}\left(\frac{r-1}{r+1}\right). \tag{45}$$

The practicality of the method depends on the distance $S/2$ that is required from the antenna to the mirror and the required dimensions of the latter. The use of the image antenna is based on the ideal situation of an infinite reflector. The criterion for the distance S is the same as previously discussed, $S \geq 2d^2/\lambda$, although experimentally distances less than $2d^2/\lambda$ have been tried without appreciable error. The mirror must

[1] E. M. Purcell, "A Method for Measuring the Absolute Gain of Microwave Antennas," RL Report No. 41-9.

be large enough to intercept most of the main beam whose width is of the order $2\lambda/d$ radians. For a square mirror of edge length h, we have then

$$\frac{2\lambda}{d} \leq \frac{2h}{S} \quad \text{or} \quad h \geq \frac{S\lambda}{d}.$$

For $S = 2d^2/\lambda$, the dimension h required is equal to or greater than $2d$. The mirror must be flat to a small fraction of a wavelength, at least $\lambda/16$.

The experimental procedure consists of setting up the matched antenna on a mount, peaking the antenna in azimuth and elevation so that the reflected power, as denoted by the maximum standing-wave ratio, is a maximum. The generator should be well padded so that it appears as a matched load to the returning energy. The next step is to measure the VSWR at a distance $S/2$ from the mirror at the position chosen, and again at a position just $\lambda/4$ nearer or farther from the mirror. The $\lambda/4$ displacement reverses the phase of the returning signal, with negligible effect on its intensity; and by taking the arithmetic mean of the gains computed for the two positions, most of the error caused by any small residual mismatch in the antenna and line is eliminated. This procedure also compensates for multiple reflections from the metallic surface. The experiment should be repeated at several distances. The difficulty in the method lies in determining VSWR accurately, since the order of magnitude of VSWR will be small. The method is not so accurate as the two-antenna method.

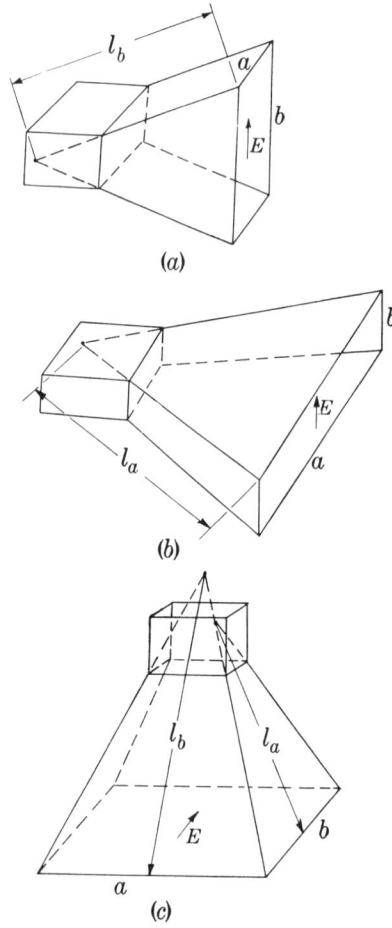

FIG. 15·19.—Electromagnetic horns: (a) E-plane sectoral horn; (b) H-plane sectoral horn; (c) pyramidal horn.

15·21. Secondary Gain Standards.—The theoretical gain has been calculated[1] for a pyramidal horn with an aperture a in the H-plane, an

[1] S. A. Schelkunoff, *Electromagnetic Waves*, Van Nostrand, New York, 1943, Chap. 9.

aperture b in the E-plane, and corresponding slant heights l_a and l_b as shown in Fig. 15·19. A few pyramidal horns have been compared with gain standards at 1, 3, and 10 cm, and the calculated values have agreed with the comparison values to within 5 per cent. Accordingly, horns may be used as secondary gain standards having a high degree of accuracy. This is particularly valuable when exploring a new wavelength region because the horns are easy to make and are fairly well matched.

The pyramidal horn may be thought of as a superposition of an E-plane sectoral horn and an H-plane sectoral horn. The gain is expressed in terms of the gains of the component horns. Defining the Fresnel integrals,

$$C(x) = \int_0^x \cos\left(\frac{\pi q^2}{2}\right) dq; \qquad S(x) = \int_0^x \sin\left(\frac{\pi q^2}{2}\right) dq, \qquad (46)$$

we have the following expressions for gain:

E-plane sectoral horn:

$$G_E = \frac{64 a l_b}{\pi \lambda b}\left[C^2\left(\frac{b}{\sqrt{2\lambda l_b}}\right) + S^2\left(\frac{b}{\sqrt{2\lambda l_b}}\right)\right]. \qquad (47)$$

H-plane sectoral horn:

$$G_H = \frac{4\pi b l_a}{\lambda a}\{[C(u) - C(v)]^2 + [S(u) - S(v)]^2\}, \qquad (48)$$

where

$$u = \frac{1}{\sqrt{2}}\left[\frac{\sqrt{\lambda l_a}}{a} + \frac{a}{\sqrt{\lambda l_a}}\right]; \qquad v = \frac{1}{\sqrt{2}}\left[\frac{\sqrt{\lambda l_a}}{a} - \frac{a}{\sqrt{\lambda l_a}}\right] \qquad (49)$$

Pyramidal horn:

$$G = \frac{\pi}{32}\left(\frac{\lambda}{b} G_H\right)\left(\frac{\lambda}{a} G_E\right). \qquad (50)$$

Curves of $(\lambda/a)G_E$ as a function of b/λ are plotted in Fig. 15·20. Corresponding curves for $(\lambda/b)G_H$ as a function of a/λ are plotted in Fig. 15·21. These curves obviate the necessity of evaluating the Fresnel integrals of Eq. (46) for most horn sizes, since the ranges of a/λ, l_a/λ, b/λ, l_b/λ are within the limits that have been found convenient for horn design.

15·22. Interaction between Antennas.—The interaction between antennas has been mentioned in the discussion of pattern and gain measurements without reference to the orders of magnitude involved. We shall now discuss the interaction between antennas such as may occur in the determination of gain. Consider, as shown in Fig. 15·22, two matched systems; one a transmitter, the antenna and generator of which are both matched to the transmission line; the other the receiver, in which the antenna and load are also both matched to their line. The voltage across the load may be considered as the superposition of component voltages generated by a series of waves arising by multiple scattering between the antennas. Also, as a result of the interaction between

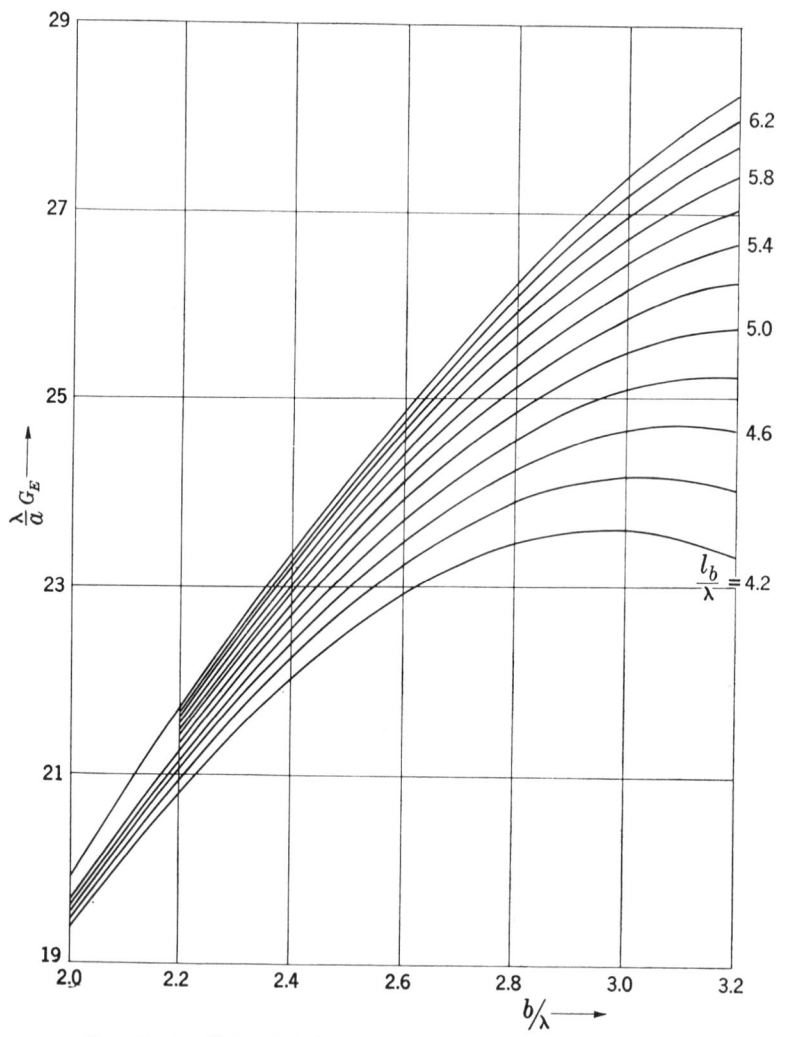

Fig. 15·20.—Gain of E-plane sectoral horns as a function of b/λ.

the antennas, a reflected wave will be observed in the transmission line of the transmitter system that may likewise be analyzed in terms of multiple scattering.

The scattering process of an antenna may be described, as in the case of absorption, in terms of an interception area, or scattering cross section, presented to a plane wave.[1] The scattered field set up by an antenna is directive and can be specified by a gain function analogous to the gain function of its transmission field. Let S be the magnitude of the Poynting vector in an incident plane wave, A_s the scattering cross section

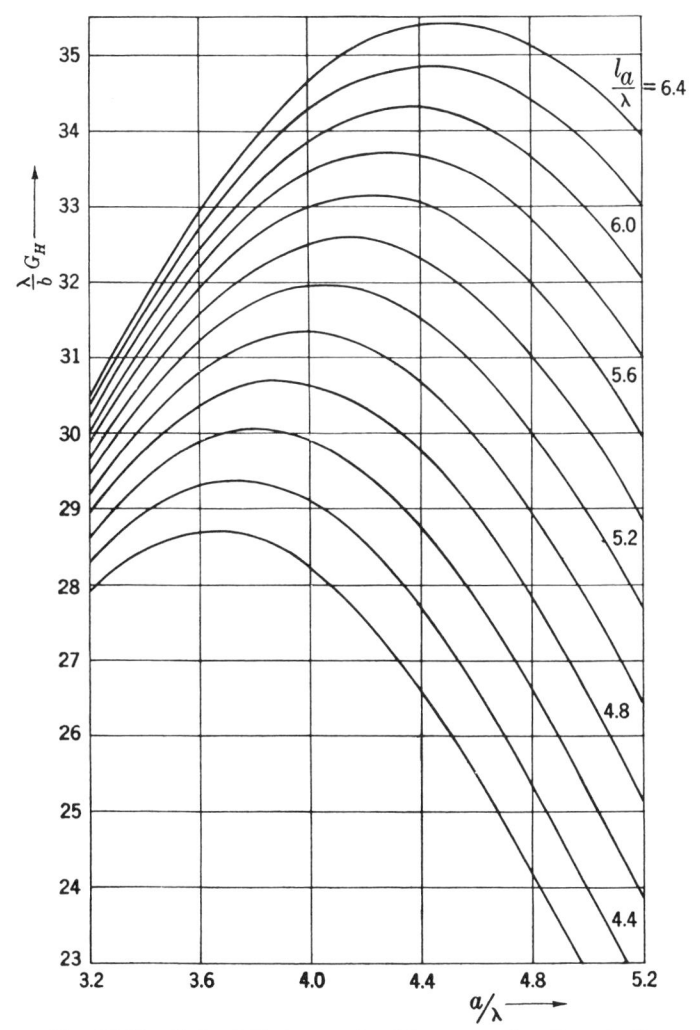

Fig. 15·21.—Gain of H-plane sectoral horns as a function of a/λ.

of the antenna; then the amplitude of the scattered field at a distance R in a given direction may be written

$$E_s = \text{const} \left(\frac{SA_s G_s}{4\pi R^2}\right)^{1/2}, \tag{51}$$

[1] *Cf.* Secs. 1·2 and 2·11.

where G_s is the scattering gain function in the given direction. The scattering cross section is a function of the aspect presented by the antenna to the incident wave; for a given direction in space G_s is likewise a function of the aspect of the antenna. It should be noted that the

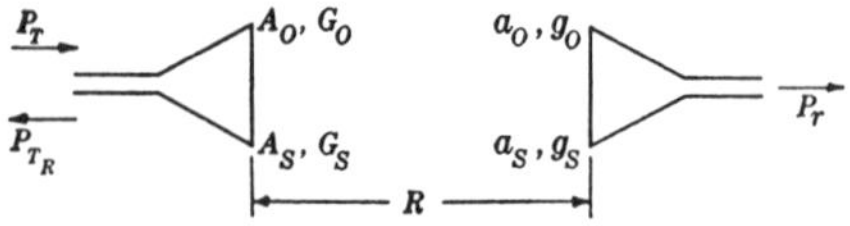

Fig. 15·22.—On the interaction of antennas.

scattered field pattern differs in general from the transmission field pattern.

Consider now the problem of the two antennas. Let A_0 be the absorption cross section and G_0 the transmission gain of the transmitting system in the direction of the line of sight between the two antennas; let A_s and G_s be respectively its scattering cross section and scattering gain for the same direction. The corresponding quantities for the receiver are a_0, g_0, a_s, g_s, respectively. We shall compute the voltage in the transmission line of the receiving system in detail. Let P_t be the total power radiated by the transmitter in the absence of interacting systems. The transmitter radiates a primary wave to the receiver with power per unit solid angle in the direction of the latter given by $G_0 P_t / 4\pi$. The receiver would extract from this wave alone the power

$$P_1 = \frac{P_t}{4} G_0 \frac{a_0}{R^2}, \qquad (52)$$

giving rise to a voltage

$$V_1 = \frac{1}{\alpha} P_1^{\frac{1}{2}} e^{j\delta} = \frac{1}{\alpha} \left(\frac{P_t G_0 a_0}{4\pi R^2} \right)^{\frac{1}{2}} e^{j\delta} \qquad (52a)$$

at a fixed reference point in the line; α and δ are constants of the receiving system, the precise values of which are not needed here. The scattering cross section of the receiver intercepts the power $G_0 P_t a_s / 4\pi R^2$ of the incident wave and sets up a scattered wave carrying power per unit solid angle

$$P_s = \frac{G_0 P_t a_s}{4\pi R^2} \frac{g_s}{4\pi}$$

in the direction of the transmitter. The latter is rescattered by the transmitter; the scattering cross section of the transmitter intercepts the power $P_s A_s / R^2$ and reradiates in the direction of the receiver the power per unit solid angle

$$P'_s = \frac{P_s A_s}{R^2} \frac{G_s}{4\pi}.$$

From this secondary wave alone the receiving antenna would abstract power

$$P_2 = P'_s \frac{a_0}{R^2} = \frac{G_0 P_t a_0}{4\pi R^2} \left(\frac{a_s g_s A_s G_s}{16\pi^2 R^4} \right),$$

corresponding to a voltage
$$V_2 = V_1\beta e^{-j[(4\pi R/\lambda)+\delta]}, \tag{53}$$
where
$$\beta = \frac{(a_s g_s A_s G_s)^{1/2}}{4\pi R^2}. \tag{54}$$

The factor $e^{-j(4\pi R/\lambda)}$ is introduced to express the phase delay introduced by the path $2R$ traversed by the scattered wave from the receiver to the transmitter and back. The secondary wave is rescattered by the receiver; and following the process through as before, it is seen that the voltage excited in the receiver line as a result of the second scattering stage is
$$V_3 = V_2\beta e^{-j[(4\pi R/\lambda)+\delta]}. \tag{55}$$

The total voltage, as a result of successive multiple-scattering processes, is then
$$\begin{aligned}V &= V_1 + V_2 + V_3 \cdots \\ &= V_1\{1 + \beta e^{-j[(4\pi R/\lambda)+\delta]} + \beta^2 e^{-2j[(4\pi R/\lambda)+\delta]} + \cdots \},\end{aligned}$$
or
$$V = \frac{1}{\alpha}\left(\frac{P_t}{4\pi}G_0\frac{a_0}{R^2}\right)^{1/2} e^{j\delta}\frac{1}{1 - \beta e^{-j[(4\pi R/\lambda)+\delta]}}. \tag{56}$$

The net power absorbed by the receiver is $P_r = \alpha^2|V|^2$, or
$$\frac{P_r}{P_t} = \frac{G_0 g_0 \lambda^2}{16\pi^2 R^2}\left[\frac{1}{1+\beta^2 - 2\beta\cos\left(\frac{4\pi R}{\lambda}+\delta\right)}\right]; \tag{57}$$

the absorption cross section of the receiving antenna has been replaced by $a_0 = g_0\lambda^2/4\pi$.

The reflected line wave voltage of the transmitter can be computed in the same way. The magnitude of the reflection coefficient in the line is then found to be
$$|\Gamma| = \frac{[G_0 A_0 g_s a_s]^{1/2}}{4\pi R^2}\left[\frac{1}{1+\beta^2 - 2\beta\cos\left(\frac{4\pi R}{\lambda}+\delta'\right)}\right]^{1/2}. \tag{58}$$

It is seen that the power absorbed by the receiver and the standing-wave ratio observed in the transmitter are periodic functions of R with a period of $\lambda/4$.

Very little information is available on the subject of the scattering cross section and gain functions. To obtain an order of magnitude of the interaction effect we shall make the *ad hoc* assumption that the scattering cross section and gain are related in the same way as the absorption cross section and transmission gain:
$$G_s = \frac{4\pi A_s}{\lambda^2}. \tag{59}$$

If the two antennas are identical, the introduction of Eq. (59) into Eqs. (57) and (58) expresses the latter in terms of only two parameters A_0 and A_s. The values of A_0 and A_s both can then be found from studies of P_r/P_t or $|\Gamma|$ as a function of R. In an experiment with a paraboloidal antenna by the mirror method it was found that $A_s = \frac{1}{2}A_0$. Taking this as a general estimate, the power received in the two-antenna experiment with identical antennas is seen to vary between limits

$$\frac{P_r}{P_t} = \left(\frac{G_0\lambda}{4\pi R}\right)^2 \left(\frac{1}{1 \pm \frac{A_0^2}{4\lambda^2 R^2}}\right)^2 \tag{60}$$

for a displacement of $\lambda/4$ in distance. If we wish to reduce this variation in power to less than $\frac{1}{4}$ db, we arrive at a distance of $R = 2d^2/\lambda$ [assuming $a_0 = 0.6(\pi d^2/4)$ which is approximately correct for paraboloidal antennas]. For a distance d^2/λ there is almost a 1-db variation. This is, therefore, another reason for the choice of $R = 2d^2/\lambda$ rather than d^2/λ in pattern and gain comparison measurements, and perhaps $R = 3d^2/\lambda$ is required for accurate gain standard measurements. The magnitude of the power reflected back into the transmitting antenna will be appreciable and results in a mismatch with respect to the generator of magnitude

$$\frac{r-1}{r-1} = \frac{G_0\lambda}{4\pi R}\sqrt{\frac{g}{4\pi}\frac{a}{R^2}}\sqrt{\frac{1}{1 \pm \beta^2}} = \frac{1}{2}\left(\frac{G_0\lambda}{4\pi R}\right)^2. \tag{61}$$

In the case of two paraboloidal antennas, assuming again that $a_0 = 0.6\left(\frac{\pi d^2}{4}\right)$, we find $r = 1.25$ for $R = d^2/\lambda$ and $r = 1.04$ for $R = 2d^2/\lambda$.

A similar argument for the mirror method leads to the relation that the power received is given by

$$\sqrt{\frac{P_R}{P_T}} = \frac{G_0\lambda}{4\pi S}\left\{\frac{1}{1 - \frac{1}{S}\sqrt{\frac{G_s A_s}{4\pi}}\,e^{-j[(2\pi S/\lambda)+\phi]}}\right\}. \tag{62}$$

It will be noted that in this case $\sqrt{P_R/P_T}$ is more sensitive to distance variations, since the correction term is proportional to the reciprocal of the distance rather than the square of the distance as in the two-antenna system. At a given distance $S/2$, taking the maximum and minimum VSWR separated by $\lambda/4$ enables one to determine $G_s A_s$ as well as G_0; experimentally of course, several distances $S/2$ are chosen for accuracy.

CHAPTER 16

ANTENNA MEASUREMENTS—EQUIPMENT

By O. A. Tyson

16·1. Survey of Equipment Requirements.—Measurements on microwave antennas differ in character from those carried out on most other radar components. A high order of amplitude stability is required of measuring equipment for the study of antennas and associated components, whereas high accuracy in timing and frequency control are the main requisites in measurements on other radar components. These requirements make the design of special equipment for antenna measurements most desirable.

The preceding chapter has mentioned briefly the chief items of equipment used in making antenna measurements. It has been pointed out that either a klystron or a magnetron is the most satisfactory source of power in the microwave region (2000 Mc/sec and higher). The reflex klystron has a definite advantage when the power requirement is 250 mw or less. The multicavity klystron, if available at the desired frequency, is useful when the power requirement is 10 watts or less; if the power requirement is greater than 10 watts, magnetrons must be used.

Mention has also been made of the demodulators or detectors commonly used, that is, bolometers and crystals. Some work has been done with diodes, but as yet they are not generally accepted for use above 1000 Mc/sec. Bolometers are especially useful because of their uniform square-law behavior. Crystals are approximately 15 db more sensitive than a bolometer but are not uniform in their behavior; they must be individually calibrated over the entire power range for which they are to be used. When a bolometer is used together with an amplitude-modulated power source, it is best to employ a tuned audio amplifier and stable vacuum-tube voltmeter for a sensitive indicating system. If a crystal or a diode is used, the amplifier-voltmeter combination may be used with an amplitude-modulated source; or with a c-w source, the detector may be connected to a microammeter or galvanometer as an indicating device. Another c-w method is to use the crystal or diode as a mixer, to amplify its output at some suitable i-f frequency, and to use as an indicator the low-current meter in the second detector circuit.

A very important instrument in antenna work is the automatic recorder, which can be used to record any r-f amplitude as a function of angle, position, or time.

16·2. Sources of R-f Power.—The discussion is here confined to a relative evaluation of various types of sources for antenna measurements. For details on the operation of these tubes, the reader is referred to *Microwave Magnetrons*, Vol. 6, and *Klystrons and Microwave Triodes*, Vol. 7, of this series.

Reflex Klystrons.—In measuring impedances, primary patterns of antennas, attenuations, etc., where the power required is between 25 and 250 mw, reflex-klystron sources may be conveniently used. This type of oscillator has several advantages, among which are

1. Wide tuning range with a single adjustment.
2. Electronic tuning for fine adjustments when precise frequency is sought.
3. Relatively small power-supply-and-modulator combination.
4. Ease of air cooling, accomplished with low-velocity quiet-operating blowers.

The circuit of a typical signal generator is shown schematically in Fig. 16·1. It consists generally of an electronically regulated anode

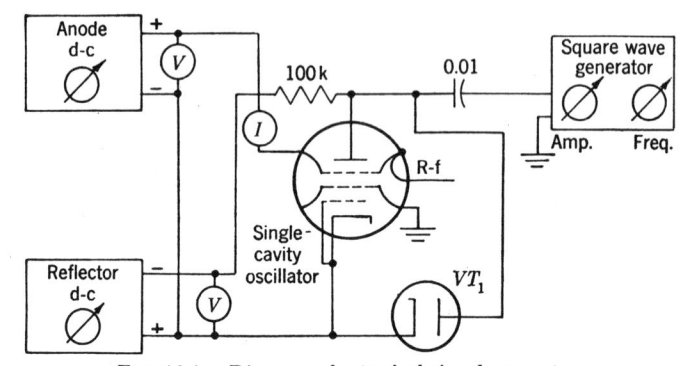

Fig. 16·1.—Diagram of a typical signal generator.

power supply, readily adjustable over a 2-to-1 voltage range, and a regulated reflector supply consisting of a string of VR tubes capable of producing the required maximum reflector voltage and a means (usually a potentiometer) for varying this reflector voltage continuously from the maximum to a very small minimum. The potentiometer in this circuit allows satisfactory adjustment because the reflector is always negative with respect to the cathode and draws no current; hence, no appreciable change of resistance as a function of applied voltage is encountered in the load element. Modulation is most suitably introduced in the reflector circuit by a square-wave voltage superimposed on the steady-state reflector voltage, which keys (i.e., switches) the oscillator off and on at the modulation rate. A limiting tube, which consists of a shunting diode, is used in this arrangement to cut off the positive half cycle of the modulation wave, thus preventing the reflector from being driven positive. Any high-vacuum diode with a low voltage drop, capable of supporting a peak voltage of several hundred volts, can serve as a limiting tube. The 6X5 is frequently used for this purpose.

This modulation and biasing procedure avoids the frequency modulation that would exist if a sinusoidal modulation wave were being used. The relation between the frequency modulation that results and the type of modulation wave used is best discussed with reference to Fig. 16·2. Figure 16·2a shows the r-f power output of a klystron as a function of reflector voltage (the numerical values indicated are merely illustrative of the order of magnitude). It is seen that power is obtained only over discrete voltage ranges corresponding to the "power modes" of the tube. The frequency of oscillation of the tube as a function of reflector voltage, within any power mode, is of the general form illustrated in Fig. 16·2b. Since the frequency is a function of reflector voltage, a sinusoidal modulation voltage would result in a wide frequency variation in the output. Further, since the r-f power is not a linear function of reflector voltage, a sinusoidal modulation will not produce a sinusoidal power envelope but one that is considerably distorted. This is avoided with on-off square-wave modulation.

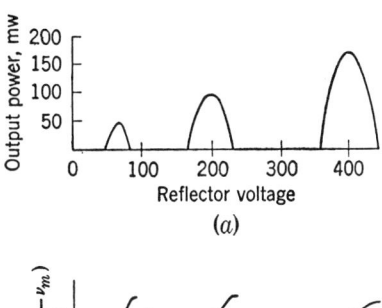

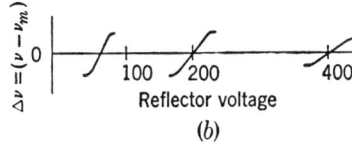

Fig. 16·2.—Operating characteristics of a reflex klystron: (a) variation in output r-f power with voltage; (b) change in output frequency with respect to the frequency ν_m of the maximum power of the mode.

It should be noted that the power output can be switched off and on with a square wave of amplitude considerably less than the maximum reflector voltage. This means that only a relatively small modulation amplitude need be superimposed on the steady-state reflector voltage to obtain complete modulation. For instance, as illustrated in Fig. 16·3, the steady-state voltage V_0 may be somewhat less than that required to activate any one of the possible modes, and the keyed voltage V added in series to this may be just equal to the interval between the nonoscillating condition and the point of maximum power. The steady-state voltage V_0 is generally so chosen that the amplitude of the keying voltage does not exceed 100 volts, axis to peak, for complete modulation. The keying is generally sustained at a regular rate of some 100 to 2000 cps.

To illustrate the operating behavior of the signal generator let us take the Sperry 419B klystron. The anode voltage would then be adjusted for 1000 volts; the square-wave generator would be set for full output of 100 volts, axis to peak, if modulation is desired. The reflector voltage can now be increased from the minimum of 25 volts until a mode of oscillation is found. At the instant the tube goes into oscillation the anode current suddenly begins to increase from its normal steady-state value of about 40 to 45 ma to a value of perhaps 43 to 48 ma at maximum power output. It will be noted that for any tuning adjustment of the tube cavity and for a fixed anode voltage, there will be from two to three modes of oscillation with the various possible reflector voltages in the range of 25 to 450 volts.

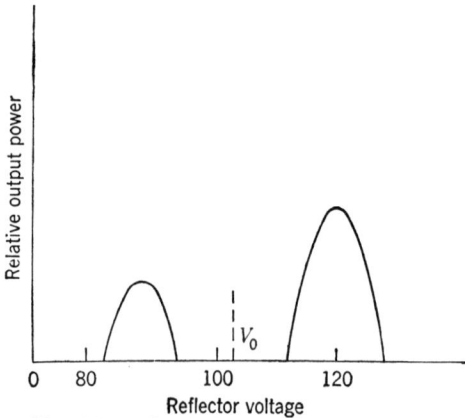

Fig. 16·3.—On the klystron operating point.

If, in coupling the klystron to the load, a rather large mismatch is unavoidable, then it is desirable to use a tuner between the klystron and load in order to obtain both maximum power and stability.

Double-cavity Klystrons.—The measurement of antenna secondary patterns, large attenuations, etc., usually requires somewhat higher power (approximately 10 watts) than is obtainable from the reflex klystron. Therefore, a different source is necessary: for example, a double-cavity klystron, such as the Sperry 410-R klystron, for the range of approximately 7 cm upward. A magnetron also may be used in this range.

A power-supply-and-modulator combination similar to that shown for the reflex tube (Fig. 16·1) may be used for the double-cavity klystron, with the elimination of the reflector supply and a modification of the method of modulation. A typical circuit for use with the 410-R klystron is shown schematically in Fig. 16·4.

The output power is a function of anode potential; if the output power is plotted as a function of applied voltage, it is found that there is a set of power modes similar to those shown in Fig. 16·2 for the reflex klystron. The double-cavity klystron differs from the reflex klystron in that there

SEC. 16·2]   SOURCES OF R-F POWER   597

is very little change in the frequency of oscillation with the variation of the anode voltage.

Since power does exist in these discrete modes, it is again necessary only to add a square wave to a constant d-c voltage when modulation is desired. This is accomplished by setting the constant d-c voltage to a value corresponding to a position just out of the desired mode on the low side. The square-wave amplitude is then chosen such that the sum of the constant d-c and the square-wave voltages causes the anode potential to rise to the optimum value for the power mode during the peak position of the wave. In a practical case the constant d-c potential

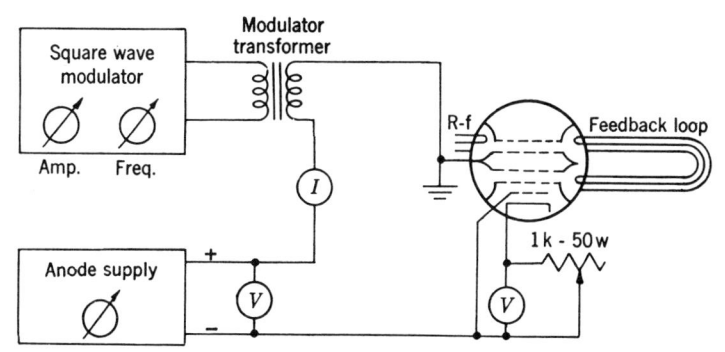

FIG. 16·4.—Schematic diagram of a signal generator employing a double-cavity klystron.

might be approximately 1700 volts, and the square wave about 200 to 300 volts, axis to peak. The modulator is required to deliver some 4 or 6 watts. The klystron anode current will be approximately 100 ma; the cathode bias voltage from 0 to 30 volts positive.

The use of the 410-R klystron as an oscillator requires that an external feedback path be provided, because this tube has been designed for use as an amplifier as well as an oscillator. This feedback path may consist of 6 in. or more of flexible coaxial cable of low loss and a coaxial-line stretcher adjustable over possibly 3 in. of length. This line stretcher is used to adjust the phase of the feedback to the optimum point. This can be done readily by setting the anode and square-wave voltages to the values previously mentioned and then moving the line stretcher over its length while observing the relative power received from the klystron. If oscillations do not occur anywhere in the range of the feedback path, a new anode voltage should be tried, approximately 200 volts higher or lower than the value originally suggested. The Sperry 410-R klystron is shown in Fig. 16.5 from two aspects. Any two of the coaxial-line output leads, one from the upper pair and one from the lower pair, may be connected to provide the feedback path. The remaining member of the upper pair is then used as a power output lead.

Double-cavity tubes will function only if the two cavities are very nearly resonant at the same frequency. Since it is often required that the tube be used at a frequency not obtainable with a narrow-range micrometer tuner, a few words pertaining to the adjustment of these cavities are in order. This adjustment can best be made by using each individual cavity as an absorption device; a reduction in power to a detector will then be noticed when the cavity is adjusted to exact reso-

FIG. 16·5.—Photograph of the Sperry 410-R klystron.

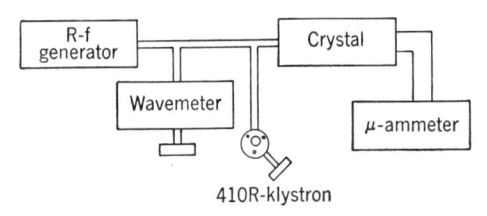

FIG. 16·6.—Schematic diagram of a circuit for adjusting a cavity to resonance.

nance. Figure 16·6 shows a suggested r-f circuit. The method is to set the generator to the exact desired wavelength as read on the wavemeter and indicated by the microammeter. The klystron cavity is connected to the circuit as shown, and the three adjustment screws are manipulated until resonance at the generated frequency is obtained (if this frequency is in the range of the klystron). It is well to note that tightening the tuning screws (clockwise motion) will cause the frequency to increase; the screws should not be tightened to the point of causing excessive bulging of the diaphragm. Also, in the tuning process care should be taken to keep the tuner frames substantially parallel. When this adjustment has been made on both cavities, they will be resonant at the same frequency and in a condition to be installed in the signal generator. It is possible that final adjustments will be needed after installation; one of the top tuner screws can then be moved back and forth slightly until satisfactory oscillation is obtained.

Magnetrons.—The investigation of the secondary pattern of large antennas requires that a long transmission path be used in order to ensure a uniform plane wave across the antenna aperture. In view of the relative insensitivity of accurate measuring devices, this generally means that an average power in excess of 10 watts is necessary. The magnetron is the source of power usually employed for these power levels; at wavelengths less than 7 cm the magnetron is the only oscillator capable of generating power high enough for the majority of secondary-pattern measurements. This type of oscillator is available for most of the ranges throughout the explored microwave regions. They are constructed in three principal types:

1. Spot frequency, pulsed operation.
2. Tunable, pulsed operation.
3. C-w, spot frequency, or tunable.

For antenna work it is generally desirable to use pulsed magnetrons because of the fairly high efficiency and small dimensions of both the tube and the driving modulators required. There are several ways in which a magnetron can be pulse modulated, but for antenna work the so-called "soft-tube" line-type modulator is preferred, except for powers in excess of 200-kw pulse peaks. This method of modulation is shown schematically in Fig. 16·7.

The modulator provides very short pulses of very large current to the magnetron, which is to have a duty ratio ($\nu\ \delta t$) of about 0.001 or less. (Here δt is the duration of the pulse, and ν is the repetition frequency, the number of pulses per unit of time.) The magnetrons used in this fashion are, of course, designed for radar use, where the short pulses of high power are needed for measurement of echo time. However, they serve very well for antenna work when a power-integrating demodulator, such as a bolometer, is used for the detecting element.

The circuit functions in the following manner. A variable d-c voltage source charges a pulse-forming network (synthetic transmission line) through a 60-henry choke during the time that the thyratron is not conducting. At regular intervals a trigger circuit drives the thyratron grid sufficiently positive to render the thyratron conducting whereupon the pulse-forming network discharges through the primary of the pulse transformer. This causes a voltage four to five times greater to appear across the magnetron, connected to the secondary of this transformer. If the original d-c voltage is adjusted properly, the amplitude of the pulse to the magnetron will be correct for operation. In this system the pulse width and shape are controlled by the pulse-forming network. The

600   ANTENNA MEASUREMENTS—EQUIPMENT   [SEC. 16·2

number of pulses per unit of time is determined by the repetition rate of the trigger.

In a typical case the d-c voltage may be adjusted to provide a 12-kv pulse across the magnetron for a period of 1 μsec, a thousand times each second. (The applied voltage is approximately the product of the d-c supply voltage and the pulse-transformer stepup ratio.) The magnetron

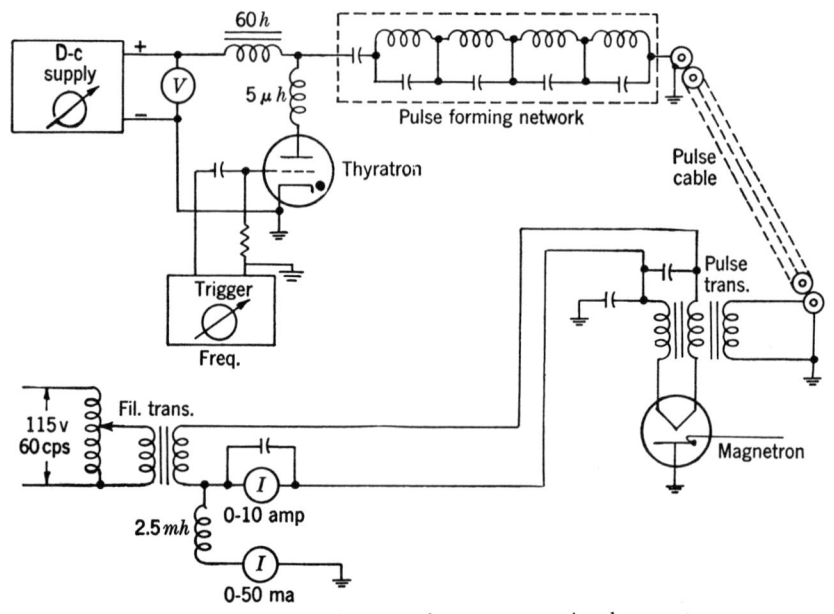

FIG. 16·7.—Schematic diagram of a magnetron signal generator.

will then pass an average current of about 10 to 12 ma. The approximate input power to the magnetron, exclusive of filament power, is 1.2×10^4 peak volts $\times 10^{-2}$ amp or 120 watts. The peak input power is given by

$$\frac{\text{Average power input}}{\text{Duty ratio}} = \frac{P}{\nu\, \delta t} = \frac{120 \text{ watts}}{10^{-3}} = 120 \text{ kw},$$

and the peak current is

$$\frac{1.2 \times 10^5 \text{ watts}}{1.2 \times 10^4 \text{ volts}} = 10 \text{ amp.}$$

The filament voltage is measured by the ammeter in the filament circuit, which has previously been calibrated for a measured potential difference at the magnetron filament of about 6 volts. This method is chosen because of the voltage drop across the pulse transformer, which is dependent on the current through it; a voltage measured at the filament transformer will not be the true voltage at the filament. The magnetron filament voltage should be reduced to about 3 volts—for a 6-volt tube—

after oscillation starts in order to maintain the cathode at a safe operating temperature. This is necessary because of the large electron bombardment of the cathode under operating conditions.

The load to which power is being supplied by the transmission line from the magnetron must be very well matched to the line if it is not to pull the magnetron frequency excessively.[1] If the mismatch is sufficient to cause a power standing-wave ratio of several to 1 with a practical length of feed line, the tube will probably not oscillate at all; at least it will have poor stability. Poor stability must particularly be avoided in antenna work, where a high order of amplitude stability, of 1 or 2 per cent, is absolutely necessary. Any of the undesirable effects just mentioned can usually be avoided by coupling the magnetron to the transmitting antenna by as short a transmission line as is mechanically feasible —2 ft or less—and maintaining the mismatch below 2 to 1 in power.

16·3. Detectors.—Bolometers and crystals are most frequently used as detectors in microwave antenna measurements. The bolometer has the decided advantage that its resistance-power curve is linear over a wide range of power, whereas crystals, although much more sensitive than bolometers, have a nonlinear response.

The crystal detector is a very small barrier layer rectifier composed of a chip of silicon in contact with a fine tungsten wire. When this element is properly introduced into an r-f circuit, a very sensitive and efficient rectification even at very high frequencies results.

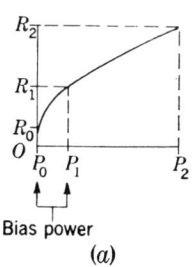

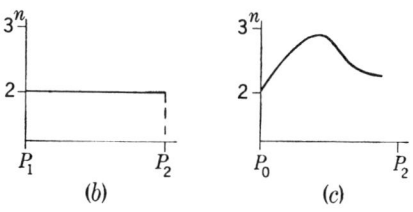

FIG. 16·8.—Characteristics of a bolometer and a silicon crystal: (a) resistance vs. power applied to bolometer; (b) "law of behavior" n vs. power above bias point; (c) "law of behavior" n vs. power for typical crystal.

In spite of their sensitivity, however, crystals find comparatively limited application in antenna measurements because of their nonlinear character.

The graph (a) in Fig. 16·8 shows the static relation between resistance and applied power in a bolometer, and it may be noted that this relation is linear in the power range P_1 to P_2. A relation exists between the power P applied to the measuring device and the output voltage V developed across the load:

$$P = KV^{2/n},$$

[1] *Microwave Magnetrons*, Vol. 6, RL Technical Series.

where K is a proportionality factor and n the law of behavior. In the range P_1 to P_2, $n = 2$ for the bolometer. In crystals for this same range of applied signal power, however, n varies considerably. This comparison for a typical case is shown graphically in (b) and (c) of Fig. 16·8. As a result of this nonlinear behavior, the crystal, while about 15 db more sensitive than a bolometer, cannot be used for measurements without being previously calibrated and the measured data corrected accordingly.

The bolometer element consists of a short platinum wire from 30 to 70 microinches in diameter. This element has an extremely low thermal capacity and because of its very small diameter possesses a very favorable surface-to-volume ratio.

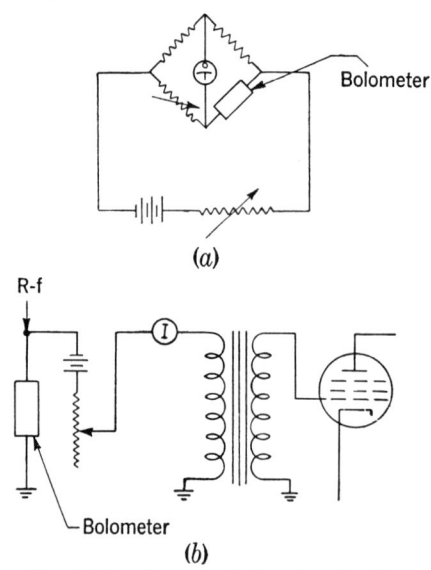

FIG. 16·9.—Common bolometer circuits.

It has been shown that the resistance of a platinum wire bolometer is linear with power above a certain minimum. Therefore, when a bolometer element is used in a circuit, it is necessary to provide a bias in order to work on the linear portion of the curve. Figure 16·9 shows the two most common circuit arrangements used with bolometers.

In circuit a the proper bias is obtained by choosing the current in the bolometer branch of the bridge which yields an initial power dissipation in the bolometer equal at least to P_1 of the curve a of Fig. 16·8. Then if the power dissipation of the bolometer is increased by coupling it to an r-f field, the consequent linear rise in resistance will cause the deflection of the linear null meter to be directly proportional to the absorbed r-f power.

Similarly in circuit b the bolometer current is controlled by adjusting the rheostat to a point where the uncoupled power dissipation is equal to or greater than P_1. When modulated r-f is coupled to the bolometer, a periodic resistance change will occur, causing a varying current that is proportional to power to flow in the transformer primary. There is then induced in the secondary a voltage that is likewise proportional to power.

The best workable range of modulation frequencies lies between 100 and 2000 cps. In Fig. 16·10 is graphed bolometer sensitivity vs. frequency for a wire 70 microinches in diameter. It is clear from this curve that greatest sensitivity appears below 100 cps but on the other hand

SEC. 16·3]   DETECTORS   603

difficulties in construction of a suitable amplifier eliminate the use of lower frequencies for this purpose.

The particular bolometer used depends on the radio frequency involved. For frequencies of less than about 4000 Mc the standard 8 AG meter fuse of $\frac{1}{100}$ or $\frac{1}{200}$ amp made by Littelfuse Company of Chicago is an excellent performer. For frequencies above 4000 Mc it becomes necessary to design the element for the particular application.[1]

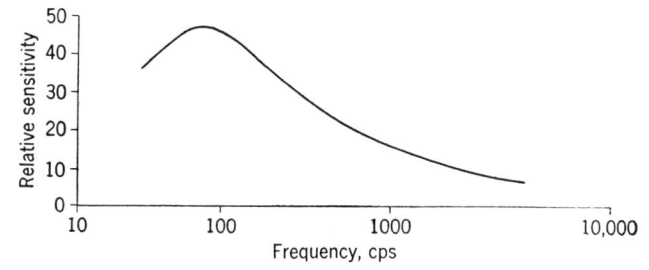

FIG. 16·10.—Bolometer sensitivity vs. modulation frequency for platinum wire (70 microinches in diameter).

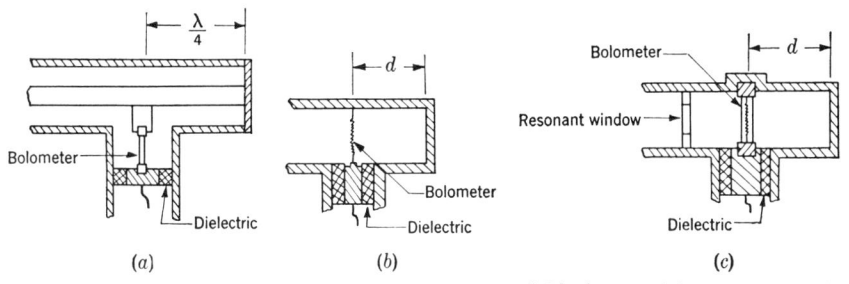

FIG. 16·11.—Methods of coupling bolometer to r-f field: (a) coaxial termination; (b) bolometer wire stretched directly across small dimension of waveguide; (c) bolometer in housing across small dimension of waveguide.

Figure 16·11 shows several methods of coupling the bolometer element to the r-f field. These methods of coupling along with the geometry and loss characteristics of the protective enclosure greatly influence the bolometer element design for use with microwaves.

With frequencies below 4000 Mc coaxial bolometer terminations are most frequently used. It is generally quite satisfactory at such frequencies to use either a $\frac{1}{200}$- or $\frac{1}{100}$-amp Littelfuse as the r-f integrating element. The choice is dictated by the operating resistance of these units, since their reactive components are quite similar. The $\frac{1}{100}$-amp fuse has an operating resistance of about 200 ohms at 10 ma, while the $\frac{1}{200}$-amp fuse operates at about 400 ohms at 5 ma.

Matching a coaxial termination to the line may be achieved by varying the length of the stub, which is nominally $\lambda/4$, and/or the distance

[1] A detailed description of the construction of Wallaston wire bolometers may be found in Vol. 11, Chap. 3, of this series.

between the center conductor and the point at which contact is made with the fuse. By adjusting these two parameters properly a reasonable match (VSWR $\sim$ 2) may be achieved on $\frac{7}{8}$-in. (46-ohm) line for about a 10 per cent bandwidth. If a better match is desired (VSWR of 1.2), it may be effected by introducing a coaxial transformer[1] of correct dimension, but the bandwidth for which this improved match will hold is reduced to about 1 per cent.

For frequencies above 4000 Mc waveguide bolometer terminations are most frequently employed. These are shown in Fig. 16·11b and c.

When used directly in waveguide the bolometer element must have axial electrodes that are parallel to the E-vector so as to achieve a reasonably large coupling. The material of the protective envelope, if used, must have a very low loss and in general should be no larger in diameter than absolutely necessary for mechanical support. It has been found that this envelope need not exceed $\frac{3}{16}$ in. in diameter and can be successfully made of polystyrene with $\frac{1}{32}$-in. wall.

The match in guide is influenced by the choice of the effective resistance of the wire and the distance d between the wire and the short circuit in Fig. 16·11b and c. With proper choice of variables very close matching (VSWR = 1.1) can be achieved for a bandwidth of 1 per cent or less. However by using a window (Fig. 16·11c) that is resonant at one frequency, the impedance match may be held for a bandwidth of, say, 8 per cent; the reactance of the window varies with frequency in the opposite sense to the variation of the reactance of the bolometer element with the short-circuited waveguide termination.

16·4. Amplifiers.—The power available at the detecting element is very small (5 mw or less) in the majority of the methods used in antenna measurements. For this reason the available voltage at the detector output terminals will likewise be small. For instance, the voltage at the terminals of a sensitive bolometer ranges from 10^{-7} to 10^{-2} volt rms over the linear part of the detection curve. Crystals also have an upper useful terminal voltage of about the same order of magnitude, but the lower limit extends down to around 10^{-8} volt for the audio region. It is evident, therefore, that some amplification is needed with these microwave detectors in order that a practical indicating meter be used for measurements.

The required performance characteristics of such an amplifier are:

1. Linearity over a range of at least 100 db.
2. An inherent noise level at least 6 db below the minimum input signal.
3. Good stability.
4. Freedom from response to outside fields.

[1] See Sec. 7·9.

Sec. 16·4]   AMPLIFIERS   605

Figure 16·12 shows the schematic diagram of a tuned audio amplifier meeting the above requirements which was designed to operate a Ballantine Model 300 voltmeter as the indicating device. This unit is linear from 0.001 to 100 volts rms output, or in other words over a range of 100 db, which corresponds to the range of the Ballantine meter.

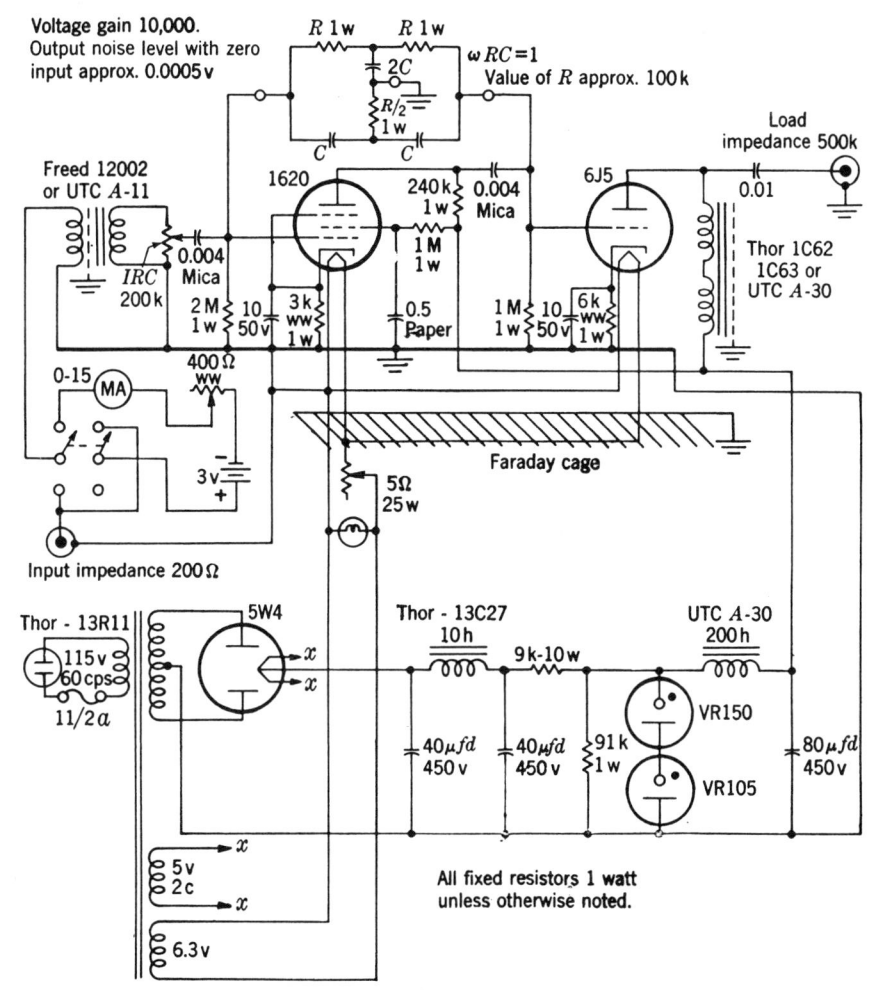

Fig. 16·12.—Tuned audio amplifier.

The voltage gain is 10,000, which permits operation of the input from 10^{-7} to 10^{-2} volt.

The tuned audio amplifier is based upon the use of a twin-T RC-bridge as a feedback element to reduce the bandwidth. This reduced bandwidth has an appreciable effect on the signal-to-noise ratio; in fact, the average peak noise voltage at the output terminal is about 5×10^{-4}.

This is equivalent to an input voltage of 5×10^{-8}, or 6 db below the minimum signal voltage. Typical performance curves are presented in Fig. 16·13. In (a) is plotted voltage gain against logarithm of the voltage output of the amplifier. This curve is clearly linear for practical purposes, over the range from 0.001 to 100 volts. The graph in (b) is a plot of relative voltage amplitude vs. frequency for the amplifier, showing the discrimination of the filter at 1000 cps.

Considerable care must be taken in selecting the components for the amplifier, also in the layout of the parts and wiring. For instance, it

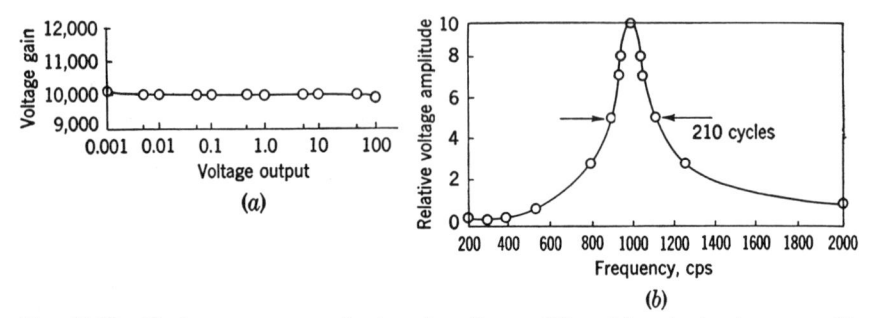

Fig. 16·13.—Performance curves for tuned audio amplifier: (a) output-gain curve; (b) curve showing discrimination of filter at 1000 cps.

is imperative to assure excellent magnetic shielding for the input transformer and the output choke; otherwise undesirable currents can be induced in the windings by surrounding fields. For a similar reason it is necessary that ground wires be short and connected to a common point. Tubes must have very low microphonic response, hence the use of the 1620 tube in the first stage. Also, the first tube should be shock-mounted. The power transformer is best packaged separately; it is not included in the amplifier cabinet but housed in a separate container and connected to the amplifier by a flexible cable about 3 ft long. This permits the transformer to be placed 2 or 3 ft away when in actual operation, a procedure that is not inconvenient, since the required transformer is quite small and light in weight.

The filters are constructed as small fixed-tuned plug-in units which are arranged in the amplifier for easy replacement. The frequency range is normally about 100 to 5000 cps.

There are many variations possible with this type of amplifier, most of which are just adaptations to special requirements. However, one modification is very desirable for impedance work. It consists essentially of the system already shown with the exception that the vacuum-tube voltmeter is built as part of the amplifier with an indicating meter placed in the front panel. This meter is generally calibrated in voltage standing-wave ratio with full-scale deflection as unity. The calibration is based on a square-law detector regardless of the type of detector used.

Another type receiver, based on a heterodyne principle, is used to some extent in antenna work, especially in the measurement of secondary patterns either where space attenuation is large or where for some reason the transmitted power is low. This method is capable of considerably greater sensitivity than is realizable with the audio method and also does not require the signal source to be modulated. The order of maximum practical power sensitivity with the heterodyne system, when using

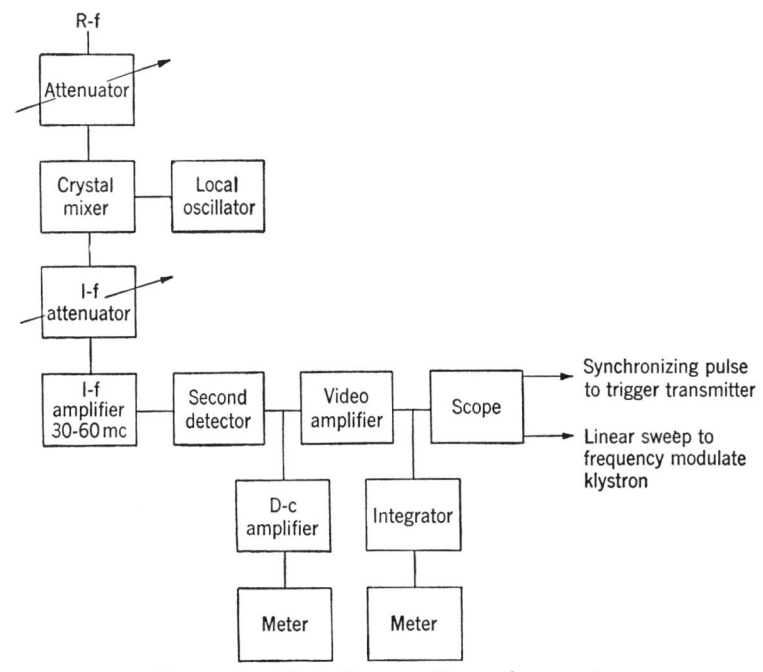

FIG. 16·14.—Block diagram of heterodyne receiver.

a crystal mixer, is about 10^{-12} watt, whereas the audio amplifier and crystal combination has a maximum power sensitivity of about 10^{-9} watt.[1] Figure 16·14 is a block diagram of a typical heterodyne receiver arranged for measurement work.

This setup makes use of either an r-f or an i-f attenuator or both and depends largely upon the attenuator as the measuring element. The attenuator, which is previously calibrated, is adjusted to have a minimum insertion in the circuit when the received signal is a minimum, and the indication on either of the meters or the oscilloscope is noted. Then for any signal of greater amplitude the attenuator is adjusted to an insertion greater than the original setting until the output indicator

[1] See *Vacuum Tube Amplifiers*, Vol. 18 of this series.

returns to the value observed for minimum signal. The change in signal amplitude then, of course, is the difference indicated by the calibrated attenuator. Actually either the c-w meter or the video meter may be calibrated for a range of, say, 5 or 10 db. Thus, signal differences can be measured by the indicating meter and the attenuator in combination. This provides a simple means of measuring the smaller variations that may be within the limits of the indicating meter without readjusting the attenuator.

The use of an i-f attenuator as the measuring element is to be preferred if the mixer can be shown to be linear over the desired range to be covered. Usually, for crystal mixers, this is true to about 30 db above the minimum detectable signal and to about 20 db further with carefully selected crystals. This method does not require a different attenuator for widely different frequencies of received signal but, on the contrary, functions equally well at any region for which a suitable mixer can be introduced. For very large power changes in the received signal (60 db or more), it may be necessary to use an r-f attenuator solely or in combination with an i-f attenuator to maintain the desired accuracy.

The heterodyne circuit contains three output indicating devices. Choice will depend upon such conditions as character of signal and flexibility of indication: (1) If the signal is appreciably modulated, either the oscilloscope or meter may be used following the video amplifier. The scope has the advantage of interval-timing, and therefore the desired signal and any spurious signal can usually be distinguished when using synchronized short-time pulse modulation on the transmitter. The undesired components are usually reflections from surrounding objects which produce an echo of different time delay from the desired signal. However, the oscilloscope is difficult to calibrate accurately over even a few decibels of power range. Thus, the attenuator must be continually readjusted to maintain a constant deflection on the scope when measuring the power changes. The meter cannot distinguish between the desired signal and an interfering signal but is capable, as has been previously pointed out, of being calibrated quite accurately over a range of possibly 10 db. A synchronized source is not used with the meter. (2) If the transmitted signal is a continuous wave, no alternating component will be available at the second detector. Therefore, a d-c amplifier and meter combination or a meter alone, if the detector signal level is high enough, is used. The oscilloscope is of little advantage in this arrangement, since, with a c-w source, interval-timing cannot be used and therefore the desired signal cannot be distinguished from those caused by reflections.

The i-f amplifier for any of the mentioned methods should be wide enough to allow for the instability of the source generator. Usually an amplifier with a 5- to 10-Mc bandwidth is used for microwave work.

A slight modification of the heterodyne circuit results in a system that is sometimes useful in c-w measurements. Here the i-f amplifier is designed to have a very narrow bandwidth, and the oscilloscope sawtooth sweep is used to frequency modulate the local oscillator. Thus, there will be an output pulse from the second detector at the instant the local oscillator passes through a frequency that is equal to the intermediate frequency above or below the frequency of the received signal. This pulse is then amplified by the video circuit and applied as a deflection voltage to the oscilloscope on which the signal appears as a sharp pip or vertical line. This pip can be controlled and measured by the attenuator as previously described. The advantage of this system for intensity measurements lies chiefly in the fact that the i-f amplifier has a quite narrow bandwidth and consequently a higher signal-to-noise ratio. In addition the video amplifier is used which is less complicated than a stable d-c amplifier. It is apparent that any number of separate pips can appear on the oscilloscope screen if a like number of slightly different signal frequencies are being produced by the signal source. Thus, a very accurate means is provided for measuring the frequency distribution and relative amplitude of the energy given off from the source. Units are built using this circuit and are known as spectrum analyzers and, as such, are frequently used for amplitude measurements with single-frequency sources.

16.5. Recorders.—The use of an automatic antenna pattern recorder has been pointed out previously in the general discussion of secondary patterns. The following is a discussion of the various methods for automatically plotting receiving power patterns.

The simplest recorder consists merely of a recording current meter. This meter is attached to the output of a linear peaked audio amplifier, the input of which is fed from the antenna under test through a square-law detector. If the angular coordinate is synchronized with the angular displacement of the antenna, it is possible to obtain a very satisfactory linear plot of power. However, for any antenna of appreciable gain, much of the desired information involving the side lobes will be lost, because such a linear recorder will not visibly resolve powers that are from 20 to 40 db below the main peak power. Also, most recording meters of this type have the stylus attached to a central pivot about which it rotates as a function of varying current. This does not yield a plot in rectangular coordinates, frequently giving rise to confusion regarding the picture portrayed. It is evident, then, that in order to record side lobes and for reasons of clarity other methods must be contrived.

Experience has shown that the most desirable scale to use with an automatic system is a decibel scale extending over a range of about 40 db. This may be accomplished in two distinct ways. (1) The linear recording current meter may be driven by a logarithmic amplifier which is fed by the square-law detector, thus obtaining the desired result by the use of the nonlinear amplifier. (2) A system can be built around a nonlinear potentiometer or attenuator with servo follow-up for stylus

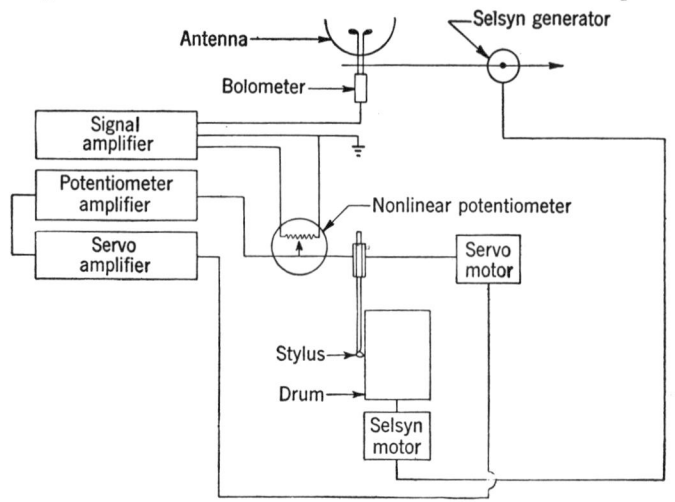

Fig. 16·15.—Audio antenna pattern recorder.

Fig. 16·16a.—Photograph of the control and recording console.

displacement. The latter method is generally preferred, since it does not require a nonlinear amplifier; accuracy of such an amplifier is usually difficult to maintain. Systems based on this second method will be described.

Figure 16·15 shows a block diagram of a typical recorder designed to operate on the audio voltage developed by a bolometer or crystal used as the antenna feed termination. It should be understood, of course, that the transmitter is modulated at the pass frequency of the audio system of the recorder.

The system operates as follows: The angular displacement of the mount to which the antenna is affixed is transmitted to the recording drum by a Selsyn generator-and-motor combination which synchronizes the angular rotation of the drum with that of the antenna.[1] The input to the signal amplifier varies with the orientation of the antenna; since this amplifier is linear, the voltage appearing across the calibrated potentiometer is directly proportional to the power picked up by the antenna. The tap-off voltage from the potentiometer is fed into the potentiometer amplifier, the output of which is compared with a constant voltage. The difference between the latter two voltages is applied to the servo amplifier; the servomechanism then drives the potentiometer tap-off to a point such as to reduce the difference voltage to zero. Since the stylus is attached to the same mechanism that drives the potentiometer, it is displaced in a like manner; consequently, the stylus displacement is proportional to the amplitude of the signal.

FIG. 16·16b.—Photograph of the electronic cabinet of an audio recorder.

An instrument based essentially on the system just described has been built and used.[2] Photographs of the control and recording console and the electronic cabinet of this instrument are shown in Fig. 16·16. With this instrument it is possible to obtain side-lobe information in fine

---

[1] Good accuracy may be achieved in this manner, since the Selsyn generator on the mount is geared to the motion through a precision gear train with a stepup ratio of about 1 to 36 and the Selsyn motor driving the drum is geared down a like amount; This means that the electrical inaccuracies of the Selsyn system are divided by 36, and by the choice of good gearing, mechanical errors may be held quite small.

[2] O. A. Tyson, "Antenna Measuring Equipment," RL Report No. 601-4, January 1945.

detail; a typical plot of power variation vs. displacement angle is shown in Fig. 16·17.

Alternative methods of supplying the information to the servomechanism make use of the r-f carrier wave rather than the audio-modulated output of a square-law detector. One method uses a calibrated i-f attenuator in a heterodyne circuit[1] in place of the a-f attenuator. This setup is shown in block diagram in Fig. 16·18. Angular coordinates

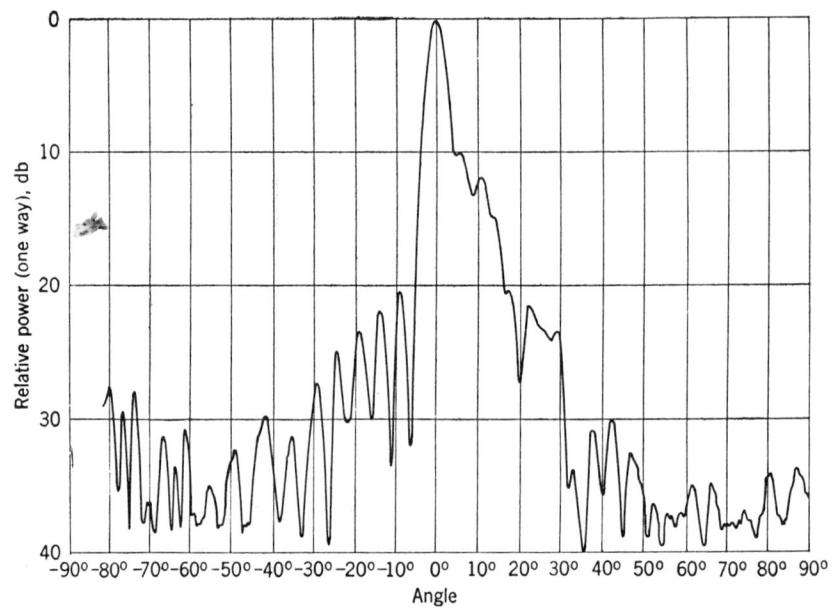

FIG. 16·17.—Antenna pattern showing detail obtainable with audio recorder.

are transmitted in a manner similar to that shown in Fig. 16·15. This consists of a Selsyn generator geared to the antenna mount and a Selsyn motor geared to the recording drum. If the antenna under test is illuminated by either a modulated or a c-w wave, an i-f voltage is developed at the output of a small mixer and local oscillator which is used to terminate the antenna feed. The i-f voltage is passed through a calibrated logarithmic attenuator and amplified; it is then rectified and used to control a servomotor just as is done in the audio system. The stylus, of course, is again geared to the servomotor, which drives the calibrated attenuator and traces a signal amplitude plot on the synchronized graph drum.

The chief advantage of the heterodyne system is that it is very much more sensitive to weak signal input than bolometers or crystals used as

[1] R. J. Symonds, "Microwave Antenna Pattern Recorder," BTL Report MM-44-170-55, Nov. 15, 1944.

rectifiers, as was indicated in Sec. 16·4. As a result it is possible to reduce transmitter power by possibly 10 db or to extend the range of power coverage from 40 to approximately 60 db under ideal conditions. The limitations to this method arise in the mixer, which is often not

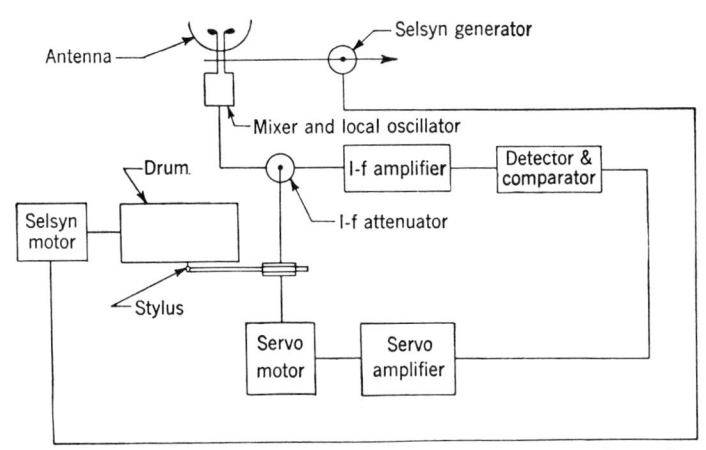

FIG. 16·18.—Antenna pattern recorder utilizing an r-f attenuator in a heterodyne circuit.

completely linear over the higher portions of the desired power range. Also, difficulty sometimes arises from the large bulk of the mixer-oscillator termination which must be attached to a relatively small antenna. Bolometers, on the other hand, are more adaptable because of their size.

Index

A

Absorbing material, 560
Achromatic doublets, 410
Admittance, characteristic, 26
  normalized, 26, 213
  terminal, of waveguide, 366
Admittance characteristics, of E-plane sectoral horns, 369–374
  of H-plane sectoral horns, 374–376
AFC, 527, 538
Air-to-surface search, 466
Aircraft antennas, 512
Aircraft installations, interference effects in, 515–519
Altar, W., 556
Altitude circle, 514
Amplifiers, 604–609
  audio, 605
Antenna feed, definition of, 12
Antenna mismatch, dependence of absorption cross section on, 51–53
Antenna pattern recorder, automatic, 609–613
Antenna system, 511
Antennas, beavertail, 453, 477
  cheese, 459
  conically scanning, 347
  ground, 510
  half-beacon, 460
  interaction between, 587–592
  pencil-beam (see Pencil-beam antennas)
  pill-box, 459–464
    impedance correction for, 463
  receiving, equivalent circuit of, 40–45
  scanning, 513–521
  shaped-beam, line sources for, 495–497
    secondary pattern of, measurement of, 579
  ship, installation problems of, 511
  shipborne, for surface search, 467
  skirt-dipole, 240–242

Antennas, strip reflector, 484
  surface, for air search, 465
  transmitting, equivalent circuit of, 37–40
Aperture, angular, 416
  circular, 192–195
  far-zone region of, 172
  gain function for, 162
  near-zone region of, 170
  rectangular, uniformly illuminated, 180–182
Aperture blocking, effects of, 190–192
Aperture efficiency, 178
Aperture-field distributions, moments of, 184
  separable, 182
Aperture-field method for reflectors, 158–160
Aperture gain, maximum, condition for, 177, 178
Aperture relations, optimum angular, 424
Aperture treatment, of radiation from horns, 357
  of radiation from waveguides, 334–336
Array designs, broadside, 318–333
  end-fire, 316–318
  nonresonant, 328–333
Array element, longitudinally polarized, 302, 309, 312
  slot-fed dipole as, 284–286
  slots as, 291–301
  streamlined, longitudinally polarized, 312
  transversely polarized, 310, 311
  tridipole radiator as, 304, 305
  waveguide radiators as, 301–303, 329
Array factor, 260
Arrays, binomial, 269
  broadside (see Broadside arrays)
  end-fire, gain of, 277–278
  horn, 319
  linear, pattern synthesis by, 279–284

Arrays, polynomial associated with, 261–264
  Tchebyscheff, 282–284
  uniform, 264–267
  uniformly illuminated, 267–269
Austin, P. M., 186
Azimuth ratio, 304

B

Babinet's principle, 167
Back-lobe interference, 428, 435
Back-scattering coefficient, 468
Baker, B. B., 108
Bandwidth, of dielectric lens, 398
  of metal-plate lens, 408–410
Barker, C. B., 139
Barrow, W. L., 349
Beacon antenna systems, 327
Beacon antennas, on aircraft, 521
Beacon elements (*see* Radiators, axially symmetrical)
Beacon pattern measurements, 573
Beacon radome designs, 540
Beam, beavertail, 477, 478
  broadside, 267–274
  end-fire, 274–278
  fanned, 7
    applications of, 450
  low-altitude, 484
  omnidirectional, 6
  pencil, 7
  sector shaped, 475
  shaped (*see* Shaped beam)
  toroidal, 6
Beam deviation factor, 488
Beam shape, variable, 508–509
Beam shaping, by obstacles in horn and wave-guide apertures, 380–383
Beamwidths vs. aperture illumination, 179, 183–187, 195
Bethe, H. A., 208
Biconical horn, 9
Bipolar charts, 33–36
Birchard, B. L., 316
Blister, 512
Bohnert, J., 442, 455, 564
Bolometer, 555, 601, 604
Bolometer element, 549
Booker, H. G., 167, 294
Born, M., 125, 166, 167, 197, 414

Boundary conditions, general formulation of, 66–68
  at infinity, 84–86
  for scattering problems, 130–132
Box horns, 377–380
Braunlich, A., 460
Breckenridge, R. G., 399
Breen, S., 251, 437
Brewster angle, 402
Brillouin, L., 42
Broadband-normal-firing arrays, 331–333
Broadside arrays, binomial illumination of, 269
  gabled illumination of, 269
  gain of, 271–274
  maximum gain conditions for, 270, 271
  resonant, 321–327
Brownlow, J. M., 399

C

Campbell's loaded-line formulas, 313–316
Carlson, J. F., 410
Center of feed, 239, 343, 562
  determination of, 564–570
Chesley, F. G., 399
Chisholm, E. B., 331, 332, 509
Choke, 242, 243, 245
Chu, L. J., 80, 146, 248, 334, 341, 349, 357, 415, 478, 497
Clapp, R. E., 299, 469
Coaxial lines, 217–226
  cascade transformers in, 221–223
  parallel stubs for, 223–225
  series reactance transformer for, 225
  TEM-mode of, 217–219
Coma lobes, 488
Condon, E. U., 415, 423
Conductance, incremental, 298
Copson, E. T., 108, 167
Coupling, between transmitter and receiver, 45–48
Coverage pattern, one-way, 4
  two-way, 5
Cross polarization, of barrel-reflector antenna, 503
  of fanned beam antennas, 457
  measurement of, 579–580
  of pencil-beam antennas, 419, 423
Cross section, absorption, 3, 42
  average, for matched system, 50, 51

INDEX

Cross section, dependence of, on antenna mismatch, 51-53
  interception, 468
  radar, 468
  receiving, 3
  scattering (see Scattering cross section)
Crystal, 555, 559, 601, 604
  calibration of law of, 552-556
$\operatorname{Csc}^2 \theta$ pattern, 466, 470, 507
Cullen, A. L., 295
Current distribution, discontinuous, 146-149
  far-zone fields of, 87-91
  over reflector, 144-149
Cut paraboloid, feed-tilt effects in, 488, 489
  offset feeding of, 454
  for shaped beams, 477-483, 487-491
  for simple fanned beams, 451-457
Cutler, C. C., 484, 497, 508
Cutoff wavelength, 205
Cylinder, parabolic, 457-459

D

De Bretteville, A. P., Jr., 399
Debye, P., 114
Detection, superheterodyne, 559, 578, 607-609
Detection system, calibration of, 552-556
Detector, bolometer, 555, 601-604
  crystal, 549, 554, 555, 601, 604
Detector response, 601-603
Dickie, R. H., 200
Dillon, R. E., 316
Dipole, electric, 92-95
  half-wave, 98
    gain of, 99
  magnetic, 95
  slot-fed, 245-248
    as array element, 284-286
  spheroidal, 249
    impedance of, 249
Dipole-disk feeds, 251-253
Dipole feeds, directive, 250
Divergence factor, 143
Dodds, J. W., 295, 299
Dolph, C. L., 282
Double-dipole system, on coaxial line, 253-254
  coaxial-line-fed, 253

Double-dipole system, theory of, 101-104
  on waveguide, 255-256
Doublet lens, 410
Dowker, Y., 523, 556
Dunbar, A. S., 480, 487

E

E-plane, principal, 103
E-plane patterns, principal, of pencil-beam antennas, 422-423, 433-437
E-plane sectoral horns, admittance characteristics of, 369-374
  modes in, 350-354
  mouth admittance of, 369
  radiation from, 357, 358-365
  tenth-power widths of, 364, 365
  throat transition, 369-371
  transmission-line equations for, 366-369
Eaton, J. E., 316
Edge diffraction, 516
Edge reflection, 518
Eisenhart, L. P., 142
Electrical length, 550-552
Elson, N., 442
Emde, F., 194, 220, 233
Error, cubic phase, 189
  linear, 188
  quadratic, 188
Everhart, E., 523
Everitt, W. L., 17, 19-21
Eyges, L. J., 316, 328, 496

F

Far-zone fields, of current distributions, 87-91
  of line-current distributions, 96-98
Far-zone region of apertures, 172
Feed, center of (see Center of feed)
  dipole-disk, 251-253
  double-dipole, 253-256, 434
  double-slot, 348-349
  extended, 473, 474, 477, 487-494
    dipole-array, 491-493
    horn-array, 493-494
  four-dipole, 256
  point-source (see Point-source feed)
  primary, 12
Feed requirements, primary, 239-240

Feed systems, front, 448
  rear, 347, 434, 448
Feed tilt, impedance changes with, 488
Feed-tilt effects, in cut paraboloids, 488, 489
  in paraboloidal reflectors, 487, 488
Feeding coefficients, 259
Fermat's principle, 122–125
Field equations for cylindrical waveguide, 201–203
Fields, dominant-mode, of sectoral horns, 352, 355
  superposition of, 99–101
  time-periodic, Maxwell's equations for, 68
Fiske, M. D., 387
Fourier integral representation of Fraunhofer region, 174–175
Frank, N. H., 173, 174
Frank, P., 110
Fraunhofer region, 160–162, 172
  criteria for, 198, 199, 561, 574
  Fourier integral representation of, 174–175
Frequency control, automatic (*see* AFC)
Frequency pulling, 538
Fresnel region, of circular aperture, 196–199
  general characteristics of, 171–172
Fresnel zones, 196
Front-feed systems, 448

G

Gain, 3, 90, 177
  absolute, 90
  of broadside arrays, 271–274
    maximum, condition for, 271
  of circular waveguides, 340–341
  dependence of, on aperture illumination, 177–178
  directive, 580–581
  effective, 583
  of end-fire arrays, 277–278
  of half-wave dipole, 99
  of parabolic cylindrical antennas, 458
  of pencil-beam antennas, 423–432
  of pyramidal horns, 587
  of rectangular waveguides, 346
  of scattering pattern, 468
  of sectoral horns, 587

Gain comparison, 581
Gain determination, reflection method for, 585–586
Gain factor, 178, 425
  vs. aperture illumination, 187, 195
Gain function, 2, 90, 94
  for aperture, 162
  scattering, 590
Gain measurements, 580–586
Gain standard determination, 582–585
Gain standards, secondary, 586
Gardner, J. H., 481, 484, 485, 508
Goldstein, H., 320
Grating, quarter-wave, 447
Grating reflectors, 449–450
Green's theorem, scalar, 108
  vector, 80
Ground antennas, 510
Ground target area, effective, 469
Guillemin, E. A., 17, 18
Guptill, E. W., 295, 299

H

H-plane, principal, 103
H-plane patterns, principal, of pencil-beam antennas, 422–423, 433–437
H-plane sectoral horns, admittance characteristics of, 374–376
  modes in, 355–357
  mouth admittances of, 375
  radiation from, 358–365
  tenth-power widths of, 365
  throat transition, 375
Half-power width, 94, 104
Hansen, W. W., 278
Hayes, W. D., 449
Hegarty, M., 523
Height finding, 467
Heins, A. E., 410
Hiatt, R., 251, 437, 442
Hill, J. F., 503
Horn arrays, 319
  as extended feeds, 481, 493–494
Horns, biconical, 9
  box, 377–380
  compound, 350, 376
  pyramidal, 587
  sectoral (*see* E-plane sectoral horns; H-plane sectoral horns)
Hull, G. F., Jr., 415

INDEX 619

Hunter, L. P., 556
Huygens-Fresnel principle, 108

I

Illumination, aperture, beamwidths vs., 179, 183–187, 195
  dependence of gain on, 177–178
  gain factor vs., 187, 195
  side lobes vs., 179, 187, 195
  binomial, of broadside arrays, 269
  gabled, of broadside arrays, 269
Images, of dipole radiators, 135–137
  principle of, 132–137
  for dipole-sources, 135–137
Impedance, characteristic, 23, 216
  normalized, 25, 213
  of spheroidal dipoles, 249
  transverse wave, 209
Impedance changes, with feed tilt, 488
Impedance correction, for pencil-beam antennas, 443–448
  for pillbox antennas, 463
Impedance effects, in normal-incidence radome, 537
Impedance-measurement apparatus, 547
Impedance measurements, probe errors in, 548, 556
Indicator, response of, 549
Intensity measurements, primary patterns, 561–564
Interference effects in aircraft installations, 515–519
Iris, trial, 552
Isotropic spherical waves, 78

J

Jahnke, E., 194, 220, 233
Junction effects, in waveguides, 214, 215

K

Keary, T. J., 455, 461, 464, 502, 508, 564
Kelvin, Lord, 121
King, A. P., 349
King, D. D., 248
King, R. W. P., 21, 248
Kingsbury, S. M., 399
Klystron, double-cavity, 596–599

Klystron, reflex, 594–596
  Sperry 410-R, 596
  Sperry 419B, 595
Kock, W. E., 402, 405, 406, 410, 442
Krutter, H., 442, 455

L

Lamont, R. L., 200
Lattice factor, 105
Lens, dielectric, 389–402
  attenuation in, 399–400
  bandwidth of, 398
  frequency sensitivity of, 398
  one surface, with elliptical contour, 392–393
    with hyperbolic contour, 390–392
  reflections from surfaces of, 401
  tolerances on, 400
  two surface, 390, 394
  zoned, 395–398
  doublet, 410
  metal-plate, 402–412
    achromatic doublets, 410
    bandwidth of, 408–410
    one surface, with elliptical contour, 403–405
    of parallel plates, 402–405
    of parallel wires, 406
    polystyrene-foam bonded, 406
    reflection from surfaces of, 410–412
    tolerances on, 407–408
  spherical, 390
Lewis, F. D., 349
Line, corrugated, 319
  lossless, 26–29
Line-current distributions, far-zone fields of, 96–98
Line-source primary pattern, 570–572
Line-source feed, and reflectors, 151–154
Line sources for shaped-beam antennas, 495–497
Littelfuse, 549, 559
Littelfuse Company, 603
Loaded-line constants, 313–316
Loaded-line formulas, Campbell's, 313–316
Lossless line, 26–29
Low-altitude $csc^2 \theta$ antennas, 484–486
Luneberg, R. K., 126

M

Macfarlane, G. G., 503
McMillan, E. B., 523
Magic T, 572
Magnetrons, 599–601
Malus, theorem of, 126
Marcuvitz, N., 200
Marshall, F. B., 556
Mason, S. J., 377, 380, 453, 564
Matching of waveguide and horn feeds, 383–387
Maximum-power theorem, 20, 21
Maxwell's equations, differential form, 64
  integral form, 64
  for time-periodic fields, 68
Mimno, H. R., 21
Mises, V., 110
Mismatch, introduced by reflector, 155–158, 439–443, 454
Moments, of aperture-field distributions, 184
Montgomery, C. G., 200
Mueller, C. E., 316

N

Nabarro, F. R. N., 415
Nacelle, 512, 513
Near-zone region of apertures, 170
Network, four-terminal, 17–19
  two-terminal-pair, 17
Network equivalent, four-terminal, of section of transmission line, 36
Network parameters, four-terminal, 17
Nowak, W. B., 253

O

Obscuration, of target signal, 520
Offset feeding technique, 453–457
Optical-Fresnel field, 171
Optical path length, 122
Orthogonality of waveguide modes, 207–209
Oster, G., 399

P

Π-section, 19
Pao, C. S., 381, 383, 453, 477, 487

Parameters, constitutive, 65
  current, 210
  voltage, 210
Pattern, apparent, 526
  primary (see Primary pattern)
  secondary (see Secondary pattern)
Pattern distortions by radomes, 524–526
Pattern synthesis by linear array, 279–284
Pencil beam, requirements for, 413, 414
Pencil-beam antennas, using circular paraboloidal reflector, aperture field of, 419
  design procedures for, 432
  cross polarization of, 419, 423
  gain of, 423–432
  impedance correction for, 443–448
  paraboloidal, 415–450
  using paraboloidal reflectors, antenna gain of, 423–433
Perry, H. A., 523
Phase constant, 23, 205
Phase determinations on point-source feeds, 564–570
Phase-error effects, on pencil beam gain, 430
  on secondary patterns, 186–192
Phase errors, defocusing, 432
Phase measurement, line-length effect in, 569
Phase-reversal techniques, in arrays, 321
Phillips, H. B., 62
Pillbox design problems, 460–464
Pippard, A. B., 442
Point-source, cone, 240
Point-source feeds, primary pattern for, 557–570
  and reflectors, 149–151
Polar diagram, 3, 90
Polarization, circular, 92
  cross (see Cross polarization)
  elliptical, 91
Polynomial, associated with array, 261–264
Poole, A. R., 461, 508
Porterfield, C. F., 478
Power modes, 595
Poynting's theorem, 69–71
Poynting vector, 69
  time-average, 70
PPI, 514

INDEX

Pressurizing, 383–387
Pressurizing device, 376
Primary feed, 12
Primary pattern, 12
   line-source, 570–572
   for point-source feeds, 557–570
Primary-pattern apparatus, siting of, 560
Probe, response of, 549
Probe errors, in impedance measurements, 556
Propagation constant, 205
Purcell, E. M., 200, 301, 330, 464, 585

R

Radar cross section, 468
Radar shadow, 515
Radiation, from E-plane sectoral horns, 357
   from H-plane sectoral horns, 358
   from waveguide, aperture treatment of, 334
Radiation conditions, 85
Radiation pattern, 90
Radiation resistance, 39, 95
Radiators, axially symmetrical, 303–309
   dipole, 92–96
      images of, 135–137
      streamlined, 310–313.
   waveguide, 301–303, 329
Radio Research Laboratory, 223
Radome, normal incidence, 537–539
   design considerations for, 538
   impedance effects on, 537
   pattern effects of, 537
   pattern distortions by, 524–526
   streamlined, 524, 535, 540–542
      design considerations for, 542
      pattern effects of, 540
Radome walls, reflection coefficients of, 529–537
   transmission coefficients of, 529–537
Range reduction, 524
Ray curvature, 111, 112
Ray velocity, 110
Rayleigh, Lord, 19
Rayleigh reciprocity theorem, 19
Rear-feed systems, 448
Receiving pattern, 4
Reciprocity, between transmitting and receiving patterns, 48–50

Reciprocity relation between transfer coefficients of network, 18
Redheffer, R. M., 523, 556
Reflection, from curved surfaces, 138–143
   from curved undersurface of fuselage, 517
   edge, 518
   laws of, 124, 134
Reflection coefficient, current, 32
   electric-field, 212
   of paraboloidal reflectors, 439–443
   of radome walls, 529–537
   voltage, 25
Reflection coefficient chart, 29–33
   (*See also* Smith charts)
Reflection method for gain determination, 585
Reflector antennas, double-curvature, 502–508
Reflector mismatch, 155–158
Reflectors, aperture-field method for, 158–160
   barrel, 480, 503
   contour cutting of, 453–457
   current distribution over, 144–149
   cut paraboloidal, 451–457, 477
      symmetrical type, 451–453
   cylindrical, for shaped beams, 494–497
   grating, 449–450
   line-source feed and, 151–155
   mismatch introduced by, 155–158, 439–443, 454
   modified, 474, 477, 479–495, 503
   paraboloidal, feed-tilt effects in, 487, 488
      structural design problems, 448–450
   point-source feed and, 149–151
   screen, 449
   shovel-, 481, 483
Refraction, laws of, 125
Responder, 327
Riblet, H., 139, 247, 282, 305, 316, 325
Rice, S. O., 247
Ridenour, L., 414
Ring source, 348
Risser, J. R., 332, 366, 369, 461, 496, 508

S

Scalar, Green's theorem, 108
Scanning antennas, 513–521

Scattering, multiple, 588
Scattering cross section, 5, 42, 468
  equivalent sphere, 5
Scattering gain function, 590
Scattering pattern, gain of, 468
Schelkunoff, S. A., 248, 261, 279, 282, 586
Schwartz inequality, 177
Screen, absorbing, 560
Screen reflectors, 449
Secondary pattern, 12, 169
  general features of, 175–180
  phase-error effects on, 186–192
Sector blanking, 517
Series reactances, on coaxial lines, 223–226
Shaped beam, applications for, 465–468
  cut paraboloid for, 477–483, 487–491
  cylindrical reflectors for, 494–497
  requirements for, 465–468
Ship antennas, 511
Shipborne antenna for surface search, 467
Sichak, W., 230, 231, 255, 301, 330, 436, 464
Side lobes, 176
  vs. illumination, 179, 187, 195
Signal generator, 594
Silver, S., 155, 157, 441, 442, 463, 487, 497, 503
Siting, of impedance-measurement apparatus, 548
  of primary-pattern apparatus, 560
  of secondary-pattern apparatus, 574–578
Skellett, A. M., 396
Skin thickness, effective, 534
Skirt-dipole antenna, 240–242
Slater, J. C., 51, 173, 174, 200, 211
Slope, 549
Slot array, axially symmetrical, 305
Slot radiators, impedance in waveguide, 287–299
Slots, dumbbell-shaped, 296
  longitudinal, 291–297
  nonradiating, 287
  probe-fed, 299–301
  resonant, 291–299
  transverse, 291, 292
Smith, P. H., 29
Smith, W. O., 497

Smith charts, 29
Snell's laws, 123–125
Sommerfeld, A., 59, 114
Southworth, G. C., 349
Space arrays, regular, 104–106
Space factors, 89
Spatial pattern measurements, 563
Spencer, R. C., 139, 184, 186, 191, 194, 473, 497, 503
Sperry 410-R klystron, 596
Sperry 419B klystron, 595
Spherical waves, isotropic, 78
Squint, 245, 438
Stabilization, 508
Standing-wave measurements, 548–550
Standing-wave ratio, 28, 545
  power, 28
  voltage, 28
Standing-wave voltage ratio, measurement of, 547–549
Stationary phase, principle of, 119–122, 157
Steele, E. R., 523
Steenland, A. M., 496
Steinberger, J., 331, 332, 496
Stergiopoulos, C. G., 479, 490
Stevenson, A. F., 292
Stratton, J. A., 57, 61, 67, 78, 80, 93, 132, 146, 159, 248
Stubs, parallel, 223–226
Suen, T. J., 523
Superposition principle, 66
Surface-to-air search, 465
Surface antenna for air search, 465
Symonds, R. J., 612
Synge, J. L., 110, 122

T

T-section, 18
Taggart, M. A., 497
Target response, directional, 468–471
Target signal, obscuration of, 520
Tchebyscheff arrays, 282–284
TE-mode, of circular waveguide, 233
  of coaxial line, 220
  of cylindrical waveguide, 204–206
  of E-plane sectoral guide, 352–353
  of H-plane sectoral guide, 355–356
  of parallel-plate waveguide, 237
  of rectangular waveguide, 226–229

INDEX

TE_{10}- to TM_{01}-mode converter, 308
TEM-mode, of coaxial lines, 217–219
  current of, 216
  of cylindrical waveguide, 203, 204
  of parallel-plate waveguides, 235–236
  voltage of, 216
Tenth-power width, 94, 104
Terman, F. E., 349, 415
Thévenin's theorem, 20
Tisza, L., 399
TM-mode, of circular waveguide, 233
  of coaxial line, 220
  of cylindrical waveguide, 206
  of parallel-plate waveguide, 236
  of rectangular waveguide, 226–229
Transfer admittance coefficients, 18
Transfer coefficients of network, reciprocity relation between, 18
Transfer impedance coefficients, 18
Transformation charts, 29–36
Transformers, cascade, in coaxial lines, 221–223
Transmission coefficients, of radome walls, 529–537
Transmission line, homogeneous, 23–26, 544–546
  three-wire, 247
  two-wire, 21–37
Transmission-line equations, for E-plane sectoral horns, 366–369
Transmitter pulling, 517, 537
Transmitter-receiver system, equivalent network of, 53–60
Transmitting pattern, 3, 557
Transponder, 327
Tridipole transverse element, 304
Truell, R., 221
Tyson, O. A., 612

V

Vallee Poussin, C. de la, 280
Values, characteristic, 204
Van Atta, L. C., 434
Variable beam shape, 508–509
Vector, Green's theorem, 80
Vertex plate, 443, 463
Vogel, B. R., 509
Voltage attenuation constant, 23
Voltage parameters, 210

Voltage reflection coefficient, 25
Von Hippel, A., 399

W

Walkinshaw, W., 503
Wall, sandwich, 532
  single, 531
Watson, G. N., 76, 194, 195, 236, 341, 353, 354, 367
Watson, W. H., 295, 299
Wave equations, 71–73
Wave velocity, 110
Waveguide, circular, TE-mode of, 233
  TM-mode of, 233
  cylindrical, field equations for, 201–203
  TEM-mode of, 203, 204
  junction effect in, 214, 215
  radiation from, aperture treatment of, 334
  rectangular, TE-mode of, 226–229
  TM-mode of, 226–229
  terminal admittance of, 366
Waveguide modes, orthogonality of, 207–209
  two-wire line representation of, 209–216
Waveguide radiation patterns, circular, 336–341
  rectangular, 341–347
Waveguide radiators, as array elements, 301–303, 329
Waves, cylindrical, 75
  plane, 73
  principal, 203
Wesson, L. G., 399
West, W. J., 478
Westphal, W. B., 399
Whelpton, J., 285
White, J. S., 523
Wind resistance, 450, 453
Window, capacitive, 230
  inductive, 230
  resonant, 231
Wing, A. H., 21
Winkler, E. D., 523
Wintner, A., 122
Wolfe, H., 461, 508
Wolff, I., 282
Wollaston wire, 559, 603
Woodyard, J. R., 278